VECTOR AND COMPLEX CALCULUS

Vector and Complex Calculus

A Textbook for Students of the Physical Sciences

Fabian Waleffe

University of Wisconsin-Madison

OXFORD
UNIVERSITY PRESS

OXFORD
UNIVERSITY PRESS

Great Clarendon Street, Oxford, OX2 6DP,
United Kingdom

Oxford University Press is a department of the University of Oxford.
It furthers the University's objective of excellence in research, scholarship,
and education by publishing worldwide. Oxford is a registered trade mark of
Oxford University Press in the UK and in certain other countries

Published in the United States of America by Oxford University Press
198 Madison Avenue, New York, NY 10016, United States of America

British Library Cataloguing in Publication Data
Data available

Library of Congress Control Number: 2024939148

ISBN 9780198927839
ISBN 9780198927822 (pbk.)

DOI: 10.1093/oso/9780198927839.001.0001

Printed and bound by
CPI Group (UK) Ltd, Croydon, CR0 4YY

To the memory of my parents, Roger Waleffe and Louise Godefroid. Born in Liège and Esneux in Belgium a few years before World War II, their formal schooling was limited, but they valued education, and encouraged and supported our studies.

Preface

This is a textbook about vector calculus and fundamentals of complex calculus. Vector calculus concepts and techniques are not too cumbersome, not too abstract, but just right for electromagnetism and Maxwell's equations, as well as mechanics, fluid dynamics and the differential geometry of curves and surfaces. This textbook aims to bridge the gap between the mathematics covered in standard multivariable calculus courses and the vector and complex calculus needed for intermediate and advanced undergraduate courses in the mathematical, physical and engineering sciences.

I have taught many courses that require vector calculus and a dash of complex calculus. Courses such as *fluid dynamics, aerodynamics, continuum mechanics, elasticity, flight dynamics, advanced dynamics* and *hydrodynamic stability and turbulence*, for example. Teaching those courses made it clear that the concepts and methods of *vector calculus* are fundamental, yet fall through the educational cracks. The standard multi-variable calculus courses do not, and cannot, go much beyond 'xyz calculus' and students do not absorb the more geometric, vector and curvilinear concepts of vector calculus. Applied mathematics, engineering and physics students repeatedly 'review' key results and formulas in various courses that do not have time to properly introduce, derive and justify those results. Mathematics students jump from elementary calculus to vector analysis in $\mathbb{R}^n$, largely skipping cross products, and applications in 3D space. Both groups of students are likely to have been accelerated through precalculus in order to race through 'xyz calculus'.

Students of this textbook will have been exposed to multi-variable calculus and learned about partial derivatives and iterated integrals. They will have heard about gradient and divergence and might remember that those involve partial derivatives with respect to x and y and z. They may have heard about the divergence and Stokes theorems, but perhaps late in the semester and those did not show up on the final exam. They will have one of the textbooks such as those of Thomas and Finney (*Calculus and Analytic Geometry*), or James Stewart (*Calculus*). Besides those and many other widely used textbooks, I would recommend the books by Marsden and Tromba (*Vector Calculus*), Paul Matthews (*Vector Calculus*), Hartley Rogers (*Multivariable Calculus with vectors*), or George Simmons (*Calculus with Analytic Geometry*) as resources for prerequisite material and their emphasis on geometrical aspects of calculus. I recommend another book by George Simmons, *Precalculus mathematics in a nutshell.* That small book provides a nice review of (euclidean) geometry, algebra and trigonometry that remain fundamental. Hartley Rogers' *Multivariable Calculus with Vectors* also reviews the geometry of two- and three-dimensional euclidean spaces, $\mathbb{E}^2$ and $\mathbb{E}^3$, and emphasizes a geometric approach similar to this textbook.

There is some overlap with the textbooks mentioned above since we start by reviewing vector algebra and geometry, but we go further and cover material discussed in more advanced textbooks that may be titled *Vector Analysis*, *Tensor Analysis*, *Complex Analysis*, and *Differential Geometry*. Although the original vector calculus textbook by E.B. Wilson, based on lectures by J.W. Gibbs, was titled *Vector Analysis*, the word *analysis* is now used to denote the field of mathematics focused on the finer details of convergence, limits, differentiation and integration of functions taken in a broad sense. That material will be found in other books, such as Walter Rudin's *Principles of Mathematical Analysis*. Here, we discuss vector and complex *calculus*, and encourage students to think of $\mathrm{d}t$ as a small time increment, $\mathrm{d}\boldsymbol{r}$ as a small displacement, and $\int$ as a sum of very many of such very small elements, as done by the original developers of calculus and still used effectively throughout applied mathematics, engineering and physics.

Vector, index and matrix notations are discussed and used throughout the book since each have their own merits and limitations and all three are used interchangeably in applications. A student should learn to use the most appropriate notation for a particular problem, not be limited to one way of expressing concepts. Good notation facilitates understanding and problem-solving. Many exercises are provided and aim at helping students to develop problem-solving skills and an ability to process information, analyze data, set up a problem and derive a solution. Many useful and interesting results are derived in the exercises and the final answers are often given explicitly, so they can be used in other problems later on, or because of their intrinsic interest.

Parts of the material in this textbook have been taught for many semesters at the University of Wisconsin-Madison in a sophomore-level class aimed at applied mathematics, engineering and physics students. UW-Madison students will have completed multi-variable calculus as well as a course on differential equations and linear algebra prior to taking that class.

The book is divided into three parts: (I) Vector Algebra and Geometry, (II) Vector Calculus, and (III) Complex Calculus. The first few chapters introduce the main concepts of vector algebra for two and three-dimensional euclidean spaces, paying particular attention to geometrical and physical applications of dot and cross products. A good understanding of the geometric properties of the dot and cross products is essential for applications, and many examples are provided in the text and the exercises. Cartesian index notation is introduced and used in the discussion of determinants, change of cartesian basis and matrix algebra. Euler angles and rotation matrices are discussed in Chap. 8.

The second part of the book, Vector Calculus, includes core material on kinematics, dynamics, curves, surfaces and fields. More advanced topics such as index notation for general coordinates (Ricci calculus), differential geometry of surfaces, and curvilinear coordinates are treated in separate chapters 14 and 15. Chapter 16 on fields discusses gradient, curl and divergence in general coordinates as this provides deeper insights into the concepts, but that core chapter is written to enable an instructor to limit the presentation to orthogonal coordinates, or even to the standard cartesian, cylindrical and spherical coordinates. This is what has generally been done in our one-semester undergraduate version of the course where we typically skip chapter 14 (curves on surfaces) and cover only selected parts of chapters 9 and 15. Chapter 17 provides a few selected applications of vector calculus to electromagnetism. This chapter is also more advanced and is not systematically covered in our one-semester course at UW-Madison but provides a few examples of vector calculus applications that may be interesting even to students taking an electromagnetism course.

The third part of the course, Complex Calculus, reviews basic concepts of algebra and calculus extended to the complex plane before discussing the Cauchy-Riemann equations, conformal mapping, Cauchy's integral theorem, contour integration and residue calculus. This is an introduction to the concepts and tools of complex calculus for use in upper level undergraduate courses in the physical sciences.

Acknowledgements

I thank the many generations of students in Math 321 especially, but also in EMA 521, 523, 542, 622 and 700 at UW-Madison who willingly or unwillingly helped me shape the content of this book. I thank my plasma physics, nuclear engineering and engineering mechanics colleagues in the former Department of Engineering Physics, and Michael Corradini and Riccardo Bonazza in particular, for their support. I am grateful for colleagues in the Department of Mathematics who have used early drafts and provided comments and suggestions: Anakewit Boonkasame, Alfredo Wetzel, Billy Jackson and especially Jean-Luc Thiffeault who taught it multiple times.

The skull surfaces in section 13.3 were provided by Laurent Rineau and the cgal project. The Mercator and Lambert projections in Chap. 13 are from Daniel Strebe of 'Mapthematics' through Wikipedia. Some surface figures were produced using GeoGebra and a few with Matlab. Most of the figures were produced by the author using MetaPost, and the text was written in LaTeX.

I thank Sonke Adlung, acquisition editor at Oxford, for his patience and encouragement. He contacted me too many years ago to write a book on Turbulence, and ended up many years later with a Vector Calculus textbook.

A significant part of this book was (re)written while I was on sabbatical at Brown University's Department of Applied Mathematics and I thank Govind Menon, in particular, for his hospitality.

I was lucky to have access to quality public education. I remember in particular my high school math teacher, Abel Vanden Broeck at the Athénée de Liège 1, and Jacques Nihoul at the University of Liège. I thank my many friends and family who encouraged me to finish this work.

Contents

Part I

Vector Geometry and Algebra

Find Your Bearings

1

Vectors in 2D and 3D euclidean spaces are introduced from a geometric point of view with reference to navigation and astronomy. Cartesian, cylindrical, and spherical representations of vectors are discussed, as well as addition and scaling operations.

1.1 Magnitude and direction

The prototypical vectors are the *displacements* in two- or three- dimensional euclidean space. We write

$$\boldsymbol{a} = |\boldsymbol{a}|\,\hat{\boldsymbol{a}} \tag{1.1}$$

for a vector $\boldsymbol{a}$ of magnitude $|\boldsymbol{a}|$ in direction $\hat{\boldsymbol{a}}$, represented by an arrow. Thus, $\boldsymbol{a}$ = '1 km heading α' is a horizontal displacement of *magnitude* $|\boldsymbol{a}| = 1$ km in *direction* $\hat{\boldsymbol{a}}$ specified by angle α clockwise from north in navigation (Fig. 1.1), and '1km heading α' is the same as 'heading α 1km,' $\boldsymbol{a} = |\boldsymbol{a}|\,\hat{\boldsymbol{a}} = \hat{\boldsymbol{a}}\,|\boldsymbol{a}|$.

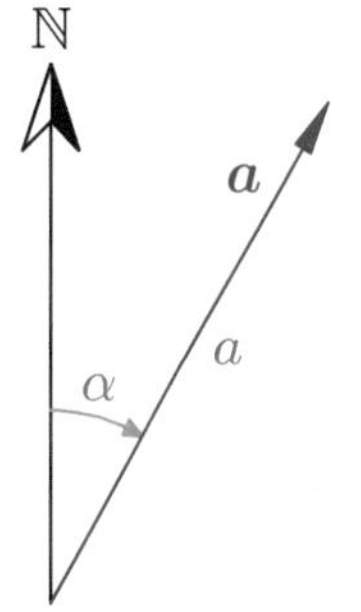

Fig. 1.1 Distance $|\boldsymbol{a}| = a$, heading α.

The *magnitude* of vector $\boldsymbol{a}$ is denoted with vertical bars $|\boldsymbol{a}|$, or in regular font italic a if there is no risk of confusion,

$$|\boldsymbol{a}| = a \geq 0. \tag{1.2}$$

The magnitude $|\boldsymbol{a}| = a \geq 0$ is a positive real number with appropriate physical units (meters, meters/second, newtons, ...).

The *direction* of vector $\boldsymbol{a}$ is denoted in boldface with a hat, $\hat{\boldsymbol{a}}$. This $\hat{\boldsymbol{a}}$ is a special vector, a *direction vector* with unit magnitude

$$|\hat{\boldsymbol{a}}| = 1, \tag{1.3}$$

and for that reason direction vectors $\hat{\boldsymbol{a}}$ are also called *unit* vectors, although these "unit" vectors do not have physical units.

Two points A and B specify a displacement vector $\boldsymbol{a} = \overrightarrow{AB}$. The same *displacement* $\boldsymbol{a}$ starting from point C leads to point D, Fig. 1.2, with

$$\overrightarrow{AB} = \boldsymbol{a} = \overrightarrow{CD},$$

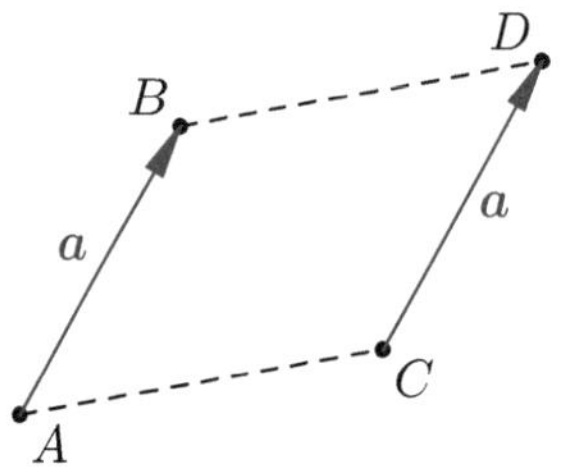

Fig. 1.2 $\overrightarrow{AB} = \overrightarrow{CD} = \boldsymbol{a}$.

and two vectors $\boldsymbol{a}$ and $\boldsymbol{b}$ are equal when their magnitudes and directions match

$$\boldsymbol{a} = \boldsymbol{b} \Leftrightarrow \begin{cases} |\boldsymbol{a}| = |\boldsymbol{b}|, \\ \hat{\boldsymbol{a}} = \hat{\boldsymbol{b}}. \end{cases} \tag{1.4}$$

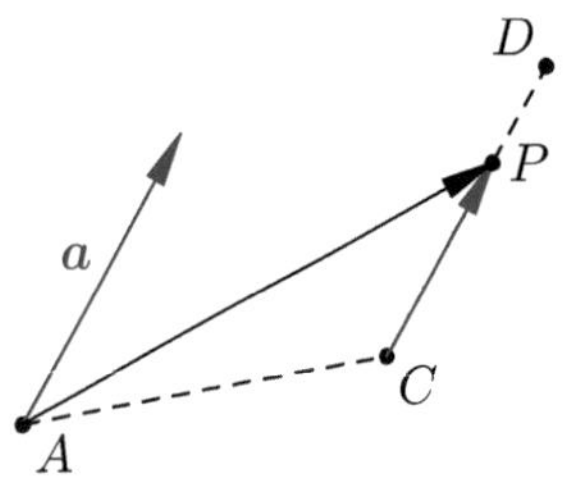

Fig. 1.3 $\overrightarrow{AP} = \overrightarrow{AC} + \frac{2}{3}a$.

Geometric vectors are often described as "oriented line segments"; however, a *line segment* is a set of points P. Points and vectors are related but distinct concepts. The oriented line segment $A \to B$ is the set of points P such that (Fig. 1.3)

$$\overrightarrow{AP} = t\,\overrightarrow{AB} = t\,\boldsymbol{a}, \qquad 0 \le t \le 1. \tag{1.5}$$

The oriented line segment $C \to D$ is a distinct set of points P such that

$$\overrightarrow{AP} = \overrightarrow{AC} + t\,\boldsymbol{a}, \qquad 0 \le t \le 1. \tag{1.6}$$

Notation We use an upper arrow $\overrightarrow{AB}$ to denote the displacement from point A to point B. The upper arrow notation $\vec{a}$ for vectors is common in introductory courses. The boldface notation $\boldsymbol{a}$, $\boldsymbol{b}$, $\boldsymbol{F}$, ... for vectors prevails in intermediate and advanced textbooks. The boldface notation for vectors was introduced by J. W. Gibbs and E. B. Wilson in the original *Vector Analysis* textbook.[1]

[1] E. B. Wilson, *Vector Analysis*, Yale University, 1901. `https://archive.org/details/117714283`

The handwritten notation for boldface is to put a squiggle $\sim$ under the character, for example

$$\underset{\sim}{a} \equiv \boldsymbol{a}, \quad \underset{\sim}{\hat{a}} \equiv \hat{\boldsymbol{a}}, \text{ etc.}$$

This notation enables distinguishing between a collection of unit vectors

$$\{\hat{\boldsymbol{a}}_1, \hat{\boldsymbol{a}}_2, \hat{\boldsymbol{a}}_3\} \equiv \{\underset{\sim}{\hat{a}}_1, \underset{\sim}{\hat{a}}_2, \underset{\sim}{\hat{a}}_3\},$$

and the cartesian components of a unit vector

$$\hat{\boldsymbol{a}} \equiv \underset{\sim}{\hat{a}} \equiv (\hat{a}_1, \hat{a}_2, \hat{a}_3).$$

That notational ability to distinguish between those two concepts, unit vector $\hat{\boldsymbol{a}}_i$ and component $\hat{a}_i$ of unit vector $\hat{\boldsymbol{a}}$, may be useful in some instances.

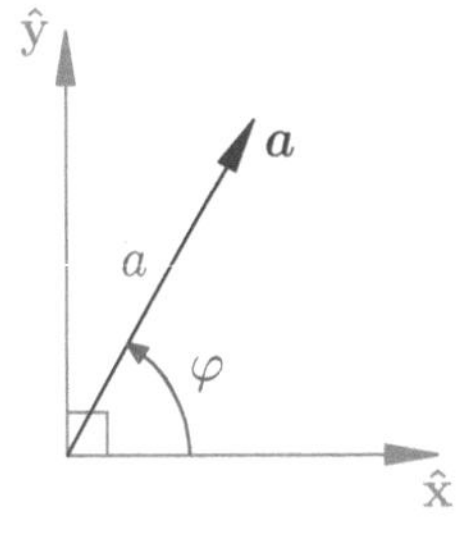

Fig. 1.4 Magnitude a, azimuth φ.

1.2 Vectors in a plane

A plane is oriented by choosing two perpendicular reference directions $\hat{\mathbf{x}}$ and $\hat{\mathbf{y}}$. A vector $\boldsymbol{a}$ in that plane can then be specified by its magnitude $a = |\boldsymbol{a}|$ with its direction $\hat{\boldsymbol{a}}$ specified by the azimuthal angle φ from $\hat{\mathbf{x}}$ to $\boldsymbol{a}$, positive toward $\hat{\mathbf{y}}$. This is the *polar representation* of $\boldsymbol{a}$ (Fig. 1.4), written

$$\boldsymbol{a} = a\,\hat{\boldsymbol{a}}(\varphi). \tag{1.7}$$

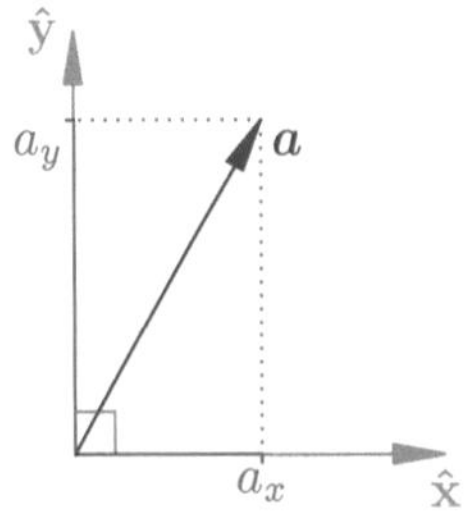

Fig. 1.5 Components a_x and a_y.

The vector $\boldsymbol{a}$ can also be represented as a sum of like vectors in the perpendicular reference directions, $\hat{\mathbf{x}}$ and $\hat{\mathbf{y}}$. That is the *cartesian representation* with (Fig. 1.5)

$$\boldsymbol{a} = a_x\,\hat{\mathbf{x}} + a_y\,\hat{\mathbf{y}}. \tag{1.8}$$

The cartesian reference directions $\hat{\mathbf{x}}$ and $\hat{\mathbf{y}}$ are fixed independently of the arbitrary vector $\boldsymbol{a}$; thus, the cartesian *components* a_x and a_y can be negative, contrary to magnitude $a = |\boldsymbol{a}| \geq 0$.

The relationships between the polar (a, φ) and cartesian (a_x, a_y) representations follow from trigonometry

$$\begin{cases} a_x = a\cos\varphi \\ a_y = a\sin\varphi \end{cases} \quad \Leftrightarrow \quad \begin{cases} a = \sqrt{a_x^2 + a_y^2}, \\ \varphi = \operatorname{atan2}(a_y, a_x), \end{cases} \tag{1.9}$$

where atan2 is the two-argument arctangent function with range in $(-\pi, \pi]$

$$-\pi < \operatorname{atan2}(a_y, a_x) \leq \pi.$$

Note the historical reversal of the arguments, $\varphi = \operatorname{atan2}(a_y, a_x)$, instead of $\varphi = \operatorname{angle}(a_x, a_y)$.

The one argument arctangent function $\arctan(a_y/a_x)$ has the range $[-\pi/2, \pi/2]$ and cannot distinguish (a_x, a_y) from $(-a_x, -a_y)$.

It is clear from (1.9) that

$$(a, \varphi) \neq (a_x, a_y),$$

in general, yet these two pairs of numbers (a, φ) and (a_x, a_y) represent the *same* 2D vector $\boldsymbol{a}$. To express equality of the representations, we need the direction vectors to write

$$\boldsymbol{a} = a\,\hat{\boldsymbol{a}}(\varphi) = a_x\,\hat{\mathbf{x}} + a_y\,\hat{\mathbf{y}}, \tag{1.10}$$

yielding (Fig. 1.6)

$$\hat{\boldsymbol{a}} = \frac{a_x}{a}\hat{\mathbf{x}} + \frac{a_y}{a}\hat{\mathbf{y}} = \cos\varphi\,\hat{\mathbf{x}} + \sin\varphi\,\hat{\mathbf{y}}. \tag{1.11}$$

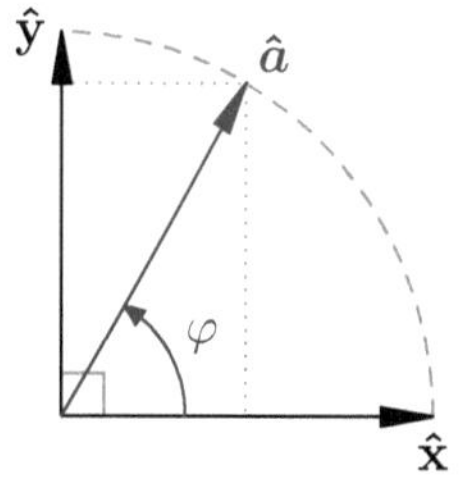

Fig. 1.6 $\hat{\boldsymbol{a}} = \cos\varphi\,\hat{\mathbf{x}} + \sin\varphi\,\hat{\mathbf{y}}$.

Thus, $\hat{\boldsymbol{a}}$ is a vector function of φ independent of magnitude a, while the cartesian direction vectors $\hat{\mathbf{x}}$ and $\hat{\mathbf{y}}$ are fixed independently of both a and φ, and independent of (a_x, a_y) as well.

The vector $\boldsymbol{a}$ is specified by the pair (a, φ) or (a_x, a_y), but the *same* physical 2D vector can be represented by *any* pair of numbers depending on the choice of reference magnitudes (physical units) and directions. This is the case even if we restrict to cartesian representations as shown in Fig. 1.7, where the same vector $\boldsymbol{a}$ has distinct cartesian components $(a_x, a_y) \neq (a'_x, a'_y)$, since the reference directions $\{\hat{\mathbf{x}}, \hat{\mathbf{y}}\}$ and $\{\hat{\mathbf{x}}', \hat{\mathbf{y}}'\}$ are distinct. To express equality, we again need the direction vectors to write

$$\boldsymbol{a} = a_x\,\hat{\mathbf{x}} + a_y\,\hat{\mathbf{y}} = a'_x\,\hat{\mathbf{x}}' + a'_y\,\hat{\mathbf{y}}'. \tag{1.12}$$

Components are useless without the directions.

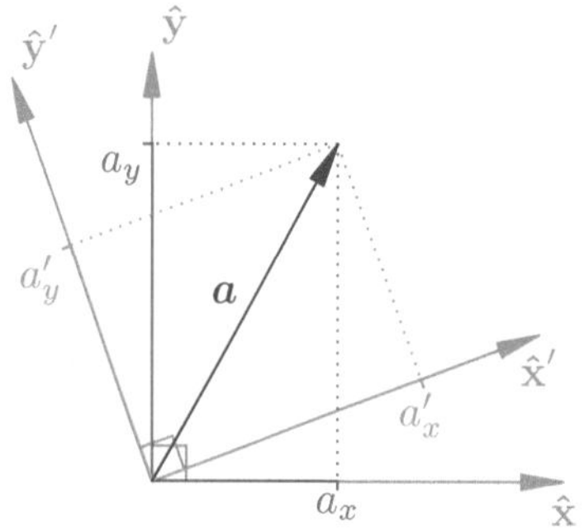

Fig. 1.7 Two cartesian decompositions of the same $\boldsymbol{a}$ but $(a_x, a_y) \neq (a'_x, a'_y)$.

Notation We use $\hat{\mathbf{x}}$, $\hat{\mathbf{y}}$, $\hat{\mathbf{z}}$ to denote the x, y, z directions, respectively, instead of the old **i**, **j**, **k** notation used in many elementary calculus and physics books. The **i**, **j**, **k** notation originates from *quaternions*, a generalization of complex numbers that preceded the concept of vector. That **i**, **j**, **k** notation clashes with the very useful index notation that we will soon study and develop. It also clashes with the complex plane where the imaginary direction **i** is the y direction, not x!

1.3 Vectors in 3D space

In our 3D world, a direction $\hat{\boldsymbol{a}}$ is specified using an *inclination* angle ζ from the upward direction, opposite to the local gravity force, and an *azimuth* angle α — the angle about the vertical (Fig. 1.8). An *elevation* angle measured up from the horizontal plane may be used instead of inclination.

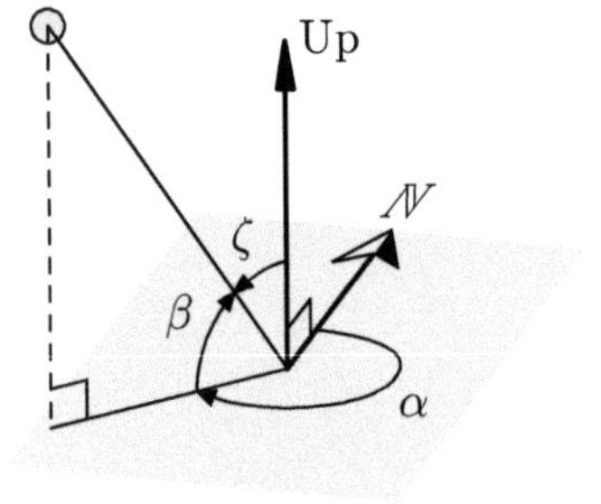

Fig. 1.8 The Sun at inclination ζ and azimuth α measured clockwise from north. Elevation $\beta = \pi/2 - \zeta$.

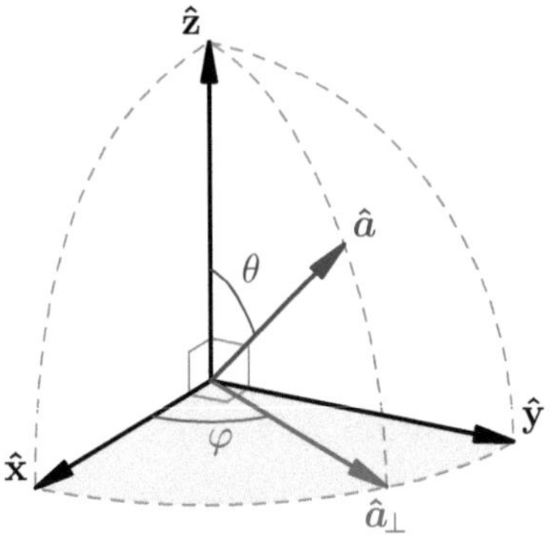

Fig. 1.9 Direction $\hat{\boldsymbol{a}}$: polar angle θ, azimuthal angle φ.

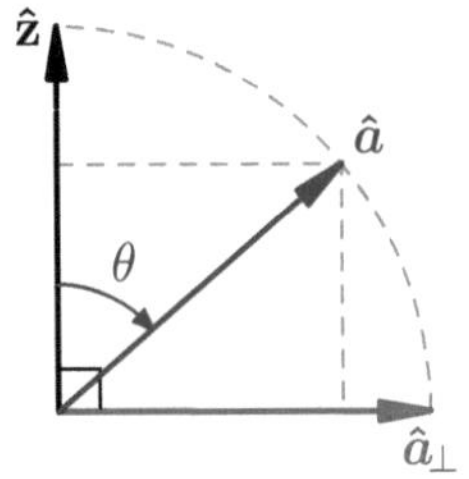

Fig. 1.10 Side view: $\hat{\mathbf{z}}, \hat{\boldsymbol{a}}, \hat{\boldsymbol{a}}_\perp$

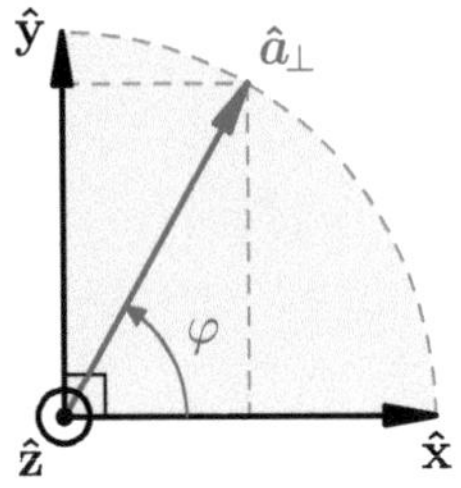

Fig. 1.11 Top view: $\hat{\mathbf{x}}, \hat{\boldsymbol{a}}_\perp, \hat{\mathbf{y}}$.

The physics and ISO[2] convention for spherical coordinates is to use the inclination or *polar* angle θ between a reference direction $\hat{\mathbf{z}}$ and the arbitrary direction $\hat{\boldsymbol{a}}$, and an *azimuth* angle φ *counterclockwise around* $\hat{\mathbf{z}}$, that is the angle between the vertical planes $(\hat{\mathbf{z}}, \hat{\mathbf{x}})$ and $(\hat{\mathbf{z}}, \hat{\boldsymbol{a}})$ (Fig. 1.9). The azimuth φ is undefined if the polar angle $\theta = 0$ or π.

The 2D side view in Fig. 1.10 and top view in Fig. 1.11, yield

$$\hat{\boldsymbol{a}} = \cos\theta\,\hat{\mathbf{z}} + \sin\theta\,\hat{\boldsymbol{a}}_\perp, \tag{1.13a}$$

$$\hat{\boldsymbol{a}}_\perp = \cos\varphi\,\hat{\mathbf{x}} + \sin\varphi\,\hat{\mathbf{y}}. \tag{1.13b}$$

Substituting (1.13b) for $\hat{\boldsymbol{a}}_\perp$ into (1.13a) gives

$$\hat{\boldsymbol{a}} = \sin\theta\,\cos\varphi\,\hat{\mathbf{x}} + \sin\theta\,\sin\varphi\,\hat{\mathbf{y}} + \cos\theta\,\hat{\mathbf{z}}, \tag{1.13c}$$

which specifies an arbitrary direction $\hat{\boldsymbol{a}}$ in 3D space in terms of the mutually orthogonal cartesian directions $\hat{\mathbf{x}}, \hat{\mathbf{y}}, \hat{\mathbf{z}}$ using two angles: a polar angle θ and an azimuth φ.

Magnitude and direction An arbitrary vector $\boldsymbol{a} = a\hat{\boldsymbol{a}}$ in 3D can then be specified by its magnitude a with its direction $\hat{\boldsymbol{a}}$ specified by the polar and azimuthal angles (θ, φ). That is the *spherical representation* $\boldsymbol{a} \equiv (a, \theta, \varphi)$. The *cylindrical* representation $(a_\perp, \varphi, a_z)$ specifies $\boldsymbol{a}$ by its horizontal magnitude $a_\perp = a\sin\theta$, azimuth φ and vertical component $a_z = a\cos\theta$. The *cartesian* representation consists of the familiar (a_x, a_y, a_z).

[2]International Standards Organization.

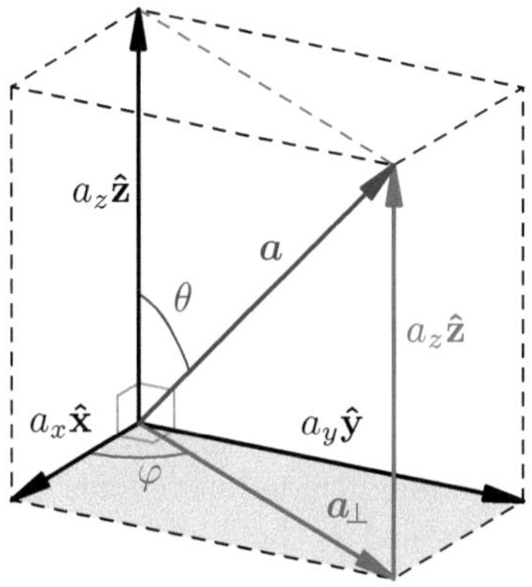

Fig. 1.12 Arbitrary vector $\boldsymbol{a}$ in 3D.

A 3D vector $\boldsymbol{a}$ can thus be represented by three real numbers, for instance (a, θ, φ) or $(a_\perp, \varphi, a_z)$ or (a_x, a_y, a_z). Each of these triplets represents the same 3D vector, yet they are not equal to each other, in general,

$$(a, \theta, \varphi) \neq (a_\perp, \varphi, a_z) \neq (a_x, a_y, a_z). \tag{1.14}$$

To express equality, we need the direction vectors to write (Fig. 1.12)

$$\boldsymbol{a} = a\, \hat{\boldsymbol{a}}(\theta, \varphi) = a_\perp \hat{\boldsymbol{a}}_\perp(\varphi) + a_z\hat{\mathbf{z}} = a_x\hat{\mathbf{x}} + a_y\hat{\mathbf{y}} + a_z\hat{\mathbf{z}}. \tag{1.15}$$

From that vector equation (1.15) and basic trigonometry (Fig. 1.13), we deduce

$$a^2 = a_\perp^2 + a_z^2, \quad a_\perp^2 = a_x^2 + a_y^2. \tag{1.16}$$

Eliminating $a_\perp^2$ yields

$$a = \sqrt{a_x^2 + a_y^2 + a_z^2} \leq |a_x| + |a_y| + |a_z|. \tag{1.17}$$

The latter inequality is a special case of the *triangle inequality* (1.21).

Eliminating the magnitudes in (1.15) yields relations between the direction vectors, already illustrated in Fig. 1.9

$$\begin{aligned} \hat{\boldsymbol{a}} &= \frac{a_\perp}{a}\, \hat{\boldsymbol{a}}_\perp + \frac{a_z}{a}\, \hat{\mathbf{z}} &&= \sin\theta\, \hat{\boldsymbol{a}}_\perp + \cos\theta\, \hat{\mathbf{z}}, \\ \hat{\boldsymbol{a}}_\perp &= \frac{a_x}{a_\perp}\, \hat{\mathbf{x}} + \frac{a_y}{a_\perp}\, \hat{\mathbf{y}} &&= \cos\varphi\, \hat{\mathbf{x}} + \sin\varphi\, \hat{\mathbf{y}}. \end{aligned} \tag{1.18}$$

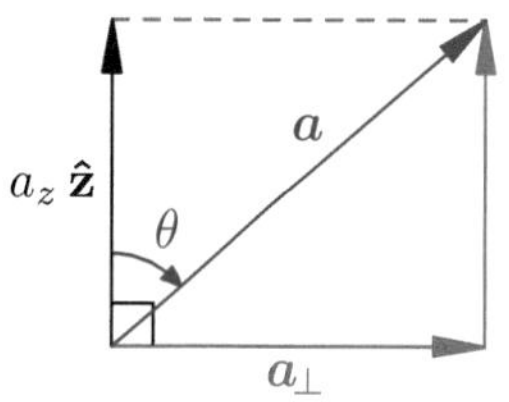

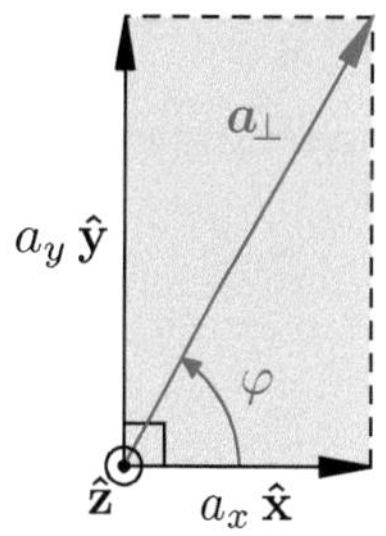

Fig. 1.13 *Top:* side view of vertical plane $(\hat{\mathbf{z}}, \boldsymbol{a}, \boldsymbol{a}_\perp)$. *Bottom:* top view of horizontal plane $(\hat{\mathbf{x}}, \hat{\mathbf{y}}, \boldsymbol{a}_\perp)$.

Unique and positive angles can be specified by restricting them to the ranges

$$0 \leq \varphi < 2\pi \quad \text{and} \quad 0 \leq \theta \leq \pi,$$

similar to the navigation convention where headings are specified as angles between 0° and 359° from north. In mathematical physics, the standard definition is

$$-\pi < \varphi \leq \pi \quad \text{and} \quad 0 \leq \theta \leq \pi,$$

then

$$\varphi = \operatorname{atan2}(a_y, a_x), \qquad \theta = \arccos(a_z/a), \tag{1.19}$$

where atan2 is the arctangent function with range in $(-\pi, \pi]$ and arccos is the arccosine function whose range is $[0, \pi]$.

1.4 Vector addition and scaling

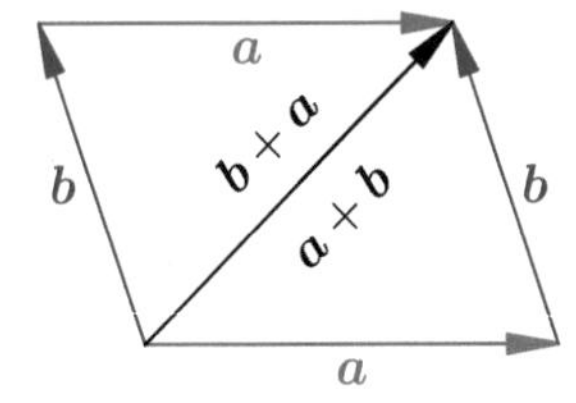

Fig. 1.14 Vector addition.

Geometric vectors such as displacements and forces can be *added* and *scaled* to compose new vectors, as we have done in (1.15), for instance. Geometric vectors add according to the *parallelogram rule* (Fig. 1.14), and addition is *commutative*,

$$\boldsymbol{a}+\boldsymbol{b}=\boldsymbol{b}+\boldsymbol{a}.$$

To any vector $\boldsymbol{b}$ we can associate an opposite vector denoted $-\boldsymbol{b}$ that is the vector with the same magnitude but reverse direction to $\boldsymbol{b}$. Vector subtraction $\boldsymbol{a}-\boldsymbol{b}$ is then defined as the addition of $\boldsymbol{a}$ and $-\boldsymbol{b}$ (Fig. 1.15),

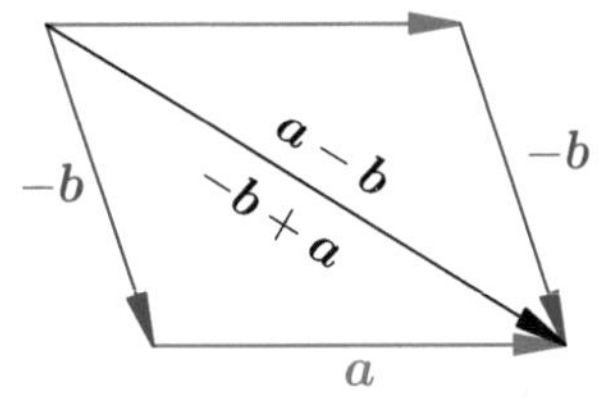

Fig. 1.15 Vector subtraction.

$$\boldsymbol{a}-\boldsymbol{b}=-\boldsymbol{b}+\boldsymbol{a}.$$

The vectors $\boldsymbol{a}\pm\boldsymbol{b}$ are the diagonals of the parallelogram with sides $\boldsymbol{a}$, $\boldsymbol{b}$.

An important difference between *points* and *displacements* is that there is no special point in 3D space, but there is one special displacement: the zero vector $\mathbf{0}$ such that

$$\boldsymbol{a}-\boldsymbol{a}=\mathbf{0}, \qquad \boldsymbol{a}+\mathbf{0}=\boldsymbol{a}.$$

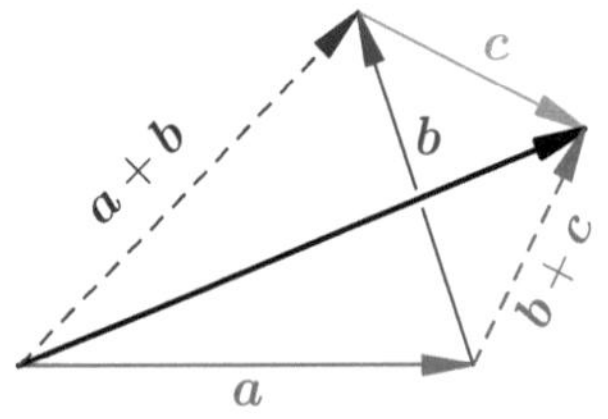

Fig. 1.16 Associativity.

Vector addition is also *associative* (Fig. 1.16)

$$(\boldsymbol{a}+\boldsymbol{b})+\boldsymbol{c}=\boldsymbol{a}+(\boldsymbol{b}+\boldsymbol{c}).$$

Vector addition is straightforward in cartesian form since it suffices to add the respective components (Fig. 1.17)

$$\boldsymbol{a}+\boldsymbol{b}=(a_x+b_x)\hat{\mathbf{x}}+(a_y+b_y)\hat{\mathbf{y}}+(a_z+b_z)\hat{\mathbf{z}}. \tag{1.20}$$

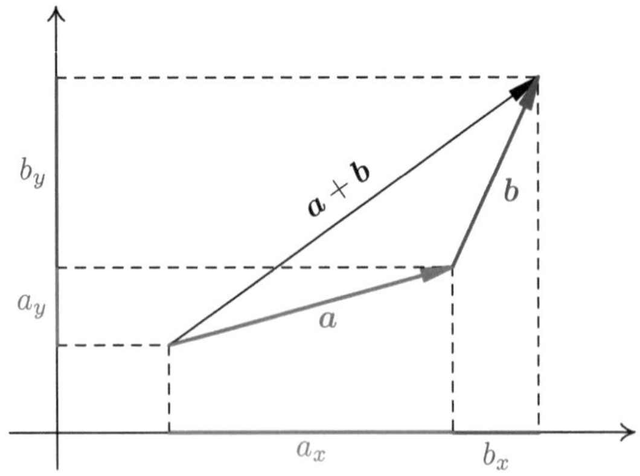

Fig. 1.17 Vector addition in cartesian form, and triangle inequality $|\boldsymbol{a}+\boldsymbol{b}|\leq|\boldsymbol{a}|+|\boldsymbol{b}|$.

That simple addition rule does *not* apply to the magnitude direction representation $\boldsymbol{a}=|\boldsymbol{a}|\hat{\boldsymbol{a}}$ and $\boldsymbol{b}=|\boldsymbol{b}|\hat{\boldsymbol{b}}$. Indeed, the magnitude of a sum is *not* the sum of the magnitudes in general. The magnitude of a sum is *less than or equal* to the sum of the magnitudes

$$|\boldsymbol{a}+\boldsymbol{b}| \leq |\boldsymbol{a}|+|\boldsymbol{b}|, \tag{1.21}$$

which is the *triangle inequality.* Likewise, the direction of a sum is *not* the sum of the directions. The direction of $\boldsymbol{v}=\boldsymbol{a}+\boldsymbol{b}$ is

$$\hat{\boldsymbol{v}}=\frac{\boldsymbol{a}+\boldsymbol{b}}{|\boldsymbol{a}+\boldsymbol{b}|} \neq \hat{\boldsymbol{a}}+\hat{\boldsymbol{b}}. \tag{1.22}$$

Vectors can also be *scaled,* that is, multiplied by a real number $\alpha \in \mathbb{R}$, called a *scalar.* Geometrically, $\alpha\boldsymbol{a}$ is a new vector parallel to $\boldsymbol{a}$ but of length $|\alpha||\boldsymbol{a}|$. The direction of $\alpha\boldsymbol{a}$ is the same as that of $\boldsymbol{a}$ if $\alpha>0$ and opposite to $\boldsymbol{a}$ if $\alpha<0$. Thus, $(-1)\boldsymbol{a}=(-\boldsymbol{a})$; multiplying $\boldsymbol{a}$ by (-1) yields the previously defined additive opposite of $\boldsymbol{a}$ (Fig. 1.18). Other geometrically immediate properties are *distributivity* with respect to addition and multiplication of scalars, and with respect to vector addition,

$$\begin{aligned}(\alpha+\beta)\boldsymbol{a}&=\alpha\boldsymbol{a}+\beta\boldsymbol{a},\\(\alpha\beta)\boldsymbol{a}&=\alpha(\beta\boldsymbol{a}),\\\alpha(\boldsymbol{a}+\boldsymbol{b})&=\alpha\boldsymbol{a}+\alpha\boldsymbol{b}.\end{aligned}$$

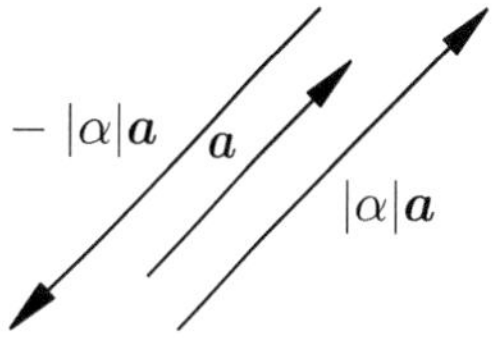

Fig. 1.18 Scaling $\boldsymbol{a}$ by $\alpha=\pm|\alpha|$.

Exercises

(1.1) What is the magnitude of a direction? the direction of a magnitude?

(1.2) What are $|\boldsymbol{v}|$ and $\hat{\boldsymbol{v}}$ for the vector $\boldsymbol{v} \equiv -5$ miles per hour heading northeast?

(1.3) What is φ for the vector whose cartesian components are $(-1,\sqrt{3})$?

(1.4) Sketch $\tan\varphi$ for $-\pi<\varphi\leq\pi$, and $\arctan x$ for $-\infty<x<\infty$.

(1.5) Let $\varphi=\operatorname{atan2}(y,x)$, defined in (1.9). Derive

$$\frac{\partial\varphi}{\partial x}=\frac{-y}{x^2+y^2}, \quad \frac{\partial\varphi}{\partial y}=\frac{x}{x^2+y^2}. \tag{1.23}$$

(Hint: work with $\sin\varphi$, $\cos\varphi$, and the chain rule.) Note that $\operatorname{atan2}(y,x)$ jumps by 2π across $x<0$, $y=0$, yet, (1.23) exist there. Discuss.

(1.6) Write `cart2polar` and `polar2cart` codes (in your favorite computer language or in pseudo-code) that transform cartesian to polar representations and polar to cartesian, respectively, for 2D vectors.

(1.7) If you move distance a heading α then distance b heading β, how far are you and in what heading from your original position? Draw a sketch and explain your *algorithm.* Headings are clockwise from north.

(1.8) An airplane travels at airspeed V, heading α (clockwise from north). The wind has speed W heading β. Make a clean sketch. Show/explain the algorithm to calculate the airplane's ground speed and heading.

(1.9) True or false: in SI units $|\hat{\boldsymbol{a}}|=1$ meter while in CGS units $|\hat{\boldsymbol{a}}|=1$ cm.

(1.10) True or false: $\hat{\mathbf{x}}+\hat{\mathbf{y}}$ is a unit vector in the northeast direction.

(1.11) *The direction of a sum is* not *the sum of the directions* (1.22), but is there anything special about the sum of arbitrary directions $\hat{\boldsymbol{a}}+\hat{\boldsymbol{b}}$? Draw a few sets of sample vectors and formulate a conjecture, then prove it using plane geometry.

(1.12) What is an *azimuth*? what are *elevation* and *inclination* in the context of this chapter?

(1.13) What are the cylindrical and spherical representations for the vectors whose cartesian components are (0,0,1) and (0,-1,0)?

(1.14) Find the azimuth φ and inclination θ (Fig. 1.9) for the vector $\boldsymbol{a}$ whose cartesian components are (1,1,1). What is the angle between $\hat{\mathbf{x}}$ and $\boldsymbol{a}$?

(1.15) Find the azimuthal and polar angles φ and θ (Fig. 1.9) for the vector $\boldsymbol{a}$ given in cartesian form as $(-1,-1,-1)$.

(1.16) A vector $\boldsymbol{v}$ is specified in cartesian components as $(-3,2,1)$. Find $\hat{\boldsymbol{v}}_\perp$ for that vector and express $\hat{\boldsymbol{v}}_\perp$ in terms of the cartesian direction vectors. Write $\boldsymbol{v}$ in cylindrical representation.

(1.17) A vector is specified in cartesian components as $(3,2,1)$. Find its magnitude and direction. Express its direction in cartesian, cylindrical and spherical representations using the *cartesian* direction vectors.

(1.18) $A \equiv (1,2,3)$ and $B \equiv (3,2,1)$ in cartesian coordinates. Find θ and φ for $\overrightarrow{AB}$.

(1.19) True or false?: $|\hat{\boldsymbol{a}}_\perp| = \cos\theta$

(1.20) True or false?: $|\boldsymbol{a}_\perp| = a\cos\theta$

(1.21) Write `cart2spher` and `spher2cart` codes (in Matlab or Python, for example, or in pseudo-code) that transform cartesian to spherical representations and spherical to cartesian, respectively, for 3D vectors.

(1.22) In (1.15), why is there just one magnitude and one direction in the first equality but three magnitudes and three directions in the third equality?

(1.23) Pick generic $\boldsymbol{a}$ and $\boldsymbol{b}$. Sketch $\boldsymbol{a}$, $\boldsymbol{b}$, $\boldsymbol{a}-\boldsymbol{b}/2$, $\boldsymbol{a}+\boldsymbol{b}/3$, $\boldsymbol{a}+\boldsymbol{b}$, $\boldsymbol{a}+3\boldsymbol{b}/2$ all together.

(1.24) The edges of a regular hexagon are a chain of vectors oriented clockwise around the hexagon. If $\boldsymbol{a}$ and $\boldsymbol{b}$ are two consecutive edge vectors, express the other four edges in terms of $\boldsymbol{a}$ and $\boldsymbol{b}$.[3]

(1.25) If $\overrightarrow{AB} = \overrightarrow{CD}$, prove that $\overrightarrow{AC} = \overrightarrow{BD}$.

(1.26) Take a blank sheet of paper and make a 3D sketch of $\hat{\mathbf{x}}, \hat{\mathbf{y}}, \hat{\mathbf{z}}$ and an arbitrary $\boldsymbol{a}$. Next, sketch the appropriate 2D projections, define angles, and derive all the formulas relating the three fundamental ways to specify $\boldsymbol{a}$ in 3D space; in particular, derive explicit expressions for $\hat{\boldsymbol{a}}$ and $\hat{\boldsymbol{a}}_\perp$ in terms of $\hat{\mathbf{x}}, \hat{\mathbf{y}}, \hat{\mathbf{z}}$. Derive all those formulas from your understanding of basic geometry and trigonometry, from scratch, all by yourself.

(1.27) What is the sum of the inner angles of a triangle with vertices A, B, C? Prove it. Next, consider a triangle ABC on a sphere. Find a lower bound and an upper bound on the sum of the inner angles of that triangle.

(1.28) Let θ be the angle between vectors $\hat{\boldsymbol{a}}$ and $\boldsymbol{v}$ (Fig. 1.19), and $\boldsymbol{v}'$ the vector obtained by rotating $\boldsymbol{v}$ about $\hat{\boldsymbol{a}}$ by angle α. Identify and use two isosceles triangles with common edge $\boldsymbol{v}'-\boldsymbol{v}$ to show that the angle β between $\boldsymbol{v}$ and $\boldsymbol{v}'$ is given by

$$\sin(\beta/2) = \sin(\alpha/2)\sin\theta. \tag{1.24}$$

Explain/justify your derivation. Illustrate.

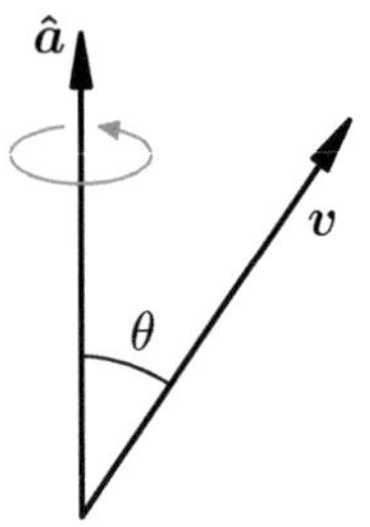

Fig. 1.19 Rotation of $\boldsymbol{v}$ about $\boldsymbol{a}$.

(1.29) Three points A, B, C lie on a circle, with AB a diameter. The inner angles of the triangle ABC are α, β, and γ, respectively (Fig. 1.20). Use isosceles triangles to prove that γ is a right angle, then use similar triangles to prove the double-angle identity

$$2\cos^2\alpha = 1 + \cos 2\alpha.$$

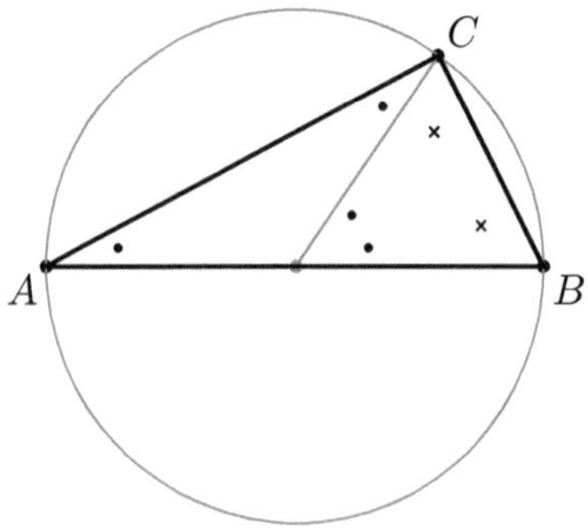

Fig. 1.20 Double angle.

[3]Hartley Rogers Jr., *Multivariable Calculus with Vectors*, Prentice-Hall, 1999.

(1.30) *Inscribed angle theorem.* Use plane geometry and isosceles triangles to prove that the *inscribed* angle is always half of the corresponding *central* angle (Fig. 1.21). Conclude that the angle ACB is invariant as C moves along a circular arc between A and B (Fig. 1.22). This geometric result is key to *bipolar coordinates.*

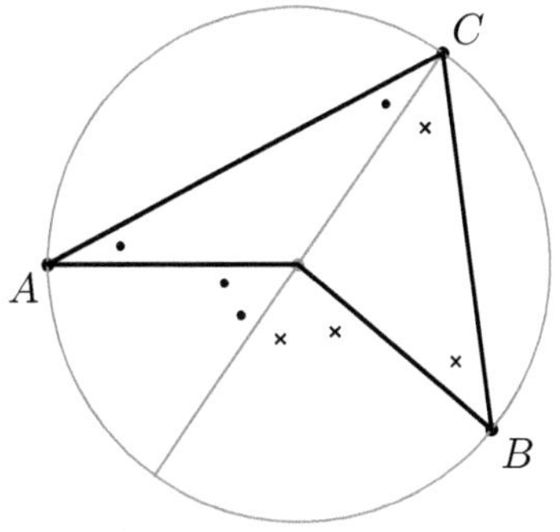

Fig. 1.21 Hints for problems 1.29, 1.30.

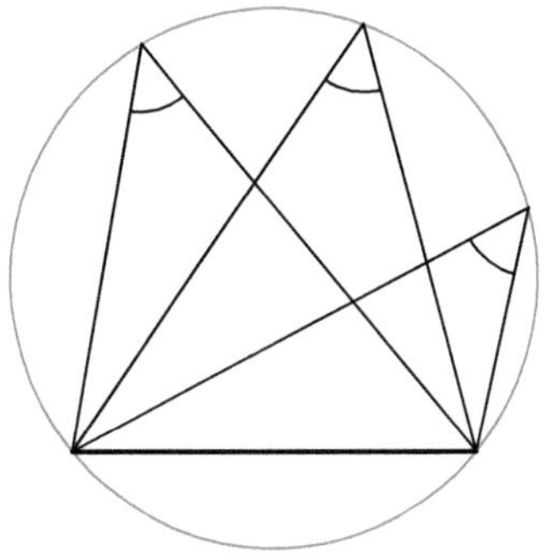

Fig. 1.22 Inscribed angle theorem.

(1.31) Specifying a point P on a spherical Earth is equivalent to specifying the direction of its *radius vector* $\boldsymbol{r} = \overrightarrow{OP}$, from the sphere center O. In that context, the reference direction $\hat{\mathbf{z}}$ is the *polar axis*, the axis of rotation of the Earth; the azimuthal angle around the polar axis is the *longitude*; and the elevation angle, measured from the equatorial plane toward the north pole, is the *latitude.* Meridians are half-circles from the north pole to the south pole at fixed longitude. Parallels are circles of fixed latitude. Longitude is measured eastward or westward from the *prime meridian* through Greenwich, UK. Figure 1.23 is a 2D projection of the Earth in a plane perpendicular to the radial direction at latitude 40° north, longitude 90° west showing meridians and parallels at 10° intervals.[4]
Let λ be the latitude and φ the longitude, consistent with our physics convention for φ (these angles are reversed in geography).
Make a 3D plot similar to Fig. 1.9 as well as 2D meridional and equatorial views similar to Figs. 1.10 and 1.11, indicating the polar axis, the prime meridian, and the longitude and latitude angles. Derive expressions for the local radial, eastward, and northward directions $\hat{\boldsymbol{r}}(\lambda,\varphi)$, $\hat{\boldsymbol{e}}(\lambda,\varphi)$, $\hat{\boldsymbol{n}}(\lambda,\varphi)$, respectively, at arbitrary latitude λ and longitude φ, in terms of the global earth cartesian frame $\hat{\mathbf{x}}, \hat{\mathbf{y}}, \hat{\mathbf{z}}$. Justify your derivation using 2D views.

Fig. 1.23 Orthographic projection of the earth showing meridians and parallels at 10° intervals.

[4]J. P. Snyder, and M. P. Voxland, *Album of Map Projections*, US Geological Survey, Professional Paper 1453, 1989. `http://pubs.usgs.gov/pp/1453/report.pdf`

Vector Spaces

2

Our main focus will be on geometric vectors in 2D and 3D euclidean space. Here we briefly review some of the language and concepts of linear algebra, in particular *linear independence* of vectors, *basis*, and *dimension* of a vector space or *subspace.*

2.1 Abstract vector spaces

The concepts of *vector* and *space* have evolved beyond geometric vectors and physical space. In its most general form, a *vector space* is a set of mathematical objects $\boldsymbol{a}, \boldsymbol{b}, \ldots$, such as *displacements* or *ordered lists of numbers,* for which addition $\boldsymbol{a} + \boldsymbol{b}$ and scalar multiplication $\alpha\boldsymbol{a}$ are defined and satisfy the eight properties, or *axioms*, listed below. *Closure* under addition and scaling is required; that is, addition $\boldsymbol{a}+\boldsymbol{b}$ and scaling $\alpha\boldsymbol{a}$ must yield objects in the same set as that of the objects $\boldsymbol{a}, \boldsymbol{b}$ that are added or scaled. Thus a sum of displacements is a displacement, and a multiple of a displacement is a displacement, for example.

Vector addition axioms

$$\boldsymbol{a} + \boldsymbol{b} = \boldsymbol{b} + \boldsymbol{a}, \tag{2.1a}$$
$$\boldsymbol{a} + (\boldsymbol{b} + \boldsymbol{c}) = (\boldsymbol{a} + \boldsymbol{b}) + \boldsymbol{c}, \tag{2.1b}$$
$$\boldsymbol{a} + \boldsymbol{0} = \boldsymbol{a}, \tag{2.1c}$$
$$\boldsymbol{a} + (-\boldsymbol{a}) = \boldsymbol{0}. \tag{2.1d}$$

Scalar multiplication axioms

$$(\alpha + \beta)\boldsymbol{a} = \alpha\boldsymbol{a} + \beta\boldsymbol{a}, \tag{2.1e}$$
$$(\alpha\beta)\boldsymbol{a} = \alpha(\beta\boldsymbol{a}), \tag{2.1f}$$
$$\alpha(\boldsymbol{a} + \boldsymbol{b}) = \alpha\boldsymbol{a} + \alpha\boldsymbol{b}, \tag{2.1g}$$
$$1\,\boldsymbol{a} = \boldsymbol{a}. \tag{2.1h}$$

A *scalar* in vector algebra is a scaling factor. A *scalar* in physics is a quantity with no direction such as mass, pressure, or temperature. Forces $\boldsymbol{F}$ and accelerations $\boldsymbol{a}$ are not in the same vector space despite $\boldsymbol{F} = m\boldsymbol{a}$, since $\boldsymbol{F}$ and $\boldsymbol{a}$ have different physical units.

2.2 The vector space $\mathbb{R}^n$

Consider the set of ordered n-tuplets of real numbers $\mathsf{x} = (x_1, x_2, \ldots, x_n)$. These could correspond to student grades on a particular exam, for instance. We will want to *add* different exam grades for *each student* and

$\triangleq$ means *equal by definition.*

$\in$ means *in* or *belonging to.*

rescale the grades. So the natural operations on these n-tuplets are *addition* defined by adding the respective components:

$$\mathbf{x} + \mathbf{y} \triangleq (x_1 + y_1, x_2 + y_2, \ldots, x_n + y_n) = \mathbf{y} + \mathbf{x}, \tag{2.2a}$$

and multiplication by a real number $\alpha \in \mathbb{R}$ defined as

$$\alpha\mathbf{x} \triangleq (\alpha x_1, \alpha x_2, \ldots, \alpha x_n). \tag{2.2b}$$

The set of n-tuplets of real numbers equipped with addition and multiplication by a real number, as just defined, is a fundamental vector space called $\mathbb{R}^n$.

The vector spaces $\mathbb{R}^2$ and $\mathbb{R}^3$ are fundamental since they correspond to the spaces of components of physical vectors, as discussed in Section 2.3. We also use $\mathbb{R}^n$ for very large n when studying systems of equations, for instance. The vector space $\mathbb{C}^n$, which consists of ordered lists of n complex numbers, is also useful in applications.

2.3 Bases and components

Addition and scaling of vectors allow us to define the concepts of *linear combination, linear (in)dependence, dimension, basis,* and *components.*

A *linear combination* of k vectors $\{\boldsymbol{a}_1, \boldsymbol{a}_2, \ldots, \boldsymbol{a}_k\}$ is an expression of the form

$$\chi_1\boldsymbol{a}_1 + \chi_2\boldsymbol{a}_2 + \cdots + \chi_k\boldsymbol{a}_k,$$

where χ_1, χ_2, ..., χ_k are arbitrary real numbers. A collection of k vectors $\{\boldsymbol{a}_1, \boldsymbol{a}_2, \ldots, \boldsymbol{a}_k\}$ are *linearly independent* if

$$\chi_1\boldsymbol{a}_1 + \chi_2\boldsymbol{a}_2 + \cdots + \chi_k\boldsymbol{a}_k = 0 \quad \Leftrightarrow \quad \chi_1 = \chi_2 = \cdots = \chi_k = 0.$$

Otherwise, the vectors are linearly dependent. For instance if $3\boldsymbol{a}_1 + 2\boldsymbol{a}_2 + \boldsymbol{a}_3 = 0$, then $\boldsymbol{a}_1, \boldsymbol{a}_2, \boldsymbol{a}_3$ are linearly *dependent.* The *dimension* of a vector space is the largest number of linearly independent vectors, n say, in that space.

A *basis* for an n-dimensional vector space is *any* collection of n linearly independent vectors. If $\{\boldsymbol{a}_1, \boldsymbol{a}_2, \ldots, \boldsymbol{a}_n\}$ is a basis for an n-dimensional vector space, then any vector $\boldsymbol{v}$ in that space can be *expanded* (or *decomposed*) as

$$\boldsymbol{v} = \eta_1\boldsymbol{a}_1 + \eta_2\boldsymbol{a}_3 + \cdots + \eta_n\boldsymbol{a}_n.$$

The n real numbers $(\eta_1, \eta_2, \ldots, \eta_n)$ are the *(scalar) components* of $\boldsymbol{v}$ in the basis $\{\boldsymbol{a}_1, \boldsymbol{a}_2, \ldots, \boldsymbol{a}_n\}$. The expansion of $\boldsymbol{v}$ is unique for that basis as the reader is asked to show in the exercises.

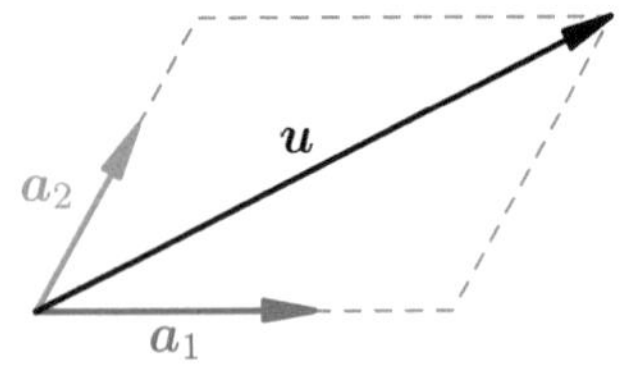

Fig. 2.1 Decomposing $\boldsymbol{u}$ into $\boldsymbol{u} = \zeta_1\,\boldsymbol{a}_1 + \zeta_2\,\boldsymbol{a}_2$ by projections *parallel* to $\boldsymbol{a}_1$ and $\boldsymbol{a}_2$ in 2D.

Basis in a plane Any two *non-parallel* vectors $\boldsymbol{a}_1$ and $\boldsymbol{a}_2$ in a plane, a horizontal plane or a vertical plane or any oblique plane, are linearly independent and form a basis for vectors in that plane. Any given vector $\boldsymbol{u}$ in the plane can be expanded as

$$\boldsymbol{u} = \zeta_1\boldsymbol{a}_1 + \zeta_2\boldsymbol{a}_2,$$

for a unique pair of real numbers (ζ_1, ζ_2). Geometrically, the components are obtained by *projections parallel* to each of the basis vectors (Fig. 2.1). Three or more vectors in a plane are necessarily *linearly dependent.* Vectors in a plane form a two-dimensional vector space.

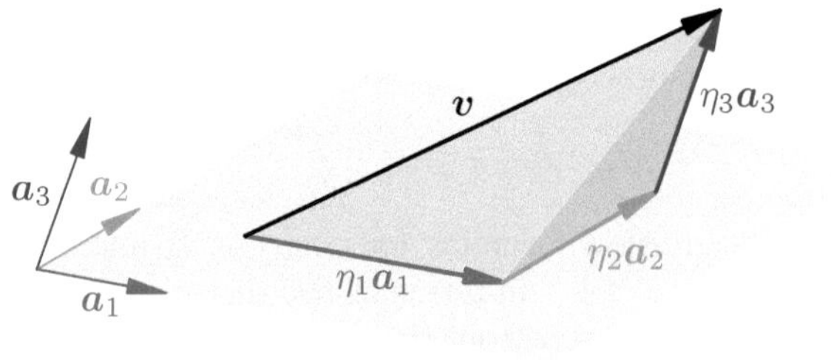

Fig. 2.2 Expanding vector $\boldsymbol{v}$ in terms of an arbitrary basis $\{\boldsymbol{a}_1, \boldsymbol{a}_2, \boldsymbol{a}_3\}$ in 3D.

Basis in 3D space Any three *non-coplanar* vectors $\boldsymbol{a}_1$, $\boldsymbol{a}_2$ and $\boldsymbol{a}_3$ in 3D space are linearly independent and those vectors form a basis for 3D space. Any given vector $\boldsymbol{v}$ can be expanded as

$$\boldsymbol{v} = \eta_1\boldsymbol{a}_1 + \eta_2\boldsymbol{a}_2 + \eta_3\boldsymbol{a}_3,$$

for a unique triplet of real numbers (η_1, η_2, η_3). Geometrically, the components are obtained by *projections parallel* to each of the basis vectors as illustrated in Fig. 2.2. Thus, any four or more vectors in 3D are necessarily *linearly dependent.* If the vectors are known in terms of cartesian components, we can calculate the components using *dot* and *cross* products without the need for geometrical constructions, as discussed in later sections.

The eight properties of addition and scalar multiplication imply that if two vectors $\boldsymbol{u}$ and $\boldsymbol{v}$ are expanded with respect to the *same* basis $\{\boldsymbol{a}_1, \boldsymbol{a}_2, \boldsymbol{a}_3\}$, that is if

$$\begin{aligned}\boldsymbol{u} &= \zeta_1\boldsymbol{a}_1 + \zeta_2\boldsymbol{a}_2 + \zeta_3\boldsymbol{a}_3,\\ \boldsymbol{v} &= \eta_1\boldsymbol{a}_1 + \eta_2\boldsymbol{a}_2 + \eta_3\boldsymbol{a}_3,\end{aligned}$$

then

$$\begin{aligned}\boldsymbol{u}+\boldsymbol{v} &= (\zeta_1+\eta_1)\boldsymbol{a}_1 + (\zeta_2+\eta_2)\boldsymbol{a}_2 + (\zeta_3+\eta_3)\boldsymbol{a}_3,\\ \alpha\boldsymbol{v} &= (\alpha\eta_1)\boldsymbol{a}_1 + (\alpha\eta_2)\boldsymbol{a}_2 + (\alpha\eta_3)\boldsymbol{a}_3.\end{aligned} \tag{2.3}$$

Thus, addition and scalar multiplication are performed component by component and the triplets of real components $(\zeta_1, \zeta_2, \zeta_3)$ and (η_1, η_2, η_3) are elements of the vector space $\mathbb{R}^3$. A basis $\{\boldsymbol{a}_1, \boldsymbol{a}_2, \boldsymbol{a}_3\}$ in 3D space thus provides a one-to-one map between *displacements* $\boldsymbol{v}$ in 3D euclidean space $\mathbb{E}^3$ and *triplets of real numbers* in $\mathbb{R}^3$

$$\boldsymbol{v} \in \mathbb{E}^3 \xleftrightarrow{\{\boldsymbol{a}_1,\boldsymbol{a}_2,\boldsymbol{a}_3\}} (\eta_1, \eta_2, \eta_3) \in \mathbb{R}^3.$$

The basis vectors $\{\boldsymbol{a}_1, \boldsymbol{a}_2, \boldsymbol{a}_3\}$ do not need to be orthogonal to each other; they only need to be *linearly independent.*

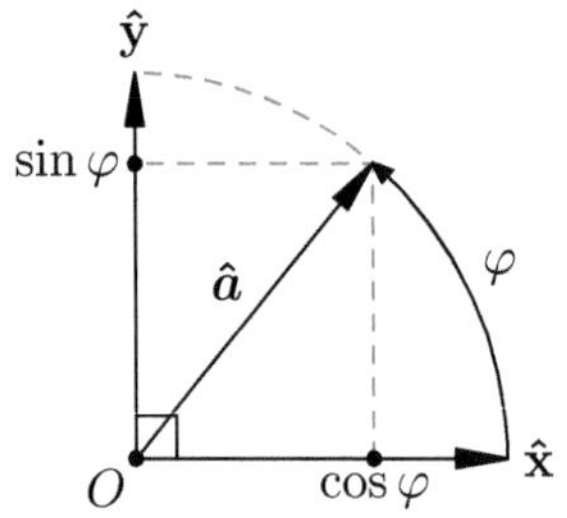

Fig. 2.3 The unit circle.

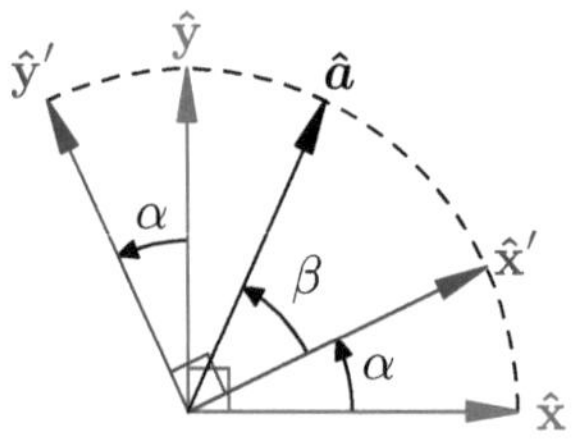

Fig. 2.4 Addition formulas.

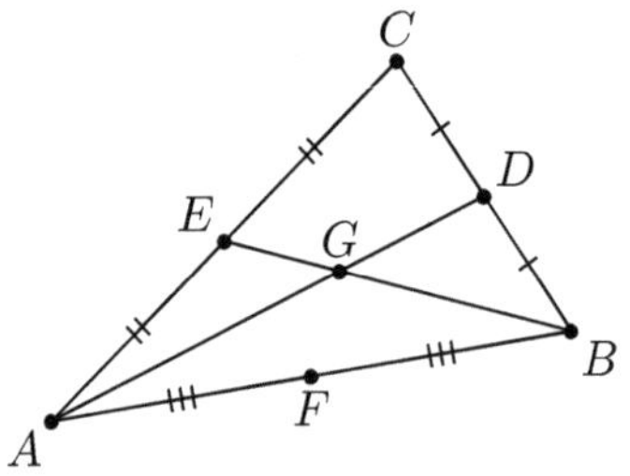

Fig. 2.5 Two medians intersect at G.

2.4 Geometric examples

Angle addition formulas A quick derivation of the angle addition formulas of trigonometry readily follows from a judicious use of *two* distinct cartesian bases. Trigonometry is about angles between radials on the *unit circle*—angles between unit vectors. If $\hat{\boldsymbol{a}}$ is any unit vector in the plane of mutually orthogonal unit vectors $\hat{\mathbf{x}}$ and $\hat{\mathbf{y}}$, then (Fig. 2.3)

$$\hat{\boldsymbol{a}} = \cos\varphi\,\hat{\mathbf{x}} + \sin\varphi\,\hat{\mathbf{y}}.$$

To derive the angle addition formulas for $\cos(\alpha+\beta)$ and $\sin(\alpha+\beta)$, write the vector $\hat{\boldsymbol{a}}$ for $\varphi = \alpha+\beta$ in terms of *two* distinct cartesian bases $\{\hat{\mathbf{x}}, \hat{\mathbf{y}}\}$ and $\{\hat{\mathbf{x}}', \hat{\mathbf{y}}'\}$ rotated by α from the former (Fig. 2.4); thus,

$$\begin{aligned}\hat{\mathbf{x}}' &= \quad\cos\alpha\,\hat{\mathbf{x}} + \sin\alpha\,\hat{\mathbf{y}}\\ \hat{\mathbf{y}}' &= -\sin\alpha\,\hat{\mathbf{x}} + \cos\alpha\,\hat{\mathbf{y}},\end{aligned}$$

and

$$\hat{\boldsymbol{a}} = \cos(\alpha+\beta)\,\hat{\mathbf{x}} + \sin(\alpha+\beta)\,\hat{\mathbf{y}} = \cos\beta\,\hat{\mathbf{x}}' + \sin\beta\,\hat{\mathbf{y}}'.$$

Substituting for $\hat{\mathbf{x}}'$ and $\hat{\mathbf{y}}'$ in terms of $\hat{\mathbf{x}}, \hat{\mathbf{y}}$ on the right hand side, *linear independence* of $\{\hat{\mathbf{x}}, \hat{\mathbf{y}}\}$ implies that the $\hat{\mathbf{x}}$ terms match on both sides of the $\hat{\boldsymbol{a}}$ equation, and likewise for the $\hat{\mathbf{y}}$ terms. This yields the addition angle formula

$$\begin{aligned}\cos(\alpha+\beta) &= \cos\alpha\cos\beta - \sin\alpha\sin\beta,\\ \sin(\alpha+\beta) &= \sin\alpha\cos\beta + \sin\beta\cos\alpha.\end{aligned} \tag{2.4}$$

Medians of a triangle As another example of the judicious choice of basis, we solve a classic geometry problem: *prove that the medians of any triangle are concurrent.*

A *median* of a triangle is a line that passes through a vertex and the midpoint of the opposite side. Let D be the midpoint of segment BC, and E that of segment AC (Fig. 2.5). It is clear that the two medians AD and BE intersect at a point, call it G. It is not obvious, however, that G lies on the third median through C and F, the midpoint of AB. That seems true on a well-drawn sketch, but in general three distinct lines do not intersect at the same point. We have to show that G is on the median CF.

Since G is on AD, there is a real number α such that

$$\overrightarrow{AG} = \alpha\,\overrightarrow{AD}. \tag{2.5}$$

Likewise, since G is on BE, there is a real number β such that

$$\overrightarrow{AG} = \overrightarrow{AB} + \beta\,\overrightarrow{BE}. \tag{2.6}$$

Intrinsic basis To avoid getting lost in the alphabet, we select a single *intrinsic system of coordinates*. That is, we select a reference point and a vector basis that *belong* to the problem at hand. Since the problem is inherently two-dimensional even if A, B, C are points in 3D space (or even in $\mathbb{R}^n$), we can specify any point in the plane A, B, C in terms of a reference point in that plane, A say, and *two* basis vectors $\{\boldsymbol{a}_1, \boldsymbol{a}_2\}$. One fine choice is the basis

$$\boldsymbol{a}_1 = \overrightarrow{AB}, \qquad \boldsymbol{a}_2 = \overrightarrow{BC}. \tag{2.7}$$

Point G in (2.5) is readily expressed in terms of $\{A, \boldsymbol{a}_1, \boldsymbol{a}_2\}$ as

$$\overrightarrow{AG} = \alpha \left(\overrightarrow{AB} + \overrightarrow{BC}/2\right) = \alpha\, \boldsymbol{a}_1 + (\alpha/2)\, \boldsymbol{a}_2 \tag{2.8}$$

$\{C, \overrightarrow{CA}, \overrightarrow{CB}\}$ is a slightly nicer choice since we are using the midpoints of those sides. Try that also!

Equation (2.6) can also be rewritten in terms of $\{A, \boldsymbol{a}_1, \boldsymbol{a}_2\}$ since

$$\overrightarrow{BE} = \overrightarrow{BA} + \tfrac{1}{2}\overrightarrow{AC} = \overrightarrow{BA} + \tfrac{1}{2}\left(\overrightarrow{AB} + \overrightarrow{BC}\right) = -\tfrac{1}{2}\,\boldsymbol{a}_1 + \tfrac{1}{2}\boldsymbol{a}_2,$$

so (2.6) becomes

$$\overrightarrow{AG} = (1 - \beta/2)\, \boldsymbol{a}_1 + (\beta/2)\, \boldsymbol{a}_2. \tag{2.9}$$

Equating (2.8) and (2.9), gives

$$\overrightarrow{AG} = \alpha\, \boldsymbol{a}_1 + (\alpha/2)\boldsymbol{a}_2 = (1 - \beta/2)\boldsymbol{a}_1 + (\beta/2)\boldsymbol{a}_2, \tag{2.10}$$

for some α and β. This is a vector equation for α and β. Since $\boldsymbol{a}_1$ and $\boldsymbol{a}_2$ are *linearly independent*, the coefficients of $\boldsymbol{a}_1$ must match on both sides of the equation, requiring $\alpha = 1 - \beta/2$. The coefficients of $\boldsymbol{a}_2$ must match also, giving $\alpha = \beta$. This yields the two equations

$$\alpha = 1 - \beta/2 = \beta$$

whose solution is

$$\alpha = \beta = 2/3.$$

Thus, $\overrightarrow{AG} = \frac{2}{3}\overrightarrow{AD}$ and $\overrightarrow{BG} = \frac{2}{3}\overrightarrow{BE}$; that is, G is *2/3 down the medians from the vertices*, and *a single median is enough to locate G*.

If we follow the same reasoning starting from the same median AD but with median CF instead of BE, we will find the same intersection point G, with $\overrightarrow{AG} = \frac{2}{3}\overrightarrow{AD}$, and the three medians have to intersect at G. Another way to verify that G is on the third median is to show that $\overrightarrow{CG}$ is parallel to $\overrightarrow{CF}$. That check is simple vector algebra:

$$\begin{aligned}\overrightarrow{CG} &= \overrightarrow{CB} + \overrightarrow{BA} + \overrightarrow{AG} = \overrightarrow{CB} + \overrightarrow{BA} + \frac{2}{3}\left(\overrightarrow{AB} + \frac{1}{2}\overrightarrow{BC}\right)\\ &= \frac{2}{3}\left(\overrightarrow{CB} + \frac{1}{2}\overrightarrow{BA}\right) = \frac{2}{3}\overrightarrow{CF}.\end{aligned}$$

Exercises

(2.1) Verify that addition and scalar multiplication of n-tuplets as defined in (2.2a) and (2.2b) satisfy the eight required vector properties.

(2.2) Show that $\mathbf{e}_1 = (1, 0, \cdots, 0)$, $\mathbf{e}_2 = (0, 1, 0, \cdots, 0)$, etc. is a basis for $\mathbb{R}^n$. It is called the *natural basis* for $\mathbb{R}^n$.

(2.3) Show that the set of polynomials $P(x)$ of degree at most n is a vector space, where $x \in \mathbb{R}$. What is the dimension of that vector space? What is a basis for that vector space?

(2.4) Show that the set of smooth functions $f(x)$ periodic of period 2π is a vector space. What is the dimension of that space? What is a basis for that space?

(2.5) Suppose you define addition of n-tuplets $\mathbf{x} = (x_1, x_2, \ldots, x_n)$ as usual but define scalar multiplication according to $\alpha\mathbf{x} = (\alpha x_1, x_2, \cdots, x_n)$, that is, only the first component is multiplied by α. Which property is violated? What if you defined $\alpha\mathbf{x} = (\alpha x_1, 0, \cdots, 0)$, which property would be violated?

(2.6) Prove that the components of any $\mathbf{v}$ with respect to a basis $\{\mathbf{a}_1, \mathbf{a}_2, \ldots, \mathbf{a}_n\}$ are unique. [Hint: assume that $\mathbf{v}$ can be expanded in two distinct ways, then use what you know about $\{\mathbf{a}_1, \mathbf{a}_2, \ldots, \mathbf{a}_n\}$.]

(2.7) Given vectors $\boldsymbol{a}$, $\boldsymbol{b}$ in $\mathbb{E}^3$, show that the set of all $\boldsymbol{v} = \alpha\boldsymbol{a} + \beta\boldsymbol{b}$, $\forall\alpha, \beta \in \mathbb{R}$ is a vector space. What is the dimension of that vector space?

(2.8) Show that the set of all vectors $\boldsymbol{v} = \alpha\boldsymbol{a} + \boldsymbol{b}$, $\forall\alpha \in \mathbb{R}$ and fixed $\boldsymbol{a}$, $\boldsymbol{b}$ is *not* a vector space, in general.

(2.9) Consider the set of all unit vectors in $\mathbb{E}^3$. Is that a vector space? Explain.

(2.10) Is $\hat{\mathbf{x}} = (1, 0, 0)$? Discuss.

(2.11) Find the components of *North* in the bases (a) $\{$*East*, *Northeast*$\}$, (b) $\{$*East*, *West*$\}$. Sketch and briefly explain your *geometrical* construction.

(2.12) Given three points P_1, P_2, P_3 in euclidean 3D space, let M be the midpoint of segment P_1P_2. What are the components of $\overrightarrow{P_2P_3}$ and $\overrightarrow{P_3M}$ in the basis $\overrightarrow{P_1P_2}$, $\overrightarrow{P_1P_3}$? Sketch.

(2.13) Given three points P_1, P_2, P_3 in euclidean 3D space, what are the coordinates of the midpoints of the triangle sides with respect to P_1 and the basis $\overrightarrow{P_1P_2}$, $\overrightarrow{P_1P_3}$?

(2.14) Pick two generic vectors $\boldsymbol{a}$, $\boldsymbol{b}$ and some arbitrary point A in the plane of your sheet of paper. If $\overrightarrow{AP} = \alpha\boldsymbol{a} + \beta\boldsymbol{b}$, sketch the region where P can be: (i) when α and β are both between 0 and 1; (ii) when $|\beta| \le |\alpha| \le 1$.

(2.15) Prove that the line segment connecting the midpoints of two sides of a triangle is parallel to and equal to half of the third side using (1) geometry, and (2) vectors.

(2.16) Prove that the diagonals of a parallelogram intersect at their midpoints using (1) geometry, and (2) vectors.

(2.17) Prove that the medians of a triangle intersect at the same point, the *centroid* G, which is 2/3 of the way down from the vertices along each median (a median is a line that connects a vertex to the middle of the opposite side). Do this in two ways: (1) using geometry, and (2) using vector methods.

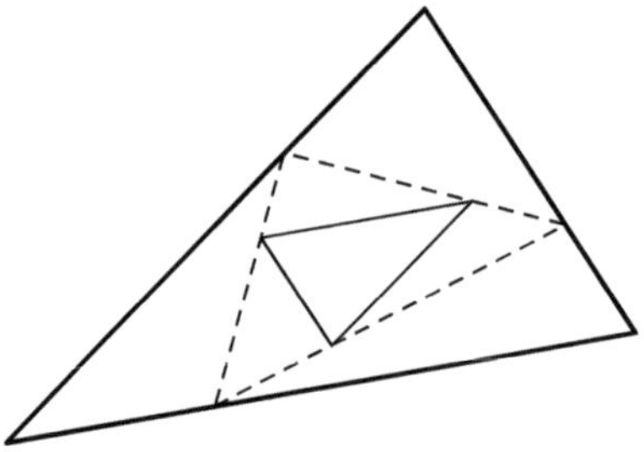

Fig. 2.6 Problem 2.18.

(2.18) From Isaacs[1]: Given an arbitrary triangle with vertices A, B, C, consider A' at 1/3 from A along edge AB, and likewise for B' along BC, and C' along CA. Similarly, consider A'' at 1/3 from A' along $A'B'$, and likewise for B'' along $B'C'$, and C'' along $C'A'$. Use vectors to prove that the triangle $A''B''C''$ is similar to the triangle ABC (Fig. 2.6).

(2.19) Find a point X such that $\overrightarrow{XA} + \overrightarrow{XB} + \overrightarrow{XC} = 0$, where A, B, C are given but arbitrary points in

[1]Martin Isaacs, *Geometry for College Students*, American Mathematical Society, 2001.

Euclidean space. If A, B, C are not colinear, show that the line through A and X cuts BC at its midpoint. Deduce similar results for the other sides of the triangle ABC and therefore that X is the point of intersection of the medians. Sketch. [Hint: since X is unknown, the vector equation has three unknown vectors; however, any two of those can be expressed in terms of the third.]

(2.20) Find X such that $\overrightarrow{XA} + \overrightarrow{XB} + \overrightarrow{XC} + \overrightarrow{XD} = 0$, where A, B, C, D are given but arbitrary points in 3D euclidean space. If A, B, C, D are not coplanar, show that the line through A and X intersects the triangle BCD at its centroid. Deduce similar results for the other faces and therefore that the medians of the tetrahedron $ABCD$, defined as the lines joining each vertex to the centroid of the opposite triangle, all intersect at the same point X, which is 3/4 of the way down from the vertices along the medians (Fig. 2.7).

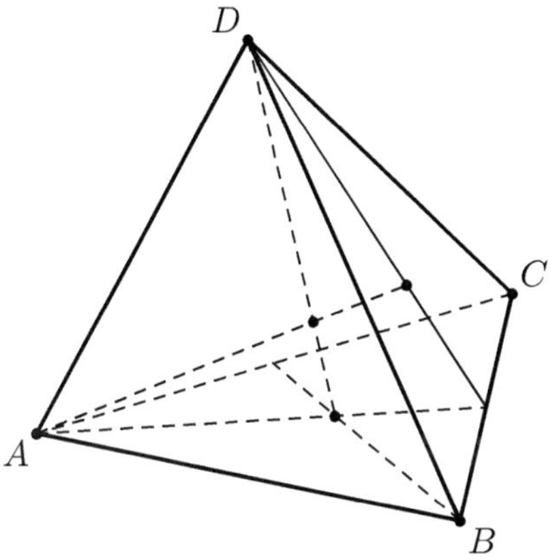

Fig. 2.7 Problem 2.20.

(2.21) Given four points A, B, C, D not coplanar, let G be the intersection of the medians of triangle A, B, C. Find the point X that is the intersection of the plane passing through A and the midpoints of BD and CD and the line through D and G (Fig. 2.8). Show that X is 3/5 down DG. Show your work.

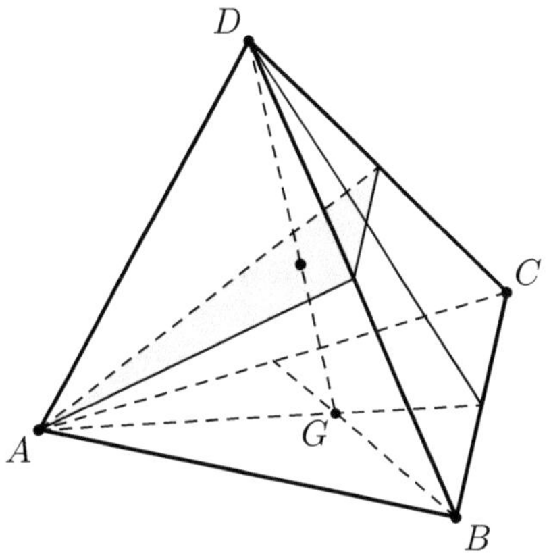

Fig. 2.8 Problem 2.21.

(2.22) Given any four distinct points P_0, P_1, P_2, P_3 in 3D space, let M_i be the midpoint of segment (P_i, P_{i+1}) with $P_4 = P_0$. Show that $M_0M_1M_2M_3$ is a parallelogram (Fig. 2.9).

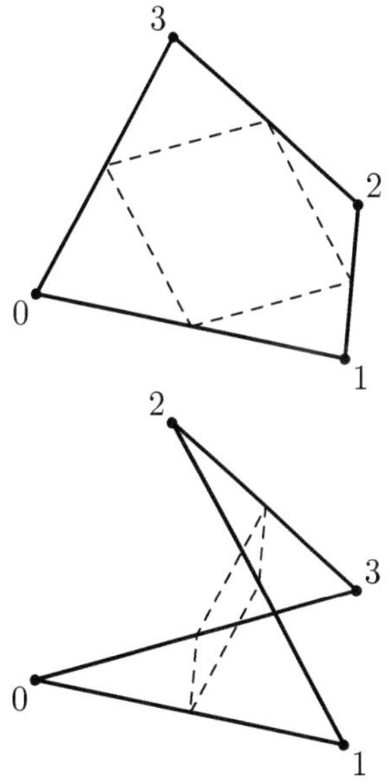

Fig. 2.9 Problem 2.22.

Dot Product

3

3.1 Geometry and algebra

The geometric definition of the dot product of vectors in 3D euclidean space is

$$\boldsymbol{a} \cdot \boldsymbol{b} \triangleq a\, b\, \cos\theta, \tag{3.1}$$

where $a = |\boldsymbol{a}|$, $b = |\boldsymbol{b}|$, and θ is the angle between the vectors $\boldsymbol{a}$ and $\boldsymbol{b}$, with $0 \leq \theta \leq \pi$ (Fig. 3.1).

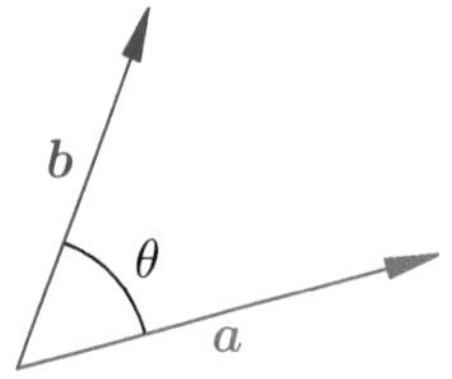

Fig. 3.1 $0 \leq \theta \leq \pi$.

The dot product is a real number such that $\boldsymbol{a} \cdot \boldsymbol{b} = 0$ iff $\boldsymbol{a}$ and $\boldsymbol{b}$ are *orthogonal*, that is when $\theta = \pi/2$ if $|\boldsymbol{a}|$ and $|\boldsymbol{b}|$ are not zero. The $\boldsymbol{0}$ vector is considered orthogonal to any vector. The dot product of any vector with itself is the square of its magnitude

$$\boldsymbol{a} \cdot \boldsymbol{a} = |\boldsymbol{a}|^2. \tag{3.2}$$

The dot product is also called the *scalar product*, since its result is a scalar, or the *inner product* in linear algebra.

Geometric properties

The dot product is directly related to the *orthogonal projections* of $\boldsymbol{a}$ onto $\boldsymbol{b}$ (Fig. 3.2)

$$\boldsymbol{a}_{\parallel} = a_{\parallel} \hat{\boldsymbol{b}} = a \cos\theta\, \hat{\boldsymbol{b}} = (\boldsymbol{a} \cdot \hat{\boldsymbol{b}})\, \hat{\boldsymbol{b}}, \tag{3.3}$$

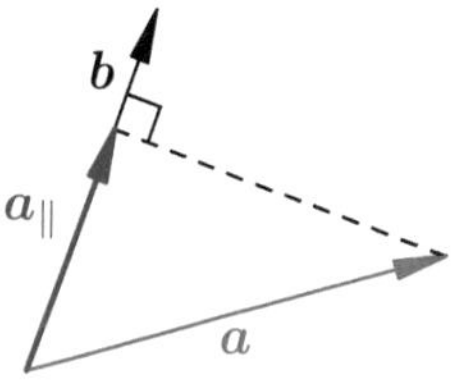

Fig. 3.2 $\boldsymbol{a}_{\parallel} = (\boldsymbol{a} \cdot \hat{\boldsymbol{b}})\hat{\boldsymbol{b}}$ Projection of $\boldsymbol{a}$ onto $\boldsymbol{b}$.

and $\boldsymbol{b}$ onto $\boldsymbol{a}$ (Fig. 3.3)

$$\boldsymbol{b}_{\parallel} = b_{\parallel} \hat{\boldsymbol{a}} = b \cos\theta\, \hat{\boldsymbol{a}} = (\hat{\boldsymbol{a}} \cdot \boldsymbol{b})\, \hat{\boldsymbol{a}}, \tag{3.4}$$

where $\hat{\boldsymbol{a}}$ and $\hat{\boldsymbol{b}}$ are the *unit* vectors in the $\boldsymbol{a}$ and $\boldsymbol{b}$ directions, respectively. The orthogonal projections are not equal in general $\boldsymbol{a}_{\parallel} \neq \boldsymbol{b}_{\parallel}$, they have different magnitudes and different directions, but the dot product has the fundamental property that

$$\boldsymbol{a} \cdot \boldsymbol{b} = \boldsymbol{a} \cdot \boldsymbol{b}_{\parallel} = \boldsymbol{a}_{\parallel} \cdot \boldsymbol{b}. \tag{3.5}$$

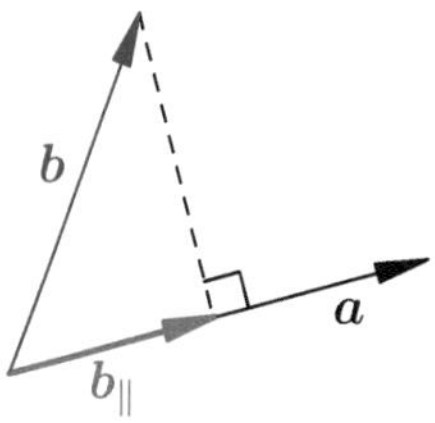

Fig. 3.3 $\boldsymbol{b}_{\parallel} = (\boldsymbol{b} \cdot \hat{\boldsymbol{a}})\hat{\boldsymbol{a}}$ Projection of $\boldsymbol{b}$ onto $\boldsymbol{a}$.

Once the parallel component $\boldsymbol{b}_{\parallel}$ has been obtained, subtracting it from $\boldsymbol{b}$ yields the perpendicular component $\boldsymbol{b}_{\perp}$,

$$\begin{cases} \boldsymbol{b}_{\parallel} = (\boldsymbol{b} \cdot \hat{\boldsymbol{a}})\, \hat{\boldsymbol{a}} \\ \boldsymbol{b}_{\perp} = \boldsymbol{b} - \boldsymbol{b}_{\parallel}, \end{cases} \tag{3.6}$$

such that $\boldsymbol{b} = \boldsymbol{b}_{\parallel} + \boldsymbol{b}_{\perp}$. This yields the decomposition of $\boldsymbol{b}$ in terms of components parallel and perpendicular to $\boldsymbol{a}$ (Fig. 3.4). Likewise, if it is $\boldsymbol{a}$ that we wish to decompose into vector components parallel and perpendicular to $\boldsymbol{b}$, such that $\boldsymbol{a} = \boldsymbol{a}_{\parallel} + \boldsymbol{a}_{\perp}$, then

$$\boldsymbol{a}_{\parallel} = (\boldsymbol{a} \cdot \hat{\boldsymbol{b}})\, \hat{\boldsymbol{b}}, \qquad \boldsymbol{a}_{\perp} = \boldsymbol{a} - \boldsymbol{a}_{\parallel}. \tag{3.7}$$

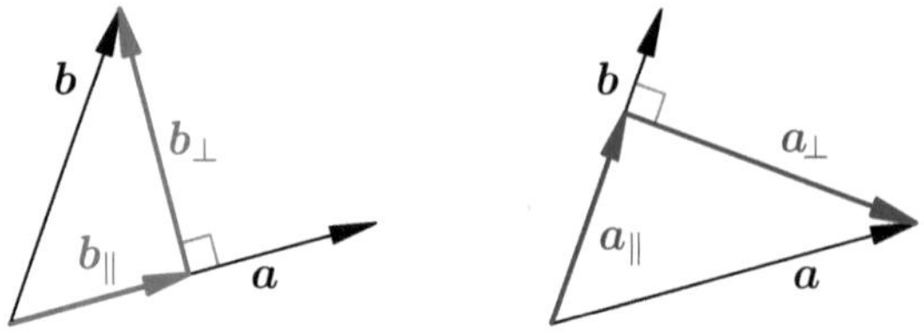

Fig. 3.4 Parallel and perpendicular components: $\boldsymbol{b} = \boldsymbol{b}_{\parallel} + \boldsymbol{b}_{\perp}$ and $\boldsymbol{a} = \boldsymbol{a}_{\parallel} + \boldsymbol{a}_{\perp}$.

Algebraic properties

The dot product has the following properties:

(1) $\boldsymbol{a} \cdot \boldsymbol{b} = \boldsymbol{b} \cdot \boldsymbol{a}$, *(commutativity)*

(2) $\boldsymbol{a} \cdot \boldsymbol{a} \geq 0, \quad \boldsymbol{a} \cdot \boldsymbol{a} = 0 \Leftrightarrow \boldsymbol{a} = \boldsymbol{0}$, *(positive definiteness)*

(3) $(\boldsymbol{a} \cdot \boldsymbol{b})^2 \leq (\boldsymbol{a} \cdot \boldsymbol{a})\, (\boldsymbol{b} \cdot \boldsymbol{b})$, *(Cauchy–Schwarz)*

(4) $\boldsymbol{a} \cdot (\beta \boldsymbol{b} + \gamma \boldsymbol{c}) = \beta(\boldsymbol{a} \cdot \boldsymbol{b}) + \gamma(\boldsymbol{a} \cdot \boldsymbol{c})$. *(distributivity)*

The first three properties follow directly from the geometric definition (3.1).

► *Proof of distributivity* Let $\boldsymbol{v} \triangleq \beta \boldsymbol{b} + \gamma \boldsymbol{c}$. All dot products are with $\boldsymbol{a}$, so decompose all vectors into components parallel and perpendicular to $\boldsymbol{a}$, that is

$$\boldsymbol{v}_{\parallel} + \boldsymbol{v}_{\perp} = \boldsymbol{v} \triangleq \beta \boldsymbol{b} + \gamma \boldsymbol{c} = \beta(\boldsymbol{b}_{\parallel} + \boldsymbol{b}_{\perp}) + \gamma(\boldsymbol{c}_{\parallel} + \boldsymbol{c}_{\perp}), \tag{3.8}$$

where

$$\boldsymbol{v}_{\parallel} = (\hat{\boldsymbol{a}} \cdot \boldsymbol{v})\, \hat{\boldsymbol{a}}, \quad \boldsymbol{b}_{\parallel} = (\hat{\boldsymbol{a}} \cdot \boldsymbol{b})\, \hat{\boldsymbol{a}}, \quad \boldsymbol{c}_{\parallel} = (\hat{\boldsymbol{a}} \cdot \boldsymbol{c})\, \hat{\boldsymbol{a}}. \tag{3.9}$$

Identity (3.8) yields

$$\boldsymbol{v}_{\parallel} - \beta \boldsymbol{b}_{\parallel} - \gamma \boldsymbol{c}_{\parallel} = \beta \boldsymbol{b}_{\perp} + \gamma \boldsymbol{c}_{\perp} - \boldsymbol{v}_{\perp},$$

but vectors parallel to $\boldsymbol{a}$ cannot add up to vectors perpendicular to $\boldsymbol{a}$, unless both combinations add up to zero; therefore

$$\boldsymbol{v}_{\parallel} = \beta \boldsymbol{b}_{\parallel} + \gamma \boldsymbol{c}_{\parallel}, \tag{3.10}$$

$$\boldsymbol{v}_{\perp} = \beta \boldsymbol{b}_{\perp} + \gamma \boldsymbol{c}_{\perp}. \tag{3.11}$$

Substituting (3.9) into (3.10) gives

$$\hat{\boldsymbol{a}} \cdot (\beta \boldsymbol{b} + \gamma \boldsymbol{c}) = \beta(\hat{\boldsymbol{a}} \cdot \boldsymbol{b}) + \gamma(\hat{\boldsymbol{a}} \cdot \boldsymbol{c}),$$

and multiplying by $|\boldsymbol{a}|$ yields the distributivity property. □

Work is a dot product

In physics, the work W done by a force $\boldsymbol{F}$ on an object undergoing the displacement $\boldsymbol{\ell}$ is (Fig. 3.5)

$$W = \boldsymbol{F} \cdot \boldsymbol{\ell} = \boldsymbol{F} \cdot \boldsymbol{\ell}_{\parallel} = \boldsymbol{F}_{\parallel} \cdot \boldsymbol{\ell},$$

where $\boldsymbol{\ell}_{\parallel} = (\boldsymbol{\ell} \cdot \hat{\boldsymbol{F}})\hat{\boldsymbol{F}}$ and $\boldsymbol{F}_{\parallel} = (\boldsymbol{F} \cdot \hat{\boldsymbol{\ell}})\hat{\boldsymbol{\ell}}$. Thus, work is the product of the force component in the direction of the displacement, $F_{\parallel} = \boldsymbol{F} \cdot \hat{\boldsymbol{\ell}}$, times the full displacement ℓ, or, equivalently, the product of the displacement in the direction of the force, $\ell_{\parallel} = \boldsymbol{\ell} \cdot \hat{\boldsymbol{F}}$, times the full force F.

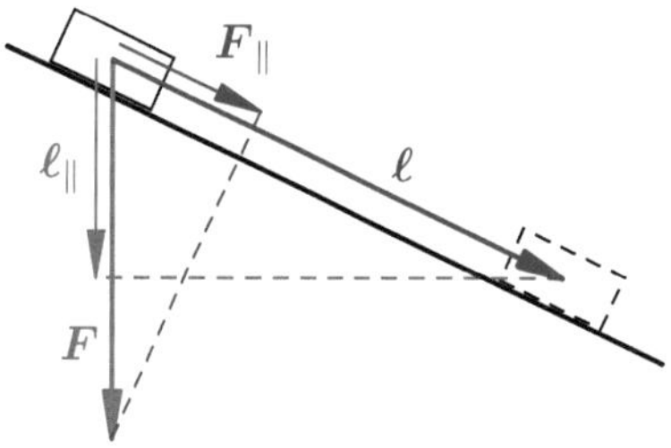

Fig. 3.5 The work $W = \boldsymbol{F} \cdot \boldsymbol{\ell}$ done by $\boldsymbol{F}$ on a box sliding $\boldsymbol{\ell}$.

Cartesian formula

The distributivity property is a fundamental algebraic property of the dot product. It yields

$$\begin{aligned} \boldsymbol{a} \cdot \boldsymbol{b} &= (a_x\hat{\mathbf{x}} + a_y\hat{\mathbf{y}} + a_z\hat{\mathbf{z}}) \cdot (b_x\hat{\mathbf{x}} + b_y\hat{\mathbf{y}} + b_z\hat{\mathbf{z}}) \\ &= a_x b_x\, \hat{\mathbf{x}} \cdot \hat{\mathbf{x}} + a_x b_y\, \hat{\mathbf{x}} \cdot \hat{\mathbf{y}} + \cdots \\ &= a_x b_x + a_y b_y + a_z b_z \end{aligned} \tag{3.12}$$

in terms of cartesian components for $\boldsymbol{a}$ and $\boldsymbol{b}$, since (Fig. 3.6)

$$\begin{aligned} \hat{\mathbf{x}} \cdot \hat{\mathbf{x}} &= 1 = \hat{\mathbf{y}} \cdot \hat{\mathbf{y}} = \hat{\mathbf{z}} \cdot \hat{\mathbf{z}}, \\ \hat{\mathbf{x}} \cdot \hat{\mathbf{y}} &= 0 = \hat{\mathbf{y}} \cdot \hat{\mathbf{z}} = \hat{\mathbf{z}} \cdot \hat{\mathbf{x}}. \end{aligned} \tag{3.13}$$

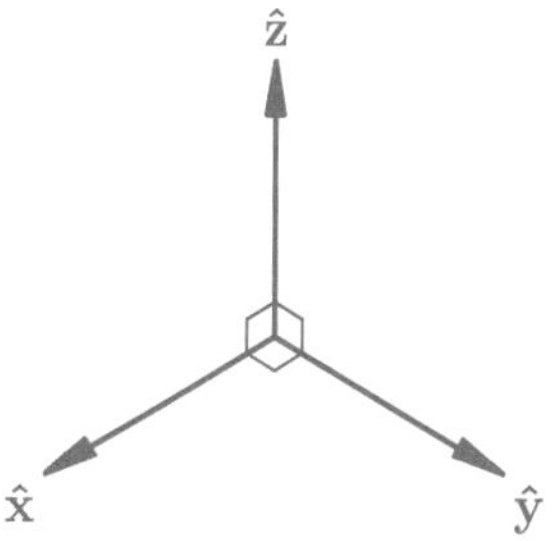

Fig. 3.6 3D cartesian basis.

Thus, computing a dot product is easy when the vectors are known in cartesian form. In particular, the angle between two cartesian vectors can be obtained from

$$\cos\theta = \frac{\boldsymbol{a} \cdot \boldsymbol{b}}{|\boldsymbol{a}||\boldsymbol{b}|} = \frac{a_x b_x + a_y b_y + a_z b_z}{\sqrt{a_x^2 + a_y^2 + a_z^2}\,\sqrt{b_x^2 + b_y^2 + b_z^2}}.$$

That result (3.12) is the standard definition of dot product in $\mathbb{R}^3$ but we deduced it from the geometric definition (3.1). The geometric definition directly indicates that other definitions of the dot product in $\mathbb{R}^3$ can and may need to be used. For example, if vectors $\boldsymbol{u}$ and $\boldsymbol{v}$ are expanded in terms of an arbitrary basis $\{\boldsymbol{a}_1, \boldsymbol{a}_2, \boldsymbol{a}_3\}$, the distributivity and commutativity of the dot product yield

$$\begin{aligned} \boldsymbol{u} \cdot \boldsymbol{v} &= (\zeta_1\boldsymbol{a}_1 + \zeta_2\boldsymbol{a}_2 + \zeta_3\boldsymbol{a}_3) \cdot (\eta_1\boldsymbol{a}_1 + \eta_2\boldsymbol{a}_2 + \eta_3\boldsymbol{a}_3) \\ &= \zeta_1\eta_1\,(\boldsymbol{a}_1 \cdot \boldsymbol{a}_1) + \zeta_2\eta_2\,(\boldsymbol{a}_2 \cdot \boldsymbol{a}_2) + \zeta_3\eta_3\,(\boldsymbol{a}_3 \cdot \boldsymbol{a}_3) \\ &\quad + (\zeta_1\eta_2 + \zeta_2\eta_1)\,(\boldsymbol{a}_1 \cdot \boldsymbol{a}_2) + (\zeta_1\eta_3 + \zeta_3\eta_1)\,(\boldsymbol{a}_1 \cdot \boldsymbol{a}_3) \\ &\quad + (\zeta_2\eta_3 + \zeta_3\eta_2)\,(\boldsymbol{a}_2 \cdot \boldsymbol{a}_3) \\ &\neq \zeta_1\eta_1 + \zeta_2\eta_2 + \zeta_3\eta_3. \end{aligned} \tag{3.14}$$

In matrix notation, this reads

$$\boldsymbol{u} \cdot \boldsymbol{v} = \begin{bmatrix} \zeta_1 & \zeta_2 & \zeta_3 \end{bmatrix} \begin{bmatrix} \boldsymbol{a}_1 \cdot \boldsymbol{a}_1 & \boldsymbol{a}_1 \cdot \boldsymbol{a}_2 & \boldsymbol{a}_1 \cdot \boldsymbol{a}_3 \\ \boldsymbol{a}_2 \cdot \boldsymbol{a}_1 & \boldsymbol{a}_2 \cdot \boldsymbol{a}_2 & \boldsymbol{a}_2 \cdot \boldsymbol{a}_3 \\ \boldsymbol{a}_3 \cdot \boldsymbol{a}_1 & \boldsymbol{a}_3 \cdot \boldsymbol{a}_2 & \boldsymbol{a}_3 \cdot \boldsymbol{a}_3 \end{bmatrix} \begin{bmatrix} \eta_1 \\ \eta_2 \\ \eta_3 \end{bmatrix}, \tag{3.15}$$

while in index notation with summation convention, this is

$$\boldsymbol{u} \cdot \boldsymbol{v} = g_{ij}\, \zeta_i\, \eta_j \tag{3.16}$$

with implicit sums over the repeated i and j indices, where $g_{ij} \triangleq \boldsymbol{a}_i \cdot \boldsymbol{a}_j = g_{ji}$ is the *metric* of the basis $\{\boldsymbol{a}_1, \boldsymbol{a}_2, \boldsymbol{a}_3\}$. Index and matrix notations are discussed in the following and in later sections.

3.2 Orthonormal bases

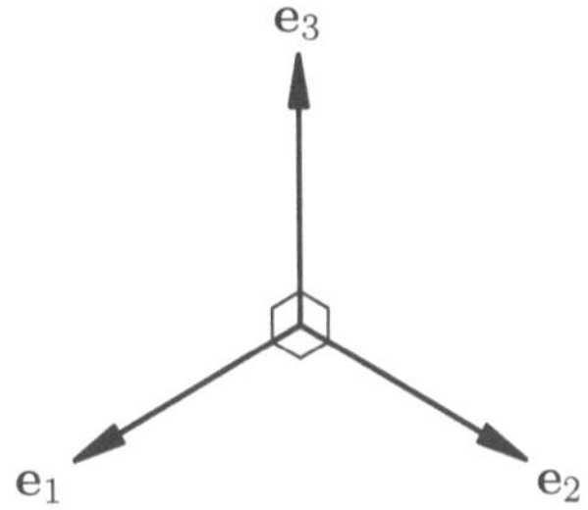

Fig. 3.7 3D orthonormal basis.

An *orthonormal* basis is a complete set of mutually orthogonal unit vectors. A cartesian basis $\{\hat{\mathbf{x}}, \hat{\mathbf{y}}, \hat{\mathbf{z}}\}$ is an *orthonormal* basis for 3D euclidean space. The vectors $\{\hat{\mathbf{x}}, \hat{\mathbf{y}}\}$ yield an orthonormal basis for the *subspace* of vectors perpendicular to $\hat{\mathbf{z}}$, but it is not a *complete* basis for 3D space. Switching to index notation, we write $\{\mathbf{e}_1, \mathbf{e}_2, \mathbf{e}_3\}$ in place of $\{\hat{\mathbf{x}}, \hat{\mathbf{y}}, \hat{\mathbf{z}}\}$ (Fig. 3.7).[1]Orthonormality of the $\{\mathbf{e}_1, \mathbf{e}_2, \mathbf{e}_3\}$ basis is then compactly expressed as

$$\mathbf{e}_i \cdot \mathbf{e}_j = \delta_{ij}, \tag{3.17}$$

where i and j can *each* be any of $1, 2, 3$ and δ_{ij} is the **Kronecker delta**[2] defined as

$$\delta_{ij} \triangleq \begin{cases} 1 & \text{if } \; i = j, \\ 0 & \text{if } \; i \neq j. \end{cases} \tag{3.18}$$

[2]Named after German mathematician Leopold Kronecker, 1823–1891.

Orthogonal expansion and projections

The components of a vector $\boldsymbol{v}$ with respect to a basis $\{\mathbf{e}_1, \mathbf{e}_2, \mathbf{e}_3\}$ are the signed scalars (v_1, v_2, v_3) such that

$$\boldsymbol{v} = v_1\mathbf{e}_1 + v_2\mathbf{e}_2 + v_3\mathbf{e}_3 = \sum_{j=1}^{3} v_j \mathbf{e}_j. \tag{3.19}$$

The vector equation (3.19) is the *expansion* of $\boldsymbol{v}$ in terms of the basis $\{\mathbf{e}_1, \mathbf{e}_2, \mathbf{e}_3\}$. Dotting (3.19) with each of the basis vectors yields

$$\left.\begin{aligned} v_1 &= \boldsymbol{v} \cdot \mathbf{e}_1, \\ v_2 &= \boldsymbol{v} \cdot \mathbf{e}_2, \\ v_3 &= \boldsymbol{v} \cdot \mathbf{e}_3 \end{aligned}\right\} \quad \Leftrightarrow \quad v_i = \boldsymbol{v} \cdot \mathbf{e}_i, \tag{3.20}$$

since $\{\mathbf{e}_1, \mathbf{e}_2, \mathbf{e}_3\}$ is orthonormal (3.17). In the expression $v_i = \boldsymbol{v} \cdot \mathbf{e}_i$, the index i is *free* to take all the possible values $i = 1, 2, 3$. The three scalar equations (3.20) are the *orthogonal projections* of $\boldsymbol{v}$ onto each of the unit vectors $\{\mathbf{e}_1, \mathbf{e}_2, \mathbf{e}_3\}$. The orthogonal projections of a vector $\boldsymbol{v}$ onto an *orthonormal* basis $\{\mathbf{e}_1, \mathbf{e}_2, \mathbf{e}_3\}$ are the components of $\boldsymbol{v}$ in that basis.

[1]We do not use **i**, **j**, **k** for cartesian unit vectors. Instead, we use $\{\hat{\mathbf{x}}, \hat{\mathbf{y}}, \hat{\mathbf{z}}\}$ or$\{\mathbf{e}_1, \mathbf{e}_2, \mathbf{e}_3\}$ or $\{\mathbf{e}_x, \mathbf{e}_y, \mathbf{e}_z\}$ to denote a set of three orthogonal direction vectors in 3D euclidean space. We use indices i, j, and k to denote *anyone* of those three directions. Those indices are positive integers that can take all the values from 1 to n, the dimension of the space. We focus on 3D space, so indices i, j, and k can *each* be 1, 2, or 3, typically. Indices i, j, k should not be confused with those old basis vectors **i**, **j**, **k** from elementary calculus.

Invariance of cartesian dot product

If two vectors $\boldsymbol{a}$ and $\boldsymbol{b}$ are expanded in terms of the orthonormal $\{\mathbf{e}_1, \mathbf{e}_2, \mathbf{e}_3\}$, that is

$$\boldsymbol{a} = a_1\mathbf{e}_1 + a_2\mathbf{e}_2 + a_3\mathbf{e}_3,$$
$$\boldsymbol{b} = b_1\mathbf{e}_1 + b_2\mathbf{e}_2 + b_3\mathbf{e}_3,$$

then the distributivity properties of the dot product and the orthonormality of the basis yield

$$\boldsymbol{a} \cdot \boldsymbol{b} = a_1 b_1 + a_2 b_2 + a_3 b_3. \tag{3.21}$$

This is (3.12) written in index notation $(1, 2, 3)$ instead of (x, y, z). One remarkable property of this cartesian formula is that its value is independent of the orthonormal basis. The dot product is a geometric property of the vectors $\boldsymbol{a}$ and $\boldsymbol{b}$, independent of the basis. This is obvious from the geometric definition (3.1) but not from its expression in terms of components (3.21). Indeed, if $\{\mathbf{e}_1, \mathbf{e}_2, \mathbf{e}_3\}$ and $\{\mathbf{e}_1', \mathbf{e}_2', \mathbf{e}_3'\}$ are two distinct orthonormal bases then

$$\boldsymbol{a} = a_1\mathbf{e}_1 + a_2\mathbf{e}_2 + a_3\mathbf{e}_3 = a_1'\mathbf{e}_1' + a_2'\mathbf{e}_2' + a_3'\mathbf{e}_3',$$
$$\boldsymbol{b} = b_1\mathbf{e}_1 + b_2\mathbf{e}_2 + b_3\mathbf{e}_3 = b_1'\mathbf{e}_1' + b_2'\mathbf{e}_2' + b_3'\mathbf{e}_3',$$

but in general

$$(a_1, a_2, a_3) \neq (a_1', a_2', a_3'),$$
$$(b_1, b_2, b_3) \neq (b_1', b_2', b_3'),$$

yet

$$\boldsymbol{a} \cdot \boldsymbol{b} = a_1 b_1 + a_2 b_2 + a_3 b_3 = a_1' b_1' + a_2' b_2' + a_3' b_3'. \tag{3.22}$$

The simple algebraic form of the dot product is *invariant under a change of orthonormal basis*, although each component may be different.

Orthogonal basis

If the basis vectors $\{\boldsymbol{a}_1, \boldsymbol{a}_2, \boldsymbol{a}_3\}$ are mutually *orthogonal* but not necessarily of unit norm, then there exist three (signed) scalars (χ_1, χ_2, χ_3) such that

$$\boldsymbol{v} = \chi_1\, \boldsymbol{a}_1 + \chi_2\, \boldsymbol{a}_2 + \chi_3\, \boldsymbol{a}_3. \tag{3.23}$$

The components (χ_1, χ_2, χ_3) can be obtained independently of one another and expressed in terms of dot products, but there are normalization factors. Dotting (3.23) with $\boldsymbol{a}_1$ yields

$$\boldsymbol{v} \cdot \boldsymbol{a}_1 = \chi_1\, \boldsymbol{a}_1 \cdot \boldsymbol{a}_1 + \chi_2\, \boldsymbol{a}_2 \cdot \boldsymbol{a}_1 + \chi_3\, \boldsymbol{a}_3 \cdot \boldsymbol{a}_1. \tag{3.24}$$

The last two terms drop out *if* $\boldsymbol{a}_2 \cdot \boldsymbol{a}_1 = 0$ and $\boldsymbol{a}_3 \cdot \boldsymbol{a}_1 = 0$ yielding

$$\chi_1 = \frac{\boldsymbol{v} \cdot \boldsymbol{a}_1}{\boldsymbol{a}_1 \cdot \boldsymbol{a}_1},$$

independent of $\boldsymbol{a}_2$ and $\boldsymbol{a}_3$. The other components χ_2 and χ_3 are obtained likewise by projecting the vector $\boldsymbol{v}$ onto the respective basis vector and using orthogonality $\boldsymbol{a}_1 \cdot \boldsymbol{a}_2 = \boldsymbol{a}_2 \cdot \boldsymbol{a}_3 = \boldsymbol{a}_3 \cdot \boldsymbol{a}_1 = 0$ to obtain

$$\chi_2 = \frac{\boldsymbol{a}_2 \cdot \boldsymbol{v}}{\boldsymbol{a}_2 \cdot \boldsymbol{a}_2}, \quad \chi_3 = \frac{\boldsymbol{a}_3 \cdot \boldsymbol{v}}{\boldsymbol{a}_3 \cdot \boldsymbol{a}_3}.$$

Arbitrary basis

If the vectors $\{\boldsymbol{a}_1, \boldsymbol{a}_2, \boldsymbol{a}_3\}$ are linearly independent but not orthogonal, then we can still expand

$$\boldsymbol{v} = \eta_1\, \boldsymbol{a}_1 + \eta_2\, \boldsymbol{a}_2 + \eta_3\, \boldsymbol{a}_3 \tag{3.25}$$

where the three unique components (η_1, η_2, η_3) can be obtained, geometrically, by *parallel* projections as illustrated in figs. 2.1 & 2.2. Algebraically, this corresponds to solving a linear system of 3 equations for the 3 unknown components (η_1, η_2, η_3). The components can still be expressed in terms of dot products by constructing a *reciprocal* basis $\{\breve{\boldsymbol{a}}_1, \breve{\boldsymbol{a}}_2, \breve{\boldsymbol{a}}_3\}$ or an orthogonal basis $\{\mathbf{q}_1, \mathbf{q}_2, \mathbf{q}_3\}$ using the *Gram-Schmidt* algorithm, as explored in exercises 3.9, 3.33, 3.38, and later in §6, problems 6.9, 6.15.

3.3 Dot product in $\mathbb{R}^n$

The geometric definition of the dot product (3.1) emphasizes the geometric aspects, but the algebraic formulas (3.12), (3.14), (3.21) are needed for calculations. Those formulas show how to define the dot product for other vector spaces where the concept of angle between vectors is not defined *a priori*; for example, what is the angle between the vectors (1,2,3,4) and (4,3,2,1) in $\mathbb{R}^4$? The dot product, a.k.a. *scalar* product or *inner* product, of the vectors $\mathbf{x}$ and $\mathbf{y}$ in $\mathbb{R}^n$ is defined as suggested by (3.21):

$$\mathbf{x} \cdot \mathbf{y} \triangleq x_1 y_1 + x_2 y_2 + \cdots + x_n y_n. \tag{3.26}$$

It can be verified that this definition satisfies the fundamental properties of the dot product (§3.1): commutativity $\mathbf{x} \cdot \mathbf{y} = \mathbf{y} \cdot \mathbf{x}$, positive definiteness $\mathbf{x} \cdot \mathbf{x} \geq 0$, and multilinearity (or distributivity):

$$(\alpha_1 \mathbf{x}_1 + \alpha_2 \mathbf{x}_2) \cdot \mathbf{y} = \alpha_1 (\mathbf{x}_1 \cdot \boldsymbol{y}) + \alpha_2 (\mathbf{x}_2 \cdot \mathbf{y}).$$

To show the Cauchy–Schwarz property, you need a bit of calculus and a classical trick: consider $\mathbf{v} = \mathbf{x} + \lambda \mathbf{y}$, then

$$F(\lambda) \triangleq \mathbf{v} \cdot \mathbf{v} = \lambda^2 \mathbf{y} \cdot \mathbf{y} + 2\lambda \mathbf{x} \cdot \mathbf{y} + \mathbf{x} \cdot \mathbf{x} \geq 0.$$

For given, but arbitrary, $\mathbf{x}$ and $\mathbf{y}$, this is a quadratic polynomial in λ. That polynomial $F(\lambda)$ has a single minimum at $\lambda_* = -(\mathbf{x} \cdot \mathbf{y})/(\mathbf{y} \cdot \mathbf{y})$. That minimum value is

$$F(\lambda_*) = (\mathbf{x} \cdot \mathbf{x}) - \frac{(\mathbf{x} \cdot \mathbf{y})^2}{(\mathbf{y} \cdot \mathbf{y})} \geq 0,$$

which must still be positive since $F \geq 0$, $\forall \lambda$, hence the Cauchy–Schwarz inequality

$$(\mathbf{x} \cdot \mathbf{y})^2 \leq (\mathbf{x} \cdot \mathbf{x})\ (\mathbf{y} \cdot \mathbf{y}).$$

Since definition (3.26) satisfies the Cauchy–Schwarz inequality, we can define the length of a vector by $|\mathbf{x}| = (\mathbf{x} \cdot \mathbf{x})^{1/2}$, called the *euclidean* length since it corresponds to length in euclidean geometry by Pythagoras's theorem. The angle θ between two vectors in $\mathbb{R}^n$ can then be defined by

$$\cos\theta = \frac{\mathbf{x}\cdot\mathbf{y}}{|\mathbf{x}|\,|\mathbf{y}|}.$$

A vector space for which a dot (or inner) product is defined is called a *Hilbert space*, or an *inner product space*. As indicated in (3.14), other definitions of the dot product in $\mathbb{R}^n$ besides (3.26) are not only possible, but sometimes required for the particular application.

Norm of a vector

The *norm* of a vector, denoted $\|\mathbf{a}\|$, is a positive real number that defines its size or "length" (but not in the sense of the number of its components). For displacement vectors in euclidean spaces, the norm is the length of the displacement, $\|\boldsymbol{a}\| = |\boldsymbol{a}|$, that is the distance between point A and B if $\overrightarrow{AB} = \boldsymbol{a}$. The following properties are geometrically straightforward for length of displacement vectors:

(1) $\|\mathbf{a}\| \geq 0$ and $\|\mathbf{a}\| = 0 \Leftrightarrow \mathbf{a} = \mathbf{0}$,

(2) $\|\alpha\mathbf{a}\| = |\alpha|\,\|\mathbf{a}\|$,

(3) $\|\mathbf{a}+\mathbf{b}\| \leq \|\mathbf{a}\| + \|\mathbf{b}\|$. (triangle inequality)

For more general vector spaces, these properties become the *defining properties* (*axioms*) that a norm must satisfy. A vector space for which a norm is defined is called a *Banach space.*

Norms for $\mathbb{R}^n$

For other types of vector space, there are many possible definitions for the norm of a vector as long as those definitions satisfy the three norm properties listed above. In $\mathbb{R}^n$, the p-norm of vector $\mathbf{x}$ is defined by the positive number

$$\|\mathbf{x}\|_p \triangleq \left(|x_1|^p + |x_2|^p + \cdots + |x_n|^p\right)^{1/p}, \tag{3.27}$$

where $p \geq 1$ is a real number. Commonly used norms are the 2-norm $\|\mathbf{x}\|_2$ which is the square root of the sum of the squares, the 1-norm $\|\mathbf{x}\|_1$ (sum of absolute values), and the infinity norm, $\|\mathbf{x}\|_\infty$, defined as the limit as $p \to \infty$ of the above expression.

Note that the 2-norm $\|\mathbf{x}\|_2 = (\mathbf{x}\cdot\mathbf{x})^{1/2}$ and for that reason is also called the *euclidean norm.* In fact, if a dot product is defined, then a norm can always be defined as the square root of the dot product. In other words, *every Hilbert space is a Banach space,* but the converse is not true.

Exercises

(3.1) A skier slides down an inclined plane with a total vertical drop of h. Show that the work done by gravity is independent of the slope.

(3.2) Discuss and sketch the solutions $\boldsymbol{u}$ of $\boldsymbol{a} \cdot \boldsymbol{u} = b$, where $\boldsymbol{a}$ and b are known.

(3.3) Discuss and sketch the solutions $\boldsymbol{u}$ of $\boldsymbol{a} \cdot \hat{\boldsymbol{u}} = b$, where $\boldsymbol{a}$ and b are known.

(3.4) Prove that $(\alpha \boldsymbol{a}) \cdot \boldsymbol{b} = \alpha(\boldsymbol{a} \cdot \boldsymbol{b})$ from the geometric definition (3.1). Discuss both $\alpha \geq 0$ and $\alpha < 0$.

(3.5) True or false: (1) $\boldsymbol{a} \cdot \boldsymbol{a}_\perp = a_\perp^2$. (2) $\boldsymbol{a} \cdot \boldsymbol{b} = a_\| b_\|$. Explain.

(3.6) Sketch $\boldsymbol{c} = \boldsymbol{a} + \boldsymbol{b}$, then calculate $\boldsymbol{c} \cdot \boldsymbol{c} = (\boldsymbol{a} + \boldsymbol{b}) \cdot (\boldsymbol{a} + \boldsymbol{b})$ and deduce the *law of cosines:* $c^2 = a^2 + b^2 - 2ab\cos\gamma$.

(3.7) $\boldsymbol{B}$ is a magnetic field and $\boldsymbol{v}$ is the velocity of a particle. Let $\boldsymbol{v} = \boldsymbol{v}_\perp + \boldsymbol{v}_\|$ where $\boldsymbol{v}_\perp$ is perpendicular to the magnetic field and $\boldsymbol{v}_\|$ is parallel to it. Derive vector expressions for $\boldsymbol{v}_\perp$ and $\boldsymbol{v}_\|$ in terms of $\boldsymbol{v}$ and $\boldsymbol{B}$.

(3.8) Use vector algebra to show that $\boldsymbol{a}' \triangleq \boldsymbol{a} - (\boldsymbol{a} \cdot \hat{\boldsymbol{n}})\hat{\boldsymbol{n}}$ is orthogonal to $\hat{\boldsymbol{n}}$, for any $\boldsymbol{a}$, $\hat{\boldsymbol{n}}$. Sketch.

(3.9) Given two nonparallel vectors $\boldsymbol{a}_1$, $\boldsymbol{a}_2$, construct two vectors $\breve{\boldsymbol{a}}_1$, $\breve{\boldsymbol{a}}_2$, such that

$$\begin{cases} \boldsymbol{a}_1 \cdot \breve{\boldsymbol{a}}_1 = 1 = \boldsymbol{a}_2 \cdot \breve{\boldsymbol{a}}_2 \\ \boldsymbol{a}_1 \cdot \breve{\boldsymbol{a}}_2 = 0 = \boldsymbol{a}_2 \cdot \breve{\boldsymbol{a}}_1. \end{cases} \tag{3.28}$$

Sketch. Write $(\breve{\boldsymbol{a}}_1, \breve{\boldsymbol{a}}_2)$ in terms of $(\boldsymbol{a}_1, \boldsymbol{a}_2)$ and dot products of the latter. Find $(\breve{\boldsymbol{a}}_1, \breve{\boldsymbol{a}}_2)$ when the cartesian components of $\boldsymbol{a}_1$ and $\boldsymbol{a}_2$ are $(2, 3, 4)$ and $(3, 2, 1)$, respectively. If $\boldsymbol{v} = v_1\boldsymbol{a}_2 + v_2\boldsymbol{a}_2$, express v_1, v_2 in terms of $\boldsymbol{v}$, $\breve{\boldsymbol{a}}_1$ and $\breve{\boldsymbol{a}}_2$.

(3.10) Derive $\cos(\alpha - \beta) = \cos\alpha\cos\beta + \sin\alpha\sin\beta$ with a dot product (Fig. 3.8). Explain.

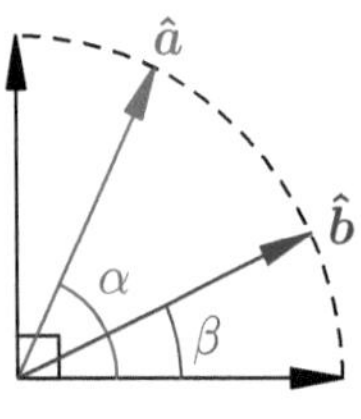

Fig. 3.8 $\cos(\alpha - \beta) = ?$

(3.11) If $\boldsymbol{a}$ and $\boldsymbol{b}$ have components (a_1, a_2) and (b_1, b_2) in the basis $\{$*North*, *Northwest*$\}$, calculate their dot product.

(3.12) Consider $\boldsymbol{v}(t) = \boldsymbol{a} + t\,\boldsymbol{b}$ where $t \in \mathbb{R}$. What is the minimum $|\boldsymbol{v}|$ and for what t? Solve two ways: (1) geometrically and (2) using calculus.

(3.13) Use vectors to show that if A, B, and C lie on a circle, with AC as a diameter, then ABC is a right angle (Fig. 3.9).

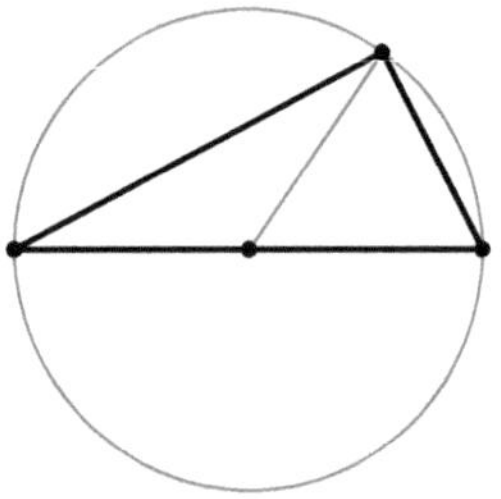

Fig. 3.9 Right angle? Why?

(3.14) *Reflection.* Given $\hat{\boldsymbol{a}}$, consider the vector transformation

$$\boldsymbol{v} \to \boldsymbol{v}' = \boldsymbol{v} - 2\,\hat{\boldsymbol{a}}\,(\hat{\boldsymbol{a}} \cdot \boldsymbol{v}). \tag{3.29}$$

(1) Show that $\boldsymbol{v}' \cdot \boldsymbol{w}' = \boldsymbol{v} \cdot \boldsymbol{w}$ for any $\boldsymbol{v}$ and $\boldsymbol{w}$, so lengths and angles are preserved by the transformation. (2) Sketch and show that the transformation corresponds to a *reflection* about a plane perpendicular to $\hat{\boldsymbol{a}}$.

(3.15) Show that the transformation $\boldsymbol{v} \to \boldsymbol{v}' = T(\boldsymbol{v})$ in (3.29) is a *linear transformation,* that is $T(\lambda\boldsymbol{v} + \mu\boldsymbol{w}) = \lambda T(\boldsymbol{v}) + \mu T(\boldsymbol{w})$ for any scalars λ, μ and vectors $\boldsymbol{v}$, $\boldsymbol{w}$.

(3.16) Show that the diagonals of the parallelogram spanned by $\boldsymbol{a}$ and $\boldsymbol{b}$ are orthogonal to each other if and only if $|\boldsymbol{a}| = |\boldsymbol{b}|$.

(3.17) Prove that $\hat{\boldsymbol{a}} + \hat{\boldsymbol{b}}$ bisects the angle between arbitrary vectors $\boldsymbol{a}$ and $\boldsymbol{b}$. Sketch.

(3.18) Prove that $a\boldsymbol{b} + b\boldsymbol{a}$ bisects the angle between arbitrary vectors $\boldsymbol{a}$ and $\boldsymbol{b}$.

(3.19) Prove that $a\boldsymbol{b} + b\boldsymbol{a}$ and $a\boldsymbol{b} - b\boldsymbol{a}$ are orthogonal to each other, for any $\boldsymbol{a}$ and $\boldsymbol{b}$.

(3.20) For a triangle with vertices A, B, C, pick D on side

BC such that $|BD|/|DC| = |BA|/|AC|$. Prove that AD bisects the angle BAC. Sketch. (Hint: express $\overrightarrow{AD}$ in terms of $\boldsymbol{a} = \overrightarrow{AB}$ and $\boldsymbol{b} = \overrightarrow{AC}$.)

(3.21) Given three arbitrary points A, B, C in 3D space, derive a vector equation for the *line* through A that intersects BC orthogonally.

(3.22) Use vectors to show that the three heights (a.k.a. *altitudes* or *normals*) dropped from the vertices of a triangle perpendicular to their opposite sides intersect at the same point, the *orthocenter* H (Fig. 3.10). (Hint: If H is defined by $\overrightarrow{AH} \cdot \overrightarrow{CB} = 0$ and $\overrightarrow{BH} \cdot \overrightarrow{CA} = 0$, show that $\overrightarrow{CH} \cdot \overrightarrow{AB} = 0$.)

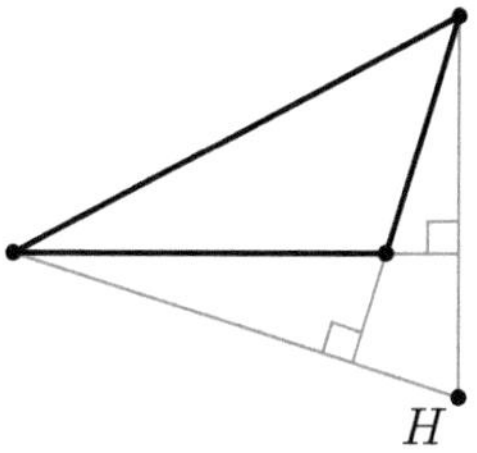

Fig. 3.10 Orthocenter H.

(3.23) Three points A, B, C in 3D space are specified by their cartesian coordinates. Show that the three equations $\overrightarrow{AB} \cdot \overrightarrow{CH} = 0$, $\overrightarrow{BC} \cdot \overrightarrow{AH} = 0$, $\overrightarrow{CA} \cdot \overrightarrow{BH} = 0$, are *not* sufficient to find the coordinates of H. Explain.

(3.24) Three points A, B, C in 3D space are specified by their cartesian coordinates. Derive an algorithm to compute the coordinates of the point H that is the intersection of the heights using linear combinations of known vectors and three dot products.

(3.25) A and B are two points on a sphere of radius R specified by their longitude and latitude. Specify the algorithm to compute the shortest distance between A and B, traveling on the sphere.

(3.26) Prove that any point P on the perpendicular bisector of AB is equidistant from A and B. Prove that the perpendicular bisectors of the sides of a triangle are concurrent and that their intersection O is the center of a circle that passes through all three vertices of the triangle. That point O is called the *circumcenter* (Fig. 3.11).

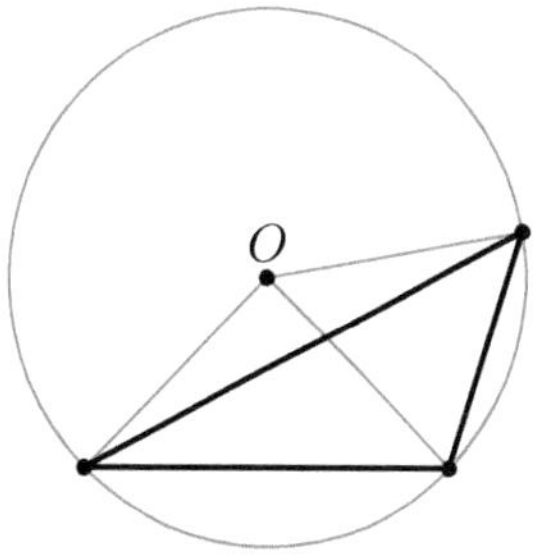

Fig. 3.11 Circumcenter O.

(3.27) Prove that any point P on an angle bisector is equidistant from the sides of the angle. Prove that the angle bisectors of a triangle are concurrent and that their intersection I is the center of a circle tangent to the three sides (the *incircle*).

(3.28) In an arbitrary triangle, let O be the circumcenter and G the centroid. Consider the point P such that $\overrightarrow{GP} = 2\,\overrightarrow{OG}$. Show that P is the orthocenter H; hence O, G, and H are on the same line called the *Euler line* (Fig. 3.12). (Hint: consider any vertex A and let D be the midpoint of the opposite side. Relate $\overrightarrow{AP}$ to $\overrightarrow{OD}$ using what you know about G and that $\overrightarrow{DO}$ is perpendicular to the BC side.)

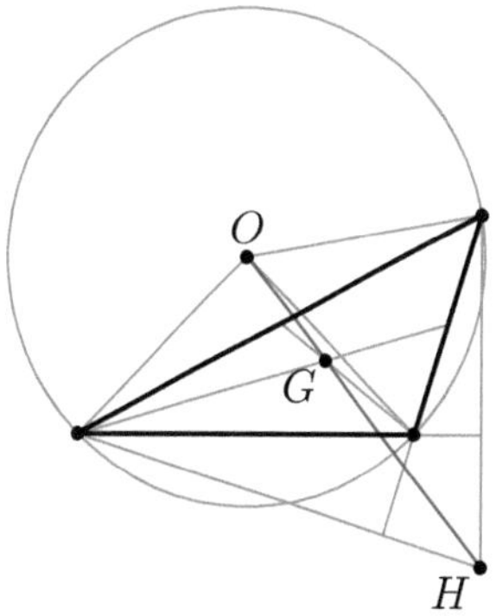

Fig. 3.12 Euler line OGH.

(3.29) In an arbitrary triangle, let H be the orthocenter and G the centroid. Consider the point P such that $\overrightarrow{HG} = 2\overrightarrow{GP}$. Show that P is the circumcenter O; hence O, G, and H are on the same line called the *Euler line*. (Hint: with vectors, consider any vertex V and let M be the midpoint of the opposite side. Relate $\overrightarrow{VH}$ to $\overrightarrow{PM}$.)

(3.30) Show that for any real α

$$\begin{cases} \hat{\mathbf{x}}' = \cos\alpha\,\hat{\mathbf{x}} + \sin\alpha\,\hat{\mathbf{y}}, \\ \hat{\mathbf{y}}' = -\sin\alpha\,\hat{\mathbf{x}} + \cos\alpha\,\hat{\mathbf{y}}, \end{cases}$$

are orthonormal if $\hat{\mathbf{x}}$, $\hat{\mathbf{y}}$ are. Sketch $\hat{\mathbf{x}}, \hat{\mathbf{y}}, \hat{\mathbf{x}}', \hat{\mathbf{y}}'$ for $\alpha \neq 0$.

(3.31) If $\boldsymbol{a} = a_x\hat{\mathbf{x}} + a_y\hat{\mathbf{y}}$, what are its components (a'_x, a'_y) in the basis $\{\hat{\mathbf{x}}', \hat{\mathbf{y}}'\}$ of the previous problem? Show and explain your work.

(3.32) For $\hat{\mathbf{x}}', \hat{\mathbf{y}}'$ as in problem 3.30, express $\hat{\mathbf{x}}$ and $\hat{\mathbf{y}}$ in terms of $\hat{\mathbf{x}}'$ and $\hat{\mathbf{y}}'$. Show your work.

(3.33) *Gram–Schmidt.* For any linearly independent vectors $\boldsymbol{a}$, $\boldsymbol{b}$ with $\boldsymbol{a}\cdot\boldsymbol{b} = ab\cos\theta$, let (Fig. 3.13)

$$\mathbf{e}_1 = \hat{\boldsymbol{a}}, \quad \boldsymbol{b}' = \boldsymbol{b} - (\boldsymbol{b}\cdot\mathbf{e}_1)\mathbf{e}_1, \quad \mathbf{e}_2 = \boldsymbol{b}'/|\boldsymbol{b}'|.$$

Prove that $\mathbf{e}_i\cdot\mathbf{e}_j = \delta_{ij}$ and express $(\boldsymbol{b}\cdot\mathbf{e}_1)$ and $|\boldsymbol{b}'|$ in terms of a, b, and θ. If $\boldsymbol{a} \equiv (2,3,4)$ and $\boldsymbol{b} \equiv (3,2,1)$ in cartesian form, find $\mathbf{e}_1$ and $\mathbf{e}_2$.

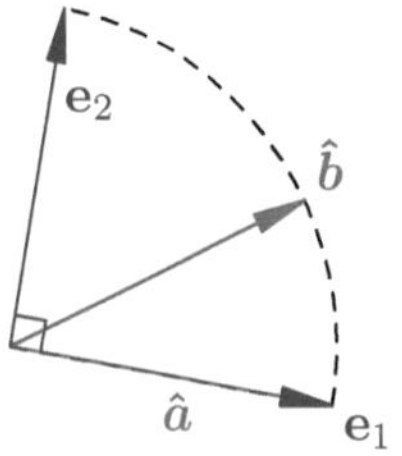

Fig. 3.13 Gram–Schmidt orthonormalization.

(3.34) For any linearly independent vectors $\boldsymbol{a}$, $\boldsymbol{b}$ with $\boldsymbol{a}\cdot\boldsymbol{b} = ab\cos\theta$, let (Fig. 3.14)

$$\mathbf{e}_1 = \frac{\hat{\boldsymbol{a}} + \hat{\boldsymbol{b}}}{|\hat{\boldsymbol{a}} + \hat{\boldsymbol{b}}|}, \quad \mathbf{e}_2 = \frac{\hat{\boldsymbol{b}} - \hat{\boldsymbol{a}}}{|\hat{\boldsymbol{b}} - \hat{\boldsymbol{a}}|}.$$

Show that $\mathbf{e}_i\cdot\mathbf{e}_j = \delta_{ij}$ and $|\hat{\boldsymbol{a}} + \hat{\boldsymbol{b}}| = 2\cos(\theta/2)$, $|\hat{\boldsymbol{a}} - \hat{\boldsymbol{b}}| = 2\sin(\theta/2)$. If $\boldsymbol{a} \equiv (2,3,4)$ and $\boldsymbol{b} \equiv (3,2,1)$ in cartesian, find $\mathbf{e}_1$ and $\mathbf{e}_2$.

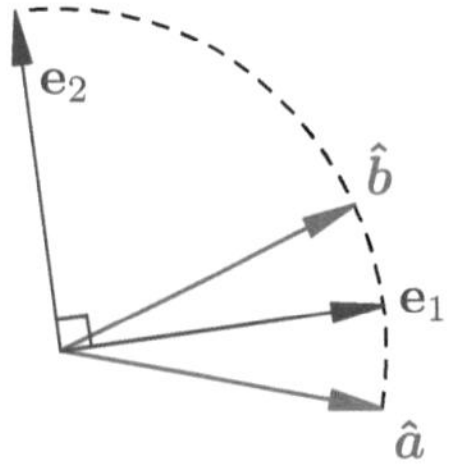

Fig. 3.14 Orthonormalization by reflection.

(3.35) Let $\hat{\boldsymbol{v}}'$ be the rotation of $\hat{\boldsymbol{v}}$ about $\hat{\boldsymbol{a}}$ by α. Derive (1.24) using suitable cartesian components (and some trigonometry). What intrinsic basis do you select?

(3.36) The orthographic projection of the earth in Fig. 1.23 is the orthogonal projection of the radius vector $\hat{\boldsymbol{r}}(\lambda, \varphi)$ on the local east–north basis

$$\hat{\mathbf{x}}' = \hat{\boldsymbol{e}}(\lambda_0, \varphi_0), \quad \hat{\mathbf{y}}' = \hat{\boldsymbol{n}}(\lambda_0, \varphi_0),$$

with $(\lambda_0, \varphi_0) = (40°, -90°)$ in Fig. 1.23. Derive

$$\begin{aligned} x' &= \hat{\boldsymbol{r}}(\lambda,\varphi)\cdot\hat{\boldsymbol{e}}(\lambda_0,\varphi_0) = \cos\lambda\,\sin(\varphi - \varphi_0), \\ y' &= \hat{\boldsymbol{r}}(\lambda,\varphi)\cdot\hat{\boldsymbol{n}}(\lambda_0,\varphi_0) \\ &= \sin\lambda\cos\lambda_0 - \cos\lambda\sin\lambda_0\cos(\varphi - \varphi_0). \end{aligned}$$

(3.37) If $\{\mathbf{e}_1, \mathbf{e}_2, \mathbf{e}_3\}$ and $\{\mathbf{e}'_1, \mathbf{e}'_2, \mathbf{e}'_3\}$ are two distinct orthogonal bases prove (3.22) but construct an explicit example that, in general,

$$a_1b_1 + 2a_2b_2 + 3a_3b_3 \neq a'_1b'_1 + 2a'_2b'_2 + 3a'_3b'_3.$$

(3.38) Rewriting (3.25) as $\boldsymbol{v} = \alpha\boldsymbol{a} + \beta\boldsymbol{b} + \gamma\boldsymbol{c}$ where $\boldsymbol{a}, \boldsymbol{b}, \boldsymbol{c}$ are not necessarily orthogonal, show/explain how to find (α, β, γ) using dot products and the following (*modified Gram–Schmidt*) orthogonalization algorithm:

(a) Construct $\boldsymbol{b}_\perp$ and $\boldsymbol{c}_\perp$ orthogonal to $\boldsymbol{a}$,
(b) Construct $\boldsymbol{c}_{\perp\perp}$ orthogonal to $\boldsymbol{a}$ and $\boldsymbol{b}_\perp$,
(c) Find $(\alpha', \beta', \gamma')$ such that $\boldsymbol{v} = \alpha'\,\boldsymbol{a} + \beta'\,\boldsymbol{b}_\perp + \gamma'\,\boldsymbol{c}_{\perp\perp}$,
(d) Find (α, β, γ) from $(\alpha', \beta', \gamma')$.

(3.39) If $\boldsymbol{w} = \sum_{i=1}^{3} w_i\mathbf{e}_i$, calculate $\mathbf{e}_j\cdot\boldsymbol{w}$ using $\sum$ notation and (3.17).

(3.40) Right or wrong? Diagnose and repair if necessary:

$$\begin{aligned} \mathbf{e}_i\cdot\sum_{i=1}^{3} w_i\mathbf{e}_i &= \sum_{i=1}^{3} w_i(\mathbf{e}_i\cdot\mathbf{e}_i) \\ &= \sum_{i=1}^{3} w_i\delta_{ii} = w_1 + w_2 + w_3. \end{aligned}$$

(3.41) If $\boldsymbol{v} = \sum_{i=1}^{3} v_i\mathbf{e}_i$ and $\boldsymbol{w} = \sum_{i=1}^{3} w_i\mathbf{e}_i$, calculate $\boldsymbol{v}\cdot\boldsymbol{w}$ using $\sum$ notation and (3.17).

(3.42) If $\boldsymbol{v} = \sum_{i=1}^{3} v_i\boldsymbol{a}_i$ and $\boldsymbol{w} = \sum_{i=1}^{3} w_i\boldsymbol{a}_i$, where the basis $\boldsymbol{a}_i$, $i = 1, 2, 3$, is *not* orthonormal, calculate $\boldsymbol{v}\cdot\boldsymbol{w}$ using $\sum$ notation.

(3.43) Calculate (i) $\sum_{j=1}^{3}\delta_{ij}a_j$, (ii) $\sum_{i=1}^{3}\sum_{j=1}^{3}\delta_{ij}a_jb_i$, (iii) $\sum_{j=1}^{3}\delta_{jj}$.

(3.44) Find all nonzero vectors orthogonal to $(1,2,3,4)$ in $\mathbb{R}^4$.

(3.45) What is the angle between $(1,2,3,4)$ and $(4,3,2,1)$?

(3.46) Decompose the vector $(4,2,1,7)$ into the sum of two vectors, one of which is parallel and the other perpendicular to $(1,2,3,4)$.

(3.47) Show that the infinity norm $\|\mathbf{x}\|_\infty = \max_i |x_i|$ for $\mathbf{x} \in \mathbb{R}^n$.

(3.48) Show that the integral $\int_a^b f(x)g(x)dx$ can be used as definition of the dot product for the vector space of real functions $f(x)$, $g(x)$ defined on $a < x < b$.

(3.49) *Orthogonal functions.* Show that $\cos nx$ with n integer, is a set of orthogonal functions for $x \in (0,\pi)$ for the dot product defined in problem 3.48. Use that orthogonality to show that the components f_n of the expansion of a function $f(x)$ in terms of that orthogonal basis,

$$f(x) = \sum_n f_n \cos nx,$$

are

$$f_n = \frac{2}{\pi\kappa_n}\int_0^\pi f(x)\cos nx\,\mathrm{d}x,$$

where $\kappa_n = 1+\delta_{n,0} = 2$ if $n=0$, and 1 otherwise. In particular, show that for $0 < x < \pi$,

$$\sin x = \frac{2}{\pi} - \frac{4}{\pi}\sum_{m=1}^{\infty}\frac{\cos 2mx}{4m^2-1},$$

yielding

$$\pi = 2 + \sum_{k=1}^{\infty}\frac{16}{(4k-3)(4k-1)(4k+1)}.$$

Cross Product

4

4.1 Geometry and algebra

The cross product, also called the *vector product* or the *area product*, is defined as

$$\boldsymbol{a} \times \boldsymbol{b} \triangleq A\hat{\boldsymbol{n}}, \tag{4.1a}$$

where $\hat{\boldsymbol{n}}$ is the unit vector perpendicular to both $\boldsymbol{a}$ and $\boldsymbol{b}$, with $\{\boldsymbol{a}, \boldsymbol{b}, \hat{\boldsymbol{n}}\}$ right-handed, and the magnitude

$$A = |\boldsymbol{a} \times \boldsymbol{b}| \triangleq |\boldsymbol{a}|\,|\boldsymbol{b}|\,\sin\theta,$$

is the area of the parallelogram spanned by $\boldsymbol{a}$ and $\boldsymbol{b}$, where θ is the angle between $\boldsymbol{a}$ and $\boldsymbol{b}$ (Fig. 4.1). The right-hand rule implies that the cross product *anti-commutes*

$$\boldsymbol{a} \times \boldsymbol{b} = -\boldsymbol{b} \times \boldsymbol{a}. \tag{4.1b}$$

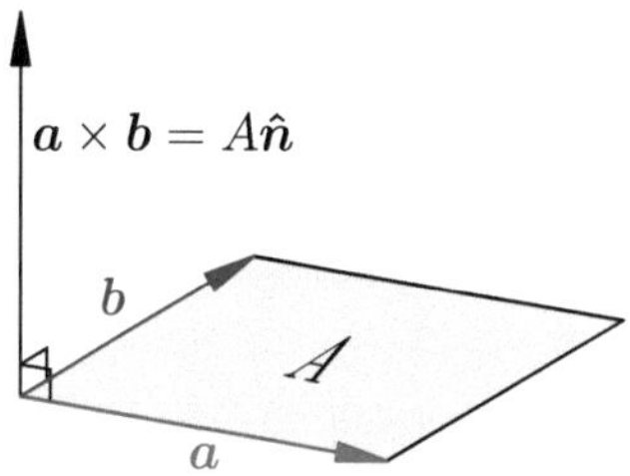

Fig. 4.1 3D view of $\boldsymbol{a} \times \boldsymbol{b}$.

Geometric properties

Since the area of a parallelogram is *base* × *height* and there are two ways to pick a base and a height, $(a, b_\perp)$ and $(a_\perp, b)$, the geometric definition (4.1a) yields the fundamental cross product property that (Fig. 4.2)

$$\boldsymbol{a} \times \boldsymbol{b} = \boldsymbol{a} \times \boldsymbol{b}_\perp = \boldsymbol{a}_\perp \times \boldsymbol{b}, \tag{4.2}$$

where $\boldsymbol{b}_\perp$ is the vector component of $\boldsymbol{b}$ perpendicular to $\boldsymbol{a}$, while $\boldsymbol{a}_\perp$ is the vector component of $\boldsymbol{a}$ perpendicular to $\boldsymbol{b}$. The magnitude of the cross product is

$$A = ab_\perp = a_\perp b = ab\sin\theta,$$

where $b_\perp = |\boldsymbol{b}_\perp| = b\sin\theta$ and $a_\perp = |\boldsymbol{a}_\perp| = a\sin\theta$.

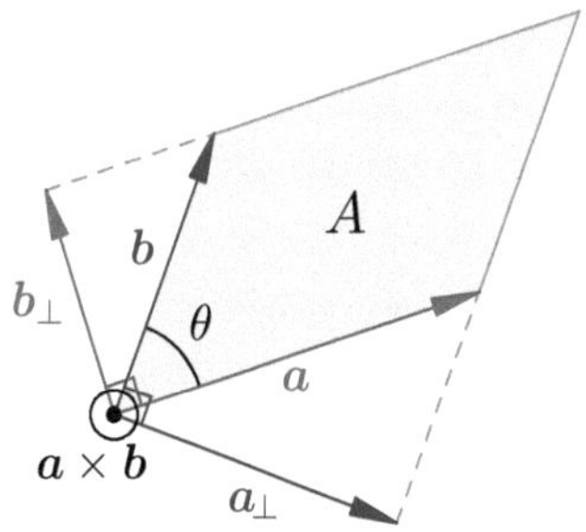

Fig. 4.2 Top view of $\boldsymbol{a} \times \boldsymbol{b}$.

Torque

In physics, the torque $\boldsymbol{T}$ about a point induced by a force $\boldsymbol{F}$ acting at a displacement $\boldsymbol{\ell}$ from that point is (Fig. 4.3)

$$\boldsymbol{T} = \boldsymbol{\ell} \times \boldsymbol{F} = \boldsymbol{\ell} \times \boldsymbol{F}_\perp = \boldsymbol{\ell}_\perp \times \boldsymbol{F}.$$

Torque $\boldsymbol{T} = \boldsymbol{\ell} \times \boldsymbol{F}$ has the same units as work $W = \boldsymbol{F} \cdot \boldsymbol{\ell}$ but torque is a vector quantity. Its direction indicates the direction of the axis of rotation that the torque would impart on a body.

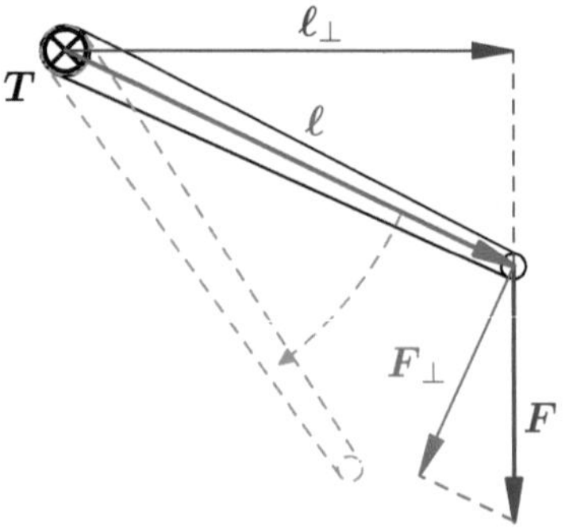

Fig. 4.3 Torque $\boldsymbol{T} = \boldsymbol{\ell} \times \boldsymbol{F}$ on a crank.

Distributive property

The cross product, like the dot product, has the distributivity property,

$$\boldsymbol{a} \times (\beta \boldsymbol{b} + \gamma \boldsymbol{c}) = \beta(\boldsymbol{a} \times \boldsymbol{b}) + \gamma(\boldsymbol{a} \times \boldsymbol{c}), \tag{4.3}$$

for any vectors $\boldsymbol{a}, \boldsymbol{b}, \boldsymbol{c}$ and scalars β, γ. This is not directly evident from the geometric definition, so a proof is in order.

▶ *Proof* Distributivity with respect to scalar multiplication, that is $\boldsymbol{a} \times (\beta \boldsymbol{b}) = \beta(\boldsymbol{a} \times \boldsymbol{b})$, is left as an exercise. For distributivity with respect to vector addition, let $\boldsymbol{v} \triangleq \boldsymbol{b} + \boldsymbol{c}$ and decompose all vectors into components parallel and perpendicular to $\boldsymbol{a}$. From (4.1a) and (4.2)

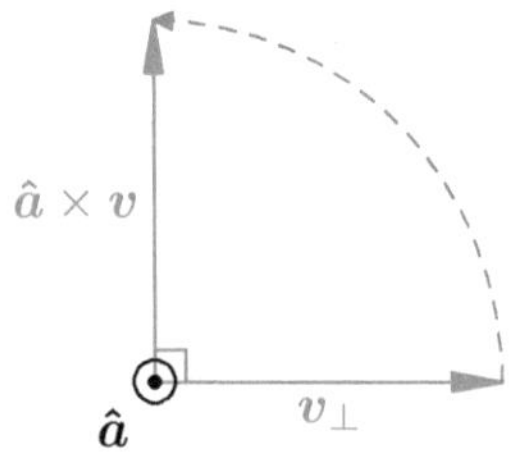

Fig. 4.4 $\hat{\boldsymbol{a}} \times \boldsymbol{v}$ is $\boldsymbol{v}_\perp$ rotated by $\pi/2$ about $\hat{\boldsymbol{a}}$.

$$\hat{\boldsymbol{a}} \times \boldsymbol{v} = \hat{\boldsymbol{a}} \times \boldsymbol{v}_\perp,$$

where $\perp$ is relative to $\boldsymbol{a}$ here. Furthermore, $\hat{\boldsymbol{a}} \times \boldsymbol{v}_\perp$ is seen to be the rotation of $\boldsymbol{v}_\perp$ by $\pi/2$ about $\hat{\boldsymbol{a}}$ (Fig. 4.4), and likewise for $\hat{\boldsymbol{a}} \times \boldsymbol{b} = \hat{\boldsymbol{a}} \times \boldsymbol{b}_\perp$ and $\hat{\boldsymbol{a}} \times \boldsymbol{c} = \hat{\boldsymbol{a}} \times \boldsymbol{c}_\perp$. Then, since $\boldsymbol{v} = \boldsymbol{b} + \boldsymbol{c}$ implies that $\boldsymbol{v}_\perp = \boldsymbol{b}_\perp + \boldsymbol{c}_\perp$ as shown in (3.11), it follows that

$$\hat{\boldsymbol{a}} \times \boldsymbol{v}_\perp = \hat{\boldsymbol{a}} \times \boldsymbol{b}_\perp + \hat{\boldsymbol{a}} \times \boldsymbol{c}_\perp,$$

because each of these cross products is simply the rotation of $\boldsymbol{v}_\perp$, $\boldsymbol{b}_\perp$, and $\boldsymbol{c}_\perp$, respectively, by a quarter-turn about $\hat{\boldsymbol{a}}$. Since the perpendicular components add up as $\boldsymbol{v}_\perp = \boldsymbol{b}_\perp + \boldsymbol{c}_\perp$, so do the rotated vectors $\hat{\boldsymbol{a}} \times \boldsymbol{v}_\perp$, $\hat{\boldsymbol{a}} \times \boldsymbol{b}_\perp$, $\hat{\boldsymbol{a}} \times \boldsymbol{c}_\perp$. This is illustrated in Fig. 4.5. □

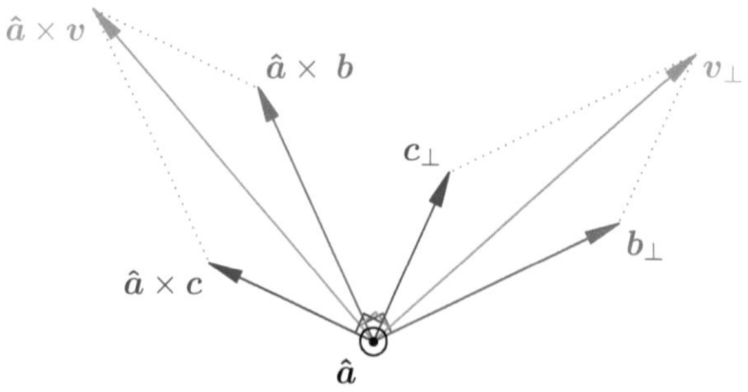

Fig. 4.5 Geometrically, the distributivity property $\hat{\boldsymbol{a}} \times \boldsymbol{v} = \hat{\boldsymbol{a}} \times \boldsymbol{b} + \hat{\boldsymbol{a}} \times \boldsymbol{c}$ is simply the $\frac{\pi}{2}$ rotation of $\boldsymbol{v}_\perp = \boldsymbol{b}_\perp + \boldsymbol{c}_\perp$.

Cartesian formula

The distributivity property (4.3) yields the cartesian formula for the cross product

$$\begin{aligned} \boldsymbol{a} \times \boldsymbol{b} &= (a_x\hat{\mathbf{x}} + a_y\hat{\mathbf{y}} + a_z\hat{\mathbf{z}}) \times (b_x\hat{\mathbf{x}} + b_y\hat{\mathbf{y}} + b_z\hat{\mathbf{z}}) \\ &= a_x b_x\, \hat{\mathbf{x}} \times \hat{\mathbf{x}} + a_x b_y\, \hat{\mathbf{x}} \times \hat{\mathbf{y}} + a_x b_z\, \hat{\mathbf{x}} \times \hat{\mathbf{z}} + \cdots \\ &= \hat{\mathbf{x}}(a_y b_z - a_z b_y) + \hat{\mathbf{y}}(a_z b_x - a_x b_z) + \hat{\mathbf{z}}(a_x b_y - a_y b_x) \end{aligned} \tag{4.4}$$

from (4.1b) and (Fig. 4.6)

$$\hat{\mathbf{x}} \times \hat{\mathbf{y}} = \hat{\mathbf{z}}, \quad \hat{\mathbf{y}} \times \hat{\mathbf{z}} = \hat{\mathbf{x}}, \quad \hat{\mathbf{z}} \times \hat{\mathbf{x}} = \hat{\mathbf{y}}, \tag{4.5}$$

that is

$$\hat{\mathbf{x}} = \hat{\mathbf{y}} \times \hat{\mathbf{z}}, \quad \hat{\mathbf{y}} = \hat{\mathbf{z}} \times \hat{\mathbf{x}}, \quad \hat{\mathbf{z}} = \hat{\mathbf{x}} \times \hat{\mathbf{y}}. \tag{4.6}$$

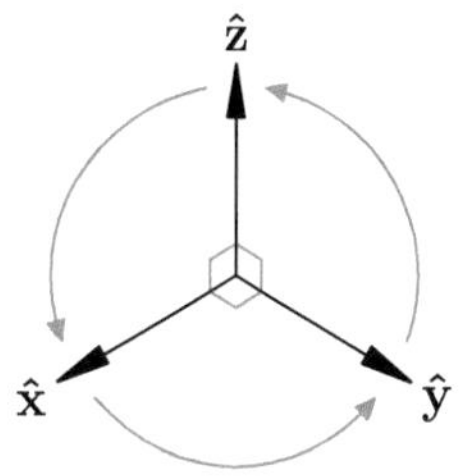

Fig. 4.6 Right-handed basis.

Each of these expressions is a cyclic permutation of the previous one

$$(x, y, z) \to (y, z, x) \to (z, x, y)$$

and (4.6) is obtained from (4.5) by switching the = and the ×, and vice versa. This cyclic property enables us to easily reconstruct formula (4.4) — or any one of its components, without having to use our right hand explicitly. We can simply reconstruct any component with that cyclic (even) or acyclic (odd) permutation rule. The cartesian expansion (4.4) of the cross product is often remembered using determinant notation

$$\begin{vmatrix} \hat{\mathbf{x}} & \hat{\mathbf{y}} & \hat{\mathbf{z}} \\ a_x & a_y & a_z \\ b_x & b_y & b_z \end{vmatrix} = \begin{vmatrix} \hat{\mathbf{x}} & a_x & b_x \\ \hat{\mathbf{y}} & a_y & b_y \\ \hat{\mathbf{z}} & a_z & b_z \end{vmatrix}$$
$$= \hat{\mathbf{x}}(a_y b_z - a_z b_y) + \hat{\mathbf{y}}(a_z b_x - a_x b_z) + \hat{\mathbf{z}}(a_x b_y - a_y b_x). \tag{4.7}$$

4.2 Double cross product

The double cross product, also known as the *vector triple product*, occurs frequently in applications directly or indirectly as a result of mirror symmetry, as discussed in section 4.3.

An important special case of double cross product is (Fig. 4.7)

$$(\boldsymbol{a} \times \boldsymbol{b}) \times \boldsymbol{a} = a^2\, \boldsymbol{b}_\perp, \tag{4.8}$$

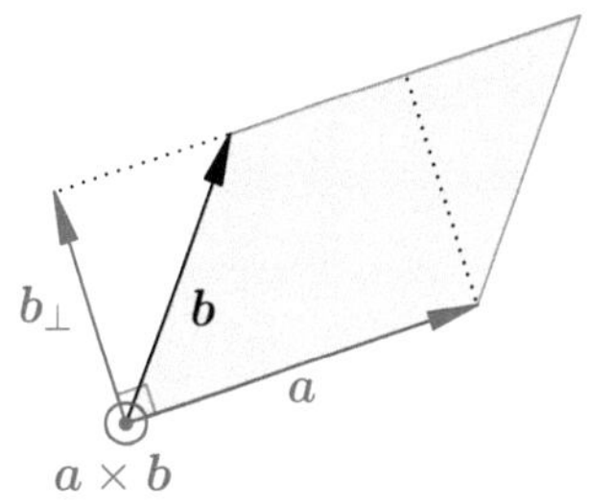

Fig. 4.7 $(\boldsymbol{a} \times \boldsymbol{b}) \times \boldsymbol{a} = a^2 \boldsymbol{b}_\perp$.

where $a^2 = \boldsymbol{a} \cdot \boldsymbol{a}$ and $\boldsymbol{b}_\perp$ is the vector component of $\boldsymbol{b}$ perpendicular to $\boldsymbol{a}$. The identity (4.8) follows from the geometric definition of the cross product since $\boldsymbol{a} \times \boldsymbol{b} = \boldsymbol{a} \times \boldsymbol{b}_\perp$, that proof is left as exercise 4.25.

The general double cross product obeys the identities

$$\begin{aligned} (\boldsymbol{a} \times \boldsymbol{b}) \times \boldsymbol{c} &= \boldsymbol{b}\,(\boldsymbol{a} \cdot \boldsymbol{c}) - \boldsymbol{a}\,(\boldsymbol{b} \cdot \boldsymbol{c}), \\ \boldsymbol{a} \times (\boldsymbol{b} \times \boldsymbol{c}) &= \boldsymbol{b}\,(\boldsymbol{a} \cdot \boldsymbol{c}) - \boldsymbol{c}\,(\boldsymbol{a} \cdot \boldsymbol{b}). \end{aligned} \tag{4.9}$$

Thus, in general

$$(\boldsymbol{a} \times \boldsymbol{b}) \times \boldsymbol{c} \neq \boldsymbol{a} \times (\boldsymbol{b} \times \boldsymbol{c}),$$

except when $\boldsymbol{c} = \alpha\,\boldsymbol{a}$ as in (4.8). The identities (4.9) follow from each other after some manipulations and renaming of vectors, but we can remember both at once as:

middle vector times dot product of the other two,
minus
other vector in parentheses times dot product of the other two.

The double cross product identity (4.9) can be verified by tediously expanding both sides in cartesian components and reducing the algebra. We give two other proofs that serve as exercises for the vector geometry and algebra of the cross product.

▶ *Vector geometry proof of* (4.9) Let $\boldsymbol{X} = (\boldsymbol{a} \times \boldsymbol{b}) \times \boldsymbol{c}$. By definition of the cross product, $\boldsymbol{X}$ must be orthogonal to $(\boldsymbol{a} \times \boldsymbol{b})$ and to $\boldsymbol{c}$. Orthogonality to $\boldsymbol{a} \times \boldsymbol{b}$, that is $\boldsymbol{X} \cdot (\boldsymbol{a} \times \boldsymbol{b}) = 0$, implies that

$$\boldsymbol{X} = x_1\,\boldsymbol{a} + x_2\,\boldsymbol{b}. \tag{4.10a}$$

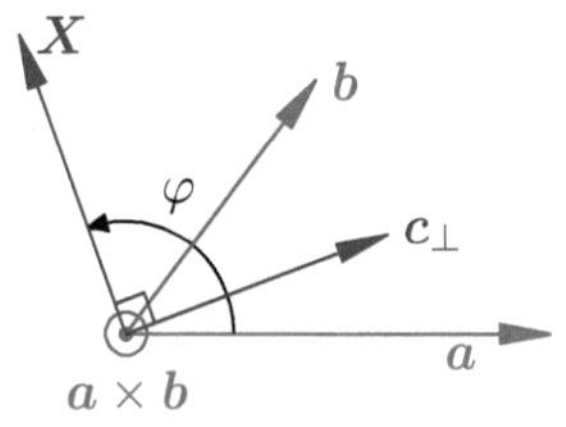

Fig. 4.8 $\boldsymbol{X} = (\boldsymbol{a} \times \boldsymbol{b}) \times \boldsymbol{c}$.

Orthogonality to $\mathbf{c}$ requires that $\boldsymbol{X} \cdot \boldsymbol{c} = x_1\,(\boldsymbol{a} \cdot \boldsymbol{c}) + x_2\,(\boldsymbol{b} \cdot \boldsymbol{c}) = 0$. Hence, $x_1 = \mu(\boldsymbol{b} \cdot \boldsymbol{c})$, $x_2 = -\mu(\boldsymbol{a} \cdot \boldsymbol{c})$, for any μ, such that (Fig. 4.8)

$$\boldsymbol{X} = \mu\,(\boldsymbol{b} \cdot \boldsymbol{c})\,\boldsymbol{a} - \mu\,(\boldsymbol{a} \cdot \boldsymbol{c})\,\boldsymbol{b}. \tag{4.10b}$$

To find the scalar μ, we take yet another cross product of (4.10b) with $\boldsymbol{a}$ (or $\boldsymbol{b}$),

$$\boldsymbol{a} \times \boldsymbol{X} = -\mu(\boldsymbol{a} \cdot \boldsymbol{c})\,(\boldsymbol{a} \times \boldsymbol{b}). \tag{4.10c}$$

The left-hand side of (4.10c) is a triple cross product, but most of the vectors are orthogonal to each other so we can figure it out from the geometric definition (4.1a). We find that

$$\boldsymbol{a} \times \boldsymbol{X} = |\boldsymbol{a}||\boldsymbol{X}|\sin\varphi\,\frac{\boldsymbol{a} \times \boldsymbol{b}}{|\boldsymbol{a} \times \boldsymbol{b}|} = |\boldsymbol{a}||\boldsymbol{c}_\perp|\sin\varphi\,(\boldsymbol{a} \times \boldsymbol{b}),$$

since $|\boldsymbol{X}| = |\boldsymbol{a} \times \boldsymbol{b}|\,|\boldsymbol{c}_\perp|$, where $\boldsymbol{c}_\perp$ is the component of $\boldsymbol{c}$ perpendicular to $\boldsymbol{a} \times \boldsymbol{b}$, and φ is the angle from $\boldsymbol{a}$ to $\boldsymbol{X}$, positive for right-hand rotation about $\boldsymbol{a} \times \boldsymbol{b}$. Now for the right-hand side of (4.10c): $\boldsymbol{a} \cdot \boldsymbol{c} = \boldsymbol{a} \cdot \boldsymbol{c}_\perp = |\boldsymbol{a}||\boldsymbol{c}_\perp|\cos(\varphi - \pi/2) = |\boldsymbol{a}||\boldsymbol{c}_\perp|\sin\varphi$. Thus, $\mu = -1$ and (4.9) is proven. □

▶ *Vector algebra proof of* (4.9) Use the intrinsic orthogonal basis $\boldsymbol{a}, \boldsymbol{b}_\perp, (\boldsymbol{a} \times \boldsymbol{b})$. In that basis

$$\boldsymbol{a} = \boldsymbol{a}, \quad \boldsymbol{b} = \beta\boldsymbol{a} + \boldsymbol{b}_\perp, \quad \boldsymbol{c} = \gamma_1\boldsymbol{a} + \gamma_2\boldsymbol{b}_\perp + \gamma_3(\boldsymbol{a} \times \boldsymbol{b}), \tag{4.11}$$

where

$$\beta = \frac{\boldsymbol{a} \cdot \boldsymbol{b}}{\boldsymbol{a} \cdot \boldsymbol{a}}, \; \gamma_1 = \frac{\boldsymbol{a} \cdot \boldsymbol{c}}{\boldsymbol{a} \cdot \boldsymbol{a}}, \; \gamma_2 = \frac{\boldsymbol{b}_\perp \cdot \boldsymbol{c}}{\boldsymbol{b}_\perp \cdot \boldsymbol{b}_\perp}, \; \gamma_3 = \frac{(\boldsymbol{a} \times \boldsymbol{b}) \cdot \boldsymbol{c}}{|\boldsymbol{a} \times \boldsymbol{b}|^2}.$$

Substituting for $\boldsymbol{b}$ and $\boldsymbol{c}$ from (4.11) and using (4.8) twice, once for $(\boldsymbol{a}\times\boldsymbol{b}_\perp)\times\boldsymbol{a} = (\boldsymbol{a}\cdot\boldsymbol{a})\boldsymbol{b}_\perp$ and another for $(\boldsymbol{a}\times\boldsymbol{b}_\perp)\times\boldsymbol{b}_\perp = -(\boldsymbol{b}_\perp\cdot\boldsymbol{b}_\perp)\boldsymbol{a}$, yields

$$(\boldsymbol{a}\times\boldsymbol{b})\times\boldsymbol{c} = (\boldsymbol{a}\cdot\boldsymbol{c})\boldsymbol{b}_\perp - (\boldsymbol{b}_\perp\cdot\boldsymbol{c})\boldsymbol{a}.$$

Similarly, using $\boldsymbol{b} = \beta\boldsymbol{a} + \boldsymbol{b}_\perp$ yields

$$(\boldsymbol{a}\cdot\boldsymbol{c})\boldsymbol{b} - (\boldsymbol{b}\cdot\boldsymbol{c})\boldsymbol{a} = (\boldsymbol{a}\cdot\boldsymbol{c})\boldsymbol{b}_\perp - (\boldsymbol{b}_\perp\cdot\boldsymbol{c})\boldsymbol{a},$$

since $(\boldsymbol{a}\cdot\boldsymbol{c})(\beta\boldsymbol{a}) - (\beta\boldsymbol{a}\cdot\boldsymbol{c})\boldsymbol{a} = 0$. Thus, (4.9) is true for all $\boldsymbol{a}, \boldsymbol{b}, \boldsymbol{c}$. □

4.3 Orientation of bases

If we pick an arbitrary unit vector $\mathbf{e}_1$, then a unit vector $\mathbf{e}_2$ orthogonal to $\mathbf{e}_1$, there are two opposite unit vectors $\mathbf{e}_3$ orthogonal to both $\mathbf{e}_1$ and $\mathbf{e}_2$. One choice gives a *right-handed basis* (i.e. $\mathbf{e}_1$ in right thumb direction, $\mathbf{e}_2$ in right index direction, and $\mathbf{e}_3$ in right major direction). The other choice gives a *left-handed basis*. These two types of bases are *mirror images* of each other as illustrated in Fig. 4.9, where $\mathbf{e}_1{}' = \mathbf{e}_1$ point straight out of the page (or screen).

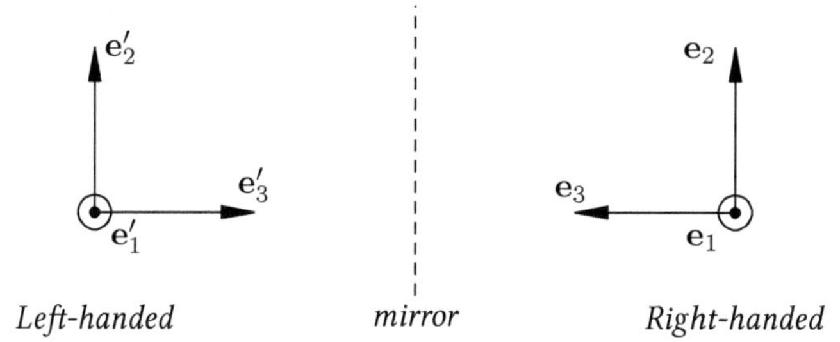

Fig. 4.9 Left- and right-handed bases.

Figure 4.9 reveals an interesting subtlety of the cross product. For this particular choice of left- and right-handed bases (other arrangements are possible, of course), $\{\mathbf{e}_1', \mathbf{e}_2', \mathbf{e}_3'\} = \{\mathbf{e}_1, \mathbf{e}_2, -\mathbf{e}_3\}$ with $\mathbf{e}_1\times\mathbf{e}_2 = \mathbf{e}_3$ but then

$$\mathbf{e}_1'\times\mathbf{e}_2' = -\mathbf{e}_3{}'.$$

Thus, the mirror image of the cross product is *not* the cross product of the mirror images. The mirror image of the cross product, $\mathbf{e}_3'$, is *minus* the cross-product of the images $\mathbf{e}_1'\times\mathbf{e}_2'$.

We showed this for a special case, but this is general. The cross-product is not invariant under reflection; it changes sign. Physical laws should not depend on the choice of basis, so this implies that they should not be expressed in terms of an *odd* number of cross products. When we write that the velocity of a particle is $\boldsymbol{v} = \boldsymbol{\omega}\times\boldsymbol{r}$, the mirror image of $\boldsymbol{v}$ must be the mirror image of $\boldsymbol{\omega}\times\boldsymbol{r}$, but this requires that $\boldsymbol{\omega}$ change sign under reflection. So $\boldsymbol{\omega}$ is not quite a vector; it is a *pseudo-vector*. It changes sign under reflection. That is because angular velocity vectors such as $\boldsymbol{\omega}$ are themselves defined according to the right-hand rule, so an expression such as $\boldsymbol{\omega}\times\boldsymbol{r}$ actually contains two applications of the right-hand rule. Likewise in the Lorentz force $\boldsymbol{F} = q\boldsymbol{v} \times \boldsymbol{B}$, $\boldsymbol{F}$ and $\boldsymbol{v}$ are true vectors, but since the definition involves a cross product, it must be that the magnetic field $\boldsymbol{B}$ is a pseudo-vector. Indeed, $\boldsymbol{B}$ itself arises from a cross product (the Biot–Savart law) so the definition of $\boldsymbol{F}$ actually contains two cross products.

The orientation (right-handed or left-handed) did not matter before, but now that we have defined the cross product with the right-hand rule, we typically choose right-handed bases. We don't have to, geometrically speaking, but we need to from an algebraic point of view; otherwise, we'd need to change the sign of the cartesian formula (4.4) to obtain the correct *right*-hand rule for *left*-handed bases.

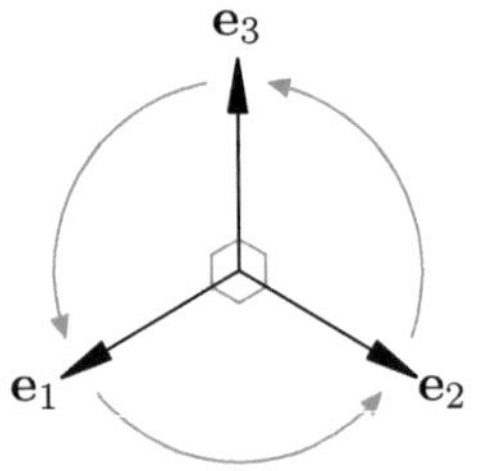

Fig. 4.10 Right-handed basis.

The right-handed cross product implies that a right-handed basis $\{\mathbf{e}_1, \mathbf{e}_2, \mathbf{e}_3\}$ satisfies (Fig. 4.10)

$$\mathbf{e}_1 \times \mathbf{e}_2 = \mathbf{e}_3, \quad \mathbf{e}_2 \times \mathbf{e}_3 = \mathbf{e}_1, \quad \mathbf{e}_3 \times \mathbf{e}_1 = \mathbf{e}_2, \tag{4.12a}$$

$$\mathbf{e}_2 \times \mathbf{e}_1 = -\mathbf{e}_3, \quad \mathbf{e}_3 \times \mathbf{e}_2 = -\mathbf{e}_1, \quad \mathbf{e}_1 \times \mathbf{e}_3 = -\mathbf{e}_2. \tag{4.12b}$$

Note that (4.12a) are *cyclic* permutations of the basis vectors

$$(1, 2, 3) \to (2, 3, 1) \to (3, 1, 2).$$

The orderings of the basis vectors in (4.12b) are *acyclic* permutations of $(1, 2, 3)$. For three elements, a *cyclic* permutation corresponds to an *even* number of permutations. For instance, we can go from $(1, 2, 3)$ to $(2, 3, 1)$ in two permutations $(1, 2, 3) \to (2, 1, 3) \to (2, 3, 1)$. The concept of *even* and *odd* number of permutations is more general than that of cyclic and acyclic permutations, as explored in the exercises. For three elements it is useful to think in terms of cyclic and acyclic permutations.

If we expand $\boldsymbol{a}$ and $\boldsymbol{b}$ in terms of the right-handed $\{\mathbf{e}_1, \mathbf{e}_2, \mathbf{e}_3\}$, then distribute the cross product, i.e., in compact summation form,

$$\boldsymbol{a} = \sum_{i=1}^{3} a_i \mathbf{e}_i, \quad \boldsymbol{b} = \sum_{j=1}^{3} b_j \mathbf{e}_j, \quad \Rightarrow \boldsymbol{a} \times \boldsymbol{b} = \sum_{i=1}^{3}\sum_{j=1}^{3} a_i b_j \, (\mathbf{e}_i \times \mathbf{e}_j);$$

we obtain

$$\boldsymbol{a} \times \boldsymbol{b} = \mathbf{e}_1(a_2 b_3 - a_3 b_2) + \mathbf{e}_2(a_3 b_1 - a_1 b_3) + \mathbf{e}_3(a_1 b_2 - a_2 b_1), \tag{4.13}$$

which can be reconstructed using the cyclic permutations of

$$+ : (1, 2, 3) \to (2, 3, 1) \to (3, 1, 2),$$
$$- : (1, 3, 2) \to (2, 1, 3) \to (3, 2, 1).$$

Exercises

(4.1) Show that $|\boldsymbol{a} \times \boldsymbol{b}|^2 + (\boldsymbol{a} \cdot \boldsymbol{b})^2 = |\boldsymbol{a}|^2\,|\boldsymbol{b}|^2,\ \forall \boldsymbol{a}, \boldsymbol{b}$.

(4.2) Prove that $\boldsymbol{\omega} \times (\alpha \boldsymbol{a}) = \alpha(\boldsymbol{\omega} \times \boldsymbol{a})$ for any vectors $\boldsymbol{\omega}$, $\boldsymbol{a}$ and scalar α. Consider both $\alpha \geq 0$ and $\alpha < 0$.

(4.3) True or false: $\boldsymbol{a} \times \boldsymbol{b} = \boldsymbol{a}_\perp \times \boldsymbol{b}_\perp$. Explain.

(4.4) A particle of charge q moving at velocity $\boldsymbol{v}$ in a magnetic field $\boldsymbol{B}$ experiences the Lorentz force $\boldsymbol{F} = q\boldsymbol{v} \times \boldsymbol{B}$. Show that there is no force in the direction of the magnetic field and that the Lorentz force does no work on the particle. What does it do to the particle?

(4.5) Derive $\sin(\alpha - \beta) = \sin\alpha \cos\beta - \sin\beta \cos\alpha$ using a cross product (Fig. 4.11). Explain.

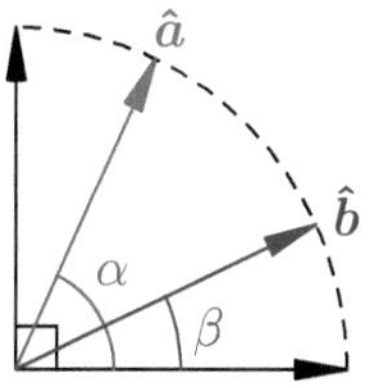

Fig. 4.11 $\sin(\alpha - \beta) = ?$

(4.6) If $\hat{\boldsymbol{n}}$ is orthogonal to both $\boldsymbol{a}$ and $\boldsymbol{b}$, prove and sketch

$$(\boldsymbol{a} \times \hat{\boldsymbol{n}}) \cdot (\boldsymbol{b} \times \hat{\boldsymbol{n}}) = \boldsymbol{a} \cdot \boldsymbol{b},$$
$$(\boldsymbol{a} \times \hat{\boldsymbol{n}}) \times (\boldsymbol{b} \times \hat{\boldsymbol{n}}) = \boldsymbol{a} \times \boldsymbol{b}.$$

(4.7) Let (a_1, a_2, a_3) and (b_1, b_2, b_3) represent the components of $\boldsymbol{a}$ and $\boldsymbol{b}$ with respect to a *left-handed* orthonormal basis $\{\mathbf{e}_1, \mathbf{e}_2, \mathbf{e}_3\}$. What is the formula corresponding to (4.13) in that basis? Explain.

(4.8) Sketch three vectors such that $\boldsymbol{a}+\boldsymbol{b}+\boldsymbol{c} = 0$. Show that $\boldsymbol{a}\times\boldsymbol{b} = \boldsymbol{b}\times\boldsymbol{c} = \boldsymbol{c}\times\boldsymbol{a}$ in two ways: (1) from the geometric definition of the cross product and (2) from the algebraic properties of the cross product. Deduce the *law of sines* (Fig. 4.12)

$$\frac{\sin\alpha}{a} = \frac{\sin\beta}{b} = \frac{\sin\gamma}{c} = \frac{1}{D},$$

where D is the diameter of the circumcircle (cf. *inscribed angle theorem*, Fig. 1.22).

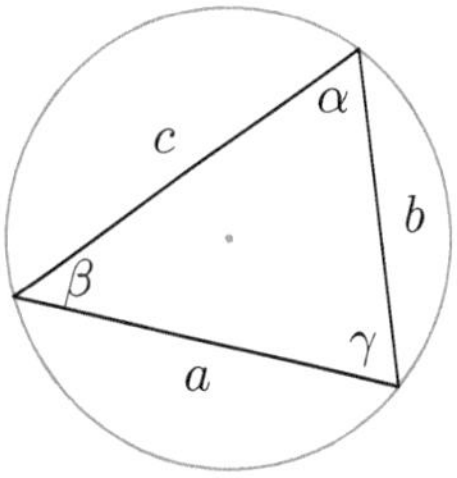

Fig. 4.12 Law of sines.

(4.9) Show that the angle between $\boldsymbol{a} = \overrightarrow{F_1P}$ and $\boldsymbol{b} = \overrightarrow{F_2P}$ is

$$\alpha = \operatorname{atan2}(\hat{\mathbf{z}} \cdot (\boldsymbol{a} \times \boldsymbol{b}), \boldsymbol{a} \cdot \boldsymbol{b}),$$

where atan2 is the two argument arctan function (1.9), and $\hat{\mathbf{z}}$ is the unit normal to the oriented plane of F_1, F_2, P (Fig. 4.13). Show that $\alpha > 0$ when P is to the left of $\overrightarrow{F_1F_2}$ and $\alpha < 0$ when P is to the right of $\overrightarrow{F_1F_2}$. What happens to α as P moves across the line segment F_1F_2?

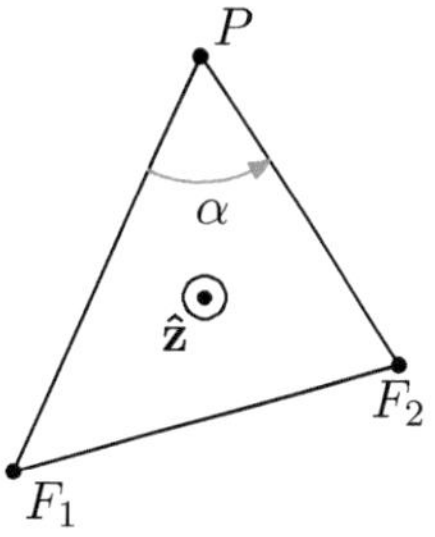

Fig. 4.13 Find α from F_1, F_2, P.

(4.10) Consider the triangle $x/a + y/b + z/c = 1$ for constant positive a, b, c and variable positive cartesian coordinates x,y,z. Show that

$$A\hat{\boldsymbol{n}} = \tfrac{1}{2}\left(\hat{\mathbf{x}}\,bc + \hat{\mathbf{y}}\,ca + \hat{\mathbf{z}}\,ab\right) \qquad (4.14)$$

is its *area vector*, that is a vector perpendicular to the triangle with a magnitude equal to its area (Fig. 4.14).

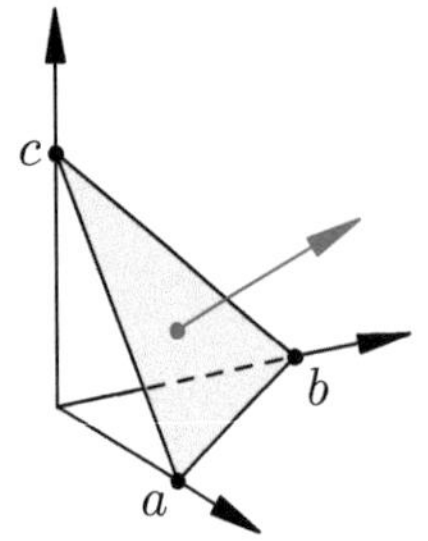

Fig. 4.14 Area vector.

(4.11) Let $\boldsymbol{r}_k = \overrightarrow{OP_k}$ denote the position vectors of points P_0, P_1, P_2 from point O in 3D space, respectively. Show that

$$A\hat{\boldsymbol{n}} \triangleq \tfrac{1}{2}(\boldsymbol{r}_0 \times \boldsymbol{r}_1 + \boldsymbol{r}_1 \times \boldsymbol{r}_2 + \boldsymbol{r}_2 \times \boldsymbol{r}_0) \qquad (4.15)$$

is a vector whose magnitude is the area of the triangle P_0, P_1, P_2 and direction is perpendicular to the triangle in a direction determined by the ordering P_0, P_1, P_2 (Fig. 4.15). Sketch the contributions of each of the three terms on the right-hand side of (4.15) for three planar cases: (i) O is one of the vertices P_0, P_1, P_2. (ii) O is inside the triangle. (iii) O is outside the triangle but in the same plane.

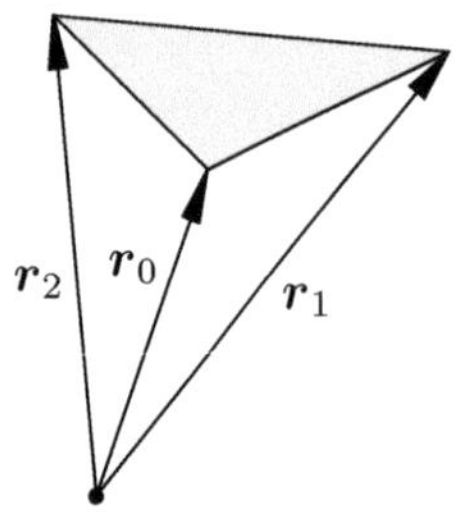

Fig. 4.15 Area vector.

(4.12) Consider an arbitrary non-self intersecting quadrilateral in the (x, y) plane with vertices P_0, P_1, P_2, P_3. Show that

$$A\hat{\boldsymbol{n}} \triangleq \tfrac{1}{2}(\boldsymbol{r}_0 \times \boldsymbol{r}_1 + \boldsymbol{r}_1 \times \boldsymbol{r}_2 + \boldsymbol{r}_2 \times \boldsymbol{r}_3 + \boldsymbol{r}_3 \times \boldsymbol{r}_0)$$

is a vector whose magnitude is the area of the quadrilateral and points in the $\pm\hat{\mathbf{z}}$ direction, depending on whether P_0, P_1, P_2, P_3 are oriented counterclockwise, or clockwise, respectively, where $\boldsymbol{r}_k = \overrightarrow{OP_k}$ is the position vector of point P_k.

(4.13) Let $P_0, P_1, \ldots, P_{n-1}$ be the vertices of a non-self intersecting polygon in the (x, y) plane (Fig. 4.16). Explain why the area of the polygon is

$$A = \frac{1}{2}\sum_{k=0}^{n-1}(x_k y_{k+1} - x_{k+1} y_k), \qquad (4.16)$$

where (x_k, y_k) are the cartesian coordinates of vertex P_k and $P_n \equiv P_0$. The area is positive if the vertices are ordered counterclockwise.

Fig. 4.16 Area of planar polygon.

(4.14) Four arbitrary points in 3D euclidean space $\mathbb{E}^3$ are the vertices of a tetrahedron. Consider the *area vectors* perpendicular, pointing outward for each of the four faces of the tetrahedron, with magnitudes equal to the area of the corresponding triangular face (Fig. 4.17). Show that the sum of these area vectors is zero:

$$\boldsymbol{A}_1 + \boldsymbol{A}_2 + \boldsymbol{A}_2 + \boldsymbol{A}_4 = 0.$$

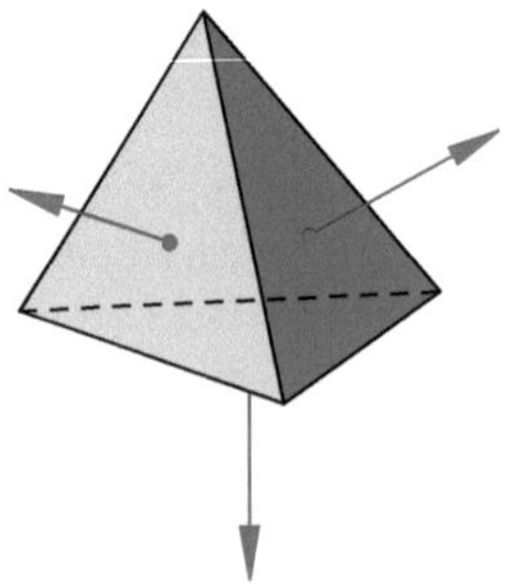

Fig. 4.17 Area vectors of a tetrahedron.

(4.15) Imagine "gluing" together two tetrahedra with a common interior face that is removed. The resulting polyhedron has six triangular faces (hexahedron). Show that the sum of the outward pointing area vectors is zero. This result and procedure extend to more complex polyhedra and to curved surfaces, as will be shown later with the *gradient theorem* in vector calculus.

(4.16) Let V be the number of vertices, F the number of faces, and E the number of edges of the polyhedron obtained by gluing together tetrahedra, as in the previous problem. Deduce Euler's formula that $V + F - E = 2$ for such polyhedra. Can you find examples where that formula breaks down?

(4.17) Consider arbitrary points $P_0, P_1, \ldots, P_{n-1}$ in 3D space. Let $P_n = P_0$. Show that

$$\boldsymbol{A} = \frac{1}{2}\sum_{k=1}^{n} \boldsymbol{r}_{k-1} \times \boldsymbol{r}_k = \frac{1}{2}\sum_{k=1}^{n} \overrightarrow{OP}_{k-1} \times \overrightarrow{OP}_k$$

is independent of the choice of origin O, and that

$$\boldsymbol{A} = \frac{1}{2}\sum_{k=2}^{n-1} (\boldsymbol{r}_{k-1} - \boldsymbol{r}_0) \times (\boldsymbol{r}_k - \boldsymbol{r}_{k-1}).$$

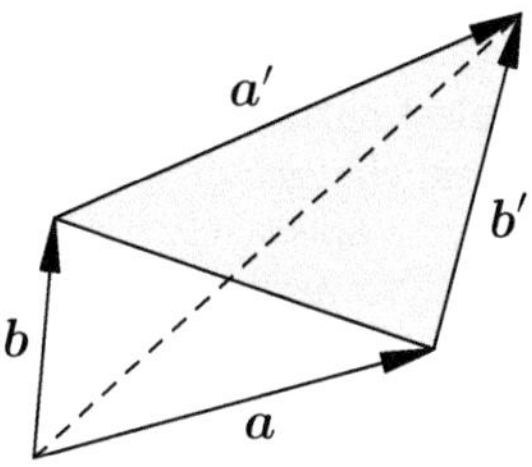

Fig. 4.18 Quadrilateral in 3D space.

(4.18) Consider an arbitrary quadrilateral with edges $\boldsymbol{a} + \boldsymbol{b}' = \boldsymbol{a}' + \boldsymbol{b}$ (Fig. 4.18). Show that

$$\begin{aligned}\boldsymbol{A} &= \tfrac{1}{2}\boldsymbol{a} \times \boldsymbol{b} + \tfrac{1}{2}\boldsymbol{a}' \times \boldsymbol{b}' \\ &= \tfrac{1}{2}\boldsymbol{a} \times \boldsymbol{b}' + \tfrac{1}{2}\boldsymbol{a}' \times \boldsymbol{b} = \bar{\boldsymbol{a}} \times \bar{\boldsymbol{b}}, \quad (4.17)\end{aligned}$$

where $\bar{\boldsymbol{a}}$ and $\bar{\boldsymbol{b}}$ are the average $\boldsymbol{a}$'s and $\boldsymbol{b}$'s,

$$\bar{\boldsymbol{a}} = \tfrac{1}{2}(\boldsymbol{a} + \boldsymbol{a}'), \quad \bar{\boldsymbol{b}} = \tfrac{1}{2}(\boldsymbol{b} + \boldsymbol{b}'),$$

or the opposite midpoint connectors

$$\bar{\boldsymbol{a}} = \boldsymbol{a} + \tfrac{1}{2}(\boldsymbol{b}' - \boldsymbol{b}), \quad \bar{\boldsymbol{b}} = \boldsymbol{b} + \tfrac{1}{2}(\boldsymbol{a}' - \boldsymbol{a}).$$

Interpret geometrically in terms of triangle and parallelogram area vectors. This result is relevant to *surface elements* $\mathrm{d}\boldsymbol{S}$ made of quadrilateral patches (13.15).

(4.19) Point P rotates at angular rate ω about the axis parallel to $\boldsymbol{a}$ passing through point A. Make a 3D sketch as well as "side" and "top" views. Derive a vector formula for the *velocity* of P.

(4.20) *Vector rotation formula.* Vector $\boldsymbol{v}'$ is the rotation of vector $\boldsymbol{v}$ about $\hat{\boldsymbol{a}}$ by angle α. Derive and explain the formula (Fig. 4.19)

$$\boldsymbol{v}' = \boldsymbol{v}_{\parallel} + \boldsymbol{v}_{\perp} \cos\alpha + (\hat{\boldsymbol{a}} \times \boldsymbol{v}_{\perp}) \sin\alpha. \quad (4.18)$$

Write a computer code or *pseudocode* that finds $\boldsymbol{v}'$ given $\boldsymbol{v}$, $\boldsymbol{a}$, α, assuming that the vectors are provided in cartesian form (and $|\boldsymbol{a}|$ is not necessarily 1). Use calls to `dot(a,b)` and `cross(a,b)` functions, as defined in Matlab, for example. Equation (4.18) is known as *Euler's*, *Rodrigues'*, or *Gibbs' formula* in the literature. Vectors were not explicitly used by Euler in 1775 or Rodrigues in 1840. The vector formula appears in §127 of Wilson's 1901 redaction of Gibbs' lectures. See also problem 4.37.

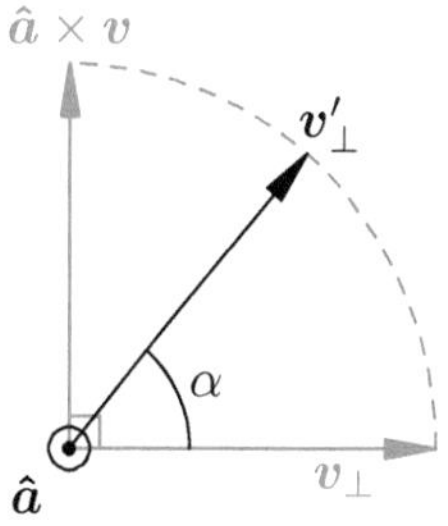

Fig. 4.19 Rotation $\boldsymbol{v} \to \boldsymbol{v}'$.

(4.21) Test the rotation code implementing (4.18) by designing simple rotation tests. Show that if $\boldsymbol{v} \equiv (2, 3, 4)$, $\boldsymbol{a} \equiv (3, 2, 1)$, $\alpha = \pi/3$, then $\boldsymbol{v}' \equiv (3.8716, 0.3283, 3.7287)$.

(4.22) *Point rotation formula.* Point P' is obtained by rotating point P by angle α about the axis parallel to $\hat{\boldsymbol{a}}$ that passes through point A. Derive and explain the formula

$$\boldsymbol{r}' = \boldsymbol{r}_A + \boldsymbol{s}_{\parallel} + \boldsymbol{s}_{\perp} \cos\alpha + (\hat{\boldsymbol{a}} \times \boldsymbol{s}_{\perp}) \sin\alpha,$$

where $\boldsymbol{r}'$, $\boldsymbol{r}$, and $\boldsymbol{r}_A$ are the positions vectors of P', P, and A from the origin O, respectively, and $\boldsymbol{s} = \boldsymbol{r} - \boldsymbol{r}_A$. Is $\boldsymbol{r}_A$ parallel to $\hat{\boldsymbol{a}}$ in general? Make clear sketches to illustrate your derivation. Modify your (problem 4.20) code for rotation of a *vector* about a *direction* $\hat{\boldsymbol{a}}$ to handle rotations of a *point* about an *axis*, $\{A, \hat{\boldsymbol{a}}\}$.

(4.23) For an arbitrary triangle, construct the outer equilateral triangles attached to each of the sides, in the plane of the original triangle (Fig. 4.20). Show that the triangle connecting the centroids of those equilateral triangles is itself equilateral. (Hint: Use $\boldsymbol{a} + \boldsymbol{b} + \boldsymbol{c} = 0$ for the original sides. Show that the vector connecting two centroids is the rotation by $2\pi/3$ of the previous such vector.)

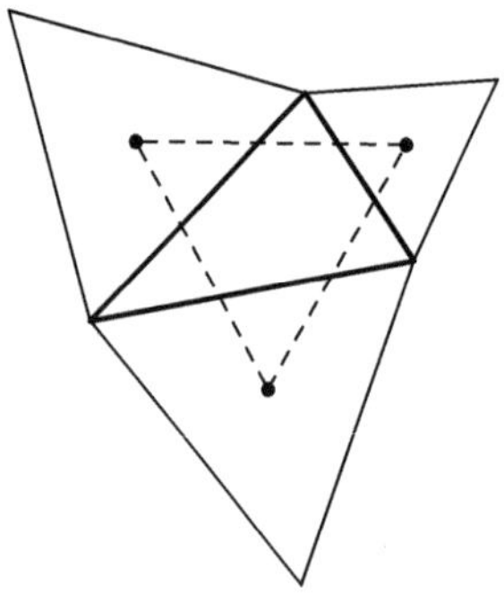

Fig. 4.20 Problem 4.23.

(4.24) True or false: $\boldsymbol{v} \perp (\boldsymbol{a} \times \boldsymbol{b}) \Leftrightarrow \boldsymbol{v} = x_1\boldsymbol{a} + x_2\boldsymbol{b}$ for some real x_1 and x_2. Explain.

(4.25) Derive formula (4.8) from the geometric definition of the cross-product (4.1a).

(4.26) Starting from the identity for $(\boldsymbol{a} \times \boldsymbol{b}) \times \boldsymbol{c}$, derive that for $\boldsymbol{a} \times (\boldsymbol{b} \times \boldsymbol{c})$.

(4.27) Prove the vector identity (4.9) for $\boldsymbol{a} \times (\boldsymbol{b} \times \boldsymbol{c})$ by expanding all vectors in cartesian form as in (4.4) and doing the algebra.

(4.28) Derive and explain the value of the coefficients β, γ_1, γ_2, γ_3 in (4.11).

(4.29) If $\boldsymbol{a} \equiv (2, 3, 4)$ and $\boldsymbol{b} \equiv (3, 2, 1)$ in cartesian components, can you find $\boldsymbol{u}$ such that $\boldsymbol{a} \times \boldsymbol{u} = \boldsymbol{b}$? Explain.

(4.30) Show by (1) cross product geometry and (2) cross product algebra that all the vectors $\boldsymbol{u}$ such that $\boldsymbol{a} \times \boldsymbol{u} = \boldsymbol{b}$ have the form

$$\boldsymbol{u} = \alpha\boldsymbol{a} + \frac{\boldsymbol{b} \times \boldsymbol{a}}{|\boldsymbol{a}|^2}, \quad \forall\alpha \in \mathbb{R} \tag{4.19}$$

as long as $\boldsymbol{a}$ and $\boldsymbol{b}$ are ....

(4.31) The net force $\boldsymbol{F}$ acting on a particle of electric charge q moving at velocity $\boldsymbol{v}$ through electric field $\boldsymbol{E}$ and magnetic field $\boldsymbol{B}$ is $\boldsymbol{F} = q(\boldsymbol{E} + \boldsymbol{v} \times \boldsymbol{B})$. What is $\boldsymbol{v}$ if $\boldsymbol{F} = 0$ but $\boldsymbol{E}$ and $\boldsymbol{B}$ are not zero?

(4.32) Forces $\boldsymbol{F}_A$ and $\boldsymbol{F}_B$ are applied at points A and B of an object, respectively. Find the center of force C such that there is no net torque about C, that is

$$\overrightarrow{CA} \times \boldsymbol{F}_A + \overrightarrow{CB} \times \boldsymbol{F}_B = 0.$$

(4.33) Prove the *Jacobi identity:*

$$\boldsymbol{a} \times (\boldsymbol{b} \times \boldsymbol{c}) + \boldsymbol{b} \times (\boldsymbol{c} \times \boldsymbol{a}) + \boldsymbol{c} \times (\boldsymbol{a} \times \boldsymbol{b}) = 0.$$

(4.34) If $\hat{\boldsymbol{a}}$ is any unit vector, show geometrically *and* vector algebraically that any vector $\boldsymbol{v}$ can be decomposed as

$$\boldsymbol{v} = (\hat{\boldsymbol{a}} \cdot \boldsymbol{v})\hat{\boldsymbol{a}} + \hat{\boldsymbol{a}} \times (\boldsymbol{v} \times \hat{\boldsymbol{a}}). \tag{4.20}$$

(4.35) For any $\boldsymbol{v}$ perpendicular to $\hat{\boldsymbol{a}}$, show that

$$\hat{\boldsymbol{a}} \times (\hat{\boldsymbol{a}} \times (\hat{\boldsymbol{a}} \times (\hat{\boldsymbol{a}} \times \boldsymbol{v}))) = \boldsymbol{v}.$$

(4.36) Magnetic fields $\boldsymbol{B}$ are created by electric currents according to the Ampère and Biot–Savart laws. The simplest current is a moving charge. Consider two electric charges q_1 and q_2 moving at velocity $\boldsymbol{v}_1$ and $\boldsymbol{v}_2$, respectively. *Assume* along the lines of the Biot–Savart law that the magnetic field induced by q_1 at q_2 is

$$\boldsymbol{B}_2 = \frac{\mu_0}{4\pi} \frac{q_1\boldsymbol{v}_1 \times (\boldsymbol{r}_2 - \boldsymbol{r}_1)}{|\boldsymbol{r}_2 - \boldsymbol{r}_1|^3},$$

where $\boldsymbol{r}_1$ and $\boldsymbol{r}_2$ are the position vectors of q_1 and q_2, respectively, and μ_0 is the magnetic constant. The Lorentz force experienced by q_2 is $\boldsymbol{F}_2 = q_2\boldsymbol{v}_2 \times \boldsymbol{B}_2$. What is the corresponding magnetic field $\boldsymbol{B}_1$ and Lorentz force $\boldsymbol{F}_1$ induced by q_2 at q_1? Show that the forces $\boldsymbol{F}_1$ and $\boldsymbol{F}_2$ do *not* satisfy Newton's action–reaction law!

(4.37) *Alternate rotation formula.* Show that formula (4.18) can be written in the equivalent forms

$$\begin{aligned}\boldsymbol{v}' &= \boldsymbol{v}\cos\alpha + (1 - \cos\alpha)\,\hat{\boldsymbol{a}}(\hat{\boldsymbol{a}} \cdot \boldsymbol{v}) \\ &\qquad + \sin\alpha\,(\hat{\boldsymbol{a}} \times \boldsymbol{v}) && (4.21a)\\ &= \boldsymbol{v} + \sin\alpha\,(\hat{\boldsymbol{a}} \times \boldsymbol{v}) + (1 - \cos\alpha)\,\boldsymbol{v}_\perp && (4.21b)\\ &= \boldsymbol{v} + 2\sin\tfrac{\alpha}{2}\cos\tfrac{\alpha}{2}\,(\hat{\boldsymbol{a}} \times \boldsymbol{v}) + 2\sin^2\tfrac{\alpha}{2}\boldsymbol{v}_\perp && (4.21c)\\ &= \boldsymbol{v} + \boldsymbol{\tau} \times (\boldsymbol{v} + \boldsymbol{v}'), && (4.21d)\end{aligned}$$

where $\boldsymbol{v}_\perp = \hat{\boldsymbol{a}} \times (\boldsymbol{v} \times \hat{\boldsymbol{a}})$, and $\boldsymbol{\tau} = \hat{\boldsymbol{a}}\tan(\alpha/2)$ is the *Rodrigues vector.* Show how the last three forms can be deduced from Fig. 4.21, proving the double angle identities in the process.

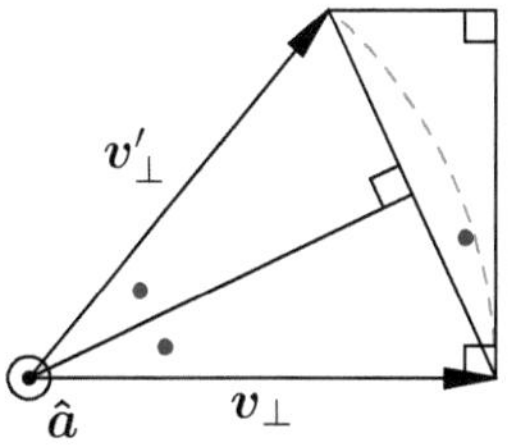

Fig. 4.21 Rotation $\boldsymbol{v} \to \boldsymbol{v}'$.

(4.38) Rewrite (4.21d) as $\boldsymbol{u}' = \boldsymbol{\tau} \times (2\boldsymbol{v} + \boldsymbol{u}')$ in terms of $\boldsymbol{u}' = \boldsymbol{v}' - \boldsymbol{v}$; then use a dot and a cross product with $\boldsymbol{\tau}$ to show that

$$\boldsymbol{v}' = \boldsymbol{v} + \frac{2}{1 + \boldsymbol{\tau} \cdot \boldsymbol{\tau}} \left(\boldsymbol{\tau} \times \boldsymbol{v} + \boldsymbol{\tau} \times (\boldsymbol{\tau} \times \boldsymbol{v}) \right). \quad (4.22)$$

(4.39) *Linear transformations.* Show that vector rotation as given by formulas (4.18) and (4.21) is a linear transformation $\boldsymbol{v} \to \boldsymbol{v}' = T(\boldsymbol{v})$, that is

$$T(\lambda \boldsymbol{v} + \mu \boldsymbol{w}) = \lambda T(\boldsymbol{v}) + \mu T(\boldsymbol{w}),$$

for any scalars λ, μ and vectors $\boldsymbol{v}$, $\boldsymbol{w}$.

(4.40) Let $\boldsymbol{u}$ be the 180° rotation of $\boldsymbol{v}$ about $\hat{\boldsymbol{n}}$ and $\boldsymbol{w}$ the reflection of $\boldsymbol{v}$ about a plane perpendicular to $\hat{\boldsymbol{n}}$. Explain and illustrate why

$$\boldsymbol{u} = 2(\hat{\boldsymbol{n}} \cdot \boldsymbol{v})\hat{\boldsymbol{n}} - \boldsymbol{v}, \quad \boldsymbol{w} = \boldsymbol{v} - 2(\hat{\boldsymbol{n}} \cdot \boldsymbol{v})\hat{\boldsymbol{n}} = -\boldsymbol{u}.$$

(4.41) *Rotation by reflections.* Show that rotation by α about $\hat{\boldsymbol{a}}$ is the same as reflection about any plane containing $\hat{\boldsymbol{a}}$ followed by reflection about the plane obtained by rotating the first one about $\hat{\boldsymbol{a}}$ by $\alpha/2$ (Fig. 4.22). Linearity $T(\boldsymbol{v}) = T(\boldsymbol{v}_\parallel) + T(\boldsymbol{v}_\perp)$ helps visualizing the geometry and doing the algebra.

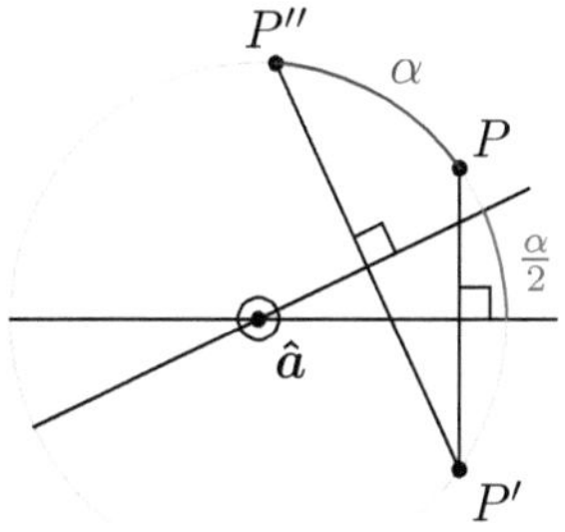

Fig. 4.22 Rotation by reflections.

(4.42) Show that rotation by α about $\hat{\boldsymbol{a}}$ is equal to rotation by π about any $\hat{\boldsymbol{p}}$ perpendicular to $\hat{\boldsymbol{a}}$, followed by rotation by π about $\hat{\boldsymbol{q}}$, where $\hat{\boldsymbol{q}}$ is the rotation of $\hat{\boldsymbol{p}}$ about $\hat{\boldsymbol{a}}$ by $\alpha/2$ (Fig. 4.23). This can be expressed symbolically for any vector $\boldsymbol{v}$ as

$$R(\alpha, \hat{\boldsymbol{a}})\boldsymbol{v} = R(\pi, \hat{\boldsymbol{q}})R(\pi, \hat{\boldsymbol{p}})\boldsymbol{v},$$

with $\hat{\boldsymbol{q}} = R(\alpha/2, \hat{\boldsymbol{a}})\,\hat{\boldsymbol{p}}$, for any $\hat{\boldsymbol{p}} \perp \hat{\boldsymbol{a}}$, where $R(\gamma, \hat{\boldsymbol{c}})$ is a vector operator that performs (right hand) rotation by angle γ about direction $\hat{\boldsymbol{c}}$. Show that

$$\hat{\boldsymbol{a}} \tan \frac{\alpha}{2} = \frac{\hat{\boldsymbol{p}} \times \hat{\boldsymbol{q}}}{\hat{\boldsymbol{p}} \cdot \hat{\boldsymbol{q}}}. \quad (4.23)$$

Linearity $T(\boldsymbol{v}) = T(\boldsymbol{v}_\parallel) + T(\boldsymbol{v}_\perp)$ helps visualizing the geometry and doing the algebra. Compare this decomposition with that in problem 4.41.

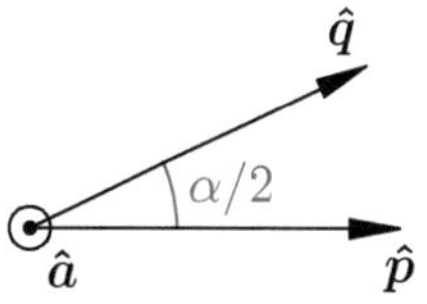

Fig. 4.23 Exercise 4.42.

(4.43) *Composing rotations in a plane.* For arbitrary points P in an oriented plane, show that rotation of $P \to P'$ by angle α about point A, then rotation of $P' \to P''$ by angle β about point B, yield a net rotation by angle γ about point C, where the rotation centers and angles are related as indicated in Fig. 4.24. Solve by purely geometric reasoning, considering transformations of C and A by the two rotations. Positive angles correspond to counterclockwise rotations.

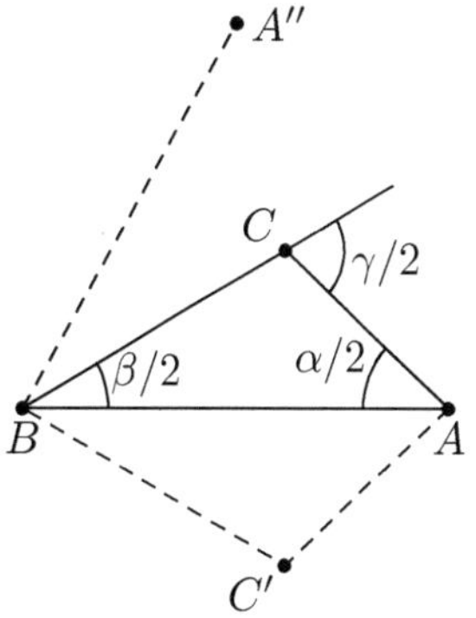

Fig. 4.24 Composition of 2D rotations.

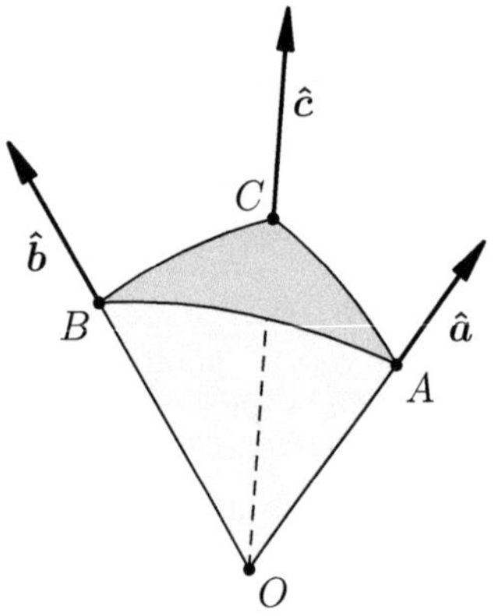

Fig. 4.25 $\hat{c}$ is the intersection of the plane $(\hat{a}, \hat{b})$ rotated by $-\alpha/2$ about $\hat{a}$ and the plane $(\hat{a}, \hat{b})$ rotated by $\beta/2$ about $\hat{b}$. The angle between those planes is $\gamma/2$.

(4.44) *Composing 3D rotations.* Exercises 4.41 and 4.42 provide ways to determine the net rotation of two successive rotations, by α about $\hat{a}$, then by β about $\hat{b}$. The first rotation about $\hat{a}$ can be decomposed into π-rotations about $\hat{p}_a$ then $\hat{q}_a$, as defined in exercise 4.42, and the rotation about $\hat{b}$ into π-rotations about $\hat{p}_b$ then $\hat{q}_b$. To combine those rotations, pick $\hat{q}_a = \hat{p}_b = \hat{n} = (\hat{a}\times\hat{b})/|\hat{a}\times\hat{b}|$, so the two successive π-rotations about $\hat{n}$ cancel out yielding

$$R(\beta, \hat{b})R(\alpha, \hat{a}) = R(\pi, \hat{q}_b)R(\pi, \hat{p}_a) = R(\gamma, \hat{c}).$$

Show that

$$\cos\frac{\gamma}{2} = \hat{p}_a \cdot \hat{q}_b, \quad \hat{c}\sin\frac{\gamma}{2} = \hat{p}_a \times \hat{q}_b, \tag{4.24a}$$

with

$$\begin{cases} \hat{p}_a = R(-\frac{1}{2}\alpha, \hat{a})\hat{n} = \cos\frac{1}{2}\alpha\, \hat{n} - \sin\frac{1}{2}\alpha\, \hat{a}\times\hat{n}, \\ \hat{q}_b = R(\frac{1}{2}\beta, \hat{b})\hat{n} = \cos\frac{1}{2}\beta\, \hat{n} + \sin\frac{1}{2}\beta\, \hat{b}\times\hat{n}, \end{cases}$$

yielding

$$\tau_c = \frac{\tau_a + \tau_b - \tau_a \times \tau_b}{1 - \tau_a \cdot \tau_b}, \tag{4.24b}$$

where $\tau_a = \hat{a}\tan(\alpha/2)$, $\tau_b = \hat{b}\tan(\beta/2)$, and $\tau_c = \hat{c}\tan(\gamma/2)$ are *Rodrigues* vectors, which fully characterize the rotations as established in (4.21), (4.22). The net rotation resulting from rotating by α about $\hat{a}$ then by β about $\hat{b}$ thus amounts to a rotation by γ about $\hat{c}$ (Fig. 4.25).
This vector derivation is essentially that in §128 of E. Wilson's 1901 redaction of Gibbs' lectures. The composition formula (4.24b) was published by Olinde Rodrigues in 1840 in cartesian coordinates. Rodrigues also described the solution geometrically (Fig. 4.25), generalizing the 2D construction in Fig. 4.24. Rotations of vectors in 3D is equivalent to rotations of points on a sphere about radial axes. That description directly yields (4.24a). Note that the composition of two rotations is *not* given by the sum of Rodrigues vectors, $\tau_c \neq \tau_a + \tau_b$, except in the limit of small rotations $\alpha \to 0$, $\beta \to 0$. See Altmann (1989)[1] for a discussion of the work of Hamilton and Rodrigues, and the connection with *quaternions*.

(4.45) Derive (4.24b) like Rodrigues but using Gibbs' vector calculus. Show that the two successive rotations $v \to v' \to v''$ can be expressed from (4.21d) as

$$u' = \tau_a \times (v + v'' - u''),$$
$$u'' = \tau_b \times (v + v'' + u'),$$

where $u' = v' - v$, $u'' = v'' - v'$, and $u' + u'' = v'' - v$. Substitute for u'' in the u' equation, and vice-versa. Reduce and combine the results to obtain $v'' - v = \tau_c \times (v + v'')$ with τ_c as in (4.24b).

[1]Simon L. Altmann, "Hamilton, Rodrigues, and the Quaternion Scandal." *Mathematics Magazine* 62.5 (Dec. 1989): 291-308.

Cartesian Index Notation

5

5.1 Levi-Civita symbol

We have used the Kronecker delta (3.18)

$$\delta_{ij} \triangleq \begin{cases} 1 & \text{if } i = j, \\ 0 & \text{if } i \neq j, \end{cases}$$

to express all nine dot products between orthonormal basis vectors in compact form as

$$\mathbf{e}_i \cdot \mathbf{e}_j = \delta_{ij}, \qquad \forall i, j,$$

There is a similar symbol, ϵ_{ijk}., the *Levi-Civita* symbol,[1] also known as the *alternating* or *permutation* symbol, defined as

$$\epsilon_{ijk} \triangleq \begin{cases} 1 & \text{if } (i,j,k) \text{ is an even permutation of } (1,2,3), \\ -1 & \text{if } (i,j,k) \text{ is an odd permutation of } (1,2,3), \\ 0 & \text{otherwise,} \end{cases} \tag{5.1}$$

or, explicitly,

$$\begin{aligned} \epsilon_{123} &= \epsilon_{231} = \epsilon_{312} = 1, \\ \epsilon_{213} &= \epsilon_{132} = \epsilon_{321} = -1, \end{aligned}$$

with all other $\epsilon_{ijk} = 0$.

For three distinct elements, (a, b, c) say, an *even* permutation is the same as a *cyclic* permutation; for example, the cyclic permutation

$$(a, b, c) \to (b, c, a)$$

is equivalent to the two permutations

$$(a, b, c) \to (b, a, c) \to (b, c, a).$$

Thus, the even permutations of $(1, 2, 3)$ are the cyclic permutations $(1, 2, 3)$, $(2, 3, 1)$, $(3, 1, 2)$ and the odd permutations are the acyclic permutations $(2, 1, 3)$, $(3, 2, 1)$, $(1, 3, 2)$. This implies that

$$\epsilon_{ijk} = \epsilon_{jki} = \epsilon_{kij}, \qquad \forall i, j, k. \tag{5.2}$$

[1] Named after Italian mathematician Tullio Levi-Civita, 1873–1941.

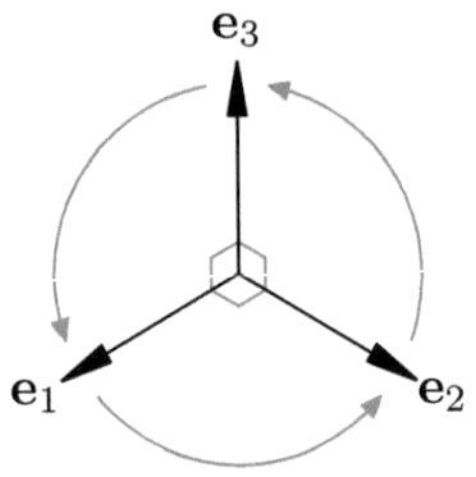

Fig. 5.1 Right-handed basis.

The ϵ_{ijk} symbol provides a compact expression for the components of the cross product of right-handed cartesian basis vectors (Fig. 5.1),

$$(\mathbf{e}_i \times \mathbf{e}_j) \cdot \mathbf{e}_k = \epsilon_{ijk}, \tag{5.3}$$

but since this is the k-component of $(\mathbf{e}_i \times \mathbf{e}_j)$ we can also write

$$\mathbf{e}_i \times \mathbf{e}_j = \sum_{k=1}^{3} \epsilon_{ijk} \mathbf{e}_k. \tag{5.4}$$

Note that there is at most one nonzero term in the latter sum.

5.2 The dummy and the free

The *expansion* of a vector $\boldsymbol{a}$ in terms of a basis $\{\mathbf{e}_1, \mathbf{e}_2, \mathbf{e}_3\}$ can be written compactly using Σ (sigma) notation

$$\boldsymbol{a} = a_1\mathbf{e}_1 + a_2\mathbf{e}_2 + a_3\mathbf{e}_3 \equiv \sum_{i=1}^{3} a_i \mathbf{e}_i. \tag{5.5}$$

We have introduced the Kronecker delta δ_{ij} and the Levi-Civita symbol ϵ_{ijk} in order to write and perform basic vector operations such as dot and cross products in compact forms, *when the basis is orthonormal and right-handed.* Using distributivity of the dot product and (3.18) yields

$$\begin{aligned} \boldsymbol{a} \cdot \boldsymbol{b} = \left(\sum_{i=1}^{3} a_i \mathbf{e}_i\right) \cdot \left(\sum_{j=1}^{3} b_j \mathbf{e}_j\right) &= \sum_{i=1}^{3}\sum_{j=1}^{3} a_i b_j \, \mathbf{e}_i \cdot \mathbf{e}_j \\ &= \sum_{i=1}^{3}\sum_{j=1}^{3} a_i b_j \delta_{ij} = \sum_{i=1}^{3} a_i b_i. \end{aligned} \tag{5.6}$$

Likewise, distributivity of the cross product with (5.4) yields

$$\boldsymbol{a} \times \boldsymbol{b} = \sum_{i=1}^{3}\sum_{j=1}^{3} a_i b_j \, \mathbf{e}_i \times \mathbf{e}_j = \sum_{i=1}^{3}\sum_{j=1}^{3}\sum_{k=1}^{3} a_i b_j \epsilon_{ijk} \, \mathbf{e}_k. \tag{5.7}$$

The indices i and j are called *dummy* or *summation* indices in the sums (5.5) and (5.6), they do not have a specific value, they have *all* the possible values in their range. It is their *place* in the particular expression and their *range* that matters, not their name

$$\boldsymbol{a} = \sum_{i=1}^{3} a_i \mathbf{e}_i = \sum_{j=1}^{3} a_j \mathbf{e}_j = \sum_{k=1}^{3} a_k \mathbf{e}_k = \cdots. \tag{5.8}$$

Indices come in two kinds, the *dummies* and the *free*. For example, j and k are dummy indices in

$$\mathbf{e}_i \cdot (\boldsymbol{a} \cdot \boldsymbol{b})\boldsymbol{c} = \left(\sum_{j=1}^{3} a_j b_j\right) c_i = \left(\sum_{k=1}^{3} a_k b_k\right) c_i, \tag{5.9}$$

j and k are summation indices, and they must take all values in their range. The index i is *free*, free to have any value in the range $1, 2, 3$. But freedom comes with constraints. If we use i on the left-hand side of the equation, then we have no choice: we must use i for c_i on the right-hand side. By convention we try to use i, j, k, l, m, n, to denote indices, which are positive integers. Greek letters are sometimes used for indices.

Mathematical operations impose some naming constraints, however. Although, we can use the same index i, writing $\boldsymbol{a} = \sum_i a_i \mathbf{e}_i$ and $\boldsymbol{b} = \sum_i b_i \mathbf{e}_i$, when they appear separately, we *cannot use the same index if we multiply them together* as in (5.6) and (5.7). Bad things will happen if you do, for instance

$$\boldsymbol{a} \times \boldsymbol{b} = \left(\sum_{i=1}^{3} a_i \mathbf{e}_i\right) \times \left(\sum_{i=1}^{3} b_i \mathbf{e}_i\right) = \sum_{i=1}^{3} a_i b_i \, \mathbf{e}_i \times \mathbf{e}_i = 0 \text{ !?!} \tag{5.10}$$

5.3 Summation convention

While he was developing the theory of general relativity, Einstein noticed that the sums that occur in vector calculations involve terms where the summation index appears *twice*. For example, i appears twice in the sums in (5.5), i and j appear twice in the double sum in (5.6), and i, j, and k each appear twice in the triple sum in (5.7). To facilitate such manipulations, Einstein dropped the Σ summation signs and introduced the

> ***summation convention:*** *a repeated index in a term implicitly denotes a sum over all values of that index.*

Thus, with the summation convention,

$$\sum_{i=1}^{3} a_i \mathbf{e}_i \longrightarrow a_i \mathbf{e}_i = a_1 \mathbf{e}_1 + a_2 \mathbf{e}_2 + a_2 \mathbf{e}_3$$

with the repeated index i on the *left-hand side* of the equality implicitly indicating a sum over all values of that index. This is a very useful notation, widely used in mathematics, physics, and engineering. It is merely a notation convention, but good notation allows us to do otherwise laborious calculations in compact form.

The name of the index does not matter if it is repeated — it is a *dummy* or *summation* index; thus,

$$a_i \mathbf{e}_i = a_j \mathbf{e}_j = \cdots = a_1 \mathbf{e}_1 + a_2 \mathbf{e}_2 + a_3 \mathbf{e}_3,$$

and any repeated index i, j, k, l, ... implies a sum over all values of that index. With the summation convention, (5.4) reads

$$\mathbf{e}_i \times \mathbf{e}_j = \epsilon_{ijk} \mathbf{e}_k \tag{5.11}$$

with an implicit sum over $k = 1, 2, 3$ on the right-hand side; that is,

$$\epsilon_{ijk} \mathbf{e}_k = \epsilon_{ij1} \mathbf{e}_1 + \epsilon_{ij2} \mathbf{e}_2 + \epsilon_{ij3} \mathbf{e}_3.$$

Likewise, the dot product (5.6) reads

$$\boldsymbol{a} \cdot \boldsymbol{b} = a_i b_i, \tag{5.12}$$

where $a_i b_i = a_1 b_1 + a_2 b_2 + a_3 b_3$ is a sum over all values of i, and the triple sum in (5.7) reduces to the very compact form

$$\boldsymbol{a} \times \boldsymbol{b} = \epsilon_{ijk} a_i b_j \mathbf{e}_k, \tag{5.13}$$

which is a sum over all values of i, j, and k and has $3^3 = 27$ terms. However, $\epsilon_{ijk} = 0$ for $3^3 - 3! = 27 - 6 = 21$ of those terms, whenever an index value is repeated in the triplet (i, j, k).

Note that the index expressions for $\boldsymbol{a} \cdot \boldsymbol{b}$ in (5.12) and $\boldsymbol{a} \times \boldsymbol{b}$ in (5.13) assume that the underlying basis, $\{\mathbf{e}_1, \mathbf{e}_2, \mathbf{e}_3\}$, is a right-handed orthonormal basis. That is where those δ_{ij}'s and ϵ_{ijk}'s come from. The summation convention can be used for nonorthonormal bases but then the dot product involves a *metric* tensor g_{ij} as in (3.16)

$$\boldsymbol{a} \cdot \boldsymbol{b} = a_i b_j g_{ij}.$$

The summation convention is a very useful and widely used notation but must be used with care — one should not write or read an i for a j or a 1 for an l, for example — and there are situations where it cannot be used because it leads to conflicts. For instance, if we have three vectors

$$\boldsymbol{a}_i \equiv \{\boldsymbol{a}_1, \boldsymbol{a}_2, \boldsymbol{a}_3\}$$

and want to write them all in magnitude direction form, $\boldsymbol{a} = a\hat{\boldsymbol{a}}$, we would write

$$\boldsymbol{a}_i = a_i \hat{\boldsymbol{a}}_i,$$

but then we *cannot* use the summation convention, since the right-hand side does *not* include a sum in this case.

The summation convention will be used hereafter, except where explicitly excluded. Two basic rules facilitate index manipulations, the *repetition rule* and the *subsitution rule.*

Repetition rule

Indices *cannot* appear more than twice *in the same term*; if they are, that is probably a mistake as in (5.10),

$$\boldsymbol{a} \times \boldsymbol{b} = (a_i \mathbf{e}_i) \times (b_i \mathbf{e}_i) = a_i b_i \; \mathbf{e}_i \times \mathbf{e}_i = 0 \; ??!$$

Terms are elements of a sum; *factors* are elements of a product.

where i appears four times in the same term. If we need to multiply two sums such as in $\boldsymbol{a} \times \boldsymbol{b}$ or $\boldsymbol{a} \cdot \boldsymbol{b}$ where $\boldsymbol{a} = a_i \mathbf{e}_i$ and $\boldsymbol{b} = b_i \mathbf{e}_i$, we need to change the name of one of the dummies, for example $i \to j$ in the $\boldsymbol{b}$ expansion, then

$$\boldsymbol{a} \times \boldsymbol{b} = (a_i \mathbf{e}_i) \times (b_j \mathbf{e}_j) = a_i b_j \; \mathbf{e}_i \times \mathbf{e}_j = \epsilon_{ijk} a_i b_j \mathbf{e}_k$$

and we did not miss any cross terms. However,

$$a_i + b_i + c_i \equiv (a_1 + b_1 + c_1,\ a_2 + b_2 + c_2,\ a_3 + b_3 + c_3)$$
$$\equiv \boldsymbol{a} + \boldsymbol{b} + \boldsymbol{c}$$

is OK since the index i appears in *different terms*, and the expression is, in fact, the index form for the vector sum $\boldsymbol{a} + \boldsymbol{b} + \boldsymbol{c}$. In contrast, the expression

$$a_i + b_j + c_k = ?!$$

does not make vector sense since i, j, k are free indices here, but there is no vector operation that adds components corresponding to different basis vectors. Free indices in different terms must match; for example,

$$\boldsymbol{v} = \boldsymbol{a} + \boldsymbol{b} \Leftrightarrow v_i = a_i + b_i \Leftrightarrow v_j = a_j + b_j,$$

but $a_i + b_j$ is a nonsensical vector operation.

Substitution rule

If one of the indices of δ_{ij} is involved in a sum, we substitute the summation index for the other index in δ_{ij} and drop δ_{ij}, for example

$$a_i\delta_{ij} \equiv a_1\delta_{1j} + a_2\delta_{2j} + a_3\delta_{3j} = a_j,$$

since i is a dummy in this example and δ_{ij} eliminates all terms in the sum except that corresponding to index j, whatever its value; thus,

$$a_i\delta_{ij} = a_j. \tag{5.14}$$

If both indices of δ_{ij} are summed over as in the double sum $a_ib_j\delta_{ij}$, it does not matter which index we substitute for; thus,

$$a_ib_j\delta_{ij} = a_ib_i = a_jb_j.$$

For example,

$$\boldsymbol{a} \cdot \boldsymbol{b} = (a_i\mathbf{e}_i) \cdot (b_j\mathbf{e}_j) = a_ib_j\, \mathbf{e}_i \cdot \mathbf{e}_j = a_ib_j\delta_{ij} = a_ib_i = a_jb_j$$

is indeed the dot product in cartesian components $\boldsymbol{a} \cdot \boldsymbol{b} = a_1b_1 + a_2b_2 + a_3b_3$. Likewise

$$\delta_{kl}\delta_{kl} = \delta_{kk} = \delta_{ll} = 3, \tag{5.15}$$

because k (and l) are repeated, so there is a sum over all values of k and

$$\delta_{kk} = \delta_{11} + \delta_{22} + \delta_{33} = 1 + 1 + 1 = 3.$$

For a similar reason, we write

$$a_1^2 + a_2^2 + a_3^2 = a_ia_i, \quad \text{not } a_i^2$$

to avoid confusion about whether there is a repeated index. There is clearly a repeated i index in a_ia_i, but not so clearly in a_i^2 that might be misinterpreted as the triplet

$$a_i^2 \stackrel{?}{=} (a_1^2, a_2^2, a_3^2)?$$

Example Computing the i component of $\boldsymbol{a} \times \boldsymbol{b}$ from (5.13) is

$$(\boldsymbol{a} \times \boldsymbol{b}) \cdot \mathbf{e}_i = \epsilon_{ljk} a_l b_j \mathbf{e}_k \cdot \mathbf{e}_i = \epsilon_{ljk} a_l b_j \delta_{ki} = \epsilon_{lji} a_l b_j = \epsilon_{ilj} a_l b_j. \quad (5.16)$$

Why? First, we switch the name of the dummy i to l in (5.13) since we want to dot with $\mathbf{e}_i$ where i is free. Second, $\mathbf{e}_k \cdot \mathbf{e}_i = \delta_{ki}$. Third, we perform the sum over k, dropping the δ_{ki} and replacing k by i. Fourth, $\epsilon_{lji} = \epsilon_{ilj}$ since (i, l, j) is a cyclic permutation of (l, j, i). Finally, we might want to rename dummy indices $l \to j$, $j \to k$ and write

$$\epsilon_{ilj} a_l b_j = \epsilon_{ijk} a_j b_k. \quad (5.17)$$

In calculations, we will want to jump directly from the vector equation to the corresponding cartesian index expression, that is

$$\boldsymbol{v} = \boldsymbol{a} \times \boldsymbol{b} \iff v_i = \epsilon_{ijk} a_j b_k. \quad (5.18)$$

The $\epsilon\epsilon = \delta\delta - \delta\delta$ identity

A mighty example that leads to a fundamental $\epsilon\delta$ identity is to write the double cross product identity

$$(\boldsymbol{a} \times \boldsymbol{b}) \times \boldsymbol{c} = \boldsymbol{b}(\boldsymbol{a} \cdot \boldsymbol{c}) - \boldsymbol{a}(\boldsymbol{b} \cdot \boldsymbol{c}) \quad (5.19)$$

in index notation. Let $\boldsymbol{v} = \boldsymbol{a} \times \boldsymbol{b}$ and

$$\boldsymbol{w} = (\boldsymbol{a} \times \boldsymbol{b}) \times \boldsymbol{c} = \boldsymbol{v} \times \boldsymbol{c}.$$

Then using (5.18) repeatedly, and making sure no index is repeated more than once, yields the i component of $\boldsymbol{w}$ as

$$w_i = \epsilon_{ijk} v_j c_k = \epsilon_{ijk} \epsilon_{jlm} a_l b_m c_k. \quad (5.20)$$

Note that j, k, l, and m are repeated, so this expression is a quadruple sum! According to the double cross product identity (5.19), the expression (5.20) should be equal to the i component of $(\boldsymbol{a} \cdot \boldsymbol{c})\boldsymbol{b} - (\boldsymbol{b} \cdot \boldsymbol{c})\boldsymbol{a}$, that is

$$(\boldsymbol{a} \cdot \boldsymbol{c}) b_i - (\boldsymbol{b} \cdot \boldsymbol{c}) a_i = (a_j c_j)\, b_i - (b_j c_j)\, a_i. \quad (5.21)$$

Since (5.20) and (5.21) are equal to each other for any $\boldsymbol{a}$, $\boldsymbol{b}$, $\boldsymbol{c}$, this should be telling us something about $\epsilon_{ijk}\epsilon_{jlm}$, but to pull out that information we need to rewrite (5.21) in the form $a_l b_m c_k$. How? By making use of our ability to rename dummy variables using δ_{ij} and the substitution rule.

Let's look at the first term in (5.21), $(a_j c_j) b_i$. We can rewrite it in the form $a_l b_m c_k$ of (5.20) as follows:

$$(a_j c_j) b_i = (a_k c_k) b_i = (\delta_{lk} a_l c_k)(\delta_{im} b_m) = \delta_{lk} \delta_{im} a_l b_m c_k. \quad (5.22)$$

Similar manipulations to the second term in (5.21) yield $(b_j c_j) a_i = \delta_{il} \delta_{km} a_l b_m c_k$. Equating (5.20) with the suitably rewritten (5.21) yields

$$\epsilon_{ijk} \epsilon_{jlm} a_l b_m c_k = (\delta_{lk} \delta_{im} - \delta_{il} \delta_{km}) a_l b_m c_k. \quad (5.23)$$

Since this equality holds *for any* a_l, b_m, c_k, we must have

$$\epsilon_{ijk}\epsilon_{jlm} = \delta_{lk}\delta_{im} - \delta_{il}\delta_{km}.$$

That's true but it is not written in a nice way so let's clean it up to a form that's easier to reconstruct. First, note that $\epsilon_{ijk} = \epsilon_{jki}$ since ϵ_{ijk} is invariant under a cyclic permutation of its indices. So our identity becomes $\epsilon_{jki}\epsilon_{jlm} = (\delta_{lk}\delta_{im} - \delta_{il}\delta_{km})$. We've done that flipping so the summation index j is in first place in both ϵ factors. Now we prefer the alphabetical order (i, j, k) to (j, k, i), so let's rename all the indices, being careful to rename the correct indices on both sides. This yields the $\epsilon\delta$ identity

$$\epsilon_{ijk}\epsilon_{ilm} = \delta_{jl}\delta_{km} - \delta_{jm}\delta_{kl}. \tag{5.24}$$

Problem 5.14 sketches an alternative derivation. The identity (5.24) can also be verified directly. The expression $\epsilon_{ijk}\epsilon_{ilm}$ is a sum over i with j, k, l, m free. That sum vanishes whenever $j = k$ or $l = m$, and the right-hand side of (5.24) also vanishes in that case. If $j \neq k$ and $l \neq m$, then only one value of i contributes to the sum, the value of i distinct from the values of j and k and l and m. The sum equals 1 if $j = l \neq k = m$, but in that case the right-hand side of (5.24) is also 1. The sum equals -1 if $j = m \neq k = l$, but in that case the right-hand side is -1 also.

Formula (5.24) has a generalization that does not include summation over any index:

$$\begin{aligned}\epsilon_{ijk}\epsilon_{lmn} =& \delta_{il}\delta_{jm}\delta_{kn} + \delta_{im}\delta_{jn}\delta_{kl} + \delta_{in}\delta_{jl}\delta_{km} \\ & - \delta_{im}\delta_{jl}\delta_{kn} - \delta_{in}\delta_{jm}\delta_{kl} - \delta_{il}\delta_{jn}\delta_{km}.\end{aligned} \tag{5.25}$$

Note that the first line corresponds to (i, j, k) matching a cyclic permutation of (l, m, n), while the second line corresponds to (i, j, k) matching an odd permutation of (l, m, n).

Exercises

(5.1) Explain why $\epsilon_{ijk} = \epsilon_{jki} = -\epsilon_{jik}$ for any integer i, j, k.

(5.2) Show that the cyclic permutations of (a, b, c) are *even* permutations. How many cyclic permutations are there?

(5.3) Show that for four distinct elements, cyclic permutations are *odd* permutations. Show that there are twenty-four permutations of (a, b, c, d) but only four cyclic permutations.

(5.4) For n distinct elements $(a_1, \ldots, a_n)$, how many permutations are there? How many cyclic permutations are there?

(5.5) Let $\boldsymbol{v} = \mathbf{e}_1$. What is v_i in compact index notation?

(5.6) If $(v_1, v_2, v_3) = (1, 2, 3)$ and $w_i = \delta_{i1}\delta_{2j}v_j$, what is (w_1, w_2, w_3)?

(5.7) If $w_i = (\delta_{i1}\delta_{2j} + \delta_{i2}\delta_{3j} + \delta_{i3}\delta_{1j})v_j$, what is (w_1, w_2, w_3)?

(5.8) If $w_i = (\delta_{j1}\delta_{i2} + \delta_{j2}\delta_{i3} + \delta_{j3}\delta_{i1})v_j$, what is (w_1, w_2, w_3)?

(5.9) Use (5.3) and Einstein's summation convention to

show that

$$\mathbf{e}_k \cdot (\boldsymbol{a} \times \boldsymbol{b}) = \epsilon_{ijk} a_i b_j = \epsilon_{kij} a_i b_j$$

$$\boldsymbol{a} \times \boldsymbol{b} = \epsilon_{ijk} a_i b_j \, \mathbf{e}_k = \mathbf{e}_i \, \epsilon_{ijk} a_j b_k.$$

(5.10) Evaluate $\epsilon_{ijk}\delta_{i1}\delta_{j2}\delta_{k3}$ and $\epsilon_{ijk}\delta_{il}\delta_{jm}\delta_{kn}$.

(5.11) Show that $\epsilon_{ijk}\epsilon_{ijk} = 6$ by (1) expanding the sums and (2) applying (5.24).

(5.12) Show that $\epsilon_{ijk}\epsilon_{ljk} = 2\delta_{il}$ by (1) expanding the sums and (2) applying (5.24).

(5.13) Deduce (5.24) from (5.25).

(5.14) Use (5.11) twice to obtain $((\mathbf{e}_i \times \mathbf{e}_j) \times \mathbf{e}_k) \cdot \mathbf{e}_l = \epsilon_{ijm}\epsilon_{mkl}$, then (5.19) to obtain another expression for $((\mathbf{e}_i \times \mathbf{e}_j) \times \mathbf{e}_k) \cdot \mathbf{e}_l$ and derive (5.24).

(5.15) True or false: $\epsilon_{ijk} a_i b_j c_k = \epsilon_{lmn} a_m b_n c_l$. Explain.

(5.16) Simplify $\epsilon_{ijk}\epsilon_{jkl}\epsilon_{lmi}$ and $\epsilon_{ijk}\epsilon_{lim}\epsilon_{nlp}$.

(5.17) Verify the identity

$$2\left(\epsilon_{1jk} a_{j2} a_{k3} + \epsilon_{i2k} a_{i1} a_{k3} + \epsilon_{ij3} a_{i1} a_{j2}\right) = a_{ii} a_{jj} - a_{ij} a_{ji}$$

(1) by explicit expansion of all the implicit sums and doing the algebra, and
(2) by showing that both sides equal $\epsilon_{ijk}\epsilon_{klm} a_{il} a_{jm}$.

(5.18) If $\boldsymbol{u}_i \times \boldsymbol{u}_j = \epsilon_{ijk}\boldsymbol{u}_k$, show that $\boldsymbol{u}_i = \frac{1}{2}\,\epsilon_{ijk}\boldsymbol{u}_j \times \boldsymbol{u}_k$ and $\boldsymbol{u}_i \cdot \boldsymbol{u}_j = \delta_{ij}$.

(5.19) For an *arbitrary* basis $\{\boldsymbol{a}_1, \boldsymbol{a}_2, \boldsymbol{a}_3\}$, consider the vectors $\boldsymbol{a}'_i = \frac{1}{2}\epsilon_{ijk}\boldsymbol{a}_j \times \boldsymbol{a}_k$. Discuss their geometric relations to the original basis. Do they form a basis? Why?

(5.20) Evaluate $((\boldsymbol{a} \times \boldsymbol{b}) \times \boldsymbol{c}) \times \boldsymbol{d}$, (1) using (4.9) and (2) in index notation, using (5.24).

(5.21) Write and evaluate $(\boldsymbol{a} \times \boldsymbol{b}) \times (\boldsymbol{c} \times \boldsymbol{d})$ using index notation.

(5.22) Show that $\boldsymbol{v}' = \boldsymbol{\omega} \times \boldsymbol{v}$ is $v'_i = A_{ij} v_j$ in cartesian index notation, with

$$A_{ij} = -\epsilon_{ijk}\omega_k \quad \Leftrightarrow \quad \omega_k = -\tfrac{1}{2}\,\epsilon_{kij} A_{ij}, \tag{5.26}$$

and $A_{ij} = -A_{ji}$ are the elements of a 3-by-3 *anti-symmetric matrix*

$$[A_{ij}] = \begin{bmatrix} 0 & -\omega_3 & \omega_2 \\ \omega_3 & 0 & -\omega_1 \\ -\omega_2 & \omega_1 & 0 \end{bmatrix}.$$

(5.23) Show that $\boldsymbol{\omega} \times \mathbf{e}_i = \epsilon_{ijk}\omega_k \, \mathbf{e}_j$ where $\mathbf{e}_i \times \mathbf{e}_j = \epsilon_{ijk}\mathbf{e}_k$ and $\omega_k = \boldsymbol{\omega} \cdot \mathbf{e}_k$.

(5.24) Let $\boldsymbol{b}_i = \boldsymbol{\omega} \times \mathbf{e}_i$ for some $\boldsymbol{\omega}$. Show that $\boldsymbol{b}_i = B_{ij}\mathbf{e}_j$ with

$$B_{ij} = \epsilon_{ijk}\omega_k \quad \Leftrightarrow \quad \omega_k = \tfrac{1}{2}\,\epsilon_{kij} B_{ij}. \tag{5.27}$$

Thus, $B_{ij} = A_{ji}$ in (5.26), so A_{ij} and B_{ij} are *transpose* of each other. B_{ij} are the components of the transformed basis vectors $\mathbf{e}_i \to \boldsymbol{b}_i$, while A_{ij} are the components of the matrix that yields the components of a transformed vector $\boldsymbol{v}'$ in terms of the original basis.

(5.25) Let $\boldsymbol{v}' = (\boldsymbol{a} \cdot \boldsymbol{v})\boldsymbol{a}$, then in index notation $v'_i = A_{ij} v_j$, where v'_i and v_j are cartesian components of $\boldsymbol{v}'$ and $\boldsymbol{v}$. Find A_{ij} and show that $A_{ij} = A_{ji}$.

(5.26) Let $\boldsymbol{v}' = \boldsymbol{v} - 2(\hat{\boldsymbol{a}} \cdot \boldsymbol{v})\hat{\boldsymbol{a}}$, then $v'_i = A_{ij} v_j$, where v'_i and v_j are cartesian components of $\boldsymbol{v}'$ and $\boldsymbol{v}$. Find A_{ij}. What is the geometric relation between $\boldsymbol{v}'$ and $\boldsymbol{v}$? Sketch.

(5.27) Let $\boldsymbol{v}' = (\hat{\boldsymbol{a}} \times \boldsymbol{v}) \times \hat{\boldsymbol{a}}$, yielding the relations $v'_i = A_{ij} v_j$ between cartesian components v'_i and v_j of $\boldsymbol{v}'$ and $\boldsymbol{v}$. Find A_{ij}. What is the geometric relation between $\boldsymbol{v}'$ and $\boldsymbol{v}$? Sketch.

Determinant

6

6.1 Mixed product

A *mixed product*, also called the *triple scalar product* or the *box product*, is a combination of a cross product and a dot product; the result is a *scalar* that is the *signed volume* V of the parallelepiped spanned by $\boldsymbol{a}, \boldsymbol{b}, \boldsymbol{c}$:

$$(\boldsymbol{a} \times \boldsymbol{b}) \cdot \boldsymbol{c} = V = \pm|V|. \tag{6.1}$$

The $+$ sign holds when $\boldsymbol{a}, \boldsymbol{b}, \boldsymbol{c}$ is right-handed, and the $-$ sign when $\boldsymbol{a}, \boldsymbol{b}, \boldsymbol{c}$ is left-handed. Thus, the mixed product $\boldsymbol{a} \times \boldsymbol{b} \cdot \boldsymbol{c}$ *determines* both the volume $|V|$ and the orientation of $\boldsymbol{a}, \boldsymbol{b}, \boldsymbol{c}$ (Fig. 6.1).

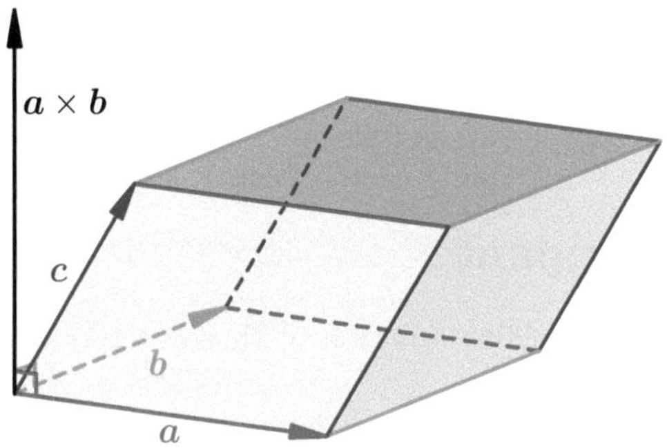

Fig. 6.1 $(\boldsymbol{a} \times \boldsymbol{b}) \cdot \boldsymbol{c}$ is the signed volume of $\boldsymbol{a}, \boldsymbol{b}, \boldsymbol{c}$.

Derivation of (6.1) The cross product $\boldsymbol{a} \times \boldsymbol{b} = A\hat{\boldsymbol{n}}$ where A is the area of the base parallelogram and $\hat{\boldsymbol{n}}$ is perpendicular to that base, with $\{\boldsymbol{a}, \boldsymbol{b}, \hat{\boldsymbol{n}}\}$ right-handed. The height h of the parallelepiped spanned by $\{\boldsymbol{a}, \boldsymbol{b}, \boldsymbol{c}\}$ is then simply $h = \hat{\boldsymbol{n}} \cdot \boldsymbol{c}$; thus, its volume is

$$V = A\,h = A\hat{\boldsymbol{n}} \cdot \boldsymbol{c} = (\boldsymbol{a} \times \boldsymbol{b}) \cdot \boldsymbol{c}.$$

Signwise, $(\boldsymbol{a} \times \boldsymbol{b}) \cdot \boldsymbol{c} > 0$ if $\{\boldsymbol{a}, \boldsymbol{b}, \boldsymbol{c}\}$ is a right-handed basis (not orthogonal in general), and $(\boldsymbol{a} \times \boldsymbol{b}) \cdot \boldsymbol{c} < 0$ if the triplet is left-handed. □

Taking $\boldsymbol{b} \times \boldsymbol{c} = A_2\hat{\boldsymbol{n}}_2$ as the area vector of the base parallelogram with $\boldsymbol{a} \cdot \hat{\boldsymbol{n}}_2 = h_2$ the corresponding height, or $\boldsymbol{c} \times \boldsymbol{a} = A_3\hat{\boldsymbol{n}}_3$ as the base with $h_3 = \boldsymbol{b} \cdot \hat{\boldsymbol{n}}_3$ as its height, yield the same volume,

$$V = Ah = A_2h_2 = A_3h_3,$$

and the same sign. Therefore the mixed product obeys the *cyclic permutation property:*

$$(\boldsymbol{a} \times \boldsymbol{b}) \cdot \boldsymbol{c} = (\boldsymbol{b} \times \boldsymbol{c}) \cdot \boldsymbol{a} = (\boldsymbol{c} \times \boldsymbol{a}) \cdot \boldsymbol{b}. \tag{6.2a}$$

Since the dot product commutes, $(\boldsymbol{b} \times \boldsymbol{c}) \cdot \boldsymbol{a} = \boldsymbol{a} \cdot (\boldsymbol{b} \times \boldsymbol{c})$, the mixed product also has the *product swap property,*

$$(\boldsymbol{a} \times \boldsymbol{b}) \cdot \boldsymbol{c} = \boldsymbol{a} \cdot (\boldsymbol{b} \times \boldsymbol{c}). \tag{6.2b}$$

That is, the mixed product does not change if we swap the dot and the cross, keeping the order of the vectors. These properties are useful in vector calculations and lead to other identities as explored in the exercises. Note that $\boldsymbol{a} \times (\boldsymbol{b} \cdot \boldsymbol{c})$ does not make sense, so the only meaningful interpretation of $\boldsymbol{a} \times \boldsymbol{b} \cdot \boldsymbol{c}$ is $(\boldsymbol{a} \times \boldsymbol{b}) \cdot \boldsymbol{c}$, and we can drop the parentheses.

Although the dot product commutes, the cross product anti-commutes, so out of the six mixed products arising from the six permutations of $\boldsymbol{a}, \boldsymbol{b}, \boldsymbol{c}$, three will have one sign and three will have the opposite sign,

$$\begin{aligned} (\boldsymbol{a} \times \boldsymbol{b}) \cdot \boldsymbol{c} &= \quad (\boldsymbol{b} \times \boldsymbol{c}) \cdot \boldsymbol{a} = \quad (\boldsymbol{c} \times \boldsymbol{a}) \cdot \boldsymbol{b} \\ = -(\boldsymbol{b} \times \boldsymbol{a}) \cdot \boldsymbol{c} &= -(\boldsymbol{c} \times \boldsymbol{b}) \cdot \boldsymbol{a} = -(\boldsymbol{a} \times \boldsymbol{c}) \cdot \boldsymbol{b}, \end{aligned}$$

and the dot and the cross could be swapped in each of these expressions using (6.2b).

Thus, there are twelve different ways to write a mixed product of the vectors $\boldsymbol{a}$, $\boldsymbol{b}$, $\boldsymbol{c}$ but they are all equal to $\pm$ the volume of the parallelepiped V; six are positive with right-handed ordering, six are negative with left-handed ordering. The mixed product also inherits distributivity properties from those of the dot and cross products. We can *abstract* all these algebraic properties as follows.

6.2 Determinant

A mixed product is a scalar function of three vectors called the *determinant*:

$$(\boldsymbol{a} \times \boldsymbol{b}) \cdot \boldsymbol{c} = \det(\boldsymbol{a}, \boldsymbol{b}, \boldsymbol{c}). \tag{6.3}$$

This determinant is a function that takes three vector inputs and returns a scalar output, that is $\det : \mathbb{E}^3 \times \mathbb{E}^3 \times \mathbb{E}^3 \to \mathbb{R}$.

The determinant has three fundamental properties that are its defining properties to extend the determinant concept to other vector spaces such as $\mathbb{R}^n$:

(1) $\det(\boldsymbol{a}, \boldsymbol{b}, \boldsymbol{c})$ changes sign if *any* two vectors are interchanged, for example,

$$\det(\boldsymbol{a}, \boldsymbol{b}, \boldsymbol{c}) = -\det(\boldsymbol{b}, \boldsymbol{a}, \boldsymbol{c}) = \det(\boldsymbol{b}, \boldsymbol{c}, \boldsymbol{a}); \tag{6.4a}$$

(2) $\det(\boldsymbol{a}, \boldsymbol{b}, \boldsymbol{c})$ is *multilinear*, that is *linear* in *each* of its vector arguments separately; for example $\forall\ \alpha_1,\ \alpha_2 \in \mathbb{R}$, and $\forall\, \boldsymbol{a}_1, \boldsymbol{a}_2, \boldsymbol{b}, \boldsymbol{c} \in \mathbb{E}^3$,

$$\det(\alpha_1 \boldsymbol{a}_1 + \alpha_2 \boldsymbol{a}_2, \boldsymbol{b}, \boldsymbol{c}) = \alpha_1 \det(\boldsymbol{a}_1, \boldsymbol{b}, \boldsymbol{c}) + \alpha_2 \det(\boldsymbol{a}_2, \boldsymbol{b}, \boldsymbol{c}); \tag{6.4b}$$

(3) $\det(\boldsymbol{a}, \boldsymbol{b}, \boldsymbol{c})$ is unity if the vectors are right-handed and orthonormal, that is

$$\det(\mathbf{e}_1, \mathbf{e}_2, \mathbf{e}_3) = 1. \tag{6.4c}$$

These three properties (6.4a), (6.4b), (6.4c) fully define the determinant, as explored in exercises 6.17, 6.18 below.

For vectors in $\mathbb{E}^3$, these properties directly follow from those of the dot and cross products and geometry. Property (6.4c) is the volume of a right-handed cubic box of unit sides. The permutation property (6.4a) is the base × height geometric property (6.2a) together with anti-commutativity of the cross-product. The multilinearity property (6.4b) is a combination of the distributivity properties of the dot and cross products,

$$\begin{aligned}\det(\alpha_1 \boldsymbol{a}_1 + \alpha_2 \boldsymbol{a}_2, \boldsymbol{b}, \boldsymbol{c}) &= (\alpha_1 \boldsymbol{a}_1 + \alpha_2 \boldsymbol{a}_2) \cdot (\boldsymbol{b} \times \boldsymbol{c}) \\ &= \alpha_1 \boldsymbol{a}_1 \cdot (\boldsymbol{b} \times \boldsymbol{c}) + \alpha_2 \boldsymbol{a}_2 \cdot (\boldsymbol{b} \times \boldsymbol{c}) \\ &= \alpha_1 \det(\boldsymbol{a}_1, \boldsymbol{b}, \boldsymbol{c}) + \alpha_2 \det(\boldsymbol{a}_2, \boldsymbol{b}, \boldsymbol{c}).\end{aligned}$$

It follows from the three defining properties (6.4) that the determinant is zero if any two vectors are identical (from (6.4a)), or if any vector is zero (from (6.4b) with $\alpha = 1$ and $\boldsymbol{u} = \boldsymbol{0}$), and that the determinant does not change if we add a multiple of one vector to another; for example,

$$\begin{aligned}\det(\boldsymbol{a}, \boldsymbol{b}, \boldsymbol{a}) &= 0, \\ \det(\boldsymbol{a}, \boldsymbol{0}, \boldsymbol{c}) &= 0, \\ \det(\boldsymbol{a}, \boldsymbol{b}, \boldsymbol{c} + \alpha\, \boldsymbol{a}) &= \det(\boldsymbol{a}, \boldsymbol{b}, \boldsymbol{c}).\end{aligned} \tag{6.5}$$

Geometrically, this last property corresponds to a *shearing* of the parallelepiped in the $\boldsymbol{a}$ direction, with no change in volume or orientation.

The *determinant* determines whether three vectors $\boldsymbol{a}, \boldsymbol{b}, \boldsymbol{c}$ are linearly independent and can be used as a basis for the vector space,

$$\det(\boldsymbol{a}, \boldsymbol{b}, \boldsymbol{c}) \neq 0 \;\Leftrightarrow\; \{\boldsymbol{a}, \boldsymbol{b}, \boldsymbol{c}\} \text{ form a basis.} \tag{6.6}$$

If $\det(\boldsymbol{a}, \boldsymbol{b}, \boldsymbol{c}) = 0$, then either one of the vectors is zero or they are co-planar and $\boldsymbol{a}$, $\boldsymbol{b}$, $\boldsymbol{c}$ cannot provide a basis for vectors in $\mathbb{E}^3$. This is how the determinant is introduced in linear algebra: it determines whether a system of linear equations has a solution. But the determinant is much more than a number that may or may not be zero; it determines the volume of the parallelepiped *and* the orientation of its edges.

6.3 Cartesian determinant

Expanding the vectors $\boldsymbol{a}$, $\boldsymbol{b}$, $\boldsymbol{c}$ in terms of a right-handed orthonormal basis $\{\mathbf{e}_1, \mathbf{e}_2, \mathbf{e}_3\}$,

$$\boldsymbol{a} = a_i \mathbf{e}_i, \quad \boldsymbol{b} = b_j \mathbf{e}_j, \quad \boldsymbol{c} = c_k \mathbf{e}_k$$

(summation convention) with $\mathbf{e}_i \times \mathbf{e}_j \cdot \mathbf{e}_k = \epsilon_{ijk}$ (5.3), the distributivity properties of dot and cross products yield

$$\det(\boldsymbol{a}, \boldsymbol{b}, \boldsymbol{c}) = (\boldsymbol{a} \times \boldsymbol{b}) \cdot \boldsymbol{c} = a_i b_j c_k (\mathbf{e}_i \times \mathbf{e}_j) \cdot \mathbf{e}_k = \epsilon_{ijk}\, a_i b_j c_k. \tag{6.7}$$

Expanding that expression

$$\epsilon_{ijk}\, a_i b_j c_k = a_1 b_2 c_3 + a_2 b_3 c_1 + a_3 b_1 c_2 \\ - a_2 b_1 c_3 - a_1 b_3 c_2 - a_3 b_2 c_1, \tag{6.8a}$$

we recover the algebraic determinant usually denoted with vertical bars,

$$\begin{vmatrix} a_1 & a_2 & a_3 \\ b_1 & b_2 & b_3 \\ c_1 & c_2 & c_3 \end{vmatrix} \triangleq \epsilon_{ijk}\, a_i b_j c_k. \tag{6.8b}$$

For 3-by-3 determinants, the determinant formula (6.8a) can be reconstructed with the graphical mnemonic (known as *rule of Sarrus*) shown in Fig. 6.2: copy the first two columns next to the third, add the products of the southeast diagonal elements, and subtract the products of the southwest diagonal elements.

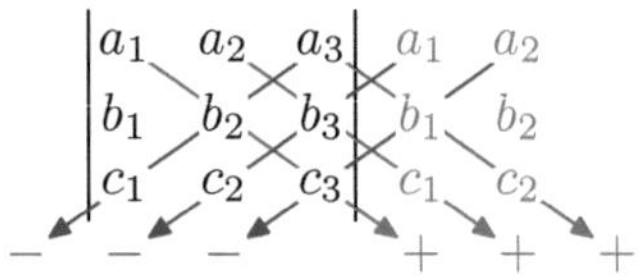

Fig. 6.2 Graphical rule for 3-by-3 determinants recovers (6.8a).

The algebraic determinant (6.8a) is a sum of row-ordered products $a_i b_j c_k$ with one and only one factor from each row and each column, and a positive (negative) sign if the column order is an even (odd) permutation of the row order. Therefore, it does not matter whether we put the vector components along rows or columns,

$$\begin{vmatrix} a_1 & a_2 & a_3 \\ b_1 & b_2 & b_3 \\ c_1 & c_2 & c_3 \end{vmatrix} = \begin{vmatrix} a_1 & b_1 & c_1 \\ a_2 & b_2 & c_2 \\ a_3 & b_3 & c_3 \end{vmatrix} = \epsilon_{ijk}\, a_i b_j c_k. \tag{6.9}$$

This is a general property that the determinant of a square matrix equals the determinant of its transpose, $\det \mathbf{A} = \det \mathbf{A}^{\mathsf{T}}$.

This familiar algebraic determinant has the same three fundamental properties (6.4a), (6.4b), (6.4c), of course:

(1) It changes sign if *any* two columns (or rows) are permuted, e.g.

$$\begin{vmatrix} a_1 & b_1 & c_1 \\ a_2 & b_2 & c_2 \\ a_3 & b_3 & c_3 \end{vmatrix} = - \begin{vmatrix} b_1 & a_1 & c_1 \\ b_2 & a_2 & c_2 \\ b_3 & a_3 & c_3 \end{vmatrix}. \tag{6.10a}$$

(2) It is multilinear, that is linear in *each* of its columns (or rows) separately, for example, $\forall \alpha$, (u_1, u_2, u_3),

$$\begin{vmatrix} \alpha a_1 + u_1 & b_1 & c_1 \\ \alpha a_2 + u_2 & b_2 & c_2 \\ \alpha a_3 + u_3 & b_3 & c_3 \end{vmatrix} = \alpha \begin{vmatrix} a_1 & b_1 & c_1 \\ a_2 & b_2 & c_2 \\ a_3 & b_3 & c_3 \end{vmatrix} + \begin{vmatrix} u_1 & b_1 & c_1 \\ u_2 & b_2 & c_2 \\ u_3 & b_3 & c_3 \end{vmatrix}. \tag{6.10b}$$

(3) Finally, the determinant of the *natural basis* is

$$\begin{vmatrix} 1 & 0 & 0 \\ 0 & 1 & 0 \\ 0 & 0 & 1 \end{vmatrix} = 1. \tag{6.10c}$$

We deduce from these three defining properties that the determinant vanishes if any column (or row) is zero or if any column (or row) is a multiple of another, and that the determinant does not change if we add to one column (row) a linear combination of the other columns (rows). These properties are the algebraic equivalents of (6.5), and enable calculating determinants by row (or column) elimination as discussed in linear algebra courses.

Cofactor expansion

Another useful formula for determinants, in addition to the $\epsilon_{ijk}a_ib_jc_k$ formula (6.7) is the *Laplace* or *cofactor* expansion in terms of 2-by-2 determinants. For example,

$$\begin{aligned}\begin{vmatrix} a_1 & b_1 & c_1 \\ a_2 & b_2 & c_2 \\ a_3 & b_3 & c_3 \end{vmatrix} &= a_1 \begin{vmatrix} b_2 & c_2 \\ b_3 & c_3 \end{vmatrix} - a_2 \begin{vmatrix} b_1 & c_1 \\ b_3 & c_3 \end{vmatrix} + a_3 \begin{vmatrix} b_1 & c_1 \\ b_2 & c_2 \end{vmatrix} \\ &= -b_1 \begin{vmatrix} a_2 & c_2 \\ a_3 & c_3 \end{vmatrix} + b_2 \begin{vmatrix} a_1 & c_1 \\ a_3 & c_3 \end{vmatrix} - b_3 \begin{vmatrix} a_1 & c_1 \\ a_2 & c_2 \end{vmatrix} \\ &= c_1 \begin{vmatrix} a_2 & b_2 \\ a_3 & b_3 \end{vmatrix} - c_2 \begin{vmatrix} a_1 & b_1 \\ a_3 & b_3 \end{vmatrix} + c_3 \begin{vmatrix} a_1 & b_1 \\ a_2 & b_2 \end{vmatrix},\end{aligned} \tag{6.11}$$

where the 2-by-2 determinants are

$$\begin{vmatrix} a_1 & b_1 \\ a_2 & b_2 \end{vmatrix} \triangleq a_1b_2 - a_2b_1 = \epsilon_{ij3}a_ib_j.$$

These Laplace formulas are simply $\boldsymbol{a} \cdot (\boldsymbol{b} \times \boldsymbol{c}) = -\boldsymbol{b} \cdot (\boldsymbol{a} \times \boldsymbol{c}) = \boldsymbol{c} \cdot (\boldsymbol{a} \times \boldsymbol{b})$ expressed with respect to a right-handed orthonormal basis and can also be derived from the full formula using index notation. Indeed,

$$\begin{aligned}\epsilon_{ijk}a_ib_jc_k &= a_1\epsilon_{1jk}b_jc_k + a_2\epsilon_{2jk}b_jc_k + a_3\epsilon_{3jk}b_jc_k \\ &= b_1\epsilon_{i1k}a_ic_k + b_2\epsilon_{i2k}a_ic_k + b_3\epsilon_{i3k}a_ic_k \\ &= c_1\epsilon_{ij1}a_ib_j + c_2\epsilon_{ij2}a_ib_j + c_3\epsilon_{ij3}a_ib_j.\end{aligned} \tag{6.12}$$

The cofactor expansion is particularly useful when one row (or column) has zeros, for example, if $b_2 = 0 = b_3$.

Exercises

(6.1) Show that

$$\begin{vmatrix} a_1 & b_1 \\ a_2 & b_2 \end{vmatrix} = a_1b_2 - a_2b_1 = A$$

is the *signed area* A of the parallelogram with sides $\boldsymbol{a} = a_1\mathbf{e}_1 + a_2\mathbf{e}_2$ and $\boldsymbol{b} = b_1\mathbf{e}_1 + b_2\mathbf{e}_2$, with $A > 0$ if $\boldsymbol{a}, \boldsymbol{b}, -\boldsymbol{a}, -\boldsymbol{b}$ is a counterclockwise cycle, and $A < 0$ if the cycle is clockwise. Sketch.

(6.2) Prove the (Lagrange) identity

$$(\boldsymbol{a}\times\boldsymbol{b})\cdot(\boldsymbol{a}\times\boldsymbol{b}) = (\boldsymbol{a}\cdot\boldsymbol{a})(\boldsymbol{b}\cdot\boldsymbol{b}) - (\boldsymbol{a}\cdot\boldsymbol{b})^2 = \begin{vmatrix}\boldsymbol{a}\cdot\boldsymbol{a} & \boldsymbol{a}\cdot\boldsymbol{b}\\ \boldsymbol{a}\cdot\boldsymbol{b} & \boldsymbol{b}\cdot\boldsymbol{b}\end{vmatrix} \quad (6.13)$$

using (1) vector identities, and (2) cartesian index notation.

(6.3) Prove the (Cauchy–Binet) identity

$$(\boldsymbol{a}\times\boldsymbol{b})\cdot(\boldsymbol{c}\times\boldsymbol{d}) = (\boldsymbol{a}\cdot\boldsymbol{c})(\boldsymbol{b}\cdot\boldsymbol{d}) - (\boldsymbol{a}\cdot\boldsymbol{d})(\boldsymbol{b}\cdot\boldsymbol{c}) = \begin{vmatrix}\boldsymbol{a}\cdot\boldsymbol{c} & \boldsymbol{a}\cdot\boldsymbol{d}\\ \boldsymbol{b}\cdot\boldsymbol{c} & \boldsymbol{b}\cdot\boldsymbol{d}\end{vmatrix} \quad (6.14)$$

using (1) vector identities, and (2) cartesian index notation.

(6.4) Show that

$$(\boldsymbol{a}\times\boldsymbol{b})\times(\boldsymbol{b}\times\boldsymbol{c}) = (\boldsymbol{a}\times\boldsymbol{b}\cdot\boldsymbol{c})\,\boldsymbol{b}.$$

(6.5) Show that

$$(\boldsymbol{a}\times\boldsymbol{b})\times(\boldsymbol{b}\times\boldsymbol{c})\cdot(\boldsymbol{c}\times\boldsymbol{a}) = (\boldsymbol{a}\times\boldsymbol{b}\cdot\boldsymbol{c})^2.$$

(6.6) True or false:

$$(\boldsymbol{a}\times\boldsymbol{b})\times(\boldsymbol{c}\times\boldsymbol{d}) = ((\boldsymbol{a}\times\boldsymbol{b})\times\boldsymbol{c})\times\boldsymbol{d}.$$

Justify geometrically.

(6.7) *Triple cross product.* Show that

$$\begin{aligned}(\boldsymbol{a}\times\boldsymbol{b})&\times(\boldsymbol{c}\times\boldsymbol{d})\\ &= (\boldsymbol{a}\times\boldsymbol{b}\cdot\boldsymbol{d})\,\boldsymbol{c} - (\boldsymbol{a}\times\boldsymbol{b}\cdot\boldsymbol{c})\,\boldsymbol{d}\\ &= -(\boldsymbol{c}\times\boldsymbol{d}\cdot\boldsymbol{b})\,\boldsymbol{a} + (\boldsymbol{c}\times\boldsymbol{d}\cdot\boldsymbol{a})\,\boldsymbol{b}. \quad (6.15)\end{aligned}$$

(6.8) Show that

$$\begin{aligned}&(\boldsymbol{a}\times\boldsymbol{b}\cdot\boldsymbol{c})\,\boldsymbol{v}\\ &= (\boldsymbol{v}\times\boldsymbol{b}\cdot\boldsymbol{c})\,\boldsymbol{a} + (\boldsymbol{a}\times\boldsymbol{v}\cdot\boldsymbol{c})\,\boldsymbol{b} + (\boldsymbol{a}\times\boldsymbol{b}\cdot\boldsymbol{v})\,\boldsymbol{c}. \quad (6.16)\end{aligned}$$

(6.9) If $\det(\boldsymbol{a},\boldsymbol{b},\boldsymbol{c}) \neq 0$, then any vector $\boldsymbol{v}$ can be expanded as $\boldsymbol{v} = \alpha\boldsymbol{a}+\beta\boldsymbol{b}+\gamma\boldsymbol{c}$. Find explicit expressions for the components α, β, γ in terms of $\boldsymbol{v}$, and $\boldsymbol{a}$, $\boldsymbol{b}$, $\boldsymbol{c}$ in the general case when the latter are *not* orthogonal. (Hint: dot $\boldsymbol{v}$ with cross products of the basis vectors $\boldsymbol{a},\boldsymbol{b},\boldsymbol{c}$, then collect the mixed products into determinants and deduce *Cramer's rule.*)

(6.10) Explain why

$$-|\boldsymbol{a}|\,|\boldsymbol{b}|\,|\boldsymbol{c}| \leq \det(\boldsymbol{a},\boldsymbol{b},\boldsymbol{c}) \leq |\boldsymbol{a}|\,|\boldsymbol{b}|\,|\boldsymbol{c}|.$$

When do the equalities apply?

(6.11) Show that

$$\det(\boldsymbol{a},\boldsymbol{b},\boldsymbol{c}) = |\boldsymbol{a}|\,|\boldsymbol{b}|\,|\boldsymbol{c}|\sin\varphi\cos\theta.$$

Specify φ and θ. Are those angles unique? Sketch.

(6.12) Consider $\boldsymbol{v}(s,t) = s\,\boldsymbol{a}+t\,\boldsymbol{b}+\boldsymbol{c}$. Find s,t that make $|\boldsymbol{v}|$ a minimum. Solve in two ways: geometrically by visualizing the set of vectors $\boldsymbol{v}(s,t)$ and using calculus by minimizing $|\boldsymbol{v}(s,t)|$. Derive a simple formula for the $\boldsymbol{v}$ of minimum magnitude.

(6.13) *Least squares.* Given arbitrary vectors $\boldsymbol{a}_1,\boldsymbol{a}_2$, $\boldsymbol{b}$, find $\boldsymbol{x} = x_1\boldsymbol{a}_1 + x_2\boldsymbol{a}_2$, such that $|\boldsymbol{x}-\boldsymbol{b}|$ is minimum. Derive an explicit formula for $\boldsymbol{x}$. Sketch.

(6.14) If $\boldsymbol{a}\times\boldsymbol{b}\cdot\boldsymbol{c} = J$, show that

$$(\boldsymbol{a}\times\boldsymbol{b})\times(\boldsymbol{b}\times\boldsymbol{c})\cdot(\boldsymbol{c}\times\boldsymbol{a}) = J^2.$$

(6.15) Given any three vectors $\boldsymbol{a}_1$, $\boldsymbol{a}_2$, $\boldsymbol{a}_3$ such that $J = \boldsymbol{a}_1\times\boldsymbol{a}_2\cdot\boldsymbol{a}_3 \neq 0$, define

$$\begin{cases}\breve{\boldsymbol{a}}_1 = J^{-1}\,\boldsymbol{a}_2\times\boldsymbol{a}_3,\\ \breve{\boldsymbol{a}}_2 = J^{-1}\,\boldsymbol{a}_3\times\boldsymbol{a}_1,\\ \breve{\boldsymbol{a}}_3 = J^{-1}\,\boldsymbol{a}_1\times\boldsymbol{a}_2.\end{cases} \quad (6.17)$$

Show that $\{\breve{\boldsymbol{a}}_1,\breve{\boldsymbol{a}}_2,\breve{\boldsymbol{a}}_3\}$ is the *reciprocal* of the basis $\{\boldsymbol{a}_1,\boldsymbol{a}_2,\boldsymbol{a}_3\}$, in the sense that

(a) $\boldsymbol{a}_i\cdot\breve{\boldsymbol{a}}_j = \delta_{ij}$,

(b) $\breve{\boldsymbol{a}}_1\times\breve{\boldsymbol{a}}_2\cdot\breve{\boldsymbol{a}}_3 = J^{-1}$,

(c) $\breve{\boldsymbol{a}}_i = \epsilon_{ijk}(\boldsymbol{a}_j\times\boldsymbol{a}_k)/(2J)$,

(d) $\boldsymbol{a}_i = \epsilon_{ijk}(\breve{\boldsymbol{a}}_j\times\breve{\boldsymbol{a}}_k)/(2J^{-1})$, so the inverse of the inverse is the original basis.

(e) If $\boldsymbol{v} = \breve{v}_i\,\boldsymbol{a}_i = v_i\,\breve{\boldsymbol{a}}_i$, then $v_i = \boldsymbol{v}\cdot\boldsymbol{a}_i$ and $\breve{v}_i = \boldsymbol{v}\cdot\breve{\boldsymbol{a}}_i$, so the components in one basis are obtained by projecting onto the *reciprocal* basis.[1]

(f) Show that, as a special case of these $\boldsymbol{v}$ expansions,

$$\begin{aligned}\breve{\boldsymbol{a}}_i &= (\breve{\boldsymbol{a}}_i\cdot\breve{\boldsymbol{a}}_j)\,\boldsymbol{a}_j = \breve{g}_{ij}\,\boldsymbol{a}_j,\\ \boldsymbol{a}_i &= (\boldsymbol{a}_i\cdot\boldsymbol{a}_j)\,\breve{\boldsymbol{a}}_j = g_{ij}\,\breve{\boldsymbol{a}}_j,\end{aligned} \quad (6.18)$$

where $g_{ij} = \boldsymbol{a}_i\cdot\boldsymbol{a}_j$ and $\breve{g}_{ij} = \breve{\boldsymbol{a}}_i\cdot\breve{\boldsymbol{a}}_j$ are the respective *metrics* (3.16).

(6.16) Show that for any $\boldsymbol{a}_1$, $\boldsymbol{a}_2$, $\boldsymbol{a}_3$,

$$\boldsymbol{a}_i\times\boldsymbol{a}_j\cdot\boldsymbol{a}_k = \epsilon_{ijk}\,\boldsymbol{a}_1\times\boldsymbol{a}_2\cdot\boldsymbol{a}_3.$$

(6.17) Use the alternating property (6.4a) to show that $\det(\mathbf{e}_i,\mathbf{e}_j,\mathbf{e}_k) = \epsilon_{ijk}$, and more generally for any vectors $\boldsymbol{a}_1,\boldsymbol{a}_2,\boldsymbol{a}_3$:

$$\det(\boldsymbol{a}_i,\boldsymbol{a}_j,\boldsymbol{a}_k) = \epsilon_{ijk}\det(\boldsymbol{a}_1,\boldsymbol{a}_2,\boldsymbol{a}_3). \quad (6.19)$$

[1] Ricci notation using *upper* indices $\boldsymbol{a}^i \equiv \breve{\boldsymbol{a}}_i$ is discussed in §14.4, 15.2, 16.4.

(6.18) Use multilinearity (6.4b) and exercise 6.17 to show that if $\boldsymbol{a} = a_i \boldsymbol{e}_i$, $\boldsymbol{b} = b_j \boldsymbol{e}_j$, $\boldsymbol{c} = c_k \boldsymbol{e}_k$ (summation convention), then $\det(\boldsymbol{a}, \boldsymbol{b}, \boldsymbol{c}) = \epsilon_{ijk} a_i b_j c_k$.

(6.19) Use multilinearity (6.4b) to show that (with summation convention),

$$\det(\alpha_i \boldsymbol{a}_i,\ \beta_j \boldsymbol{a}_j,\ \gamma_k \boldsymbol{a}_k) = \epsilon_{ijk} \alpha_i \beta_j \gamma_k \det(\boldsymbol{a}_1, \boldsymbol{a}_2, \boldsymbol{a}_3). \tag{6.20}$$

(6.20) If $V = \epsilon_{ijk} a_{i1} a_{j2} a_{k3}$, explain/show why

$$\epsilon_{ijk}\, a_{il} a_{jm} a_{kn} = \epsilon_{lmn} V.$$

(6.21) Use (6.4) to show that if $\boldsymbol{c}_i = a_{ji} \boldsymbol{b}_j$ (sum over j), then

$$\det(\boldsymbol{c}_1, \boldsymbol{c}_2, \boldsymbol{c}_3) = \begin{vmatrix} a_{11} & a_{12} & a_{13} \\ a_{21} & a_{22} & a_{23} \\ a_{31} & a_{32} & a_{33} \end{vmatrix} \det(\boldsymbol{b}_1, \boldsymbol{b}_2, \boldsymbol{b}_3).$$

(6.22) Use (6.10) to show that if $c_{ij} = a_{ik} b_{kj}$ (sum over k), then

$$\begin{vmatrix} c_{11} & c_{12} & c_{13} \\ c_{21} & c_{22} & c_{23} \\ c_{31} & c_{32} & c_{33} \end{vmatrix} = \begin{vmatrix} a_{11} & a_{12} & a_{13} \\ a_{21} & a_{22} & a_{23} \\ a_{31} & a_{32} & a_{33} \end{vmatrix} \begin{vmatrix} b_{11} & b_{12} & b_{13} \\ b_{21} & b_{22} & b_{23} \\ b_{31} & b_{32} & b_{33} \end{vmatrix}.$$

(6.23) Prove the (Cauchy–Binet) identity

$$(\boldsymbol{a} \times \boldsymbol{b} \cdot \boldsymbol{c})\, (\boldsymbol{u} \times \boldsymbol{v} \cdot \boldsymbol{w}) = \begin{vmatrix} \boldsymbol{a} \cdot \boldsymbol{u} & \boldsymbol{a} \cdot \boldsymbol{v} & \boldsymbol{a} \cdot \boldsymbol{w} \\ \boldsymbol{b} \cdot \boldsymbol{u} & \boldsymbol{b} \cdot \boldsymbol{v} & \boldsymbol{b} \cdot \boldsymbol{w} \\ \boldsymbol{c} \cdot \boldsymbol{u} & \boldsymbol{c} \cdot \boldsymbol{v} & \boldsymbol{c} \cdot \boldsymbol{w} \end{vmatrix}.$$

(6.24) If A is the area the parallelogram with sides $\boldsymbol{a}$ and $\boldsymbol{b}$, show the identities

$$A^2 = \begin{vmatrix} \boldsymbol{a} \cdot \boldsymbol{a} & \boldsymbol{a} \cdot \boldsymbol{b} \\ \boldsymbol{a} \cdot \boldsymbol{b} & \boldsymbol{b} \cdot \boldsymbol{b} \end{vmatrix} = \begin{vmatrix} \boldsymbol{a} \cdot \boldsymbol{a} & \boldsymbol{a} \cdot \boldsymbol{b} \\ 0 & \boldsymbol{b}_\perp \cdot \boldsymbol{b}_\perp \end{vmatrix} = a^2 b_\perp^2.$$

(6.25) Consider the successive orthogonalizations (3.6)

$$(\boldsymbol{a}, \boldsymbol{b}, \boldsymbol{c}) \to (\boldsymbol{a}, \boldsymbol{b}_\perp, \boldsymbol{c}_\perp) \to (\boldsymbol{a}, \boldsymbol{b}_\perp, \boldsymbol{c}_{\perp\perp}), \tag{6.21}$$

where $\boldsymbol{b}_\perp$ and $\boldsymbol{c}_\perp$ are perpendicular to $\boldsymbol{a}$, while $\boldsymbol{c}_{\perp\perp}$ is perpendicular to both $\boldsymbol{a}$ and $\boldsymbol{b}_\perp$. Explain *geometrically* why these transformations do not change the *volume*. Explain *algebraically* why these transformations do not change the *determinant*. If a, $b_\perp$, $c_{\perp\perp}$ are the magnitudes of the corresponding vectors, show that $\boldsymbol{a} \times \boldsymbol{b} \cdot \boldsymbol{c} = \pm a\, b_\perp c_{\perp\perp}$, but the sign cannot be determined from dot products alone.

(6.26) If V is the volume of the parallelepiped $\boldsymbol{a}$, $\boldsymbol{b}$, $\boldsymbol{c}$ show that

$$V^2 = \begin{vmatrix} \boldsymbol{a} \cdot \boldsymbol{a} & \boldsymbol{a} \cdot \boldsymbol{b} & \boldsymbol{a} \cdot \boldsymbol{c} \\ \boldsymbol{b} \cdot \boldsymbol{a} & \boldsymbol{b} \cdot \boldsymbol{b} & \boldsymbol{b} \cdot \boldsymbol{c} \\ \boldsymbol{c} \cdot \boldsymbol{a} & \boldsymbol{c} \cdot \boldsymbol{b} & \boldsymbol{c} \cdot \boldsymbol{c} \end{vmatrix} = \begin{vmatrix} \boldsymbol{a} \cdot \boldsymbol{a} & \boldsymbol{a} \cdot \boldsymbol{b} & \boldsymbol{a} \cdot \boldsymbol{c} \\ 0 & \boldsymbol{b}_\perp \cdot \boldsymbol{b}_\perp & \boldsymbol{b}_\perp \cdot \boldsymbol{c}_\perp \\ 0 & 0 & \boldsymbol{c}_{\perp\perp} \cdot \boldsymbol{c}_{\perp\perp} \end{vmatrix} = a^2 b_\perp^2 c_{\perp\perp}^2$$

where $\boldsymbol{b}_\perp$ and $\boldsymbol{c}_\perp$ are $\perp$ to $\boldsymbol{a}$, while $\boldsymbol{c}_{\perp\perp}$ is $\perp$ to both $\boldsymbol{a}$ and $\boldsymbol{b}_\perp$. (Hint: use row elimination, and interpret geometrically.)

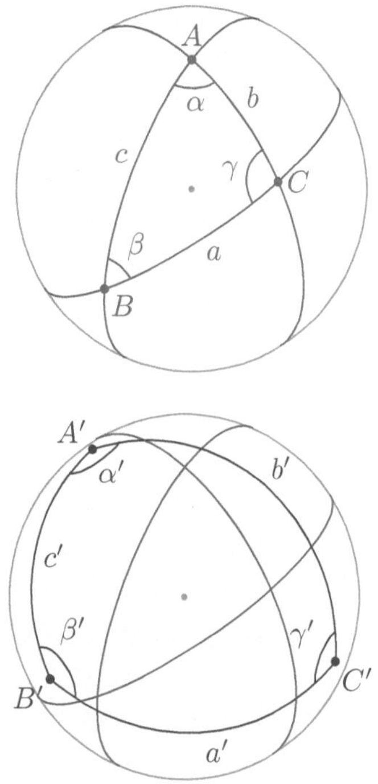

Fig. 6.3 Spherical triangles ABC and $A'B'C'$.

(6.27) *Spherical triangle.* Three points A, B, C on a unit sphere, connected by arcs of great circles, form a *geodesic triangle* (Fig. 6.3). A *great circle* is a circle on the sphere whose center is that of the sphere. Let a, b, c be the lengths of arcs BC, CA, AB, equal to their radian measure on a unit sphere, and α, β, γ the inner angles between the (tangents to the) arcs at A, B, and C, respectively. The center of the sphere is the origin O and define

$$\boldsymbol{A} = \overrightarrow{OA}, \qquad \boldsymbol{B} = \overrightarrow{OB}, \qquad \boldsymbol{C} = \overrightarrow{OB}.$$

Take a, b, c as the smallest arcs joining the vertices, so all angles are between 0 and π, and ABC is a *proper* geodesic triangle with each circular arc the shortest distance between the vertices on the sphere. Assume furthermore, that vertices are labeled such that $\boldsymbol{A}, \boldsymbol{B}, \boldsymbol{C}$ is right-handed, that is $\boldsymbol{A} \times \boldsymbol{B} \cdot \boldsymbol{C} \geq 0$.

(a) Calculate $(\boldsymbol{A} \times \boldsymbol{B}) \cdot (\boldsymbol{A} \times \boldsymbol{C})$ in two ways to obtain the spherical law of cosines for sides,

$$\cos a = \cos b \cos c + \sin b \sin c \, \cos \alpha, \qquad (6.22)$$

with similar identities for $\cos b$ and $\cos c$ by cyclic rotations of the symbols. (Hint: use (1) Fig. 6.4 for a direct calculation, and (2) vector identities.)

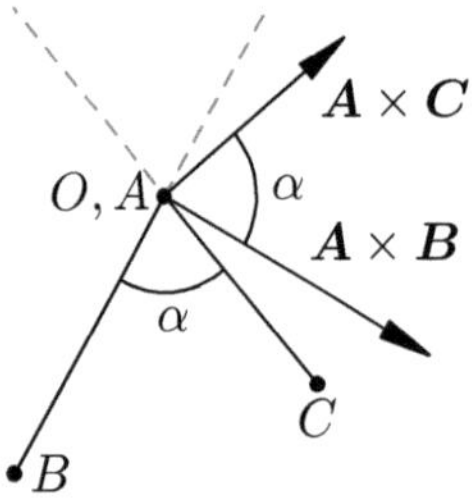

Fig. 6.4 Angles between planes AOB and AOC.

(b) Derive the spherical law of sines

$$\frac{\sin \alpha}{\sin a} = \frac{\sin \beta}{\sin b} = \frac{\sin \gamma}{\sin c} \qquad (6.23)$$

using spherical coordinates with polar axis $\hat{\mathbf{z}} = \boldsymbol{A}$ and $\boldsymbol{B}$ in the xz-plane, to show that $\boldsymbol{A} \times \boldsymbol{B} \cdot \boldsymbol{C} = \sin c \,(\sin b \sin \alpha)$; then cyclic rotations that do not change the mixed product.

(c) For a sphere of radius R, where a, b, c need to be replaced by their radian measures $a/R, b/R, c/R$ in the above formulas (6.22), (6.23), show that $R \to \infty$, with a, b, c fixed yields the law of cosines and the law of sines for flat triangles:

$$a^2 = b^2 + c^2 - 2bc \cos \alpha,$$
$$bc \sin \alpha = ca \sin \beta = ab \sin \gamma.$$

(6.28) *Spherical polar triangle.* Let A, B, C be three points on a unit sphere, with position vectors $\boldsymbol{A}, \boldsymbol{B}, \boldsymbol{C}$ from the center of the sphere at O. As in problem 6.27, a, b, c are the lengths of the great circle arcs connecting BC, CA, AB, equal to their radian measures on a unit sphere, and α, β, γ are the angles between the arcs at A, B, C. Restrict to *proper* geodesic triangles formed by the shortest arcs joining the vertices, so all angles are less or equal to π, then

$$\begin{aligned} \boldsymbol{A} \cdot \boldsymbol{B} &= \cos c, & \boldsymbol{A} \times \boldsymbol{B} &= \sin c \, \boldsymbol{C}', \\ \boldsymbol{B} \cdot \boldsymbol{C} &= \cos a, & \boldsymbol{B} \times \boldsymbol{C} &= \sin a \, \boldsymbol{A}', \\ \boldsymbol{C} \cdot \boldsymbol{A} &= \cos b, & \boldsymbol{C} \times \boldsymbol{A} &= \sin b \, \boldsymbol{B}', \end{aligned} \qquad (6.24)$$

where $\boldsymbol{A}', \boldsymbol{B}', \boldsymbol{C}'$ are unit vectors, with $\boldsymbol{A}'$ the polar axis to the BC equator, and similarly for $\boldsymbol{B}'$ and $\boldsymbol{C}'$ (Fig. 6.5).

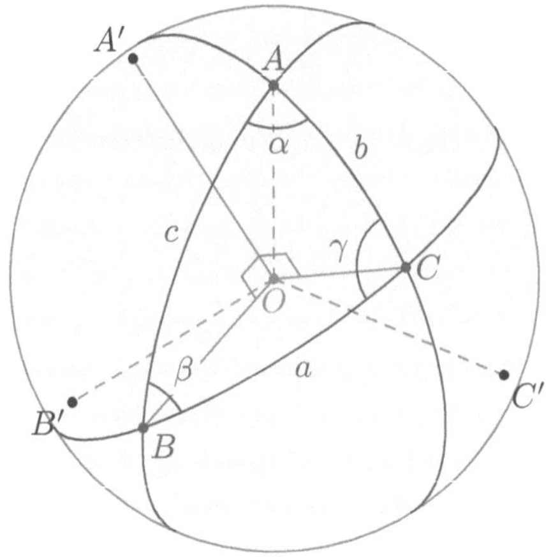

Fig. 6.5 Relative poles A', B', C'.

(a) From (6.24) and vector algebra, deduce

$$\begin{aligned} \boldsymbol{A}' \cdot \boldsymbol{B}' &= -\cos \gamma, & \boldsymbol{A}' \times \boldsymbol{B}' &= \sin \gamma \, \boldsymbol{C}, \\ \boldsymbol{B}' \cdot \boldsymbol{C}' &= -\cos \alpha, & \boldsymbol{B}' \times \boldsymbol{C}' &= \sin \alpha \, \boldsymbol{A}, \\ \boldsymbol{C}' \cdot \boldsymbol{A}' &= -\cos \beta, & \boldsymbol{C}' \times \boldsymbol{A}' &= \sin \beta \, \boldsymbol{B}. \end{aligned} \qquad (6.25)$$

(b) Use $(\boldsymbol{A}' \times \boldsymbol{B}') \cdot (\boldsymbol{A}' \times \boldsymbol{C}')$ to derive the *law of cosines for angles*

$$\cos \alpha = -\cos \beta \cos \gamma + \cos a \sin \beta \sin \gamma, \qquad (6.26)$$

with similar equations for $\cos \beta$ and $\cos \gamma$ by cyclic rotations of the symbols.

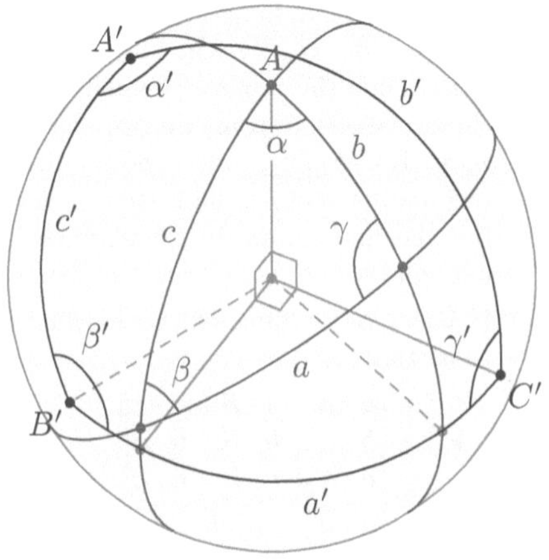

Fig. 6.6 Geodesic polar triangle $A'B'C'$.

(c) For the geodesic polar triangle $A'B'C'$ (Fig. 6.6), define the lengths of those arcs $B'C', C'A', A'B'$ as a', b', c' and the angles between those arcs at A', B', C' as α', β', γ', respectively. Justify geometrically why

$$\begin{aligned} a' &= \pi - \alpha, & \alpha' &= \pi - a, \\ b' &= \pi - \beta, & \beta' &= \pi - b, \\ c' &= \pi - \gamma, & \gamma' &= \pi - c. \end{aligned} \tag{6.27}$$

Deduce (6.25) from these.

(d) From the law of sines (6.23), show that

$$\boldsymbol{A}' \cdot \boldsymbol{A} = \sin\beta \sin c = \sin b \sin\gamma, \tag{6.28}$$

and similarly for $\boldsymbol{B}' \cdot \boldsymbol{B}$ and $\boldsymbol{C}' \cdot \boldsymbol{C}$ by cyclic rotations of symbols.

(e) Explain why $\{\boldsymbol{C}', \boldsymbol{A}, \boldsymbol{C}' \times \boldsymbol{A}\}$ is a right handed orthonormal basis, and deduce from (6.28), (6.24), (6.25), or geometrically, that

$$\begin{aligned} \boldsymbol{A}' &= \sin\beta \sin c\, \boldsymbol{A} - \sin\beta \cos c\, \boldsymbol{C}' \times \boldsymbol{A} - \cos\beta\, \boldsymbol{C}', \\ \boldsymbol{B} &= \cos c\, \boldsymbol{A} + \sin c\, \boldsymbol{C}' \times \boldsymbol{A}, \\ \boldsymbol{C} &= \cos b\, \boldsymbol{A} + \sin b \left(\cos\alpha\, \boldsymbol{C}' \times \boldsymbol{A} + \sin\alpha\, \boldsymbol{C}'\right). \end{aligned} \tag{6.29}$$

From $\boldsymbol{B} \times \boldsymbol{C} = \sin a\, \boldsymbol{A}'$, deduce the identity

$$\cos\alpha \sin b \cos c - \cos b \sin c = -\cos\beta \sin a. \tag{6.30}$$

Similarly, find $\boldsymbol{A}', \boldsymbol{B}, \boldsymbol{C}$ in terms of the basis $\{\boldsymbol{A}, \boldsymbol{B}', \boldsymbol{A} \times \boldsymbol{B}'\}$, and derive the identity

$$\cos\alpha \cos b \sin c - \sin b \cos c = -\cos\gamma \sin a. \tag{6.31}$$

(f) Derive identity (6.30) by substituting the expression for $\cos a$ in (6.22) into the corresponding one for $\cos b$. Likewise, derive (6.31) from (6.22) by substituting the expression for $\cos a$ into that for $\cos c$.
Further spherical trigonometry results and insights can be found in Isaac Todhunter's textbook.[2]

[2] I. Todhunter, *Spherical Trigonometry*, Macmillan, 1886. `https://www.gutenberg.org/ebooks/19770`

Points, Lines, Planes, etc.

7

7.1 Radius vector

In multivariable calculus and linear algebra it is expedient to confuse points and vectors. In $\mathbb{R}^3$ for instance, a point P is defined as a triplet (x_1, x_2, x_3) but a vector $\boldsymbol{a}$ is also defined by a real number triplet (a_1, a_2, a_3); yet in physical space, points and displacements are different objects. Two points define a displacement, and a line. Displacements can be added and scaled, but points are not added or scaled.

The confusion arises from the fundamental way to locate points in euclidean space by specifying displacements from a *reference point* called the *origin* and denoted O. A point P is then specified by providing the displacement $\boldsymbol{r} = \overrightarrow{OP}$ from the reference point O. That vector $\boldsymbol{r}$ is called the *radius* or *position*[1] vector of point P (Fig. 7.1).

A *cartesian system of coordinates* consists of a reference point O and three mutually orthogonal directions $\hat{\mathbf{x}}$, $\hat{\mathbf{y}}$, $\hat{\mathbf{z}}$ that provide a basis for displacements in 3D euclidean space $\mathbb{E}^3$. The radius vector $\boldsymbol{r} = \overrightarrow{OP}$ of point P can then be specified in spherical, cylindrical, or cartesian form as in §1.3, now for the radius vector $\boldsymbol{r}$ instead of the arbitrary vector $\boldsymbol{a}$,

$$\overrightarrow{OP} = \boldsymbol{r} = r\,\hat{\boldsymbol{r}} = \rho\,\hat{\boldsymbol{\rho}} + z\,\hat{\mathbf{z}} = x\,\hat{\mathbf{x}} + y\,\hat{\mathbf{y}} + z\,\hat{\mathbf{z}}. \tag{7.1}$$

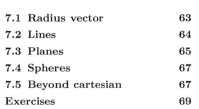

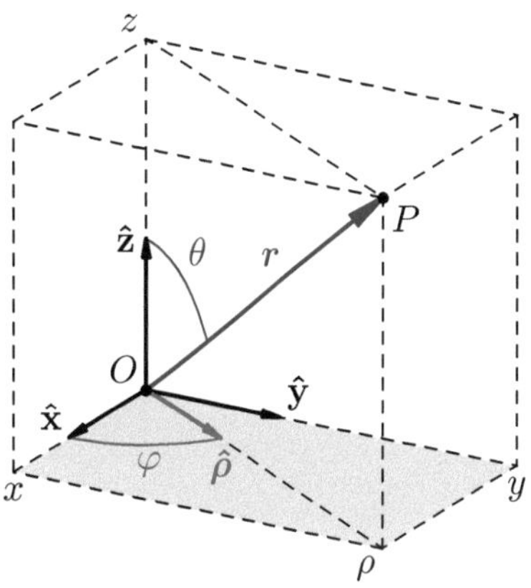

Fig. 7.1 *Radius* vector $\boldsymbol{r} = \overrightarrow{OP}$, a.k.a. the *position* vector, in spherical, cylindrical, and cartesian coordinates, eqn (7.1). In the physics and ISO conventions, θ is the angle between $\hat{\mathbf{z}}$ and $\boldsymbol{r}$.

The meridional plane equality $r\hat{\boldsymbol{r}} = \rho\hat{\boldsymbol{\rho}} + z\hat{\mathbf{z}}$ with $\hat{\boldsymbol{\rho}} \cdot \hat{\mathbf{z}} = 0$ yields

$$\left.\begin{aligned} r &= \sqrt{\rho^2 + z^2} \\ \theta &= \arccos(z/r) \end{aligned}\right\} \quad \Leftrightarrow \quad \left\{\begin{aligned} \rho &= r\sin\theta \\ z &= r\cos\theta. \end{aligned}\right. \tag{7.2}$$

Similarly, the equatorial plane equality $\rho\,\hat{\boldsymbol{\rho}} = x\,\hat{\mathbf{x}} + y\,\hat{\mathbf{y}}$ with $\hat{\mathbf{x}} \cdot \hat{\mathbf{y}} = 0$ gives

$$\left.\begin{aligned} \rho &= \sqrt{x^2 + y^2} \\ \varphi &= \operatorname{atan2}(y, x) \end{aligned}\right\} \quad \Leftrightarrow \quad \left\{\begin{aligned} x &= \rho\cos\varphi \\ y &= \rho\sin\varphi. \end{aligned}\right. \tag{7.3}$$

Eliminating ρ from (7.2) and (7.3) gives

$$\left.\begin{aligned} r &= \sqrt{x^2 + y^2 + z^2} \\ \theta &= \arccos(z/r) \\ \varphi &= \operatorname{atan2}(y, x) \end{aligned}\right\} \quad \Leftrightarrow \quad \left\{\begin{aligned} x &= r\sin\theta\cos\varphi \\ y &= r\sin\theta\sin\varphi \\ z &= r\cos\theta \end{aligned}\right. \tag{7.4}$$

with $0 \le \theta \le \pi$ and $-\pi < \varphi \le \pi$.

[1]Using $\boldsymbol{p}$ for position vector is avoided since it is used for momentum $\boldsymbol{p} = m\boldsymbol{v}$ in physics.

We can also deduce expressions for the direction vectors $\hat{\boldsymbol{\rho}}$ and $\hat{\boldsymbol{r}}$ in terms of the cartesian directions $\hat{\mathbf{x}}$, $\hat{\mathbf{y}}$, $\hat{\mathbf{z}}$,

$$\hat{\boldsymbol{\rho}} = \cos\varphi\,\hat{\mathbf{x}} + \sin\varphi\,\hat{\mathbf{y}}, \qquad \hat{\boldsymbol{r}} = \sin\theta\,\hat{\boldsymbol{\rho}} + \cos\theta\,\hat{\mathbf{z}}. \tag{7.5}$$

We refer to these as *hybrid* representations using spherical coordinates (θ, φ) with cartesian direction vectors $\hat{\mathbf{x}}$, $\hat{\mathbf{y}}$, $\hat{\mathbf{z}}$, to distinguish them from the full cartesian expressions

$$\hat{\boldsymbol{\rho}} = \frac{x\,\hat{\mathbf{x}} + y\,\hat{\mathbf{y}}}{\sqrt{x^2+y^2}}, \qquad \hat{\boldsymbol{r}} = \frac{x\,\hat{\mathbf{x}} + y\,\hat{\mathbf{y}} + z\,\hat{\mathbf{z}}}{\sqrt{x^2+y^2+z^2}}. \tag{7.6}$$

Note that θ is the angle between $\hat{\mathbf{z}}$ and $\boldsymbol{r}$ (called the *polar* or *zenith* or *inclination* angle, depending on the context) while φ is the azimuthal angle *about* the z axis, the angle between the $(\hat{\mathbf{z}}, \hat{\mathbf{x}})$ and the $(\hat{\mathbf{z}}, \boldsymbol{r})$ planes. The distance to the origin is r while $\rho = r_\perp$ is the distance to the z-axis. This is the physics convention in ISO 80000-2 (International Standards Organization) widely used in mathematical physics. See for example Fig. 10-10 in Volume III of *The Feynman Lectures on Physics.*[2] American calculus teachers often reverse the definitions of θ and φ, creating confusion for students in the physical sciences.

The unit vectors $\hat{\mathbf{x}}$, $\hat{\mathbf{y}}$, $\hat{\mathbf{z}}$ form a basis for 3D euclidean vector space, but $\hat{\boldsymbol{\rho}}$ and $\hat{\mathbf{z}}$ do not, and $\hat{\boldsymbol{r}}$ does not either. The cartesian basis $\hat{\mathbf{x}}$, $\hat{\mathbf{y}}$, $\hat{\mathbf{z}}$ consists of three fixed and mutually orthogonal directions independent of P, but $\hat{\boldsymbol{\rho}}$ and $\hat{\boldsymbol{r}}$ depend on the location of P with respect to O; each point has its own $\hat{\boldsymbol{\rho}}$ and $\hat{\boldsymbol{r}}$. We will construct full cylindrical and spherical orthogonal bases, $(\hat{\boldsymbol{\rho}}, \hat{\boldsymbol{\varphi}}, \hat{\mathbf{z}})$ and $(\hat{\boldsymbol{r}}, \hat{\boldsymbol{\theta}}, \hat{\boldsymbol{\varphi}})$, later in vector calculus. These cylindrical and spherical basis vectors vary with position P. They yield local bases for vector functions of position P.

Once a cartesian system of coordinates, O, $\hat{\mathbf{x}}$, $\hat{\mathbf{y}}$, $\hat{\mathbf{z}}$, has been chosen, the cartesian *coordinates* (x, y, z) of P are the cartesian *components* of $\boldsymbol{r} = x\,\hat{\mathbf{x}} + y\,\hat{\mathbf{y}} + z\hat{\mathbf{z}}$. The cylindrical coordinates of P are (ρ, φ, z) and its *spherical coordinates* are (r, θ, φ), but cylindrical and spherical coordinates are *not* vector components. They are the cylindrical and spherical representations of $\boldsymbol{r}$.

7.2 Lines

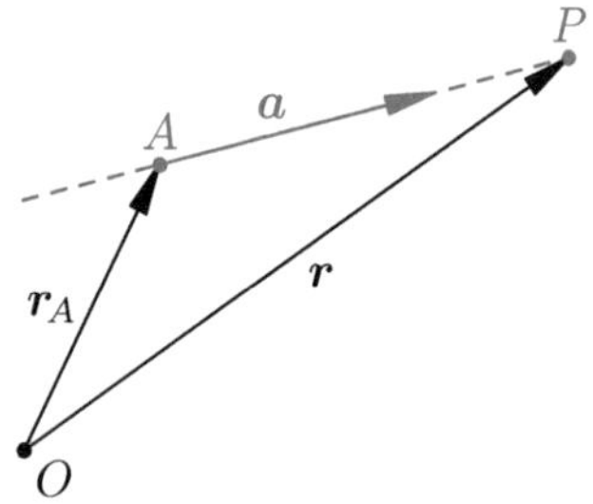

Fig. 7.2 Line through A, parallel to $\boldsymbol{a}$.

The line through point A that is parallel to the vector $\boldsymbol{a}$ consists of all points P such that (Fig. 7.2)

$$\overrightarrow{AP} = t\,\boldsymbol{a}, \quad \forall t \in \mathbb{R}. \tag{7.7a}$$

This vector equation expresses that the vector $\overrightarrow{AP}$ is parallel to $\boldsymbol{a}$. In terms of an origin O we have $\boldsymbol{r} = \overrightarrow{OP} = \overrightarrow{OA} + \overrightarrow{AP} = \boldsymbol{r}_A + \overrightarrow{AP}$, that is

$$\boldsymbol{r} = \boldsymbol{r}_A + t\,\boldsymbol{a}. \tag{7.7b}$$

[2] http://www.feynmanlectures.caltech.edu/III_10.html#Ch10-F10

This is the *parametric vector equation* of the line. The real number t is the parameter; it is the coordinate of P in the system of coordinates $(A, \boldsymbol{a})$ with reference point A and basis vector $\boldsymbol{a}$. We can eliminate t by taking the cross product of (7.7b) with $\boldsymbol{a}$ to obtain the implicit equation

$$(\boldsymbol{r} - \boldsymbol{r}_A) \times \boldsymbol{a} = 0. \tag{7.7c}$$

In cartesian coordinates, $\boldsymbol{r} = x\hat{\mathbf{x}} + y\hat{\mathbf{y}} + z\hat{\mathbf{z}}$, $\boldsymbol{r}_A = x_A\hat{\mathbf{x}} + y_A\hat{\mathbf{y}} + z_A\hat{\mathbf{z}}$, and $\boldsymbol{a} = a_x\hat{\mathbf{x}} + a_y\hat{\mathbf{y}} + a_z\hat{\mathbf{z}}$; thus, (7.7b) yields

$$\begin{cases} x = x_A + t\, a_x, \\ y = y_A + t\, a_y, \\ z = z_A + t\, a_z, \end{cases} \tag{7.7d}$$

which provides a map from $t \in \mathbb{R} \to (x, y, z) \in \mathbb{R}^3$. Eliminating t yields the implicit equations for the line

$$\frac{x - x_A}{a_x} = \frac{y - y_A}{a_y} = \frac{z - z_A}{a_z}. \tag{7.7e}$$

Notation We use $\boldsymbol{r}$ for the radius vector of a variable *point* P with cartesian coordinates (x, y, z). We use $\boldsymbol{r}_A$ for the radius vector of a specific point A with cartesian coordinates (x_A, y_A, z_A), and (a_x, a_y, a_z) for the cartesian *components* of vector $\boldsymbol{a}$, to help distinguish between points and vectors.

7.3 Planes

The equation of a plane passing through point A and parallel to the vectors $\boldsymbol{a}$ and $\boldsymbol{b}$ consists of all points P such that (Fig. 7.3)

$$\overrightarrow{AP} = u\,\boldsymbol{a} + v\,\boldsymbol{b}, \quad \forall u, v \in \mathbb{R}. \tag{7.8a}$$

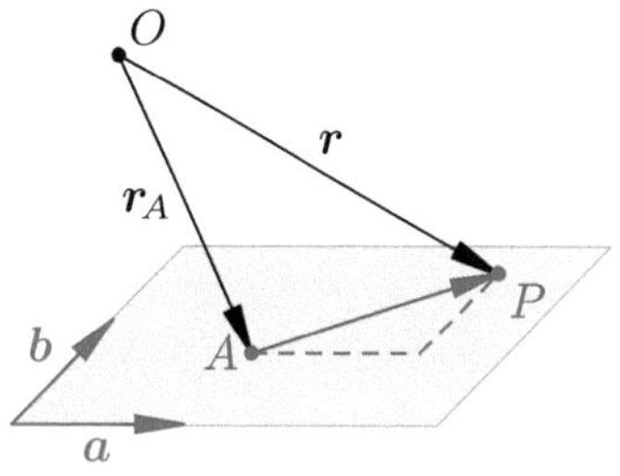

Fig. 7.3 Plane through point A parallel to $\boldsymbol{a}$ and $\boldsymbol{b}$.

In terms of an origin O, with $\boldsymbol{r} = \overrightarrow{OP} = \overrightarrow{OA} + \overrightarrow{AP}$, that is

$$\boldsymbol{r} = \boldsymbol{r}_A + u\,\boldsymbol{a} + v\,\boldsymbol{b}. \tag{7.8b}$$

This is the parametric vector equation of that plane with real parameters u, v, that are the coordinates of P in the system of coordinates specified by A, $\boldsymbol{a}$, $\boldsymbol{b}$. The parameters u and v can be eliminated by dotting the parametric equation (7.8b) with $\boldsymbol{n} = \boldsymbol{a} \times \boldsymbol{b}$, yielding

$$\overrightarrow{AP} \cdot \boldsymbol{n} = (\boldsymbol{r} - \boldsymbol{r}_A) \cdot \boldsymbol{n} = 0. \tag{7.8c}$$

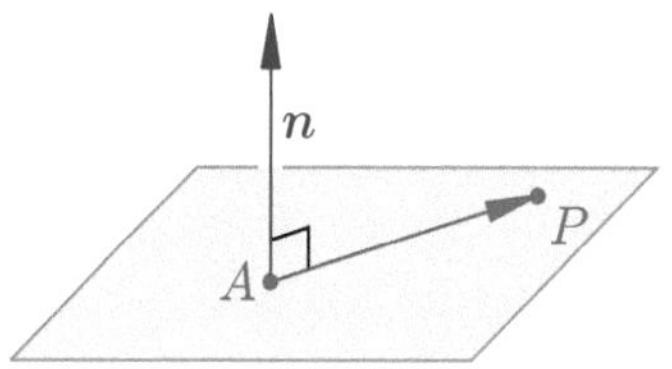

Fig. 7.4 Plane through point A perpendicular to $\boldsymbol{n}$.

This is the implicit equation of the plane passing through A and perpendicular to $\boldsymbol{n}$ (Fig. 7.4).

In cartesian coordinates, the vector equation $\boldsymbol{r} = \boldsymbol{r}_A + u\boldsymbol{a} + v\boldsymbol{b}$ reads

$$\begin{cases} x = x_A + u\, a_x + v\, b_x, \\ y = y_A + u\, a_y + v\, b_y, \\ z = z_A + u\, a_z + v\, b_z, \end{cases} \tag{7.8d}$$

that provide a map from $(u, v) \in \mathbb{R}^2 \to (x, y, z) \in \mathbb{R}^3$, while the implicit equation $(\boldsymbol{r} - \boldsymbol{r}_A) \cdot \boldsymbol{n} = 0$ reads

$$n_x(x - x_A) + n_y(y - y_A) + n_z(z - z_A) = 0, \tag{7.8e}$$

where $n_x = (a_y b_z - a_z b_y)$, $n_y = (a_z b_x - a_x b_z)$, $n_z = (a_x b_y - a_y b_x)$.

Line and plane

To illustrate the judicious use of these various parametric and implicit equations of lines and planes, consider finding the point of intersection of a line and a plane. There is a unique point of intersection, in general, unless the line is parallel to the plane (no intersection), or the line is in the plane (all points are in the plane).

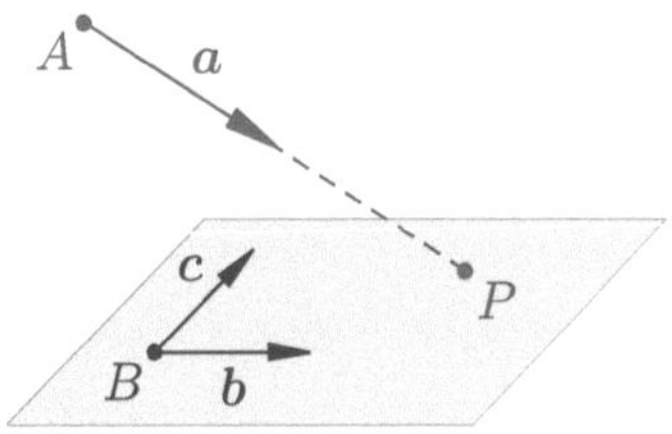

Fig. 7.5 Line and plane.

Let the line be that passing through point A, parallel to vector $\boldsymbol{a}$ so its parametric and implicit equations are (7.7b) and (7.7c), and let the plane pass through point B and parallel to vectors $\boldsymbol{b}$ and $\boldsymbol{c}$ (Fig. 7.5), so its parametric and implicit equations are

$$\boldsymbol{r} = \boldsymbol{r}_B + u\,\boldsymbol{b} + v\,\boldsymbol{c} \quad \Leftrightarrow \quad (\boldsymbol{r} - \boldsymbol{r}_B) \cdot (\boldsymbol{b} \times \boldsymbol{c}) = 0,$$

in lieu of (7.8b) and (7.8c), since the reference point A and direction $\boldsymbol{a}$ for the line and the plane are distinct, in general. There are four distinct formulations, depending on the choice of implicit or parametric equations for the line and the plane.

(a) *Implicit–implicit:* Find $\boldsymbol{r}$ such that

$$(\boldsymbol{r} - \boldsymbol{r}_A) \times \boldsymbol{a} = 0 \quad \text{and} \quad (\boldsymbol{r} - \boldsymbol{r}_B) \cdot (\boldsymbol{b} \times \boldsymbol{c}) = 0. \tag{7.9a}$$

In cartesian coordinates, this yields a 4-by-3 linear system of equations for $\boldsymbol{r} \equiv (x, y, z)$ that can be solved by elimination (exercise 7.6).

(b) *Parametric–parametric:* If $\boldsymbol{r}$ is on the line, then $\boldsymbol{r} = \boldsymbol{r}_A + t\boldsymbol{a}$ for some t. If it is on the plane then $\boldsymbol{r} = \boldsymbol{r}_B + u\boldsymbol{b} + v\boldsymbol{c}$, for some u, v. Eliminating $\boldsymbol{r}$ gives

$$\boldsymbol{r}_A + t\,\boldsymbol{a} = \boldsymbol{r}_B + u\,\boldsymbol{b} + v\,\boldsymbol{c}, \tag{7.9b}$$

which is a 3-by-3 linear system for the parameters (t, u, v).

(c) *Implicit–parametric:* A third solution method is to use implicit equation for the line $(\boldsymbol{r} - \boldsymbol{r}_A) \times \boldsymbol{a} = 0$ together with the parametric equation for the plane, $\boldsymbol{r} = \boldsymbol{r}_B + u\boldsymbol{b} + v\boldsymbol{c}$. Substituting for $\boldsymbol{r}$ yields a 3-by-2 linear system for u and v,

$$(\boldsymbol{a} \times \boldsymbol{b})\,u + (\boldsymbol{a} \times \boldsymbol{c})\,v = \boldsymbol{a} \times (\boldsymbol{r}_A - \boldsymbol{r}_B). \tag{7.9c}$$

(d) *Parametric–implicit:* The winning combination is to use the parametric equation for the line together with the implicit equation for the plane since, to locate the intersection, we just need to find t along the line. Thus, the intersection is given by the value of t such that $\boldsymbol{r} = \boldsymbol{r}_A + t\boldsymbol{a}$ satisfies $(\boldsymbol{r} - \boldsymbol{r}_B) \cdot (\boldsymbol{b} \times \boldsymbol{c}) = 0$. Substituting for $\boldsymbol{r}$ and solving for t gives

$$t = \frac{(\boldsymbol{r}_B - \boldsymbol{r}_A) \cdot (\boldsymbol{b} \times \boldsymbol{c})}{\boldsymbol{a} \cdot (\boldsymbol{b} \times \boldsymbol{c})}. \tag{7.9d}$$

7.4 Spheres

The implicit equation of a sphere of radius R centered at C is $|\boldsymbol{r}-\boldsymbol{r}_C| = R$, that is

$$(\boldsymbol{r}-\boldsymbol{r}_C)\cdot(\boldsymbol{r}-\boldsymbol{r}_C) = R^2. \tag{7.10}$$

The parametric equation of that sphere is (Fig. 7.6)

$$\boldsymbol{r} = \boldsymbol{r}_C + R\,\hat{\mathbf{a}}, \tag{7.11}$$

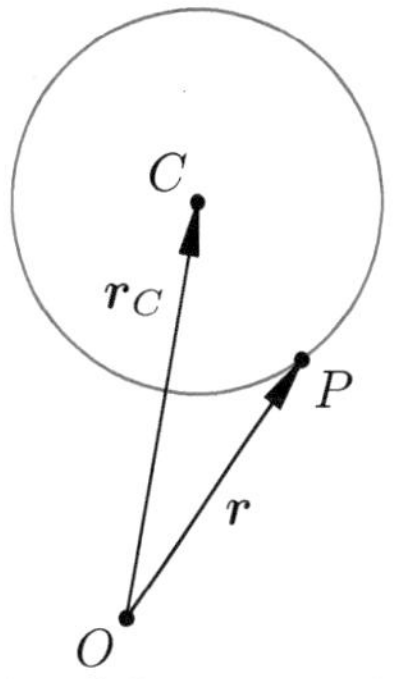

Fig. 7.6 Sphere centered at C.

where $\hat{\mathbf{a}}$ is any direction in 3D space. We have seen in eqn (1.13c) that such a direction can be expressed as

$$\hat{\mathbf{a}}(\theta,\varphi) = \cos\varphi\sin\theta\,\hat{\mathbf{x}} + \sin\varphi\sin\theta\,\hat{\mathbf{y}} + \cos\theta\,\hat{\mathbf{z}},$$

where $\hat{\mathbf{x}}, \hat{\mathbf{y}}, \hat{\mathbf{z}}$ are any set of mutually orthogonal unit vectors. The angles θ and φ are the two parameters for the sphere.

In cartesian coordinates, the implicit vector equation (7.10) becomes

$$(x-x_C)^2 + (y-y_C)^2 + (z-z_C)^2 = R^2,$$

and the parametric equation (7.11) reads

$$\begin{cases} x = x_C + R\sin\theta\cos\varphi, \\ y = y_C + R\sin\theta\sin\varphi, \\ z = z_C + R\cos\theta. \end{cases}$$

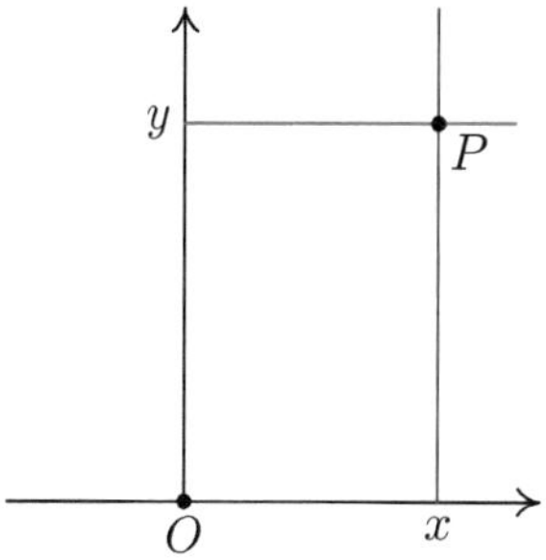

Fig. 7.7 Cartesian coordinates (x, y).

7.5 Beyond cartesian

Cartesian coordinates (x, y) specify the position of point P with respect to an origin O by specifying signed distances in two orthogonal directions (Fig. 7.7). Cartesian coordinates are fundamental for euclidean space, but there are many other useful ways to specify the coordinates of P. *Polar coordinates* (r, φ) allow us to go straight to the point by specifying distance r in direction φ (Fig. 7.8).

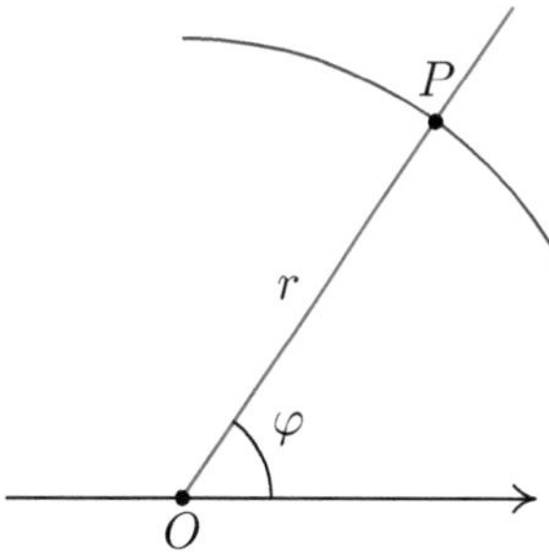

Fig. 7.8 Polar coordinates (r, φ).

Biradial coordinates (α_1, α_2) allow us to *triangulate* our position P from, say, two lighthouses located at points F_1 and F_2, by measuring the angles between the lines F_1P, F_2P and the northern direction, thus locating P as the intersection of two *radials*, one from F_1 and one from F_2 (Fig. 7.9).

Bicircular coordinates locate P using the *distances* $s_1 = |F_1P|$ and $s_2 = |F_2P|$ as in the global positioning system (GPS) that measures distance from satellites in 3D space. The point P is the intersection of two circles in 2D, and three spheres in 3D (Fig. 7.10). The GPS system uses redundancy and least squares to determine points uniquely and robustly.

Apollonian coordinates locate P using the ratio of distances s_1/s_2 and the difference of the angles $u = \alpha_2 - \alpha_1$. These are *orthogonal coordinates*, since P is then the intersection of *Apollonian circles* that meet orthogonally (Fig. 7.11).

Bipolar coordinates are even better. They use the natural log of the distance ratio, $v = \ln(s_1/s_2)$ instead of s_1/s_2, together with the angle difference $u = \alpha_2 - \alpha_1$. Bipolar coordinates (u, v) are not only orthogonal but also *conformal*; that is, they preserve all angles as will be discussed later in vector and complex calculus (Fig. 7.12). Bipolar coordinates are the natural coordinates if there are two *electric line charges* of opposite sign at F_1 and F_2 in electrostatics, two opposite *line currents* in magnetostatics, a line *source* and a line *sink* in fluid dynamics, or counter-rotating line *vortices* in aerodynamics, such as the trailing vortices behind an airplane. These will be revisited in §16.2 and 16.3.

Of all these useful coordinate systems, only (x, y) correspond to *components* of the position vector $\boldsymbol{r}$ in the basis $(\hat{\mathbf{x}}, \hat{\mathbf{y}})$. *Coordinates* of points are usually *not* the *components* of vectors. We will study these and other *curvilinear coordinates* later in vector calculus. But before going beyond cartesian, we study how to switch from one cartesian basis to another in chapter 8.

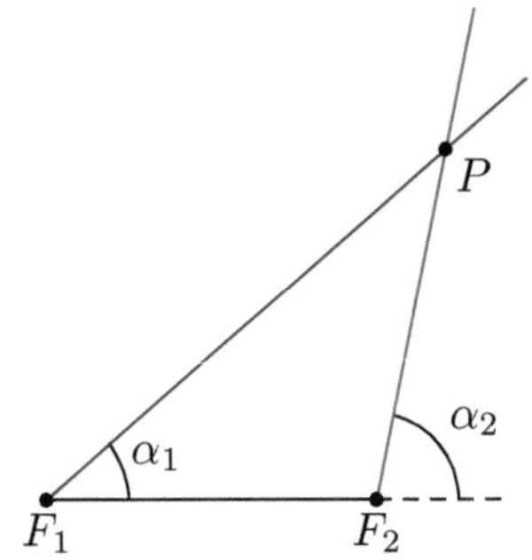

Fig. 7.9 Biradials (α_1, α_2).

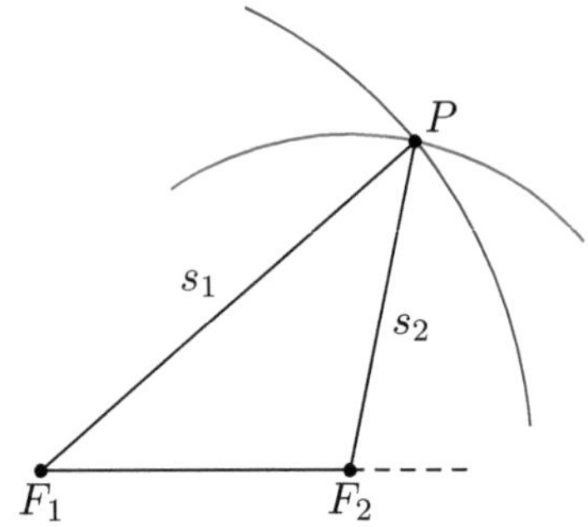

Fig. 7.10 Bicircles (s_1, s_2).

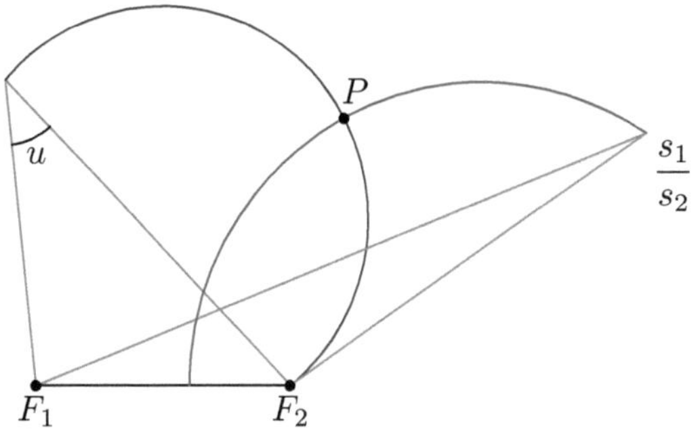

Fig. 7.11 Apollonian coordinates $u = \alpha_2 - \alpha_1$ and s_1/s_2.

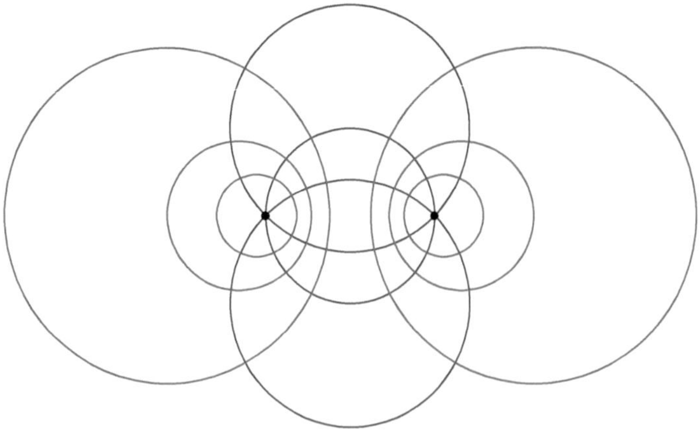
Fig. 7.12 Bipolar coordinate curves $u = \pm(\pi/4, 2\pi/4, 3\pi/4)$ and $v = \pm(0.5, 1, 1.5)$ are Apollonian circles, with $(u, v) = (\alpha_2 - \alpha_1, \ln(s_1/s_2))$ for $\alpha_1, \alpha_2, s_1, s_2$, as in Figs. 7.9 and 7.10.

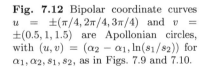

Exercises

(7.1) Consider the point A whose position vector $\boldsymbol{r}_A = \overrightarrow{OA}$ is the average of the position vectors of three arbitrary points P_1, P_2, P_3 in 3D euclidean space,

$$\boldsymbol{r}_A \triangleq \frac{\boldsymbol{r}_1 + \boldsymbol{r}_2 + \boldsymbol{r}_3}{3}.$$

Show that A lies on each of the medians of the triangle P_1, P_2, P_3 and is therefore independent of the origin O.

(7.2) *Center of mass.* The center of mass of a system of N particles of mass m_i located at position $\boldsymbol{r}_i$, $i = 1, \dots, N$, is the *mass-averaged position* defined by

$$\boldsymbol{r}_c \triangleq \frac{1}{M} \sum_{i=1}^{N} m_i \boldsymbol{r}_i$$

where $M = \sum_{i=1}^{N} m_i$ is the total mass. Show that $\boldsymbol{r}_c$ is independent of the origin. (Consider $O' \neq O$ and define $\boldsymbol{r}_i'$ as the position vector of P_i from O'. Show that the mass-averaged position from O is the same *point* as the mass-averaged position from O'.)

(7.3) Find vector equations for the line passing through the two points with radius vectors $\boldsymbol{r}_1$, $\boldsymbol{r}_2$, and for the plane through the three points $\boldsymbol{r}_1$, $\boldsymbol{r}_2$, $\boldsymbol{r}_3$.

(7.4) Let $\boldsymbol{r}_A$ and $\boldsymbol{r}_B$ be the position vectors of points A and B, respectively, and consider all points $\boldsymbol{r} = (\alpha \boldsymbol{r}_A + \beta \boldsymbol{r}_B)/(\alpha + \beta)$ for all real α and β with $\alpha + \beta \neq 0$. Do these points lie along a line or a plane? Explain.

(7.5) Let $\boldsymbol{r}_A$, $\boldsymbol{r}_B$, $\boldsymbol{r}_C$ be the position vectors of points A, B and C in 3D space, respectively, and consider all points $\boldsymbol{r} = (\alpha \boldsymbol{r}_A + \beta \boldsymbol{r}_B + \gamma \boldsymbol{r}_C))/(\alpha + \beta + \gamma)$ for all real α, β, γ with $\alpha + \beta + \gamma \neq 0$. Show that these points lie on a plane. Which plane is it?

(7.6) Express (7.9a) in cartesian coordinates. When is the 4-by-3 linear system solvable? When is there a unique solution? Infinitely many solutions? Answer using elimination (row reduction) as well as using vectors and geometry.

(7.7) Derive (7.9d) by dotting (7.9b) with a suitable vector. Which vector is that?

(7.8) Find u by dotting (7.9c) with a suitable vector, then find v in a similar way.

(7.9) What is the distance between point A and the plane that passes through B and is perpendicular to $\boldsymbol{n}$? Sketch and explain.

(7.10) What is the distance between point B and the line that passes through A and is parallel to $\boldsymbol{a}$? Sketch and explain.

(7.11) Find the point on plane A, B, C that is closest to point D. Sketch and explain.

(7.12) A laser pointer is located at point L, its beam pointing in the direction of $\boldsymbol{b}$. Where does it hit the plane passing through points A, B, C? Sketch and explain.

(7.13) What is the distance between the line parallel to $\boldsymbol{a}$ that passes through point A and the line parallel to $\boldsymbol{b}$ that passes through point B? Sketch and explain.

(7.14) A particle was at point P_1 at time t_1 and is moving at the constant velocity $\boldsymbol{v}_1$. Another particle was at P_2 at t_2 and is moving at the constant velocity $\boldsymbol{v}_2$. How close did the particles get to each other and at what time? What conditions are needed for a collision?

(7.15) Show that all $\boldsymbol{r}$'s such that $\boldsymbol{a} \cdot \hat{\boldsymbol{r}} = b$ is a cone, not a plane. Which cone is it?

(7.16) Find all $\boldsymbol{r}$'s such that $(\boldsymbol{r} - \boldsymbol{r}_A) \cdot \hat{\boldsymbol{a}} = 0$ and $(\boldsymbol{r} - \boldsymbol{r}_B) \cdot \hat{\boldsymbol{b}} = 0$, by expressing $\boldsymbol{r}$ in terms of the intrinsic basis $\{\hat{\boldsymbol{a}}, \hat{\boldsymbol{a}} \times \hat{\boldsymbol{b}}, (\hat{\boldsymbol{a}} \times \hat{\boldsymbol{b}}) \times \hat{\boldsymbol{a}}\}$. Sketch and discuss the geometry.

(7.17) Find the intersections of a line passing through point A in the direction of $\boldsymbol{a}$ and the sphere of radius R centered at point C. Illustrate.

(7.18) Construct a vector parametrization for the intersection of the plane $\boldsymbol{a} \cdot \boldsymbol{r} = b$ and the sphere of radius R centered at C. Illustrate.

(7.19) For the geodesic triangle ABC on a unit sphere of center O (Fig. 6.3), use spherical coordinates (θ, φ) with OA as polar axis, and AB as prime meridian with $\varphi = 0$ such that $(\theta_B, \varphi_B) = (c, 0)$ and $(\theta_C, \varphi_C) = (b, \alpha)$, and AC is the $\varphi = \alpha$ meridian. In that underlying cartesian basis,

$$\boldsymbol{A} = \hat{\mathbf{z}}, \qquad \boldsymbol{B} = \hat{\mathbf{x}} \sin c + \hat{\mathbf{z}} \cos c,$$
$$\boldsymbol{C} = (\hat{\mathbf{x}} \cos \alpha + \hat{\mathbf{y}} \sin \alpha) \sin b + \hat{\mathbf{z}} \cos b.$$

Assume that ABC is a *proper* geodesic triangle and the labeling of B and C insures that all an-

gles are between 0 and π. The BC geodesic is the intersection of the sphere $\boldsymbol{r}(\theta, \varphi)$ with the plane OBC such that $\boldsymbol{r} \cdot \boldsymbol{B} \times \boldsymbol{C} = \sin a\, \boldsymbol{r} \cdot \boldsymbol{A}' = 0$.
Show that the BC geodesic is given by

$$\cot\theta = -\tan\theta_* \cos(\varphi - \varphi_*). \qquad (7.12a)$$

Characterize the point with coordinates (θ_*, φ_*).
Use (6.29) to obtain

$$\begin{cases} \cos\theta_* = \sin\beta \sin c \geq 0, \\ \sin\theta_* \cos\varphi_* = -\sin\beta\cos c, \\ \sin\theta_* \sin\varphi_* = -\cos\beta. \end{cases} \qquad (7.12b)$$

That will be used later to compute the area of a spherical triangle (13.25).

(7.20) Find an explicit formula for all $\boldsymbol{r}$'s such that $|\boldsymbol{r} - \boldsymbol{r}_A| = a$ and $|\boldsymbol{r} - \boldsymbol{r}_B| = b$. Sketch and interpret geometrically.

(7.21) Find an explicit formula for all $\boldsymbol{r}$'s such that $|\boldsymbol{r}| = a$ and $\hat{\boldsymbol{\omega}} \cdot \hat{\boldsymbol{r}} = b$. Interpret geometrically.

(7.22) Given three points P_1, P_2, P_3 in 3D space, specified by their cartesian coordinates ($P_1 \equiv (x_1, y_1, z_1)$ etc.), derive an algorithm to find the center C of the circle through the three points (the *circumcircle*) using vector algebra. Write a code in Matlab or Python, or your favorite language.

(7.23) *Delaunay triangulation.* Given four points P_0, P_1, P_2, P_3 in 3D space, write an algorithm to assemble them into two distinct triangles such that no vertex is inside the smallest circumscribed sphere of any triangle formed by the other vertices.

(7.24) Given three points P_1, P_2, P_3 in 3D space, specified by their cartesian coordinates ($P_1 \equiv (x_1, y_1, z_1)$ etc.), derive an algorithm to find the center I of the circle tangent to the three sides (the *inner circle*) using vector algebra. Write a code in Matlab or Python, for example.

(7.25) Point C is obtained by rotating point B about the axis passing through point A, with direction $\boldsymbol{a}$, by angle α (right-hand rotation by α about $\boldsymbol{a}$). Find an explicit vector expression for $\overrightarrow{OC}$ in terms of $\overrightarrow{OB}$, $\overrightarrow{OA}$, $\boldsymbol{a}$ and α. Make clean sketches. Express your vector result in cartesian form.

(7.26) A ray emanating from point L propagating in direction $\boldsymbol{a}$ hits a triangle with vertices A, B, C and bounces from it according to the equal angles law of reflection. Construct the vector equation of the reflected ray in terms of the data. How do you check the data to verify that the ray does hit the triangle?

(7.27) The skull surface is specified by a database of N triangles (Fig. 7.13). Each triangle has three vertices P_0, P_1, P_2 ordered such that $\overrightarrow{P_0P_1} \times \overrightarrow{P_1P_2}$ is pointing outward. Each vertex is specified in cartesian coordinates. To visualize the skull surface as seen from direction $\hat{\boldsymbol{a}}$, the triangles on the 'backside' are not drawn. Construct an algorithm to: (1) decide which triangles to draw, and (2) decide on the brightness of the triangles.

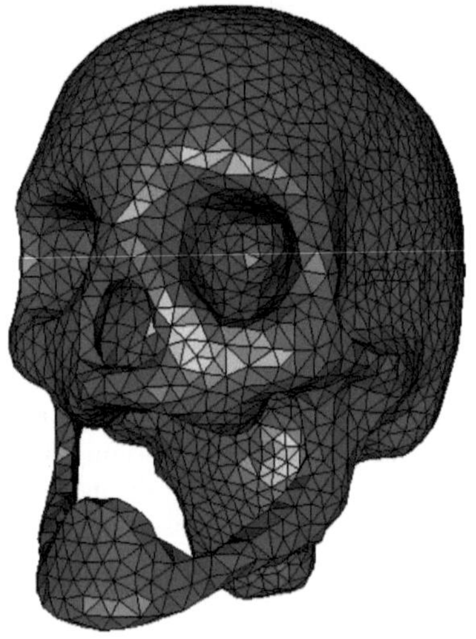

Fig. 7.13 Skull surface. Courtesy of Rineau and Yvinec, *The Computational Geometry Algorithms Library*, `https://www.cgal.org`.

(7.28) Prove that all points P with a fixed angle between the lines from P to two fixed points F_1, F_2, lie on a circle (in 2D, Fig. 7.11).

(7.29) Prove that all points P with a fixed ratio of distances to two fixed points F_1, F_2, lie on a circle (in 2D, Fig. 7.11).

(7.30) Derive the cartesian (x, y) to biangular (α_1, α_2) equations

$$\begin{cases} \alpha_1 = \operatorname{atan2}(y, x + a) \\ \alpha_2 = \operatorname{atan2}(y, x - a) \end{cases}$$

$$\begin{cases} x = a(t_2 + t_1)/(t_2 - t_1) \\ y = 2at_1t_2/(t_2 - t_1), \end{cases}$$

where $t_1 = \tan\alpha_1$, $t_2 = \tan\alpha_2$, and the (x, y) axes are such that F_1, F_2 are at $(\pm a, 0)$, respectively.

(7.31) Derive the cartesian (x, y) to bipolar (u, v) coordinates equations (Fig. 7.11)

$$\begin{cases} u = \operatorname{atan2}(2ay, x^2 + y^2 - a^2), \\ v = \frac{1}{2}\ln\left(\frac{(x+a)^2 + y^2}{(x-a)^2 + y^2}\right), \end{cases} \tag{7.13}$$

where $u = \alpha_2 - \alpha_1$, $v = \ln(s_1/s_2)$, and the (x, y) axes are such that F_1, F_2 are at $(\pm a, 0)$, respectively. Deduce that

$$\begin{cases} \tan u = \dfrac{2ay}{x^2 + y^2 - a^2}, \\ \tanh v = \dfrac{2ax}{x^2 + y^2 + a^2}, \end{cases}$$

and derive the inverse relations $(u, v) \to (x, y)$,

$$\begin{cases} x = a\dfrac{\sinh v}{\cosh v - \cos u}, \\ y = a\dfrac{\sin u}{\cosh v - \cos u} \end{cases} \tag{7.14}$$

(an important constraint in the full derivation is that $y/\sin(u) \geq 0$), where

$$\cosh v = \frac{e^v + e^{-v}}{2}, \quad \sinh v = \frac{e^v - e^{-v}}{2},$$
$$\tanh v = \frac{\sinh v}{\cosh v}$$

are the hyperbolic cosine, sine, and tangent functions, respectively, that satisfy the identity

$$\cosh^2 v - \sinh^2 v = 1.$$

Orthogonal Transformations

8

An *orthogonal transformation* is a linear transformation of vectors that preserves all dot products, and thus all magnitudes and angles. Our first example is the transformation of cartesian components of a vector when changing from one cartesian basis to another. We learn about orthogonal matrices and Euler angles. The chapter ends with discussion of body-fixed and space-fixed rotations.

8.1 Change of cartesian basis

A vector $\boldsymbol{v}$ is represented by two distinct sets of cartesian components (v_1, v_2) and (v_1', v_2') in 2D as shown in Fig. 8.1. The relations between the components (v_1, v_2) and (v_1', v_2') can be found using geometry and trigonometry,

$$\begin{cases} v_1' = v_1 \cos\alpha + v_2 \sin\alpha, \\ v_2' = -v_1 \sin\alpha + v_2 \cos\alpha. \end{cases} \tag{8.1}$$

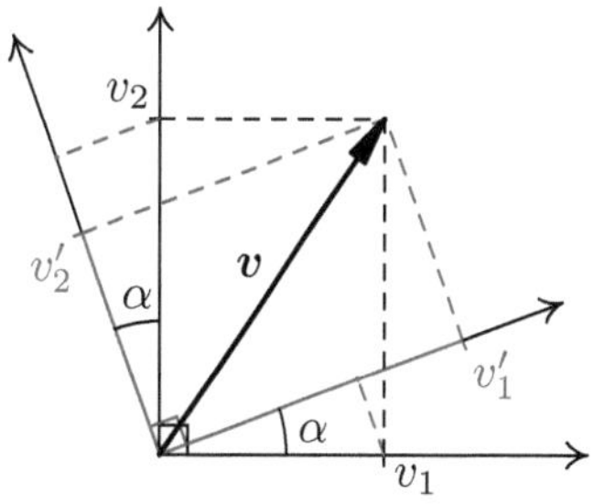

Fig. 8.1 Two cartesian representations of $\boldsymbol{v}$.

These transformation equations can be written in *matrix–vector* form as

$$\begin{bmatrix} v_1' \\ v_2' \end{bmatrix} = \begin{bmatrix} \cos\alpha & \sin\alpha \\ -\sin\alpha & \cos\alpha \end{bmatrix} \begin{bmatrix} v_1 \\ v_2 \end{bmatrix}, \tag{8.2}$$

where the matrix–vector product is performed *row by column* as given in (8.1).

Deriving eqn (8.1) from Fig. 8.1 is a good exercise in geometry and trigonometry, but it is easier to obtain the transformation rules by using *basis vectors* explicitly (Fig. 8.2). The starting point is the vector identity

$$\boldsymbol{v} = v_1\mathbf{e}_1 + v_2\mathbf{e}_2 = v_1'\mathbf{e}_1' + v_2'\mathbf{e}_2'. \tag{8.3a}$$

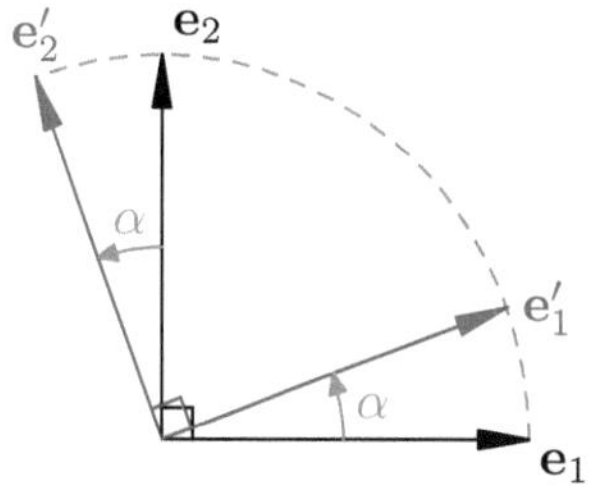

Fig. 8.2 Change of cartesian bases in 2D.

Dotting (8.3a) with $\mathbf{e}_1'$ and $\mathbf{e}_2'$, using orthonormality $\mathbf{e}_i' \cdot \mathbf{e}_j' = \delta_{ij}$, yields

$$\begin{cases} v_1' = \mathbf{e}_1' \cdot \boldsymbol{v} = \mathbf{e}_1' \cdot \mathbf{e}_1\, v_1 + \mathbf{e}_1' \cdot \mathbf{e}_2\, v_2, \\ v_2' = \mathbf{e}_2' \cdot \boldsymbol{v} = \mathbf{e}_2' \cdot \mathbf{e}_1\, v_1 + \mathbf{e}_2' \cdot \mathbf{e}_2\, v_2, \end{cases}$$

which is clearer in matrix–vector form

$$\begin{bmatrix} v_1' \\ v_2' \end{bmatrix} = \begin{bmatrix} \mathbf{e}_1' \cdot \mathbf{e}_1 & \mathbf{e}_1' \cdot \mathbf{e}_2 \\ \mathbf{e}_2' \cdot \mathbf{e}_1 & \mathbf{e}_2' \cdot \mathbf{e}_2 \end{bmatrix} \begin{bmatrix} v_1 \\ v_2 \end{bmatrix} = \begin{bmatrix} \cos\alpha & \sin\alpha \\ -\sin\alpha & \cos\alpha \end{bmatrix} \begin{bmatrix} v_1 \\ v_2 \end{bmatrix}, \tag{8.3b}$$

recovering (8.2) since, reading off Fig. 8.2,

$$\begin{cases} \mathbf{e}_1' \cdot \mathbf{e}_1 = \cos\alpha, & \mathbf{e}_1' \cdot \mathbf{e}_2 = \sin\alpha, \\ \mathbf{e}_2' \cdot \mathbf{e}_1 = -\sin\alpha, & \mathbf{e}_2' \cdot \mathbf{e}_2 = \cos\alpha. \end{cases}$$

The inverse transformation is obtained by dotting (8.3a) with $\mathbf{e}_1$ and $\mathbf{e}_2$, using $\mathbf{e}_i \cdot \mathbf{e}_j = \delta_{ij}$, yielding

$$\begin{bmatrix} v_1 \\ v_2 \end{bmatrix} = \begin{bmatrix} \mathbf{e}_1 \cdot \mathbf{e}_1' & \mathbf{e}_1 \cdot \mathbf{e}_2' \\ \mathbf{e}_2 \cdot \mathbf{e}_1' & \mathbf{e}_2 \cdot \mathbf{e}_2' \end{bmatrix} \begin{bmatrix} v_1' \\ v_2' \end{bmatrix} = \begin{bmatrix} \cos\alpha & -\sin\alpha \\ \sin\alpha & \cos\alpha \end{bmatrix} \begin{bmatrix} v_1' \\ v_2' \end{bmatrix}. \tag{8.3c}$$

A transformation $(v_1, v_2) \to (v_1', v_2')$ followed by its inverse $(v_1', v_2') \to (v_1, v_2)$ should return the original components. Indeed, substituting (8.2) for (v_1', v_2') into (8.3c) gives

$$\begin{aligned} \begin{bmatrix} v_1 \\ v_2 \end{bmatrix} &= \begin{bmatrix} \cos\alpha & -\sin\alpha \\ \sin\alpha & \cos\alpha \end{bmatrix} \begin{bmatrix} \cos\alpha & \sin\alpha \\ -\sin\alpha & \cos\alpha \end{bmatrix} \begin{bmatrix} v_1 \\ v_2 \end{bmatrix} \\ &= \begin{bmatrix} \cos^2\alpha + \sin^2\alpha & 0 \\ 0 & \cos^2\alpha + \sin^2\alpha \end{bmatrix} \begin{bmatrix} v_1 \\ v_2 \end{bmatrix} = \begin{bmatrix} v_1 \\ v_2 \end{bmatrix}, \end{aligned}$$

where the *matrix-matrix* product is performed *row by column.*

For 3D space, let $\{\mathbf{e}_1, \mathbf{e}_2, \mathbf{e}_3\}$ and $\{\mathbf{e}_1', \mathbf{e}_2', \mathbf{e}_3'\}$ denote two distinct orthonormal bases, such that

$$\mathbf{e}_i \cdot \mathbf{e}_j = \delta_{ij} = \mathbf{e}_i' \cdot \mathbf{e}_j'. \tag{8.4}$$

The corresponding components (v_1, v_2, v_3) and (v_1', v_2', v_3') of a vector $\boldsymbol{v}$ will be distinct in general, $(v_1, v_2, v_3) \neq (v_1', v_2', v_3')$, but equality of the representations can be expressed with the basis vectors

$$\boldsymbol{v} = v_1\mathbf{e}_1 + v_2\mathbf{e}_2 + v_3\mathbf{e}_3 = v_1'\mathbf{e}_1' + v_2'\mathbf{e}_2' + v_3'\mathbf{e}_3',$$

that is,

$$\boldsymbol{v} = v_j\mathbf{e}_j = v_j'\mathbf{e}_j', \tag{8.5a}$$

with summation over repeated indices. The relationships between the two sets of components follow from (8.5a) and simple dot products

$$v_i' = \mathbf{e}_i' \cdot \boldsymbol{v} = (\mathbf{e}_i' \cdot \mathbf{e}_j)\, v_j = Q_{ij}\, v_j, \tag{8.5b}$$

$$v_i = \mathbf{e}_i \cdot \boldsymbol{v} = (\mathbf{e}_i \cdot \mathbf{e}_j')\, v_j' = Q_{ji}\, v_j', \tag{8.5c}$$

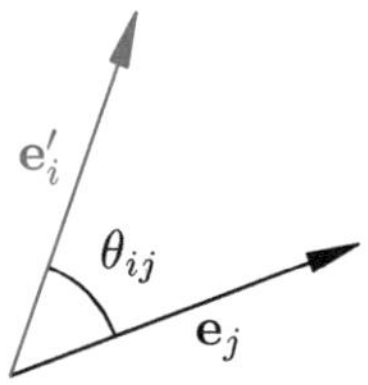

Fig. 8.3 $Q_{ij} = \mathbf{e}_i' \cdot \mathbf{e}_j$.

with

$$Q_{ij} \triangleq \mathbf{e}_i' \cdot \mathbf{e}_j = \cos\theta_{ij}, \tag{8.6a}$$

where θ_{ij} is the angle between the unit vectors $\mathbf{e}_i'$ and $\mathbf{e}_j$ (Fig. 8.3). The Q_{ij}'s are the *direction cosines* yielding the components of one basis in terms of the other

$$\mathbf{e}_i' = Q_{ij}\,\mathbf{e}_j, \qquad \mathbf{e}_j = Q_{ij}\,\mathbf{e}_i'. \tag{8.6b}$$

8.2 Direction cosine matrix

It is helpful to view those nine $Q_{ij} = \mathbf{e}'_i \cdot \mathbf{e}_j$ as the elements of a 3-by-3 *matrix*

$$\mathsf{Q} = \begin{bmatrix} Q_{11} & Q_{12} & Q_{13} \\ Q_{21} & Q_{22} & Q_{23} \\ Q_{31} & Q_{32} & Q_{33} \end{bmatrix} \equiv [Q_{ij}], \tag{8.7}$$

that is, a 3-by-3 table with the first index corresponding to *rows* and the second index to *columns*, so Q_{23} is the element on row 2, column 3, for example. The matrix Q whose elements are (8.6) is the *direction cosine matrix*

$$\mathsf{Q} = \begin{bmatrix} \mathbf{e}'_1 \cdot \mathbf{e}_1 & \mathbf{e}'_1 \cdot \mathbf{e}_2 & \mathbf{e}'_1 \cdot \mathbf{e}_3 \\ \mathbf{e}'_2 \cdot \mathbf{e}_1 & \mathbf{e}'_2 \cdot \mathbf{e}_2 & \mathbf{e}'_2 \cdot \mathbf{e}_3 \\ \mathbf{e}'_3 \cdot \mathbf{e}_1 & \mathbf{e}'_3 \cdot \mathbf{e}_2 & \mathbf{e}'_3 \cdot \mathbf{e}_3 \end{bmatrix} = \begin{bmatrix} \cos\theta_{11} & \cos\theta_{12} & \cos\theta_{13} \\ \cos\theta_{21} & \cos\theta_{22} & \cos\theta_{23} \\ \cos\theta_{31} & \cos\theta_{32} & \cos\theta_{33} \end{bmatrix}. \tag{8.8}$$

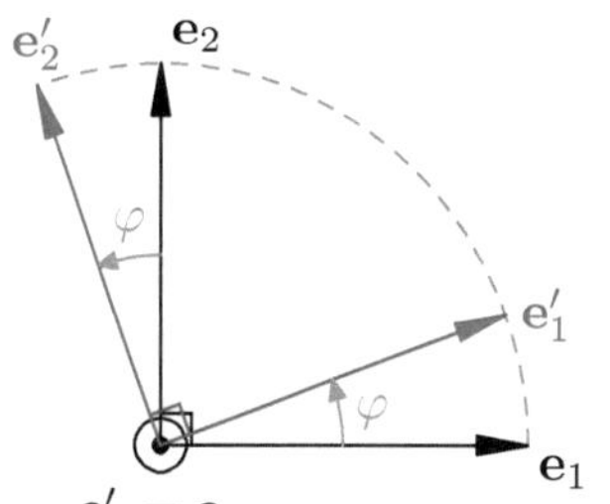

Fig. 8.4 Rotation about $\mathbf{e}_3$ by φ.

Example 1 If $\{\mathbf{e}'_1, \mathbf{e}'_2, \mathbf{e}'_3\}$ is the rotation of $\{\mathbf{e}_1, \mathbf{e}_2, \mathbf{e}_3\}$ about[1] $\mathbf{e}_3$ by φ, then

$$\mathsf{Q} = \begin{bmatrix} \cos\varphi & \sin\varphi & 0 \\ -\sin\varphi & \cos\varphi & 0 \\ 0 & 0 & 1 \end{bmatrix}, \tag{8.9}$$

with

$$\theta_{11} = \varphi, \quad \theta_{12} = \frac{\pi}{2} - \varphi, \quad \theta_{21} = \varphi + \frac{\pi}{2}, \quad \theta_{22} = \varphi,$$

but (8.9) is derived from Fig. 8.4 and basic trigonometry, yielding directly

$$\mathbf{e}'_2 \cdot \mathbf{e}_1 = -\sin\varphi,$$

for example, and avoiding the correct but unnecessary steps:

$$\mathbf{e}'_2 \cdot \mathbf{e}_1 = \cos\theta_{21} = \cos(\varphi + \pi/2) = -\sin\varphi.$$

Example 2 *Global to local Earth basis.* Let $\{\mathbf{e}_1, \mathbf{e}_2, \mathbf{e}_3\}$ denote an Earth basis with $\mathbf{e}_3$ in the *polar axis* direction, $(\mathbf{e}_1, \mathbf{e}_3)$ in the plane of the *prime meridian* and a local basis $\{\mathbf{e}'_1, \mathbf{e}'_2, \mathbf{e}'_3\}$ at longitude φ, polar angle θ (latitude $\lambda = \pi/2 - \theta$) with $\mathbf{e}'_1$ south, $\mathbf{e}'_2$ east and $\mathbf{e}'_3$ up (Fig. 8.5). Most of the required vector analysis has already been done in section 1.3, where an arbitrary 3D direction $\hat{\mathbf{a}}$ is first written in terms of $(\hat{\mathbf{z}}, \hat{\mathbf{a}}_\perp)$ (1.13a) in the vertical plane $(\hat{\mathbf{z}}, \hat{\mathbf{a}})$, with $\hat{\mathbf{a}}_\perp$ then expressed in terms of $(\hat{\mathbf{x}}, \hat{\mathbf{y}})$ (1.13b) in that horizontal plane. In fact, (1.13c) gives $\mathbf{e}'_3 = \hat{\mathbf{a}}$. A similar procedure yields $\mathbf{e}'_1$ and $\mathbf{e}'_2$ with the help of the intermediate horizontal direction vector $\hat{\boldsymbol{\rho}} = \cos\varphi\,\mathbf{e}_1 + \sin\varphi\,\mathbf{e}_2 = \hat{\mathbf{a}}_\perp$ shown in Fig. 8.5. One finds that (exercise 8.8)

$$\begin{cases} \mathbf{e}'_1 = \cos\theta\cos\varphi\,\mathbf{e}_1 + \cos\theta\sin\varphi\,\mathbf{e}_2 - \sin\theta\,\mathbf{e}_3, \\ \mathbf{e}'_2 = -\sin\varphi\,\mathbf{e}_1 + \cos\varphi\,\mathbf{e}_2, \\ \mathbf{e}'_3 = \sin\theta\cos\varphi\,\mathbf{e}_1 + \sin\theta\sin\varphi\,\mathbf{e}_2 + \cos\theta\,\mathbf{e}_3, \end{cases} \tag{8.10a}$$

[1]Positive angle directions are defined by the right-hand rule: for rotation by α about $\hat{\mathbf{a}}$, imagine your right thumb in the $\hat{\mathbf{a}}$ direction, then positive α is the direction of your curling fingers.

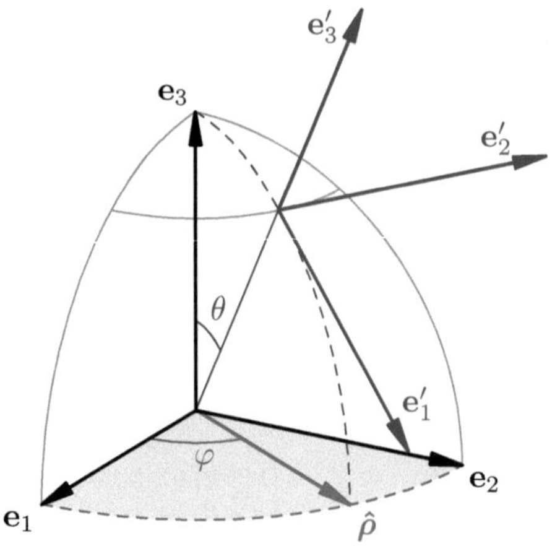

Fig. 8.5 3D view of global Earth basis $\{\mathbf{e}_1, \mathbf{e}_2, \mathbf{e}_3\}$ and local basis $\{\mathbf{e}_1', \mathbf{e}_2', \mathbf{e}_3'\}$ at longitude φ, polar angle θ. $\mathbf{e}_1'$ is south, $\mathbf{e}_2'$ east, and $\mathbf{e}_3'$ up.

yielding the transformation matrix

$$\mathbf{Q} = \begin{bmatrix} \cos\theta\cos\varphi & \cos\theta\sin\varphi & -\sin\theta \\ -\sin\varphi & \cos\varphi & 0 \\ \sin\theta\cos\varphi & \sin\theta\sin\varphi & \cos\theta \end{bmatrix}. \tag{8.10b}$$

The two angles φ and θ specify this direction cosine matrix. In general, three angles are needed, two to specify any one vector, the third for rotation about that vector. The azimuthal and polar angles, (φ, θ), specify $\mathbf{e}_3'$, and a third angle, ψ say, specifies the rotation of $(\mathbf{e}_1', \mathbf{e}_2')$ about $\mathbf{e}_3'$; here, $\psi = 0$. This is discussed in section 8.5, on *Euler angles.*

8.3 Vector, index, and matrix notations

The transformation rule (8.5b) can be viewed as a *row-by-column* matrix–vector product

$$v_i' = Q_{ij} v_j \;\Leftrightarrow\; \begin{bmatrix} v_1' \\ v_2' \\ v_3' \end{bmatrix} = \begin{bmatrix} Q_{11} & Q_{12} & Q_{13} \\ Q_{21} & Q_{22} & Q_{23} \\ Q_{31} & Q_{32} & Q_{33} \end{bmatrix} \begin{bmatrix} v_1 \\ v_2 \\ v_3 \end{bmatrix} \;\Leftrightarrow\; \mathsf{v}' = \mathsf{Q}\mathsf{v} \tag{8.11}$$

where v' and v are the *columns* of cartesian components (v_1', v_2', v_3') and (v_1, v_2, v_3), respectively. The inverse transformation (8.5c) is

$$v_i = Q_{ji} v_j' \;\Leftrightarrow\; \begin{bmatrix} v_1 \\ v_2 \\ v_3 \end{bmatrix} = \begin{bmatrix} Q_{11} & Q_{21} & Q_{31} \\ Q_{12} & Q_{22} & Q_{32} \\ Q_{13} & Q_{23} & Q_{33} \end{bmatrix} \begin{bmatrix} v_1' \\ v_2' \\ v_3' \end{bmatrix} \;\Leftrightarrow\; \mathsf{v} = \mathsf{Q}^{\mathsf{T}}\mathsf{v}' \tag{8.12}$$

where Q^{T} is the *transpose* of Q, the matrix whose (i, j) element is the (j, i) element of Q

$$Q^{\mathsf{T}}_{ij} \triangleq Q_{ji}.$$

The *rows* of Q^{T} are the corresponding *columns* of Q, and vice versa.

We use bold italic $\boldsymbol{v}$ for a geometric vector, a quantity with magnitude and direction independent of the system of coordinates. The cartesian components of that vector with respect to any orthonormal basis $\{\mathbf{e}_1, \mathbf{e}_2, \mathbf{e}_3\}$ are written compactly as v_i in *index notation*, with $i = 1, 2, 3$. Those components (v_1, v_2, v_3) form a vector in $\mathbb{R}^3$ that is considered to be a *column* vector v in *matrix notation.* Column vectors will be written in bold sans serif font, such as v, w. The vector, index, and matrix are equivalent but distinct representations for the vector $\boldsymbol{v}$

$$\boldsymbol{v} \equiv v_i \equiv \mathsf{v}.$$

Index notation does not in itself have the concept of "rows" and "columns," but it distinguishes between first and second indices, thus

$$Q_{ij}\, v_j \neq Q_{ji}\, v_j.$$

The sum $Q_{ij}v_j$ is three dot products between the *rows* of Q and the column v

$$Q_{ij}\, v_j \equiv \mathsf{Qv},$$

which is the standard *row-by-column* product of matrix algebra. However, the sum $Q_{ji}v_j$ is three dot products between the *columns* of Q and the column v. Thus, the matrix transpose concept is needed to write $Q_{ji}v_j$ in matrix notation, but there are still two distinct matrix ways of interpreting $Q_{ji}v_j$ as either

$$Q_{ji}\, v_j \equiv \mathsf{v}^\mathsf{T}\mathsf{Q}$$

or

$$Q_{ji}\, v_j \equiv \mathsf{Q}^\mathsf{T}\mathsf{v},$$

where v^T and $\mathsf{v}^\mathsf{T}\mathsf{Q}$ are *row vectors* while $\mathsf{Q}^\mathsf{T}\mathsf{v}$ and v are *column vectors,*

$$\mathsf{v}^\mathsf{T} = \begin{bmatrix} v_1 & v_2 & v_3 \end{bmatrix}, \qquad \mathsf{v} = \begin{bmatrix} v_1 \\ v_2 \\ v_3 \end{bmatrix}.$$

In matrix notation, a row cannot equal a column, unless they are mere 1-by-1; thus, $\mathsf{v}^\mathsf{T} \neq \mathsf{v}$, although both v^T and v are equivalent to $v_i \equiv (v_1, v_2, v_3)$. Hence,

$$w_i = Q_{ji}v_j = v_jQ_{ji} = Q^\mathsf{T}_{ij}v_j$$

is written in matrix notation as

$$\mathsf{w}^\mathsf{T} = \mathsf{v}^\mathsf{T}\mathsf{Q} \quad \text{or} \quad \mathsf{w} = \mathsf{Q}^\mathsf{T}\mathsf{v}$$

but $\mathsf{w} \neq \mathsf{v}^\mathsf{T}\mathsf{Q}$ since w is a column but $\mathsf{v}^\mathsf{T}\mathsf{Q}$ is a row vector.

8.4 Orthogonal matrices

There are nine direction cosines $Q_{ij} = \cos\theta_{ij}$ but orthonormality of both $\{\mathbf{e}_1, \mathbf{e}_2, \mathbf{e}_3\}$ and $\{\mathbf{e}'_1, \mathbf{e}'_2, \mathbf{e}'_3\}$ imply several constraints, so these Q matrices are special and the nine angles θ_{ij} are not independent. The constraints follow from (8.5b), (8.5c) which must hold for any (v_1, v_2, v_3) and (v'_1, v'_2, v'_3). Substituting (8.5c) into (8.5b) yields

$$v'_i = Q_{ik} v_k = Q_{ik} Q_{jk}\, v'_j,$$

and since this must be true for *any* (v'_1, v'_2, v'_3) this implies (exercises 8.15 and 8.16)

$$Q_{ik} Q_{jk} = \delta_{ij}.$$

The expression $Q_{ik}Q_{jk}$ is a sum over k with i, j free that corresponds to the dot product between the i and j *rows* of $\mathbf{Q}$. In *row-by-column* matrix notation this is

$$Q_{ik}Q_{jk} = Q_{ik}Q^{\mathsf{T}}_{kj} = \delta_{ij} \;\Leftrightarrow\; \mathbf{Q}\mathbf{Q}^{\mathsf{T}} = \begin{bmatrix} 1 & 0 & 0 \\ 0 & 1 & 0 \\ 0 & 0 & 1 \end{bmatrix} \triangleq \mathbf{I} \tag{8.13}$$

and the *rows* of matrix $\mathbf{Q}$ (8.7) are *orthonormal*, where $\mathbf{I}$ is the *identity matrix.*

Likewise, substituting (8.5b) into (8.5c) gives

$$v_i = Q_{ki} v'_k = Q_{ki} Q_{kj}\, v_j,$$

which holds for any (v_1, v_2, v_3), implying that

$$Q_{ki} Q_{kj} = \delta_{ij}.$$

Now $Q_{ki}Q_{kj}$ is a sum over k with i, j free and this is the dot product between the i and j *columns* of $\mathbf{Q}$. In *row-by-column* matrix notation this is

$$Q_{ki}Q_{kj} = Q^{\mathsf{T}}_{ik}Q_{kj} = \delta_{ij} \;\Leftrightarrow\; \mathbf{Q}^{\mathsf{T}}\mathbf{Q} = \begin{bmatrix} 1 & 0 & 0 \\ 0 & 1 & 0 \\ 0 & 0 & 1 \end{bmatrix} = \mathbf{I}, \tag{8.14}$$

and the *columns* of matrix $\mathbf{Q}$ (8.7) are also *orthonormal.*

Geometric interpretation The two sets of orthogonal relationships (8.13), (8.14) can be derived more geometrically as follows. The coefficient $Q_{ij} = \mathbf{e}'_i \cdot \mathbf{e}_j$ is both the j component of $\mathbf{e}'_i$ in the $\{\mathbf{e}_1, \mathbf{e}_2, \mathbf{e}_3\}$ basis, and the i component of $\mathbf{e}_j$ in the $\{\mathbf{e}'_1, \mathbf{e}'_2, \mathbf{e}'_3\}$ basis. Therefore we can write

$$\mathbf{e}'_i = (\mathbf{e}'_i \cdot \mathbf{e}_j)\, \mathbf{e}_j = Q_{ij}\, \mathbf{e}_j. \tag{8.15}$$

In other words, the ith *row* of matrix $\mathbf{Q}$ in (8.7) contains the components of $\mathbf{e}'_i$ in the basis $\{\mathbf{e}_1, \mathbf{e}_2, \mathbf{e}_3\}$:

$$\mathbf{e}'_i \equiv \begin{bmatrix} Q_{i1} & Q_{i2} & Q_{i3} \end{bmatrix}.$$

Thus, just like $\boldsymbol{a} \cdot \boldsymbol{b} = a_k b_k$, we have

$$\mathbf{e}'_i \cdot \mathbf{e}'_j = Q_{ik} Q_{jk} = Q_{ik} Q^{\mathsf{T}}_{kj} = \delta_{ij}, \tag{8.16}$$

since $\mathbf{e}'_i \cdot \mathbf{e}'_j = \delta_{ij}$. So the *rows* of $\mathbf{Q}$ are orthonormal because they are the components of $\{\mathbf{e}'_1, \mathbf{e}'_2, \mathbf{e}'_3\}$ in the basis $\{\mathbf{e}_1, \mathbf{e}_2, \mathbf{e}_3\}$ (Fig. 8.6).

Likewise,

$$\mathbf{e}_i = (\mathbf{e}_i \cdot \mathbf{e}'_j)\, \mathbf{e}'_j = Q_{ji}\, \mathbf{e}'_j, \tag{8.17}$$

and the ith *column* of matrix $\mathbf{Q}$ in (8.7) contains the components of $\mathbf{e}_i$ in the basis $\{\mathbf{e}'_1, \mathbf{e}'_2, \mathbf{e}'_3\}$:

$$\mathbf{e}_i \equiv \begin{bmatrix} Q_{1i} \\ Q_{2i} \\ Q_{3i} \end{bmatrix}.$$

Then

$$\mathbf{e}_i \cdot \mathbf{e}_j = Q_{ki} Q_{kj} = Q^{\mathsf{T}}_{ik} Q_{kj} = \delta_{ij}, \tag{8.18}$$

and the columns of $\mathbf{Q}$ are orthonormal because they are the components of $\{\mathbf{e}_1, \mathbf{e}_2, \mathbf{e}_3\}$ in the basis $\{\mathbf{e}'_1, \mathbf{e}'_2, \mathbf{e}'_3\}$ (Fig. 8.6).

This explains why a matrix with orthonormal *rows* automatically has orthonormal *columns*, and vice versa, as long as the matrix is *square*. Indeed, orthonormal *rows* can be interpreted as the components of the orthonormal $\{\mathbf{e}'_1, \mathbf{e}'_2, \mathbf{e}'_3\}$ in terms of $\{\mathbf{e}_1, \mathbf{e}_2, \mathbf{e}_3\}$, but then the *columns* are the components of $\{\mathbf{e}_1, \mathbf{e}_2, \mathbf{e}_3\}$ in terms of $\{\mathbf{e}'_1, \mathbf{e}'_2, \mathbf{e}'_3\}$,

$$\mathbf{Q}\mathbf{Q}^{\mathsf{T}} = \mathbf{I} \quad \Leftrightarrow \quad \mathbf{Q}^{\mathsf{T}}\mathbf{Q} = \mathbf{I} \tag{8.19}$$

as long as $\mathbf{Q}$ is square, where $\mathbf{I}$ is the identity matrix defined in (8.13). Such a matrix $\mathbf{Q}$ is called an *orthogonal matrix.*

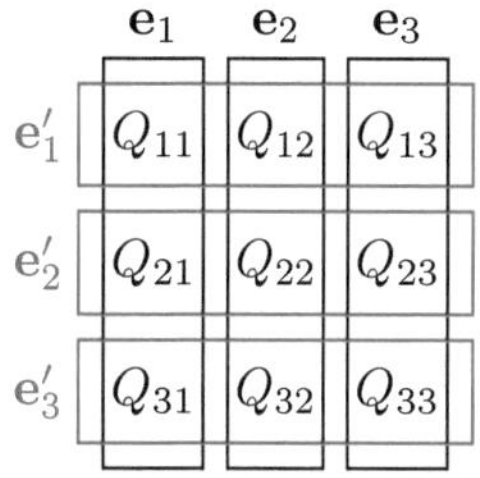

Fig. 8.6 $Q_{ij} \triangleq \mathbf{e}'_i \cdot \mathbf{e}_j$.

A 3-by-3 matrix $\mathbf{Q}$ has nine elements Q_{ij}. Orthonormality of the rows imposes six constraints on those Q_{ij}'s. Orthonormality of the columns imposes six constraints as well; however, these are redundant with row orthonormality. Thus there are only six independent constraints and the nine elements of a 3-by-3 orthogonal matrix depend on only *three* independent parameters. Parametrizations of 3-by-3 orthogonal matrices by three Euler angles are discussed in §8.5 and problems 8.30, 8.33, 8.34.

Successive transformations

A fundamental property of orthogonal matrices is that their *product* is an orthogonal matrix.[2] This is straightforward to prove with matrix algebra (exercise 9.5) and is readily understood geometrically as the result of successive orthogonal transformations since

$$\left.\begin{aligned} v'_k &= Q^{(1)}_{kj}\, v_j \\ v''_i &= Q^{(2)}_{ik}\, v'_k \end{aligned}\right\} \Rightarrow v''_i = Q^{(2)}_{ik} Q^{(1)}_{kj} v_j = Q_{ij} v_j, \tag{8.20a}$$

that is,

$$\mathsf{v}'' = \mathbf{Q}^{(2)}\, \mathsf{v}' = \mathbf{Q}^{(2)}\, \mathbf{Q}^{(1)} \mathsf{v}, \tag{8.20b}$$

where v, v' and v'' are the *column vectors* of $\boldsymbol{v}$ components in the cartesian bases $\{\mathbf{e}_1, \mathbf{e}_2, \mathbf{e}_3\}$, $\{\mathbf{e}'_1, \mathbf{e}'_2, \mathbf{e}'_3\}$, and $\{\mathbf{e}''_1, \mathbf{e}''_2, \mathbf{e}''_3\}$, respectively.

Thus, if $\mathbf{Q}^{(1)}$ and $\mathbf{Q}^{(2)}$ correspond to two successive orthogonal transformations

$$\{\mathbf{e}_1, \mathbf{e}_2, \mathbf{e}_3\} \xrightarrow{\mathbf{Q}^{(1)}} \{\mathbf{e}'_1, \mathbf{e}'_2, \mathbf{e}'_3\}$$

[2] So n-by-n orthogonal matrices form a *group*, the orthogonal group $O(n)$.

$$\{\mathbf{e}'_1, \mathbf{e}'_2, \mathbf{e}'_3\} \xrightarrow{\mathsf{Q}^{(2)}} \{\mathbf{e}''_1, \mathbf{e}''_2, \mathbf{e}''_3\}$$

with $Q^{(1)}_{ij} \triangleq \mathbf{e}'_i \cdot \mathbf{e}_j$ and $Q^{(2)}_{ij} \triangleq \mathbf{e}''_i \cdot \mathbf{e}'_j$, then the transformation

$$\{\mathbf{e}_1, \mathbf{e}_2, \mathbf{e}_3\} \xrightarrow{\mathsf{Q}} \{\mathbf{e}''_1, \mathbf{e}''_2, \mathbf{e}''_3\}$$

is itself an orthogonal transformation with

$$\mathsf{Q} = \mathsf{Q}^{(2)}\,\mathsf{Q}^{(1)} \tag{8.21}$$

and

$$\mathbf{e}''_i \cdot \mathbf{e}_j = Q_{ij} = Q^{(2)}_{ik}\,Q^{(1)}_{kj}. \tag{8.22}$$

This result that the product of two orthogonal matrices is an orthogonal matrix can be shown purely algebraically (exercise 8.20). Note that the successive transformation matrices appear in *right-to-left* order when applied to column vectors of components (8.20b).

8.5 Euler angles

The elements Q_{ij} of a 3-by-3 orthogonal matrix can be interpreted as the dot products of unit vectors of two distinct orthonormal bases, $Q_{ij} = \mathbf{e}'_i \cdot \mathbf{e}_j = \cos\theta_{ij}$, where θ_{ij} is the angle between $\mathbf{e}'_i$ and $\mathbf{e}_j$. In 3D, there are 9 such angles but these angles are not independent since both bases $\{\mathbf{e}'_1, \mathbf{e}'_2, \mathbf{e}'_3\}$ and $\{\mathbf{e}_1, \mathbf{e}_2, \mathbf{e}_3\}$ consist of mutually orthogonal unit vectors. If both bases have the same handedness (both right- or left-handed), each basis can be transformed into the other through only *three* elementary rotations, in general. "Elementary" in the sense that the rotation is about one of the basis vectors. The three angles corresponding to those three elementary rotations are called *Euler angles.*[3]

[3] A distinct three-angle representation of rotation matrices is explored in exercises 8.30, 8.33, 8.34.

That only three angles are needed is not a surprise since an arbitrary direction $\hat{\boldsymbol{a}}$ in 3D space can be specified by two angles (1.13c). Thus, θ and φ are sufficient to specify $\mathbf{e}'_3$, say, in terms of $\{\mathbf{e}_1, \mathbf{e}_2, \mathbf{e}_3\}$, and we only need one extra angle, ψ say, to specify the orientation of $\{\mathbf{e}'_1, \mathbf{e}'_2\}$ about the direction $\mathbf{e}'_3$. That is spherical coordinates θ, φ, with an extra twist ψ as illustrated in Fig. 8.7.

Thus, we can construct (or *decompose*, or *factorize*) any orthogonal matrix $\mathbf{Q}$ corresponding to the transformation from $\{\mathbf{e}_1, \mathbf{e}_2, \mathbf{e}_3\}$ to $\{\mathbf{e}'_1, \mathbf{e}'_2, \mathbf{e}'_3\}$ in the following three successive elementary (right-handed) rotations.

First rotation about $\mathbf{e}_3$

Rotate from the original basis about $\mathbf{e}_3$ by φ to obtain the intermediate basis (Fig. 8.8)

$$\begin{cases} \mathbf{e}^{(1)}_1 = \cos\varphi\,\mathbf{e}_1 + \sin\varphi\,\mathbf{e}_2 \\ \mathbf{e}^{(1)}_2 = -\sin\varphi\,\mathbf{e}_1 + \cos\varphi\,\mathbf{e}_2 \\ \mathbf{e}^{(1)}_3 = \mathbf{e}_3, \end{cases} \tag{8.23}$$

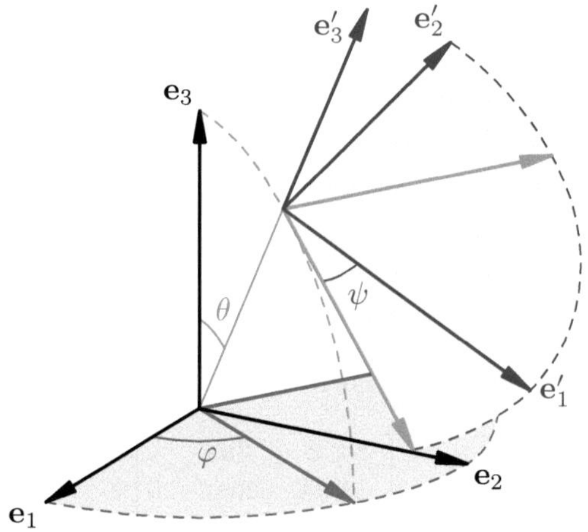

Fig. 8.7 3D view of the zyz Euler angles φ, θ, ψ, relating two right handed orthonormal bases $\{\mathbf{e}_1, \mathbf{e}_2, \mathbf{e}_3\}$ to $\{\mathbf{e}'_1, \mathbf{e}'_2, \mathbf{e}'_3\}$.

defining $Q^{(1)}_{ij} = \mathbf{e}^{(1)}_i \cdot \mathbf{e}_j$, yields the rotation matrix

$$\mathbf{Q}^{(1)} = \begin{bmatrix} \cos\varphi & \sin\varphi & 0 \\ -\sin\varphi & \cos\varphi & 0 \\ 0 & 0 & 1 \end{bmatrix}. \tag{8.24}$$

The goal of this rotation is to put $\mathbf{e}^{(1)}_1$ in the plane of $\mathbf{e}_3$ and the target $\mathbf{e}'_3$. This sets $\mathbf{e}^{(1)}_2 = (\mathbf{e}_3 \times \mathbf{e}'_3)/|\mathbf{e}_3 \times \mathbf{e}'_3|$.

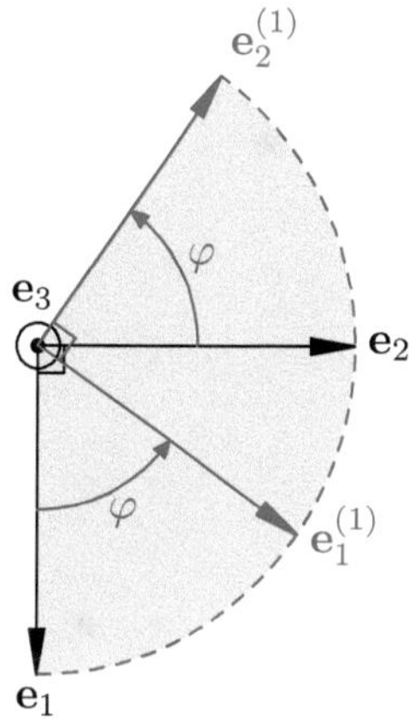

Fig. 8.8 Rotation about $\mathbf{e}_3$.

Second rotation about $\mathbf{e}^{(1)}_2$

Rotate from the intermediate basis $\{\mathbf{e}^{(1)}_1, \mathbf{e}^{(1)}_2, \mathbf{e}^{(1)}_3\}$ about $\mathbf{e}^{(1)}_2$ by θ to obtain the basis (Fig. 8.9)

$$\begin{cases} \mathbf{e}^{(2)}_1 = \cos\theta\, \mathbf{e}^{(1)}_1 - \sin\theta\, \mathbf{e}^{(1)}_3 \\ \mathbf{e}^{(2)}_2 = \mathbf{e}^{(1)}_2 \\ \mathbf{e}^{(2)}_3 = \sin\theta\, \mathbf{e}^{(1)}_1 + \cos\theta\, \mathbf{e}^{(1)}_3 \end{cases} \tag{8.25}$$

with $Q^{(2)}_{ij} = \mathbf{e}^{(2)}_i \cdot \mathbf{e}^{(1)}_j$, this corresponds to the rotation matrix

$$\mathbf{Q}^{(2)} = \begin{bmatrix} \cos\theta & 0 & -\sin\theta \\ 0 & 1 & 0 \\ \sin\theta & 0 & \cos\theta \end{bmatrix}. \tag{8.26}$$

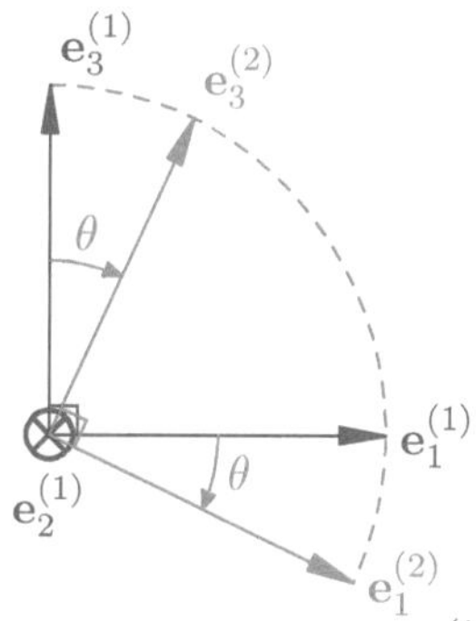

Fig. 8.9 Rotation about $\mathbf{e}^{(1)}_2$.

The goal of this second rotation is to align $\mathbf{e}^{(1)}_3 = \mathbf{e}_3$ with the target $\mathbf{e}'_3$, but the vectors $(\mathbf{e}^{(2)}_1, \mathbf{e}^{(2)}_2)$ do not yet match the target $(\mathbf{e}'_1, \mathbf{e}'_2)$, in general.

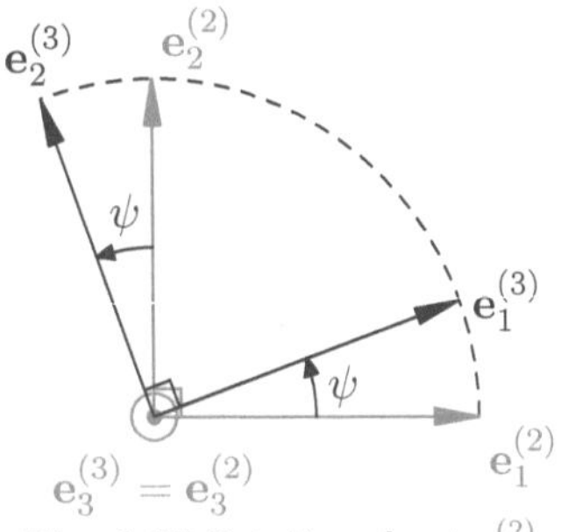

Fig. 8.10 Rotation about $\mathbf{e}_3^{(2)}$.

Third rotation about $\mathbf{e}_3^{(2)}$

A final elementary rotation about $\mathbf{e}_3^{(2)} = \mathbf{e}'_3$ by an angle ψ is required in general to align $(\mathbf{e}_1^{(3)}, \mathbf{e}_2^{(3)})$ with the target $(\mathbf{e}'_1, \mathbf{e}'_2)$ (Fig. 8.10),

$$\left\{ \begin{array}{rcrcr} \mathbf{e}_1^{(3)} & = & \cos\psi\, \mathbf{e}_1^{(2)} & + & \sin\psi\, \mathbf{e}_2^{(2)} \\ \mathbf{e}_2^{(3)} & = & -\sin\psi\, \mathbf{e}_1^{(2)} & + & \cos\psi\, \mathbf{e}_2^{(2)} \\ \mathbf{e}_3^{(3)} & = & & & \mathbf{e}_3^{(2)} \end{array} \right. \tag{8.27}$$

defining $Q_{ij}^{(3)} = \mathbf{e}_i^{(3)} \cdot \mathbf{e}_j^{(2)}$; this corresponds to the rotation matrix

$$\mathbf{Q}^{(3)} = \begin{bmatrix} \cos\psi & \sin\psi & 0 \\ -\sin\psi & \cos\psi & 0 \\ 0 & 0 & 1 \end{bmatrix}. \tag{8.28}$$

Combined rotation

The orthogonal transformation $\mathbf{Q}$ from $\{\mathbf{e}_1, \mathbf{e}_2, \mathbf{e}_3\}$ to $\{\mathbf{e}_1^{(3)}, \mathbf{e}_2^{(3)}, \mathbf{e}_3^{(3)}\}$ is then obtained by taking the product of elementary rotation matrices in *right-to-left order*,

$$\mathbf{Q} = \mathbf{Q}^{(3)}\mathbf{Q}^{(2)}\mathbf{Q}^{(1)} \tag{8.29}$$

$$= \begin{bmatrix} \cos\psi & \sin\psi & 0 \\ -\sin\psi & \cos\psi & 0 \\ 0 & 0 & 1 \end{bmatrix} \begin{bmatrix} \cos\theta & 0 & -\sin\theta \\ 0 & 1 & 0 \\ \sin\theta & 0 & \cos\theta \end{bmatrix} \begin{bmatrix} \cos\varphi & \sin\varphi & 0 \\ -\sin\varphi & \cos\varphi & 0 \\ 0 & 0 & 1 \end{bmatrix}.$$

The right-hand rule fixes the positive angle directions, and a unique angle representation is obtained, for example, by restricting

$$-\pi < \varphi \le \pi, \quad 0 \le \theta \le \pi, \quad -\pi < \psi \le \pi,$$

and setting $\varphi = 0$ if $\theta = 0$ or π to settle the degeneracy in those cases.

If $\{\mathbf{e}_1, \mathbf{e}_2, \mathbf{e}_3\}$ and $\{\mathbf{e}'_1, \mathbf{e}'_2, \mathbf{e}'_3\}$ do not have the same handedness and ψ was determined so that $\mathbf{e}_1^{(3)} = \mathbf{e}'_1$, then $\mathbf{e}_2^{(3)}$ will be opposite to $\mathbf{e}'_2$. In that case, a final sign change of the second row of $\mathbf{Q}$ will be required and the transformation from $\{\mathbf{e}_1, \mathbf{e}_2, \mathbf{e}_3\}$ to $\{\mathbf{e}'_1, \mathbf{e}'_2, \mathbf{e}'_3\}$ consists of three elementary rotations and one reflection, in general. If no reflection is needed then $\mathbf{Q}$ is a *proper orthogonal* matrix, it corresponds to a pure rotation.

Any 3-by-3 proper orthogonal matrix $\mathbf{Q}$ can thus be represented with only three angles. There are multiple ways to choose those angles. The (φ, θ, ψ) choice made here fits with spherical coordinates (φ, θ) and is labeled $zy'z''$ (or $z_0y_1z_2$), since we rotated about the original z axis, then the new $y \equiv y'$, then the new $z \equiv z''$ again.

A $zx'z''$ decomposition would perform a rotation about the original $\mathbf{e}_3$ so that $\mathbf{e}_1^{(1)}$ is in the $\mathbf{e}_3 \times \mathbf{e}'_3$ direction. Then a rotation about $\mathbf{e}_1^{(1)}$ would be performed to align $\mathbf{e}_3^{(2)} = \mathbf{e}'_3$, followed by a rotation about $\mathbf{e}_3^{(2)}$ to align $(\mathbf{e}_1^{(2)}, \mathbf{e}_2^{(2)})$. There are four other such Euler angle decompositions: $xy'x''$, $xz'x''$, $yx'y''$, $yz'y''$, where the first and third rotations are about the same intrinsic direction.

Yaw, pitch, and roll

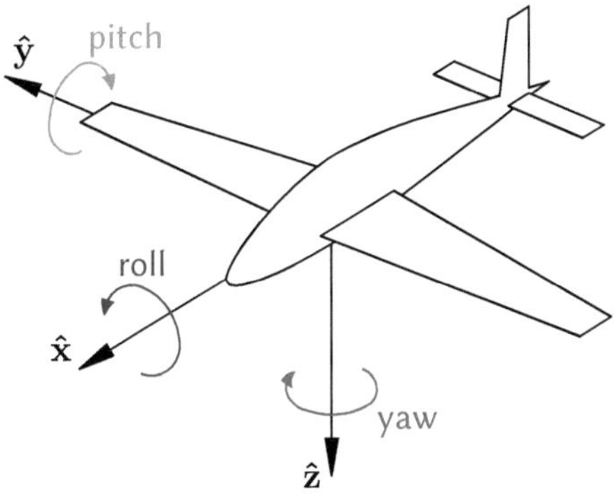

Fig. 8.11 Airplane attitude: yaw, pitch, and roll.

In aircraft dynamics and control, three angles about three distinct axes are used (Fig. 8.11). Imagine a frame $\hat{\mathbf{x}}$, $\hat{\mathbf{y}}$, $\hat{\mathbf{z}}$ attached to an airplane, with $\hat{\mathbf{x}}$ pointing from tail to nose (roll axis), $\hat{\mathbf{y}}$ pointing from one wingtip to the other (pitch axis) and $\hat{\mathbf{z}}$ perpendicular to the plane of the airplane downward (yaw axis). The orientation of the airplane with respect to a fixed reference frame can be specified by the following

- the *yaw* (or *heading*)—the angle α around $\hat{\mathbf{z}}$ to align $\hat{\mathbf{x}}$ with the desired horizontal direction;[4]
- the *pitch* (or *elevation*)—the angle β about the new $\hat{\mathbf{y}}$ to pitch the nose up or down to align $\hat{\mathbf{x}}$ with the desired direction in the vertical plane; and
- the *roll* (or *bank*)—the angle γ about the new $\hat{\mathbf{x}}$ to rotate the wings around the axis of the airplane to achieve the desired *bank* angle.

The yaw–pitch–roll decomposition is a zyx (or $zy'x''$ to emphasize that the axes rotate with the airplane) *factorization* of the orthogonal matrix $\mathbf{Q}$. Such choices of Euler angles are sometimes called *Bryan* angles, after G. H. Bryan, a pioneer in the mathematical analysis of airplane stability.[5] A $zy'x''$ factorization of an orthogonal $\mathbf{Q}$ reads

$$\mathbf{Q} = \begin{bmatrix} 1 & 0 & 0 \\ 0 & \cos\gamma & \sin\gamma \\ 0 & -\sin\gamma & \cos\gamma \end{bmatrix} \begin{bmatrix} \cos\beta & 0 & -\sin\beta \\ 0 & 1 & 0 \\ \sin\beta & 0 & \cos\beta \end{bmatrix} \begin{bmatrix} \cos\alpha & \sin\alpha & 0 \\ -\sin\alpha & \cos\alpha & 0 \\ 0 & 0 & 1 \end{bmatrix}, \tag{8.30}$$

such that

$$\begin{bmatrix} \hat{\mathbf{x}} \\ \hat{\mathbf{y}} \\ \hat{\mathbf{z}} \end{bmatrix} = \mathbf{Q}_\gamma \begin{bmatrix} \hat{\mathbf{x}}_2 \\ \hat{\mathbf{y}}_2 \\ \hat{\mathbf{z}}_2 \end{bmatrix} = \mathbf{Q}_\gamma \mathbf{Q}_\beta \begin{bmatrix} \hat{\mathbf{x}}_1 \\ \hat{\mathbf{y}}_1 \\ \hat{\mathbf{z}}_1 \end{bmatrix} = \mathbf{Q}_\gamma \mathbf{Q}_\beta \mathbf{Q}_\alpha \begin{bmatrix} \hat{\mathbf{x}}_0 \\ \hat{\mathbf{y}}_0 \\ \hat{\mathbf{z}}_0 \end{bmatrix} = \mathbf{Q} \begin{bmatrix} \hat{\mathbf{x}}_0 \\ \hat{\mathbf{y}}_0 \\ \hat{\mathbf{z}}_0 \end{bmatrix}, \tag{8.31}$$

[4] In aerodynamics and flight dynamics, the yaw, pitch, and roll angles are often denoted ψ, θ, ϕ, while α and β denote angles of attack and sideslip, respectively. Here we use α, β, γ for yaw, pitch, and roll to distinguish from the φ, θ, ψ Euler angles used in the previous pages.

[5] G. H. Bryan, *Stability in Aviation*, Macmillan, 1911 (p. 20). See T. J. M. Boyd, *Journal of Aeronautical History* 2011, 97-115 for an historical review of G. H. Bryan and his contributions.

where α is the *yaw*, β the *pitch*, and γ the *roll* angles, and $\{\hat{\mathbf{x}}_0, \hat{\mathbf{y}}_0, \hat{\mathbf{z}}_0\}$ is the fixed reference frame. The first transformation by α about $\hat{\mathbf{z}}_0 = \hat{\mathbf{z}}_1$ is the rightmost matrix $\mathbf{Q}_\alpha$ in (8.30), the second by β about $\hat{\mathbf{y}}_1 = \hat{\mathbf{y}}_2$ is the middle matrix $\mathbf{Q}_\beta$, and the last one by γ about $\hat{\mathbf{x}}_2 = \hat{\mathbf{x}}$ is the leftmost matrix $\mathbf{Q}_\gamma$.

Degeneracy and "gimbal lock"

Euler angles inherit a singularity from spherical coordinates where the azimuth φ is undetermined when the polar angle $\theta = 0$ or π, and φ jumps by $\pm\pi$ as the direction vector $\hat{\boldsymbol{a}}$ passes through the polar directions $\pm\hat{\mathbf{z}}$ if $0 \le \theta \le \pi$.

Likewise in (8.29) when $\theta = 0$ or π, the first and third rotations are about the same actual direction. In those cases, the orientation of $(\mathbf{e}'_1, \mathbf{e}'_2)$ depends only on the sum $\varphi + \psi$, and φ and ψ are not uniquely determined. If $\mathbf{e}'_3$ moves along a meridian ($\theta \to 0$ or π) with φ and ψ fixed in Fig. 8.7, both φ and ψ will jump by $\pm\pi$ as $\mathbf{e}'_3$ passes through $\pm\mathbf{e}_3$. The zyx decomposition (8.30) also has a degeneracy but for $\beta = \pm\pi/2$, corresponding to the airplane pitching straight up or down.

This degeneracy is known as *gimbal lock* in mechanical engineering and may have real mechanical consequences depending on the nature and limitations of the mechanical device (telescope, gyroscope, robotic arm, etc.) and its controls (Fig. 8.12). Imagine tracking an airplane flying straight towards you; your head tilts backward as the airplane approaches but you need a quick 180° full-body rotation followed by a downwards tilt of your head to keep following the airplane beyond straight overhead.

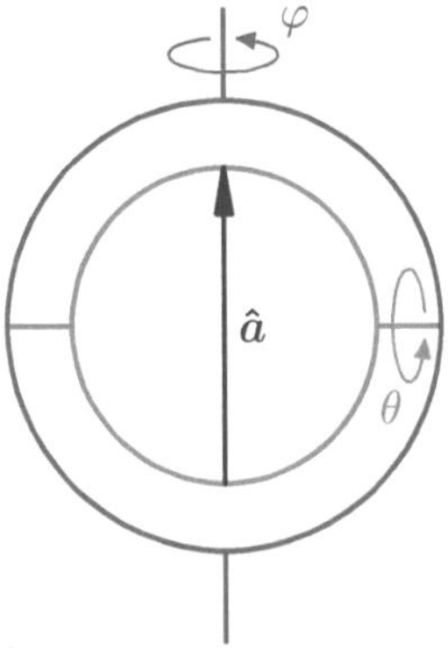

Fig. 8.12 Two-axis gimbals.

8.6 Rotations and reflections

This chapter began with orthogonal transformations of the *components* of a vector $\boldsymbol{v}$ from one cartesian basis to another. The starting point was the identity (8.5a)

$$\boldsymbol{v} = v_i \mathbf{e}_i = v'_i \mathbf{e}'_i \tag{8.32}$$

stating that the *same* vector $\boldsymbol{v}$ is expressed in two different cartesian bases (Fig. 8.13). Dotting (8.32) with $\mathbf{e}'_i$ yields (8.5b)

$$v'_i = Q_{ij} v_j \tag{8.33}$$

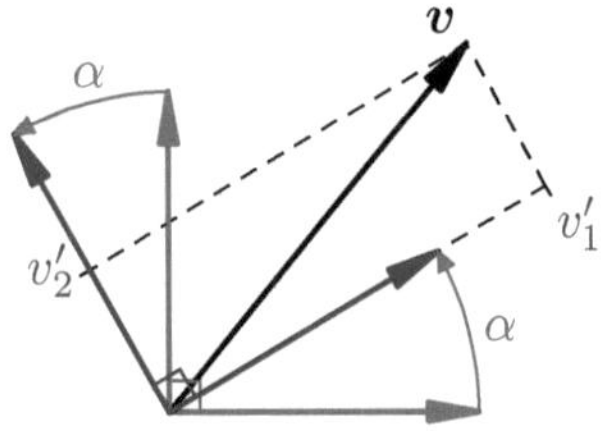

Fig. 8.13 Rotate basis, fixed $\boldsymbol{v}$.

where $Q_{ij} \triangleq \mathbf{e}'_i \cdot \mathbf{e}_j$ are the components of an orthogonal matrix.

A vector $\boldsymbol{v}$ can also be orthogonally transformed into *another* vector $\boldsymbol{v}'$ through reflections and rotations. These geometric operations can be expressed in vector form as done in (3.29) and (4.18), without reference to a cartesian basis. If we do introduce a cartesian basis $\{\mathbf{e}_1, \mathbf{e}_2, \mathbf{e}_3\}$, then the orthogonal transformation can be applied to that basis to yield a transformed cartesian basis $\{\mathbf{e}'_1, \mathbf{e}'_2, \mathbf{e}'_3\}$. The linear transformation $\boldsymbol{v} \to \boldsymbol{v}'$ then reads

$$\boldsymbol{v} = v_i \mathbf{e}_i \to \boldsymbol{v}' = v_i \mathbf{e}'_i. \tag{8.34a}$$

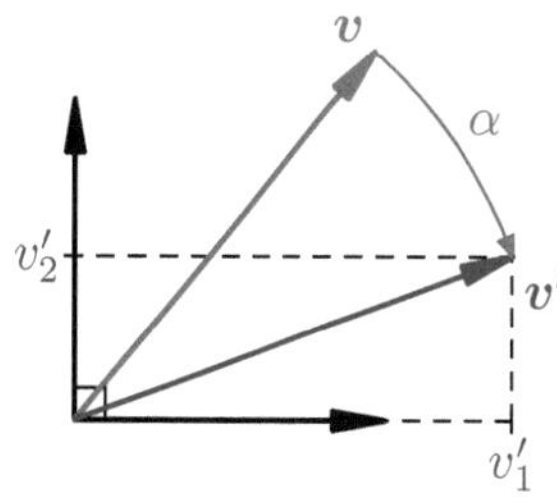

Fig. 8.14 *Reverse* rotate $\boldsymbol{v}$, fixed basis.

The vector $\boldsymbol{v}$ has been transformed into $\boldsymbol{v}'$ just as $\mathbf{e}_i$ has been transformed into $\mathbf{e}'_i$, but the components (v_1, v_2, v_3) have not changed. Of course, we are interested then in the components (v'_1, v'_2, v'_3) of the new $\boldsymbol{v}'$ in the original basis $\{\mathbf{e}_1, \mathbf{e}_2, \mathbf{e}_3\}$ (Fig. 8.14). These are now defined by the identity

$$\boldsymbol{v}' = v_i \mathbf{e}'_i = v'_i \mathbf{e}_i, \tag{8.34b}$$

instead of (8.32). Dotting (8.34b) with $\mathbf{e}_i$ yields

$$v'_i = Q_{ji}\, v_j = Q^{\mathsf{T}}_{ij}\, v_j, \tag{8.34c}$$

which is the inverse of the transformation in (8.33), when $Q_{ij} \triangleq \mathbf{e}'_i \cdot \mathbf{e}_j$ as before.

The meaning of v'_i in (8.33) is of course different from its meaning in (8.34c). Equation (8.33) gives the components of the original vector $\boldsymbol{v}$ in the transformed basis, while (8.34c) gives the components of the transformed vector $\boldsymbol{v}'$ in the original basis.

Those operations are inverses of one another: rotating a basis one way with respect to a fixed vector is equivalent to rotating the vector the *opposite* way relative to the fixed basis, as far as the cartesian components are concerned, which is illustrated in Figs. 8.13 and 8.14. The former transformation involves matrix $\mathbf{Q}$; the latter involves its transpose, $\mathbf{Q}^{\mathsf{T}}$. If $\mathbf{Q}$ is factored into a product of orthogonal matrices as in (8.29) or (8.30), then $\mathbf{Q}^{\mathsf{T}}$ is the product of the transposed matrices in *reverse order* (exercise 8.20). Rotation of vectors with fixed basis is sometimes called an *active rotation* (8.34c), while rotation of a basis with fixed vector is a *passive rotation* (8.33).

Linear combinations and transformations

The definition $Q_{ij} \triangleq \mathbf{e}'_i \cdot \mathbf{e}_j$ implies (8.6b)

$$\mathbf{e}'_i = Q_{ij}\, \mathbf{e}_j, \tag{8.35a}$$

which gives the vectors $\{\mathbf{e}'_1, \mathbf{e}'_2, \mathbf{e}'_3\}$ as *linear combinations* of $\{\mathbf{e}_1, \mathbf{e}_2, \mathbf{e}_3\}$.

It is useful to unpack (8.35a) using matrix notation, for example in the case of (8.10a) with its $\mathbf{Q}$ (8.10b) decomposed as in (8.29) with $\psi = 0$,

$$\begin{bmatrix} \mathbf{e}'_1 \\ \mathbf{e}'_2 \\ \mathbf{e}'_3 \end{bmatrix} = \mathbf{Q} \begin{bmatrix} \mathbf{e}_1 \\ \mathbf{e}_2 \\ \mathbf{e}_3 \end{bmatrix} = \begin{bmatrix} \cos\theta & 0 & -\sin\theta \\ 0 & 1 & 0 \\ \sin\theta & 0 & \cos\theta \end{bmatrix} \begin{bmatrix} \cos\varphi & \sin\varphi & 0 \\ -\sin\varphi & \cos\varphi & 0 \\ 0 & 0 & 1 \end{bmatrix} \begin{bmatrix} \mathbf{e}_1 \\ \mathbf{e}_2 \\ \mathbf{e}_3 \end{bmatrix}, \tag{8.35b}$$

but the columns containing the vectors $\mathbf{e}'_i$ and $\mathbf{e}_j$ are not "vectors." They are columns of vectors, and not any vectors; these are specific *linear combinations* expressing directions $\{\mathbf{e}'_1, \mathbf{e}'_2, \mathbf{e}'_3\}$ in terms of $\{\mathbf{e}_1, \mathbf{e}_2, \mathbf{e}_3\}$. If these are vectors in $\mathbb{R}^3$, then they would have to be *row* vectors in (8.35b). Equations (8.35b) can also be written as

$$\begin{aligned} \begin{bmatrix} \mathbf{e}'_1 & \mathbf{e}'_2 & \mathbf{e}'_3 \end{bmatrix} &= \begin{bmatrix} \mathbf{e}_1 & \mathbf{e}_2 & \mathbf{e}_3 \end{bmatrix} \mathbf{Q}^{\mathsf{T}} \\ &= \begin{bmatrix} \mathbf{e}_1 & \mathbf{e}_2 & \mathbf{e}_3 \end{bmatrix} \begin{bmatrix} \cos\varphi & -\sin\varphi & 0 \\ \sin\varphi & \cos\varphi & 0 \\ 0 & 0 & 1 \end{bmatrix} \begin{bmatrix} \cos\theta & 0 & \sin\theta \\ 0 & 1 & 0 \\ -\sin\theta & 0 & \cos\theta \end{bmatrix}, \end{aligned} \tag{8.35c}$$

which is appropriate for *column* vectors $\mathbf{e}'_1, \mathbf{e}'_2, \mathbf{e}'_3$ and $\mathbf{e}_1, \mathbf{e}_2, \mathbf{e}_3$, in $\mathbb{R}^3$.

In contrast to the linear combinations (8.35), the equations $v'_i = Q_{ij} v_j$ in (8.33) and $v'_i = Q_{ji}\, v_j = Q^{\mathsf{T}}_{ij} v_j$ in (8.34c), are linear transformations, where the inputs v_j are the components of any vector $\boldsymbol{v}$. Transformation (8.34c) is the "active"' transformation with respect to the same original basis

$$\boldsymbol{v} = v_i \mathbf{e}_i \to \boldsymbol{v}' = v_i \mathbf{e}'_i = v'_i \mathbf{e}_i = \mathbf{e}_i\, Q^{\mathsf{T}}_{ij} v_j. \tag{8.36a}$$

In matrix form, for (8.35b), this is

$$\begin{bmatrix} v'_1 \\ v'_2 \\ v'_3 \end{bmatrix} = \begin{bmatrix} \cos\varphi & -\sin\varphi & 0 \\ \sin\varphi & \cos\varphi & 0 \\ 0 & 0 & 1 \end{bmatrix} \begin{bmatrix} \cos\theta & 0 & \sin\theta \\ 0 & 1 & 0 \\ -\sin\theta & 0 & \cos\theta \end{bmatrix} \begin{bmatrix} v_1 \\ v_2 \\ v_2 \end{bmatrix}, \tag{8.36b}$$

where (v_1, v_2, v_3) are the components of the original vector and (v'_1, v'_2, v'_3) the components of the rotated vector, both in the $\{\mathbf{e}_1, \mathbf{e}_2, \mathbf{e}_3\}$ basis.

That "active" transformation can be applied to any vector and in particular to the basis vectors. For example, $\boldsymbol{v} = \mathbf{e}_1$ has components $v_j = \delta_{j1} \equiv (1, 0, 0)$ and the active transformation $\boldsymbol{v} = \mathbf{e}_1 \to \boldsymbol{v}' = \mathbf{e}'_1 = v'_i \mathbf{e}_i$ is, from (8.34c),

$$v'_i = Q^{\mathsf{T}}_{ij} \delta_{j1} = Q^{\mathsf{T}}_{i1} = \mathbf{e}'_1 \cdot \mathbf{e}_i. \tag{8.37a}$$

This is the first column of $\mathbf{Q}^T$, which is the first *row* of $\mathbf{Q}$. For (8.36b), this yields $(v'_1, v'_2, v'_3) = (\cos\varphi\cos\theta, \sin\varphi\cos\theta, -\sin\theta)$, and

$$\mathbf{e}'_1 = v'_i \mathbf{e}_i = \cos\theta\cos\varphi\, \mathbf{e}_1 + \cos\theta\sin\varphi\, \mathbf{e}_2 - \sin\theta\, \mathbf{e}_3, \tag{8.37b}$$

which is (8.35b) *left* multiplied by the *row* vector $\begin{bmatrix} 1 & 0 & 0 \end{bmatrix}$, or (8.35c) *right* multiplied by the *column* vector $(1, 0, 0)$. Thus, if the matrix $\mathbf{Q} = [Q_{ij}]$ in the linear combinations (8.35) defines an active transformation, the matrix that performs that transformation of column vectors is $\mathbf{Q}^{\mathsf{T}}$.

Body-fixed and space-fixed rotations

In a sequence of active rotations such as (8.36b), the inputs and outputs of each transformation are components in the original cartesian basis $\{\mathbf{e}_1, \mathbf{e}_2, \mathbf{e}_3\}$. Hence, each successive elementary rotation is about those original direction vectors. In other words, those rotations are *extrinsic* or *space-fixed.* The rotations in the linear combinations (8.35b), (8.35c), apply to direction vectors, yielding new direction vectors after each elementary rotation, and the next elementary rotation is therefore with respect to those new directions. Hence, the rotations expressed there are *intrinsic* or *body-fixed.* In (8.35b) and (8.35c), for example, a cartesian basis $\{\mathbf{e}_1, \mathbf{e}_2, \mathbf{e}_3\}$ undergoes two successive *intrinsic* (or *body-fixed*) orthogonal transformations (Fig. 8.15),

$$\mathbf{e}_k \to \mathbf{e}^{(1)}_j = Q^{(1)}_{jk} \mathbf{e}_k \to \mathbf{e}^{(2)}_i = Q^{(2)}_{ij} \mathbf{e}^{(1)}_j = \mathbf{e}'_i. \tag{8.38}$$

This corresponds to the (active) transformation $\boldsymbol{v} = v_i \mathbf{e}_i \to \boldsymbol{v}' = v_i \mathbf{e}'_i$ with

$$\boldsymbol{v}' = v_i \mathbf{e}'_i = v_i\, Q^{(2)}_{ij} Q^{(1)}_{jk} \mathbf{e}_k. \tag{8.39}$$

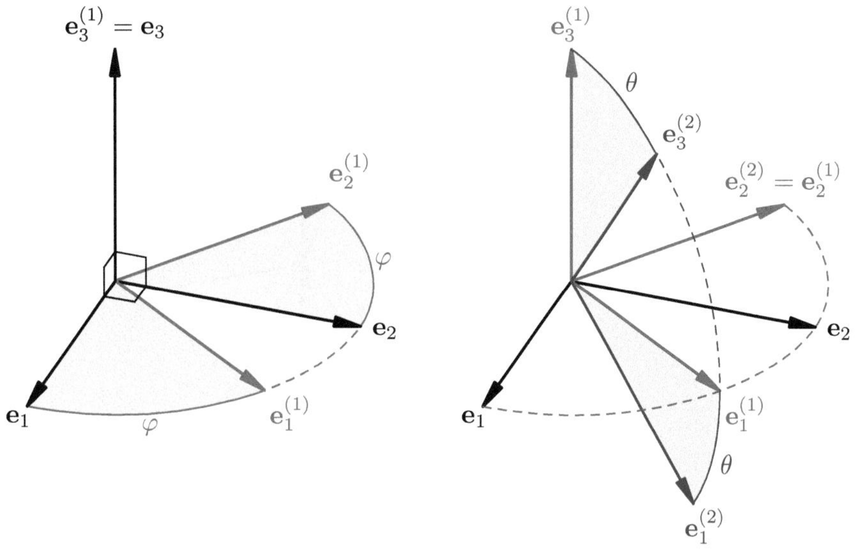

Fig. 8.15 Body-fixed rotations $\{\mathbf{e}_1, \mathbf{e}_2, \mathbf{e}_3\} \to \{\mathbf{e}_1^{(1)}, \mathbf{e}_2^{(1)}, \mathbf{e}_3^{(1)}\} \to \{\mathbf{e}_1^{(2)}, \mathbf{e}_2^{(2)}, \mathbf{e}_3^{(2)}\}$.

This final product can be interpreted from right to left, as a sequence of intrinsic rotations of the direction vectors as in (8.38), or from left to right as a sequence of *extrinsic*, or *space-fixed*, transformations of components about the original directions

$$v_i \to v_j^{(1)} = v_i\, Q_{ij}^{(2)} \to v_k^{(2)} = v_j^{(1)} Q_{jk}^{(1)} = v_k'. \tag{8.40}$$

These components correspond to the sequence of transformed vectors

$$\boldsymbol{v} = v_i \mathbf{e}_i \to \boldsymbol{v}^{(1)} = v_i\, Q_{ij}^{(2)} \mathbf{e}_j \to \boldsymbol{v}^{(2)} = v_i\, Q_{ij}^{(2)} Q_{jk}^{(1)} \mathbf{e}_k = \boldsymbol{v}', \tag{8.41}$$

and thus to the sequence of extrinsic (space-fixed) transformed bases

$$\mathbf{e}_i \to \boldsymbol{\epsilon}_i^{(1)} = Q_{ij}^{(2)} \mathbf{e}_j \to \boldsymbol{\epsilon}_i^{(2)} = Q_{ij}^{(2)} Q_{jk}^{(1)} \mathbf{e}_k = \mathbf{e}_i', \tag{8.42}$$

illustrated in Fig. 8.16. The final basis $\boldsymbol{\epsilon}_i^{(2)} = \mathbf{e}_i^{(2)} = \mathbf{e}_i'$ is the same as (8.38) but the intermediate bases are not, $\mathbf{e}_i^{(1)} \neq \boldsymbol{\epsilon}_i^{(1)}$, in general, and the sequence of extrinsic transformations are applied in *reverse* order to that of the sequence of intrinsic transformations. For example, rotating a basis by φ about the original z, then by θ about the new y, yields the same result as rotating first about the original y by θ, then about the *original* z by φ, as illustrated in Figs. 8.15 and 8.16. This applies to general orthogonal transformations, to any orthogonal matrices Q_1, Q_2, not just elementary rotation leaving one axis fixed, and to more than two transformations.

In a sequence of body-fixed orthogonal transformations B_1, B_2, B_3, $\ldots$, the successive transformations apply to the current basis $\mathbf{e}_i^{(k)}$ yielding the sequence

$$\mathbf{e}_i \to \mathbf{e}_i^{(1)} = B_{ij}^{(1)} \mathbf{e}_j \to \mathbf{e}_i^{(2)} = B_{ij}^{(2)} \mathbf{e}_j^{(1)} \to \mathbf{e}_i^{(3)} = B_{ij}^{(3)} \mathbf{e}_j^{(2)}, \ldots, \tag{8.43}$$

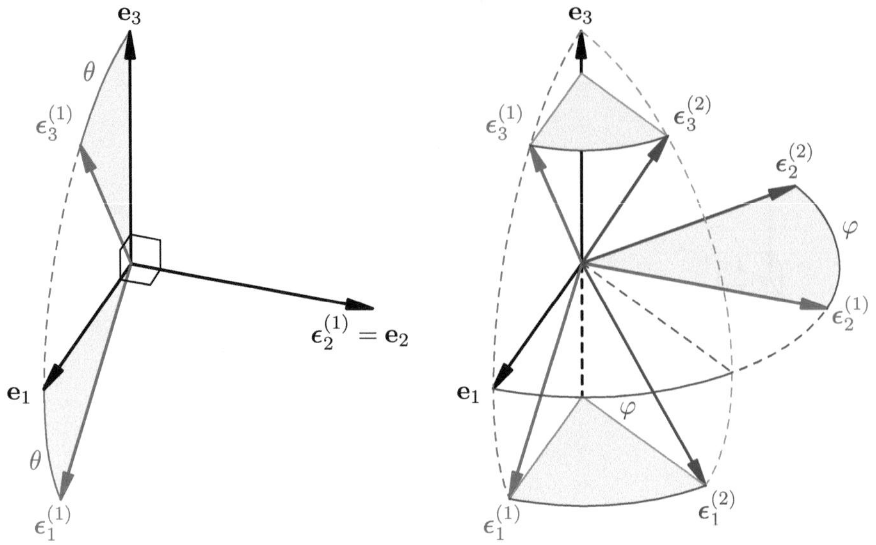

Fig. 8.16 Space-fixed rotations $\{\mathbf{e}_1, \mathbf{e}_2, \mathbf{e}_3\} \to \{\epsilon_1^{(1)}, \epsilon_2^{(1)}, \epsilon_3^{(1)}\} \to \{\epsilon_1^{(2)}, \epsilon_2^{(2)}, \epsilon_3^{(2)}\}$.

such that

$$\mathbf{e}_i^{(3)} = B_{ij}^{(3)} B_{jk}^{(2)} B_{kl}^{(1)} \mathbf{e}_l, \tag{8.44}$$

resulting in a combined orthogonal transformation matrix

$$\mathbf{Q} = \ldots \mathbf{B}_3 \mathbf{B}_2 \mathbf{B}_1, \tag{8.45}$$

with the matrices occurring in right-to-left order, for $\mathbf{Q}$ such that $Q_{ij} = \mathbf{e}'_i \cdot \mathbf{e}_j$.

In a sequence of space-fixed orthogonal transformations $\mathbf{S}_1$, $\mathbf{S}_2$, $\mathbf{S}_3$, ..., the successive transformations apply to the original basis $\mathbf{e}_i$ yielding

$$\mathbf{e}_i \to \epsilon_i^{(1)} = S_{ij}^{(1)} \mathbf{e}_j \to \epsilon_i^{(2)} = S_{ij}^{(1)} S_{jk}^{(2)} \mathbf{e}_k \to \epsilon_i^{(3)} = S_{ij}^{(1)} S_{jk}^{(2)} S_{kl}^{(3)} \mathbf{e}_l, \tag{8.46}$$

resulting in a combined orthogonal transformation matrix

$$\mathbf{Q} = \mathbf{S}_1 \mathbf{S}_2 \mathbf{S}_3 \ldots \tag{8.47}$$

with the matrices now appearing in left-to-right order.

If a basis $\mathbf{e}_i$ is transformed into $\mathbf{e}'_i = Q_{ij} \mathbf{e}_j$ through a series of body-fixed transformations $\mathbf{B}_1$, $\mathbf{B}_2$, $\mathbf{B}_3$, ..., and space-fixed transformations $\mathbf{S}_1$, $\mathbf{S}_2$, ..., the combined orthogonal transformation matrix is

$$\mathbf{Q} = \ldots \mathbf{B}_3 \mathbf{B}_2 \mathbf{B}_1 \mathbf{S}_1 \mathbf{S}_2 \ldots. \tag{8.48}$$

Linear transformations of vectors are further discussed in section §9.6.

Exercises

(8.1) What is the most general explicit form of a 2-by-2 orthogonal matrix? Justify

(8.2) Find $\mathbf{Q}$ if $\{\hat{\mathbf{x}}', \hat{\mathbf{y}}', \hat{\mathbf{z}}'\}$ is the right hand rotation of $\{\hat{\mathbf{x}}, \hat{\mathbf{y}}, \hat{\mathbf{z}}\}$ about $\hat{\mathbf{z}}$ by angle α. Sketch. Verify (8.16) and (8.18) for your $\mathbf{Q}$. If $\boldsymbol{v}$ has components (a, b, c) in the basis $\{\hat{\mathbf{x}}, \hat{\mathbf{y}}, \hat{\mathbf{z}}\}$, provide explicit formula for its components (a', b', c') in the basis $\{\hat{\mathbf{x}}', \hat{\mathbf{y}}', \hat{\mathbf{z}}'\}$.

(8.3) Find $\mathbf{Q}$ if $\{\hat{\mathbf{x}}', \hat{\mathbf{y}}', \hat{\mathbf{z}}'\}$ is the right-hand rotation of $\{\hat{\mathbf{x}}, \hat{\mathbf{y}}, \hat{\mathbf{z}}\}$ about $\hat{\mathbf{y}}$ by angle β. Sketch. Verify (8.16) and (8.18) for your $\mathbf{Q}$.

(8.4) Find $\mathbf{Q}$ if $\{\hat{\mathbf{x}}', \hat{\mathbf{y}}', \hat{\mathbf{z}}'\}$ is the right-hand rotation of $\{\hat{\mathbf{x}}, \hat{\mathbf{y}}, \hat{\mathbf{z}}\}$ about $\hat{\mathbf{x}}$ by angle γ. Sketch. Verify (8.16) and (8.18) for your $\mathbf{Q}$.

(8.5) Find $\mathbf{Q}$ if $\mathbf{e}_1' = -\mathbf{e}_2$, $\mathbf{e}_2' = \mathbf{e}_3$, $\mathbf{e}_3' = \mathbf{e}_1$ and verify (8.16) and (8.18) for it.

(8.6) Find $\mathbf{Q}$ if $\mathbf{e}_1' = \mathbf{e}_2$, $\mathbf{e}_2' = \mathbf{e}_1$, $\mathbf{e}_3' = \mathbf{e}_3$ and verify (8.16) and (8.18) for it. This $\mathbf{Q}$ is a special kind of orthogonal matrix called a *permutation matrix*.

(8.7) Verify by explicit calculations that the rows of (8.9) and (8.10b) are orthonormal, and likewise for the columns.

(8.8) Derive (8.10) using 2D projections in meridional planes containing $(\hat{\boldsymbol{\rho}}, \mathbf{e}_3)$ and $(\mathbf{e}_1', \mathbf{e}_3')$, and equatorial planes containing $(\mathbf{e}_1, \mathbf{e}_2)$ and $(\hat{\boldsymbol{\rho}}, \mathbf{e}_2')$, as in §1.3.

(8.9) Having obtained $\mathbf{e}_3'$ as $\hat{\boldsymbol{a}}$ in §1.3, calculate $\mathbf{e}_2'$ and $\mathbf{e}_1'$ in (8.10) from

$$\mathbf{e}_2' = \frac{\mathbf{e}_3 \times \mathbf{e}_3'}{|\mathbf{e}_3 \times \mathbf{e}_3'|}, \quad \mathbf{e}_1' = \mathbf{e}_2' \times \mathbf{e}_3' = \frac{(\mathbf{e}_3 \times \mathbf{e}_3') \times \mathbf{e}_3'}{|\mathbf{e}_3 \times \mathbf{e}_3'|}.$$

(8.10) Determine the angles θ_{12} and θ_{21} in terms of φ and θ in (8.10b).

(8.11) Let $\{\mathbf{e}_1, \mathbf{e}_2, \mathbf{e}_3\}$ denote an Earth basis as defined in Fig. 8.5, and $\{\mathbf{e}_1', \mathbf{e}_2', \mathbf{e}_3'\}$ a local basis at longitude φ and latitude $\lambda = \pi/2 - \theta$ with $\mathbf{e}_1'$ east, $\mathbf{e}_2'$ north and $\mathbf{e}_3'$ up. Show that the transformation matrix $\mathbf{Q}$ is

$$\mathbf{Q} = \begin{bmatrix} -\sin\varphi & \cos\varphi & 0 \\ -\cos\varphi\sin\lambda & -\sin\varphi\sin\lambda & \cos\lambda \\ \cos\varphi\cos\lambda & \sin\varphi\cos\lambda & \sin\lambda \end{bmatrix},$$

(i) from meridional and equatorial plane projections as in §1.3, (ii) from (8.10b), and (iii) from (8.29). Explain/justify your derivations.

(8.12) *Orthogonal projection of a 3D figure* (Fig. 8.17). One way to make a 2D drawing of a 3D figure is to plot the orthogonal projection of the 3D data specified in an $\{\mathbf{e}_1, \mathbf{e}_2, \mathbf{e}_3\}$ basis onto a plane perpendicular to the viewpoint in direction $\mathbf{e}_3'$ at azimuth φ and elevation λ. Find the relevant $\mathbf{Q}$ in terms of φ and λ and specify how to obtain the 2D plotting data from the 3D data. Almost all 3D pictures in this book are orthogonal projections.

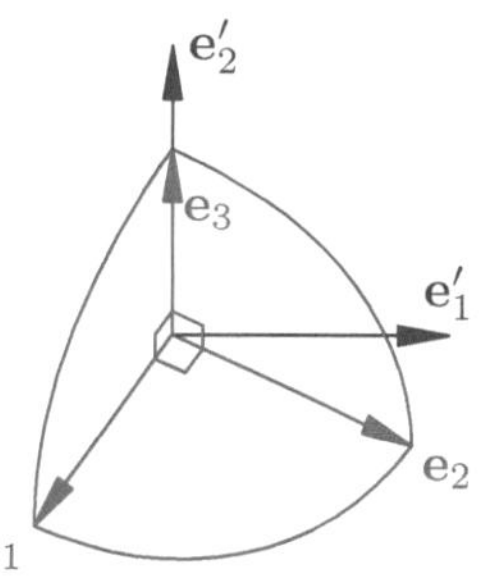

Fig. 8.17 Orthogonal projection of $\mathbf{e}_1, \mathbf{e}_2, \mathbf{e}_3$ on the $\mathbf{e}_1', \mathbf{e}_2'$ plane for $\varphi = 30°$, $\lambda = 50°$.

(8.13) The velocity of a satellite is v_1 east, v_2 north, v_3 up as measured from a cartesian basis located at longitude φ, latitude λ. What are the corresponding velocity components with respect to the Earth basis?

(8.14) The velocity of a satellite is v_1 east, v_2 north, and v_3 up as measured from a cartesian basis located at longitude φ_1, latitude λ_1. What are the corresponding velocity components with respect to a local basis at longitude φ_2, latitude λ_2? Derive and explain an algorithm to compute those components.

(8.15) Prove that if $u_i = A_{ij}u_j$ for any (u_1, u_2, u_3), then $A_{ij} = \delta_{ij}$.

(8.16) Find *all* the A_{ij}'s for which $u_i = A_{ij}u_j$ for a given (u_1, u_2, u_3). For example, what are all the A_{ij}'s when $(u_1, u_2, u_3) = (2, 3, 4)$? Explain.

(8.17) What is the orthogonal matrix corresponding to a reflection about the (x, z) plane? What is its determinant?

(8.18) Explain why the determinant of any orthogonal matrix $\mathbf{Q}$ is ± 1. When is $\det(\mathbf{Q}) = +1$ and when is it -1, in general? Give explicit examples.

(8.19) If $\boldsymbol{a}$ and $\boldsymbol{b}$ are two arbitrary vectors and $\{\mathbf{e}_1', \mathbf{e}_2', \mathbf{e}_3'\}$ and $\{\mathbf{e}_1, \mathbf{e}_2, \mathbf{e}_3\}$ are two distinct orthonormal bases, we have shown that $a_i b_i = a_i' b_i'$ (eqn (3.22), here with summation over repeated indices). Verify this invariance directly from the transformation rule (8.5b), $v_i' = Q_{ij} v_j$, showing your mastery of index notation.

(8.20) If $\mathbf{Q} = \mathbf{Q}_1 \mathbf{Q}_2$, show that $\mathbf{Q}^\mathsf{T} = \mathbf{Q}_2^\mathsf{T} \mathbf{Q}_1^\mathsf{T}$, so the transpose of a product is the product of the transposes in reverse order.

(8.21) Prove that the product of two orthogonal matrices is an orthogonal matrix but that their sum is not, in general.

(8.22) Find the $zy'z''$ and $zx'z''$ factorizations of (8.10b). Specify each angle in both cases.

(8.23) The orthonormal basis $\{\mathbf{e}_1'', \mathbf{e}_2'', \mathbf{e}_3''\}$ is the rotation of $\{\mathbf{e}_1', \mathbf{e}_2', \mathbf{e}_3'\}$ by ψ about $\mathbf{e}_3'$. The transformation from $\{\mathbf{e}_1, \mathbf{e}_2, \mathbf{e}_3\}$ to $\{\mathbf{e}_1', \mathbf{e}_2', \mathbf{e}_3'\}$ is given in (8.10b). What is the transformation matrix for $\{\mathbf{e}_1, \mathbf{e}_2, \mathbf{e}_3\}$ to $\{\mathbf{e}_1'', \mathbf{e}_2'', \mathbf{e}_3''\}$? Find an elementary rotation factorization of that matrix.

(8.24) Let $\{\mathbf{e}_1', \mathbf{e}_2', \mathbf{e}_3'\}$ be the rotation of $\{\mathbf{e}_1, \mathbf{e}_2, \mathbf{e}_3\}$ about $\mathbf{e}_1$ by α. Find $\mathbf{Q}$. Find the $zy'x''$ and $zy'z''$ factorizations of $\mathbf{Q}$ specifying all angles in terms of α.

(8.25) Let $\mathbf{Q}$ be any orthogonal matrix. If its zyz factorization is specified by angles φ, θ, and ψ, construct its $zx'z''$ and $zy'x''$ factorizations. Write the explicit forms of the matrix factorizations and determine the angles of the latter factorizations in terms of the original φ, θ, and ψ.

(8.26) The right-handed orthonormal basis $\{\mathbf{e}_1', \mathbf{e}_2', \mathbf{e}_3'\}$ is such that $\mathbf{e}_1'$ has azimuth φ and polar angle θ relative to the basis $\{\mathbf{e}_1, \mathbf{e}_2, \mathbf{e}_3\}$ with $\mathbf{e}_3' = (\mathbf{e}_3 \times \mathbf{e}_1')/\sin\theta$. Make a 3D sketch of the bases, similar to Fig. 8.5. Find the transformation matrix $\mathbf{Q} = [Q_{ij}]$ such that $\mathbf{e}_i' = Q_{ij} \mathbf{e}_j$ as well as zyz and zyx factorizations for that matrix. Specify all angles in terms of φ and θ.

(8.27) Constructing a $zy'z''$ Euler angle representation of a given orthogonal matrix $\mathbf{Q}$ consists of finding the angles φ, θ, ψ given the Q_{ij} coefficients. Show that

$$\begin{aligned} \theta &= \arccos Q_{33}, \\ \varphi &= \operatorname{atan2}(Q_{32}, Q_{31}), \\ \psi &= \operatorname{atan2}(Q_{23}, -Q_{13}). \end{aligned}$$

Show how this fails when $\theta = 0$ or π. Setting $\varphi = 0$ in those cases, find ψ.

(8.28) *Givens rotations.* Show that an arbitrary 3-by-3 matrix $\mathbf{A}$ can be transformed into an upper triangular matrix $\mathbf{R}$ through a sequence of elementary rotations:

$$\begin{bmatrix} 1 & 0 & 0 \\ 0 & c_3 & s_3 \\ 0 & -s_3 & c_3 \end{bmatrix} \begin{bmatrix} c_2 & 0 & s_2 \\ 0 & 1 & 0 \\ -s_2 & 0 & c_2 \end{bmatrix} \begin{bmatrix} c_1 & s_1 & 0 \\ -s_1 & c_1 & 0 \\ 0 & 0 & 1 \end{bmatrix} \mathbf{A} = \begin{bmatrix} r_{11} & r_{12} & r_{13} \\ 0 & r_{22} & r_{23} \\ 0 & 0 & r_{33} \end{bmatrix}, \tag{8.49}$$

where $c_k^2 + s_k^2 = 1$ such that $(c_k, s_k) = (\cos\theta_k, \sin\theta_k)$ for some angle θ_k.
To eliminate the a_{21} element of $\mathbf{A}$ with $c_1^2 + s_1^2 = 1$ requires

$$c_1 = \frac{a_{11}}{\sqrt{a_{11}^2 + a_{21}^2}}, \qquad s_1 = \frac{a_{21}}{\sqrt{a_{11}^2 + a_{21}^2}}.$$

Next, (c_2, s_2) eliminate the a_{31}' element, then (c_3, s_3) eliminate the a_{32}'' element. This is $\mathbf{Q}^\mathsf{T}\mathbf{A} = \mathbf{R}$, where $\mathbf{Q}$ is orthogonal, yielding $\mathbf{A} = \mathbf{QR}$, the QR decomposition of $\mathbf{A}$. Show that when this process is applied to an orthogonal matrix $\mathbf{A} = \mathbf{Q}$, the upper triangular matrix $\mathbf{R}$ is the identity matrix and thus the product of the elementary rotation matrices is $\mathbf{Q}^\mathsf{T}$, thereby yielding a decomposition of $\mathbf{Q}$ into a product of elementary rotation matrices. This approach can be extended to orthogonal matrices of any size. How many elementary rotations (Euler angles) are needed to factor a 4-by-4 orthogonal matrix? An n-by-n orthogonal matrix? Givens rotations are thus the algebraic version of Euler angle factorizations of rotation matrices. Givens rotations are defined to align vectors in $\mathbb{R}^2$ with the lower row index, for example $(a_{11}, a_{13}) \to (a_{11}', 0)$, not by the right-hand rule.

(8.29) Let $\boldsymbol{v}' = \boldsymbol{v} - 2(\hat{\boldsymbol{a}} \cdot \boldsymbol{v})\hat{\boldsymbol{a}}$, yielding $v_i' = A_{ij} v_j$ in a cartesian basis. Find A_{ij}. Show that $[A_{ij}]$ is an orthogonal matrix. Note that this is *not* a change of basis.

(8.30) If $\boldsymbol{v}'$ is the rotation of $\boldsymbol{v}$ by α about $\hat{\boldsymbol{a}}$, as seen in (4.18) and (4.21), then $v_i' = A_{ij} v_j$ in a cartesian basis with $\hat{\boldsymbol{a}} = a_i \mathbf{e}_i$, $\boldsymbol{v}' = v_i' \mathbf{e}_i$, $\boldsymbol{v} = v_i \mathbf{e}_i$. Show that

$$A_{ij} = \cos\alpha\, \delta_{ij} + (1 - \cos\alpha)\, a_i a_j - \sin\alpha\, \epsilon_{ijk} a_k. \tag{8.50}$$

Prove that $\mathbf{A}$ is orthogonal using index notation. Note that this is *not* a change of basis; it is a rotation of vectors $\boldsymbol{v} \to \boldsymbol{v}'$. The matrix $\mathbf{Q}$, such that $Q_{ij} \triangleq \mathbf{e}_i' \cdot \mathbf{e}_j$ when $\{\mathbf{e}_1', \mathbf{e}_2', \mathbf{e}_3'\}$ is the rotation of $\{\mathbf{e}_1, \mathbf{e}_2, \mathbf{e}_3\}$ by α about $\hat{\boldsymbol{a}}$, is $\mathbf{Q} = \mathbf{A}^\mathsf{T}$, as shown in (8.34c), that is,

$$Q_{ij} = \cos\alpha\, \delta_{ij} + (1 - \cos\alpha)\, a_i a_j + \sin\alpha\, \epsilon_{ijk} a_k. \quad (8.51)$$

(8.31) Since $\hat{\boldsymbol{a}} = a_i \mathbf{e}_i$ is a unit vector in (8.50), the components of $\hat{\boldsymbol{a}}$ can be written in spherical form as $(a_1, a_2, a_3) = (\sin\theta\cos\varphi, \sin\theta\sin\varphi, \cos\theta)$ as in (1.13c). Show that the (active) rotation matrix $\mathbf{A} = [A_{ij}]$ in (8.50) can be written as

$$\mathbf{A} = \mathbf{Q}_1^\mathsf{T}\mathbf{Q}_2^\mathsf{T} \begin{bmatrix} \cos\alpha & -\sin\alpha & 0 \\ \sin\alpha & \cos\alpha & 0 \\ 0 & 0 & 1 \end{bmatrix} \mathbf{Q}_2\mathbf{Q}_1 = \mathbf{Q}^\mathsf{T}\mathbf{R}\mathbf{Q}, \quad (8.52)$$

with $\mathbf{Q}_1$, $\mathbf{Q}_2$ as in (8.24) and (8.26), respectively, and $\mathbf{Q}_2\mathbf{Q}_1 = \mathbf{Q}$ as in (8.10b), with θ and φ determined by $\hat{\boldsymbol{a}}$ here. The active rotation of a column vector $\mathbf{v}$ is $\mathbf{v}' = \mathbf{A}\mathbf{v}$. The matrices $\mathbf{Q}_1$ and $\mathbf{Q}_2$ transform the components $\mathbf{v}$ into a basis where $\mathbf{e}_3' = \hat{\boldsymbol{a}}$, the middle matrix $\mathbf{R}$ performs the active rotation by α about $\mathbf{e}_3'$ in that basis, then $\mathbf{Q}^\mathsf{T} = \mathbf{Q}_1^\mathsf{T}\mathbf{Q}_2^\mathsf{T}$ transforms the components back to the original basis.
Hint: direct calculation of (8.50) and (8.52) is mind-numbing. A better approach is to split the middle matrix $\mathbf{R}$ in (8.52) as

$$\mathbf{R} = \cos\alpha \begin{bmatrix} 1 & 0 & 0 \\ 0 & 1 & 0 \\ 0 & 0 & 1 \end{bmatrix} + (1 - \cos\alpha) \begin{bmatrix} 0 & 0 & 0 \\ 0 & 0 & 0 \\ 0 & 0 & 1 \end{bmatrix} + \sin\alpha \begin{bmatrix} 0 & -1 & 0 \\ 1 & 0 & 0 \\ 0 & 0 & 0 \end{bmatrix}.$$

In index notation, this is

$$R_{ij} = \cos\alpha\, \delta_{ij} + (1 - \cos\alpha)\, \delta_{i3}\delta_{j3} - \sin\alpha\, (\delta_{i1}\delta_{j2} - \delta_{i2}\delta_{j1}).$$

Distributing and using $\mathbf{Q} = \mathbf{Q}_2\mathbf{Q}_1$ as in (8.10b) whose last row is (a_1, a_2, a_3), simplifies the calculations.

(8.32) The orthonormal basis $\{\mathbf{e}_1', \mathbf{e}_2', \mathbf{e}_3'\}$ is the rotation of $\{\mathbf{e}_1, \mathbf{e}_2, \mathbf{e}_3\}$ by $\alpha = 2\pi/3$ about $\boldsymbol{a} = \mathbf{e}_1 + \mathbf{e}_2 + \mathbf{e}_3$. Visualize this rotation and deduce the matrix $\mathbf{Q}$ such that $Q_{ij} = \mathbf{e}_i' \cdot \mathbf{e}_j$ without calculations. Verify that the matrix $\mathbf{Q}$ is indeed that given by (8.51) for this α and $\hat{\boldsymbol{a}}$. Find the angles φ, θ, ψ that yield the ZYZ Euler angle decomposition of this matrix $\mathbf{Q}$ as specified in (8.29). Find the spherical angles (θ_a, φ_a) that specify $\hat{\boldsymbol{a}}$ in the $\{\mathbf{e}_1, \mathbf{e}_2, \mathbf{e}_3\}$ basis. Show that those two sets of angles $(\alpha, \theta_a, \varphi_a)$ and (φ, θ, ψ) that characterize the same $\mathbf{Q}$ are distinct.

(8.33) Equation (8.51) provides the matrix $\mathbf{Q}$ when $\{\mathbf{e}_1', \mathbf{e}_2', \mathbf{e}_3'\}$ is the rotation of $\{\mathbf{e}_1, \mathbf{e}_2, \mathbf{e}_3\}$ by angle α about direction $\hat{\boldsymbol{a}}$, with $Q_{ij} \triangleq \mathbf{e}_i' \cdot \mathbf{e}_j$. Here we investigate the inverse construction: find $\hat{\boldsymbol{a}}$ and α given $\mathbf{Q}$. This is called the *axis-angle* representation of a rotation matrix. Since $\hat{\boldsymbol{a}} = a_l \mathbf{e}_l$ can be specified by two angles, this is another three-angle representation of $\mathbf{Q}$, distinct from the Euler angle factorization into elementary rotation matrices, in general. Show that

$$\cos\alpha = \frac{Q_{ii} - 1}{2}, \quad a_l = \frac{\epsilon_{ijl}\,(Q_{ij} - Q_{ji})}{4\sin\alpha}, \quad (8.53)$$

where $Q_{ii} = Tr(\mathbf{Q})$ is the *trace* of $\mathbf{Q}$, the sum of its diagonal elements (and of its *eigenvalues*). Note that this provides $0 \le \alpha \le \pi$ and the formula for a_l is indeterminate when $\alpha = \pi$.

(8.34) Explain and implement as a computer program, or pseudo-code, the following more geometric approach to finding the axis-angle parameters, $\hat{\boldsymbol{a}}$ and α, given the orthogonal matrix $\mathbf{Q}$. Recall (8.15) that $\mathbf{e}_i'$ and $\mathbf{e}_i$ are rows i of $\mathbf{Q}$ and $\mathbf{I}$, respectively, in the $\{\mathbf{e}_1, \mathbf{e}_2, \mathbf{e}_3\}$ basis.
(i) Compute $(\mathbf{e}_1' - \mathbf{e}_1)$, $(\mathbf{e}_2' - \mathbf{e}_2)$, $(\mathbf{e}_3' - \mathbf{e}_3)$ that are orthogonal to $\hat{\boldsymbol{a}}$ since $\mathbf{e}_i'$ is the rotation of $\mathbf{e}_i$ about $\hat{\boldsymbol{a}}$. Pick the two largest in magnitude, $\boldsymbol{p}$ and $\boldsymbol{q}$ say.
(ii) Let $\boldsymbol{a} = \boldsymbol{p} \times \boldsymbol{q}$ then $\hat{\boldsymbol{a}} = \boldsymbol{a}/|\boldsymbol{a}|$.
(iii) Let $\mathbf{e}_i$ be the original basis vector corresponding to $\boldsymbol{p}$. Compute $\boldsymbol{b} = \mathbf{e}_i - \hat{\boldsymbol{a}}\,\hat{\boldsymbol{a}} \cdot \mathbf{e}_i$ and $\boldsymbol{b}' = \boldsymbol{b} + \boldsymbol{p}$ then

$$\alpha = \operatorname{atan2}(\hat{\boldsymbol{a}} \cdot \boldsymbol{b} \times \boldsymbol{b}', \boldsymbol{b} \cdot \boldsymbol{b}').$$

Show that $\hat{\boldsymbol{a}}$ is determined even when $\alpha = \pi$. Note that $-\pi < \alpha \le \pi$.

(8.35) Demonstrate that a sequence of body-fixed rotations as defined in (8.29) and (8.30) yields the same result as the *reverse* sequence of space-fixed rotations using a physical object (a book, a model airplane, etc.).

Matrices and Tensors

9

The basic concepts, notations, and results of matrix and vector algebra are briefly reviewed in this chapter, then expanded to the coordinate-free geometric vectors. Linear algebra and matrices are typically introduced in the solution of systems of linear equations. Our first encounter with matrices in this book arose from rotations of vectors in Chapter 8.

9.1 Matrices

The matrix Q in Chapter 8 is a very special matrix, an *orthogonal* matrix. An arbitrary 3-by-3 real matrix A is any array of nine independent real numbers

$$\mathsf{A} \equiv \begin{bmatrix} a_{11} & a_{12} & a_{13} \\ a_{21} & a_{22} & a_{23} \\ a_{31} & a_{32} & a_{33} \end{bmatrix}. \tag{9.1}$$

More generally, an M-by-N matrix will have M rows and N columns and M does not have to be equal to N; that is, matrices can be *rectangular*.

Matrices are denoted in bold upper case sans serif fonts such as A and Q, and by square brackets [ ]. The (i,j) element of A is denoted A_{ij} or a_{ij}. The matrix of elements may be defined in compact form as[1]

$$\mathsf{A} = [A_{ij}] \quad \text{or} \quad \mathsf{A} = [a_{ij}].$$

[1] a_{ij} is preferred when the full matrix is displayed as in (9.1), but we prefer A_{ij} for compact index expressions with summation convention as in (9.2).

By convention, vectors in $\mathbb{R}^3$ are defined as 3-by-1 matrices

$$\mathsf{x} = \begin{bmatrix} x_1 \\ x_1 \\ x_3 \end{bmatrix},$$

which we may write as $\mathsf{x} = (x_1, x_2, x_3)$ but not $[x_1, x_2, x_3]$, which denotes a 1-by-3 matrix, or *row* vector that is the transpose of the column vector: $[x_1, x_2, x_3] = \mathsf{x}^\mathsf{T}$.

Matrix–vector multiplication

Equation (8.5b) shows how matrix–vector multiplication should be defined. The matrix–vector product Ax, where A is 3-by-3 and x 3-by-1, is a 3-by-1 vector b in $\mathbb{R}^3$ whose ith component is the dot product of row i of matrix A with the column x,

$$\mathsf{Ax} = \mathsf{b} \quad \Leftrightarrow \quad A_{ij}x_j = b_i \tag{9.2}$$

where

$$A_{ij}x_j \equiv A_{i1}x_1 + A_{i2}x_2 + A_{i3}x_3$$

Summation over repeated indices is used in this chapter.

in the summation convention and $i = 1, 2, 3$. The product of a matrix with a (column) vector is performed *row-by-column*. This product is defined only if the number of columns of $\mathbf{A}$ is equal to the number of rows of $\mathbf{x}$. A 3-by-3 matrix cannot multiply a 2-by-1 vector.

Since $\mathbf{Ax}$ and $\mathbf{Bx}$ are vectors, it follows from the laws of vector addition and scaling that

$$\mathbf{v} = \mathbf{Ax} + \mathbf{Bx} \Leftrightarrow v_i = A_{ij}x_j + B_{ij}x_j = (A_{ij} + B_{ij})x_j = C_{ij}x_j,$$
$$\mathbf{v} = \alpha\mathbf{Ax} \Leftrightarrow v_i = \alpha A_{ij}x_j = (\alpha A_{ij})x_j = C_{ij}x_j.$$

Thus, matrices should be added and scaled *term by term*

$$\begin{aligned} \mathbf{C} = \mathbf{A} + \mathbf{B} &\Leftrightarrow C_{ij} = A_{ij} + B_{ij}, \\ \mathbf{C} = \alpha\mathbf{A} &\Leftrightarrow C_{ij} = \alpha A_{ij}. \end{aligned} \tag{9.3}$$

This is valid only when $\mathbf{A}$, $\mathbf{B}$, and $\mathbf{C}$ are matrices of the same M-by-N size, of course.

Matrix–Matrix multiplication

Two successive linear transformation of components, that is,

$$x'_i = A_{ij}\,x_j, \quad \text{then} \quad x''_i = B_{ij}\,x'_j,$$

can be combined into one transformation from x_j to x''_i

$$x''_i = B_{ik}x'_k = B_{ik}A_{kj}\,x_j = C_{ij}\,x_j.$$

Thus, the product $\mathbf{BA}$ of two matrices $\mathbf{B}$ and $\mathbf{A}$ is a matrix $\mathbf{C}$ whose element C_{ij} is the dot product of row i of $\mathbf{B}$ with column j of $\mathbf{A}$:

$$\mathbf{C} = \mathbf{BA} \quad \Leftrightarrow \quad C_{ij} = B_{ik}A_{kj}. \tag{9.4}$$

The matrix product is done *row-by-column*, as for matrix–vector multiplication. This requires that the *number of columns of the first matrix* in the product ($\mathbf{B}$) equals the *number of rows of the second matrix* ($\mathbf{A}$). We can only multiply an M-by-N by an N-by-P, that is, the *inner dimensions must match.*

In general,

$$\mathbf{AB} \neq \mathbf{BA};$$

matrix multiplication does not commute since $a_ib_j \neq a_jb_i$ unless $i = j$. For example,

$$\begin{aligned} &\begin{bmatrix} a_1 & a_2 \\ a_3 & a_4 \end{bmatrix} \begin{bmatrix} b_1 & b_2 \\ b_3 & b_4 \end{bmatrix} = \begin{bmatrix} a_1b_1 + a_2b_3 & a_1b_2 + a_2b_4 \\ a_3b_1 + a_4b_3 & a_3b_2 + a_4b_4 \end{bmatrix} \\ \neq &\begin{bmatrix} b_1 & b_2 \\ b_3 & b_4 \end{bmatrix} \begin{bmatrix} a_1 & a_2 \\ a_3 & a_4 \end{bmatrix} = \begin{bmatrix} b_1a_1 + b_2a_3 & b_1a_2 + b_2a_4 \\ b_3a_1 + b_4a_3 & b_3a_2 + b_4a_4 \end{bmatrix}. \end{aligned}$$

Matrix transpose

The transformation (8.5c) involves the sum $Q_{ji}x'_j$, which is similar to the matrix–vector multiplication except that the multiplication is column-by-column. To write this as a *row-by-column* matrix–vector multiplication, we defined the *transpose matrix* A^T such that

$$A^\mathsf{T}_{ij} = A_{ji}, \tag{9.5}$$

that is, the columns of A become the rows of A^T and vice versa. Then

$$x_i = A_{ji}x'_j = A^\mathsf{T}_{ij}x'_j \quad \Leftrightarrow \quad \mathsf{x} = \mathsf{A}^\mathsf{T}\mathsf{x}'.$$

It is left as an exercise in index notation to show that the transpose of a product is equal to the product of the transposes in *reverse order*,

$$(\mathsf{AB})^\mathsf{T} = \mathsf{B}^\mathsf{T}\mathsf{A}^\mathsf{T}. \tag{9.6}$$

9.2 Some special matrices

Identity matrix

There is a unique matrix such that $\mathbf{I}\mathsf{x} = \mathsf{x}$, $\forall \mathsf{x}$. For $\mathsf{x} \in \mathbb{R}^3$, that is

The symbol $\forall$ means *for all*, or *for any*.

$$\mathbf{I} = \begin{bmatrix} 1 & 0 & 0 \\ 0 & 1 & 0 \\ 0 & 0 & 1 \end{bmatrix}. \tag{9.7}$$

Thus, the (i, j) element of the identity matrix is the Kronecker delta,

Kronecker delta:

$$\delta_{ij} = \begin{cases} 1 \text{ if } i = j \\ 0 \text{ if } i \neq j. \end{cases}$$

$$I_{ij} = \delta_{ij}. \tag{9.8}$$

The proof of uniqueness follows from picking $\mathsf{x} = \mathbf{e}_1 = (1, 0, 0)$; then $\mathbf{I}\mathbf{e}_1 = \mathbf{e}_1$ implies that the first column of $\mathbf{I}$ is $\mathbf{e}_1$. Picking $\mathsf{x} = \mathbf{e}_2 = (0, 1, 0)$ shows that the second column of $\mathbf{I}$ is $\mathbf{e}_2$; then $\mathsf{x} = \mathbf{e}_3 = (0, 0, 1)$ implies that its third column is $\mathbf{e}_3$ (see also prob. 8.15 in §8).

Symmetric and antisymmetric matrices

A *symmetric matrix* A is equal to its transpose,

$$\mathsf{A} = \mathsf{A}^\mathsf{T} \quad \Leftrightarrow A_{ij} = A_{ji}. \tag{9.9}$$

For example,

$$\mathsf{A} = \begin{bmatrix} a & b & c \\ b & d & e \\ c & e & f \end{bmatrix} = \mathsf{A}^\mathsf{T}$$

is symmetric. An *antisymmetric matrix* (or *skew-symmetric*) A is the opposite of its transpose

$$\mathsf{A} = -\mathsf{A}^\mathsf{T} \quad \Leftrightarrow A_{ij} = -A_{ji}. \tag{9.10}$$

For example,

$$\mathbf{A} = \begin{bmatrix} 0 & -a_3 & a_2 \\ a_3 & 0 & -a_1 \\ -a_2 & a_1 & 0 \end{bmatrix} = -\mathbf{A}^\mathsf{T}$$

is antisymmetric.

$A_{ij} = -\epsilon_{ijk} a_k,$
$a_i = -\frac{1}{2}\epsilon_{ikl} A_{kl}.$

That matrix is the matrix version of the vector operation $\boldsymbol{a} \times (\cdot)$ in cartesian components; indeed, if $\boldsymbol{v}' = \boldsymbol{a} \times \boldsymbol{v}$, then

$$v_i' = \epsilon_{ikj} a_k v_j = A_{ij} v_j = -A_{ji} v_j.$$

Thus, any 3-by-3 antisymmetric matrix corresponds to a cross product. The diagonal elements of an antisymmetric matrix must be zero.

Orthogonal and permutation matrices

In Chapter 8, we encountered square *orthogonal* matrices $\mathbf{Q}$ such that

$$\mathbf{Q}^\mathsf{T}\mathbf{Q} = \mathbf{I} = \mathbf{Q}\mathbf{Q}^\mathsf{T}.$$

A special kind of orthogonal matrices is the *permutation* matrices $\mathbf{P}$ that perform a permutation of the components of a vector. For example

$$\mathbf{P}\mathbf{v} = \begin{bmatrix} 0 & 1 & 0 \\ 0 & 0 & 1 \\ 1 & 0 & 0 \end{bmatrix} \begin{bmatrix} v_1 \\ v_2 \\ v_3 \end{bmatrix} = \begin{bmatrix} v_2 \\ v_3 \\ v_1 \end{bmatrix} \tag{9.11}$$

is a matrix that performs the $(2,3,1)$ permutation of $(1,2,3)$. Its inverse is $\mathbf{P}^\mathsf{T}$; indeed,

$$\mathbf{P}^\mathsf{T}\mathbf{P} = \begin{bmatrix} 0 & 0 & 1 \\ 1 & 0 & 0 \\ 0 & 1 & 0 \end{bmatrix} \begin{bmatrix} 0 & 1 & 0 \\ 0 & 0 & 1 \\ 1 & 0 & 0 \end{bmatrix} = \mathbf{I}.$$

By itself, $\mathbf{P}^\mathsf{T}$ yields the $(3,1,2)$ permutation of $(1,2,3)$

$$\mathbf{P}^\mathsf{T}\mathbf{v} = \begin{bmatrix} 0 & 0 & 1 \\ 1 & 0 & 0 \\ 0 & 1 & 0 \end{bmatrix} \begin{bmatrix} v_1 \\ v_2 \\ v_3 \end{bmatrix} = \begin{bmatrix} v_3 \\ v_1 \\ v_2 \end{bmatrix}.$$

The elements of a permutation matrix are all zero except for a single 1 somewhere along each row and each column.

9.3 Determinant

The mixed product of three vectors *determines* the signed volume of the parallelepiped spanned by $\boldsymbol{a}, \boldsymbol{b}, \boldsymbol{c}$. The connection with algebraic determinants arises from writing the mixed product in cartesian components

$$\boldsymbol{a} \times \boldsymbol{b} \cdot \boldsymbol{c} = \epsilon_{ijk} a_i b_j c_k = \begin{vmatrix} a_1 & a_2 & a_3 \\ b_1 & b_2 & b_3 \\ c_1 & c_2 & c_3 \end{vmatrix} = \begin{vmatrix} a_1 & b_1 & c_1 \\ a_2 & b_2 & c_2 \\ a_3 & b_3 & c_3 \end{vmatrix}$$

as discussed in Chapter 6, where the components of $\boldsymbol{a}$, $\boldsymbol{b}$, $\boldsymbol{c}$ can be arranged as rows or columns. For a 3-by-3 matrix $\mathbf{A}$ with elements A_{ij} this means that

$$\det(\mathbf{A}) = \epsilon_{ijk}A_{i1}A_{j2}A_{k3} = \epsilon_{lmn}A_{1l}A_{2m}A_{3n} = \det(\mathbf{A}^\mathsf{T}). \tag{9.12}$$

This can be verified by explicit expansions of the sums for a 3-by-3 matrix, or by the following more abstract argument. We can reorder any column-ordered term into a row-ordered term,

$$A_{i1}A_{j2}A_{k3} \to A_{1l}A_{2m}A_{3n},$$

by performing the permutations that led from $(1,2,3) \to (i,j,k)$ in *reverse order*. Thus, the *number* of permutations is preserved and (l,m,n) is even if (i,j,k) is even, and odd if (i,j,k) is odd. Thus,

$$\epsilon_{ijk}A_{i1}A_{j2}A_{k3} = \epsilon_{lmn}A_{1l}A_{2m}A_{3n},$$

and (9.12) follows. For example, if $(i,j,k) = (3,1,2)$, then

$$A_{31}A_{12}A_{23} = A_{12}A_{23}A_{31}$$

and $(l,m,n) = (2,3,1) \neq (i,j,k) = (3,1,2)$; however, $\epsilon_{312} = \epsilon_{231} = 1$.

Another useful identity is

$$\epsilon_{ijk}A_{il}A_{jm}A_{kn} = \epsilon_{lmn}\det\mathbf{A} = \epsilon_{lmn}\epsilon_{ijk}A_{i1}A_{j2}A_{k3} \tag{9.13}$$

since the left-hand side is the determinant of the matrix $\mathbf{A}$ whose columns $(1,2,3)$ have been replaced by columns (l,m,n). That property then yields

$$\begin{aligned}\det(\mathbf{AB}) &= \epsilon_{ijk}\,(A_{il}B_{l1})\,(A_{jm}B_{m2})\,(A_{kn}B_{n3})\\ &= \epsilon_{ijk}\,(A_{il}A_{jm}A_{kn})\,(B_{l1}B_{m2}B_{n3})\\ &= \epsilon_{ijk}\epsilon_{lmn}\,A_{i1}A_{j2}A_{k3}\,B_{l1}B_{m2}B_{n3}\\ &= \det(\mathbf{A})\det(\mathbf{B}).\end{aligned} \tag{9.14}$$

These more abstract arguments are useful to extend the results to N-by-N matrices, indeed if $\mathbf{A}$ is N-by-N, then its determinant is the N-tuple sums over repeated indices

$$\det(\mathbf{A}) = \epsilon_{j_1j_2\dots j_N}\,A_{1j_1}A_{2j_2}\cdots A_{Nj_N} = \det(\mathbf{A}^\mathsf{T}), \tag{9.15}$$

where

$$\epsilon_{j_1j_2\dots j_N} \equiv \operatorname{sgn}(j_1,j_2,\dots,j_N) = \pm 1,$$

depending on whether $(j_1,j_2,\dots,j_N)$ is an even or odd permutation of $(1,2,\dots,N)$ and 0 otherwise.

9.4 Matrix inverse and Ax=b

If $\det(\mathbf{A}) \neq 0$, then there exists a unique matrix denoted $\mathbf{A}^{-1}$ called the *inverse* of $\mathbf{A}$ such that

$$\mathbf{A}^{-1}\mathbf{A} = \mathbf{I}. \tag{9.16}$$

If it exists, then

$$\mathbf{Ax} = \mathbf{b} \Leftrightarrow \mathbf{x} = \mathbf{A}^{-1}\mathbf{b}.$$

This implies that

$$\mathbf{AA}^{-1}\mathbf{b} = \mathbf{b} \Leftrightarrow \mathbf{AA}^{-1} = \mathbf{I}$$

since the left equality holds for any $\mathbf{b}$, and $\mathbf{I}$ is unique (9.7). Hence, the left inverse is the same as the right inverse

$$\mathbf{A}^{-1}\mathbf{A} = \mathbf{I} = \mathbf{AA}^{-1},$$

and uniqueness of the inverse follows (exercise 9.7). It is also left as an exercise (9.6) to prove that the inverse of a product is the product of the inverses *in reverse order*,

$$(\mathbf{AB})^{-1} = \mathbf{B}^{-1}\mathbf{A}^{-1}. \tag{9.17}$$

The matrix $\mathbf{A}^{-1}$ can be computed column by column by solving the linear systems of equation $\mathbf{Ax}_j = \mathbf{e}_j$ where $\mathbf{e}_j$ is the jth column of $\mathbf{I}$, although in general one is interested in finding $\mathbf{x}$ for a single $\mathbf{b}$. That $\mathbf{Ax} = \mathbf{b}$ problem is a system of linear equations for $\mathbf{x}$ given the matrix $\mathbf{A}$ and the vector $\mathbf{b}$. There are two interpretations of that problem: (1) as a search for the components $(x_1, \ldots, x_n)$ of $\mathbf{b}$ in the basis given by the columns of $\mathbf{A}$ and (2) as a search for a point $\mathbf{x}$ that is the intersection of certain planes perpendicular to the rows of $\mathbf{A}$.

Linear combination of columns

Consider $\mathbf{A}$ as a row of columns, $\mathbf{A} = [\mathbf{a}_1, \mathbf{a}_2, \mathbf{a}_3]$, where $\mathbf{a}_j$ is the jth column of $\mathbf{A}$, then

$$\mathbf{Ax} = \begin{bmatrix} \mathbf{a}_1 & \mathbf{a}_2 & \mathbf{a}_3 \end{bmatrix} \begin{bmatrix} x_1 \\ x_2 \\ x_3 \end{bmatrix} = x_1\mathbf{a}_1 + x_2\mathbf{a}_2 + x_3\mathbf{a}_3 = \mathbf{b}$$

and $\mathbf{x} = (x_1, x_2, x_3)$ are the components of $\mathbf{b}$ in the basis $\{\mathbf{a}_1, \mathbf{a}_2, \mathbf{a}_3\}$.

Finding these components is possible in general only if $\{\mathbf{a}_1, \mathbf{a}_2, \mathbf{a}_3\}$ are linearly independent, that is iff $\det(\mathbf{a}_1, \mathbf{a}_2, \mathbf{a}_3) = \det(\mathbf{A}) \neq 0$. If $\det(\mathbf{A}) = 0$, then the three columns are in the same plane and the system will have a solution only if $\mathbf{b}$ is also in that plane. Actually, there will be an infinite number of solutions in that case since there is a nonzero combination $\alpha_1\mathbf{a}_1 + \alpha_2\mathbf{a}_2 + \alpha_3\mathbf{a}_3 = 0$, and any multiple of $(\alpha_1, \alpha_1, \alpha_3)$ can be added to (x_1, x_2, x_3) while preserving $\mathbf{Ax} = \mathbf{b}$.

We can find the components (x_1, x_2, x_3) when $\det \mathbf{A} \neq 0$ by thinking geometrically and projecting $\mathbf{b}$ on the *reciprocal basis* (6.17),

$$\mathbf{r}_1 = \frac{\mathbf{a}_2 \times \mathbf{a}_3}{\det \mathbf{A}}, \quad \mathbf{r}_2 = \frac{\mathbf{a}_3 \times \mathbf{a}_1}{\det \mathbf{A}}, \quad \mathbf{r}_3 = \frac{\mathbf{a}_1 \times \mathbf{a}_2}{\det \mathbf{A}}, \tag{9.18a}$$

yielding

$$x_i = \mathbf{r}_i \cdot \mathbf{b} = \mathbf{r}_i^\mathsf{T}\mathbf{b} \tag{9.18b}$$

and

$$\mathbf{A}^{-1} = \begin{bmatrix} \mathbf{r}_1 & \mathbf{r}_2 & \mathbf{r}_3 \end{bmatrix}^\mathsf{T} \tag{9.18c}$$

since $\mathbf{r}_i^\mathsf{T}\mathbf{a}_j = \delta_{ij}$. Thus the reciprocal vectors (9.18a) are the *rows* of $\mathbf{A}^{-1}$.

Cramer's rule

Formula (9.18b) for the solution of $\mathbf{Ax} = \mathbf{b}$ can be written in terms of determinants since $\mathbf{a} \cdot (\mathbf{b} \times \mathbf{c}) = \det(\mathbf{a}, \mathbf{b}, \mathbf{c})$, and thereby generalized to higher dimension. We find that

$$x_1 = \frac{\det(\mathbf{b}, \mathbf{a}_2, \mathbf{a}_3)}{\det \mathbf{A}}, \quad x_2 = \frac{\det(\mathbf{a}_1, \mathbf{b}, \mathbf{a}_3)}{\det \mathbf{A}}, \quad x_3 = \frac{\det(\mathbf{a}_1, \mathbf{a}_2, \mathbf{b})}{\det \mathbf{A}},$$

and this is *Cramer's rule*: the ith component of the solution $\mathbf{x}$ of $\mathbf{Ax} = \mathbf{b}$ is

$$x_i = \frac{\det \mathbf{A}^{(i)}}{\det \mathbf{A}}, \tag{9.19}$$

where the matrix $\mathbf{A}^{(i)}$ is the matrix $\mathbf{A}$ but with column i replaced by $\mathbf{b}$.

Cramer's rule can deduced directly from the algebraic properties of determinants, for example,

$$\begin{aligned}\det(\mathbf{b}, \mathbf{a}_2, \mathbf{a}_3) &= \det(x_1\mathbf{a}_1 + x_2\mathbf{a}_2 + x_3\mathbf{a}_3,\ \mathbf{a}_2, \mathbf{a}_3) \\ &= x_1 \det(\mathbf{a}_1, \mathbf{a}_2, \mathbf{a}_3) + x_2 \det(\mathbf{a}_2, \mathbf{a}_2, \mathbf{a}_3) + x_3 \det(\mathbf{a}_3, \mathbf{a}_2, \mathbf{a}_3) \\ &= x_1 \det(\mathbf{a}_1, \mathbf{a}_2, \mathbf{a}_3) = x_1 \det \mathbf{A},\end{aligned}$$

using properties (6.4a) and (6.4b) of determinants. Cramer's formula (9.19) applies to N-by-N matrices.

Formula for matrix inverse

Carmer's rule yields a formula for the matrix inverse $\mathbf{A}^{-1}$ in terms of determinants. This generalizes the cross-product formula (9.18c) to n-by-n matrices.

Since $\mathbf{A}^{-1}$ is the solution $\mathbf{X}$ of $\mathbf{AX} = \mathbf{I}$, column j of $\mathbf{A}^{-1}$ is the solution $\mathbf{x}_j$ of

$$\mathbf{Ax}_j = \mathbf{e}_j,$$

where $\mathbf{e}_j$ is column j of $\mathbf{I}$. The (i, j) component of $\mathbf{A}^{-1}$ is thus the i component of $\mathbf{x}_j$. Using (9.19) that is

$$A^{-1}_{ij} = \frac{\det \mathbf{A}^{(ij)}}{\det \mathbf{A}}, \tag{9.20}$$

where $\mathbf{A}^{(ij)}$ is the matrix $\mathbf{A}$ but with column i replaced by $\mathbf{e}_j$, such that column i of $\mathbf{A}^{(ij)}$ is 0, except for a 1 on row j. That special determinant in the numerator,

$$\det \mathbf{A}^{(ij)} \triangleq (-1)^{i+j} M_{ji},$$

is the *cofactor* of A_{ij}, where M_{ji} is the determinant of the matrix obtained by deleting *row* j and *column* i of $\mathbf{A}$.

Cramer's formula (9.20) holds in any dimension, but computing determinants in dimension n can be impossibly costly. Formula (9.15) contains $n! = n(n-1)\cdots 2 \cdot 1$ terms, all the permutations of n distinct elements. The factorial function grows very quickly with n,

$$3! = 6, \quad 6! = 720, \quad 12! = 479\,001\,600, \quad 24! \approx 6.2\ 10^{23}, \quad \ldots,$$

so 24! is about equal to Avogadro's constant! There is a famous asymptotic formula, *Stirling's formula*, that gives

$$n! \sim \sqrt{2\pi n}\left(\frac{n}{e}\right)^n \tag{9.21}$$

where $f(n) \sim g(n)$ means $\lim_{n\to\infty}(f(n)/g(n)) = 1$. In any case, the direct formula (9.15) for determinants is nice, but uses excessive calculations for $n > 4$. Determinants of large matrices, if needed, can be calculated after reduction to triangular form or orthogonalization, as discussed below.

Intersection of planes

The linear system of equations, $\mathbf{Ax} = \mathbf{b}$, can also be interpreted as an intersection of planes. Here, we view $\mathbf{A}$ as a *column of rows*, but it may be simpler then to transpose the problem and work with the columns of $\mathbf{A}^\mathsf{T}$ that are the rows of $\mathbf{A}$. Thus,

$$\mathbf{Ax} = \mathbf{b} \Leftrightarrow \mathbf{x}^\mathsf{T}\mathbf{A}^\mathsf{T} = \mathbf{b}^\mathsf{T}$$

where $\mathbf{x}^\mathsf{T}$ and $\mathbf{b}^\mathsf{T}$ are now *rows* and the columns of $\mathbf{A}^\mathsf{T} = [\mathbf{n}_1, \mathbf{n}_2, \mathbf{n}_3]$ are the *rows* of $\mathbf{A}$. The linear system is

$$\mathbf{x}^\mathsf{T}\mathbf{A}^\mathsf{T} = \mathbf{x}^\mathsf{T}[\mathbf{n}_1, \mathbf{n}_2, \mathbf{n}_3] = \mathbf{b}^\mathsf{T} \Leftrightarrow \begin{cases} \mathbf{n}_1 \cdot \mathbf{x} = b_1 \\ \mathbf{n}_2 \cdot \mathbf{x} = b_2 \\ \mathbf{n}_3 \cdot \mathbf{x} = b_3 \end{cases}$$

and $\mathbf{x}$ is seen as the position vector of a point at the intersection of three planes.[2] To find $\mathbf{x}$, the standard procedure is *elimination* (row reduction) where we combine the equations

[2]Recall that $\mathbf{n} \cdot \mathbf{x} = b$ is the equation of a plane perpendicular to $\mathbf{n}$ and passing through a point $\mathbf{x}_0$ such that $\mathbf{n} \cdot \mathbf{x}_0 = b$, for instance the point $\mathbf{x}_0 = b\mathbf{n}/|\mathbf{n}|^2$, which is, in fact, the closest to the origin.

$$\begin{cases} \mathbf{n}_1 \cdot \mathbf{x} = b_1 \\ \mathbf{n}_2 \cdot \mathbf{x} = b_2 \\ \mathbf{n}_3 \cdot \mathbf{x} = b_3 \end{cases} \longrightarrow \begin{cases} \mathbf{n}_1 \cdot \mathbf{x} &= b_1 \\ (\mathbf{n}_2 - L_{21}\mathbf{n}_1) \cdot \mathbf{x} &= b_2 - L_{21}b_1 \\ (\mathbf{n}_3 - L_{31}\mathbf{n}_1) \cdot \mathbf{x} &= b_3 - L_{31}b_1 \end{cases} \tag{9.22}$$

picking the coefficients L_{21} and L_{31} such that $\mathbf{n}'_2 = \mathbf{n}_2 - L_{21}\mathbf{n}_1$ and $\mathbf{n}'_3 = \mathbf{n}_3 - L_{31}\mathbf{n}_1$ are orthogonal to $\mathbf{e}_1$,

$$\mathbf{e}_1 \cdot \mathbf{n}'_2 = 0 = \mathbf{e}_1 \cdot \mathbf{n}'_3,$$

that is, $\mathbf{n}'_2$ and $\mathbf{n}'_3$ have a zero first component. At the next step, one combines the last two equations defining $\mathbf{n}''_3 = \mathbf{n}'_3 - L_{32}\mathbf{n}'_2$ with L_{32} such that

$$\mathbf{e}_2 \cdot \mathbf{n}''_3 = 0.$$

Thus, the first and second components of $\mathbf{n}''_3$ are zero. The final system of equations is then upper triangular and easy to solve by *backward substitution.* This elimination requires permutations of equations, in general, in case $\mathbf{e}_1 \cdot \mathbf{n}_1$ is zero or small compared to $\mathbf{e}_1 \cdot \mathbf{n}_2$ or $\mathbf{e}_1 \cdot \mathbf{n}_3$, for instance. That introduces a permutation matrix $\mathbf{P}$. The final decomposition is known as the LU or PLU decomposition of $\mathbf{A}$,

$$\mathbf{A} = \mathbf{P}^\mathsf{T}\mathbf{LU},$$

where $\mathbf{P}$ is a permutation matrix, $\mathbf{L}$ is a lower triangular matrix, and $\mathbf{U}$ is an upper triangular matrix. The determinant of $\mathbf{A}$ is then the product of the diagonal elements of $\mathbf{U}$ times the *sign* of the permutation $\mathbf{P}$.

9.5 Gram–Schmidt and QR

A better approach than Cramer's rule for finding the components of $\mathbf{b}$ in the basis $\{\mathbf{a}_1, \mathbf{a}_2, \ldots\}$, and for computing determinants, is the *Gram–Schmidt* orthogonalization that has already been used in exercises 3.38 and 6.25. Starting from any $\mathbf{A} = [\mathbf{a}_1, \mathbf{a}_2, \mathbf{a}_3]$, we orthonormalize the columns $\mathbf{a}_j$ by subtracting the parallel projections as in §3 where $\boldsymbol{a}_\perp = \boldsymbol{a} - (\boldsymbol{a} \cdot \hat{\boldsymbol{n}})\hat{\boldsymbol{n}}$ is the component of $\boldsymbol{a}$ perpendicular to unit vector $\hat{\boldsymbol{n}}$.

For three vectors $[\mathbf{a}_1, \mathbf{a}_2, \mathbf{a}_3]$, the Gram–Schmidt algorithm proceeds in three steps

$$\begin{array}{llll} (1) & \mathbf{q}_1 = \dfrac{\mathbf{a}_1}{|\mathbf{a}_1|}, & \mathbf{a}_2' = \mathbf{a}_2 - \mathbf{a}_2 \cdot \mathbf{q}_1 \mathbf{q}_1, & \mathbf{a}_3' = \mathbf{a}_3 - \mathbf{a}_3 \cdot \mathbf{q}_1 \mathbf{q}_1, \\ (2) & & \mathbf{q}_2 = \dfrac{\mathbf{a}_2'}{|\mathbf{a}_2'|}, & \mathbf{a}_3'' = \mathbf{a}_3' - \mathbf{a}_3' \cdot \mathbf{q}_2 \mathbf{q}_2, \\ (3) & & & \mathbf{q}_3 = \dfrac{\mathbf{a}_3''}{|\mathbf{a}_3''|}, \end{array} \qquad (9.23)$$

where $\mathbf{u} \cdot \mathbf{v} = \mathbf{u}^\mathsf{T}\mathbf{v}$ is the dot (or *inner*) product of vector $\mathbf{u}$ and $\mathbf{v}$.

The vectors $\mathbf{q}_1$, $\mathbf{q}_2$, $\mathbf{q}_3$ thus obtained form an orthonormal basis. If we keep track of the coefficients in an *upper triangular* matrix $\mathbf{R}$ where

$$\begin{array}{llll} (1) & r_{11} = |\mathbf{a}_1|, & r_{12} = \mathbf{a}_2 \cdot \mathbf{q}_1, & r_{13} = \mathbf{a}_3 \cdot \mathbf{q}_1, \\ (2) & & r_{22} = |\mathbf{a}_2'|, & r_{23} = \mathbf{a}_3' \cdot \mathbf{q}_2, \\ (3) & & & r_{33} = |\mathbf{a}_3''|, \end{array} \qquad (9.24)$$

we obtain the so-called $\mathbf{QR}$ decomposition of $\mathbf{A}$,

$$\mathbf{A} = \begin{bmatrix} \mathbf{a}_1 & \mathbf{a}_2 & \mathbf{a}_3 \end{bmatrix} = \begin{bmatrix} \mathbf{q}_1 & \mathbf{q}_2 & \mathbf{q}_3 \end{bmatrix} \begin{bmatrix} r_{11} & r_{12} & r_{13} \\ 0 & r_{22} & r_{23} \\ 0 & 0 & r_{23} \end{bmatrix} = \mathbf{QR}.$$

To solve $\mathbf{Ax} = \mathbf{b}$ using $\mathbf{A} = \mathbf{QR}$, project $\mathbf{b}$ onto the orthogonal basis $\{\mathbf{q}_1, \mathbf{q}_2, \mathbf{q}_3\}$, that is, $\mathbf{b}' = \mathbf{Q}^\mathsf{T}\mathbf{b}$ in matrix notation, then *backsolve* the upper triangular linear system $\mathbf{Rx} = \mathbf{b}'$, that is,

$$\mathbf{Ax} = (\mathbf{QR})\mathbf{x} = \mathbf{b} \Rightarrow \mathbf{Rx} = \mathbf{Q}^\mathsf{T}\mathbf{b} = \mathbf{b}'.$$

Then

$$\begin{cases} x_3 = & b_3'/r_{33}, \\ x_2 = & (b_2' - r_{23}x_3)/r_{22}, \\ x_1 = & (b_1' - r_{12}x_2 - r_{13}x_3)/r_{11}. \end{cases} \qquad (9.25)$$

The $\mathbf{QR}$ decomposition can also be applied to the *rows* of $\mathbf{A}$, which are the columns of $\mathbf{A}^\mathsf{T}$. Thus, $\mathbf{A}^\mathsf{T} = \mathbf{QR}$, where the matrices $\mathbf{Q}$ and $\mathbf{R}$ will be different from those for the columns of A, in general. The solution to $\mathsf{Ax} = \mathsf{b} \Leftrightarrow \mathsf{x}^\mathsf{T}\mathsf{A}^\mathsf{T} = \mathsf{b}^\mathsf{T}$ can be obtained as follows

$$\mathsf{x}^\mathsf{T}\mathsf{A}^\mathsf{T} = \mathsf{x}^\mathsf{T}\mathsf{QR} = \mathsf{b}^\mathsf{T} \Rightarrow \mathsf{R}^\mathsf{T}\mathsf{Q}^\mathsf{T}\mathsf{x} = \mathsf{R}^\mathsf{T}\mathsf{y} = \mathsf{b}.$$

Thus, one starts with a *forward solve* for y, since R^T is *lower triangular*, then $\mathsf{x} = \mathsf{Qy}$:

$$\mathsf{R}^\mathsf{T}\mathsf{y} = \mathsf{b} \Rightarrow \mathsf{x} = \mathsf{Qy}.$$

The reader should consult Trefethen and Bau,[3] for example, for more insights and information about numerical linear algebra.

9.6 Tensors

A *tensor* is an operator that yields a basis-invariant linear transformation of vectors. We begin with several fundamental examples of tensors that have already been encountered implicitly in earlier vector operations.

Projection tensors

The vector projections of a vector $\boldsymbol{v}$ parallel or perpendicular to a vector $\boldsymbol{a}$, that is, the vector transformations

$$\boldsymbol{v} \to \boldsymbol{v}' = \boldsymbol{v}_\parallel = \hat{\boldsymbol{a}}(\hat{\boldsymbol{a}} \cdot \boldsymbol{v})$$

or

$$\boldsymbol{v} \to \boldsymbol{v}' = \boldsymbol{v}_\perp = \boldsymbol{v} - \boldsymbol{v}_\parallel,$$

are linear transformations of $\boldsymbol{v}$. To separate the parallel and perpendicular projection operations from the vector $\boldsymbol{v}$ on which they apply, we write

$$\begin{aligned} \boldsymbol{v}_\parallel &= \hat{\boldsymbol{a}}(\hat{\boldsymbol{a}} \cdot \boldsymbol{v}) = \boldsymbol{\mathcal{P}}_\parallel \cdot \boldsymbol{v}, \\ \boldsymbol{v}_\perp &= \boldsymbol{v} - \hat{\boldsymbol{a}}(\hat{\boldsymbol{a}} \cdot \boldsymbol{v}) = \boldsymbol{\mathcal{P}}_\perp \cdot \boldsymbol{v}, \end{aligned}$$

thereby defining the parallel and perpendicular projection operators

$$\begin{aligned} \boldsymbol{\mathcal{P}}_\parallel &= \hat{\boldsymbol{a}}\hat{\boldsymbol{a}} = \frac{\boldsymbol{a}\boldsymbol{a}}{\boldsymbol{a} \cdot \boldsymbol{a}}, \\ \boldsymbol{\mathcal{P}}_\perp &= \boldsymbol{I} - \hat{\boldsymbol{a}}\hat{\boldsymbol{a}} = \boldsymbol{I} - \frac{\boldsymbol{a}\boldsymbol{a}}{\boldsymbol{a} \cdot \boldsymbol{a}}. \end{aligned} \tag{9.26}$$

Those $\boldsymbol{\mathcal{P}}_\parallel$, $\boldsymbol{\mathcal{P}}_\perp$, and $\boldsymbol{I}$ operators are not vectors since their dot product with a vector $\boldsymbol{v}$ yields a vector not a scalar. Those operators are *tensors*. The tensor $\boldsymbol{I}$ is the *identity tensor* such that for any $\boldsymbol{v}$

$$\boldsymbol{I} \cdot \boldsymbol{v} = \boldsymbol{v}. \tag{9.27}$$

[3] L. N. Trefethen and D. Bau, *Numerical Linear Algebra*, SIAM, 1997.

In a cartesian basis $\{\mathbf{e}_1, \mathbf{e}_2, \mathbf{e}_3\}$ with $\boldsymbol{a} = a_j \mathbf{e}_j$, these tensors have the cartesian components

$$\begin{aligned} I_{ij} = \delta_{ij}, \qquad P^{\parallel}_{ij} = \hat{a}_i \hat{a}_j = \frac{a_i a_j}{a_k a_k}, \\ P^{\perp}_{ij} = \delta_{ij} - \hat{a}_i \hat{a}_j = \delta_{ij} - \frac{a_i a_j}{a_k a_k}. \end{aligned} \tag{9.28}$$

Thus, a *tensor* is represented by a *matrix* with respect to a specific basis.

Reflection and Rotation tensors

Other examples of tensors are *reflections* about a plane perpendicular to $\boldsymbol{a}$,

$$\boldsymbol{v}' = \boldsymbol{v} - 2\hat{\boldsymbol{a}}(\hat{\boldsymbol{a}} \cdot \boldsymbol{v}) = \left(\boldsymbol{\mathcal{P}}_{\perp} - \boldsymbol{\mathcal{P}}_{\parallel}\right) \cdot \boldsymbol{v} = (\boldsymbol{I} - 2\hat{\boldsymbol{a}}\hat{\boldsymbol{a}}) \cdot \boldsymbol{v}, \tag{9.29}$$

and *rotation* by angle α about a direction $\boldsymbol{a}$, from (4.18),

$$\begin{aligned} \boldsymbol{v}' &= \boldsymbol{v}_{\parallel} + \cos\alpha\, \boldsymbol{v}_{\perp} + \sin\alpha\, \hat{\boldsymbol{a}} \times \boldsymbol{v} \\ &= \left(\cos\alpha\, \boldsymbol{I} + (1 - \cos\alpha)\, \hat{\boldsymbol{a}}\hat{\boldsymbol{a}} + \sin\alpha\, \hat{\boldsymbol{a}} \times \boldsymbol{I}\right) \cdot \boldsymbol{v} = \boldsymbol{\mathcal{R}} \cdot \boldsymbol{v}. \end{aligned} \tag{9.30}$$

In a cartesian basis, $\hat{\boldsymbol{a}} \times \boldsymbol{I}$ is $\epsilon_{ikl} \hat{a}_k \delta_{lj} = \epsilon_{ikj} \hat{a}_k$, and

$$\boldsymbol{v}' = \boldsymbol{\mathcal{R}} \cdot \boldsymbol{v} \quad \Leftrightarrow \quad v'_i = R_{ij} v_j,$$

where

$$R_{ij} = \cos\alpha\, \delta_{ij} + (1 - \cos\alpha)\, \hat{a}_i \hat{a}_j + \sin\alpha\, \epsilon_{ikj} \hat{a}_k \tag{9.31}$$

is the cartesian matrix of the rotation tensor $\boldsymbol{\mathcal{R}}$, already encountered in (8.50).

Tensor of inertia

The first tensor that students encounter in physics is usually the *tensor of inertia* $\boldsymbol{\mathcal{I}}$ that arises in the expression for the angular momentum of a rigid system. Consider N particles with masses m_l, position vectors $\boldsymbol{r}_l$, and velocities $\boldsymbol{v}_l$, with $l = 1, \ldots, N$. The total angular momentum of the system is[4]

$$\boldsymbol{L} = \sum_{l=1}^{N} \boldsymbol{r}_l \times m_l \boldsymbol{v}_l.$$

If the system is rigid and rotating about a point that we picked as the origin, then $\boldsymbol{v}_l = \boldsymbol{\omega} \times \boldsymbol{r}_l$ where $\boldsymbol{\omega}$ may vary in time but is the same for all rigidly connected particles. The angular momentum then reduces to

$$\boldsymbol{L} = \sum_{l=1}^{N} \boldsymbol{r}_l \times m_l (\boldsymbol{\omega} \times \boldsymbol{r}_l) = \sum_{l=1}^{N} m_l \big((\boldsymbol{r}_l \cdot \boldsymbol{r}_l) \boldsymbol{\omega} - (\boldsymbol{r}_l \cdot \boldsymbol{\omega}) \boldsymbol{r}_l \big).$$

Since $\boldsymbol{\omega}$ is the same for all l, we would like to pull it out from the sum and write $\boldsymbol{L} = \boldsymbol{\mathcal{I}} \cdot \boldsymbol{\omega}$, but that requires introducing the concept of *tensor of inertia*,

$$\boldsymbol{\mathcal{I}} = \sum_{l=1}^{N} m_l \big((\boldsymbol{r}_l \cdot \boldsymbol{r}_l)\, \boldsymbol{I} - \boldsymbol{r}_l \boldsymbol{r}_l \big). \tag{9.32}$$

[4]Note that l is a particle index here, not a cartesian component index. It appears *three* times in each term so we write the sum explicitly using $\sum$ and cannot use the summation convention for those sums over particles.

Dyadic product

The tensors $\boldsymbol{aa}$ and $\boldsymbol{r}_l\boldsymbol{r}_l$ are special cases of the *dyadic product* $\boldsymbol{ab}$ of two vectors $\boldsymbol{a}$ and $\boldsymbol{b}$. In a coordinate-free form, the dyadic product $\boldsymbol{ab}$ is defined by how it operates on an arbitrary vector $\boldsymbol{v}$,

The symbol $\triangleq$ means *equals by definition.*

$$(\boldsymbol{ab}) \cdot \boldsymbol{v} \triangleq \boldsymbol{a}(\boldsymbol{b} \cdot \boldsymbol{v}), \tag{9.33}$$

and is naturally written $\boldsymbol{ab}$ in applied mathematics, physics, and engineering. The dyadic product is also called the *tensor product* and denoted $\boldsymbol{a} \otimes \boldsymbol{b}$ in the mathematical literature. In a cartesian basis with $\boldsymbol{a} = a_i \mathbf{e}_i$ and $\boldsymbol{b} = b_j \mathbf{e}_j$, we have

$$\boldsymbol{a}(\boldsymbol{b} \cdot \boldsymbol{v}) \to a_i(b_j v_j) = (a_i b_j) v_j,$$

so the dyadic product $\boldsymbol{ab}$ has cartesian components $a_i b_j$ and is represented by the matrix

$$\boldsymbol{ab} \equiv \begin{bmatrix} a_1 b_1 & a_1 b_2 & a_1 b_3 \\ a_2 b_1 & a_2 b_2 & a_2 b_3 \\ a_3 b_1 & a_3 b_2 & a_3 b_3 \end{bmatrix} = \begin{bmatrix} a_1 \\ a_2 \\ a_3 \end{bmatrix} \begin{bmatrix} b_1 & b_2 & b_3 \end{bmatrix} = \mathsf{ab}^\mathsf{T}. \tag{9.34}$$

For example,

$$\mathbf{e}_1\mathbf{e}_2 \equiv \begin{bmatrix} 1 \\ 0 \\ 0 \end{bmatrix} \begin{bmatrix} 0 & 1 & 0 \end{bmatrix} = \begin{bmatrix} 0 & 1 & 0 \\ 0 & 0 & 0 \\ 0 & 0 & 0 \end{bmatrix} \tag{9.35a}$$

in the cartesian basis $\{\mathbf{e}_1, \mathbf{e}_2, \mathbf{e}_3\}$. For another cartesian basis $\{\mathbf{e}'_1, \mathbf{e}'_2, \mathbf{e}'_3\}$ such that $\mathbf{e}'_i = q_{ij}\mathbf{e}_j$ and thus $\mathbf{e}_i = q_{ji}\mathbf{e}'_j$,

$$\mathbf{e}_1\mathbf{e}_2 \equiv \begin{bmatrix} q_{11} \\ q_{21} \\ q_{31} \end{bmatrix} \begin{bmatrix} q_{12} & q_{22} & q_{32} \end{bmatrix} = \begin{bmatrix} q_{11}q_{12} & q_{11}q_{22} & q_{11}q_{32} \\ q_{21}q_{12} & q_{21}q_{22} & q_{21}q_{32} \\ q_{31}q_{12} & q_{31}q_{22} & q_{31}q_{32} \end{bmatrix} \tag{9.35b}$$

and the matrix representation of the tensor $\mathbf{e}_1\mathbf{e}_2$ depends on the basis, just like the components (v_1, v_2, v_3) of a vector $\boldsymbol{v}$ depend on the basis.

A dyad $\boldsymbol{ab}$ can also be *left* multiplied by a vector and the result is naturally[5]

$$\boldsymbol{v} \cdot (\boldsymbol{ab}) = (\boldsymbol{v} \cdot \boldsymbol{a})\boldsymbol{b} = (\boldsymbol{ba}) \cdot \boldsymbol{v}. \tag{9.36}$$

That is not equal to (9.33) for all $\boldsymbol{v}$, unless $\boldsymbol{a}$ and $\boldsymbol{b}$ are parallel, so the dyadic product does not commute

$$\boldsymbol{ab} \neq \boldsymbol{ba}.$$

The difference

$$\boldsymbol{ab} - \boldsymbol{ba}$$

is the *commutator*; it is itself a tensor whose action on any vector $\boldsymbol{v}$ is equal to the cross product of $\boldsymbol{v}$ with $\boldsymbol{a} \times \boldsymbol{b}$ since

$$\boldsymbol{v} \times (\boldsymbol{a} \times \boldsymbol{b}) = \boldsymbol{a}(\boldsymbol{b} \cdot \boldsymbol{v}) - \boldsymbol{b}(\boldsymbol{a} \cdot \boldsymbol{v}) = (\boldsymbol{ab} - \boldsymbol{ba}) \cdot \boldsymbol{v},$$

but note the position of $\boldsymbol{v}$ in the cross product and in the tensor expressions.

[5] Recall that in vector (or *dyadic*) notation there is no concept of rows and columns as in matrix notation in which (9.36) has to be written

$$\mathsf{v}^\mathsf{T}(\mathsf{ab}^\mathsf{T}) = (\mathsf{v}^\mathsf{T}\mathsf{a})\mathsf{b}^\mathsf{T} = ((\mathsf{ba}^\mathsf{T})\mathsf{v})^\mathsf{T}.$$

Linear transformations

A vector transformation $\boldsymbol{v} \to \boldsymbol{w} = T(\boldsymbol{v})$ is *linear* if

$$T(\lambda \boldsymbol{u} + \mu \boldsymbol{v}) = \lambda T(\boldsymbol{u}) + \mu T(\boldsymbol{v}) \tag{9.37}$$

for any scalars λ, μ and vectors $\boldsymbol{u}$, $\boldsymbol{v}$. If a transformation $T(\boldsymbol{v})$ is linear it can be expressed as a *tensor* $\boldsymbol{\mathcal{T}}$ dotted with $\boldsymbol{v}$,

$$T(\boldsymbol{v}) = \boldsymbol{\mathcal{T}} \cdot \boldsymbol{v}, \tag{9.38}$$

as in examples (9.26), (9.29), (9.30).

A linear transformation is fully specified by its action on a *basis*. Let $\{\boldsymbol{a}_1, \boldsymbol{a}_2, \boldsymbol{a}_3\}$ be an arbitrary basis and define

$$T(\boldsymbol{a}_j) \triangleq T_{ij}\, \boldsymbol{a}_i, \tag{9.39}$$

so column j of $[T_{ij}]$ contains the components of the transformed vector $\boldsymbol{b}_j = T(\boldsymbol{a}_j)$ in the chosen basis. Then, for any $\boldsymbol{v} = v_j \boldsymbol{a}_j$,

$$\boldsymbol{w} = T(\boldsymbol{v}) = T(v_j \boldsymbol{a}_j) = v_j T(\boldsymbol{a}_j) = T_{ij} v_j \boldsymbol{a}_i,$$

using linearity of $T(\cdot)$ and the summation convention. Thus, $\boldsymbol{w} = w_i \boldsymbol{a}_i$ with

$$w_i = T_{ij} v_j,$$

and the nine T_{ij}'s indeed fully specify the linear transformation $\boldsymbol{w} = T(\boldsymbol{v})$ for any $\boldsymbol{v}$ with respect to the basis $\{\boldsymbol{a}_1, \boldsymbol{a}_2, \boldsymbol{a}_3\}$.

Dyadic representations

The tensor $\boldsymbol{\mathcal{T}}$ such that $\boldsymbol{\mathcal{T}} \cdot \boldsymbol{v} = T(\boldsymbol{v})$ can be written explicitly in terms of T_{ij} and dyadic products between the basis vectors $\{\boldsymbol{a}_1, \boldsymbol{a}_2, \boldsymbol{a}_3\}$ and their *reciprocals* $\{\breve{\boldsymbol{a}}_1, \breve{\boldsymbol{a}}_2, \breve{\boldsymbol{a}}_3\}$ (6.17), as

$$\boldsymbol{\mathcal{T}} = T_{ij}\, \boldsymbol{a}_i \breve{\boldsymbol{a}}_j. \tag{9.40a}$$

Indeed, for any $\boldsymbol{v} = v_j \boldsymbol{a}_j$, using (9.37), (9.39), and $\boldsymbol{a}_i \cdot \breve{\boldsymbol{a}}_j = \delta_{ij}$, so $v_j = \breve{\boldsymbol{a}}_j \cdot \boldsymbol{v}$,

$$\boldsymbol{\mathcal{T}} \cdot \boldsymbol{v} = T_{ij} \boldsymbol{a}_i \breve{\boldsymbol{a}}_j \cdot \boldsymbol{v} = T_{ij} \boldsymbol{a}_i\, v_j = T(\boldsymbol{v}).$$

Equation (9.40a) also gives

$$T_{ij} = \breve{\boldsymbol{a}}_i \cdot \boldsymbol{\mathcal{T}} \cdot \boldsymbol{a}_j. \tag{9.40b}$$

An effective discussion of general bases requires introducing Ricci index notation (Ricci calculus) where upper indices are used to denote reciprocal vectors $\breve{\boldsymbol{a}}_i \equiv \boldsymbol{a}^i$ and (9.40a), (9.40b) are written as

$$\boldsymbol{\mathcal{T}} = T^i_j\, \boldsymbol{a}_i \boldsymbol{a}^j \;\Leftrightarrow\; T^i_j = \boldsymbol{a}^i \cdot \boldsymbol{\mathcal{T}} \cdot \boldsymbol{a}_j.$$

This will not be pursued here and we will restrict the following to cartesian bases such that $\mathbf{e}_i \cdot \mathbf{e}_j = \delta_{ij}$.

For a cartesian basis $\{\mathbf{e}_1, \mathbf{e}_2, \mathbf{e}_3\}$, that is equal to its reciprocal, there is no need to distinguish between upper and lower indices. In a cartesian basis,

$$\boldsymbol{\mathcal{T}} = T_{ij}\,\mathbf{e}_i\mathbf{e}_j \;\Leftrightarrow\; T_{ij} = \mathbf{e}_i \cdot \boldsymbol{\mathcal{T}} \cdot \mathbf{e}_j, \tag{9.41a}$$

and for any $\boldsymbol{v} = v_k\mathbf{e}_k$

$$\boldsymbol{\mathcal{T}} \cdot \boldsymbol{v} = T_{ij}\,\mathbf{e}_i\mathbf{e}_j \cdot v_k\mathbf{e}_k = \mathbf{e}_i T_{ij}\delta_{jk}v_k = \mathbf{e}_i T_{ij} v_j. \tag{9.41b}$$

Symmetric and antisymmetric tensors

As for single dyads, left multiplication of a tensor by a vector is not equal to right multiplication, in general,

$$\boldsymbol{v} \cdot \boldsymbol{\mathcal{T}} \neq \boldsymbol{\mathcal{T}} \cdot \boldsymbol{v},$$

but left multiplication defines the *transpose* $\boldsymbol{\mathcal{T}}^{\mathsf{T}}$ as the tensor such that, for any $\boldsymbol{v}$,

$$\boldsymbol{\mathcal{T}}^{\mathsf{T}} \cdot \boldsymbol{v} \triangleq \boldsymbol{v} \cdot \boldsymbol{\mathcal{T}}. \tag{9.42}$$

A tensor is *symmetric* if $\boldsymbol{\mathcal{T}} = \boldsymbol{\mathcal{T}}^{\mathsf{T}}$, that is, if

$$\boldsymbol{u} \cdot \boldsymbol{\mathcal{T}} \cdot \boldsymbol{v} = \boldsymbol{v} \cdot \boldsymbol{\mathcal{T}} \cdot \boldsymbol{u}, \tag{9.43a}$$

for any $\boldsymbol{u}$ and $\boldsymbol{v}$. In a cartesian basis, $\boldsymbol{\mathcal{T}} = T_{ij}\mathbf{e}_i\mathbf{e}_j$, this implies that

$$T_{ij} = T_{ji}. \tag{9.43b}$$

A tensor is *antisymmetric* or *skew-symmetric* if $\boldsymbol{\mathcal{T}} = -\boldsymbol{\mathcal{T}}^{\mathsf{T}}$, that is, if

$$\boldsymbol{v} \cdot \boldsymbol{\mathcal{T}} \cdot \boldsymbol{u} = -\boldsymbol{u} \cdot \boldsymbol{\mathcal{T}} \cdot \boldsymbol{v}, \tag{9.44a}$$

for any $\boldsymbol{u}$ and $\boldsymbol{v}$. In a cartesian basis, $\boldsymbol{\mathcal{T}} = T_{ij}\mathbf{e}_i\mathbf{e}_j$, this implies that

$$T_{ij} = -T_{ji}. \tag{9.44b}$$

An antisymmetric tensor $\boldsymbol{\mathcal{A}}$ is equivalent to a cross product with a (pseudo) vector $\boldsymbol{\omega}$:

$$\mathsf{A} = \begin{bmatrix} 0 & -\omega_3 & \omega_2 \\ \omega_3 & 0 & -\omega_1 \\ -\omega_2 & \omega_1 & 0 \end{bmatrix}$$

$$\boldsymbol{\mathcal{A}}^{\mathsf{T}} = -\boldsymbol{\mathcal{A}} \;\Leftrightarrow\; \boldsymbol{\mathcal{A}} \cdot \boldsymbol{v} = \boldsymbol{\omega} \times \boldsymbol{v}. \tag{9.45}$$

In a cartesian basis, $A_{ij} = \epsilon_{ikj}\omega_k$ so the cartesian matrix $\mathsf{A} = -\mathsf{A}^{\mathsf{T}}$ as well.

Change of cartesian basis

Equation (9.40a) shows how to transform tensor components from one basis to another. We restrict the discussion to cartesian bases. Given two distinct cartesian bases $\{\mathbf{e}_1, \mathbf{e}_2, \mathbf{e}_3\}$ and $\{\mathbf{e}'_1, \mathbf{e}'_2, \mathbf{e}'_3\}$, the starting point are the identities

$$\boldsymbol{v} = v_i\mathbf{e}_i = v'_i\mathbf{e}'_i, \tag{9.46}$$

(8.5a) for vectors, and

$$\boldsymbol{\mathcal{T}} = T_{ij}\,\mathbf{e}_i\mathbf{e}_j = T'_{ij}\,\mathbf{e}'_i\mathbf{e}'_j, \tag{9.47}$$

for tensors, where the bases are distinct but both cartesian so

$$\mathbf{e}_i \cdot \mathbf{e}_j = \delta_{ij} = \mathbf{e}'_i \cdot \mathbf{e}'_j.$$

The vector component transformations follow from dotting (9.46) with $\mathbf{e}'_i$ and $\mathbf{e}_i$ to obtain

$$\begin{cases} v'_i = (\mathbf{e}'_i \cdot \mathbf{e}_j)\, v_j = Q_{ij}\, v_j, \\ v_i = (\mathbf{e}_i \cdot \mathbf{e}'_j)\, v'_j = Q_{ji}\, v'_j, \end{cases} \tag{9.48}$$

where $\mathsf{Q} = [Q_{ij}]$ is an orthogonal matrix as discussed in §8.4. The tensor component transformations follow from dotting (9.47) with $\mathbf{e}'_i$ on the *left* and $\mathbf{e}'_j$ on the *right* to obtain

$$T'_{ij} = (\mathbf{e}'_i \cdot \mathbf{e}_k)\, T_{kl}\, (\mathbf{e}_l \cdot \mathbf{e}'_j) = Q_{ik}\, T_{kl}\, Q^{\mathsf{T}}_{lj}. \tag{9.49}$$

In matrix notation, these component transformations read

$$\begin{aligned} \mathsf{v}' &= \mathsf{Q}\mathsf{v}, \\ \mathsf{T}' &= \mathsf{Q}\mathsf{T}\mathsf{Q}^{\mathsf{T}}, \end{aligned} \tag{9.50}$$

where v and v' are column vectors and T', Q, T, $\mathsf{Q}^{\mathsf{T}} = \mathsf{Q}^{-1}$ are matrices. The matrix transformation $\mathsf{T}' = \mathsf{Q}\mathsf{T}\mathsf{Q}^{-1}$ is called a *similarity transformation* in matrix algebra. Similarity transformations do not change the eigenvalues of a matrix.

Notation, notation, notation

We write the operation of a tensor $\boldsymbol{\mathcal{T}}$ on a vector $\boldsymbol{v}$ in vector, index, and matrix notations as

$$\boldsymbol{\mathcal{T}} \cdot \boldsymbol{v} \equiv T_{ij} v_j \equiv \mathsf{T}\mathsf{v}$$

and

$$\boldsymbol{v} \cdot \boldsymbol{\mathcal{T}} \equiv T_{ji} v_j \equiv \mathsf{v}^{\mathsf{T}}\mathsf{T}.$$

Some authors drop the dot and write $\boldsymbol{\mathcal{T}}\boldsymbol{v}$ for the action of a tensor $\boldsymbol{\mathcal{T}}$ on a vector $\boldsymbol{v}$ but that is a confusing mix of dyadic and matrix notations. Indeed, if $\boldsymbol{\mathcal{T}}$ is a single dyad $\boldsymbol{\mathcal{T}} = \boldsymbol{ab}$, then

$$\boldsymbol{\mathcal{T}}\boldsymbol{c} = \boldsymbol{abc}$$

is a *triad* or a *third-order tensor* whose components are $a_i b_j c_k$, that is, a tensor whose dot product with a vector yields a *dyad* $(\boldsymbol{abc}) \cdot \boldsymbol{v} = (\boldsymbol{ab})(\boldsymbol{c} \cdot \boldsymbol{v})$. Thus, in dyadic notation $\boldsymbol{\mathcal{T}}\boldsymbol{c} \neq \boldsymbol{\mathcal{T}} \cdot \boldsymbol{c}$.

The vector and index notations do not have rows and columns. Thus, the commutative dot product is written

$$\begin{aligned} \boldsymbol{a} \cdot \boldsymbol{b} &= \boldsymbol{b} \cdot \boldsymbol{a}, \\ a_i b_i &= b_i a_i, \\ \mathsf{a}^{\mathsf{T}}\mathsf{b} &= \mathsf{b}^{\mathsf{T}}\mathsf{a}, \end{aligned}$$

in vector, index, and matrix notations, the latter two assuming that the basis is cartesian. If the basis is not cartesian then

$$\boldsymbol{a} \cdot \boldsymbol{b} = a_i g_{ij} b_j = \mathsf{a}^\mathsf{T} \mathsf{G} \mathsf{b},$$

where $\mathsf{G} = [g_{ij}] = \mathsf{G}^\mathsf{T}$ is the symmetric matrix of dot products between the basis vectors (the *metric* of the basis), $g_{ij} = \boldsymbol{a}_i \cdot \boldsymbol{a}_j$. In matrix notation, the fundamental data structure is the matrix and the fundamental operations are the row-by-column matrix multiplications and the transpose operation that switches rows and columns. Vectors are N-by-1 column matrices by default.

Students will also encounter the *double dot product* that was originally defined by Gibbs/Wilson[6] as

$$(\boldsymbol{ab}) : (\boldsymbol{cd}) = (\boldsymbol{a} \cdot \boldsymbol{c})(\boldsymbol{b} \cdot \boldsymbol{d}) \tag{9.51a}$$

for dyads. For second-order tensors $\boldsymbol{\mathcal{A}} = A_{ij} \mathbf{e}_i \mathbf{e}_j$ and $\boldsymbol{\mathcal{B}} = B_{kl} \mathbf{e}_k \mathbf{e}_l$ in a cartesian basis, that is,

$$\boldsymbol{\mathcal{A}} : \boldsymbol{\mathcal{B}} = A_{ij} B_{kl} (\mathbf{e}_i \cdot \mathbf{e}_k)(\mathbf{e}_j \cdot \mathbf{e}_l) = A_{ij} B_{ij} = \mathrm{tr}(\mathsf{A}\mathsf{B}^\mathsf{T}),$$

where $\mathrm{tr}(\mathsf{A}) = A_{ii}$ is the *trace* of the matrix, the sum of its diagonal elements. Some authors prefer to define the double dot product as

$$(\boldsymbol{ab}) : (\boldsymbol{cd}) = (\boldsymbol{a} \cdot \boldsymbol{d})(\boldsymbol{b} \cdot \boldsymbol{c}), \tag{9.51b}$$

yielding

$$\boldsymbol{\mathcal{A}} : \boldsymbol{\mathcal{B}} = A_{ij} B_{kl} (\mathbf{e}_i \cdot \mathbf{e}_l)(\mathbf{e}_j \cdot \mathbf{e}_k) = A_{ij} B_{ji} = \mathrm{tr}(\mathsf{A}\mathsf{B}).$$

[6] E. B. Wilson, *Vector Analysis* §117, Yale University Press, 1901.

9.7 Eigenvectors

Transformations $\boldsymbol{v} \to \boldsymbol{w} = T(\boldsymbol{v})$ naturally lead to the question of whether there exist vectors $\boldsymbol{v}$ that are invariant under the transformation. Strict invariance $T(\boldsymbol{v}) = \boldsymbol{v}$ is too restrictive to be generally useful, but invariance up to a scaling factor λ is an important concept with general implications. Thus, given a linear transformation $\boldsymbol{v} \to \boldsymbol{w} = T(\boldsymbol{v})$, we search for special vectors $\boldsymbol{v}$ such that

$$T(\boldsymbol{v}) = \lambda \boldsymbol{v},$$

for some scalar λ. Such vectors, if they exist, are called *eigenvectors* of the transformation and the scaling factors λ are the corresponding *eigenvalues.*

Geometric examples

Eigenvectors and eigenvalues for the parallel and perpendicular projections in (9.26) can be found by inspection. Clearly

$$(\hat{\boldsymbol{a}}\hat{\boldsymbol{a}}) \cdot \boldsymbol{v} = \lambda \boldsymbol{v}$$

for $\boldsymbol{v} = \hat{\boldsymbol{a}}$ with $\lambda = 1$, and for any $\boldsymbol{v}\cdot\hat{\boldsymbol{a}} = 0$ with $\lambda = 0$. The set of vectors perpendicular to $\hat{\boldsymbol{a}}$ is an *eigen-subspace* for that tensor $\boldsymbol{\mathcal{T}} = \hat{\boldsymbol{a}}\hat{\boldsymbol{a}}$. We can pick any two orthonormal vectors $\boldsymbol{q}_1$ and $\boldsymbol{q}_2$ in that plane and thus define an orthonormal *eigenbasis* $\{\boldsymbol{q}_1, \boldsymbol{q}_2, \hat{\boldsymbol{a}}\}$, that is, a basis consisting of orthonormal eigenvectors of the linear transformation. The $(\lambda, \boldsymbol{v})$ eigenvalue–eigenvector pairs are thus

$$(\lambda, \boldsymbol{v}) = (\lambda_1, \boldsymbol{q}_1),\ (\lambda_2, \boldsymbol{q}_2),\ (\lambda_3, \hat{\boldsymbol{a}})$$

with

$$(\lambda_1, \lambda_2, \lambda_3) = \begin{cases} (0,0,1) & \text{for } \boldsymbol{\mathcal{T}} = \hat{\boldsymbol{a}}\hat{\boldsymbol{a}}, \\ (1,1,0) & \text{for } \boldsymbol{\mathcal{T}} = \boldsymbol{I} - \hat{\boldsymbol{a}}\hat{\boldsymbol{a}}, \\ (1,1,-1) & \text{for } \boldsymbol{\mathcal{T}} = \boldsymbol{I} - 2\hat{\boldsymbol{a}}\hat{\boldsymbol{a}}. \end{cases} \tag{9.52}$$

For the rotation tensor (9.30), it is geometrically and algebraically evident that

$$\boldsymbol{\mathcal{R}} \cdot \hat{\boldsymbol{a}} = \hat{\boldsymbol{a}}.$$

Thus, $\lambda = 1$ for $\boldsymbol{v} = \hat{\boldsymbol{a}}$ since vectors parallel to $\hat{\boldsymbol{a}}$ are unaffected by rotation about $\hat{\boldsymbol{a}}$. Vectors *perpendicular* to $\hat{\boldsymbol{a}}$ remain perpendicular to $\hat{\boldsymbol{a}}$ after rotation, but no such vector can be parallel to itself after rotation unless the rotation angle $\alpha = 0$ or π.

Let $\boldsymbol{q}_1$ be any vector perpendicular to $\hat{\boldsymbol{a}}$ and $\boldsymbol{q}_2 = \hat{\boldsymbol{a}}\times\boldsymbol{q}_1$, so $\{\boldsymbol{q}_1, \boldsymbol{q}_2, \hat{\boldsymbol{a}}\}$ form a right-handed orthogonal basis, and consider rotation by $\alpha = \pi/2$; in that case,

$$\begin{cases} \boldsymbol{\mathcal{R}}_{\frac{\pi}{2}} \cdot \boldsymbol{q}_1 = \boldsymbol{q}_2, \\ \boldsymbol{\mathcal{R}}_{\frac{\pi}{2}} \cdot \boldsymbol{q}_2 = -\boldsymbol{q}_1. \end{cases}$$

Eigenvectors for this $\pi/2$ rotation operator, $\boldsymbol{\mathcal{R}}_{\frac{\pi}{2}}$, require a complex leap:

$$\boldsymbol{\mathcal{R}}_{\frac{\pi}{2}} \cdot (\boldsymbol{q}_1 + \imath\boldsymbol{q}_2) = -\imath(\boldsymbol{q}_1 + \imath\boldsymbol{q}_2),$$

where $\imath\imath = \imath^2 = -1$. Then, for rotation at arbitrary α,

We write $\imath$ for the imaginary unit here, such that $\imath^2 = -1$, since we have been using indices i and j in this section.

$$\boldsymbol{\mathcal{R}} \cdot (\boldsymbol{q}_1 \pm \imath\boldsymbol{q}_2) = e^{\mp\imath\alpha}(\boldsymbol{q}_1 \pm \imath\boldsymbol{q}_2),$$

where $e^{\pm\imath\alpha} = \cos\alpha \pm \imath\sin\alpha$. Thus, the $(\lambda, \boldsymbol{v})$ eigenvalue–eigenvector pairs for rotation by α about $\hat{\boldsymbol{a}}$ are

$$(\lambda, \boldsymbol{v}) = (e^{-\imath\alpha}, \boldsymbol{q}_1 + \imath\boldsymbol{q}_2), \quad (e^{\imath\alpha}, \boldsymbol{q}_1 - \imath\boldsymbol{q}_2), \quad (1, \hat{\boldsymbol{a}}) \tag{9.53}$$

where $\boldsymbol{q}_1$ is any real vector orthogonal to $\hat{\boldsymbol{a}}$ and $\boldsymbol{q}_2 = \hat{\boldsymbol{a}} \times \boldsymbol{q}_1$. The eigenvector corresponding to $\lambda = 1$ is the rotation axis $\hat{\boldsymbol{a}}$, and the other eigenvalues $\lambda = e^{\pm\imath\alpha}$ yield the rotation angle α.

Eigenvalue problem

The eigenvalue problem for an arbitrary linear transformation $\boldsymbol{v} \to \boldsymbol{\mathcal{A}}\cdot\boldsymbol{v}$ is to find λ and $\boldsymbol{v} \neq 0$ such that

$$\boldsymbol{\mathcal{A}} \cdot \boldsymbol{v} = \lambda\boldsymbol{v}.$$

If the eigenvectors form a complete basis, then they and their eigenvalues fully characterize $\boldsymbol{\mathcal{A}}$. The eigenvalue problem is nonlinear in the eigenvalue λ but linear in the eigenvector $\boldsymbol{v}$ that is thus determined up to an arbitrary scaling factor.

Let $\mathbf{A}$ be the matrix of $\boldsymbol{\mathcal{A}}$ in some basis, and $\mathbf{v}$ the components of $\boldsymbol{v}$ in that same basis. The search for eigenvectors then become the *matrix eigenvalue problem*

$$\mathbf{A}\mathbf{v} = \lambda\mathbf{v}, \tag{9.54}$$

which is a system of 3-by-3 (or N-by-N) equations for three (or N) unknowns λ and $\mathbf{v}/|\mathbf{v}|$, since the amplitude of $\mathbf{v}$ drops out. This is not quite a linear system since λ and $\mathbf{v}$ depend on each other. However, $\mathbf{v}$ can be eliminated,

$$(\mathbf{A} - \lambda\mathbf{I})\mathbf{v} = 0 \Leftrightarrow \det(\mathbf{A} - \lambda\mathbf{I}) = 0. \tag{9.55}$$

Indeed since $\mathbf{v} \neq 0$, the eigenvalue λ must be such that the columns (and rows) of the matrix $\mathbf{A} - \lambda\mathbf{I}$ are linearly dependent, so the determinant must vanish.

For a 3-by-3 matrix $\mathbf{A}$, using lower case a_{ij} for the coefficients of $\mathbf{A}$,

$$\begin{aligned}\det(\lambda\mathbf{I} - \mathbf{A}) &= \epsilon_{ijk}(\lambda\delta_{i1} - a_{i1})(\lambda\delta_{j2} - a_{j2})(\lambda\delta_{k3} - a_{k3}) \\ &= \lambda^3 - \lambda^2(\epsilon_{i23}a_{i1} + \epsilon_{1j3}a_{j2} + \epsilon_{12k}a_{k3}) \\ &\quad + \lambda(\epsilon_{ij3}a_{i1}a_{j2} + \epsilon_{i2k}a_{i1}a_{k3} + \epsilon_{1jk}a_{j2}a_{k3}) - \det\mathbf{A} \\ &= \lambda^3 - \lambda^2\operatorname{tr}\mathbf{A} + \lambda\frac{1}{2}\left((\operatorname{tr}\mathbf{A})^2 - \operatorname{tr}\mathbf{A}^2\right) - \det\mathbf{A},\end{aligned} \tag{9.56}$$

where $\operatorname{tr}\mathbf{A} = a_{ii}$ and $\operatorname{tr}\mathbf{A}^2 = a_{ik}a_{ki}$ (see Chapter 5, exercise 5.17). Those traces and $\det\mathbf{A}$ are invariants under similarity transformations $\mathbf{A} \to \mathbf{S}\mathbf{A}\mathbf{S}^{-1}$. The coefficient of λ in (9.56) can also be recognized as the sum of the *cofactors* of the diagonal elements.

If $\mathbf{A}$ is N-by-N, $\det(\mathbf{A} - \lambda\mathbf{I})$ is a polynomial of degree N in λ called the *characteristic polynomial* and $\det(\mathbf{A} - \lambda\mathbf{I}) = 0$ is the *characteristic equation*. The fundamental theorem of algebra states that a polynomial of degree N has N roots; however, some can be repeated and some can be complex. If the matrix is real, then the complex roots come in complex conjugate pairs; see (9.52) and (9.53) for example.

In theory—but in practice only for 2-by-2 matrices or matrices with special structure—the eigenvector problem reduces to first finding the eigenvalues λ, then going back to the known $\mathbf{A} - \lambda\mathbf{I}$ matrix to find $\mathbf{v}$, for instance by row reduction. For more general matrices, the eigenvectors and eigenvalues are obtained simultaneously through an iterative algorithm based on repeated $\mathbf{QR}$ decompositions and similarity transformations $\mathbf{Q}^{\mathsf{T}}(\mathbf{QR})\mathbf{Q} = \mathbf{RQ} = \mathbf{A}'$, with some shifty accelerations.[7]

[7] See for example, L. N. Trefethen and D. Bau, *Numerical Linear Algebra*, SIAM, 1997.

Diagonalization of symmetric matrices

Symmetric tensors and matrices feature prominently in many applications. The parallel and perpendicular projections tensors (9.26), the reflection tensor (9.29), and the tensor of inertia (9.32) are symmetric, for instance. A fundamental property of those *symmetric linear operators* is that they have *real eigenvalues* λ and a *complete set of orthogonal eigenvectors*.

First, we will show that λ is real, then prove orthogonality and completeness of the eigenvectors. The concept of *complex conjugates* is needed. The *complex conjugate* of a scalar $\lambda = x + \imath y$, where x and y are real and $\imath^2 = -1$, is denoted with an asterisk λ^* and defined as

$$\lambda^* = x - \imath y$$

such that

$$\lambda \lambda^* = x^2 + y^2 \geq 0.$$

Likewise, a complex vector $\mathbf{v} \in \mathbb{C}^N$ is $\mathbf{v} = \mathbf{x} + \imath \mathbf{y}$ where $\mathbf{x}$, $\mathbf{y} \in \mathbb{R}^N$ are real, and its complex conjugate is

$$\mathbf{v}^* = \mathbf{x} - \imath \mathbf{y},$$

such that

$$\mathbf{v}^* \cdot \mathbf{v} = \mathbf{x} \cdot \mathbf{x} + \mathbf{y} \cdot \mathbf{y} \geq 0.$$

In matrix notation, this complex conjugate dot product is written

$$\mathbf{v}^{\mathsf{H}} \mathbf{v} = \mathbf{v}^* \cdot \mathbf{v} \geq 0,$$

where $\mathbf{v}^{\mathsf{H}}$ is the *transpose conjugate* of $\mathbf{v}$; thus, $(\mathbf{x} + \imath \mathbf{y})^{\mathsf{H}} = \mathbf{x}^{\mathsf{T}} - \imath \mathbf{y}^{\mathsf{T}}$.

We can now prove that λ is real for a *real symmetric* matrix $\mathbf{A} = \mathbf{A}^{\mathsf{T}}$, and more generally for a *complex Hermitian* matrix $\mathbf{A} = \mathbf{A}^{\mathsf{H}}$. Indeed, if $\mathbf{A}\mathbf{v} = \lambda \mathbf{v}$ with $\mathbf{v} \neq 0$, then $\mathbf{v}^{\mathsf{H}} \mathbf{A} \mathbf{v} = \lambda \mathbf{v}^{\mathsf{H}} \mathbf{v}$ and thus

$$\lambda = \frac{\mathbf{v}^{\mathsf{H}} \mathbf{A} \mathbf{v}}{\mathbf{v}^{\mathsf{H}} \mathbf{v}}. \tag{9.57}$$

Now, if $\mathbf{A}^{\mathsf{H}} = \mathbf{A}$,

$$(\mathbf{v}^{\mathsf{H}} \mathbf{A} \mathbf{v})^{\mathsf{H}} = \mathbf{v}^{\mathsf{H}} \mathbf{A} \mathbf{v}$$

since $(\mathbf{u}^{\mathsf{H}} \mathbf{A} \mathbf{v})^{\mathsf{H}} = \mathbf{v}^{\mathsf{H}} \mathbf{A}^{\mathsf{H}} \mathbf{u}$ for any $\mathbf{u}$ and $\mathbf{v}$, and we assume that $\mathbf{A}^{\mathsf{H}} = \mathbf{A}$. Hence, $\mathbf{v}^{\mathsf{H}} \mathbf{A} \mathbf{v}$ is real since it is a number equal to its conjugate. It follows from (9.57) that λ is real since $\mathbf{v}^{\mathsf{H}} \mathbf{v}$ is also real. That was a quick proof!

Showing that a symmetric matrix always has a *complete* set of *orthogonal* eigenvectors, even if some eigenvalues are repeated, is a bit more involved, but the proof is instructive. The idea is to start from one eigenvector and transform the matrix into an orthonormal basis that includes the known eigenvector. This effectively projects the N-by-N matrix onto an orthogonal $(N-1)$-by-$(N-1)$ subspace, and the projected matrix is still symmetric, so the argument can be repeated on a matrix of size $(N-1)$-by-$(N-1)$.

We start from the fact that any matrix $\mathbf{A}$ has at least one λ_1, since the fundamental theorem of algebra[8] guarantees that a (characteristic)

[8] §18.4.

polynomial such as (9.56) has at least one root λ_1, that has at least one corresponding eigenvector $\mathbf{v}_1 \neq 0$ since $\det(\mathbf{A} - \lambda_1 \mathbf{I}) = 0$,

$$\mathbf{A}\mathbf{v}_1 = \lambda_1 \mathbf{v}_1.$$

The root λ_1 could be repeated as for the identity matrix $\mathbf{I}$ for which all eigenvalues $\lambda = 1$ since $\mathbf{I}\mathbf{v} = \mathbf{v}$ for any $\mathbf{v}$. We assume that $\mathbf{A}$ is real and symmetric in the following, so λ_1 and $\mathbf{v}_1$ are real, but the proof carries through for Hermitian matrices by replacing the transpose $()^{\mathsf{T}}$ by transpose-conjugate $()^{\mathsf{H}}$. Since $\mathbf{v}_1 \neq 0$, it can be normalized so $\mathbf{v}_1^{\mathsf{T}} \mathbf{v}_1 = 1$.

Let $\mathbf{q}_1 = \mathbf{v}_1$ and pick an orthonormal basis $\mathbf{Q} = \begin{bmatrix} \mathbf{q}_1 & \mathbf{q}_2 & \mathbf{q}_3 \end{bmatrix}$. The eigenvalue problem can be orthogonally transformed

$$\mathbf{A}\mathbf{v} = \lambda \mathbf{v} \rightarrow (\mathbf{Q}^{\mathsf{T}}\mathbf{A}\mathbf{Q})(\mathbf{Q}^{\mathsf{T}}\mathbf{v}) = \lambda(\mathbf{Q}^{\mathsf{T}}\mathbf{v}) \equiv \mathbf{A}'\mathbf{v}' = \lambda \mathbf{v}'. \tag{9.58}$$

The transformed matrix $\mathbf{A}' = \mathbf{Q}^{\mathsf{T}}\mathbf{A}\mathbf{Q}$ is still real and symmetric since

$$(\mathbf{A}')^{\mathsf{T}} = (\mathbf{Q}^{\mathsf{T}}\mathbf{A}\mathbf{Q})^{\mathsf{T}} = \mathbf{Q}^{\mathsf{T}}\mathbf{A}^{\mathsf{T}}\mathbf{Q} = \mathbf{Q}^{\mathsf{T}}\mathbf{A}\mathbf{Q},$$

and it has a special structure. The elements a'_{i1} of its first column are

$$a'_{i1} = \mathbf{q}_i^{\mathsf{T}}\mathbf{A}\mathbf{q}_1 = \lambda_1 \mathbf{q}_i^{\mathsf{T}}\mathbf{q}_1 = \lambda_1 \delta_{i1}$$

since $\mathbf{q}_1 = \mathbf{v}_1$ is a λ_1 eigenvector and the $\mathbf{q}_i$ are orthogonal. The first row has the same structure, by symmetry; thus,

$$\mathbf{A}' = \mathbf{Q}^{\mathsf{T}}\mathbf{A}\mathbf{Q} = \begin{bmatrix} \lambda_1 & 0 & 0 \\ 0 & a'_{22} & a'_{23} \\ 0 & a'_{32} & a'_{33} \end{bmatrix}, \quad \mathbf{v}'_1 = \mathbf{Q}^{\mathsf{T}}\mathbf{v}_1 = \begin{bmatrix} 1 \\ 0 \\ 0 \end{bmatrix}, \tag{9.59}$$

where

$$a'_{ij} = \mathbf{q}_i^{\mathsf{T}}\mathbf{A}\mathbf{q}_j = \mathbf{q}_j^{\mathsf{T}}\mathbf{A}\mathbf{q}_i = a'_{ji}.$$

So the 2-by-2 submatrix of $\mathbf{A}'$ is symmetric, $a'_{23} = a'_{32}$, and therefore has at least one real eigenvalue λ_2 and eigenvector

$$\begin{bmatrix} a'_{22} & a'_{23} \\ a'_{32} & a'_{33} \end{bmatrix} \begin{bmatrix} v'_{22} \\ v'_{32} \end{bmatrix} = \lambda_2 \begin{bmatrix} v'_{22} \\ v'_{32} \end{bmatrix},$$

yielding

$$\mathbf{A}'\mathbf{v}'_2 = \lambda_2 \mathbf{v}'_2, \quad \mathbf{v}'_2 = \begin{bmatrix} 0 \\ v'_{22} \\ v'_{32} \end{bmatrix}.$$

The eigenvalue λ_2 may equal λ_1 but $\mathbf{v}'_2$ is obviously orthogonal to $\mathbf{v}'_1 = (1, 0, 0)$.

The procedure can be continued. Normalize $\mathbf{v}'_2$, let $\mathbf{q}'_1 = \mathbf{q}_1$, $\mathbf{q}'_2 = \mathbf{v}'_2$ and pick a $\mathbf{q}'_3$ orthogonal to $\mathbf{q}'_1$ and $\mathbf{q}'_2$. Transform the eigenvalue problem again, now with $\mathbf{Q}' = \begin{bmatrix} \mathbf{q}'_1 & \mathbf{q}'_2 & \mathbf{q}'_3 \end{bmatrix}$,

$$\mathbf{A}'\mathbf{v}' = \lambda \mathbf{v}' \rightarrow \mathbf{A}''\mathbf{v}'' = \lambda \mathbf{v}'' \tag{9.60}$$

where $\mathsf{A}'' = \mathsf{Q}'^{\mathsf{T}}\mathsf{A}'\mathsf{Q}'$, $\mathsf{v}'' = \mathsf{Q}'^{\mathsf{T}}\mathsf{v}'$, and

$$\mathsf{A}'' = \begin{bmatrix} \lambda_1 & 0 & 0 \\ 0 & \lambda_2 & 0 \\ 0 & 0 & a''_{33} \end{bmatrix}, \quad \begin{bmatrix} \mathsf{v}''_1 & \mathsf{v}''_2 \end{bmatrix} = \mathsf{Q}'^{\mathsf{T}} \begin{bmatrix} \mathsf{v}'_1 & \mathsf{v}'_2 \end{bmatrix} = \begin{bmatrix} 1 & 0 \\ 0 & 1 \\ 0 & 0 \end{bmatrix}. \quad (9.61)$$

In this 3-by-3 case, the procedure terminates here; $\lambda_3 = a''_{33}$ is the last eigenvalue with eigenvector $\mathsf{v}''_3 = (0, 0, 1)$ that is clearly orthogonal to v''_1 and v''_2. In that basis, $\mathsf{V}'' = \begin{bmatrix} \mathsf{v}''_1 & \mathsf{v}''_2 & \mathsf{v}''_3 \end{bmatrix} = \mathbf{I}$. The eigenvectors of the original matrix A are $\mathsf{v}_i = \mathsf{Q}\mathsf{Q}'\mathsf{v}''_i$; they are the columns of the matrix

$$\mathsf{V} = \begin{bmatrix} \mathsf{v}_1 & \mathsf{v}_2 & \mathsf{v}_3 \end{bmatrix} = \mathsf{Q}\mathsf{Q}'\mathsf{V}'' = \mathsf{Q}\mathsf{Q}'$$

corresponding to eigenvalues λ_1, λ_2, λ_3, respectively, and

$$\begin{aligned} \mathsf{A}\mathsf{V} &= \begin{bmatrix} \mathsf{A}\mathsf{v}_1 & \mathsf{A}\mathsf{v}_2 & \mathsf{A}\mathsf{v}_3 \end{bmatrix} \\ &= \begin{bmatrix} \lambda_1\mathsf{v}_1 & \lambda_2\mathsf{v}_2 & \lambda_3\mathsf{v}_3 \end{bmatrix} = \mathsf{V}\mathsf{\Lambda}, \end{aligned} \quad (9.62)$$

where

$$\mathsf{\Lambda} = \begin{bmatrix} \lambda_1 & 0 & 0 \\ 0 & \lambda_2 & 0 \\ 0 & 0 & \lambda_3 \end{bmatrix}$$

is a diagonal matrix and the matrix of eigenvectors V is orthogonal

$$\mathsf{V}^{\mathsf{T}}\mathsf{V} = \mathbf{I} = \mathsf{V}\mathsf{V}^{\mathsf{T}}.$$

Equation (9.62) can be written

$$\mathsf{V}^{\mathsf{T}}\mathsf{A}\mathsf{V} = \mathsf{\Lambda} \Leftrightarrow \mathsf{A} = \mathsf{V}\mathsf{\Lambda}\mathsf{V}^{\mathsf{T}}.$$

Thus, *a real symmetric matrix can be diagonalized by an orthogonal matrix.* The decomposition $\mathsf{A} = \mathsf{V}\mathsf{\Lambda}\mathsf{V}^{\mathsf{T}}$ is the eigenvalue decomposition (EVD) or *spectral decomposition* of A.

If symmetric matrix A corresponds to tensor $\boldsymbol{\mathcal{A}} = A_{ij}\mathbf{e}_i\mathbf{e}_j$ in an orthonormal basis $\mathbf{e}_1, \mathbf{e}_2, \mathbf{e}_3$, then its orthonormal eigenvectors

$$\hat{\boldsymbol{v}}_i = V_{ji}\mathbf{e}_j, \qquad \mathbf{e}_i = V_{ij}\hat{\boldsymbol{v}}_j \quad (9.63)$$

since the eigenvectors are the *columns* of V. Then its *rows* are the components of $\mathbf{e}_i$ in the eigenvector basis.[9] The symmetric tensor is

$$\boldsymbol{\mathcal{A}} = A_{ij}\mathbf{e}_i\mathbf{e}_j = \lambda_1\hat{\boldsymbol{v}}_1\hat{\boldsymbol{v}}_1 + \lambda_2\hat{\boldsymbol{v}}_2\hat{\boldsymbol{v}}_2 + \lambda_3\hat{\boldsymbol{v}}_3\hat{\boldsymbol{v}}_3, \quad (9.64)$$

where the eigenvector expansion cannot be written using the summation convention since the index appears three times in each term.

An important application of this result is that it is always possible to find a cartesian basis in which the symmetric *tensor of inertia* (9.32) is diagonal. The eigendirections corresponding to that basis are called *principal axes* and the eigenvalues the *principal moments of inertia* in that context. This is explored in this chapter's exercises. The principal axes are those about which the corresponding rigid body can rotate freely without wobbling. That is discussed later in §11.4. Another application is to the classification of quadratics.

[9] This is opposite to the row–column convention in (8.6) where

$$\mathbf{e}'_i = Q_{ij}\mathbf{e}_j \Leftrightarrow \mathbf{e}_i = Q_{ji}\mathbf{e}'_j.$$

Classification of quadratic curves

A quadratic curve is the set of solutions (x, y) of a two-variable quadratic polynomial

$$ax^2 + bxy + cy^2 + dx + ey + f = 0, \tag{9.65}$$

where a, b, c, d, e, and f are real constants. The sum of the quadratic terms, called a *quadratic form*, can be written in matrix vector notation as

$$ax^2 + bxy + cy^2 = \begin{bmatrix} x & y \end{bmatrix} \begin{bmatrix} a & \frac{1}{2}b \\ \frac{1}{2}b & c \end{bmatrix} \begin{bmatrix} x \\ y \end{bmatrix}, \tag{9.66}$$

and invariably involves a real symmetric matrix.

Let $\lambda_1 \leq \lambda_2$ be the eigenvalues of that symmetric matrix, and $\mathbf{q}_1$ and $\mathbf{q}_2$ be the corresponding orthonormal eigenvectors. In that basis, the (x, y) are orthogonally transformed to

$$\begin{bmatrix} x \\ y \end{bmatrix} = \begin{bmatrix} \mathbf{q}_1 & \mathbf{q}_2 \end{bmatrix} \begin{bmatrix} x_1 \\ x_2 \end{bmatrix}$$

and the quadratic form becomes

$$ax^2 + bxy + cy^2 = \lambda_1 x_1^2 + \lambda_2 x_2^2. \tag{9.67}$$

The linear and constant terms are modified according to the transformation of coordinates, $dx + ey = \gamma_1 x_1 + \gamma_2 x_2$ for some new constants γ_1, γ_2, and the quadratic equation reduces to

$$\lambda_1 x_1^2 + \lambda_2 x_2^2 + \gamma_1 x_1 + \gamma_2 x_2 + f = 0, \tag{9.68}$$

in the new cartesian coordinates (x_1, x_2).

The quadratic is therefore an *ellipse* if λ_1 and λ_2 have the same sign, a *hyperbola* when the eigenvalues have opposite sign, and a *parabola* when one of the eigenvalues vanishes. Since

$$\lambda_1 \lambda_2 = \begin{vmatrix} a & \frac{1}{2}b \\ \frac{1}{2}b & c \end{vmatrix} = ac - b^2/4,$$

the quadratic is an *ellipse* when $b^2 - 4ac < 0$, a *hyperbola* when $b^2 - 4ac > 0$, and a *parabola* when $b^2 - 4ac = 0$.

Epilogue

The proof developed for real 3-by-3 symmetric matrices generalizes to N-by-N matrices and to complex *Hermitian matrices* for which $\mathbf{A}^\mathsf{H} = \mathbf{A}$. In that case the eigenvalues λ are still real, but the eigenvectors $\mathbf{V}$ are complex orthogonal, that is, $\mathbf{V}^\mathsf{H}\mathbf{V} = \mathbf{I} = \mathbf{V}\mathbf{V}^\mathsf{H}$. All the steps of the proof hold with $\mathbf{Q}^\mathsf{T}$ replaced by $\mathbf{Q}^\mathsf{H}$. The proof can also be adapted to *anti-Hermitian* matrices with $\mathbf{A}^\mathsf{H} = -\mathbf{A}$, in which case the eigenvalues are pure imaginary (plus a zero eigenvalue for odd N) and the eigenvectors are complex orthogonal. Rotation matrices (9.53) are neither symmetric nor antisymmetric in general, but have complex orthogonal eigenvectors.

For arbitrary matrices, there is no guarantee that there will be a complete set of eigenvectors, a fundamental example is $\mathbf{A} = \left[\begin{smallmatrix} a & 1 \\ 0 & a \end{smallmatrix}\right]$, known as a *Jordan block*, that has only one eigenvector, although a *random* matrix will *almost certainly* have a complete set of eigenvectors. An arbitrary real matrix will have complex eigenvalues and eigenvectors in general. A useful generalization of the eigenvector decomposition of a matrix is the *singular value decomposition* (SVD). For an arbitrary 3-by-3 tensor $\boldsymbol{\mathcal{A}}$, the SVD reads

$$\boldsymbol{\mathcal{A}} = \sigma_1 \hat{\boldsymbol{u}}_1 \hat{\boldsymbol{v}}_1 + \sigma_2 \hat{\boldsymbol{u}}_2 \hat{\boldsymbol{v}}_2 + \sigma_3 \hat{\boldsymbol{u}}_3 \hat{\boldsymbol{v}}_3, \tag{9.69}$$

and

$$\boldsymbol{\mathcal{A}}^{\mathsf{T}} = \sigma_1 \hat{\boldsymbol{v}}_1 \hat{\boldsymbol{u}}_1 + \sigma_2 \hat{\boldsymbol{v}}_2 \hat{\boldsymbol{u}}_2 + \sigma_3 \hat{\boldsymbol{v}}_3 \hat{\boldsymbol{u}}_3,$$

where $\hat{\boldsymbol{u}}_i \cdot \hat{\boldsymbol{u}}_j = \delta_{ij} = \hat{\boldsymbol{v}}_i \cdot \hat{\boldsymbol{v}}_j$. Thus, $\boldsymbol{\mathcal{A}} \cdot \hat{\boldsymbol{v}}_1 = \sigma_1 \hat{\boldsymbol{u}}_1$, and

$$\boldsymbol{\mathcal{A}}^{\mathsf{T}} \cdot \boldsymbol{\mathcal{A}} \cdot \hat{\boldsymbol{v}}_1 = \sigma_1 \boldsymbol{\mathcal{A}}^{\mathsf{T}} \cdot \hat{\boldsymbol{u}}_1 = \sigma_1^2 \hat{\boldsymbol{v}}_1.$$

Likewise

$$\boldsymbol{\mathcal{A}} \cdot \boldsymbol{\mathcal{A}}^{\mathsf{T}} \cdot \hat{\boldsymbol{u}}_1 = \sigma_1 \boldsymbol{\mathcal{A}} \cdot \hat{\boldsymbol{v}}_1 = \sigma_1^2 \hat{\boldsymbol{u}}_1.$$

The *singular values* $\sigma_i \geq 0$ are real and positive, they are the local extrema of $|\boldsymbol{\mathcal{A}} \cdot \hat{\boldsymbol{v}}|$, and thus the square roots of the eigenvalues of the symmetric tensor $(\boldsymbol{\mathcal{A}}^{\mathsf{T}} \cdot \boldsymbol{\mathcal{A}})$. The right eigenvectors of $\boldsymbol{\mathcal{A}}^{\mathsf{T}} \cdot \boldsymbol{\mathcal{A}}$ are the right singular vectors $\hat{\boldsymbol{v}}_i$, and the left eigenvectors are the left singular vectors $\hat{\boldsymbol{u}}_i$.

Exercises

(9.1) Give examples of 2-by-2 and 3-by-3 symmetric and antisymmetric matrices.

(9.2) If $\mathbf{x}^{\mathsf{T}} = [x_1, x_2, x_3]$, calculate $\mathbf{x}^{\mathsf{T}}\mathbf{x}$ and $\mathbf{x}\mathbf{x}^{\mathsf{T}}$.

(9.3) For $\mathbf{x} \in \mathbb{R}^n$, show that $\mathbf{x}^{\mathsf{T}}\mathbf{x}$ and $\mathbf{x}\mathbf{x}^{\mathsf{T}}$ are symmetric, (i) explicitly using indices, (ii) by matrix manipulations.

(9.4) Prove (9.6) that $(\mathbf{AB})^{\mathsf{T}} = \mathbf{B}^{\mathsf{T}}\mathbf{A}^{\mathsf{T}}$ using index notation.

(9.5) Use (9.6) to prove that the product of two orthogonal matrices is an orthogonal matrix.

(9.6) Prove that $(\mathbf{AB})^{-1} = \mathbf{B}^{-1}\mathbf{A}^{-1}$ using matrix notation (this is one easy line).

(9.7) Prove that the inverse of a matrix, if it exists, is unique by showing that if $\mathbf{B}$ and $\mathbf{C}$ are two matrices such that $\mathbf{AB} = \mathbf{I} = \mathbf{AC}$, then $\mathbf{B} = \mathbf{C}$ (this is one easy line).

(9.8) Show that $\det \mathbf{I} = 1$ and

$$\det \mathbf{A}^{-1} = \frac{1}{\det \mathbf{A}}.$$

(9.9) Verify that (9.20) correctly gives the inverse of $\mathbf{A} = \left[\begin{smallmatrix} a & b \\ c & d \end{smallmatrix}\right]$ as

$$\mathbf{A}^{-1} = \frac{1}{ad - bc} \begin{bmatrix} d & -b \\ -c & a \end{bmatrix}.$$

Specify the matrices $\mathbf{A}^{(11)}$, $\mathbf{A}^{(12)}$, $\mathbf{A}^{(21)}$ and $\mathbf{A}^{(22)}$.

(9.10) Verify that (9.20) correctly gives the inverse of

$$\mathbf{A} = \begin{bmatrix} 1 & 0 & 0 \\ -\alpha_3 & 1 & 0 \\ \alpha_2 & -\alpha_1 & 1 \end{bmatrix}$$

as

$$\mathbf{A}^{-1} = \begin{bmatrix} 1 & 0 & 0 \\ \alpha_3 & 1 & 0 \\ \alpha_1\alpha_3 - \alpha_2 & -\alpha_1 & 1 \end{bmatrix}.$$

Specify the matrices $\mathbf{A}^{(21)}$, $\mathbf{A}^{(31)}$.

(9.11) Let $\mathsf{A} = [\mathsf{a}_1, \mathsf{a}_2]$ with a_1, a_2 two arbitrary non-parallel column vectors in $\mathbb{R}^3$. Construct matrix B whose two *rows* b_1, b_2 are such that $\mathsf{b}_i \cdot \mathsf{a}_j = \delta_{ij}$. Show that B is not unique. Construct an explicit example where $\mathsf{BA} = \mathsf{I}$ but $\mathsf{AB} \neq \mathsf{I}$.

(9.12) Give a nontrivial example of a 3-by-3 orthogonal *and* symmetric matrix (a trivial example is I). What is the geometric meaning of such matrices?

(9.13) Show that the permutation matrix that encodes $(1,2,3) \to (2,3,1)$ has the index representation $P_{ij} = \delta_{i1}\delta_{2j} + \delta_{i2}\delta_{3j} + \delta_{i3}\delta_{1j}$.

(9.14) What is the matrix P encoding the permutation $(1,2,3,4) \to (2,3,4,1)$? What is an explicit index expression for P_{ij}?

(9.15) Pick three nontrivial but arbitrary vectors in $\mathbb{R}^3$ (using Matlab's `randn(3)` for instance, or by choosing random vectors yourself); then construct an orthonormal basis q_1, q_2, q_3 using the Gram–Schmidt procedure. Verify that the matrix $\mathsf{Q} = [\mathsf{q}_1, \mathsf{q}_2, \mathsf{q}_3]$ is orthogonal. Note in particular that the rows are orthogonal eventhough you orthogonalized the columns only.

(9.16) Pick *two* arbitrary vectors a_1, a_2 in $\mathbb{R}^3$ and orthogonalize them to construct q_1, q_2. Consider the 3-by-2 matrix $\mathsf{Q} = [\mathsf{q}_1, \mathsf{q}_2]$ and compute $\mathsf{Q}\mathsf{Q}^\mathsf{T}$ and $\mathsf{Q}^\mathsf{T}\mathsf{Q}$. Explain.

(9.17) Let $\boldsymbol{\mathcal{P}} = \boldsymbol{I} - \hat{\boldsymbol{a}}\hat{\boldsymbol{a}}$ and $\boldsymbol{\mathcal{H}} = \boldsymbol{I} - 2\hat{\boldsymbol{a}}\hat{\boldsymbol{a}}$. Show that $\boldsymbol{v}' = \boldsymbol{\mathcal{P}} \cdot \boldsymbol{v}$ and $\boldsymbol{v}' = \boldsymbol{\mathcal{H}} \cdot \boldsymbol{v}$ are linear transformations. Show that $\boldsymbol{\mathcal{H}}$ preserves all dot products but $\boldsymbol{\mathcal{P}}$ does not. Show this in two ways: using index notation, and using matrix notation.

(9.18) Let $\boldsymbol{a}$ and $\boldsymbol{b}$ be arbitrary vectors and $\boldsymbol{v}_\perp$ the perpendicular projection of $\boldsymbol{v}$ into the $(\boldsymbol{a}, \boldsymbol{b})$-plane. Show that the tensor $\boldsymbol{\mathcal{T}} = \hat{\boldsymbol{a}}\hat{\boldsymbol{b}} + \hat{\boldsymbol{b}}\hat{\boldsymbol{a}}$ produces a vector $\boldsymbol{\mathcal{T}} \cdot \boldsymbol{v}$ that is the reflection of $\boldsymbol{v}_\perp$ about the bisectrix of angle $(\boldsymbol{a}, \boldsymbol{b})$.

(9.19) Let $\{\mathbf{e}'_1, \mathbf{e}'_2, \mathbf{e}'_3\}$ be a cartesian basis obtained by rotating $\{\mathbf{e}_1, \mathbf{e}_2, \mathbf{e}_3\}$ about $\mathbf{e}_3$ by α. Sketch. Find the matrix representations of the dyadic products $\mathbf{e}_1\mathbf{e}_2$ and $\mathbf{e}'_1\mathbf{e}'_2$ in terms of both bases. Show and briefly explain your work. Display the four 3-by-3 matrices.

(9.20) Let $\{\hat{\mathbf{x}}, \hat{\mathbf{y}}, \hat{\mathbf{z}}\}$ be the usual cartesian basis. Show that the identity tensor $\boldsymbol{I}$, such that $\boldsymbol{I} \cdot \boldsymbol{v} = \boldsymbol{v}$ for any vector $\boldsymbol{v}$, is $\boldsymbol{I} = \hat{\mathbf{x}}\hat{\mathbf{x}} + \hat{\mathbf{y}}\hat{\mathbf{y}} + \hat{\mathbf{z}}\hat{\mathbf{z}}$.

(9.21) Let $\{\boldsymbol{a}_1, \boldsymbol{a}_2, \boldsymbol{a}_3\}$ be an arbitrary basis and $\{\breve{\boldsymbol{a}}_1, \breve{\boldsymbol{a}}_2, \breve{\boldsymbol{a}}_3\}$ its reciprocal basis (6.17). (i) Show that the identity tensor

$$\boldsymbol{I} = \boldsymbol{a}_1\breve{\boldsymbol{a}}_1 + \boldsymbol{a}_2\breve{\boldsymbol{a}}_2 + \boldsymbol{a}_3\breve{\boldsymbol{a}}_3 \\ = \breve{\boldsymbol{a}}_1\boldsymbol{a}_1 + \breve{\boldsymbol{a}}_2\boldsymbol{a}_2 + \breve{\boldsymbol{a}}_3\boldsymbol{a}_3.$$

(ii) What is the geometric transformation corresponding to $\boldsymbol{a}_1\breve{\boldsymbol{a}}_1 + \boldsymbol{a}_2\breve{\boldsymbol{a}}_2$?

(9.22) Let $\{\boldsymbol{a}_1, \boldsymbol{a}_2, \boldsymbol{a}_3\}$ be an arbitrary basis and $\{\boldsymbol{a}^1, \boldsymbol{a}^2, \boldsymbol{a}^3\}$ its reciprocal basis. Show that the identity tensor is $\boldsymbol{I} = g^{ij}\boldsymbol{a}_i\boldsymbol{a}_j = g_{ij}\boldsymbol{a}^i\boldsymbol{a}^j$, with summation over repeated indices in one upper and one lower position, where $g^{ij} = \boldsymbol{a}^i \cdot \boldsymbol{a}^j$ and $g_{ij} = \boldsymbol{a}_i \cdot \boldsymbol{a}_j$ are the *metrics* such that $\boldsymbol{a}_i = g_{ij}\boldsymbol{a}^j$, $\boldsymbol{a}^i = g^{ij}\boldsymbol{a}_j$, and $g^{ij}g_{jk} = \delta^i_k = g_{kj}g^{ji}$.

(9.23) Show that (right-hand) rotation of $\boldsymbol{v} \to \boldsymbol{v}' = \boldsymbol{\mathcal{R}} \cdot \boldsymbol{v}$ by φ about $\hat{\mathbf{z}}$ corresponds to the tensor

$$\boldsymbol{\mathcal{R}} = \hat{\mathbf{z}}\hat{\mathbf{z}} + \cos\varphi\,(\hat{\mathbf{x}}\hat{\mathbf{x}} + \hat{\mathbf{y}}\hat{\mathbf{y}}) + \sin\varphi\,(\hat{\mathbf{y}}\hat{\mathbf{x}} - \hat{\mathbf{x}}\hat{\mathbf{y}}) \\ = \begin{bmatrix} \hat{\mathbf{x}} & \hat{\mathbf{y}} & \hat{\mathbf{z}} \end{bmatrix} \begin{bmatrix} \cos\varphi & -\sin\varphi & 0 \\ \sin\varphi & \cos\varphi & 0 \\ 0 & 0 & 1 \end{bmatrix} \begin{bmatrix} \hat{\mathbf{x}} \\ \hat{\mathbf{y}} \\ \hat{\mathbf{z}} \end{bmatrix}, \qquad (9.70)$$

where $\{\hat{\mathbf{x}}, \hat{\mathbf{y}}, \hat{\mathbf{z}}\}$ is the usual right-handed cartesian basis.

(9.24) Show that the tensor $\boldsymbol{\mathcal{R}}$ that rotates the cartesian basis vectors $\mathbf{e}_i \to \mathbf{e}'_i = \boldsymbol{\mathcal{R}} \cdot \mathbf{e}_i = Q_{ij}\mathbf{e}_j$, for Q as in (8.35b), is (with the matrix product continued on the second line)

$$\boldsymbol{\mathcal{R}} = \begin{bmatrix} \mathbf{e}_1 & \mathbf{e}_2 & \mathbf{e}_3 \end{bmatrix} \begin{bmatrix} \cos\varphi & -\sin\varphi & 0 \\ \sin\varphi & \cos\varphi & 0 \\ 0 & 0 & 1 \end{bmatrix} \cdots \\ \begin{bmatrix} \cos\theta & 0 & \sin\theta \\ 0 & 1 & 0 \\ -\sin\theta & 0 & \cos\theta \end{bmatrix} \begin{bmatrix} \mathbf{e}_1 \\ \mathbf{e}_2 \\ \mathbf{e}_3 \end{bmatrix}. \qquad (9.71)$$

(9.25) Show that $\boldsymbol{u} \cdot (\boldsymbol{\mathcal{T}} \cdot \boldsymbol{v}) = (\boldsymbol{u} \cdot \boldsymbol{\mathcal{T}}) \cdot \boldsymbol{v} = \boldsymbol{u} \cdot \boldsymbol{\mathcal{T}} \cdot \boldsymbol{v} \neq \boldsymbol{v} \cdot \boldsymbol{\mathcal{T}} \cdot \boldsymbol{u}$, in general.

(9.26) If $\boldsymbol{\mathcal{A}}$ and $\boldsymbol{\mathcal{B}}$ are arbitrary tensors, show that $(\boldsymbol{\mathcal{A}} \cdot \boldsymbol{\mathcal{B}})^\mathsf{T} = \boldsymbol{\mathcal{B}}^\mathsf{T} \cdot \boldsymbol{\mathcal{A}}^\mathsf{T}$.

(9.27) Let $\boldsymbol{\mathcal{T}} = \boldsymbol{a}_1\boldsymbol{b}_1 + \cdots + \boldsymbol{a}_N\boldsymbol{b}_N$, where the $\boldsymbol{a}_l$, $\boldsymbol{b}_l$ are arbitrary vectors. Show that its trace $\mathrm{tr}(\boldsymbol{\mathcal{T}}) \triangleq \mathbf{e}_k \cdot \boldsymbol{\mathcal{T}} \cdot \mathbf{e}_k = \sum_l \boldsymbol{a}_l \cdot \boldsymbol{b}_l$, and its transpose $\boldsymbol{\mathcal{T}}^\mathsf{T} = \sum_l \boldsymbol{b}_l\boldsymbol{a}_l$.

(9.28) Let $\boldsymbol{\mathcal{T}} = T_{ij}\mathbf{e}_i\mathbf{e}_j$ in a cartesian basis $\{\mathbf{e}_1, \mathbf{e}_2, \mathbf{e}_3\}$. Show that $T_{ij} = T_{ji}$ if and only if $\boldsymbol{u} \cdot \boldsymbol{\mathcal{T}} \cdot \boldsymbol{v} = \boldsymbol{v} \cdot \boldsymbol{\mathcal{T}} \cdot \boldsymbol{u}$ for any vectors $\boldsymbol{u}$, $\boldsymbol{v}$.

(9.29) Let $\hat{\boldsymbol{a}}$ and $\hat{\boldsymbol{b}}$ be orthogonal unit vectors, $\hat{\boldsymbol{a}} \cdot \hat{\boldsymbol{b}} = 0$, and $\hat{\boldsymbol{c}}$ obtained by rotating $\hat{\boldsymbol{b}}$ about $\hat{\boldsymbol{a}}$ by $\alpha/2 \neq 0$. Let

$$\boldsymbol{\mathcal{R}}_b = 2\hat{\boldsymbol{b}}\hat{\boldsymbol{b}} - \boldsymbol{I}, \\ \boldsymbol{\mathcal{R}}_c = 2\hat{\boldsymbol{c}}\hat{\boldsymbol{c}} - \boldsymbol{I},$$

and consider $\mathcal{R} = \mathcal{R}_c \cdot \mathcal{R}_b$.

(i) What geometric operations do $\mathcal{R}_b$ and $\mathcal{R}_c$ represent? Show and explain.

(ii) Show that $\mathcal{R}^\mathsf{T} \cdot \mathcal{R} = \boldsymbol{I}$.

(iii) Show that $\mathcal{R}$ is the tensor for rotation about $\hat{\boldsymbol{a}}$ by α by showing that applying $\mathcal{R}$ to the intrinsic basis $\hat{\boldsymbol{a}}$, $\hat{\boldsymbol{b}}$, $\hat{\boldsymbol{c}}$ rotates each vector by α about $\hat{\boldsymbol{a}}$.

(iv) Show that

$$\mathcal{R} = \hat{\boldsymbol{a}}\hat{\boldsymbol{a}} + \cos\alpha\,(\hat{\boldsymbol{b}}\hat{\boldsymbol{b}} + \hat{\boldsymbol{n}}\hat{\boldsymbol{n}}) + \sin\alpha\,(\hat{\boldsymbol{n}}\hat{\boldsymbol{b}} - \hat{\boldsymbol{b}}\hat{\boldsymbol{n}}), \quad (9.72a)$$

in terms of the intrinsic orthonormal basis $\{\hat{\boldsymbol{a}}, \hat{\boldsymbol{b}}, \hat{\boldsymbol{n}}\}$, where $\hat{\boldsymbol{n}} = \hat{\boldsymbol{a}} \times \hat{\boldsymbol{b}}$, so that its matrix $\mathbf{R}$ in *that* basis is

$$\mathbf{R} = \begin{bmatrix} 1 & 0 & 0 \\ 0 & \cos\alpha & -\sin\alpha \\ 0 & \sin\alpha & \cos\alpha \end{bmatrix}. \quad (9.72b)$$

What are its components R_{ij} in an arbitrary cartesian basis $\{\mathbf{e}_1, \mathbf{e}_2, \mathbf{e}_3\}$?

(9.30) Equations (8.15) express one cartesian basis in terms of another; define a tensor $\mathcal{T}$ that transforms $\mathbf{e}_i$ into $\mathbf{e}'_i$, such that

$$\mathcal{T} \cdot \mathbf{e}_i = \mathbf{e}'_i = Q_{ij}\mathbf{e}_j. \quad (9.73a)$$

Show that $\mathcal{T} = T_{ij}\mathbf{e}_i\mathbf{e}_j = T'_{ij}\mathbf{e}'_i\mathbf{e}'_j$, with

$$T_{ij} = Q_{ji} = Q^\mathsf{T}_{ij} = T'_{ij}. \quad (9.73b)$$

Discuss why $T_{ij} = T'_{ij}$.

(9.31) *Tensors for body-fixed rotations.* Let $\mathbf{e}'_i = Q_{ij}\,\mathbf{e}_j$ with $[Q_{ij}] = \mathbf{Q} = \mathbf{Q}_3\mathbf{Q}_2\mathbf{Q}_1$ such that

$$\mathbf{e}'_i = Q^{(3)}_{ij}\mathbf{e}^{(2)}_j, \quad \mathbf{e}^{(2)}_j = Q^{(2)}_{jk}\mathbf{e}^{(1)}_k, \quad \mathbf{e}^{(1)}_k = Q^{(1)}_{kl}\mathbf{e}_l,$$

as in (8.29) and (8.30), for example. These $\mathbf{Q}$ matrices correspond to the active rotation matrices $\mathbf{R}_1 = \mathbf{Q}^\mathsf{T}_1$, $\mathbf{R}_2 = \mathbf{Q}^\mathsf{T}_2$, $\mathbf{R}_3 = \mathbf{Q}^\mathsf{T}_3$, (8.34c), (9.73b). Define the *(body-fixed)* rotation tensors

$$\begin{aligned} \mathcal{B}^{(1)} &= R^{(1)}_{ij}\mathbf{e}_i\mathbf{e}_j, \\ \mathcal{B}^{(2)} &= R^{(2)}_{ij}\mathbf{e}^{(1)}_i\mathbf{e}^{(1)}_j, \\ \mathcal{B}^{(3)} &= R^{(3)}_{ij}\mathbf{e}^{(2)}_i\mathbf{e}^{(2)}_j. \end{aligned} \quad (9.74a)$$

Show that

(i) $\mathcal{B}^{(1)}$, $\mathcal{B}^{(2)}$, $\mathcal{B}^{(3)}$ perform the successive *body-fixed* rotations corresponding to $\mathbf{Q}_1$, $\mathbf{Q}_2$, $\mathbf{Q}_3$, respectively. That is, show that

$$\begin{aligned} \mathcal{B}^{(1)} \cdot \mathbf{e}_i &= \mathbf{e}^{(1)}_i, \\ \mathcal{B}^{(2)} \cdot \mathbf{e}^{(1)}_i &= \mathbf{e}^{(2)}_i, \\ \mathcal{B}^{(3)} \cdot \mathbf{e}^{(2)}_i &= \mathbf{e}^{(3)}_i. \end{aligned}$$

(ii) the tensor $\mathcal{R} = \mathcal{B}^{(3)} \cdot \mathcal{B}^{(2)} \cdot \mathcal{B}^{(1)}$ yields

$$\mathcal{R} \cdot \mathbf{e}_i = \mathbf{e}^{(3)}_i = \mathbf{e}'_i.$$

(iii) in the original basis $\{\mathbf{e}_1, \mathbf{e}_2, \mathbf{e}_3\}$, the matrices of these tensors are

$$\begin{aligned} &\mathbf{B}_1 = \mathbf{R}_1, \; \mathbf{B}_2 = \mathbf{Q}^\mathsf{T}_1\mathbf{R}_2\mathbf{Q}_1, \; \mathbf{B}_3 = \mathbf{Q}^\mathsf{T}_1\mathbf{Q}^\mathsf{T}_2\mathbf{R}_3\mathbf{Q}_2\mathbf{Q}_1, \\ &\mathbf{R} = \mathbf{B}_3\mathbf{B}_2\mathbf{B}_1 = \mathbf{R}_1\mathbf{R}_2\mathbf{R}_3. \end{aligned} \quad (9.74b)$$

A vector $\boldsymbol{v} = v_i\mathbf{e}_i$ is transformed into $\mathcal{R} \cdot \boldsymbol{v} = \boldsymbol{v}' = v'_i\mathbf{e}_i$ with $\mathbf{v}' = \mathbf{R}\mathbf{v}$.

(9.32) *Tensors for space-fixed rotations.* Let $\mathbf{e}'_i = Q_{ij}\,\mathbf{e}_j$ with $\mathbf{Q} = \mathbf{Q}_3\mathbf{Q}_2\mathbf{Q}_1$ as in (8.29) and (8.30), for example, corresponding to the active rotation matrices $\mathbf{R}_1 = \mathbf{Q}^\mathsf{T}_1$, $\mathbf{R}_2 = \mathbf{Q}^\mathsf{T}_2$, $\mathbf{R}_3 = \mathbf{Q}^\mathsf{T}_3$, (8.34c), (9.73b). Define the *(space-fixed)* rotation tensors

$$\begin{aligned} \mathcal{S}^{(1)} &= R^{(1)}_{ij}\mathbf{e}_i\mathbf{e}_j, \\ \mathcal{S}^{(2)} &= R^{(2)}_{ij}\mathbf{e}_i\mathbf{e}_j, \\ \mathcal{S}^{(3)} &= R^{(3)}_{ij}\mathbf{e}_i\mathbf{e}_j. \end{aligned} \quad (9.75a)$$

Show that the matrix of the successive rotation tensor $\mathcal{S} = \mathcal{S}^{(1)} \cdot \mathcal{S}^{(2)} \cdot \mathcal{S}^{(3)}$ in the basis $\{\mathbf{e}_1, \mathbf{e}_2, \mathbf{e}_3\}$ is

$$\mathbf{S} = \mathbf{R}_1\mathbf{R}_2\mathbf{R}_3. \quad (9.75b)$$

A vector $\boldsymbol{v} = v_i\mathbf{e}_i$ is transformed into $\mathcal{S} \cdot \boldsymbol{v} = \boldsymbol{v}' = v'_i\mathbf{e}_i$ with $\mathbf{v}' = \mathbf{S}\mathbf{v}$.

Matrix $\mathbf{S}$ equals the matrix $\mathbf{R}$ for the body-fixed rotations (9.74b), but the sequence of space-fixed rotations is performed in *reverse order* from that of the body-fixed rotations (9.74b), that is

$$\mathcal{B}^{(3)} \cdot \mathcal{B}^{(2)} \cdot \mathcal{B}^{(1)} = \mathcal{S}^{(1)} \cdot \mathcal{S}^{(2)} \cdot \mathcal{S}^{(3)}. \quad (9.75c)$$

(9.33) Find the eigenvectors of $\boldsymbol{ab}$ and $\boldsymbol{I} - \alpha\boldsymbol{ab}$ and their eigenvalues, where $\boldsymbol{a}$, $\boldsymbol{b}$ are arbitrary vectors. Show that the eigenvectors are not orthogonal in general, and that the eigenvectors do not form a complete basis if $\boldsymbol{a} \cdot \boldsymbol{b} = 0$.

(9.34) Show that the eigenvectors of $\boldsymbol{ab} + \boldsymbol{ba}$ are $\hat{\boldsymbol{a}} \pm \hat{\boldsymbol{b}}$ and $\boldsymbol{a} \times \boldsymbol{b}$ and that they are orthogonal, for arbitrary nonparallel $\boldsymbol{a}$ and $\boldsymbol{b}$. What are the eigenvalues?

(9.35) Show that the rotation eigenvectors $\boldsymbol{v}_i$ in (9.53) are *not* orthogonal in the sense $\boldsymbol{v}_i \cdot \boldsymbol{v}_j = \delta_{ij}$. However, they are *complex orthogonal* in the sense $\boldsymbol{v}^*_i \cdot \boldsymbol{v}_j \propto \delta_{ij}$, where $()^*$ denotes *complex conjugate* obtained by replacing $\imath$ by $-\imath$ so $(\boldsymbol{q}_1 \pm \imath\boldsymbol{q}_2)^* = \boldsymbol{q}_1 \mp \imath\boldsymbol{q}_2$. Normalize $\boldsymbol{v}_i$ so that $\boldsymbol{v}^*_i \cdot \boldsymbol{v}_j = \delta_{ij}$.

(9.36) Find the eigenvectors of $\boldsymbol{ab} - \boldsymbol{ba}$ and their eigenvalues for arbitrary $\boldsymbol{a}$ and $\boldsymbol{b}$. Are the eigenvectors orthogonal? Are they complex orthogonal?

(9.37) If $\boldsymbol{\mathcal{A}}$ has matrix $\mathbf{A}$ in some basis, its matrix in some other basis is $\mathbf{A}' = \mathbf{SAS}^{-1}$ as shown in (9.50) for cartesian bases. Show that the eigenvalues of $\mathbf{A}'$ are the same as those of $\mathbf{A}$. How are the eigenvectors related?

(9.38) Find all the eigenvalues and eigenvectors of

$$\mathbf{A} = \begin{bmatrix} a & 1 & 0 \\ 0 & b & 1 \\ 0 & 0 & c \end{bmatrix}.$$

Show that there is only eigenvector when $a = b = c$, that is when $\mathbf{A}$ is a *Jordan block*.

(9.39) Find the eigenvectors and the singular vectors of

$$\mathbf{A} = \begin{bmatrix} a & 1 \\ 0 & b \end{bmatrix}.$$

Sketch the eigenvectors and the singular vectors for $a \neq b$ and for $a = b$.

(9.40) Verify the eigenvalues (9.52) and (9.53) and their corresponding eigenvectors by picking a suitable orthonormal basis and finding the eigenvalues and eigenvectors of the resulting 3-by-3 matrix.

(9.41) Calculate the SVD of the cartesian matrices of (9.52) and (9.53).

(9.42) Explain and justify how $\mathbf{Av} = \lambda\mathbf{v}$ is transformed to $\mathbf{A}'\mathbf{v}' = \lambda\mathbf{v}'$ in (9.58).

(9.43) Let $\boldsymbol{\mathcal{A}}$ be symmetric and $\boldsymbol{\mathcal{A}} \cdot \boldsymbol{v} = \lambda \boldsymbol{v}$, $\boldsymbol{\mathcal{A}} \cdot \boldsymbol{w} = \mu \boldsymbol{w}$ with $\lambda \neq \mu$. Show that $\boldsymbol{v} \cdot \boldsymbol{w} = 0$.

(9.44) Show that the tensor of inertia is symmetric and *positive definite*, that is, $\boldsymbol{v} \cdot \boldsymbol{\mathcal{I}} \cdot \boldsymbol{v} \geq 0$ for any $\boldsymbol{v} \neq 0$. Conclude that its eigenvalues are real and positive.

(9.45) Show that the principal directions of the tensor of inertia $\boldsymbol{\mathcal{I}}$ are the same as those of the tensor $\boldsymbol{\mathcal{P}} = \sum_l m_l \boldsymbol{r}_l \boldsymbol{r}_l$. How are the eigenvalues related?

(9.46) Mass m_1 is at $\boldsymbol{r}_1 = x_1\hat{\mathbf{x}} + y_1\hat{\mathbf{y}} + z_1\hat{\mathbf{z}}$ and mass m_2 is at $\boldsymbol{r}_2 = a\boldsymbol{r}_1$ for some a. Write the tensor of inertia in vector form. Find its matrix with respect to the cartesian basis $\hat{\mathbf{x}}$, $\hat{\mathbf{y}}$, $\hat{\mathbf{z}}$. Find the principal axes and the principal moments of inertia.

(9.47) Show that $\boldsymbol{aa} + \boldsymbol{bb}$ is symmetric. Hence, that tensor has orthonormal eigenvectors $\boldsymbol{e}_1', \boldsymbol{e}_2', \boldsymbol{e}_3'$. If $\boldsymbol{a}$ and $\boldsymbol{b}$ are nonparallel, show that $\boldsymbol{e}_3' = \boldsymbol{a} \times \boldsymbol{b} / |\boldsymbol{a} \times \boldsymbol{b}|$ is an eigenvector. Thus, $\boldsymbol{e}_1'$ and $\boldsymbol{e}_2'$ are in the plane of $\boldsymbol{a}$ and $\boldsymbol{b}$. If α is the angle from $\boldsymbol{a}$ to $\boldsymbol{e}_1'$, right-handed about $\boldsymbol{e}_3'$, and θ that between $\boldsymbol{a}$ and $\boldsymbol{b}$, show that

$$\tan 2\alpha = \frac{b^2 \sin 2\theta}{a^2 + b^2 \cos 2\theta},$$

and the eigenvalues are

$$\begin{aligned} \lambda_1 &= a^2 \cos^2 \alpha + b^2 \cos^2(\theta - \alpha), \\ \lambda_2 &= a^2 \sin^2 \alpha + b^2 \sin^2(\theta - \alpha), \\ \lambda_3 &= 0. \end{aligned}$$

What is α in the limits $a \ll b$, $b \ll a$, and $a = b$? Do the results make sense to you in those limits? Discuss. Considering limits provides checks on the validity of the results. (Hint: $\boldsymbol{e}_1' \cdot (\boldsymbol{aa} + \boldsymbol{bb}) \cdot \boldsymbol{e}_2' = 0$.)

Part II

Vector Calculus

Kinematics

10

10.1 Vector functions and their derivatives

Vector calculus deals with vector functions. We begin with vector functions $\boldsymbol{a}(t)$ of one real variable t. This $\boldsymbol{a}(t)$ is *any* vector function of t. It could be a position vector at time t, a velocity at time t, an acceleration, or a time-dependent force, for example. The tip of vector $\boldsymbol{a}(t)$ traces a curve $\mathcal{C}$ with respect to its tail. The difference $\Delta\boldsymbol{a} = \boldsymbol{a}(t') - \boldsymbol{a}(t)$ is a secant vector for that curve (Fig. 10.1).

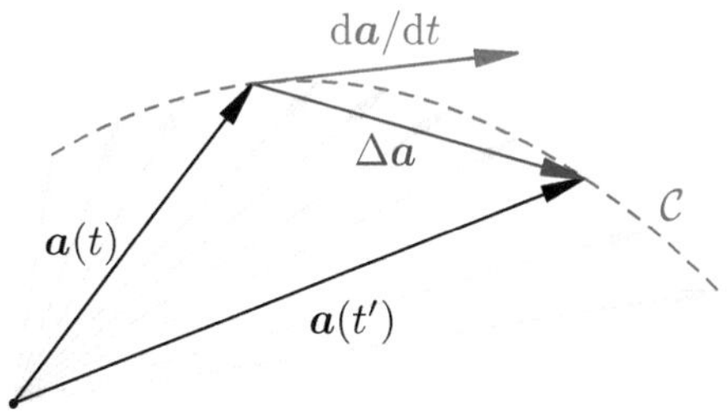

Fig. 10.1 Curve $\mathcal{C}$ of $\boldsymbol{a}(t)$ with secant $\Delta\boldsymbol{a} = \boldsymbol{a}(t') - \boldsymbol{a}(t)$ and derivative $\mathrm{d}\boldsymbol{a}/\mathrm{d}t$ at t.

A vector function $\boldsymbol{a}(t)$ is *continuous* at t if

$$\lim_{t'\to t} \boldsymbol{a}(t') = \boldsymbol{a}(t), \tag{10.1}$$

then the secant vector $\Delta\boldsymbol{a} = \boldsymbol{a}(t') - \boldsymbol{a}(t)$ goes to zero as $\Delta t = t' - t \to 0$. The limit of the ratio $\Delta\boldsymbol{a}/\Delta t$, if it exists, is the derivative of $\boldsymbol{a}(t)$ at t:

$$\frac{\mathrm{d}\boldsymbol{a}}{\mathrm{d}t} \triangleq \lim_{t'\to t} \frac{\boldsymbol{a}(t') - \boldsymbol{a}(t)}{t' - t} = \lim_{\Delta t\to 0} \frac{\Delta\boldsymbol{a}}{\Delta t}. \tag{10.2}$$

The vector derivative $\mathrm{d}\boldsymbol{a}/\mathrm{d}t$ is tangent to the curve at $\boldsymbol{a}(t)$ since it is the limit of rescaled secant vectors $\Delta\boldsymbol{a}/\Delta t$.

Differentiation rules

Rules for derivatives of vector functions are similar to those of scalar functions. The derivative of a sum of vector functions is the sum of the derivatives,

$$\frac{\mathrm{d}}{\mathrm{d}t}(\boldsymbol{a} + \boldsymbol{b}) = \frac{\mathrm{d}\boldsymbol{a}}{\mathrm{d}t} + \frac{\mathrm{d}\boldsymbol{b}}{\mathrm{d}t}. \tag{10.3}$$

There are several kinds of products involving vectors, but the *product rules* are all similar,

$$\frac{\mathrm{d}}{\mathrm{d}t}(\alpha\boldsymbol{a}) = \frac{\mathrm{d}\alpha}{\mathrm{d}t}\boldsymbol{a} + \alpha\frac{\mathrm{d}\boldsymbol{a}}{\mathrm{d}t}, \tag{10.4}$$

$$\frac{\mathrm{d}}{\mathrm{d}t}(\boldsymbol{a}\cdot\boldsymbol{b}) = \frac{\mathrm{d}\boldsymbol{a}}{\mathrm{d}t}\cdot\boldsymbol{b} + \boldsymbol{a}\cdot\frac{\mathrm{d}\boldsymbol{b}}{\mathrm{d}t}, \tag{10.5}$$

$$\frac{\mathrm{d}}{\mathrm{d}t}(\boldsymbol{a}\times\boldsymbol{b}) = \frac{\mathrm{d}\boldsymbol{a}}{\mathrm{d}t}\times\boldsymbol{b} + \boldsymbol{a}\times\frac{\mathrm{d}\boldsymbol{b}}{\mathrm{d}t}. \tag{10.6}$$

The *chain rule* also applies. If $\boldsymbol{a}$ is a function of φ that is a function of t, then

$$\frac{\mathrm{d}\boldsymbol{a}}{\mathrm{d}t} = \frac{\mathrm{d}\boldsymbol{a}}{\mathrm{d}\varphi}\frac{\mathrm{d}\varphi}{\mathrm{d}t}, \tag{10.7}$$

and if $\boldsymbol{a}$ is a function of $\theta_1, \theta_2, \cdots, \theta_n$ that are functions of t then

$$\frac{\mathrm{d}\boldsymbol{a}}{\mathrm{d}t} = \frac{\partial\boldsymbol{a}}{\partial\theta_1}\frac{\mathrm{d}\theta_1}{\mathrm{d}t} + \frac{\partial\boldsymbol{a}}{\partial\theta_1}\frac{\mathrm{d}\theta_2}{\mathrm{d}t} + \cdots + \frac{\partial\boldsymbol{a}}{\partial\theta_n}\frac{\mathrm{d}\theta_n}{\mathrm{d}t}. \tag{10.8}$$

The proofs of these differentiation rules are similar to those for scalar functions in elementary calculus. For example,

$$\begin{aligned}
\frac{\mathrm{d}}{\mathrm{d}t}(\boldsymbol{a}\cdot\boldsymbol{b}) &= \lim_{t'\to t}\frac{\boldsymbol{a}(t')\cdot\boldsymbol{b}(t') - \boldsymbol{a}(t)\cdot\boldsymbol{b}(t)}{t'-t} \\
&= \lim_{t'\to t}\frac{\boldsymbol{a}(t')\cdot\boldsymbol{b}(t') - a(t)\cdot b(t') + a(t)\cdot b(t') - \boldsymbol{a}(t)\cdot\boldsymbol{b}(t)}{t'-t} \\
&= \lim_{t'\to t}\left(\frac{\boldsymbol{a}(t') - \boldsymbol{a}(t)}{t'-t}\cdot\boldsymbol{b}(t') + \boldsymbol{a}(t)\cdot\frac{\boldsymbol{b}(t') - \boldsymbol{b}(t)}{t'-t}\right) \\
&= \frac{\mathrm{d}\boldsymbol{a}}{\mathrm{d}t}\cdot\boldsymbol{b} + \boldsymbol{a}\cdot\frac{\mathrm{d}\boldsymbol{b}}{\mathrm{d}t},
\end{aligned}$$

if $\boldsymbol{a}(t)$ and $\boldsymbol{b}(t)$ are differentiable at t. Proof of the chain rule (10.8) requires existence *and* continuity of the partial derivatives $\partial\boldsymbol{a}/\partial\theta_i$.

10.2 Magnitude and direction

The derivative of a magnitude $|\boldsymbol{a}|$ is related to the vector derivative $\mathrm{d}\boldsymbol{a}/\mathrm{d}t$ as

$$\frac{\mathrm{d}|\boldsymbol{a}|}{\mathrm{d}t} = \hat{\boldsymbol{a}}\cdot\frac{\mathrm{d}\boldsymbol{a}}{\mathrm{d}t}, \tag{10.9}$$

since the chain rule and the product rule applied to $|\boldsymbol{a}| = \sqrt{\boldsymbol{a}\cdot\boldsymbol{a}}$ yield

$$\frac{\mathrm{d}\sqrt{\boldsymbol{a}\cdot\boldsymbol{a}}}{\mathrm{d}t} = \frac{1}{2\sqrt{\boldsymbol{a}\cdot\boldsymbol{a}}}\left(2\boldsymbol{a}\cdot\frac{\mathrm{d}\boldsymbol{a}}{\mathrm{d}t}\right) = \hat{\boldsymbol{a}}\cdot\frac{\mathrm{d}\boldsymbol{a}}{\mathrm{d}t}.$$

Thus, the rate of change of the magnitude $|\boldsymbol{a}|$ is the projection of $\mathrm{d}\boldsymbol{a}/\mathrm{d}t$ onto the direction $\hat{\boldsymbol{a}}$ (Fig. 10.2).

The derivative $\mathrm{d}\hat{\boldsymbol{a}}/\mathrm{d}t$ of a unit vector is always orthogonal to $\hat{\boldsymbol{a}}$ since $|\hat{\boldsymbol{a}}| = 1$ for all t. This follows from (10.9) for $\boldsymbol{a} = \hat{\boldsymbol{a}}$, or directly from the product rule,

$$\hat{\boldsymbol{a}}\cdot\frac{\mathrm{d}\hat{\boldsymbol{a}}}{\mathrm{d}t} = \frac{\mathrm{d}}{\mathrm{d}t}\left(\frac{\hat{\boldsymbol{a}}\cdot\hat{\boldsymbol{a}}}{2}\right) = 0. \tag{10.10}$$

Thus, $\mathrm{d}\hat{\boldsymbol{a}}/\mathrm{d}t$ is orthogonal to $\hat{\boldsymbol{a}}$ and there exists a vector $\boldsymbol{\omega}(t)$, defined up to a multiple of $\boldsymbol{a}(t)$, such that

$$\frac{\mathrm{d}\hat{\boldsymbol{a}}}{\mathrm{d}t} = \boldsymbol{\omega}(t) \times \hat{\boldsymbol{a}}. \tag{10.11}$$

The vector $\boldsymbol{\omega}(t)$ is the *angular velocity* of $\hat{\boldsymbol{a}}$. The vector $\boldsymbol{\omega}(t)$ varies in magnitude and direction, in general. Angular velocities are discussed in §10.3 and §10.5.

The rate of change of an arbitrary vector $\boldsymbol{a}(t)$ arises from changes in both its magnitude $a = |\boldsymbol{a}|$ and its direction $\hat{\boldsymbol{a}}$. The product rule gives

$$\frac{\mathrm{d}\boldsymbol{a}}{\mathrm{d}t} = \frac{\mathrm{d}}{\mathrm{d}t}(a\,\hat{\boldsymbol{a}}) = \frac{\mathrm{d}a}{\mathrm{d}t}\,\hat{\boldsymbol{a}} + a\,\frac{\mathrm{d}\hat{\boldsymbol{a}}}{\mathrm{d}t}, \tag{10.12}$$

and it follows from (10.10) that these two vector components are orthogonal to each other (Fig. 10.2). Using (10.11), the general vector derivative (10.12) reads

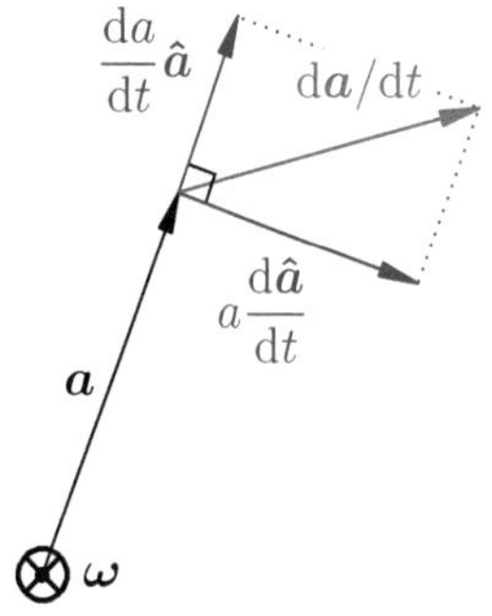

Fig. 10.2 Orthogonal components of $\mathrm{d}\boldsymbol{a}/\mathrm{d}t$.

$$\frac{\mathrm{d}\boldsymbol{a}}{\mathrm{d}t} = \frac{\mathrm{d}}{\mathrm{d}t}(a\,\hat{\boldsymbol{a}}) = \frac{\mathrm{d}a}{\mathrm{d}t}\,\hat{\boldsymbol{a}} + \boldsymbol{\omega}(t) \times \boldsymbol{a}, \tag{10.13}$$

highlighting that the rate of change of $\boldsymbol{a}(t)$ arises from two orthogonal contributions: one in the direction of $\boldsymbol{a}$ due to a change in its magnitude $a = |\boldsymbol{a}|$, and one orthogonal to $\boldsymbol{a}$ due to a rotation with angular velocity $\boldsymbol{\omega}(t)$.

10.3 Cylindrical and spherical directions

The radius vector $\boldsymbol{r} = \overrightarrow{OP}$ of an arbitrary point P with respect to origin O, is

$$\boldsymbol{r} = r\hat{\boldsymbol{r}} = \rho\hat{\boldsymbol{\rho}} + z\hat{\mathbf{z}} = x\hat{\mathbf{x}} + y\hat{\mathbf{y}} + z\hat{\mathbf{z}} \tag{10.14}$$

in spherical, cylindrical, and cartesian coordinates, respectively §7.1. For a moving particle, $\boldsymbol{r}$ is a function of time t but the cartesian direction vectors $\hat{\mathbf{x}}, \hat{\mathbf{y}}, \hat{\mathbf{z}}$ are (typically) fixed independent of the particle motion; however, the radial directions $\hat{\boldsymbol{r}}$ and $\hat{\boldsymbol{\rho}}$ depend on the particle position and, thus, vary with t in general. The horizontal radial direction $\hat{\boldsymbol{\rho}}$ depends only on the azimuth φ while the radial direction $\hat{\boldsymbol{r}}$ is a function of the polar angle θ and the azimuth φ (Fig. 10.3). Changes in $\hat{\boldsymbol{\rho}}$ and $\hat{\boldsymbol{r}}$ thus arise from θ and φ varying with t. Hence, the time derivative of (10.14) is

$$\begin{aligned}\frac{\mathrm{d}\boldsymbol{r}}{\mathrm{d}t} = \dot{r}\hat{\boldsymbol{r}} + r\frac{\mathrm{d}\hat{\boldsymbol{r}}}{\mathrm{d}t} &= \dot{\rho}\hat{\boldsymbol{\rho}} + \rho\frac{\mathrm{d}\hat{\boldsymbol{\rho}}}{\mathrm{d}t} + \dot{z}\hat{\mathbf{z}}, \\ &= \dot{x}\hat{\mathbf{x}} + \dot{y}\hat{\mathbf{y}} + \dot{z}\hat{\mathbf{z}},\end{aligned} \tag{10.15}$$

where Newton's compact dot notation has been used with

$$\dot{r} = \frac{\mathrm{d}r}{\mathrm{d}t} = \frac{\mathrm{d}|\boldsymbol{r}|}{\mathrm{d}t}, \quad \dot{\boldsymbol{r}} = \frac{\mathrm{d}\boldsymbol{r}}{\mathrm{d}t}, \quad \dot{\rho} = \frac{\mathrm{d}\rho}{\mathrm{d}t}, \quad \ddot{x} = \frac{\mathrm{d}^2 x}{\mathrm{d}t^2}, \quad \cdots.$$

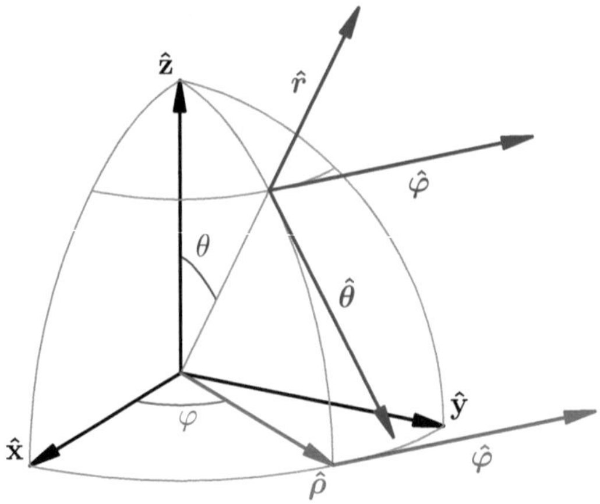

Fig. 10.3 3D view of cartesian $\{\hat{\mathbf{x}}, \hat{\mathbf{y}}, \hat{\mathbf{z}}\}$, cylindrical $\{\hat{\boldsymbol{\rho}}, \hat{\boldsymbol{\varphi}}, \hat{\mathbf{z}}\}$ and spherical $\{\hat{\boldsymbol{r}}, \hat{\boldsymbol{\theta}}, \hat{\boldsymbol{\varphi}}\}$ bases. The vectors $\{\hat{\mathbf{x}}, \hat{\mathbf{y}}, \hat{\mathbf{z}}\}$ do not depend on position, in particular they do not depend on θ, φ. The vectors $\{\hat{\boldsymbol{\rho}}, \hat{\boldsymbol{\varphi}}\}$ are orthogonal to $\hat{\mathbf{z}}$ and depend only on φ.

Cylindrical directions

To evaluate $\mathrm{d}\hat{\boldsymbol{\rho}}/\mathrm{d}t$, define

$$\hat{\boldsymbol{\varphi}} = \hat{\mathbf{z}} \times \hat{\boldsymbol{\rho}}. \tag{10.16}$$

Thus, $\{\hat{\mathbf{z}}, \hat{\boldsymbol{\rho}}, \hat{\boldsymbol{\varphi}}\}$ is a right-handed orthonormal basis (Fig. 10.4). If φ is a function of t then $\hat{\boldsymbol{\rho}}$ and $\hat{\boldsymbol{\varphi}}$ rotate about $\hat{\mathbf{z}}$ at angular rotation rate $\dot{\varphi}$; that is, $\hat{\boldsymbol{\rho}}$ and $\hat{\boldsymbol{\varphi}}$ both satisfy (10.11) with angular velocity

$$\boldsymbol{\omega}(t) = \dot{\varphi}\,\hat{\mathbf{z}}. \tag{10.17}$$

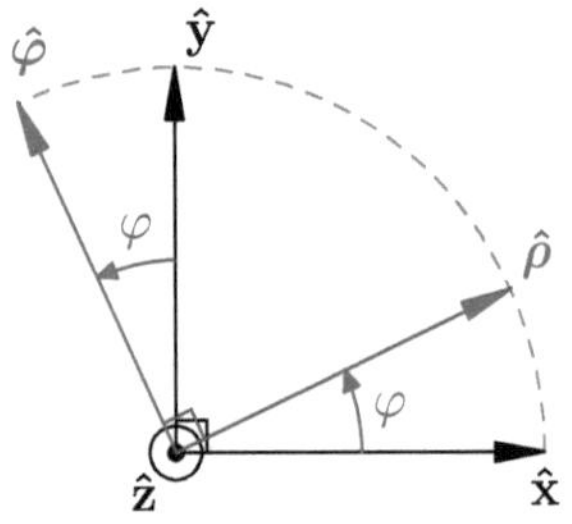

Fig. 10.4 Cartesian and cylindrical directions. Top view.

Thus,

$$\begin{aligned} \frac{\mathrm{d}\hat{\boldsymbol{\rho}}}{\mathrm{d}t} &= \dot{\varphi}\,\hat{\mathbf{z}} \times \hat{\boldsymbol{\rho}} = \dot{\varphi}\,\hat{\boldsymbol{\varphi}}, \\ \frac{\mathrm{d}\hat{\boldsymbol{\varphi}}}{\mathrm{d}t} &= \dot{\varphi}\,\hat{\mathbf{z}} \times \hat{\boldsymbol{\varphi}} = -\dot{\varphi}\,\hat{\boldsymbol{\rho}}. \end{aligned} \tag{10.18}$$

Since $\hat{\boldsymbol{\rho}} = \hat{\boldsymbol{\rho}}(\varphi(t))$ and $\hat{\boldsymbol{\varphi}} = \hat{\boldsymbol{\varphi}}(\varphi(t))$, the chain rule gives

$$\frac{\mathrm{d}\hat{\boldsymbol{\rho}}}{\mathrm{d}t} = \frac{\mathrm{d}\hat{\boldsymbol{\rho}}}{\mathrm{d}\varphi}\frac{\mathrm{d}\varphi}{\mathrm{d}t}, \qquad \frac{\mathrm{d}\hat{\boldsymbol{\varphi}}}{\mathrm{d}t} = \frac{\mathrm{d}\hat{\boldsymbol{\varphi}}}{\mathrm{d}\varphi}\frac{\mathrm{d}\varphi}{\mathrm{d}t}.$$

Then, from (10.18),

$$\begin{aligned} \frac{\mathrm{d}\hat{\boldsymbol{\rho}}}{\mathrm{d}\varphi} &= \hat{\mathbf{z}} \times \hat{\boldsymbol{\rho}} = \hat{\boldsymbol{\varphi}}, \\ \frac{\mathrm{d}\hat{\boldsymbol{\varphi}}}{\mathrm{d}\varphi} &= \hat{\mathbf{z}} \times \hat{\boldsymbol{\varphi}} = -\hat{\boldsymbol{\rho}}. \end{aligned} \tag{10.19}$$

These results can also be obtained by differentiating

$$\begin{aligned} \hat{\boldsymbol{\rho}} &= \cos\varphi\,\hat{\mathbf{x}} + \sin\varphi\,\hat{\mathbf{y}}, \\ \hat{\boldsymbol{\varphi}} &= -\sin\varphi\,\hat{\mathbf{x}} + \cos\varphi\,\hat{\mathbf{y}}. \end{aligned} \tag{10.20}$$

Spherical directions

To evaluate $\mathrm{d}\hat{\boldsymbol{r}}/\mathrm{d}t$, define

$$\hat{\boldsymbol{\theta}} = \hat{\boldsymbol{\varphi}} \times \hat{\boldsymbol{r}} \tag{10.21}$$

with $\hat{\boldsymbol{\varphi}} = \hat{\mathbf{z}}\times\hat{\boldsymbol{\rho}} = (\hat{\mathbf{z}}\times\hat{\boldsymbol{r}})/\sin\theta$, so $\{\hat{\boldsymbol{r}}, \hat{\boldsymbol{\theta}}, \hat{\boldsymbol{\varphi}}\}$ is a right-handed orthonormal basis. The relationship between those direction vectors is illustrated in Figs. 10.3 and 10.5. The radial $\hat{\boldsymbol{r}}$, polar $\hat{\boldsymbol{\theta}}$, and azimuthal $\hat{\boldsymbol{\varphi}}$ direction vectors rotate rigidly since their magnitudes are fixed and the angles between them are fixed. There are two sources of rotation now, $\dot{\varphi}\,\hat{\mathbf{z}}$ and $\dot{\theta}\,\hat{\boldsymbol{\varphi}}$, and the angular velocity vector is

$$\boldsymbol{\omega}(t) = \dot{\varphi}\,\hat{\mathbf{z}} + \dot{\theta}\,\hat{\boldsymbol{\varphi}}. \tag{10.22}$$

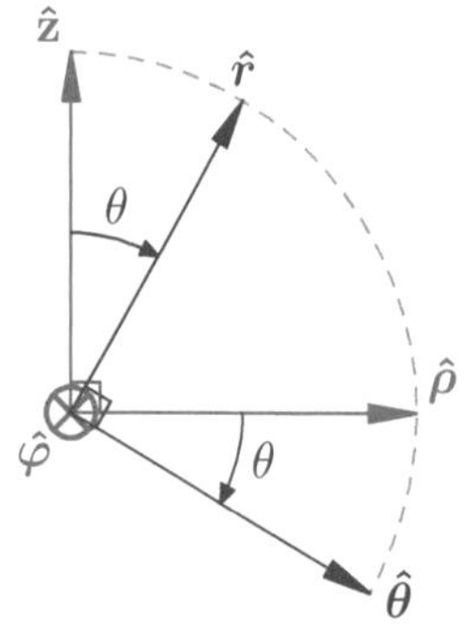

Fig. 10.5 Cylindrical and spherical directions. Side view.

Each of the basis vectors $\hat{\boldsymbol{r}}, \hat{\boldsymbol{\theta}}, \hat{\boldsymbol{\varphi}}$, evolves according to the rotation equation (10.11), with $\boldsymbol{\omega}(t)$ given by (10.22), yielding

$$\begin{aligned}
\frac{\mathrm{d}\hat{\boldsymbol{r}}}{\mathrm{d}t} &= \left(\dot{\varphi}\,\hat{\mathbf{z}} + \dot{\theta}\,\hat{\boldsymbol{\varphi}}\right) \times \hat{\boldsymbol{r}} = \dot{\varphi}\sin\theta\,\hat{\boldsymbol{\varphi}} + \dot{\theta}\,\hat{\boldsymbol{\theta}},\\
\frac{\mathrm{d}\hat{\boldsymbol{\theta}}}{\mathrm{d}t} &= \left(\dot{\varphi}\,\hat{\mathbf{z}} + \dot{\theta}\,\hat{\boldsymbol{\varphi}}\right) \times \hat{\boldsymbol{\theta}} = \dot{\varphi}\cos\theta\,\hat{\boldsymbol{\varphi}} - \dot{\theta}\,\hat{\boldsymbol{r}},\\
\frac{\mathrm{d}\hat{\boldsymbol{\varphi}}}{\mathrm{d}t} &= \left(\dot{\varphi}\,\hat{\mathbf{z}} + \dot{\theta}\,\hat{\boldsymbol{\varphi}}\right) \times \hat{\boldsymbol{\varphi}} = -\dot{\varphi}\,\hat{\boldsymbol{\rho}} = -\dot{\varphi}\,\sin\theta\,\hat{\boldsymbol{r}} - \dot{\varphi}\,\cos\theta\,\hat{\boldsymbol{\theta}}.
\end{aligned} \tag{10.23}$$

Partial derivatives of the direction vectors with respect to angles can be interpreted as rotations at unit angular rates, with $(\dot{\varphi}, \dot{\theta}) = (1, 0)$ and $\boldsymbol{\omega} = \hat{\mathbf{z}}$ for $\partial/\partial\varphi$, and $(\dot{\varphi}, \dot{\theta}) = (0, 1)$ and $\boldsymbol{\omega} = \hat{\boldsymbol{\varphi}}$ for $\partial/\partial\theta$, yielding

$$\begin{aligned}
\frac{\partial\hat{\boldsymbol{r}}}{\partial\varphi} &= \hat{\mathbf{z}} \times \hat{\boldsymbol{r}} = \sin\theta\,\hat{\boldsymbol{\varphi}},\\
\frac{\partial\hat{\boldsymbol{\theta}}}{\partial\varphi} &= \hat{\mathbf{z}} \times \hat{\boldsymbol{\theta}} = \cos\theta\,\hat{\boldsymbol{\varphi}},\\
\frac{\partial\hat{\boldsymbol{\varphi}}}{\partial\varphi} &= \hat{\mathbf{z}} \times \hat{\boldsymbol{\varphi}} = -\hat{\boldsymbol{\rho}} = -(\sin\theta\,\hat{\boldsymbol{r}} + \cos\theta\,\hat{\boldsymbol{\theta}}),
\end{aligned} \tag{10.24a}$$

and

$$\begin{aligned}
\frac{\partial\hat{\boldsymbol{r}}}{\partial\theta} &= \hat{\boldsymbol{\varphi}} \times \hat{\boldsymbol{r}} = \hat{\boldsymbol{\theta}},\\
\frac{\partial\hat{\boldsymbol{\theta}}}{\partial\theta} &= \hat{\boldsymbol{\varphi}} \times \hat{\boldsymbol{\theta}} = -\hat{\boldsymbol{r}},\\
\frac{\partial\hat{\boldsymbol{\varphi}}}{\partial\theta} &= \hat{\boldsymbol{\varphi}} \times \hat{\boldsymbol{\varphi}} = 0.
\end{aligned} \tag{10.24b}$$

These results can also be obtained from the chain rule

$$\frac{\mathrm{d}\hat{\boldsymbol{r}}}{\mathrm{d}t} = \frac{\partial\hat{\boldsymbol{r}}}{\partial\theta}\dot{\theta} + \frac{\partial\hat{\boldsymbol{r}}}{\partial\varphi}\dot{\varphi}, \qquad \frac{\mathrm{d}\hat{\boldsymbol{\theta}}}{\mathrm{d}t} = \frac{\partial\hat{\boldsymbol{\theta}}}{\partial\theta}\dot{\theta} + \frac{\partial\hat{\boldsymbol{\theta}}}{\partial\varphi}\dot{\varphi}, \qquad \frac{\mathrm{d}\hat{\boldsymbol{\varphi}}}{\mathrm{d}t} = \frac{\mathrm{d}\hat{\boldsymbol{\varphi}}}{\mathrm{d}\varphi}\dot{\varphi},$$

applied to

$$\begin{aligned}
\hat{\boldsymbol{r}} &= \cos\theta\,\hat{\mathbf{z}} + \sin\theta\,\hat{\boldsymbol{\rho}},\\
\hat{\boldsymbol{\theta}} &= -\sin\theta\,\hat{\mathbf{z}} + \cos\theta\,\hat{\boldsymbol{\rho}},
\end{aligned} \tag{10.25}$$

with $\hat{\boldsymbol{\rho}}$ and $\hat{\boldsymbol{\varphi}}$ defined in (10.20).

10.4 Velocity and acceleration

The velocity of a particle with radius vector $\boldsymbol{r}(t) = x(t)\hat{\mathbf{x}} + y(t)\hat{\mathbf{y}} + z(t)\hat{\mathbf{z}}$ is

$$\frac{\mathrm{d}\boldsymbol{r}}{\mathrm{d}t} = \dot{x}\hat{\mathbf{x}} + \dot{y}\hat{\mathbf{y}} + \dot{z}\hat{\mathbf{z}} \tag{10.26}$$

in cartesian coordinates assuming that $\hat{\mathbf{x}}, \hat{\mathbf{y}}, \hat{\mathbf{z}}$ are fixed, but in cylindrical coordinates the radius vector is $\boldsymbol{r} = \rho(t)\hat{\boldsymbol{\rho}}(t) + z(t)\hat{\mathbf{z}}$ and using (10.18)

$$\frac{\mathrm{d}\boldsymbol{r}}{\mathrm{d}t} = \dot{\rho}\hat{\boldsymbol{\rho}} + \rho\frac{\mathrm{d}\hat{\boldsymbol{\rho}}}{\mathrm{d}t} + \dot{z}\hat{\mathbf{z}} = \dot{\rho}\,\hat{\boldsymbol{\rho}} + \rho\dot{\varphi}\,\hat{\boldsymbol{\varphi}} + \dot{z}\hat{\mathbf{z}}. \tag{10.27}$$

In spherical coordinates $\boldsymbol{r} = r(t)\hat{\boldsymbol{r}}(t)$ and (10.23) gives

$$\frac{\mathrm{d}\boldsymbol{r}}{\mathrm{d}t} = \dot{r}\hat{\boldsymbol{r}} + r\frac{\mathrm{d}\hat{\boldsymbol{r}}}{\mathrm{d}t} = \dot{r}\hat{\boldsymbol{r}} + r\dot{\varphi}\sin\theta\,\hat{\boldsymbol{\varphi}} + r\dot{\theta}\,\hat{\boldsymbol{\theta}}. \tag{10.28}$$

The acceleration $\mathrm{d}^2\boldsymbol{r}/\mathrm{d}t^2$ of a particle with radius vector $\boldsymbol{r}(t)$ has the cartesian expression

$$\frac{\mathrm{d}^2\boldsymbol{r}}{\mathrm{d}t^2} = \ddot{x}\,\hat{\mathbf{x}} + \ddot{y}\,\hat{\mathbf{y}} + \ddot{z}\,\hat{\mathbf{z}}, \tag{10.29}$$

but in cylindrical coordinates, the derivative of (10.27) using (10.18) is

$$\frac{\mathrm{d}^2\boldsymbol{r}}{\mathrm{d}t^2} = \left(\ddot{\rho} - \rho\dot{\varphi}^2\right)\hat{\boldsymbol{\rho}} + \left(\rho\ddot{\varphi} + 2\dot{\rho}\dot{\varphi}\right)\hat{\boldsymbol{\varphi}} + \ddot{z}\,\hat{\mathbf{z}}. \tag{10.30}$$

In spherical coordinates, the derivative of (10.28) using (10.23) yields

$$\begin{aligned}\frac{\mathrm{d}^2\boldsymbol{r}}{\mathrm{d}t^2} &= \left(\ddot{r} - r\dot{\theta}^2 - r\dot{\varphi}^2\sin^2\theta\right)\hat{\boldsymbol{r}} \\ &\quad + \left(r\ddot{\theta} + 2\dot{r}\dot{\theta} - r\dot{\varphi}^2\sin\theta\cos\theta\right)\hat{\boldsymbol{\theta}} \\ &\quad + \left(r\ddot{\varphi}\sin\theta + 2r\dot{\theta}\dot{\varphi}\cos\theta + 2\dot{r}\dot{\varphi}\sin\theta\right)\hat{\boldsymbol{\varphi}}.\end{aligned} \tag{10.31}$$

10.5 Angular velocities

Precession, nutation, and spin

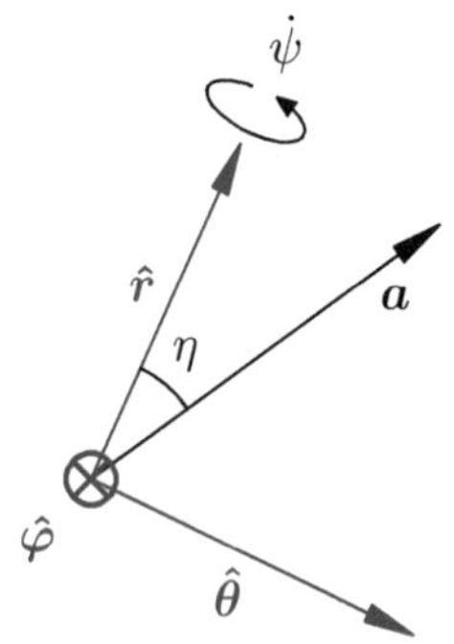

Fig. 10.6 Rotation of $\boldsymbol{a}$ about $\hat{\boldsymbol{r}}$.

Angular velocities add up, as seen in (10.22). As another example, consider a vector $\boldsymbol{a}$ rotating about $\hat{\boldsymbol{r}}(t)$ at angular rate $\dot{\psi}$ (Fig. 10.6), then

$$\frac{\mathrm{d}\boldsymbol{a}}{\mathrm{d}t} = \boldsymbol{\omega}(t) \times \boldsymbol{a} \tag{10.32}$$

with

$$\boldsymbol{\omega}(t) = \dot{\varphi}\,\hat{\mathbf{z}} + \dot{\theta}\,\hat{\boldsymbol{\varphi}} + \dot{\psi}\,\hat{\boldsymbol{r}}. \tag{10.33}$$

If $\hat{\boldsymbol{r}}(t)$ is the direction of the axis of a spinning top, then $\dot{\psi}$ is the *spin rate* of the top about its axis, while $\dot{\varphi}$ and $\dot{\theta}$ are the *precession* and *nutation* rates of its axis, respectively, and $\boldsymbol{a}(t)$ would then be any material vector moving with the spinning top. Such a vector $\boldsymbol{a}(t)$ has the explicit form

$$\boldsymbol{a}(t) = a\cos\eta\,\hat{\boldsymbol{r}}(t) + a\sin\eta\,(\cos\psi(t)\,\hat{\boldsymbol{\theta}}(t) + \sin\psi(t)\,\hat{\boldsymbol{\varphi}}(t)), \tag{10.34}$$

where $a = |\boldsymbol{a}|$ is a fixed magnitude, η is the angle between $\boldsymbol{a}$ and $\hat{\boldsymbol{r}}$, and $\psi(t)$ is the varying azimuthal angle of $\hat{\boldsymbol{a}}$ in the $(\hat{\boldsymbol{\theta}}, \hat{\boldsymbol{\varphi}})$ plane. The angles η and ψ are the polar and azimuthal angles of $\boldsymbol{a}$ in the $\{\hat{\boldsymbol{\theta}}, \hat{\boldsymbol{\varphi}}, \hat{\boldsymbol{r}}\}$ basis, that is itself rotating with angular velocity (10.22).

The angle ψ and the unit vectors $\hat{\boldsymbol{r}}, \hat{\boldsymbol{\theta}}, \hat{\boldsymbol{\varphi}}$ vary with time, but a and η are considered fixed in this example. The simple addition of angular velocities in (10.33) can be understood as a result of the chain rule since

$$\frac{\mathrm{d}\boldsymbol{a}}{\mathrm{d}t} = \frac{\partial \boldsymbol{a}}{\partial \varphi}\,\dot{\varphi} + \frac{\partial \boldsymbol{a}}{\partial \theta}\,\dot{\theta} + \frac{\partial \boldsymbol{a}}{\partial \psi}\,\dot{\psi}, \tag{10.35a}$$

and the angle partial derivatives are rotations at unit angular rate *with all other angles fixed.* Thus,

$$\frac{\partial \boldsymbol{a}}{\partial \varphi} = \hat{\mathbf{z}} \times \boldsymbol{a}, \quad \frac{\partial \boldsymbol{a}}{\partial \theta} = \hat{\boldsymbol{\varphi}} \times \boldsymbol{a}, \quad \frac{\partial \boldsymbol{a}}{\partial \psi} = \hat{\boldsymbol{r}} \times \boldsymbol{a}, \tag{10.35b}$$

substituting into (10.35a) yields (10.32) with $\boldsymbol{\omega}(t)$ as in (10.33).

Additional rotations

Of course, the polar angle η of arbitrary vector $\boldsymbol{a}(t)$ with respect to $\hat{\boldsymbol{r}}(t)$ could also vary with time, $\eta = \eta(t)$, adding yet another rotation, in which case the angular velocity for $\boldsymbol{a}(t)$ would be

$$\boldsymbol{\omega}(t) = \dot{\varphi}\,\hat{\mathbf{z}} + \dot{\theta}\,\hat{\boldsymbol{\varphi}} + \dot{\psi}\,\hat{\boldsymbol{r}} + \dot{\eta}\,\hat{\boldsymbol{\psi}}, \tag{10.36}$$

where $\hat{\boldsymbol{\psi}} = -\sin\psi\,\hat{\boldsymbol{\theta}} + \cos\psi\,\hat{\boldsymbol{\varphi}}$ is the azimuthal direction vector in the $(\hat{\boldsymbol{\theta}}, \hat{\boldsymbol{\varphi}})$ plane (Fig. 10.7). The vector $\boldsymbol{a}(t)$ still has the explicit expression (10.34) but now η, ψ, $\hat{\boldsymbol{r}}, \hat{\boldsymbol{\theta}}, \hat{\boldsymbol{\varphi}}$ vary with time t. The angular velocity (10.36) can again be justified with the chain rule since $\hat{\boldsymbol{a}}$ is now a function of $(\varphi, \theta, \psi, \eta)$ that all depend on time t. Thus,

$$\frac{\mathrm{d}\boldsymbol{a}}{\mathrm{d}t} = \frac{\partial \boldsymbol{a}}{\partial \varphi}\dot{\varphi} + \frac{\partial \boldsymbol{a}}{\partial \theta}\dot{\theta} + \frac{\partial \boldsymbol{a}}{\partial \psi}\dot{\psi} + \frac{\partial \boldsymbol{a}}{\partial \eta}\dot{\eta}.$$

Expressing the angle partial derivatives as unit angular rate rotations about specific axes as in (10.35b) yields confirmation of (10.36).

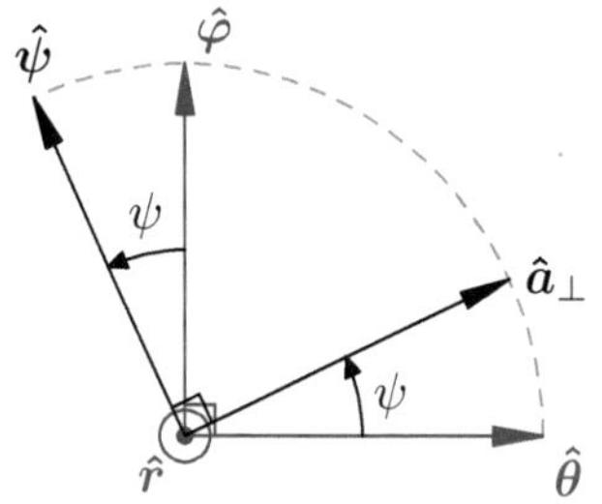

Fig. 10.7 Rotation by ψ about $\hat{\boldsymbol{r}}$.

Connection with rotation tensors

If a vector $\boldsymbol{a}(t)$ is rotating, then there exists a rotation tensor $\boldsymbol{\mathcal{R}}(t)$ such that

$$\boldsymbol{a}(t) = \boldsymbol{\mathcal{R}}(t) \cdot \boldsymbol{a}_0, \tag{10.37a}$$

where $\boldsymbol{a}_0 = \boldsymbol{a}(t_0)$, $\boldsymbol{\mathcal{R}}(t_0) = \mathbf{I}$, and t_0 is an arbitrary initial time. The tensor $\boldsymbol{\mathcal{R}}$ is proper orthogonal; that is, $\det\boldsymbol{\mathcal{R}} = 1$ and $\boldsymbol{\mathcal{R}}^{\mathsf{T}} \cdot \boldsymbol{\mathcal{R}} = \mathbf{I} = \boldsymbol{\mathcal{R}} \cdot \boldsymbol{\mathcal{R}}^{\mathsf{T}}$ for all t. Thus,

$$\frac{\mathrm{d}}{\mathrm{d}t}\left(\boldsymbol{\mathcal{R}} \cdot \boldsymbol{\mathcal{R}}^{\mathsf{T}}\right) = \dot{\boldsymbol{\mathcal{R}}} \cdot \boldsymbol{\mathcal{R}}^{\mathsf{T}} + \boldsymbol{\mathcal{R}} \cdot \dot{\boldsymbol{\mathcal{R}}}^{\mathsf{T}} = 0,$$

so $\dot{\mathcal{R}} \cdot \mathcal{R}^\mathsf{T}$ is antisymmetric. Then

$$\frac{\mathrm{d}\boldsymbol{a}}{\mathrm{d}t} = \dot{\mathcal{R}} \cdot \boldsymbol{a}_0 = \dot{\mathcal{R}} \cdot \mathcal{R}^\mathsf{T} \cdot \boldsymbol{a}(t) = \boldsymbol{\omega} \times \boldsymbol{a}, \tag{10.37b}$$

since the action of an antisymmetric tensor in 3D equals a cross product (9.45). If the rotation tensor is a product $\mathcal{R} = \mathcal{R}_2 \cdot \mathcal{R}_1$, for example, then

$$\dot{\mathcal{R}} = \dot{\mathcal{R}}_2 \cdot \mathcal{R}_1 + \mathcal{R}_2 \cdot \dot{\mathcal{R}}_1,$$

and

$$\dot{\mathcal{R}} \cdot \mathcal{R}^\mathsf{T} = \dot{\mathcal{R}}_2 \cdot \mathcal{R}_2^\mathsf{T} + \mathcal{R}_2 \cdot \dot{\mathcal{R}}_1 \cdot \mathcal{R}_1^\mathsf{T} \cdot \mathcal{R}_2^\mathsf{T}.$$

Additivity of angular velocities results from the product rule in this point of view (exercise 10.18).

Exercises

(10.1) Show that for any (sufficiently differentiable) vector function $\boldsymbol{u}(t)$

(a) $\boldsymbol{u} \cdot \dfrac{\mathrm{d}\boldsymbol{u}}{\mathrm{d}t} = \dfrac{\mathrm{d}}{\mathrm{d}t}\left(\dfrac{\boldsymbol{u} \cdot \boldsymbol{u}}{2}\right)$,

(b) $\boldsymbol{u} \times \dfrac{\mathrm{d}^2\boldsymbol{u}}{\mathrm{d}t^2} = \dfrac{\mathrm{d}}{\mathrm{d}t}\left(\boldsymbol{u} \times \dfrac{\mathrm{d}\boldsymbol{u}}{\mathrm{d}t}\right)$, and

(c) $\boldsymbol{u} \times \dfrac{\mathrm{d}\boldsymbol{u}}{\mathrm{d}t} \cdot \dfrac{\mathrm{d}^3\boldsymbol{u}}{\mathrm{d}t^3} = \dfrac{\mathrm{d}}{\mathrm{d}t}\left(\boldsymbol{u} \times \dfrac{\mathrm{d}\boldsymbol{u}}{\mathrm{d}t} \cdot \dfrac{\mathrm{d}^2\boldsymbol{u}}{\mathrm{d}t^2}\right)$.

(10.2) Show that if $\boldsymbol{u}(t)$ is any vector with constant magnitude, then $\boldsymbol{u} \cdot \dfrac{\mathrm{d}\boldsymbol{u}}{\mathrm{d}t} = 0,\ \forall t$.

(10.3) If $\boldsymbol{r} = r\hat{\boldsymbol{r}}$, show that

$$\hat{\boldsymbol{r}} \cdot \frac{\mathrm{d}\boldsymbol{r}}{\mathrm{d}t} = \frac{\mathrm{d}r}{\mathrm{d}t} \quad \text{and} \quad \boldsymbol{r} \cdot \frac{\mathrm{d}\boldsymbol{r}}{\mathrm{d}t} = r\frac{\mathrm{d}r}{\mathrm{d}t}.$$

(10.4) Show that

$$\frac{\mathrm{d}}{\mathrm{d}t}\begin{vmatrix} a_1 & b_1 & c_1 \\ a_2 & b_2 & c_2 \\ a_3 & b_3 & c_3 \end{vmatrix} = \begin{vmatrix} \dot{a}_1 & b_1 & c_1 \\ \dot{a}_2 & b_2 & c_2 \\ \dot{a}_3 & b_3 & c_3 \end{vmatrix} + \begin{vmatrix} a_1 & \dot{b}_1 & c_1 \\ a_2 & \dot{b}_2 & c_2 \\ a_3 & \dot{b}_3 & c_3 \end{vmatrix} + \begin{vmatrix} a_1 & b_1 & \dot{c}_1 \\ a_2 & b_2 & \dot{c}_2 \\ a_3 & b_3 & \dot{c}_3 \end{vmatrix}.$$

(10.5) Show that if $\boldsymbol{a}(t)$, $\boldsymbol{b}(t)$, $\boldsymbol{c}(t)$ are three distinct solutions $\boldsymbol{u}(t)$ of

$$\frac{\mathrm{d}\boldsymbol{u}}{\mathrm{d}t} = \boldsymbol{\omega}(t) \times \boldsymbol{u}$$

for the same $\boldsymbol{\omega}(t)$, then $\boldsymbol{a} \times \boldsymbol{b} \cdot \boldsymbol{c}$ is constant. Interpret geometrically.

(10.6) If $\boldsymbol{v}(t) = \mathrm{d}\boldsymbol{r}/\mathrm{d}t$ and $\boldsymbol{r}_c$ is any constant vector, show that

$$\frac{\mathrm{d}}{\mathrm{d}t}\big((\boldsymbol{r} - \boldsymbol{r}_c) \times m\boldsymbol{v}\big) = (\boldsymbol{r} - \boldsymbol{r}_c) \times \frac{\mathrm{d}}{\mathrm{d}t}(m\boldsymbol{v}).$$

In mechanics, $\boldsymbol{L} = (\boldsymbol{r} - \boldsymbol{r}_c) \times m\boldsymbol{v}$ is the *angular momentum* with respect to the point $\boldsymbol{r}_c$, of the particle of mass m with velocity $\boldsymbol{v}$ at position $\boldsymbol{r}$.

(10.7) Consider $\boldsymbol{r}(t) = \boldsymbol{a}\cos t + \boldsymbol{b}\sin t + \boldsymbol{c}$, where $\boldsymbol{a}$, $\boldsymbol{b}$, $\boldsymbol{c}$ are arbitrary constant vectors in 3D space. Sketch $\boldsymbol{r}(t)$ and indicate all points where $\boldsymbol{r} \cdot \mathrm{d}\boldsymbol{r}/\mathrm{d}t = 0$ for $\boldsymbol{c}$ zero, and also when $\boldsymbol{c}$ is nonzero.

(10.8) The position of a particle at time t is given by $\boldsymbol{r}(t) = \boldsymbol{a}\cos\theta(t) + \boldsymbol{b}\sin\theta(t)$, with $\theta(t) = -\pi/2 + (\pi/4)\cos t$ and $\boldsymbol{a}$, $\boldsymbol{b}$ arbitrary constants. What are the velocity and the acceleration of the particle? Sketch the particle motion and a few representative position, velocity and acceleration vectors.

(10.9) Verify (10.23) using (10.25).

(10.10) Close this book. Derive from scratch the partial derivatives of $\hat{\boldsymbol{r}}$, $\hat{\boldsymbol{\theta}}$, and $\hat{\boldsymbol{\varphi}}$ with respect to θ and φ.

(10.11) True or false:

$$\text{(i)}\ \frac{\partial\hat{\boldsymbol{\rho}}}{\partial\theta} = \hat{\boldsymbol{\varphi}} \times \hat{\boldsymbol{\rho}} = -\hat{\mathbf{z}}, \quad \text{(ii)}\ \frac{\partial\hat{\mathbf{z}}}{\partial\theta} = \hat{\boldsymbol{\varphi}} \times \hat{\mathbf{z}} = -\hat{\boldsymbol{\rho}},$$

where the unit vectors are as in Fig. 10.3. Explain.

(10.12) Verify (10.30) and (10.31) for the acceleration in cylindrical and spherical coordinates.

(10.13) A bead is rotating at constant angular velocity ω about a circular hoop. The hoop rotates about one of its diameters at constant angular velocity Ω. What are the bead velocity and acceleration? Sketch and show/explain your work.

(10.14) If $\mathrm{d}\boldsymbol{a}/\mathrm{d}t = \boldsymbol{\omega}(t) \times \boldsymbol{a}$ with $\hat{\boldsymbol{\omega}}$ fixed, show that $|\boldsymbol{a}|$ and $\hat{\boldsymbol{\omega}} \cdot \boldsymbol{a}$ are constant, *then* pick a suitable basis to show that

$$\boldsymbol{a}(t) = \boldsymbol{a}_0^{\parallel} + \boldsymbol{a}_0^{\perp} \cos\alpha(t) + \hat{\boldsymbol{\omega}} \times \boldsymbol{a}_0^{\perp} \sin\alpha(t)$$

where $\alpha(t) = \int_{t_0}^{t} \omega(s)\, ds$ and $\boldsymbol{a}_0 = \boldsymbol{a}(t_0)$.

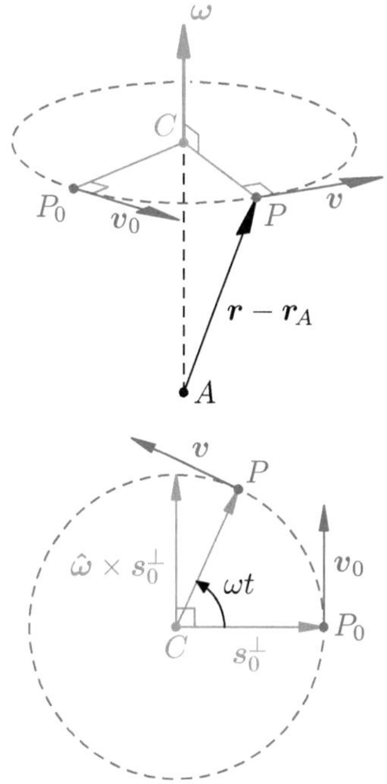

Fig. 10.8 Rotation of point P about fixed axis $(A, \boldsymbol{\omega})$. *Top:* 3D view; *bottom:* 2D view $\perp \boldsymbol{\omega}$.

(10.15) *Uniform rotation.* Point P rotates at constant angular rate ω about an *axis* that passes through the fixed point A and is parallel to the constant unit vector $\hat{\boldsymbol{\omega}}$ (Fig. 10.8). Show that its position vector $\boldsymbol{r}(t) = \overrightarrow{OP}$ with respect to a fixed origin O satisfies

$$\frac{\mathrm{d}\boldsymbol{r}}{\mathrm{d}t} = \boldsymbol{\omega} \times (\boldsymbol{r} - \boldsymbol{r}_A), \qquad (10.38)$$

where $\boldsymbol{\omega} = \omega\hat{\boldsymbol{\omega}}$ and $\boldsymbol{r}_A = \overrightarrow{OA}$. If P_0 is the position at $t = 0$, show that

$$\boldsymbol{r}(t) = \boldsymbol{r}_A + \boldsymbol{s}_0^{\parallel} + \boldsymbol{s}_0^{\perp} \cos\omega t + (\hat{\boldsymbol{\omega}} \times \boldsymbol{s}_0^{\perp}) \sin\omega t. \qquad (10.39)$$

Specify the center C, $\boldsymbol{s}_0^{\parallel}$, and $\boldsymbol{s}_0^{\perp}$ in terms of O, A, P_0, and $\hat{\boldsymbol{\omega}}$.

(10.16) Show that (10.34) with a and η fixed, solves (10.32) with angular velocity (10.33).

(10.17) Specify the tensor $\boldsymbol{\mathcal{R}}$ for rotation about a fixed $\hat{\mathbf{z}}$ by $\varphi(t)$, and the matrix of $\boldsymbol{\mathcal{R}}$ and that of $\dot{\boldsymbol{\mathcal{R}}}$ in the cartesian basis $\{\hat{\mathbf{x}}, \hat{\mathbf{y}}, \hat{\mathbf{z}}\}$. Verify that for any vector $\boldsymbol{a}$

$$\dot{\boldsymbol{\mathcal{R}}} \cdot \boldsymbol{\mathcal{R}}^{\mathsf{T}} \cdot \boldsymbol{a} = \dot{\varphi}\, \hat{\mathbf{z}} \times \boldsymbol{a}.$$

(10.18) The spherical basis $\{\hat{\boldsymbol{\theta}}, \hat{\boldsymbol{\varphi}}, \hat{\boldsymbol{r}}\}$ is related to the cartesian basis $\{\hat{\mathbf{x}}, \hat{\mathbf{y}}, \hat{\mathbf{z}}\}$ as $\{\mathbf{e}_1', \mathbf{e}_2', \mathbf{e}_3'\}$ is to $\{\mathbf{e}_1, \mathbf{e}_2, \mathbf{e}_3\}$ in (8.35b). Use that matrix formulation to calculate the time derivatives of the spherical direction vectors for arbitrary $\varphi(t)$, $\theta(t)$ with fixed $\{\hat{\mathbf{x}}, \hat{\mathbf{y}}, \hat{\mathbf{z}}\}$. Verify that your results recover (10.22), (10.23).

(10.19) *Uniqueness of* $\boldsymbol{\omega}$ *for a rigid frame.* Let $\mathbf{e}_i(t)$ denote a moving cartesian basis such that $\mathbf{e}_i \cdot \mathbf{e}_j = \delta_{ij}$ for all t. Show that $\dot{\mathbf{e}}_i = \boldsymbol{\omega}(t) \times \mathbf{e}_i$, for a unique $\boldsymbol{\omega}(t)$. One compact way to do so is to define $\dot{\mathbf{e}}_i = A_{ij}(t)\mathbf{e}_j$, show that A_{ij} is antisymmetric, then use (5.27).

(10.20) Let $\mathbf{e}_i(t)$ denote a moving cartesian basis with $\mathbf{e}_i(t) = Q_{ij}(t)\mathbf{e}_j(0)$. Show that its angular velocity is $\boldsymbol{\omega}(t) = \omega_i(t)\mathbf{e}_i(t)$ with $\omega_i = \frac{1}{2}\epsilon_{ijk}\dot{Q}_{jl}Q_{kl}$, that is

$$\dot{\mathsf{Q}}\mathsf{Q}^{\mathsf{T}} = \begin{bmatrix} 0 & \omega_3 & -\omega_2 \\ -\omega_3 & 0 & \omega_1 \\ \omega_2 & -\omega_1 & 0 \end{bmatrix}. \qquad (10.40)$$

Note the transpose compared to (9.45), and recall (5.27), (5.26) and (8.35).

(10.21) If M is any m-by-3 matrix, $\{\hat{\mathbf{x}}, \hat{\mathbf{y}}, \hat{\mathbf{z}}\}$ is a cartesian basis, and $\boldsymbol{a} = a_1\hat{\mathbf{x}} + a_2\hat{\mathbf{y}} + a_3\hat{\mathbf{z}}$ is an arbitrary vector, show/explain why

$$\mathsf{M} \begin{bmatrix} 0 & a_3 & -a_2 \\ -a_3 & 0 & a_1 \\ a_2 & -a_1 & 0 \end{bmatrix} \begin{bmatrix} \hat{\mathbf{x}} \\ \hat{\mathbf{y}} \\ \hat{\mathbf{z}} \end{bmatrix} = \mathsf{M} \begin{bmatrix} \boldsymbol{a} \times \hat{\mathbf{x}} \\ \boldsymbol{a} \times \hat{\mathbf{y}} \\ \boldsymbol{a} \times \hat{\mathbf{z}} \end{bmatrix} = \boldsymbol{a} \times \mathsf{M} \begin{bmatrix} \hat{\mathbf{x}} \\ \hat{\mathbf{y}} \\ \hat{\mathbf{z}} \end{bmatrix}. \qquad (10.41)$$

(10.22) Let $\{\hat{\mathbf{x}}(t), \hat{\mathbf{y}}(t), \hat{\mathbf{z}}(t)\}$ be a cartesian basis attached to an airplane (Fig. 10.9) with angular velocity $\boldsymbol{\omega}(t) = \omega_x\hat{\mathbf{x}} + \omega_y\hat{\mathbf{y}} + \omega_z\hat{\mathbf{z}}$ in the moving airplane frame. If $\alpha(t)$, $\beta(t)$, and $\gamma(t)$ are the Bryan angles (8.30) specifying the attitude of the airplane at time t with respect to a fixed cartesian basis $\{\hat{\mathbf{x}}_0, \hat{\mathbf{y}}_0, \hat{\mathbf{z}}_0\}$, argue that

$$\boldsymbol{\omega}(t) = \dot{\alpha}\, \hat{\mathbf{z}}_0 + \dot{\beta}\, \hat{\mathbf{y}}_1(t) + \dot{\gamma}\, \hat{\mathbf{x}}_2(t) \qquad (10.42a)$$

and specify $\hat{\mathbf{z}}_0$, $\hat{\mathbf{y}}_1(t)$, and $\hat{\mathbf{x}}_2(t)$ in terms of

$\{\hat{\mathbf{x}}, \hat{\mathbf{y}}, \hat{\mathbf{z}}\}$ at time t. Deduce that

$$\begin{bmatrix} \omega_x \\ \omega_y \\ \omega_z \end{bmatrix} = \begin{bmatrix} -\sin\beta & 0 & 1 \\ \cos\beta\sin\gamma & \cos\gamma & 0 \\ \cos\beta\cos\gamma & -\sin\gamma & 0 \end{bmatrix} \begin{bmatrix} \dot{\alpha} \\ \dot{\beta} \\ \dot{\gamma} \end{bmatrix}. \tag{10.42b}$$

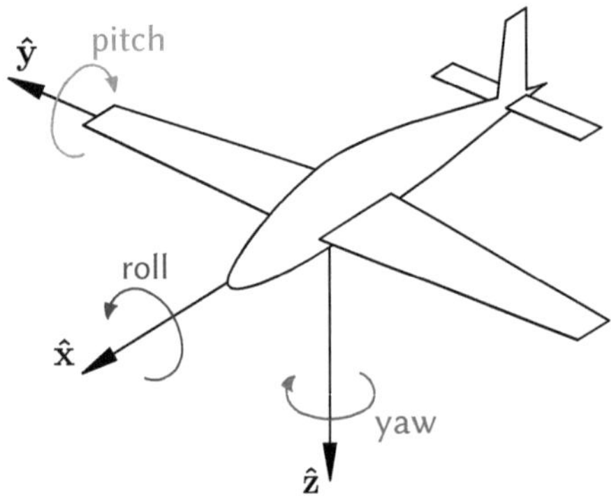

Fig. 10.9 Yaw, pitch and roll.

(10.23) Differentiate (8.31) for varying α, β, γ with fixed cartesian basis $\{\hat{\mathbf{x}}_0, \hat{\mathbf{y}}_0, \hat{\mathbf{z}}_0\}$, and use (10.41) to show that

$$\frac{\mathrm{d}}{\mathrm{d}t}\{\hat{\mathbf{x}}, \hat{\mathbf{y}}, \hat{\mathbf{z}}\} = \boldsymbol{\omega}(t) \times \{\hat{\mathbf{x}}, \hat{\mathbf{y}}, \hat{\mathbf{z}}\}$$

with $\boldsymbol{\omega}(t)$ as in (10.42a).

(10.24) Equation (10.42b) shows that finding (α, β, γ) from $(\omega_x, \omega_y, \omega_z)$ involves solving *three coupled nonlinear* ODEs that, furthermore, are singular when $\cos\beta = 0$ (the "gimbal lock" issue). In that case a direct numerical integration of

$$\frac{\mathrm{d}\boldsymbol{a}}{\mathrm{d}t} = \boldsymbol{\omega}(t) \times \boldsymbol{a},$$

for $\boldsymbol{a}(t) = \hat{\mathbf{x}}(t)$, $\hat{\mathbf{y}}(t)$, $\hat{\mathbf{z}}(t)$, may be more straightforward. *Euler's method* approximates this vector differential equation by a time marching scheme: knowing $\boldsymbol{a} = \boldsymbol{a}(t)$ and $\boldsymbol{\omega} = \boldsymbol{\omega}(t)$, we approximate $\boldsymbol{a}(t + \Delta t) \approx \boldsymbol{a}'$ with

$$\boldsymbol{a}' = \boldsymbol{a} + (\boldsymbol{\omega} \times \boldsymbol{a})\,\Delta t, \tag{10.43}$$

for small but finite Δt, such that $0 < |\boldsymbol{\omega}|\Delta t \ll 1$, then restart from $\boldsymbol{a}'$ to find an approximation to $\boldsymbol{a}(t+2\Delta t)$. A better method is *semi-implicit Euler*

$$\begin{cases} a_1' = a_1 + (\omega_2 a_3 - \omega_3 a_2)\,\Delta t, \\ a_2' = a_2 + \left(\omega_3 a_1' - \omega_1 a_3\right)\Delta t, \\ a_3' = a_3 + \left(\omega_1 a_2' - \omega_2 a_1'\right)\Delta t, \end{cases} \tag{10.44}$$

with respect to a *fixed* cartesian basis.

(a) Show that (10.42b) is singular for $\beta = \pm\pi/2$.

(b) Show that (10.43) is unstable in the sense that $|\boldsymbol{a}'| > |\boldsymbol{a}|$, so the magnitude increases at each step.

(c) Show that (10.44) preserves

$$a_1^2 + a_2^2 + a_3^2 - \Delta t(\omega_1 a_2 a_3 - \omega_2 a_3 a_1 + \omega_3 a_1 a_2)$$

by rewriting the system (10.44) in matrix form as

$$\mathsf{U}^\mathsf{T}\mathsf{a}' = \mathsf{U}\mathsf{a},$$

for a certain upper triangular matrix U; then, showing that

$$\mathsf{a}^\mathsf{T}\mathsf{U}^\mathsf{T}\mathsf{a}' = \mathsf{a}^\mathsf{T}\mathsf{U}\mathsf{a} = (\mathsf{a}')^\mathsf{T}\mathsf{U}\mathsf{a}'.$$

(d) *Optional.* Implement (10.43) and (10.44) and apply to (1) $\boldsymbol{\omega}(t) = \Omega\hat{\mathbf{z}}$ and (2) $\boldsymbol{\omega}(t) = \Omega\hat{\mathbf{z}} + \omega\hat{\boldsymbol{\varphi}}$, where $\hat{\boldsymbol{\varphi}} = (-\sin(\Omega t)\,\hat{\mathbf{x}} + \cos(\Omega t)\hat{\mathbf{y}})$, for constant ω and Ω. Find the exact solution in those cases. Compare exact and numerical solutions.

Dynamics

11

11.1 Single particle

In classical mechanics, the motion of a particle of constant mass m is governed by Newton's law

$$\boldsymbol{F} = m\boldsymbol{a}, \tag{11.1}$$

where $\boldsymbol{F}$ is the resultant of the forces acting on the particle and

$$\boldsymbol{a} = \frac{\mathrm{d}\boldsymbol{v}}{\mathrm{d}t} = \frac{\mathrm{d}^2\boldsymbol{r}}{\mathrm{d}t^2}$$

is its acceleration, with $\boldsymbol{r}$ its position vector. Newton's law is a *vector differential equation* for the particle position $\boldsymbol{r}(t)$ at time t given the mass m and the force $\boldsymbol{F}(t)$. We cover two elementary examples here, then some fundamental examples in section 11.2, and the exercises.

Constant velocity

If $\boldsymbol{F} = 0$, then

$$\boldsymbol{a} = \frac{\mathrm{d}\boldsymbol{v}}{\mathrm{d}t} = 0,$$

so the velocity of the particle is constant, $\boldsymbol{v}(t) = \boldsymbol{v}_0$ say, and its position is given by the vector differential equation $\mathrm{d}\boldsymbol{r}/\mathrm{d}t = \boldsymbol{v}(t) = \boldsymbol{v}_0$ whose solution is

$$\boldsymbol{r}(t) = \boldsymbol{r}_0 + t\boldsymbol{v}_0, \tag{11.2}$$

where $\boldsymbol{r}_0$ is a constant of integration whose meaning is clear: it is the position of the particle at time $t = 0$. If the particle is at P_0 at time t_0 then

$$\boldsymbol{r}(t) = \boldsymbol{r}_0 + (t - t_0)\boldsymbol{v}_0. \tag{11.3}$$

The trajectory of the particle is a straight line parallel to $\boldsymbol{v}_0$ (Fig. 11.1).

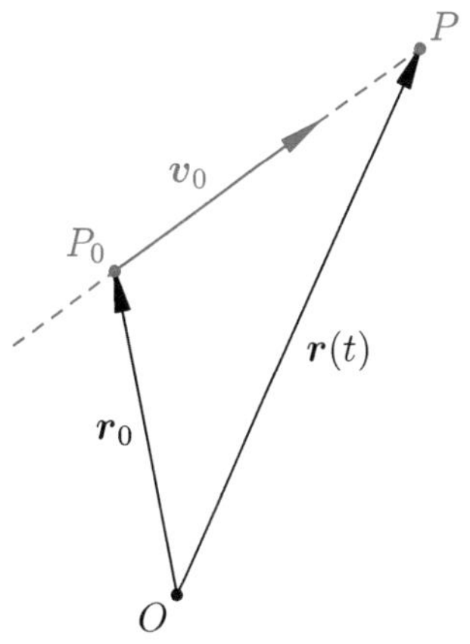

Fig. 11.1 Straight line (11.3).

Constant acceleration

If $\boldsymbol{F} = \boldsymbol{F}_0$ is constant, then the acceleration $\boldsymbol{a}_0 = \boldsymbol{F}_0/m$ is constant and

$$\frac{\mathrm{d}^2\boldsymbol{r}}{\mathrm{d}t^2} = \frac{\mathrm{d}\boldsymbol{v}}{\mathrm{d}t} = \boldsymbol{a}_0. \tag{11.4}$$

Integrating once gives

$$\dot{\boldsymbol{r}} = \boldsymbol{v} = \boldsymbol{v}_0 + t\boldsymbol{a}_0,$$

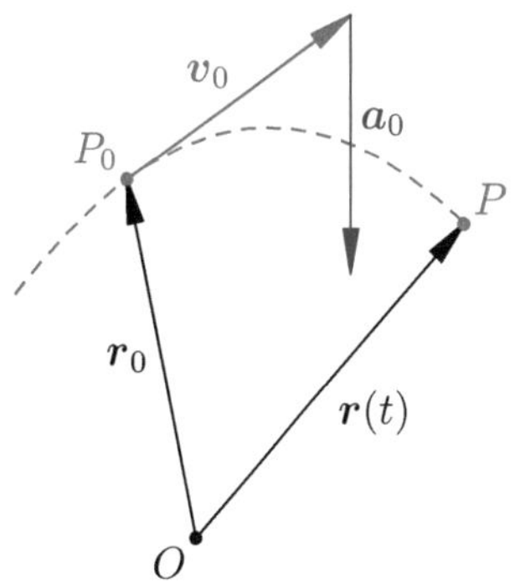

Fig. 11.2 Parabola (11.5).

and another integration yields

$$\boldsymbol{r}(t) = \boldsymbol{r}_0 + t\boldsymbol{v}_0 + \tfrac{1}{2}t^2\boldsymbol{a}_0, \tag{11.5}$$

where $\boldsymbol{v}_0$ and $\boldsymbol{r}_0$ are *vector* constants of integration. They are the velocity and position at $t = 0$, respectively. If the initial data $\boldsymbol{r}_0$ and $\boldsymbol{v}_0$ are given at t_0, it suffices to replace t by $t - t_0$ in (11.5).

The trajectory is a parabola passing through P_0 parallel to $\boldsymbol{v}_0$ at $t = 0$. The parabolic motion is in the plane through $\boldsymbol{r}_0$ that is parallel to $\boldsymbol{v}_0$ and $\boldsymbol{a}_0$ but the origin O may not be in that plane (Fig. 11.2). We can write this parabola in standard form by selecting cartesian axes such that $\boldsymbol{r}_0 = \boldsymbol{0}$, $\boldsymbol{a}_0 = -g\hat{\mathbf{y}}$ and $\boldsymbol{v}_0 = v_0(\cos\alpha\,\hat{\mathbf{x}} + \sin\alpha\,\hat{\mathbf{y}})$ (Fig. 11.3). Then (11.5) becomes

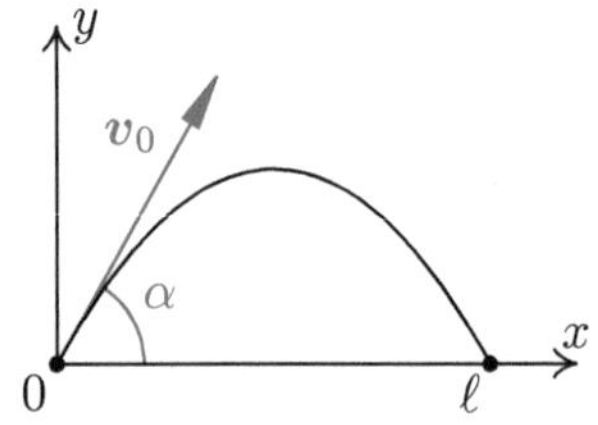

Fig. 11.3 $\ell = (v_0^2/g)\sin 2\alpha$.

$$\boldsymbol{r} = x\hat{\mathbf{x}} + y\hat{\mathbf{y}} + z\hat{\mathbf{z}} = -\tfrac{1}{2}\,g\,t^2\hat{\mathbf{y}} + v_0(\cos\alpha\hat{\mathbf{x}} + \sin\alpha\hat{\mathbf{y}})t,$$

yielding the parametric equations

$$x = v_0\,t\cos\alpha, \qquad y = v_0\,t\sin\alpha - gt^2/2.$$

Eliminating t when $\cos\alpha \neq 0$ yields

$$y = x\,\tan\alpha - \frac{g}{2v_0^2\cos^2\alpha}\,x^2. \tag{11.6}$$

11.2 Central force motion

A force $\boldsymbol{F} = F(r)\,\hat{\boldsymbol{r}}$, acting at point $\boldsymbol{r} = r\hat{\boldsymbol{r}}$, that always points in the radial direction $\hat{\boldsymbol{r}}$ and whose magnitude $F(r)$ depends only on the distance $r = |\boldsymbol{r}|$ to the origin, is called a *central force*. The origin is not an arbitrary point here: it is the location of the source of the force, the Sun for planetary motion. Newton's law for a particle of mass m submitted to such a force is

$$m\frac{\mathrm{d}\boldsymbol{v}}{\mathrm{d}t} = F(r)\,\hat{\boldsymbol{r}}, \tag{11.7}$$

where $\boldsymbol{v} = \mathrm{d}\boldsymbol{r}/\mathrm{d}t$ and $\boldsymbol{r} = r\hat{\boldsymbol{r}}$ is the position vector of the particle. Hence both r and $\hat{\boldsymbol{r}}$ are functions of time t, in general. The motion resulting from such a force *conserves* two fundamental quantities, *angular momentum* and *energy*, as we show thanks to a simple cross product, then a dot product.

Conservation of angular momentum

The cross product of (11.7) with $\boldsymbol{r}$ yields

$$\boldsymbol{r} \times m\frac{\mathrm{d}\boldsymbol{v}}{\mathrm{d}t} = \boldsymbol{r} \times F(r)\hat{\boldsymbol{r}} = 0,$$

since $\boldsymbol{r} \times \hat{\boldsymbol{r}} = r\hat{\boldsymbol{r}} \times \hat{\boldsymbol{r}} = 0$. Using that result and the product rule yields

$$\frac{\mathrm{d}}{\mathrm{d}t}\left(\boldsymbol{r} \times m\boldsymbol{v}\right) = \boldsymbol{v} \times m\boldsymbol{v} + \boldsymbol{r} \times m\frac{\mathrm{d}\boldsymbol{v}}{\mathrm{d}t} = 0.$$

Therefore

$$\boldsymbol{r} \times m\boldsymbol{v} = \boldsymbol{L}_0, \tag{11.8}$$

where $\boldsymbol{L}_0 = L_0\,\hat{\boldsymbol{L}}_0 = \boldsymbol{r}_0 \times m\boldsymbol{v}_0$ is a constant vector. The vector $\boldsymbol{L} = \boldsymbol{r} \times m\boldsymbol{v}$ is called *angular momentum* in physics. The fact that $\boldsymbol{r} \times m\boldsymbol{v}$ is constant implies that the motion is in the plane that passes through the origin O and is perpendicular to $\boldsymbol{L}_0$; the motion is *planar.* Constant $\boldsymbol{L}$ also implies that "the radius vector sweeps equal areas in equal times." Indeed, $\boldsymbol{v}\mathrm{d}t = \mathrm{d}\boldsymbol{r}$ is the displacement during the infinitesimal time span $\mathrm{d}t$. Thus,

$$\boldsymbol{r} \times \boldsymbol{v}\,\mathrm{d}t = \boldsymbol{r} \times \mathrm{d}\boldsymbol{r} = \frac{\boldsymbol{L}_0}{m}\,\mathrm{d}t,$$

but

$$\mathrm{d}A(t) = \frac{1}{2}\,|\boldsymbol{r}(t) \times \mathrm{d}\boldsymbol{r}(t)| = \frac{L_0}{2m}\,\mathrm{d}t$$

is the infinitesimal triangular area swept by $\boldsymbol{r}(t)$ in time $\mathrm{d}t$ (Fig. 11.4). This is Kepler's law: the area swept by $\boldsymbol{r}(t)$ in time Δt is independent of the start time t_1 (Fig. 11.5):

$$\int_{t_1}^{t_1+\Delta t} \mathrm{d}A(t) = \int_{t_1}^{t_1+\Delta t} \frac{L_0}{2m}\,\mathrm{d}t = \frac{L_0}{2m}\,\Delta t.$$

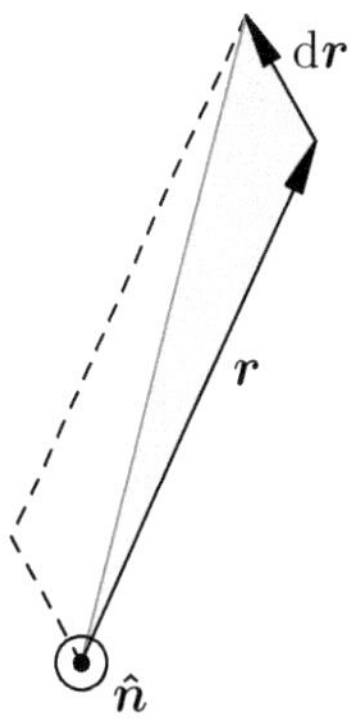

Fig. 11.4 $\mathrm{d}A = \frac{1}{2}|\boldsymbol{r} \times \mathrm{d}\boldsymbol{r}|$.

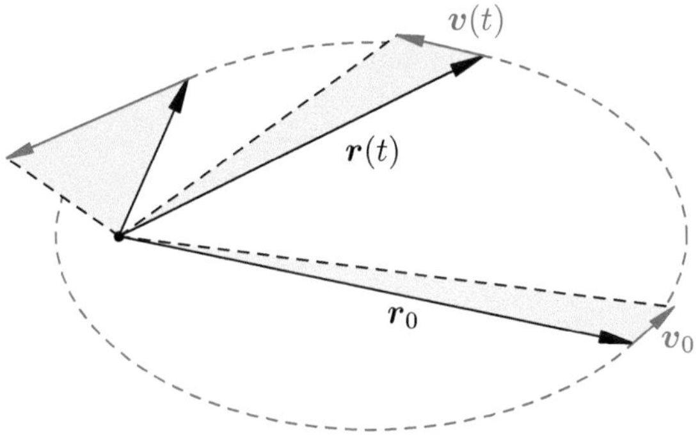

Fig. 11.5 Kepler's (2nd) law: "The radius vector sweeps equal areas in equal times" follows from conservation of the "area sweep rate": $\boldsymbol{r}(t) \times \boldsymbol{v}(t) = 2\hat{\boldsymbol{z}}\,\mathrm{d}A/\mathrm{d}t = \boldsymbol{L}_0/m$.

Conservation of energy

The dot product of (11.7) with $\boldsymbol{v}$

$$m\frac{\mathrm{d}\boldsymbol{v}}{\mathrm{d}t} \cdot \boldsymbol{v} - F(r)\,\hat{\boldsymbol{r}} \cdot \boldsymbol{v} = 0$$

yields

$$\frac{\mathrm{d}}{\mathrm{d}t}\left(m\,\frac{\boldsymbol{v} \cdot \boldsymbol{v}}{2} + V(r)\right) = 0. \tag{11.9}$$

Equation (11.9) follows from the product rules

$$\boldsymbol{v} \cdot \frac{\mathrm{d}\boldsymbol{v}}{\mathrm{d}t} = \frac{\mathrm{d}}{\mathrm{d}t}\left(\frac{\boldsymbol{v} \cdot \boldsymbol{v}}{2}\right), \qquad \hat{\boldsymbol{r}} \cdot \frac{\mathrm{d}\boldsymbol{r}}{\mathrm{d}t} = \frac{\mathrm{d}r}{\mathrm{d}t}$$

from (10.5), (10.9), and the chain rule

$$\frac{\mathrm{d}V}{\mathrm{d}t} = \frac{\mathrm{d}V}{\mathrm{d}r}\frac{\mathrm{d}r}{\mathrm{d}t} = \frac{\mathrm{d}V}{\mathrm{d}r}\,\hat{\boldsymbol{r}}\cdot\boldsymbol{v} = -F(r)\,\hat{\boldsymbol{r}}\cdot\boldsymbol{v},$$

where $V(r)$ is an antiderivative of $-F(r)$ such that

$$\frac{\mathrm{d}V}{\mathrm{d}r} = -F(r). \tag{11.10}$$

For example, $V(r) = -1/r$ for the classic inverse square law $F(r) = -1/r^2$. The function $V(r)$ is called the *potential*, and the minus sign in (11.10) is a physics convention so that the force points toward *decreasing* potential.[1]

[1] Balls roll from high to low elevation. Heat flows from hot to cold temperature.

Equation (11.9) implies that

$$\tfrac{1}{2}m|\boldsymbol{v}|^2 + V(r) = E_0 \tag{11.11}$$

is a constant called the *total energy.* The first term, $m|\boldsymbol{v}|^2/2$, is the *kinetic energy* and the second term, $V(r)$, is the *potential energy* which is defined up to an arbitrary constant. $V(r)$ and E_0 can be negative but $m|\boldsymbol{v}|^2/2 \geq 0$, so the physically admissible r domain is that where $V(r) \leq E_0$.

Effective potential

Conservation of angular momentum (11.8) implies that the motion is planar. Picking $\hat{\mathbf{z}} = \hat{\boldsymbol{L}}_0$ as the angular momentum direction, and using cylindrical coordinates with $z = 0$ and $\rho = r$ in (10.27), or spherical coordinates with $\theta = \pi/2$ in (10.28), gives

$$L_0 = |\boldsymbol{r} \times m\boldsymbol{v}| = m\,r^2\dot{\varphi} \tag{11.12}$$

and

$$|\boldsymbol{v}|^2 = \dot{r}^2 + r^2\dot{\varphi}^2 = \dot{r}^2 + \frac{L_0^2}{m^2r^2},$$

where the angular speed $\dot{\varphi}$ has been eliminated using (11.12). Substituting that expression for $|\boldsymbol{v}|^2$ in (11.11) gives

$$\tfrac{1}{2}m\dot{r}^2 + \frac{L_0^2}{2mr^2} + V(r) = E_0, \tag{11.13}$$

which is a first order, nonlinear differential equation for $r(t)$.

Conservation of angular momentum and energy has led to the reduction of the 3D, second order differential equation (11.7) for the vector $\boldsymbol{r}(t)$ to the 1D, first order differential equation (11.13) for the scalar $r(t) = |\boldsymbol{r}(t)|$. The angular momentum and total energy constants are determined by the initial position $\boldsymbol{r}_0$ and velocity $\boldsymbol{v}_0$,

$$L_0 = |\boldsymbol{r}_0 \times m\boldsymbol{v}_0|, \quad E_0 = \tfrac{1}{2}m|\boldsymbol{v}_0|^2 + V(r_0). \tag{11.14}$$

Equation (11.13) expresses conservation of energy for an equivalent *one-dimensional* radial motion

$$\tfrac{1}{2}m\dot{r}^2 + U(r) = E_0, \tag{11.15}$$

with *effective potential*

$$U(r) = \frac{L_0^2}{2mr^2} + V(r) \le E_0, \tag{11.16}$$

since the kinetic energy $\frac{1}{2}m\dot{r}^2 \ge 0$ in (11.15) implies that $U(r) \le E_0$ for all time t. Angular momentum L_0 thus provides a repulsive centrifugal potential, $L_0^2/(2mr^2)$.

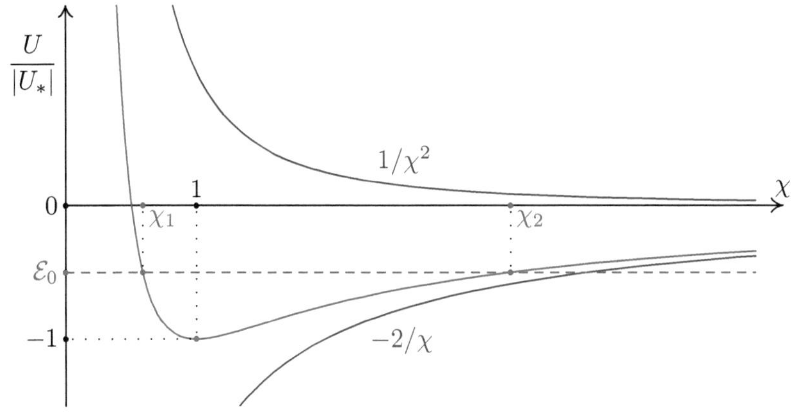

Fig. 11.6 Effective potential $U/|U_*|$ (11.21) as a function of $\chi = r/r_*$.

Planetary motion

Planetary motion corresponds to $F(r) = -k/r^2$ for some constant $k \ge 0$.[2] The potential is $V(r) = -k/r$ in this case, and the effective potential (11.16)

$$U(r) = \frac{L_0^2}{2mr^2} - \frac{k}{r} \tag{11.17}$$

tends to $+\infty$ as $r \to 0^+$ if $L_0 \ne 0$, and to 0^- as $r \to +\infty$. It has a single minimum

$$U_* = -\frac{mk^2}{2L_0^2} \quad \text{at} \quad r = r_* = \frac{L_0^2}{m\,k}. \tag{11.18}$$

We can *nondimensionalize* (11.13), with $V(r) = -k/r$, by dividing it by $|U_*|$, to obtain

$$\left(\frac{\mathrm{d}\chi}{\mathrm{d}\tau}\right)^2 + \frac{1}{\chi^2} - \frac{2}{\chi} = \mathcal{E}_0, \tag{11.19}$$

where

$$\chi = \frac{r}{r_*} = \frac{mk}{L_0^2}r, \quad \tau = \frac{L_0 t}{mr_*^2} = \frac{mk^2}{L_0^3}t, \quad \mathcal{E}_0 = \frac{E_0}{|U_*|} = \frac{2E_0L_0^2}{mk^2}, \tag{11.20}$$

are a *nondimensional* radius χ, time τ, and total energy $\mathcal{E}_0$, respectively. The non-dimensional effective potential (11.17) is

$$\frac{U(r)}{|U_*|} = \frac{1}{\chi^2} - \frac{2}{\chi} \ge -1. \tag{11.21}$$

Analysis of Fig. 11.6 and eqn (11.19) shows that $\mathcal{E}_0 \ge -1$, and

[2] $k = Gm(M+m)$, where G is the gravitational constant and M is the mass of the Sun.

- $-1 = \mathcal{E}_0$ is uniform circular rotation at $r = r_* = L_0^2/(mk)$, $\dot{\varphi} = mk^2/L_0^3$.
- $-1 < \mathcal{E}_0 < 0$ is an oscillation between $\chi_1 < 1 < \chi_2$ where $U(r_1) = U(r_2) = E_0$; and
- $0 < \mathcal{E}_0$ is an escape to infinity, perhaps after dropping to χ_1 where $U(r_1) = E_0$.

The planet position can be found by solving the ordinary differential equation (ODE) (11.19) for $r(t) = r_*\,\chi(\tau)$. The angular velocity follows from (11.12), that reads

$$\frac{\mathrm{d}\varphi}{\mathrm{d}\tau} = \frac{1}{\chi^2} \tag{11.22}$$

in terms of the nondimensional radius χ and time τ. The planet orbit can be found by eliminating $\mathrm{d}\tau$ between (11.19) and (11.22). The substitution $u = 1/\chi$ yields

$$\mathrm{d}\varphi = \frac{\pm\mathrm{d}u}{\sqrt{\mathcal{E}_0 + 2u - u^2}}. \tag{11.23}$$

A second substitution, $u = \epsilon\, v + 1$ with $\epsilon = \sqrt{1 + \mathcal{E}_0}$, transforms $\mathcal{E}_0 + 2u - u^2$ into $\epsilon^2(1 - v^2)$; then $\mathrm{d}\varphi = \pm\mathrm{d}v/\sqrt{1 - v^2}$ integrates to $\varphi - \varphi_0 = \arcsin v$, yielding

$$\chi = \frac{r}{r_*} = \frac{1}{1 - \epsilon\,\cos\varphi}, \tag{11.24}$$

where $\varphi_0 = \pi/2$ was chosen so $\varphi = \pi$ corresponds to the minimum $\chi = \chi_1$. A few orbits are plotted in Fig. 11.7 for $\mathcal{E}_0 = -1$, -0.5, 0, and 0.5. It is left as an exercise to show that these correspond to a circle, an ellipse, a parabola, and a hyperbola, respectively.

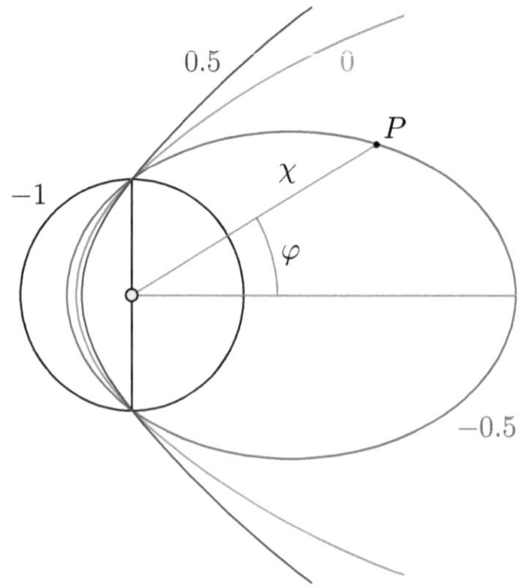

Fig. 11.7 Planet orbits for $\mathcal{E}_0 = -1$, -0.5, 0, 0.5, with $\chi = r/r_*$.

LRL vector

There is another conserved quantity for planetary motion, in addition to angular momentum and energy. It is the vector

$$\boldsymbol{A}(t) = \boldsymbol{p} \times \boldsymbol{L} - mk\,\boldsymbol{r}/r = \boldsymbol{A}, \tag{11.25}$$

where $\boldsymbol{p} = m\boldsymbol{v}$ is the momentum, $\boldsymbol{L} = \boldsymbol{r} \times m\boldsymbol{v}$ the angular momentum and k is the central force constant $\boldsymbol{F} = -k\,\hat{\boldsymbol{r}}/r^2 = \dot{\boldsymbol{p}}$. The vector $\boldsymbol{A}$ is called the LRL vector after Laplace, Runge, and Lenz.[3] It is a dynamical invariant since

$$\begin{aligned}\frac{\mathrm{d}\boldsymbol{A}}{\mathrm{d}t} &= \frac{\mathrm{d}\boldsymbol{p}}{\mathrm{d}t} \times \boldsymbol{L} + \boldsymbol{p} \times \frac{\mathrm{d}\boldsymbol{L}}{\mathrm{d}t} - mk\frac{\mathrm{d}(\boldsymbol{r}/r)}{\mathrm{d}t},\\ &= -\frac{k}{r^2}\hat{\boldsymbol{r}} \times (\boldsymbol{r} \times m\boldsymbol{v}) - \frac{mk}{r}\boldsymbol{v} + \frac{mk}{r^2}\dot{r}\,\boldsymbol{r},\\ &= -\frac{mk}{r^2}(\hat{\boldsymbol{r}} \cdot \boldsymbol{v})\,\boldsymbol{r} + \frac{mk}{r}\boldsymbol{v} - \frac{mk}{r}\boldsymbol{v} + \frac{mk}{r^2}\dot{r}\,\boldsymbol{r} = 0,\end{aligned}$$

using $\dot{\boldsymbol{p}} = -k\,\hat{\boldsymbol{r}}/r^2$, $\dot{\boldsymbol{L}} = 0$, $\dot{\boldsymbol{r}} = \boldsymbol{v}$, and $\hat{\boldsymbol{r}} \cdot \boldsymbol{v} = \dot{r}$.

Vector $\boldsymbol{A}$ points from the attracting center (the Sun at O) to the orbit's *perihelion*, the point of nearest approach to the Sun where $|\boldsymbol{r}|$ is minimum, and thus where $\boldsymbol{r} \cdot \boldsymbol{v} = 0$. Indeed, let $\boldsymbol{r} = \boldsymbol{r}_1$ at the perihelion, then $\boldsymbol{L} = \boldsymbol{r}_1 \times m\boldsymbol{v}_1$ and

$$\boldsymbol{A} = m\boldsymbol{v}_1 \times (\boldsymbol{r}_1 \times m\boldsymbol{v}_1) - \frac{mk}{r_1}\boldsymbol{r}_1 = m\left(m|\boldsymbol{v}_1|^2 - \frac{k}{r_1}\right)\boldsymbol{r}_1,$$

since $\boldsymbol{r}_1 \cdot \boldsymbol{v}_1 = 0$ at the perihelion P_1 (Fig. 11.8). The perihelion radius r_1 and speed $|\boldsymbol{v}_1|$ can be expressed in terms of the total energy $E_0 = \frac{1}{2}m|\boldsymbol{v}_1|^2 - k/r_1$ and angular momentum $L_0 = mr_1|\boldsymbol{v}_1|$, (11.51). Substituting those r_1 and $|\boldsymbol{v}_1|$ values into $\boldsymbol{A}$ gives

$$\boldsymbol{A} = mk\,\epsilon\,\hat{\boldsymbol{r}}_1, \tag{11.26}$$

where $\epsilon = \sqrt{1 + \mathcal{E}_0}$ with $\mathcal{E}_0 = (2E_0L_0^2)/(mk^2)$ (11.20).

The planetary orbit follows from a dot product of (11.25) with $\boldsymbol{r}$,

$$\begin{aligned}\boldsymbol{r} \cdot \boldsymbol{A} &= \boldsymbol{r} \cdot (m\boldsymbol{v} \times \boldsymbol{L}_0) - mkr = (\boldsymbol{r} \times m\boldsymbol{v}) \cdot \boldsymbol{L}_0 - mkr\\ &= L_0^2 - mk\,r.\end{aligned}$$

Substituting for $\boldsymbol{A}$ from (11.26), and dividing by mk, yields

$$\epsilon\, r\,(\hat{\boldsymbol{r}} \cdot \hat{\boldsymbol{r}}_1) = r_* - r, \tag{11.27}$$

where $r_* = L_0^2/(mk)$. Since $\hat{\boldsymbol{r}} \cdot \hat{\boldsymbol{r}}_1 = -\cos\varphi$, this is the orbit equation (11.24).

Planetary motion is a significant problem in the history of mathematics and physics. It was solved by Newton using his new calculus methods, as well as elaborate classical geometry arguments. *Vector calculus* yields the main results with a few dot, cross, and double cross products, and the vector differentiation rules. There is more, of course, orbit transfers and Lagrangian points for instance, but those must be left for later studies in astrodynamics and celestial mechanics.

[3]See E. B. Wilson, *Vector Analysis*, §61 (1901), for probably the first vector calculus proof. See H. Goldstein, *Am. J. Phys.* **44** (1976) for some history on the many rediscoveries of $\boldsymbol{A}$.

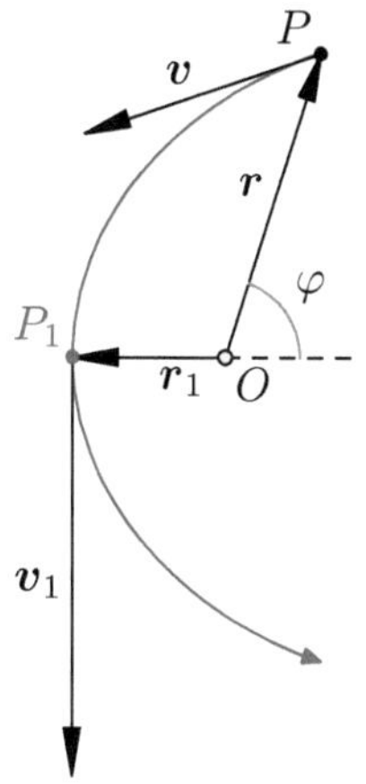

Fig. 11.8 Perihelion: $\boldsymbol{r}_1 \cdot \boldsymbol{v}_1 = 0$ at P_1.

11.3 System of particles

Important results can be derived for the dynamics of several particles from the center of mass concept, and the action–reaction law.

Consider N particles of constant mass m_l at positions $\boldsymbol{r}_l$, $l = 1, \dots, N$. The net force acting on particle number l is $\boldsymbol{F}_l$, and Newton's law for each particle reads $m_l \ddot{\boldsymbol{r}}_l = \boldsymbol{F}_l$. Summing over all l's yields[4]

$$\sum_{l=1}^{N} m_l \ddot{\boldsymbol{r}}_l = \sum_{l=1}^{N} \boldsymbol{F}_l. \tag{11.28}$$

Great cancellations occur on both sides.

On the left side of (11.28), let $\boldsymbol{r}_l = \boldsymbol{r}_c + \boldsymbol{s}_l$, where $\boldsymbol{r}_c$ is the center of mass and $\boldsymbol{s}_l$ is the position vector of particle l with respect to the center of mass. Then by definition of the center of mass $\boldsymbol{r}_c$,

$$M\,\boldsymbol{r}_c = \sum_l m_l \boldsymbol{r}_l = \sum_l m_l\,(\boldsymbol{r}_c + \boldsymbol{s}_l) = M\boldsymbol{r}_c + \sum_l m_l \boldsymbol{s}_l,$$

where $M = \sum_l m_l$ is the total mass, and thus $\sum_l m_l \boldsymbol{s}_l = 0$. Then, since the masses m_l are constants

$$\sum_l m_l \ddot{\boldsymbol{r}}_l = \frac{\mathrm{d}^2}{\mathrm{d}t^2}\left(\sum_l m_l \boldsymbol{r}_l\right) = M\ddot{\boldsymbol{r}}_c.$$

On the right side of (11.28), by Newton's third law,[5] all internal forces cancel out and the resultant is therefore the sum of all external forces only $\sum_l \boldsymbol{F}_l = \sum_l \boldsymbol{F}_l^{(e)} = \boldsymbol{F}^{(e)}$. Therefore (11.28) reduces to

[5] Also known as "action–reaction."

$$M\ddot{\boldsymbol{r}}_c = \boldsymbol{F}^{(e)}, \tag{11.29}$$

where M is the total mass and $\boldsymbol{F}^{(e)}$ is the resultant of all external forces acting on all the particles. The motion of the center of mass of a system of particles is that of a single particle of mass M with position vector $\boldsymbol{r}_c$ under the action of the sum of all external forces. This is a fundamental theorem of mechanics.

There are also great cancellations for the motion about the center of mass. This involves considering the net angular momentum and torque about the center of mass. Taking the cross-product of Newton's law, $m_l \ddot{\boldsymbol{r}}_l = \boldsymbol{F}_l$, with $\boldsymbol{s}_l$ for each particle and summing over all particles gives

$$\sum_l \boldsymbol{s}_l \times m_l \ddot{\boldsymbol{r}}_l = \sum_l \boldsymbol{s}_l \times \boldsymbol{F}_l. \tag{11.30}$$

On the left side, $\boldsymbol{r}_l = \boldsymbol{r}_c + \boldsymbol{s}_l$ with $\sum_l m_l \boldsymbol{s}_l = 0$. Therefore

$$\sum_l \boldsymbol{s}_l \times m_l \ddot{\boldsymbol{r}}_l = \sum_l m_l \boldsymbol{s}_l \times (\ddot{\boldsymbol{r}}_c + \ddot{\boldsymbol{s}}_l) = \sum_l m_l \boldsymbol{s}_l \times \ddot{\boldsymbol{s}}_l = \frac{\mathrm{d}}{\mathrm{d}t}\left(\sum_l \boldsymbol{s}_l \times m_l \dot{\boldsymbol{s}}_l\right).$$

[4] Recall that summing over particles is not a vector operation, so we do not and often cannot use summation convention for sums over particles. For example, (11.28) has a repeated summation index on the left but not on the right!

This last expression is the rate of change of the total angular momentum about the center of mass,

$$\boldsymbol{L}_c = \sum_l (\boldsymbol{s}_l \times m_l \dot{\boldsymbol{s}}_l)\,. \tag{11.31}$$

On the right side of (11.30), one can argue that the (internal) force on particle k exerted by particle l is in the direction of the relative position of l with respect to k, that is,

$$\boldsymbol{f}_{kl} = \alpha_{kl}(\boldsymbol{r}_l - \boldsymbol{r}_k).$$

By action–reaction the force from k on l is $\boldsymbol{f}_{lk} = -\boldsymbol{f}_{kl}$, and the net contribution to the torque from the internal forces will cancel out:

$$\boldsymbol{r}_k \times \boldsymbol{f}_{kl} + \boldsymbol{r}_l \times \boldsymbol{f}_{lk} = (\boldsymbol{r}_k - \boldsymbol{r}_l) \times \boldsymbol{f}_{kl} = 0.$$

This is true with respect to any point, and in particular with respect to the center of mass. Hence, for the motion about the center of mass we have

$$\frac{\mathrm{d}\boldsymbol{L}_c}{\mathrm{d}t} = \boldsymbol{T}_c^{(e)} \tag{11.32}$$

where $\boldsymbol{T}_c^{(e)} = \sum_l \boldsymbol{s}_l \times \boldsymbol{F}_l^{(e)}$ is the net torque about the center of mass due to external forces only. This is another fundamental theorem of mechanics, where the rate of change of the total angular momentum about the center of mass is equal to the total torque due to the external forces only.

11.4 Rigid body dynamics

The two vector differential equations for motion of the center of mass, (11.29), and evolution of the angular momentum about the center of mass, (11.32), are sufficient to determine the motion of a rigid body.

A rigid body is such that all lengths and angles are preserved within the body. If P_0, P_1 and P_2 are any three points of the rigid body, and $\boldsymbol{a} = \overrightarrow{P_0P_1}$, $\boldsymbol{b} = \overrightarrow{P_0P_2}$, then $\boldsymbol{a} \cdot \boldsymbol{b}$ is constant. The evolution of any vector $\boldsymbol{a}$ tied to the body is thus a rotation and

$$\frac{\mathrm{d}\boldsymbol{a}}{\mathrm{d}t} = \boldsymbol{\omega}(t) \times \boldsymbol{a},$$

for the same angular velocity $\boldsymbol{\omega}(t)$ independent of $\boldsymbol{a}$, as shown in exercise 10.19.

The center of mass of a rigid body moves according to the sum of the external forces, as for a system of particles. A continuous rigid body, such as a rock, can be considered as a continuous distribution of mass, with mass $\mathrm{d}m(\boldsymbol{s})$ at point $\boldsymbol{s}$, and the sum over all $\boldsymbol{s}$ becomes a three-dimensional integral over all points $\boldsymbol{s}$ in the domain V of the body,

$$\sum_l m_l \boldsymbol{s}_l \longrightarrow \int_{\mathcal{V}} \boldsymbol{s}\,\mathrm{d}m(\boldsymbol{s}),$$

where $\mathrm{d}m(\boldsymbol{s})$ is the mass *measure* of the infinitesimal volume element $\mathrm{d}V(\boldsymbol{s})$ at point $\boldsymbol{s}$, or in other words $\mathrm{d}m = \rho\,\mathrm{d}V$, where $\rho(\boldsymbol{s})$ is the mass density at point $\boldsymbol{s}$.

For the motion about the center of mass, the position vectors $\boldsymbol{s}_l$ are frozen into the body; hence, $\dot{\boldsymbol{s}}_l = \boldsymbol{\omega} \times \boldsymbol{s}_l$ for any point of the body. The total angular momentum about the center of mass (11.31) for a rigid system of particles then reads

$$\boldsymbol{L}_c = \sum_l m_l \boldsymbol{s}_l \times (\boldsymbol{\omega} \times \boldsymbol{s}_l) = \sum_l m_l \left(|\boldsymbol{s}_l|^2 \boldsymbol{\omega} - \boldsymbol{s}_l \left(\boldsymbol{s}_l \cdot \boldsymbol{\omega} \right) \right), \tag{11.33}$$

or, for a continuous rigid body,

$$\boldsymbol{L}_c = \int_{\mathcal{V}} \left(|\boldsymbol{s}|^2 \boldsymbol{\omega} - \boldsymbol{s} \left(\boldsymbol{s} \cdot \boldsymbol{\omega} \right) \right) \mathrm{d}m(\boldsymbol{s}). \tag{11.34}$$

Since the angular velocity $\boldsymbol{\omega}$ is unique for any vector $\boldsymbol{s}$ tied to the body, it does not depend on $\boldsymbol{s}$ and we can pull it out from the sum or integral. This requires the concept of *tensor of inertia* discussed in §9.6, yielding

$$\boldsymbol{L}_c = \boldsymbol{\mathcal{I}}_c \cdot \boldsymbol{\omega}, \tag{11.35}$$

where

$$\boldsymbol{\mathcal{I}}_c = \sum_l m_l \big((\boldsymbol{s}_l \cdot \boldsymbol{s}_l)\, \boldsymbol{I} - \boldsymbol{s}_l \boldsymbol{s}_l \big), \tag{11.36}$$

or

$$\boldsymbol{\mathcal{I}}_c = \int_{\mathcal{V}} \big((\boldsymbol{s} \cdot \boldsymbol{s})\, \boldsymbol{I} - \boldsymbol{s}\boldsymbol{s} \big)\, dm(\boldsymbol{s}), \tag{11.37}$$

for a rigidly connected system of point masses, or a continuous rigid body, respectively.

The tensor $\boldsymbol{\mathcal{I}}_c$ is symmetric positive definite; thus, it has real positive eigenvalues and orthogonal eigenvectors (9.64). Let $\mathcal{I}_1, \mathcal{I}_2, \mathcal{I}_3$ be the eigenvalues with $\mathbf{e}_1, \mathbf{e}_2, \mathbf{e}_3$ the corresponding eigenvectors. Then

$$\boldsymbol{\mathcal{I}}_c = \mathcal{I}_1\, \mathbf{e}_1 \mathbf{e}_1 + \mathcal{I}_2\, \mathbf{e}_2 \mathbf{e}_2 + \mathcal{I}_3\, \mathbf{e}_3 \mathbf{e}_3. \tag{11.38}$$

The eigenvalues $\mathcal{I}_1, \mathcal{I}_2, \mathcal{I}_3$ are called the *principal moments of inertia*, and they are constants since they describe the distribution of mass about axes fixed in the body.[6] The eigenvectors $\mathbf{e}_1, \mathbf{e}_2, \mathbf{e}_3$ are fixed in the body; thus, they vary with time t if the body is rotating.

If there is no net torque about the center of mass, then $\boldsymbol{L}_c = \boldsymbol{\mathcal{I}}_c \cdot \boldsymbol{\omega}$ is constant; that is,

$$\frac{\mathrm{d}}{\mathrm{d}t} (\boldsymbol{\mathcal{I}}_c \cdot \boldsymbol{\omega}) = \frac{\mathrm{d}}{\mathrm{d}t} (\mathcal{I}_1 \omega_1 \mathbf{e}_1 + \mathcal{I}_2 \omega_2 \mathbf{e}_2 + \mathcal{I}_3 \omega_3 \mathbf{e}_3) = 0, \tag{11.39}$$

where $\boldsymbol{\omega} = \omega_1 \mathbf{e}_1 + \omega_2 \mathbf{e}_2 + \omega_3 \mathbf{e}_3$ is the angular velocity in the eigenvector frame. The angular velocity $\boldsymbol{\omega}$ of the rigid body is a function of time, in general, and its rate of change arises from the changes of the components

[6] If $\{\mathbf{e}_1, \mathbf{e}_2, \mathbf{e}_3\}$ is any orthonormal basis moving with the rigid body, then the components of the inertia tensor $\mathcal{I}_{ij} = \mathbf{e}_i \cdot \boldsymbol{\mathcal{I}}_c \cdot \mathbf{e}_j = \mathcal{I}_{ji}$ are constants in that basis. Thus, its eigenvalues are constants also.

$(\omega_1, \omega_2, \omega_3)$ in the body frame, as well as the rotation of the body frame $\{\mathbf{e}_1, \mathbf{e}_2, \mathbf{e}_3\}$. The rates of change of the latter are simply $\dot{\mathbf{e}}_i = \boldsymbol{\omega} \times \mathbf{e}_i$. Equation (11.39) thus expands to

$$\mathcal{I}_1(\dot{\omega}_1 \mathbf{e}_1 + \omega_1 \dot{\mathbf{e}}_1) + \mathcal{I}_2(\dot{\omega}_2 \mathbf{e}_2 + \omega_2 \dot{\mathbf{e}}_2) + \mathcal{I}_3(\dot{\omega}_3 \mathbf{e}_3 + \omega_3 \dot{\mathbf{e}}_3) = 0,$$

since the principal moments of inertia are constants. Projecting onto $\{\mathbf{e}_1, \mathbf{e}_2, \mathbf{e}_3\}$, and using $\dot{\mathbf{e}}_i \cdot \mathbf{e}_j = (\boldsymbol{\omega} \times \mathbf{e}_i) \cdot \mathbf{e}_j = \epsilon_{ijk}\omega_k$, yields

$$\begin{cases} \mathcal{I}_1 \dot{\omega}_1 = (\mathcal{I}_2 - \mathcal{I}_3)\, \omega_2 \omega_3, \\ \mathcal{I}_2 \dot{\omega}_2 = (\mathcal{I}_3 - \mathcal{I}_1)\, \omega_3 \omega_1, \\ \mathcal{I}_3 \dot{\omega}_3 = (\mathcal{I}_1 - \mathcal{I}_2)\, \omega_1 \omega_2. \end{cases} \tag{11.40}$$

These are *Euler's equations* for the torque-free motion of a rigid body about its center of mass, as a rock or any solid object thrown into the air. The dynamics conserves angular kinetic energy, $E_c = \frac{1}{2}\boldsymbol{\omega} \cdot \boldsymbol{\mathcal{I}}_c \cdot \boldsymbol{\omega}$,

$$\frac{\mathrm{d}}{\mathrm{d}t}\left(\mathcal{I}_1 \omega_1^2 + \mathcal{I}_2 \omega_2^2 + \mathcal{I}_3 \omega_3^2\right) = 0 \tag{11.41}$$

and angular momentum, $\boldsymbol{L}_c \cdot \boldsymbol{L}_c = (\boldsymbol{\omega} \cdot \boldsymbol{\mathcal{I}}_c) \cdot (\boldsymbol{\mathcal{I}}_c \cdot \boldsymbol{\omega})$,

$$\frac{\mathrm{d}}{\mathrm{d}t}\left(\mathcal{I}_1^2 \omega_1^2 + \mathcal{I}_2^2 \omega_2^2 + \mathcal{I}_3^2 \omega_3^2\right) = 0. \tag{11.42}$$

Euler's equations (11.40) show that rotation about any principal axis is a fixed point; that is,

$$(\omega_1, \omega_2, \omega_3) = (\Omega, 0, 0),\ (0, \Omega, 0),\ \text{and}\ (0, 0, \Omega), \tag{11.43}$$

are the steady solutions, for any constant Ω. However rotation about the axis corresponding to the middle moment of inertia is unstable. Indeed, if we label the principal moments of inertia such that $0 \le \mathcal{I}_1 \le \mathcal{I}_2 \le \mathcal{I}_3$, then linearization of (11.40) about $(0, \Omega, 0)$ yields

$$\begin{cases} \mathcal{I}_1 \dot{\omega}_1 = (\mathcal{I}_2 - \mathcal{I}_3)\, \Omega\, \omega_3, \\ \mathcal{I}_3 \dot{\omega}_3 = (\mathcal{I}_1 - \mathcal{I}_2)\, \omega_1 \Omega, \end{cases} \tag{11.44}$$

that reduces to the second order ODE

$$\mathcal{I}_1 \mathcal{I}_3\, \ddot{\omega}_1 = (\mathcal{I}_1 - \mathcal{I}_2)\,(\mathcal{I}_2 - \mathcal{I}_3)\, \Omega^2\, \omega_1, \tag{11.45}$$

which has exponentially growing solutions whenever $(\mathcal{I}_1 - \mathcal{I}_2)\,(\mathcal{I}_2 - \mathcal{I}_3) > 0$ which is the case if $\mathcal{I}_2$ is the middle principal moment. The linearization (11.44) and the exponential growth of ω_1 and ω_3 are only valid as long as $(\mathcal{I}_3 - \mathcal{I}_1) \int_0^t \omega_1(t')\omega_3(t')\, \mathrm{d}t'$ is small compared to $\mathcal{I}_2 \Omega$. Otherwise, ω_2 cannot be assumed to be fixed at Ω and the full nonlinear system (11.40) needs to be considered.

Solutions of the full nonlinear system are typically periodic solutions. In fact, the two quadratic invariants (11.41) and (11.42) imply that, in the cartesian space with coordinates $x_i = L_i/|\boldsymbol{L}| = \mathbf{e}_i \cdot \boldsymbol{L}_c/|\boldsymbol{L}_c|$, that is

$$(x_1, x_2, x_3) = \frac{(\mathcal{I}_1 \omega_1,\ \mathcal{I}_2 \omega_2,\ \mathcal{I}_3 \omega_3)}{\sqrt{\mathcal{I}_1^2 \omega_1^2 + \mathcal{I}_2^2 \omega_2^2 + \mathcal{I}_3^2 \omega_3^2}},$$

the orbits consists of the intersection of the unit sphere (11.42)

$$x_1^2 + x_2^2 + x_3^2 = 1, \tag{11.46}$$

with the ellipsoid (11.41)

$$\frac{x_1^2}{\mathcal{I}_1} + \frac{x_2^2}{\mathcal{I}_2} + \frac{x_3^2}{\mathcal{I}_3} = C, \tag{11.47}$$

such that $\mathcal{I}_3^{-1} \le C \le \mathcal{I}_1^{-1}$. Thus, the ellipsoid always intersects the sphere.

Exercises

(11.1) A particle of mass m is thrown at speed v_O from a height h_O above the ground. At what angle α should the particle be thrown to maximize the horizontal distance traveled when it hits the ground? Show your reasoning.

(11.2) Give a closed book, step-by-step derivation of (11.11) from (11.7).

(11.3) Give a closed book, step-by-step derivation of (11.23) from (11.8) and (11.11).

(11.4) Our study of planetary motion has implicitly assumed that $L_O \neq 0$. Repeat the analysis when $L_O = 0$. Find and describe the orbits.

(11.5) A particle of mass m at position $\boldsymbol{r}$ moves under the action of a force $\boldsymbol{F} = -k\boldsymbol{r}/r^3$. If $\boldsymbol{r}(0) = \boldsymbol{r}_O$, what is the velocity $\boldsymbol{v}(0)$ of minimum magnitude that yields *escape to infinity*? Show your derivation. (Ans.: $|\boldsymbol{v}_O| = \sqrt{2k/(mr_O)}$.)

(11.6) A particle of mass m at position $\boldsymbol{r}$ moves under the action of a force $\boldsymbol{F} = -k\boldsymbol{r}/r^3$. If $\boldsymbol{r}(0) = \boldsymbol{r}_O$, what is the magnitude and direction of $\boldsymbol{v}(0)$ that will lead to the particle traveling (i) in a *circular orbit*? (ii) in a *parabolic orbit*? Show your reasoning.

(11.7) Radius $r_* = L_O^2/(mk)$ and frequency $\Omega_* = L_O/(mr_*^2) = mk^2/L_O^3$ are defined in (11.19). The radius r_* is that of a circular orbit with angular momentum L_O. Show that Ω_* is the angular frequency for that circular orbit. Thus, $T_* = 2\pi/\Omega_*$ is its period, and $v_* = \Omega_* r_* = k/L_O$ is the planet speed in its circular orbit.

(11.8) In cartesian coordinates, $x = r\cos\varphi$, $y = r\sin\varphi$, show that (11.24) is

$$(1-\epsilon^2)x^2 - 2\epsilon r_* x + y^2 = r_*^2. \tag{11.48a}$$

For $\epsilon = 1$, this is a *parabola* with

$$2r_*\, x = y^2 - r_*^2. \tag{11.48b}$$

For $\epsilon \neq 1$, completing the square yields

$$(1-\epsilon^2)\left(x - \frac{\epsilon r_*}{1-\epsilon^2}\right)^2 + y^2 = \frac{r_*^2}{1-\epsilon^2}. \tag{11.48c}$$

Show that (11.48c) is an *ellipse* when $0 \le \epsilon < 1$ with

$$\frac{(x-c)^2}{a^2} + \frac{y^2}{b^2} = 1, \tag{11.48d}$$

and a *hyperbola* when $\epsilon > 1$ with

$$\frac{(x-c)^2}{a^2} - \frac{y^2}{b^2} = 1, \tag{11.48e}$$

where $a = r_*/|1-\epsilon^2|$ and $b = r_*/\sqrt{|1-\epsilon^2|}$ are the principal radii, and $c = \epsilon r_*/(1-\epsilon^2)$ is the location of the center C from the focus F in the major direction. Note that $c < 0$ for $\epsilon > 1$.
Here $(0,0)$ is a *focus* and the *centers* of the ellipses and hyperbolas are at $(c, 0)$. The *vertices* are at $y = 0$, $x = c \pm a = r_*/(\pm 1 - \epsilon)$. Thus, a vertex has a nearest focus at a distance $r_*/(1+\epsilon)$ in the major direction. The shape of the conic is set by the *eccentricity* ϵ, with $\epsilon = \sqrt{1 - b^2/a^2}$ for an ellipse, and $\epsilon = \sqrt{1 + b^2/a^2}$ for a hyperbola. Its size is set by $r_* = b^2/a$, the *semi-latus rectum*, or *focal halfwidth* (Figs. 11.7 and 11.9).

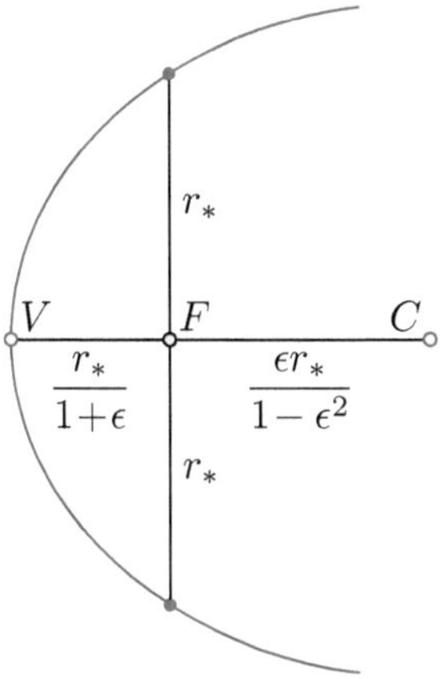

Fig. 11.9 Vertex V, focus F, center C, eccentricity ϵ, and focal halfwidth r_*.

(11.9) The area of an ellipse of major radius a and minor radius b is πab. Use the previous problems and Kepler's law of areas 11.5 to deduce Kepler's thirrd law that the square of the period T of an elliptic orbit is proportional to the cube of its major radius

$$T^2 = \frac{4\pi^2 m}{k} a^3 = \frac{4\pi^2}{G(M+m)} a^3, \qquad (11.49)$$

since $k = Gm(M+m)$. So T is (almost) independent of the planet mass $m \ll M$.

(11.10) For planetary motion with $\boldsymbol{r}(t) = r_* \chi \hat{\boldsymbol{r}}$ as defined in (11.20) and (11.24), show that

$$\boldsymbol{v}/v_* = -\epsilon \sin\varphi\, \hat{\boldsymbol{r}} + (1 - \epsilon \cos\varphi)\, \hat{\boldsymbol{\varphi}}, \qquad (11.50)$$

where $\epsilon = \sqrt{1 + \mathcal{E}_0}$ is the *eccentricity* and $v_* = k/L_0$.

(11.11) Show that the *perihelion* radius r_1 and speed v_1 are

$$r_1 = r_*/(1+\epsilon), \qquad v_1 = (1+\epsilon)\, v_*, \qquad (11.51)$$

where $r_* = L_0^2/(mk)$, $v_* = k/L_0$, and $\epsilon = \sqrt{1+\mathcal{E}_0}$, with $\mathcal{E}_0 = 2E_0 L_0^2/(mk^2)$. In the elliptic case $0 \le \epsilon < 1$, show that the *aphelion* radius r_2 and speed v_2 are

$$r_2 = r_*/(1-\epsilon), \qquad v_2 = (1-\epsilon)\, v_*,$$

and the ellipse *diameter* $r_1 + r_2 = 2a = -2k/E_0$ is independent of angular momentum L_0.

(11.12) The Coulomb force between electrically charged particles also follows the inverse square law, but the force is *repulsive* $\boldsymbol{F} = k\hat{\boldsymbol{r}}/r^2$, with $k > 0$ if the charges are of the same sign. Find the potential $V(r)$ and draw an effective potential diagram. Find the orbits. (One can still nondimensionalize (11.13) by multiplying it by $2L_0^2/(mk^2)$ and use $L_0^2/(mk)$ as a length scale, and $L_0^3/(mk^2)$ as a time scale, when $L_0 \neq 0$.)

(11.13) What are constants of motion for $\mathrm{d}\boldsymbol{r}/\mathrm{d}t = \boldsymbol{\omega} \times \boldsymbol{r}$ with $\boldsymbol{\omega}$ constant? Show that $\mathrm{d}^2\boldsymbol{r}_\perp/\mathrm{d}t^2 = -\omega^2 \boldsymbol{r}_\perp$, where $\boldsymbol{r}_\perp$ is the component of $\boldsymbol{r}$ perpendicular to $\boldsymbol{\omega}$ and $\omega = |\boldsymbol{\omega}|$. Find the force $\boldsymbol{F}$ that sustains that motion for a particle of mass m, according to Newton's law. Is $\boldsymbol{F}$ a central force?

(11.14) What are constants of motion for $\mathrm{d}^2\boldsymbol{r}/\mathrm{d}t^2 = -\omega^2 \boldsymbol{r}$ with ω real and constant? How does this differ from exercise 11.13? Find the general solution $\boldsymbol{r}(t)$.

(11.15) What are constants of motion for $\mathrm{d}^2\boldsymbol{r}/\mathrm{d}t^2 = \omega^2 \boldsymbol{r}$ with ω real and constant? Find the general solution $\boldsymbol{r}(t)$.

(11.16) A particle of mass m and electric charge q moving at velocity $\boldsymbol{v}$ in a magnetic field $\boldsymbol{B}$ experiences the Lorentz force $\boldsymbol{F} = q\boldsymbol{v} \times \boldsymbol{B}$, and Newton's law yields the $\boldsymbol{v}(t)$ evolution equation as

$$m\frac{\mathrm{d}\boldsymbol{v}}{\mathrm{d}t} = q\,\boldsymbol{v} \times \boldsymbol{B}. \qquad (11.52)$$

Show that kinetic energy is conserved but angular momentum is not, in general. If m, q and $\boldsymbol{B}$ are constants, solve for $\boldsymbol{v}(t)$ then for the particle position $\boldsymbol{r}(t)$. Express your answer in terms of the initial conditions $\boldsymbol{r}(0) = \boldsymbol{r}_0$ and $\boldsymbol{v}(0) = \boldsymbol{v}_0$. Show that the particle trajectories are *helices*, and that the particle angular rotation rate about the magnetic field $\boldsymbol{B}$ is $\omega = |q\boldsymbol{B}|/m$.

(11.17) Investigate $m\,\mathrm{d}\boldsymbol{v}/\mathrm{d}t = \boldsymbol{F} = -k(r - \ell_0)\hat{\boldsymbol{r}}$ for constant m and arbitrary initial conditions $\boldsymbol{r}(0) = \boldsymbol{r}_0$ and $\boldsymbol{v}(0) = \boldsymbol{v}_0$, where $\boldsymbol{v} = \mathrm{d}\boldsymbol{r}/\mathrm{d}t$, $\boldsymbol{r} = r\hat{\boldsymbol{r}}$, with k, ℓ_0 constant. This $\boldsymbol{F}$ corresponds to a spring of stiffness k and rest length ℓ_0. What are two constants of motion? Can you define a potential $V(r)$ for this problem? What types of motion do you expect? Sketch a few representative orbits.

(11.18) A bead of constant mass m is sliding without friction along a circular hoop of radius R that rotates about the vertical diameter at constant angular velocity Ω. The only external force acting on the particle is gravity $-mg\hat{\mathbf{z}}$. If $\theta(t)$ specifies the angular position of the particle around the hoop, use Newton's law to derive the second order differential equation for $\theta(t)$. What is the force from the hoop on the particle? Find all the possible equilibrium positions for which θ remains constant.

(11.19) If $\boldsymbol{f}_{kl} = \alpha_{kl}(\boldsymbol{r}_l - \boldsymbol{r}_k)$, where $\alpha_{kl} = \alpha_{lk}$ can be a function of the distance $|\boldsymbol{r}_l - \boldsymbol{r}_k|$, show that $\boldsymbol{s}_k \times \boldsymbol{f}_{kl} + \boldsymbol{s}_l \times \boldsymbol{f}_{lk} = 0$, where $\boldsymbol{s}$ is the position vector from the center of mass such that $\boldsymbol{r} = \boldsymbol{r}_c + \boldsymbol{s}$.

(11.20) For a two-particle system with force $\boldsymbol{f}_{12} = F(r)\hat{\boldsymbol{r}} = -\boldsymbol{f}_{21}$, where $\boldsymbol{r} = r\hat{\boldsymbol{r}} = \boldsymbol{r}_2 - \boldsymbol{r}_1$, show that (1) the center of mass moves in a straight line, (2) the angular momentum about the center of mass $\boldsymbol{L}_c$ is conserved, and (3) $\boldsymbol{r}(t)$ evolves according to

$$m\frac{\mathrm{d}^2\boldsymbol{r}}{\mathrm{d}t^2} = F(r)\hat{\boldsymbol{r}} \quad \text{with} \quad m = \frac{m_1 m_2}{m_1 + m_2}.$$

(11.21) Consider a discrete rigid system of constant masses m_l at positions $\boldsymbol{r}_l$, $l = 1, \ldots, N$. Show that the gravity forces, $\boldsymbol{F}_l = m_l\boldsymbol{g}$, with $\boldsymbol{g}$ the constant acceleration of gravity, do not produce any net torque about the center of mass.

(11.22) Verify conservation of angular kinetic energy and momentum (11.41) and (11.42).

(11.23) Show why $\mathcal{I}_3^{-1} \le C \le \mathcal{I}_1^{-1}$ in (11.47), and why the ellipsoid always intersects the unit sphere.

(11.24) Use the quadratic invariants (11.46) and (11.47), to show that (11.43) are steady states.

(11.25) Show that there are only two points on the unit sphere and the ellipsoid (11.47) when $C = \mathcal{I}_3^{-1}$, and when $C = \mathcal{I}_1^{-1}$, but that there are two great circles of points when $C = \mathcal{I}_2^{-1}$. Sketch the corresponding orbits locally near $(0, 1, 0)$, projected onto the tangent plane to the sphere at that point, and indicate the direction of motion along those orbits, deduced using Euler's equations (11.40) near that point.

Curves

12

12.1 Representations

The simplest curve is a straight line that can be defined *implicitly* as the intersection of two *planes* (Fig. 12.1)

$$\begin{cases} \hat{\boldsymbol{n}}_1 \cdot \boldsymbol{r} = h_1, \\ \hat{\boldsymbol{n}}_2 \cdot \boldsymbol{r} = h_2, \end{cases} \tag{12.1}$$

where $\hat{\boldsymbol{n}}$ is a unit normal to the plane and h is the distance from the origin to the plane in the direction $\hat{\boldsymbol{n}}$. Definition (12.1) is *implicit*; it does not specify the position vector $\boldsymbol{r}$, only equations that $\boldsymbol{r}$ must satisfy.

The *parametric* equation of a line is (Fig. 12.2)

$$\boldsymbol{r}(t) = \boldsymbol{r}_0 + t\,\boldsymbol{v}_0, \tag{12.2}$$

where $\boldsymbol{r}_0$ is the position vector of any reference point P_0 on the line, $\boldsymbol{v}_0$ is a vector parallel to the line, and t is a real variable in $-\infty < t < \infty$. A parametric representation is *explicit*; selecting any t value yields the position vector $\boldsymbol{r}(t)$ of a point on the line, for example, $\boldsymbol{r}(0) = \boldsymbol{r}_0$ and $\boldsymbol{r}(1.5) = \boldsymbol{r}_0 + 1.5\,\boldsymbol{v}_0$. The speed $|\boldsymbol{v}_0| \neq 0$ is arbitrary, and the same line can be parametrized as

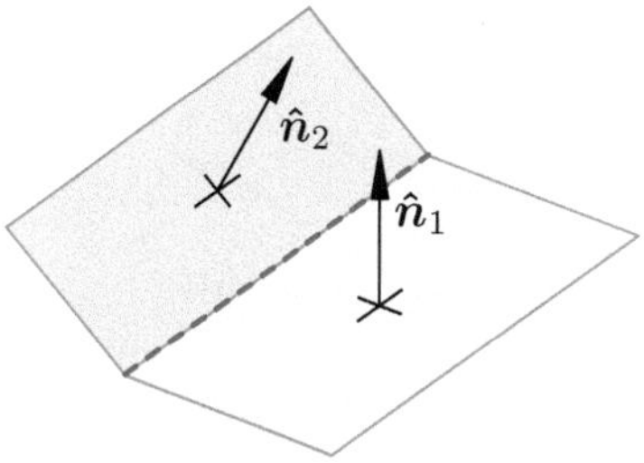

Fig. 12.1 Two planes, one line (12.1).

$$\boldsymbol{r}(s) = \boldsymbol{r}_0 + s\,\hat{\boldsymbol{v}}_0, \tag{12.3}$$

where s is now the *signed distance* along the line, with $s = 0$ at P_0 and $s > 0$ in the direction $\hat{\boldsymbol{v}}_0$. Exercise 12.1 asks to find the relation between the two representations (12.1) and (12.2).

In general, a curve $\mathcal{C}$ can be defined by a continuous vector function $\boldsymbol{r}(t)$ of a real variable t in an interval $I = (t_0, t_f)$,

$$\mathcal{C} = \left\{ P \in \mathbb{E}^3;\ \overrightarrow{OP} = \boldsymbol{r}(t),\ t \in I \subset \mathbb{R} \right\},$$

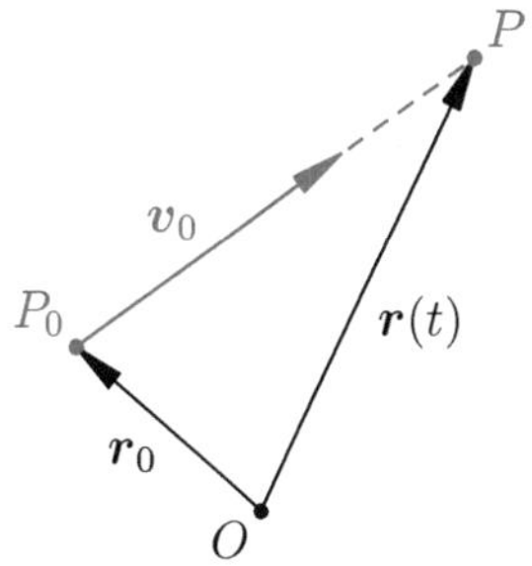

Fig. 12.2 Parametric line (12.2).

as in example (12.2). It is useful to think of t as time and $\boldsymbol{r}(t)$ as the position of a particle at time t, although the parameter may not have units of time. The vector function $\boldsymbol{r}(t)$ is a *parametric representation* of curve $\mathcal{C}$, and t is then the *coordinate* of point P on $\mathcal{C}$ in that representation. An implicit definition, $f_1(\boldsymbol{r}) = 0 = f_2(\boldsymbol{r})$, may have several disconnected sets as solutions; for example, the intersections between two ellipsoids can be the empty set, one tangent point, or one or two closed curves.

The use of the word *parameter* to describe the real independent variable t in $\boldsymbol{r}(t)$ originates from the fact that if a curve is defined implicitly by $f_1(\boldsymbol{r}) = 0 = f_2(\boldsymbol{r})$, then $\boldsymbol{r}(t)$ with $t \in I$ is a *one-parameter* set of solutions $\boldsymbol{r}$ to those equations, that is,

$$f_1(\boldsymbol{r}(t)) = 0 = f_2(\boldsymbol{r}(t)), \qquad \forall t \in I.$$

Parametrizing a curve is constructing an explicit vector function $\boldsymbol{r}(t)$ that represents it. Thus, (12.2) with (12.74) is a parametrization of the line defined implicitly by (12.1).

The word "parameter" is also used to refer to the constants that specify a particular object in a given representation; thus, $\boldsymbol{r}_0$ and $\boldsymbol{v}_0$ in (12.2) are the parameters that determine a specific line, and the variable t is then the *coordinate* of a point on that line in that representation (12.2).

Secant and tangent vectors

Given a parametrization $\boldsymbol{r}(t)$, the displacement vector

$$\Delta\boldsymbol{r} = \boldsymbol{r}(t+\Delta t) - \boldsymbol{r}(t)$$

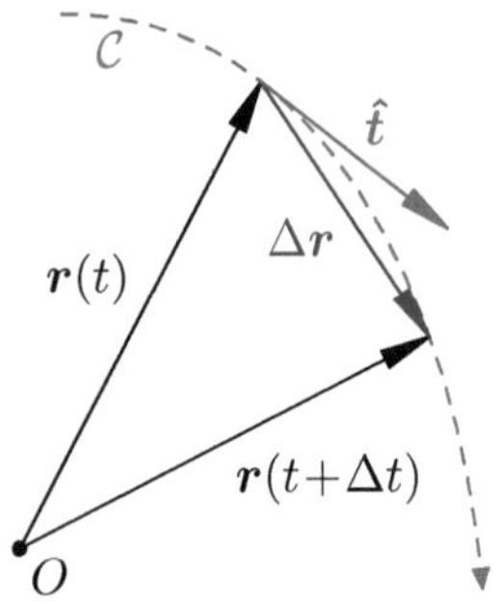

Fig. 12.3 $\boldsymbol{r}(t), \Delta\boldsymbol{r}, \hat{\boldsymbol{t}}$.

is a *secant vector* connecting two points on the curve (Fig. 12.3), and $\Delta\boldsymbol{r} \to 0$ when $\Delta t \to 0$, if the curve is *continuous*. Dividing $\Delta\boldsymbol{r}$ by its magnitude $|\Delta\boldsymbol{r}|$ gives a unit vector, whose limit,

$$\lim_{\Delta t \to 0^+} \frac{\Delta\boldsymbol{r}}{|\Delta\boldsymbol{r}|} = \hat{\boldsymbol{t}}, \tag{12.4}$$

is the *unit tangent* to the curve at $\boldsymbol{r}(t)$, with $\hat{\boldsymbol{t}}$ pointing in the direction of increasing t. Dividing $\Delta\boldsymbol{r}(t)$ by Δt, yields the *derivative*

$$\frac{\mathrm{d}\boldsymbol{r}}{\mathrm{d}t} = \lim_{\Delta t \to 0} \frac{\Delta\boldsymbol{r}}{\Delta t} = v\,\hat{\boldsymbol{t}}, \tag{12.5}$$

that is, a vector whose direction is the unit tangent $\hat{\boldsymbol{t}}$ to the curve at that $\boldsymbol{r}(t)$. Its magnitude is the "speed,"

$$v = \lim_{\Delta t \to 0} \frac{|\Delta\boldsymbol{r}|}{|\Delta t|}, \tag{12.6}$$

whose value depends on how fast we are traveling along the curve at that point. It is an actual speed only if t is time, not just *time-like*. If those limits exist, the curve is *differentiable* at that point.

Smooth, piecewise smooth, and fractal curves

Newtonian particle trajectories defined by the vector differential equation

$$m\ddot{\boldsymbol{r}} = \boldsymbol{F}(\boldsymbol{r}, \dot{\boldsymbol{r}})$$

have an $\boldsymbol{r}(t)$ that is C^2, *twice continuously differentiable*, if the force function $\boldsymbol{F}(\boldsymbol{r}, \boldsymbol{v})$ is continuous with respect to its arguments, so $\ddot{\boldsymbol{r}}$ exists and is continuous. Collisions are modeled by impulses—*Dirac delta functions* $\boldsymbol{F} = \Delta \boldsymbol{p}\, \delta(t - t_c)$ at the time and place of impact, where $\Delta \boldsymbol{p}$ is the *impulse* with units of momentum. The acceleration is nominally infinite at the collision time. The momentum $\boldsymbol{p} = m\dot{\boldsymbol{r}}$ changes discontinuously by $\Delta \boldsymbol{p}$ across the collision, but $\boldsymbol{r}(t)$ is continuous. In vector calculus, we consider curves that are at least *piecewise* C^2. A few collisions once in a while are OK, but not constantly!

Brownian motion is the motion of a particle small enough to feel the jarring effect of *relentless collisions* with fluid molecules. A mathematical model of such motion is a *random walk* of step $d\ell = \sqrt{\sigma \mathrm{d}t}$ in time $\mathrm{d}t$, such that

$$\boldsymbol{r}(t + \mathrm{d}t) - \boldsymbol{r}(t) = \hat{\boldsymbol{a}}\, \sqrt{\sigma\, \mathrm{d}t},$$

where $\hat{\boldsymbol{a}}$ is a random direction and σ has units of length square over time, L^2/T. The resulting particle motion $\boldsymbol{r}(t)$ is continuous but nowhere differentiable since $\mathrm{d}\boldsymbol{r}(t) \sim \sqrt{\mathrm{d}t}$. Such individual curves are modeled and studied with *stochastic differential equations.* The coefficient $\sigma > 0$ is a *diffusion constant.*[1]

The *Koch curve* is a deterministic example of a *fractal* curve, continuous everywhere yet nowhere differentiable (Fig. 12.4). There is even the *Peano* curve (Fig. 12.5) and the *Hilbert* curve that fill the *entire surface* of a square! These examples show that mere continuity of the parametric function $\boldsymbol{r}(t)$ is not sufficient for our intuitive notion of "curve."

[1]The diffusion constant is defined as $\nu = \sigma/(2d)$ in a space of dimension d, so that the *diffusion equation* that measures the probability of observing a particle at a certain point at a certain time, has the same diffusion constant irrespective of dimension.

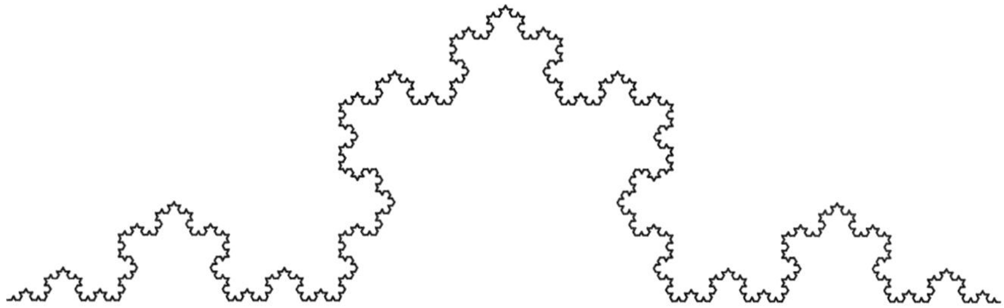

Fig. 12.4 Five iterations toward the Koch curve (based on code by Martin Geisler, *Fractals with MetaPost*, `http://mgeisler.net/downloads/metafractals/metafractals.pdf`).

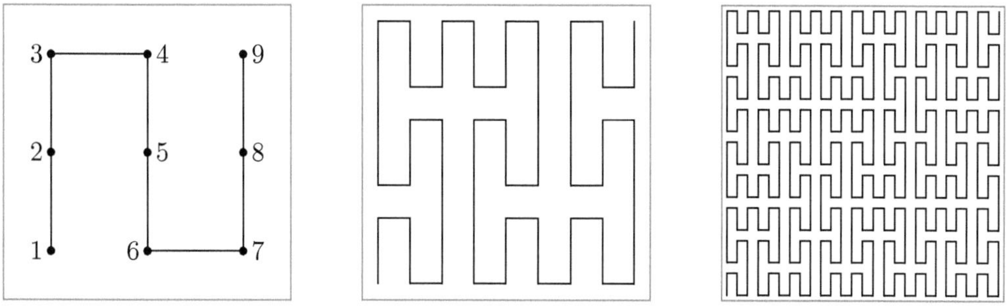

Fig. 12.5 Peano map of $t = \left[1 - \frac{1}{2}, 2 - \frac{1}{2}, \cdots, 3^{2k} - \frac{1}{2}\right]/3^{2k}$ for $k = 1, 2, 3$. The Peano curve is a continuous map of the interval $0 \le t \le 1$ to the unit square $[0, 1] \times [0, 1]$.

12.2 Classic curves

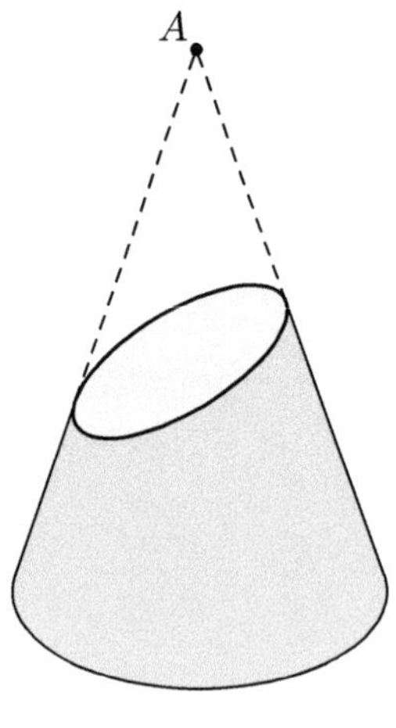

Fig. 12.6 Conic section.

A classic family of curves arises from all possible intersections of a cone with a plane (Fig. 12.6). These are the circle, ellipse, parabola, and hyperbola, collectively referred to as the *conics*, or *conic sections*. The implicit equations for these conics are thus

$$\begin{cases} \hat{\boldsymbol{\omega}} \cdot (\boldsymbol{r} - \boldsymbol{r}_A) = \cos\theta\,|\boldsymbol{r} - \boldsymbol{r}_A|, \\ \hat{\boldsymbol{n}} \cdot \boldsymbol{r} = h_0, \end{cases} \tag{12.7a}$$

where $\hat{\boldsymbol{\omega}}$ is the direction of the cone axis, A is the cone vertex (or *apex*), and θ is the cone angle (or *aperture*), with $0 \le \theta < \pi/2$. The plane is normal to $\hat{\boldsymbol{n}}$ and at distance h_0 from the origin, that is, distance $h = h_0 - \hat{\boldsymbol{n}} \cdot \boldsymbol{r}_A$ from A in the $\hat{\boldsymbol{n}}$ direction.

In cartesian coordinates with the origin at A so $\boldsymbol{r} - \boldsymbol{r}_A = x\hat{\mathbf{x}} + y\hat{\mathbf{y}} + z\hat{\mathbf{z}}$, picking $\hat{\mathbf{z}} = \hat{\boldsymbol{n}}$ and $\hat{\mathbf{y}} \cdot \hat{\boldsymbol{\omega}} = 0$, so $\hat{\boldsymbol{\omega}} = \cos\alpha\,\hat{\mathbf{x}} + \sin\alpha\,\hat{\mathbf{z}}$, for some α in $[-\pi/2, \pi/2]$ (Fig. 12.7), eqns (12.7a) reduce to $z = h$, and

$$x\cos\alpha + h\sin\alpha = \cos\theta\sqrt{x^2 + y^2 + h^2}. \tag{12.7b}$$

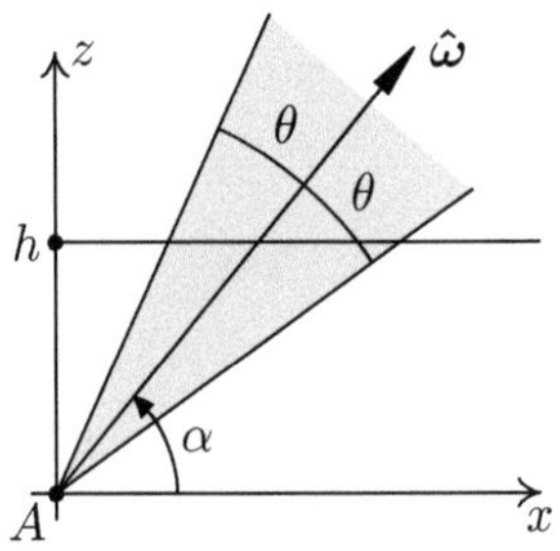

Fig. 12.7 Conic side view.

Squaring both sides,[2] and completing the square yields

$$(1 - \epsilon^2)\left(x - \frac{\epsilon^2 h\tan\alpha}{1 - \epsilon^2}\right)^2 + y^2 = \frac{r_*^2}{1 - \epsilon^2}, \tag{12.7c}$$

where $\epsilon = \cos\alpha/\cos\theta$, and $r_* = h\tan\theta$. This is (11.48c), shifted in x so $(0, 0)$ is at the cone apex not the focus, and ϵ is the *eccentricity* and r_* the *focal halfwidth.* It is an ellipse when $\alpha > \theta$ $(0 \le \epsilon < 1)$, and a hyperbola when $\alpha < \theta$ $(\epsilon > 1)$. For $\alpha = \theta$ $(\epsilon = 1)$, (12.7b) yields the parabola

$$2r_*(x - x_F) = (y^2 - r_*^2), \tag{12.7d}$$

which is in form (11.48b) with $r_* = h\tan\theta$ and focus at $x_F = h/(2\tan\theta)$.

We now consider convenient coordinate independent parameterizations for each of the conics, in addition to the universal polar parameterization (11.24)

$$\boldsymbol{r}(\varphi) = \boldsymbol{r}_F + \frac{r_*\cos\varphi}{1 - \epsilon\cos\varphi}\hat{\boldsymbol{a}} + \frac{r_*\sin\varphi}{1 - \epsilon\cos\varphi}\hat{\boldsymbol{b}}, \tag{12.8}$$

where $\boldsymbol{r}_F$ is the position vector of the focus F, $\hat{\boldsymbol{a}}$ the major principal direction, and $\hat{\boldsymbol{b}}$ the minor principal direction, orthogonal to $\hat{\boldsymbol{a}}$.

Circle

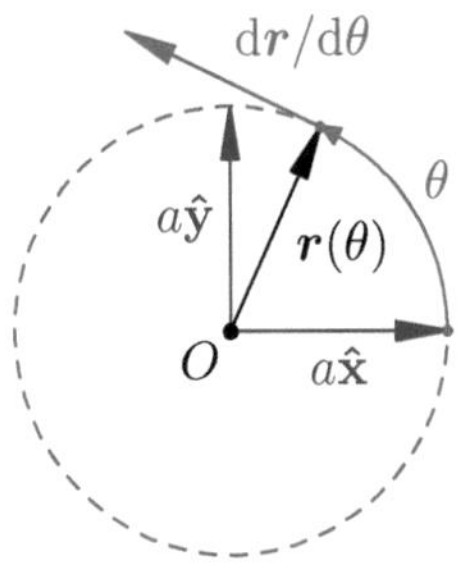

Fig. 12.8 Circle (12.9).

The circle of radius a centered at O has the implicit equation $x^2 + y^2 = a^2$, and can be parameterized by $x = a\cos\theta$, $y = a\sin\theta$, giving the position vector $\boldsymbol{r} = x\hat{\mathbf{x}} + y\hat{\mathbf{y}}$ as (Fig. 12.8)

[2]So $\hat{\omega} \to \pm\hat{\omega}$ and the cone becomes a double cone, and a hyperbola has two branches of solutions.

$$\boldsymbol{r}(\theta) = a\hat{\mathbf{x}}\cos\theta + a\hat{\mathbf{y}}\sin\theta, \tag{12.9}$$

where θ is the angle between $\hat{\mathbf{x}}$ and $\boldsymbol{r}$. The "velocity"

$$\frac{\mathrm{d}\boldsymbol{r}}{\mathrm{d}\theta} = -a\hat{\mathbf{x}}\sin\theta + a\hat{\mathbf{y}}\cos\theta,$$

is the tangent vector to the circle at $\boldsymbol{r}(\theta)$, with $\boldsymbol{r}\cdot\mathrm{d}\boldsymbol{r}/\mathrm{d}\theta = 0$ for any θ, since $\boldsymbol{r}\cdot\boldsymbol{r} = a^2$ is constant.

Ellipse

The circle parameterization is easily extended to an arbitrary ellipse centered at C with major radius vector $\boldsymbol{a}$ and minor radius vector $\boldsymbol{b}$, perpendicular to $\boldsymbol{a}$ (Fig. 12.9),

$$\boldsymbol{r}(\theta) = \boldsymbol{r}_C + \boldsymbol{a}\cos\theta + \boldsymbol{b}\sin\theta. \tag{12.10}$$

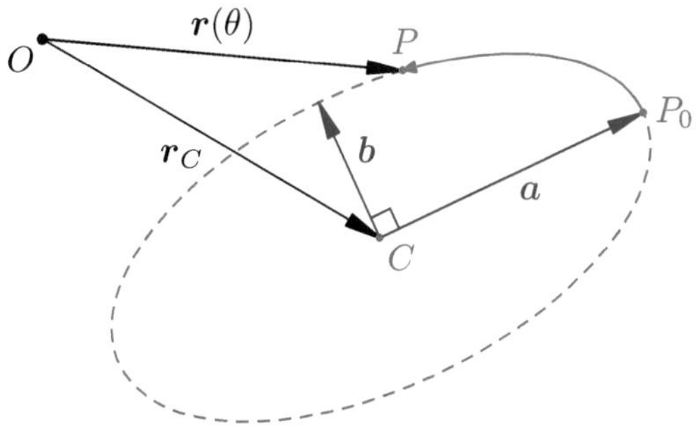

Fig. 12.9 An ellipse centered at C with principal radii $\boldsymbol{a}$ and $\boldsymbol{b}$ (12.10).

The parameter θ is an angle, but it is *not* the angle between $\boldsymbol{a}$ and $(\boldsymbol{r} - \boldsymbol{r}_C)$, unless $|\boldsymbol{a}| = |\boldsymbol{b}|$ (see exercise 12.4). If we pick cartesian coordinates with C as the origin and $\hat{\mathbf{x}} = \hat{\boldsymbol{a}}$, $\hat{\mathbf{y}} = \hat{\boldsymbol{b}}$, then $\boldsymbol{r}_C = 0$, $\boldsymbol{r} = x\hat{\mathbf{x}} + y\hat{\mathbf{y}}$, and (12.10) becomes

$$\begin{cases} x = a\cos\theta \\ y = b\sin\theta \end{cases} \quad\Leftrightarrow\quad \frac{x^2}{a^2} + \frac{y^2}{b^2} = 1, \tag{12.11}$$

which is the standard cartesian equation of an ellipse.

The vector equation (12.10) is coordinate independent. In cartesian coordinates, with (Fig. 12.10)

$$\boldsymbol{r}_c = x_c\,\hat{\mathbf{x}} + y_c\,\hat{\mathbf{y}}, \quad \boldsymbol{a} = a\cos\alpha\,\hat{\mathbf{x}} + a\sin\alpha\,\hat{\mathbf{y}}, \quad \boldsymbol{b} = -b\sin\alpha\,\hat{\mathbf{x}} + b\cos\alpha\,\hat{\mathbf{y}},$$

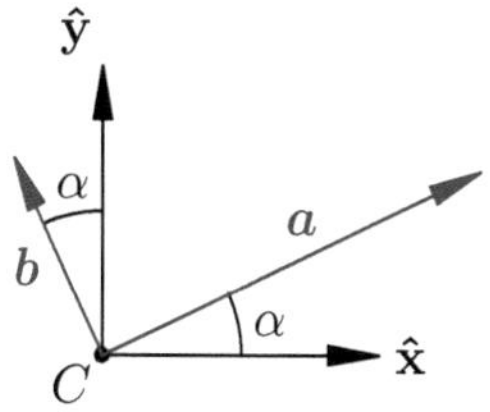

Fig. 12.10 Orientation of major and minor radius vectors $\boldsymbol{a}$ and $\boldsymbol{b}$.

separating the $\hat{\mathbf{x}}$ and $\hat{\mathbf{y}}$ components of (12.10) yields

$$\begin{cases} x = x_c + a\cos\alpha\cos\theta - b\sin\alpha\sin\theta, \\ y = y_c + a\sin\alpha\cos\theta + b\cos\alpha\sin\theta, \end{cases} \tag{12.12}$$

which can be written in matrix form as

$$\begin{bmatrix} \cos\alpha & -\sin\alpha \\ \sin\alpha & \cos\alpha \end{bmatrix} \begin{bmatrix} a & 0 \\ 0 & b \end{bmatrix} \begin{bmatrix} \cos\theta \\ \sin\theta \end{bmatrix} = \begin{bmatrix} x - x_c \\ y - y_c \end{bmatrix}. \tag{12.13}$$

We recognize the first matrix as a rotation matrix whose inverse is its transpose. The second matrix is invertible when a and b are nonzero. Multiplying by the inverses of those matrices gives $(\cos\theta, \sin\theta)$, then eliminating θ using $\cos^2\theta + \sin^2\theta = 1$ yields the general implicit equation for an ellipse in the (x, y) plane

$$(x - x_c)^2\left(\frac{\cos^2\alpha}{a^2} + \frac{\sin^2\alpha}{b^2}\right) + (y - y_c)^2\left(\frac{\sin^2\alpha}{a^2} + \frac{\cos^2\alpha}{b^2}\right) + (x - x_c)(y - y_c)\sin 2\alpha\left(\frac{1}{a^2} - \frac{1}{b^2}\right) = 1. \tag{12.14}$$

The vector equation (12.10) is even more general since it applies to ellipses in 3D (and even in $\mathbb{R}^n$), and $\boldsymbol{a}$ and $\boldsymbol{b}$ do not even have to be orthogonal for the curve to be an ellipse (exercise 12.13), but then P_0 is not a vertex.

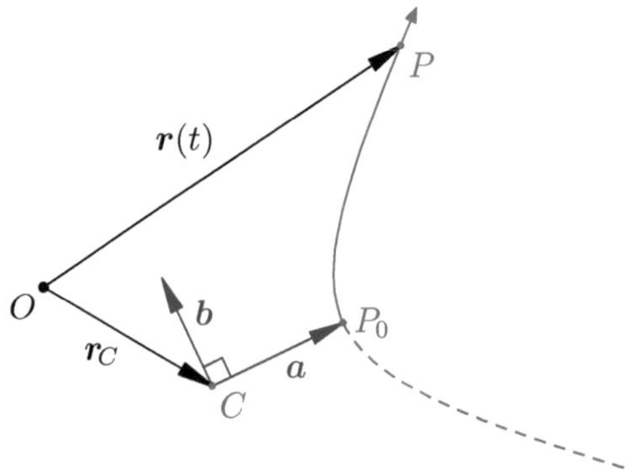

Fig. 12.11 Hyperbolic branch with center C and principal directions $\boldsymbol{a}$ and $\boldsymbol{b}$ (12.15).

Hyperbola

Replacing cos and sin by cosh and sinh in (12.10)

$$\boldsymbol{r}(t) = \boldsymbol{r}_C + \boldsymbol{a}\cosh t + \boldsymbol{b}\sinh t \tag{12.15}$$

yields a hyperbolic branch in the $C, \boldsymbol{a}, \boldsymbol{b}$ plane (Fig. 12.11), where the *hyperbolic* cosine and sine are defined as

$$\cosh t = \frac{e^t + e^{-t}}{2}, \qquad \sinh t = \frac{e^t - e^{-t}}{2}, \tag{12.16}$$

thus

$$\cosh^2 t - \sinh^2 t = 1$$

and

$$\frac{\mathrm{d}}{\mathrm{d}t}\cosh t = \sinh t, \quad \frac{\mathrm{d}}{\mathrm{d}t}\sinh t = \cosh t,$$

for any t, as is easily verified. The reduction of (12.15) to the standard implicit equation of a hyperbola is left as an exercise. As for the ellipse, $\boldsymbol{a}$ and $\boldsymbol{b}$ do not have to be perpendicular for (12.15) to be a hyperbola, but then P_0 is not a vertex.

Parabola

The parabola has already been discussed in (11.5). If P_0 is its vertex, as it was for the ellipse and hyperbola above, and $\boldsymbol{a}$ is its "acceleration" direction, then it can be written (Fig. 12.12)

$$\boldsymbol{r}(t) = \boldsymbol{r}_0 + \tfrac{1}{2}\, t^2\, \boldsymbol{a} + t\, \boldsymbol{b}, \tag{12.17}$$

with $\boldsymbol{a} \cdot \boldsymbol{b} = 0$. Again, $\boldsymbol{a}$ and $\boldsymbol{b}$ do not have to be perpendicular for (12.17) to be a parabola, but then P_0 is not the vertex (Fig. 11.2).

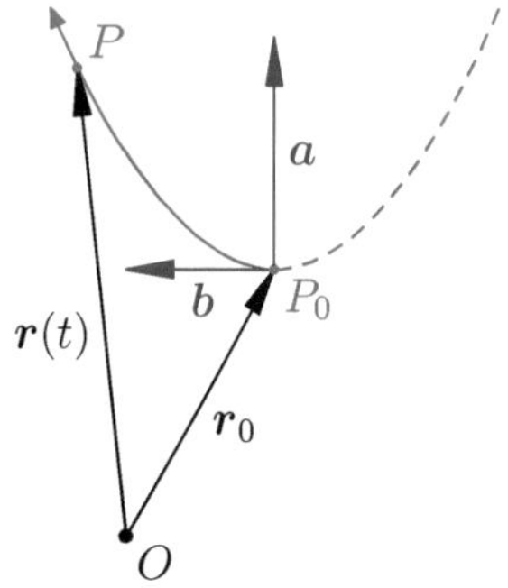

Fig. 12.12 Parabola (12.17).

12.3 Flexible curves

The conics are true classics. They have been studied for millennia from centuries BC to the Renaissance, and beyond. They turned out to be the orbits of planets and comets. Their optical properties are exploited in mirrors, telescopes, and antennas. Many other beautiful curves defined by explicit $\boldsymbol{r}(t)$ functions have been studied, but ship, airplane, and automobile design and construction, then computer graphics, required more flexible curves, easier to assemble and reproduce.

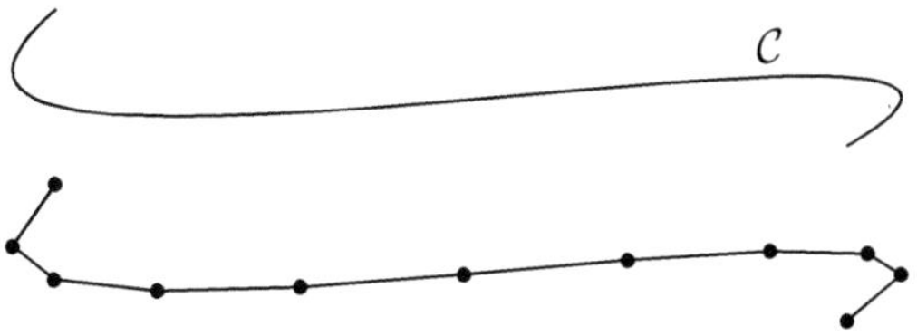

Fig. 12.13 Piecewise linear approximation to a smooth curve $\mathcal{C}$.

To represent an arbitrary curve $\mathcal{C}$, we can sample it at points $P_0, \ldots, P_N$ and simply connect successive points by straight lines, as in Fig. 12.13. This is a *piecewise linear approximation* to the curve, and naturally we would take more sampling points where the curve changes direction more rapidly. It is straightforward to construct an analytical representation for the piecewise linear curve. Since a straight line segment from point A to point B is most simply parametrized as

$$\boldsymbol{r}(t) = \boldsymbol{r}_A + t(\boldsymbol{r}_B - \boldsymbol{r}_A) = (1-t)\boldsymbol{r}_A + t\boldsymbol{r}_B,$$

with $t = 0$ at A and $t = 1$ at B, we can parametrize the full piecewise linear curve as

$$\boldsymbol{r}(t) = (1 - t_k)\boldsymbol{r}_{k-1} + t_k \boldsymbol{r}_k, \tag{12.18}$$

where $t_k = t - \lfloor t \rfloor = t \bmod 1$, with $0 \le t \le N$ for N segments and $\lfloor t \rfloor$ the integer part of t, so t_k runs from 0 to 1 in segment k. This is a simple approach but a satisfactory representation of the curve will typically require a lot of sampling data. The curve (12.18) is continuous but the tangent vectors jump at the nodes $\boldsymbol{r}_k$.

A smoother approach is to represent the curve with higher order piecewise polynomials so we can ensure continuity of $\boldsymbol{r}(t)$ and $\mathrm{d}\boldsymbol{r}/\mathrm{d}t$ at the nodes. Clearly this requires *two* more (vector) degrees of freedom and *piecewise cubic* polynomials. For example, curve $\mathcal{C}$ in Fig. 12.13 can be represented more smoothly *and* with less data as the *cubic Bézier* curve

Pierre Bézier was an engineer working for French automobile company Renault. He used and popularized the curves named after him. The piecewise polynomial curve (12.21) is essentially cubic Hermite interpolation.

$$\boldsymbol{r}(t) = (1-t)^3\boldsymbol{r}_A + 3(1-t)^2 t\,\boldsymbol{r}_A^+ + 3(1-t)t^2\boldsymbol{r}_B^- + t^3\boldsymbol{r}_B, \qquad (12.19)$$

shown in Fig. 12.14, where $\boldsymbol{r}(0) = \boldsymbol{r}_A$, $\boldsymbol{r}(1) = \boldsymbol{r}_B$ are the start and end points, and $\boldsymbol{r}_A^+$, $\boldsymbol{r}_B^-$ are interior control points that determine the tangents at $\boldsymbol{r}_A$ and $\boldsymbol{r}_B$, respectively. Indeed, d/dt of (12.19) yields

$$\dot{\boldsymbol{r}}(0) = 3(\boldsymbol{r}_A^+ - \boldsymbol{r}_A), \qquad \dot{\boldsymbol{r}}(1) = 3(\boldsymbol{r}_B - \boldsymbol{r}_B^-). \qquad (12.20)$$

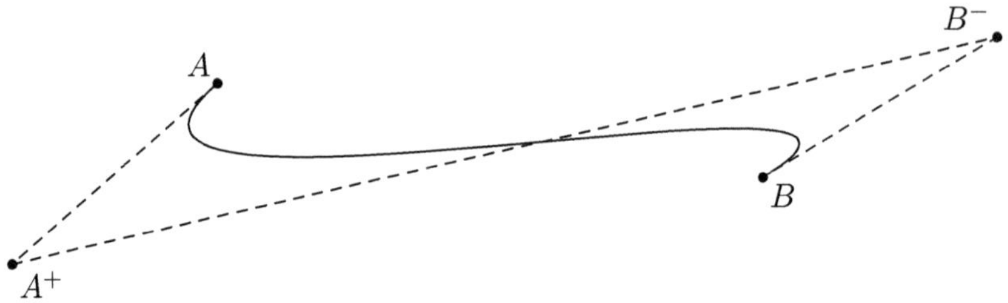

Fig. 12.14 Cubic Bézier representation of $\mathcal{C}$, with its four control points (12.19).

Fig. 12.15 Piecewise cubic Bézier.

The cubic Bézier curves can be combined to create *piecewise cubic* polynomials for more complex curves, with the explicit formula

$$\boldsymbol{r}(t) = (1-t_k)^3\boldsymbol{r}_{k-1} + 3(1-t_k)^2 t_k\,\boldsymbol{r}_{k-1}^+ + 3(1-t_k)t_k^2\boldsymbol{r}_k^- + t^3\boldsymbol{r}_k, \quad (12.21)$$

where $t_k = t - \lfloor t \rfloor = t \bmod 1$, with $0 \le t \le N$ for N cubic Bézier curves and $\lfloor t \rfloor$ the integer part of t, as in (12.18). The piecewise cubic curve (12.21) is continuous at the nodes $\boldsymbol{r}_k$ and has continuous tangents there if $\boldsymbol{r}_k$ and its *pre-* and *post-*controls points, $\boldsymbol{r}_k^-$, $\boldsymbol{r}_k^+$, are on the same line, but this is not required to allow corners and cusps, as in Fig. 12.15 which is composed of four cubic Bézier curves.

Bernstein polynomials

The cubic t-polynomials in (12.19) are cubic *Bernstein polynomials* $B_{k,3}(t)$, with

$$B_{k,n}(t) = \binom{n}{k}(1-t)^{n-k}t^k = \frac{n!}{k!(n-k)!}(1-t)^{n-k}t^k, \qquad (12.22)$$

where $k! = k(k-1)\dots 1$ is the factorial, $\binom{n}{k}$ are the binomial coefficients, and $0! = 1$. Their origin is clear: if all the control points are identical $\boldsymbol{r}_A = \boldsymbol{r}_A^+ = \boldsymbol{r}_B^- = \boldsymbol{r}_B$, then we want $\boldsymbol{r}(t) = \boldsymbol{r}_A$ for all t, so all the polynomials should add up to 1, and that is guaranteed by the binomial expansion

$$1 = ((1-t)+t)^n = \sum_{k=0}^{n}\binom{n}{k}(1-t)^{n-k}t^k. \qquad (12.23)$$

The Bernstein polynomials obey the recurrence relation

$$B_{k,n}(t) = (1-t)\, B_{k,n-1}(t) + t\, B_{k-1,n-1}(t), \qquad (12.24)$$

which is the basis of *De Casteljau*'s recursive algorithm.They also satisfy a simple recurrence for derivatives

$$\frac{\mathrm{d}}{\mathrm{d}t} B_{k,n}(t) = n\big(B_{k-1,n-1}(t) - B_{k,n-1}(t)\big). \qquad (12.25)$$

To facilitate use of these recurrences, the Bernstein polynomial $B_{k,n}(t)$ is defined to be 0 if $k < 0$ or $n < 0$ or $k > n$; then for example, (12.24) and (12.25) correctly yield

$$B_{0,n} = (1-t)B_{0,n-1}, \qquad \frac{\mathrm{d}}{\mathrm{d}t} B_{0,n} = -nB_{0,n-1}.$$

De Casteljau's algorithm

The cubic Bézier curve (12.19) can be written

$$\boldsymbol{r}(t) = \boldsymbol{r}_0\, B_{0,3}(t) + \boldsymbol{r}_1\, B_{1,3}(t) + \boldsymbol{r}_2\, B_{2,3}(t) + \boldsymbol{r}_3\, B_{3,3}(t), \qquad (12.26)$$

in terms of the cubic Bernstein polynomials $B_{k,3}(t)$, where the four control points A, A^+, B^-, B are now labelled 0, 1, 2, 3 to match the Bernstein index k and facilitate the following discussion.

The recurrence (12.24) allows (12.26) to be written as

$$\boldsymbol{r}(t) = \boldsymbol{r}_{0,3}(t) = (1-t)\, \boldsymbol{r}_{0,2}(t) + t\, \boldsymbol{r}_{1,2}(t), \qquad (12.27a)$$

where

$$\begin{aligned} \boldsymbol{r}_{0,2}(t) &= \boldsymbol{r}_0 B_{0,2}(t) + \boldsymbol{r}_1 B_{1,2}(t) + \boldsymbol{r}_2 B_{2,2}(t), \\ \boldsymbol{r}_{1,2}(t) &= \boldsymbol{r}_1 B_{0,2}(t) + \boldsymbol{r}_2 B_{1,2}(t) + \boldsymbol{r}_3 B_{2,2}(t), \end{aligned}$$

are two quadratic Bézier curves with three control points, $(\boldsymbol{r}_0, \boldsymbol{r}_1, \boldsymbol{r}_2)$ for $\boldsymbol{r}_{0,2}(t)$, and $(\boldsymbol{r}_1, \boldsymbol{r}_2, \boldsymbol{r}_3)$ for $\boldsymbol{r}_{1,2}(t)$. Here, the first index of $\boldsymbol{r}_{k,n}(t)$ is the starting point $\boldsymbol{r}_k$, and the second is the polynomial degree n, with $\boldsymbol{r}_{0,3}(t) = \boldsymbol{r}(t)$ in (12.26). The quadratic Bézier curves can themselves be expressed in terms of linear Bézier curves as

$$\begin{aligned} \boldsymbol{r}_{0,2}(t) &= (1-t)\, \boldsymbol{r}_{0,1}(t) + t\, \boldsymbol{r}_{1,1}(t), \\ \boldsymbol{r}_{1,2}(t) &= (1-t)\, \boldsymbol{r}_{1,1}(t) + t\, \boldsymbol{r}_{2,1}(t), \end{aligned} \qquad (12.27b)$$

where

$$\begin{aligned} \boldsymbol{r}_{0,1}(t) &= \boldsymbol{r}_0 B_{0,1}(t) + \boldsymbol{r}_1 B_{1,1}(t) = (1-t)\, \boldsymbol{r}_0 + t\, \boldsymbol{r}_1, \\ \boldsymbol{r}_{1,1}(t) &= \boldsymbol{r}_1 B_{0,1}(t) + \boldsymbol{r}_2 B_{1,1}(t) = (1-t)\, \boldsymbol{r}_1 + t\, \boldsymbol{r}_2, \\ \boldsymbol{r}_{2,1}(t) &= \boldsymbol{r}_2 B_{0,1}(t) + \boldsymbol{r}_3 B_{1,1}(t) = (1-t)\, \boldsymbol{r}_2 + t\, \boldsymbol{r}_3. \end{aligned} \qquad (12.27c)$$

Equations (12.27) yield the cubic Bézier curve as successive linear approximations. The algorithm is illustrated in Fig. 12.16, and can also be visualized geometrically as done in Fig. 12.17.

Paul De Casteljau was a mathematician who worked for French automobile company Citroën. He is credited with developing the concept and algorithms for the curves named after Bézier. Others developed similar concepts at about the same time, including work on *splines* by Schoenberg in the 1940s and de Boor in the late 50s, then at General Motors.

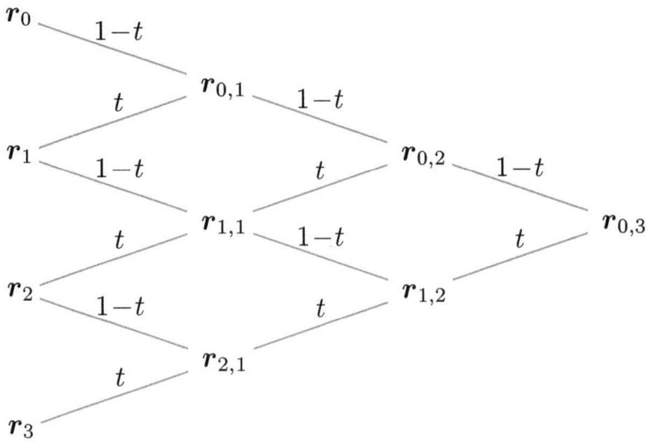

Fig. 12.16 Recurrence (12.27) to evaluate $\boldsymbol{r}_{0,3}(t)$.

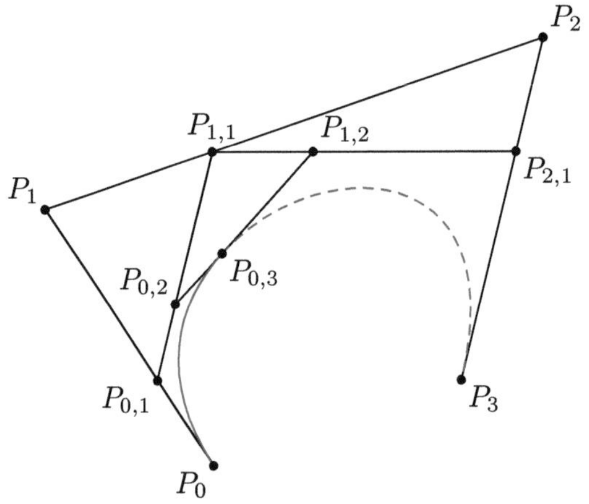

Fig. 12.17 Recurrence (12.27) to evaluate $\boldsymbol{r}_{0,3}(t)$, here for $t = 1/3$, with $P_{k,n}$ at relative distance t along line segment $[P_{k,n-1}, P_{k+1,n-1}]$.

The recursion has other benefits. One benefit is that the velocity and acceleration at $\boldsymbol{r}_{0,3}(t)$ are simply

$$\begin{aligned} \dot{\boldsymbol{r}}_{0,3}(t) &= 3\,(\boldsymbol{r}_{1,2}(t) - \boldsymbol{r}_{0,2}(t)), \\ \ddot{\boldsymbol{r}}_{0,3}(t) &= 6\,(\boldsymbol{r}_{2,1}(t) - 2\boldsymbol{r}_{1,1}(t) + \boldsymbol{r}_{0,1}(t)). \end{aligned} \tag{12.28}$$

Another is that to split a cubic Bézier curve with control points P_0, P_1, P_2, P_3, at any one of its points $P_{0,3}(t)$, the control points for the split cubics are simply P_0, $P_{0,1}$, $P_{0,2}$, $P_{0,3}$ for the first cubic, and $P_{0,3}$, $P_{1,2}$, $P_{2,1}$, P_3 for the second. That is how Fig. 12.17 was plotted with the first cubic solid, and the second dashed.

Piecewise cubic Bézier curves are an example of cubic *splines*, widely used in computer graphics. A broader class of such functions are *B-splines* that are also defined by recurrence, a special case of which is the Bernstein polynomial recurrence (12.24). B-splines can be constructed to automatically ensure continuity of the second derivatives (accelerations) at the connection betweens polynomial pieces. The reader is referred to de Boor's book.[3]

[3]Carl de Boor, *A Practical Guide to Splines*, Springer-Verlag, 2001.

12.4 Speeding through curves

The position vector $\boldsymbol{r}(t)$ of a particle P at time t contains information about its trajectory, that is, the *curve* along which it is traveling, as well as information about speed and acceleration that are proper to the particle. Different cars traveling on the same road will have different $\boldsymbol{r}(t)$, but the underlying curve is the same. Here, we separate those particle-specific speed and acceleration from the intrinsic characteristics of the underlying curve. *Arclength*, the distance along the curve, is the fundamental quantity that allows separation of speed from geometry.

Speed and arclength

Arclength s can be defined in terms of the speed $v(t) = |\boldsymbol{v}(t)|$ by the scalar differential equation

$$\frac{\mathrm{d}s}{\mathrm{d}t} = v = \left|\frac{\mathrm{d}\boldsymbol{r}}{\mathrm{d}t}\right| > 0. \tag{12.29}$$

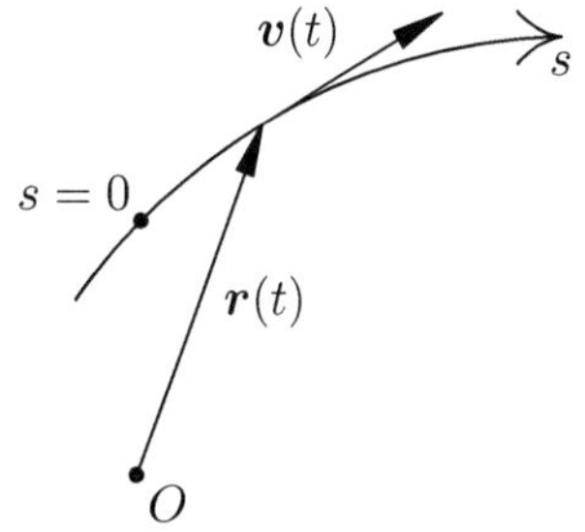

Fig. 12.18 Arclength s is distance along the curve, from an arbitrary starting point. Distance over time is speed, $\mathrm{d}s/\mathrm{d}t = |\boldsymbol{v}(t)|$.

This definition picks the direction of increasing s as the direction of travel as t increases. Thus arclength s is a monotonically increasing function of t and there is a one-to-one correspondence between s and t (Fig. 12.18), assuming that $\boldsymbol{r}(t)$ is monotonically moving along the curve; no stopping and backtracking allowed when curve mapping! Forward we go, assuming that you are up to speed on your *chain rule.*

The successive time derivatives of the position vector $\boldsymbol{r}(t)$ are well known: $\boldsymbol{v} = \dot{\boldsymbol{r}} = \mathrm{d}\boldsymbol{r}/\mathrm{d}t$ is the *velocity*, $\boldsymbol{a} = \ddot{\boldsymbol{r}}$ is the *acceleration*, and $\boldsymbol{j} = \dddot{\boldsymbol{r}}$ is the *jerk*. The *velocity* $\boldsymbol{v} = \dot{\boldsymbol{r}}$ can be written in terms of speed $v = |\dot{\boldsymbol{r}}|$ and tangent direction to the curve $\hat{\boldsymbol{t}}$,

$$\boldsymbol{v} = \frac{\mathrm{d}\boldsymbol{r}}{\mathrm{d}t} = \frac{\mathrm{d}s}{\mathrm{d}t}\frac{\mathrm{d}\boldsymbol{r}}{\mathrm{d}s} = v\,\hat{\boldsymbol{t}}, \tag{12.30}$$

where

$$\hat{\boldsymbol{t}} = \frac{\mathrm{d}\boldsymbol{r}}{\mathrm{d}s} = \frac{\mathrm{d}\boldsymbol{r}/\mathrm{d}t}{|\mathrm{d}\boldsymbol{r}/\mathrm{d}t|} \tag{12.31}$$

points in the direction of increasing s and t.

Acceleration and curvature

The *acceleration* $\boldsymbol{a} = \ddot{\boldsymbol{r}}$ is the time rate of change of velocity, and the chain rule gives

$$\begin{aligned}\boldsymbol{a} = \frac{\mathrm{d}\boldsymbol{v}}{\mathrm{d}t} = \frac{\mathrm{d}s}{\mathrm{d}t}\frac{\mathrm{d}\boldsymbol{v}}{\mathrm{d}s} &= v\frac{\mathrm{d}(v\,\hat{\boldsymbol{t}})}{\mathrm{d}s} = v\frac{\mathrm{d}v}{\mathrm{d}s}\hat{\boldsymbol{t}} + v^2\frac{\mathrm{d}\hat{\boldsymbol{t}}}{\mathrm{d}s}\\ &= \tfrac{1}{2}\frac{\mathrm{d}(v^2)}{\mathrm{d}s}\,\hat{\boldsymbol{t}} + \kappa v^2\,\hat{\boldsymbol{n}},\end{aligned} \tag{12.32}$$

where

$$\frac{\mathrm{d}\hat{\boldsymbol{t}}}{\mathrm{d}s} \triangleq \kappa\,\hat{\boldsymbol{n}} = \frac{1}{R}\,\hat{\boldsymbol{n}}. \tag{12.33a}$$

Equation (12.33a) defines the magnitude κ and direction $\hat{\boldsymbol{n}}$ of the vector $\mathrm{d}\hat{\boldsymbol{t}}/\mathrm{d}s$. Since $\hat{\boldsymbol{t}} \cdot \hat{\boldsymbol{t}} = 1$, we have $\hat{\boldsymbol{t}} \cdot \mathrm{d}\hat{\boldsymbol{t}}/\mathrm{d}s = 0$; thus, $\hat{\boldsymbol{n}}$ is perpendicular to $\hat{\boldsymbol{t}}$. Indeed $\hat{\boldsymbol{n}}$ points in the direction of the turn and is a unit vector that is *normal* (or orthogonal, or perpendicular) to the curve at $\boldsymbol{r}(t)$. The *curvature*

$$\kappa = \left| \frac{\mathrm{d}\hat{\boldsymbol{t}}}{\mathrm{d}s} \right|, \tag{12.33b}$$

has units of inverse length, and can thus be written as $\kappa = 1/R$, where $R = R(s)$ is the local *radius of curvature*. Curvature is the turning rate of the tangent per unit arclength (Fig. 12.19),

$$\mathrm{d}\hat{\boldsymbol{t}} = \hat{\boldsymbol{n}}\, \kappa\, \mathrm{d}s = \hat{\boldsymbol{n}}\, \mathrm{d}\beta. \tag{12.33c}$$

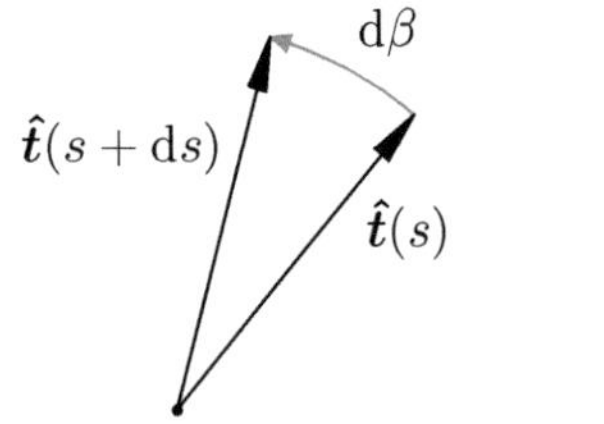

Fig. 12.19 Curvature κ is the rate of rotation of the tangent with respect to distance along the curve $\mathrm{d}\beta = \kappa\, \mathrm{d}s$.

Figure 12.20 illustrates $\boldsymbol{v}$, $\boldsymbol{a}$, $\hat{\boldsymbol{t}}$ and $\hat{\boldsymbol{n}}$ for a curve $\boldsymbol{r}(t)$ defined as a cubic Bézier (12.19). Note that $\hat{\boldsymbol{n}}$ reverses abruptly at an inflection point, but $\kappa\hat{\boldsymbol{n}}$ is continuous since $\kappa = 0$ at an inflection.

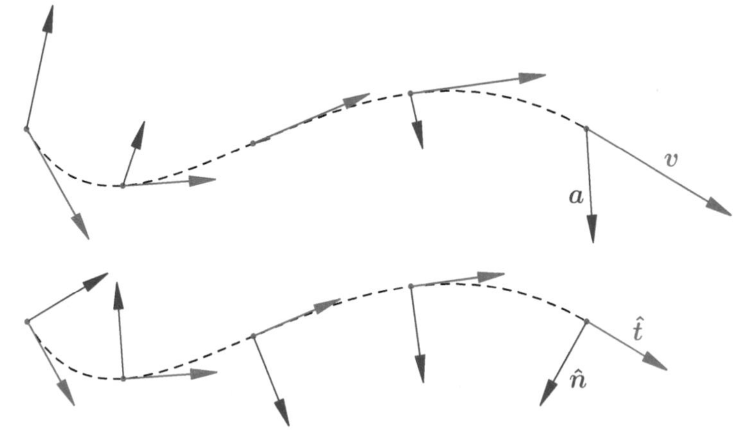

Fig. 12.20 Velocity $\boldsymbol{v}$ and acceleration $\boldsymbol{a}$ (top). Unit tangent $\hat{\boldsymbol{t}}$ and normal $\hat{\boldsymbol{n}}$ (bottom).

With these definitions, expression (12.32) yields a decomposition of the acceleration $\boldsymbol{a}$ in terms of a component in the curve direction $\hat{\boldsymbol{t}}$ and a component in the turn direction $\hat{\boldsymbol{n}}$. Physics students will recognize the $\hat{\boldsymbol{n}}$ component, $\kappa v^2 = v^2/R$, as the *centripetal acceleration* of the particle. The $\hat{\boldsymbol{t}}$ component foreshadows the *work energy theorem*; the $\hat{\boldsymbol{t}}$ component of the applied force $\boldsymbol{F} = m\boldsymbol{a}$ is the rate of change of the kinetic energy $\mathrm{d}(\frac{1}{2}mv^2)/\mathrm{d}s$. Expression (12.32) is general, holding for any C^2 curve, not just for uniform circular motion where $\boldsymbol{a} = \hat{\boldsymbol{n}}\, v^2/R = \hat{\boldsymbol{n}}\, \omega^2 R$ is purely centripetal and of constant magnitude. Expression (12.32) holds for curvature $\kappa = \kappa(s)$, and speed $v = v(s)$ that vary along the curve, in general.

Jerk and torsion

Next comes the *jerk* $\boldsymbol{j} = \dddot{\boldsymbol{r}}$, the rate of change of acceleration. The chain rule, the previous results (12.30) for $\boldsymbol{v}$ and (12.32) for $\boldsymbol{a}$, and some perseverance, yield

$$
\begin{aligned}
\boldsymbol{j} = \frac{\mathrm{d}\boldsymbol{a}}{\mathrm{d}t} &= \frac{\mathrm{d}s}{\mathrm{d}t}\frac{\mathrm{d}\boldsymbol{a}}{\mathrm{d}s} \\
&= \frac{v}{2}\frac{\mathrm{d}^2(v^2)}{\mathrm{d}s^2}\hat{\boldsymbol{t}} + v^2\frac{\mathrm{d}v}{\mathrm{d}s}\frac{\mathrm{d}\hat{\boldsymbol{t}}}{\mathrm{d}s} + v\frac{\mathrm{d}(\kappa v^2)}{\mathrm{d}s}\hat{\boldsymbol{n}} + \kappa v^3\frac{\mathrm{d}\hat{\boldsymbol{n}}}{\mathrm{d}s} \\
&= \frac{v}{2}\frac{\mathrm{d}^2(v^2)}{\mathrm{d}s^2}\hat{\boldsymbol{t}} + \frac{\mathrm{d}(\kappa v^3)}{\mathrm{d}s}\hat{\boldsymbol{n}} + \kappa v^3\frac{\mathrm{d}\hat{\boldsymbol{n}}}{\mathrm{d}s} \\
&= \left(\frac{v}{2}\frac{\mathrm{d}^2(v^2)}{\mathrm{d}s^2} - \kappa^2 v^3\right)\hat{\boldsymbol{t}} + \frac{\mathrm{d}(\kappa v^3)}{\mathrm{d}s}\hat{\boldsymbol{n}} + \kappa\tau v^3\,\hat{\boldsymbol{b}}
\end{aligned}
\tag{12.34}
$$

with

$$
\frac{\mathrm{d}\hat{\boldsymbol{n}}}{\mathrm{d}s} = \tau\,\hat{\boldsymbol{b}} - \kappa\,\hat{\boldsymbol{t}} \tag{12.35}
$$

in terms of the *binormal* $\hat{\boldsymbol{b}} = \hat{\boldsymbol{t}} \times \hat{\boldsymbol{n}}$ and the *torsion* $\tau(s)$. Equation (12.35) defines τ and $\hat{\boldsymbol{b}}$. Since $\hat{\boldsymbol{n}}$ is a unit vector, it must be orthogonal to its derivative; thus, $\mathrm{d}\hat{\boldsymbol{n}}/\mathrm{d}s$ cannot have any $\hat{\boldsymbol{n}}$ component, but it can have a $\hat{\boldsymbol{t}}$ component, and a $\hat{\boldsymbol{b}}$ component. The $\hat{\boldsymbol{t}}$ component of $\hat{\boldsymbol{n}}$ has to be $(-\kappa)$ since $\hat{\boldsymbol{t}} \cdot \hat{\boldsymbol{n}} = 0$ always, thus the derivative of $\hat{\boldsymbol{t}} \cdot \hat{\boldsymbol{n}}$ is also zero, and by the product rule that imposes

$$
\frac{\mathrm{d}\hat{\boldsymbol{t}}}{\mathrm{d}s} \cdot \hat{\boldsymbol{n}} + \hat{\boldsymbol{t}} \cdot \frac{\mathrm{d}\hat{\boldsymbol{n}}}{\mathrm{d}s} = \kappa + \hat{\boldsymbol{t}} \cdot \frac{\mathrm{d}\hat{\boldsymbol{n}}}{\mathrm{d}s} = 0.
$$

In general, $\mathrm{d}\hat{\boldsymbol{n}}/\mathrm{d}s$ can have a $\hat{\boldsymbol{b}}$ component as well, orthogonal to both $\hat{\boldsymbol{t}}$ and $\hat{\boldsymbol{n}}$. We call that component τ, the *torsion*. The binormal $\hat{\boldsymbol{b}}$ is constant and the torsion $\tau = 0$ for a planar curve, such that $\hat{\boldsymbol{t}}$ and $\hat{\boldsymbol{n}}$ stay in a plane that is, in fact, orthogonal to $\hat{\boldsymbol{b}}$.

Curvature and torsion shortcuts

We can stop before we snap.[4] Three derivatives of $\boldsymbol{r}(t)$ is enough. The first derivative, $\boldsymbol{v} = \dot{\boldsymbol{r}}$, gives the speed $v = |\dot{\boldsymbol{r}}|$ and unit tangent $\hat{\boldsymbol{t}} = \dot{\boldsymbol{r}}/v$. The second derivative, $\boldsymbol{a} = \ddot{\boldsymbol{r}}$, then gives the curvature κ and binormal $\hat{\boldsymbol{b}}$ through a cross product with $\boldsymbol{v}$. Indeed, their expressions in terms of $\hat{\boldsymbol{t}}$ and $\hat{\boldsymbol{n}}$, (12.30) and (12.32), yield

$$
\boldsymbol{v} \times \boldsymbol{a} = \dot{\boldsymbol{r}} \times \ddot{\boldsymbol{r}} = \kappa\, v^3\, \hat{\boldsymbol{b}}, \tag{12.36}
$$

and thus provides the curvature κ and binormal $\hat{\boldsymbol{b}}$, directly, without explicit use of arclength that invariably involves square roots since $\mathrm{d}s = \sqrt{\mathrm{d}\boldsymbol{r} \cdot \mathrm{d}\boldsymbol{r}}$. The normal then follows from $\hat{\boldsymbol{n}} = \hat{\boldsymbol{b}} \times \hat{\boldsymbol{t}}$. The dot product of (12.36) with the jerk (12.34), then yields the torsion τ,

$$
(\boldsymbol{v} \times \boldsymbol{a}) \cdot \boldsymbol{j} = (\dot{\boldsymbol{r}} \times \ddot{\boldsymbol{r}}) \cdot \dddot{\boldsymbol{r}} = (\kappa v^3)^2\, \tau. \tag{12.37}
$$

We need to compute one square root, to find v from $v^2 = \dot{\boldsymbol{r}} \cdot \dot{\boldsymbol{r}}$, but we do not need to labor through repeated derivatives of square roots. A cross product (12.36), and a mixed product (12.37), avoid that.

[4] *Snap* is the time derivative of *jerk*, that is $\mathrm{d}^4\boldsymbol{r}/\mathrm{d}t^4$.

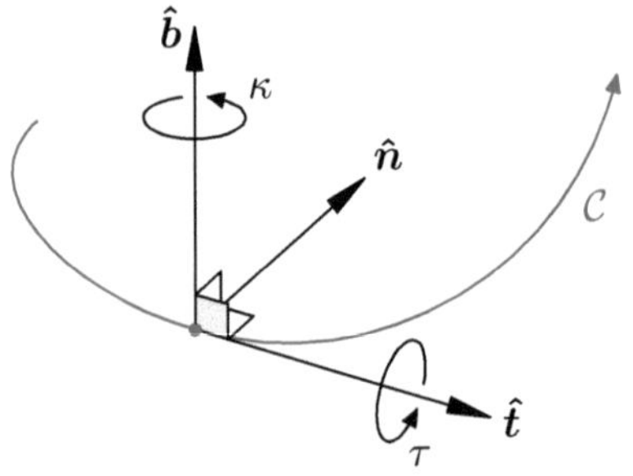

Fig. 12.21 Frenet frame rotates about $\hat{t}$ at rate τ, and $\hat{b}$ at rate κ.

Frenet–Serret frame and formula

The triad $\{\hat{t}, \hat{n}, \hat{b}\}$ forms a right-handed orthonormal basis at $r(s)$, called the *Frenet frame* (Fig, 12.21). The Frenet frame rotates as $r(s)$ moves along the curve; its basis vectors evolve according to the *Frenet–Serret formulas*, (12.33a), (12.35), best written in matrix form as

$$\frac{\mathrm{d}}{\mathrm{d}s}\begin{bmatrix}\hat{t}\\ \hat{n}\\ \hat{b}\end{bmatrix} = \begin{bmatrix}0 & \kappa & 0\\ -\kappa & 0 & \tau\\ 0 & -\tau & 0\end{bmatrix}\begin{bmatrix}\hat{t}\\ \hat{n}\\ \hat{b}\end{bmatrix}, \tag{12.38}$$

where $\mathrm{d}\hat{b}/\mathrm{d}s$ follows from the product rule applied to the definition $\hat{b} = \hat{t}\times\hat{n}$. The matrix in (12.38) is antisymmetric, and therefore equivalent to a cross product with the angular rotation rate vector $\boldsymbol{\omega}(s) = \tau\hat{t} + \kappa\hat{b}$. Indeed, each of the basis vectors, $\hat{t}(s)$, $\hat{n}(s)$, $\hat{b}(s)$, evolves according to the differential equation

Jean Frenet and *Joseph Serret*, French mathematicians who discovered the formula named after them around 1850.

$$\frac{\mathrm{d}\boldsymbol{u}}{\mathrm{d}s} = \boldsymbol{\omega}(s)\times\boldsymbol{u} = \left(\tau\hat{t} + \kappa\hat{b}\right)\times\boldsymbol{u}, \tag{12.39}$$

which corresponds to rotation about $\hat{t}$ at rate τ, and about $\hat{b}$ at rate κ. The angular rotation rate $\boldsymbol{\omega}$ has units of inverse length here; it is a rate of rotation per unit arclength, not time.

The representation of a curve $\mathcal{C}$ in terms of its arclength, $\boldsymbol{r} = \boldsymbol{r}(s)$, is unique up to a choice of origin for s, positive direction, and unit of length. Think, for instance, of the coordinates of points along US Highway 90, from Boston to Seattle. If $s = 0$ is Madison, Wisconsin, with $s > 0$ heading west, then 435 miles is a unique location.[5] But $\boldsymbol{r}(s)$ is usually not a nice explicit function, except for the simplest curves such as straight lines and circles. By contrast, an infinite number of $\boldsymbol{r}(t)$ functions can describe the same curve $\mathcal{C}$; for example, $\boldsymbol{r}(t)$ could represent the positions of different cars and trucks driving along the same highway at different speeds $v(t)$.

[5] *Sioux Falls, ND.*

12.5 Line integrals

A line integral, also called a *path integral*, is an integral along a curve $\mathcal{C}$. This is confusing to many students who have been drilled to think of an integral as the *area under a curve.* To avoid conceptual confusion, it is best to adopt the more general view of an integral as the *limit of a sum.*

A standard line integral over a curve $\mathcal{C}$ is

$$\int_{\mathcal{C}} \boldsymbol{F}(\boldsymbol{r}) \cdot \mathrm{d}\boldsymbol{r}, \tag{12.40}$$

where $\boldsymbol{F}(\boldsymbol{r})$ is a *vector field*, a vector function of position vector $\boldsymbol{r}$, such as the gravity field $\boldsymbol{F}(\boldsymbol{r}) = -\mathrm{k}\,\boldsymbol{r}/r^3$, or some other force field. In those forceful cases, the line integral is a *scalar* interpreted as the *work* done by the force on a particle moving along that curve.

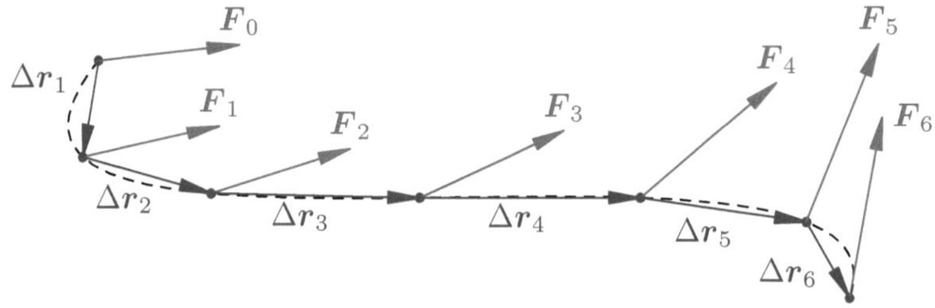

Fig. 12.22 Line integral as limit of a sum (12.41a), here with $N = 6$.

The integral is readily understood as a *Riemann sum.* Sampling the curve at successive points, or *nodes*, $P_0, \ldots, P_N$, with position vectors $\boldsymbol{r}_0, \ldots, \boldsymbol{r}_N$, along the curve (Fig. 12.22), we define the integral as the limit of a sum

$$\int_{\mathcal{C}} \boldsymbol{F}(\boldsymbol{r}) \cdot \mathrm{d}\boldsymbol{r} = \lim_{|\Delta\boldsymbol{r}_n|\to 0} \sum_{n=1}^{N} \overline{\boldsymbol{F}}_n \cdot \Delta\boldsymbol{r}_n, \tag{12.41a}$$

where $\Delta\boldsymbol{r}_n = \boldsymbol{r}_n - \boldsymbol{r}_{n-1}$ is the secant vector from P_{n-1} to P_n, and $\overline{\boldsymbol{F}}_n$ is an estimate of the average value of $\boldsymbol{F}(\boldsymbol{r})$ along the arc from P_{n-1} to P_n. A simple choice is $\overline{\boldsymbol{F}}_n = \boldsymbol{F}_{n-1}$, or $\overline{\boldsymbol{F}}_n = \boldsymbol{F}_n$, where $\boldsymbol{F}_j = \boldsymbol{F}(\boldsymbol{r}_j)$, but better choices are the *trapezoidal rule* that uses the average nodal values,

$$\overline{\boldsymbol{F}}_n = \tfrac{1}{2}\left(\boldsymbol{F}_{n-1} + \boldsymbol{F}_n\right), \tag{12.41b}$$

or the *midpoint rule* that uses the function evaluated at the secant midpoint,

$$\overline{\boldsymbol{F}}_n = \boldsymbol{F}\left(\tfrac{1}{2}(\boldsymbol{r}_{n-1} + \boldsymbol{r}_n)\right). \tag{12.41c}$$

These different choices for $\overline{\boldsymbol{F}}_n$ give finite sums that converge to the same limit, but the trapezoidal and midpoint rules will converge faster for sufficiently smooth curves and functions, and give more accurate finite sum approximations, as studied in numerical analysis.

The Riemann sum definition only requires being able to sample the curve, and the function values, at nodes P_n. If an explicit representation $\boldsymbol{r}(t)$ is known for the curve, then the sampling can be done by selecting a sequence $t_0 < t_1 < \cdots < t_N = t_f$, and the limit $|\Delta\boldsymbol{r}_n| \to 0$ becomes a limit $\Delta t_n \to 0$, assuming that $\boldsymbol{r}(t)$ is differentiable. In the limit, this reduces the line integral to a regular scalar integral over t,

$$\int_{\mathcal{C}} \boldsymbol{F}(\boldsymbol{r}) \cdot \mathrm{d}\boldsymbol{r} = \int_{t_0}^{t_f} f(t)\,\mathrm{d}t, \tag{12.42a}$$

where

$$f(t) = \boldsymbol{F}(\boldsymbol{r}(t)) \cdot \frac{\mathrm{d}\boldsymbol{r}}{\mathrm{d}t}, \tag{12.42b}$$

with $\boldsymbol{r}(t_0)$ the starting point of curve $\mathcal{C}$, and $\boldsymbol{r}(t_f)$ its end point. This is a change of variable, from $\boldsymbol{r}$ along $\mathcal{C}$ to t in the interval $[t_0, t_f]$, with

$$\mathrm{d}\boldsymbol{r} = \frac{\mathrm{d}\boldsymbol{r}}{\mathrm{d}t}\,\mathrm{d}t. \tag{12.43}$$

It is useful to think of the differential $\mathrm{d}\boldsymbol{r}$, called the *line element*, as an infinitesimal secant vector, tangent to the curve at point $\boldsymbol{r}$ and of magnitude $|\mathrm{d}\boldsymbol{r}| = \mathrm{d}s$, the arclength element with

$$\mathrm{d}s = |\mathrm{d}\boldsymbol{r}| = \sqrt{\mathrm{d}\boldsymbol{r} \cdot \mathrm{d}\boldsymbol{r}} = \sqrt{\mathrm{d}x^2 + \mathrm{d}y^2 + \mathrm{d}z^2}. \tag{12.44}$$

The arclength element $|\mathrm{d}\boldsymbol{r}| = \mathrm{d}s$ should not be confused with the differential of distance to the origin r, dr, which is

$$\mathrm{d}r = \mathrm{d}|\boldsymbol{r}| = \mathrm{d}(\sqrt{x^2+y^2+z^2}) = \frac{x\,\mathrm{d}x + y\,\mathrm{d}y + z\,\mathrm{d}z}{\sqrt{x^2+y^2+z^2}} = \hat{\boldsymbol{r}} \cdot \mathrm{d}\boldsymbol{r}, \tag{12.45}$$

which is the differential form of (10.9).

Other types of line integrals

Other types of line integrals arise in applications, besides the "work integral" (12.40). For example,

$$\int_{\mathcal{C}} f(\boldsymbol{r})\,|\mathrm{d}\boldsymbol{r}|, \quad \int_{\mathcal{C}} \boldsymbol{F}(\boldsymbol{r})\,|\mathrm{d}\boldsymbol{r}|, \quad \int_{\mathcal{C}} f(\boldsymbol{r})\,\mathrm{d}\boldsymbol{r}, \quad \int_{\mathcal{C}} \boldsymbol{F}(\boldsymbol{r}) \times \mathrm{d}\boldsymbol{r}.$$

The first one gives a *scalar* result; one example is *arclength* discussed hereafter. The latter three integrals would give *vector* results. We can make sense of these integrals from the *limit of a sum* definition. They will show up in some examples later, and in applications in this textbook and in other fields.

Examples of line integrals

The line integral equivalent of $\int_a^b x\,\mathrm{d}x = (b^2 - a^2)/2$ is (12.40) with $\boldsymbol{F}(\boldsymbol{r}) = \boldsymbol{r}$ along the straight line from point A to point B, with position vectors $\boldsymbol{r}_A = \boldsymbol{a}$ and $\boldsymbol{r}_B = \boldsymbol{b}$; then

$$\mathcal{C} : \boldsymbol{r}(t) = \boldsymbol{a} + t\,(\boldsymbol{b} - \boldsymbol{a}), \qquad \mathrm{d}\boldsymbol{r}(t) = (\boldsymbol{b} - \boldsymbol{a})\,\mathrm{d}t,$$

with $t = 0 \to 1$, and (12.42) yields

$$\begin{aligned} \int_{\mathcal{C}} \boldsymbol{r} \cdot \mathrm{d}\boldsymbol{r} &= \int_0^1 \Big(\boldsymbol{a} + t\,(\boldsymbol{b} - \boldsymbol{a})\Big) \cdot (\boldsymbol{b} - \boldsymbol{a})\,\mathrm{d}t \\ &= \boldsymbol{a} \cdot (\boldsymbol{b} - \boldsymbol{a}) + \tfrac{1}{2}(\boldsymbol{b} - \boldsymbol{a}) \cdot (\boldsymbol{b} - \boldsymbol{a}) = \tfrac{1}{2}\,(\boldsymbol{b} \cdot \boldsymbol{b} - \boldsymbol{a} \cdot \boldsymbol{a}). \end{aligned} \tag{12.46}$$

That is straightforward, yet more algebra than necessary in this case. This is a very special $\boldsymbol{F}(\boldsymbol{r}) = \boldsymbol{r}$, so special that this line integral is actually *independent of the path* from A to B. Indeed, for *any* curve $\mathcal{C}$ from A to B

$$\int_{\mathcal{C}} \boldsymbol{r} \cdot \mathrm{d}\boldsymbol{r} = \tfrac{1}{2} \int_{\mathcal{C}} \mathrm{d}(\boldsymbol{r} \cdot \boldsymbol{r}) = \tfrac{1}{2} \left(\boldsymbol{r}_B \cdot \boldsymbol{r}_B - \boldsymbol{r}_A \cdot \boldsymbol{r}_A\right) \tag{12.47}$$

using differentials as in (12.45), This is a special case of the general result that

$$\int_{\mathcal{C}} \mathrm{d}f(P) = f(B) - f(A), \tag{12.48}$$

for a curve $\mathcal{C}$ that starts at point A and ends at point B, where $f(P)$ is the value of a function at point P of $\mathcal{C}$. That integral is a sum of differences along the curve, that is,

$$\begin{aligned} \int_{\mathcal{C}} \mathrm{d}f(P) &= \lim_{N \to \infty} \Big((f_1 - f(A)) + (f_2 - f_1) + \cdots + (f(B) - f_{N-1}) \Big) \\ &= \lim_{N \to \infty} \Big(f(B) - f(A) \Big) = f(B) - f(A), \end{aligned}$$

with $f_k = f(P_k)$. The collapsing sum of successive differences is known as a *telescoping sum*, and we are essentially rehashing the *fundamental theorem of calculus.*

We can also calculate the integral by calling on whatever $\boldsymbol{r}(t)$ might specify the curve, without the need for knowing $\boldsymbol{r}(t)$ explicitly. Indeed,

$$\begin{aligned} \int_{\mathcal{C}} \boldsymbol{r} \cdot \mathrm{d}\boldsymbol{r} = \int_{t_A}^{t_B} \boldsymbol{r}(t) \cdot \frac{\mathrm{d}\boldsymbol{r}(t)}{\mathrm{d}t}\,\mathrm{d}t &= \tfrac{1}{2} \int_{t_A}^{t_B} \frac{\mathrm{d}(\boldsymbol{r} \cdot \boldsymbol{r})}{\mathrm{d}t}\,\mathrm{d}t \\ &= \tfrac{1}{2} \left(\boldsymbol{r}_B \cdot \boldsymbol{r}_B - \boldsymbol{r}_A \cdot \boldsymbol{r}_A\right), \end{aligned} \tag{12.49}$$

with a more explicit use of the *fundamental theorem of calculus*

$$\int_a^b \frac{\mathrm{d}f}{\mathrm{d}t}\mathrm{d}t = \int_a^b \mathrm{d}f(t) = f(b) - f(a). \tag{12.50}$$

Either way, we recover (12.46) *whatever the curve from* A to B. But, this *path independence* only holds for very special vector fields $\boldsymbol{F}(\boldsymbol{r})$ called *conservative vector fields.*

Work of a central force

Central force fields, $\boldsymbol{F}(\boldsymbol{r}) = F(r)\hat{\boldsymbol{r}}$ where $\boldsymbol{r} = r\hat{\boldsymbol{r}}$, are a class of conservative fields. The previous example $\boldsymbol{F}(\boldsymbol{r}) = \boldsymbol{r} = r\hat{\boldsymbol{r}}$ is in that class, as is the inverse square law $\boldsymbol{F}(\boldsymbol{r}) = -\hat{\boldsymbol{r}}/r^2$ arising in Newton's law of gravitation, and Coulomb's law in electrostatics.

As in (11.10), we can introduce a *potential* $V(r)$ such that $\mathrm{d}V/\mathrm{d}r = -F(r)$, a scalar function of one scalar variable. For such a central force field, the work integral then evaluates to the potential difference. Using (12.45),

$$\begin{aligned} \int_{\mathcal{C}} \boldsymbol{F}(\boldsymbol{r}) \cdot \mathrm{d}\boldsymbol{r} &= \int_{\mathcal{C}} F(r)\hat{\boldsymbol{r}} \cdot \mathrm{d}\boldsymbol{r} = \int_{\mathcal{C}} F(r)\mathrm{d}r \\ &= -\int_{\mathcal{C}} \frac{\mathrm{d}V}{\mathrm{d}r}\mathrm{d}r = V(r_A) - V(r_B), \end{aligned} \tag{12.51}$$

and the integral is *path independent*; it has the same value for any curve $\mathcal{C}$ from point A to point B. Conservative vector fields $\boldsymbol{F}(\boldsymbol{r})$, and path independence, are characterized by zero curl and Stokes' theorem (16.61). Next, we consider a nonconservative vector field.

Rigid body circulation

The vector field $\boldsymbol{v}(\boldsymbol{r}) = \boldsymbol{\omega} \times \boldsymbol{r}$, with $\boldsymbol{\omega}$ independent of $\boldsymbol{r}$, is a simple example of a nonconservative vector field. We recognize it as the velocity field of uniform rotation, also called *rigid (or solid) body rotation*, about an axis passing through the origin in the direction $\hat{\boldsymbol{\omega}}$. Thus, we denote it $\boldsymbol{v}$ for *velocity,* instead of $\boldsymbol{F}$ for *force*, or *field.* The standard line integral (12.40) for $\boldsymbol{v}$ instead of $\boldsymbol{F}$ is called the *circulation* of $\boldsymbol{v}$ about the curve $\mathcal{C}$. Velocities *circulate* along curves, they do not "work".

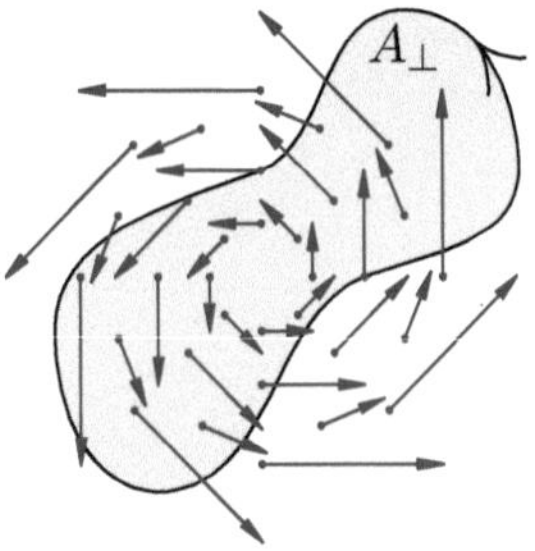

Fig. 12.23 $\oint_{\mathcal{C}} (\boldsymbol{\omega} \times \boldsymbol{r}) \cdot \mathrm{d}\boldsymbol{r} = 2\omega A_\perp$.

In cylindrical coordinates with $\hat{\mathbf{z}}$ in the direction of $\boldsymbol{\omega} = \omega\hat{\mathbf{z}}$ and, from (10.27),

$$\boldsymbol{r} = \rho\hat{\boldsymbol{\rho}}(\varphi) + z\hat{\mathbf{z}}, \qquad \mathrm{d}\boldsymbol{r} = \hat{\boldsymbol{\rho}}\,\mathrm{d}\rho + \hat{\boldsymbol{\varphi}}\rho\,\mathrm{d}\varphi + \hat{\mathbf{z}}\,\mathrm{d}z,$$

thus, $\boldsymbol{v} = \boldsymbol{\omega} \times \boldsymbol{r} = \omega\rho\hat{\boldsymbol{\varphi}}$ and (Fig. 12.23)

$$\int_{\mathcal{C}} \boldsymbol{v} \cdot \mathrm{d}\boldsymbol{r} = \int_{\mathcal{C}} (\boldsymbol{\omega} \times \boldsymbol{r}) \cdot \mathrm{d}\boldsymbol{r} = \omega \int_{\mathcal{C}} \rho^2 \mathrm{d}\varphi = 2\omega A_\perp. \tag{12.52}$$

If $\mathcal{C}$ is a full circle of radius R perpendicular to $\hat{\mathbf{z}}$ and centered on the z axis, for example, then the integral is $2\pi R^2\omega$, and thus not zero as it would be for a conservative vector field. For a more general curve $\mathcal{C}$, (12.52) is $2\omega A_\perp$, where $A_\perp$ is the *signed* area swept by the radius vector projected onto the plane perpendicular to $\boldsymbol{\omega}$, that is, the area swept by the vector $\boldsymbol{\rho} = \boldsymbol{r}_\perp = \boldsymbol{r} - \hat{\boldsymbol{\omega}}(\boldsymbol{r} \cdot \hat{\boldsymbol{\omega}})$, as discussed hereafter.

Area swept by radius vector

Since $(\boldsymbol{\omega} \times \boldsymbol{r}) \cdot \mathrm{d}\boldsymbol{r} = \boldsymbol{\omega} \cdot (\boldsymbol{r} \times \mathrm{d}\boldsymbol{r})$, from the mixed product identity (6.2b), and $\boldsymbol{\omega}$ was assumed constant, (12.52) is equal to the dot product of $\boldsymbol{\omega}$ with the integral

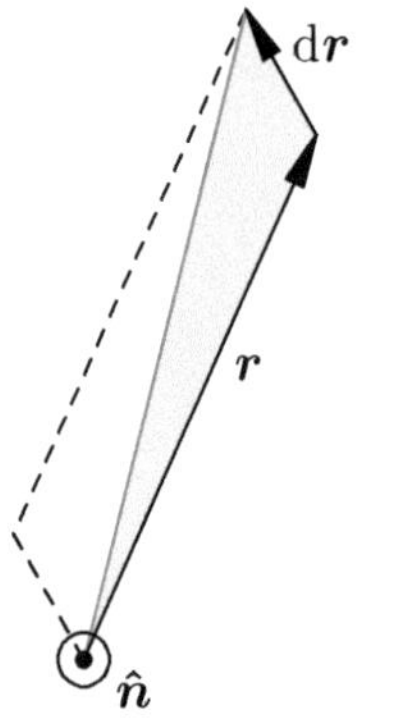

Fig. 12.24 $\hat{\boldsymbol{n}}\mathrm{d}A = \frac{1}{2}\boldsymbol{r} \times \mathrm{d}\boldsymbol{r}$.

$$\int_{\mathcal{C}} \boldsymbol{r} \times \mathrm{d}\boldsymbol{r} = \lim_{|\Delta \boldsymbol{r}_l| \to 0} \sum_{l=1}^{N} \boldsymbol{r}_l \times \Delta\boldsymbol{r}_l = \int_{t_0}^{t_1} \left(\boldsymbol{r}(t) \times \frac{\mathrm{d}\boldsymbol{r}(t)}{\mathrm{d}t} \right) \mathrm{d}t. \tag{12.53}$$

Now

$$\boldsymbol{r}_l \times \Delta\boldsymbol{r}_l = 2\Delta A_l\, \hat{\boldsymbol{n}}_l,$$

where ΔA_l is the area of the *triangle* with sides $\boldsymbol{r}_k$ and $\Delta\boldsymbol{r}_l = \boldsymbol{r}_{l+1} - \boldsymbol{r}_l$, and $\hat{\boldsymbol{n}}_l$ is the unit normal to that triangle (Fig. 12.24).

If $\boldsymbol{r}$ stays in a plane perpendicular to $\hat{\boldsymbol{n}}$, then $\boldsymbol{r} \cdot \hat{\boldsymbol{n}} = 0$ and all the $\hat{\boldsymbol{n}}_l = \pm\hat{\boldsymbol{n}}$, so the integral (12.53) is $2A\hat{\boldsymbol{n}}$, where A is the sum of the signed areas swept by $\boldsymbol{r}$, Fig. 12.25 (cf. Kepler's law, Fig. 11.5). The area is positive when $\boldsymbol{r}$ sweeps counterclockwise about $\hat{\boldsymbol{n}}$, and negative

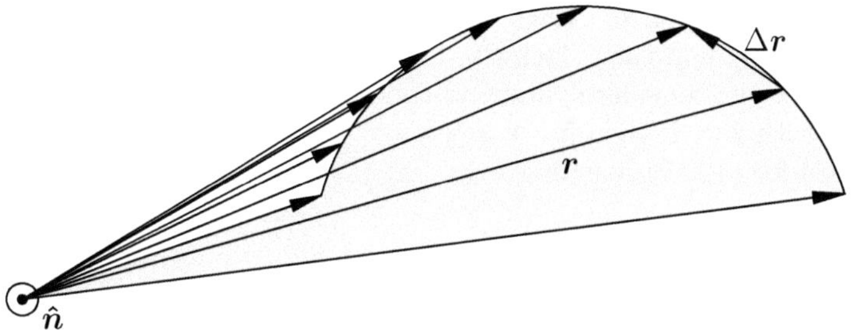

Fig. 12.25 Area swept by radius vector.

when it sweeps clockwise. More generally, if the curve is planar, then $\boldsymbol{r}\cdot\hat{\boldsymbol{n}} = h$ for some constant unit vector $\hat{\boldsymbol{n}}$ and real number h, and

$$\int_{\mathcal{C}} \boldsymbol{r} \times \mathrm{d}\boldsymbol{r} = h\hat{\boldsymbol{n}} \times (\boldsymbol{r}_1 - \boldsymbol{r}_0) + 2A\,\hat{\boldsymbol{n}}, \tag{12.54}$$

where A is the area swept by the vector $\boldsymbol{r}_\perp = \boldsymbol{r} - h\hat{\boldsymbol{n}}$, and $\boldsymbol{r}_1 - \boldsymbol{r}_0$ is the secant vector from the starting point to the end point of $\mathcal{C}$ (exercise 12.53).

If the curve $\mathcal{C}$ is not planar, then

$$\hat{\boldsymbol{a}} \cdot \int_{\mathcal{C}} \boldsymbol{r} \times \mathrm{d}\boldsymbol{r} = \hat{\boldsymbol{a}} \cdot \int_{\mathcal{C}} \boldsymbol{r}_\perp \times \mathrm{d}\boldsymbol{r}_\perp = 2A_\perp \tag{12.55}$$

for *any* constant unit vector $\hat{\boldsymbol{a}}$, where $A_\perp$ is the area swept by the radius vector projected onto a plane perpendicular to $\hat{\boldsymbol{a}}$, that is, the vector $\boldsymbol{\rho} = \boldsymbol{r}_\perp = \boldsymbol{r} - \hat{\boldsymbol{a}}(\boldsymbol{r}\cdot\hat{\boldsymbol{a}})$. The area is positive if $\boldsymbol{r}_\perp$ sweeps counterclockwise about $\hat{\boldsymbol{a}}$, and negative if it sweeps clockwise (Fig. 12.26). This is (12.52) for $\boldsymbol{v} = \hat{\boldsymbol{a}} \times \boldsymbol{r}$ since $\hat{\boldsymbol{a}} \times \boldsymbol{r} \cdot \mathrm{d}\boldsymbol{r} = \hat{\boldsymbol{a}} \cdot \boldsymbol{r} \times \mathrm{d}\boldsymbol{r}$.

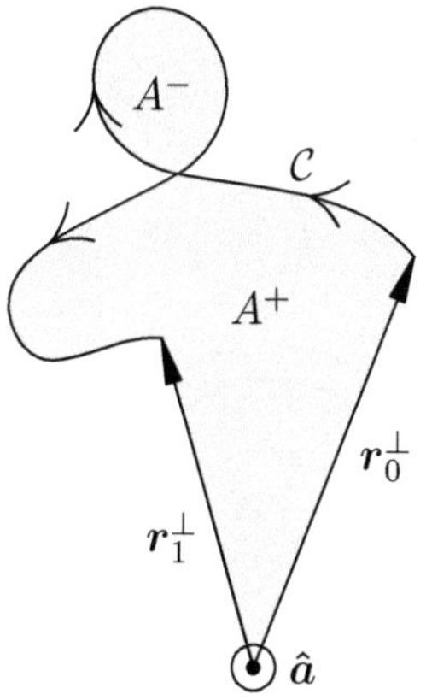

Fig. 12.26 $A_\perp = A^+ - A^-$.

If a curve $\mathcal{C}$ loops around the $\hat{\boldsymbol{a}}$ axis through the origin several times, then $\boldsymbol{r}_\perp$ will sweep the same area multiple times and these must be counted with the correct multiplicity. For example, for the helix with radius R and height H

$$\mathcal{C} : \boldsymbol{r}(t) = R\,(\cos(2k\pi t)\,\hat{\mathbf{x}} + \sin(2k\pi t)\,\hat{\mathbf{y}}) + Ht\,\hat{\mathbf{z}}, \quad t = 0 \to 1.$$

The vector $\boldsymbol{r}_\perp = R\,(\cos 2k\pi t\,\hat{\mathbf{x}} + \sin 2k\pi t\,\hat{\mathbf{y}})$ will sweep the disk of radius R, k times and the area of that disk must be counted k times; thus, for that helical $\mathcal{C}$,

$$\hat{\mathbf{z}} \cdot \int_{\mathcal{C}} \boldsymbol{r} \times \mathrm{d}\boldsymbol{r} = 2k\pi R^2.$$

Winding number and angle functions

The line integral (12.55) is related to another integral that yields the *winding number* k of a curve $\mathcal{C}$ about an axis passing through the origin in direction $\hat{\boldsymbol{a}}$,

$$k \triangleq \frac{1}{2\pi} \int_{\mathcal{C}} \frac{\hat{\boldsymbol{a}} \times \boldsymbol{r}}{|\hat{\boldsymbol{a}} \times \boldsymbol{r}|^2} \cdot \mathrm{d}\boldsymbol{r}. \tag{12.56}$$

If the curve $\mathcal{C}$ is closed, then k is an integer counting the number of counterclockwise turns of $\mathcal{C}$ about the axis $(O, \hat{\boldsymbol{a}})$ passing through the origin O in direction $\hat{\boldsymbol{a}}$. Each counterclockwise loop around the axis contributes $+1$, and each clockwise loop -1 to the winding number k.

To show that, consider cylindrical coordinates (ρ, φ, z) and directions $(\hat{\boldsymbol{\rho}}, \hat{\boldsymbol{\varphi}}, \hat{\mathbf{z}})$ with $\hat{\mathbf{z}} = \hat{\boldsymbol{a}}$; then $\hat{\boldsymbol{a}} \times \boldsymbol{r} = \hat{\mathbf{z}} \times \boldsymbol{r} = \rho\hat{\boldsymbol{\varphi}}$ and $(\hat{\boldsymbol{a}} \times \boldsymbol{r}) \cdot \mathrm{d}\boldsymbol{r} = \rho^2 \mathrm{d}\varphi$, as derived for (12.52), yielding

$$\int_{\mathcal{C}} \frac{\hat{\boldsymbol{a}} \times \boldsymbol{r}}{|\hat{\boldsymbol{a}} \times \boldsymbol{r}|^2} \cdot \mathrm{d}\boldsymbol{r} = \int_{\mathcal{C}} \mathrm{d}\varphi.$$

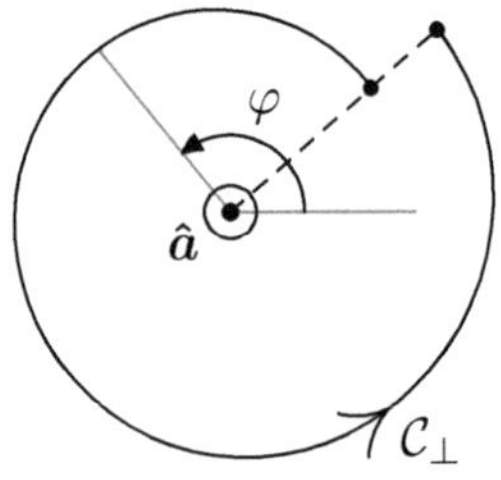

Fig. 12.27 $\int_{\mathcal{C}} \mathrm{d}\varphi = 2\pi.$

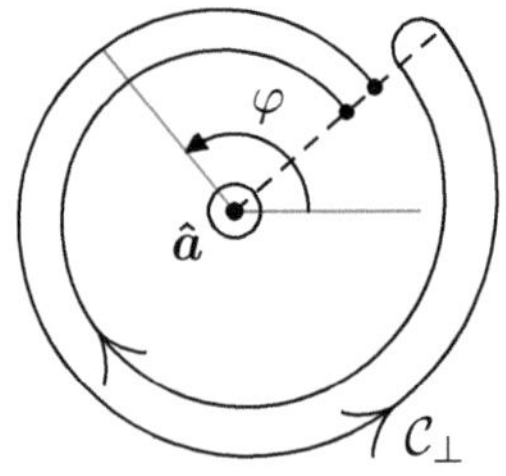

Fig. 12.28 $\int_{\mathcal{C}} \mathrm{d}\varphi = 0.$

The right hand side is the sum of the azimuthal angle increments $\mathrm{d}\varphi$ along the curve $\mathcal{C}$. This is similar to (12.48); however, here the azimuthal angle φ has to be considered as a continuous function of arclength along the curve $\mathcal{C}$, and thus a *multivalued* function of point P, defined up to a multiple of 2π. Thus if a curve comes back to the same point P, or the same radial from O more generally, after looping around the axis $(O, \hat{\boldsymbol{a}})$, then the sum of the $\mathrm{d}\varphi$'s is clearly 2π, not 0 (Fig. 12.27). Each counterclockwise loop contributes $+2\pi$, and each clockwise loop -2π (Fig. 12.28).

The integrand $(\hat{\boldsymbol{a}} \times \boldsymbol{r})/|\hat{\boldsymbol{a}} \times \boldsymbol{r}|^2$ is a vector field, uniquely defined, continuous and differentiable at all $\boldsymbol{r}$ such that $|\hat{\boldsymbol{a}} \times \boldsymbol{r}| \neq 0$, but it is not (quite) a conservative vector field.[6] For a conservative field, the line integral (12.41) depends only on the end points, and is thus invariant to deformations of the curve $\mathcal{C}$ that preserve the end points, and vanishes if the end points coincide. The integral (12.56) is invariant under deformations of the curve $\mathcal{C}$ that preserve the number of loops around the axis $(O, \hat{\boldsymbol{a}})$.

In magnetostatics, a steady line current J flowing in direction $\hat{\boldsymbol{a}}$, from $-\infty$ to $+\infty$ through the origin, induces a magnetic field $\boldsymbol{B}(\boldsymbol{r}) = \mu_0 J/(2\pi)(\hat{\boldsymbol{a}} \times \boldsymbol{r})/|\hat{\boldsymbol{a}} \times \boldsymbol{r}|^2$. The circulation of that magnetic field about a simple closed loop $\mathcal{C}$ enclosing the line current is

$$\oint_{\mathcal{C}} \boldsymbol{B} \cdot \mathrm{d}\boldsymbol{r} = \frac{\mu_0 J}{2\pi} \oint_{\mathcal{C}} \frac{\hat{\boldsymbol{a}} \times \boldsymbol{r}}{|\hat{\boldsymbol{a}} \times \boldsymbol{r}|^2} \cdot \mathrm{d}\boldsymbol{r} = \mu_0 J. \tag{12.57}$$

This is *Ampère's law*, and integral (12.56).

Turning number and total curvature

Another related integral is the *turning number* m of the tangent to a curve about a fixed direction $\hat{\boldsymbol{a}}$,

$$m \triangleq \frac{1}{2\pi} \int_{\mathcal{C}} \frac{\hat{\boldsymbol{a}} \times \hat{\boldsymbol{t}}}{|\hat{\boldsymbol{a}} \times \hat{\boldsymbol{t}}|^2} \cdot \mathrm{d}\hat{\boldsymbol{t}}, \tag{12.58}$$

where $\hat{\boldsymbol{t}} = \mathrm{d}\boldsymbol{r}/\mathrm{d}s$ is the unit tangent to the curve $\mathcal{C}$ at point $\boldsymbol{r}(s)$ of the curve, with arclength s parametrization. This is the same integral as (12.56) but with $\hat{\boldsymbol{t}}$ replacing $\boldsymbol{r}$; thus, this integral measures the winding number about $\hat{\boldsymbol{a}}$ of the hodograph of $\hat{\boldsymbol{t}}$, that is, the curve traced by the

[6] Because of the singularity at $|\hat{\boldsymbol{a}} \times \boldsymbol{r}| = \rho = 0$, see Stokes' theorem.

tip of $\hat{\boldsymbol{t}}$ when its tail is fixed at the origin. This may be enough to explain (12.58) (exercise 12.56).

We can justify it explicitly in a manner similar to that used for (12.52). Let $\hat{\boldsymbol{a}}, \hat{\boldsymbol{b}}, \hat{\boldsymbol{c}}$ denote a fixed right-handed orthonormal basis; then (Fig. 12.29)

$$\hat{\boldsymbol{t}} = \cos\alpha\,\hat{\boldsymbol{a}} + \sin\alpha\,\hat{\boldsymbol{t}}_\perp, \qquad \hat{\boldsymbol{t}}_\perp = \cos\beta\,\hat{\boldsymbol{b}} + \sin\beta\,\hat{\boldsymbol{c}}. \tag{12.59a}$$

This is a spherical representation for $\hat{\boldsymbol{t}}$ with polar angle α from $\hat{\boldsymbol{a}}$ and azimuthal angle β about $\hat{\boldsymbol{a}}$, to distinguish them from the angles θ and φ used for $\boldsymbol{r}$. The azimuthal direction is $\hat{\boldsymbol{\beta}} = \hat{\boldsymbol{a}} \times \hat{\boldsymbol{t}}_\perp$ and the polar direction is $\hat{\boldsymbol{\alpha}} = \hat{\boldsymbol{\beta}} \times \hat{\boldsymbol{t}}$. Hence, in that representation,

$$\hat{\boldsymbol{a}} \times \hat{\boldsymbol{t}} = \hat{\boldsymbol{\beta}}\,\sin\alpha, \quad \mathrm{d}\hat{\boldsymbol{t}} = \hat{\boldsymbol{\alpha}}\,\mathrm{d}\alpha + \sin\alpha\,\mathrm{d}\hat{\boldsymbol{t}}_\perp, \quad \mathrm{d}\hat{\boldsymbol{t}}_\perp = \hat{\boldsymbol{\beta}}\,\mathrm{d}\beta, \tag{12.59b}$$

and

$$\int_{\mathcal{C}} \frac{\hat{\boldsymbol{a}} \times \hat{\boldsymbol{t}} \cdot \mathrm{d}\hat{\boldsymbol{t}}}{|\hat{\boldsymbol{a}} \times \hat{\boldsymbol{t}}|^2} = \int_{\mathcal{C}} \mathrm{d}\beta = 2m\pi, \tag{12.59c}$$

yielding (12.58). This is the sum of the azimuthal increments $\mathrm{d}\beta$ for the unit tangent $\hat{\boldsymbol{t}}$, and the turning number m is an integer if the curve is closed.[7] Although the integral is again similar to (12.48), the sum of the azimuthal increments $\mathrm{d}\beta$ depends on the evolution of $\hat{\boldsymbol{t}}$ along the entire curve $\mathcal{C}$, not just at the end points. The integrand $(\hat{\boldsymbol{a}} \times \hat{\boldsymbol{t}})/|\hat{\boldsymbol{a}} \times \hat{\boldsymbol{t}}|^2$ is not a conservative vector field; it is not even a vector field since its value at $\boldsymbol{r}$ depends on the direction of the curve at that point, not just on $\boldsymbol{r}$.

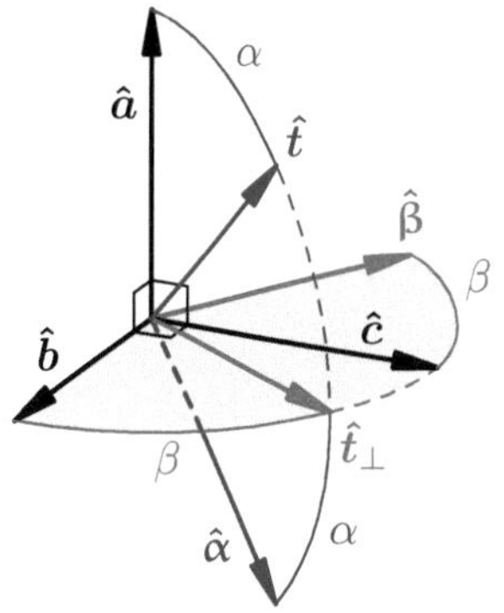

Fig. 12.29 Spherical representation of $\hat{\boldsymbol{t}}$ based on an orthonormal basis $\{\hat{\boldsymbol{a}}, \hat{\boldsymbol{b}}, \hat{\boldsymbol{c}}\}$ with $\hat{\boldsymbol{a}}$ as polar axis.

[7] In which case, m is also called the *index* of the curve.

The turning number (12.58) can be expressed in terms of the signed curvature of the curve $\mathcal{C}$ projected onto a plane perpendicular to $\hat{\boldsymbol{a}}$. That is the planar curve $\mathcal{C}_\perp$ traced out by $\boldsymbol{r}_\perp = \boldsymbol{r} - \hat{\boldsymbol{a}}(\boldsymbol{r} \cdot \hat{\boldsymbol{a}})$ and the unit tangent to that planar curve is $\hat{\boldsymbol{t}}_\perp$ with $\mathrm{d}\hat{\boldsymbol{t}}_\perp = \hat{\boldsymbol{\beta}}\,\mathrm{d}\beta$, as in (12.59), such that

$$\int_{\mathcal{C}} \frac{\hat{\boldsymbol{a}} \times \hat{\boldsymbol{t}} \cdot \mathrm{d}\hat{\boldsymbol{t}}}{|\hat{\boldsymbol{a}} \times \hat{\boldsymbol{t}}|^2} = \int_{\mathcal{C}_\perp} \hat{\boldsymbol{a}} \times \hat{\boldsymbol{t}}_\perp \cdot \mathrm{d}\hat{\boldsymbol{t}}_\perp. \tag{12.60a}$$

The latter integral can be written in terms of arclength $s_\perp$ along the planar curve $\mathcal{C}_\perp$, not arclength s along $\mathcal{C}$, as

$$\int_{\mathcal{C}_\perp} \hat{\boldsymbol{a}} \times \hat{\boldsymbol{t}}_\perp \cdot \frac{\mathrm{d}\hat{\boldsymbol{t}}_\perp}{\mathrm{d}s_\perp}\,\mathrm{d}s_\perp = \int_{\mathcal{C}_\perp} \kappa_\perp\,\mathrm{d}s_\perp \tag{12.60b}$$

where

$$\hat{\boldsymbol{a}} \times \hat{\boldsymbol{t}}_\perp \cdot \frac{\mathrm{d}\hat{\boldsymbol{t}}_\perp}{\mathrm{d}s_\perp} = \hat{\boldsymbol{a}} \times \hat{\boldsymbol{t}}_\perp \cdot \hat{\boldsymbol{n}}\,|\kappa_\perp| = \pm|\kappa_\perp| \triangleq \kappa_\perp,$$

is the *signed* curvature (12.33a) of the curve $\mathcal{C}_\perp$ in a plane perpendicular to $\hat{\boldsymbol{a}}$, with $\hat{\boldsymbol{n}}$ the turning direction and $\hat{\boldsymbol{a}} \times \hat{\boldsymbol{t}}_\perp \cdot \hat{\boldsymbol{n}} = \pm 1$. It is positive when $\hat{\boldsymbol{t}}_\perp$ turns left and negative when $\hat{\boldsymbol{t}}_\perp$ turns right, since $\hat{\boldsymbol{a}} \times \hat{\boldsymbol{t}}_\perp$ is always 90° left of $\hat{\boldsymbol{t}}_\perp$. The integral

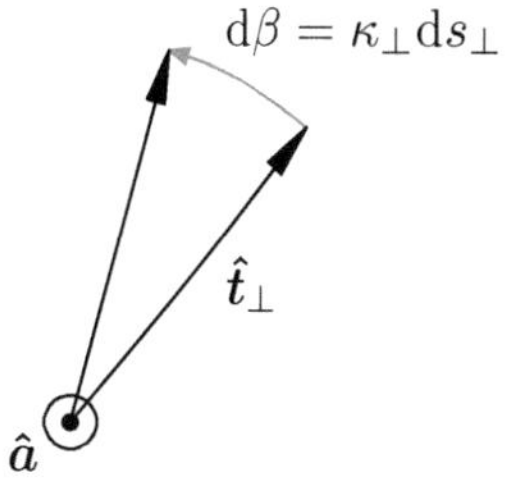

Fig. 12.30 $\hat{\boldsymbol{a}} \times \hat{\boldsymbol{t}}_\perp \cdot \mathrm{d}\hat{\boldsymbol{t}}_\perp = \mathrm{d}\beta$.

$$\int_{\mathcal{C}_\perp} \hat{\boldsymbol{a}} \times \hat{\boldsymbol{t}}_\perp \cdot \mathrm{d}\hat{\boldsymbol{t}}_\perp = \int_{\mathcal{C}_\perp} \kappa_\perp\,\mathrm{d}s_\perp = \int_{\mathcal{C}_\perp} \mathrm{d}\beta \tag{12.60c}$$

is the *total curvature* of the planar curve $\mathcal{C}_\perp$ about $\hat{\boldsymbol{a}}$ (Fig. 12.30). That is the difference between the final and initial azimuthal angle β of the tangent to the curve $\mathcal{C}_\perp$, considered as a continuous function along the curve. A curve that starts up with $\beta = \pi/2$ and ends left after a simple continuous left turn such that $\mathrm{d}\beta > 0$ ends at $\beta = \pi$ with $\int_{\mathcal{C}_\perp} \mathrm{d}\beta = \pi/2$, but a curve that starts up at $\beta = \pi/2$ and ends left after a continuous right turn such that $\mathrm{d}\beta < 0$ has $\int_{\mathcal{C}_\perp} \mathrm{d}\beta = -3\pi/2$, ending at $\beta = \pi/2 - 3\pi/2 = -\pi = \pi - 2\pi$, for example.

Since

$$\mathrm{d}\beta = \hat{\boldsymbol{a}} \times \hat{\boldsymbol{t}}_\perp \cdot \mathrm{d}\hat{\boldsymbol{t}}_\perp = \hat{\boldsymbol{a}} \cdot \hat{\boldsymbol{t}}_\perp \times \mathrm{d}\hat{\boldsymbol{t}}_\perp,$$

the total signed curvature of a plane curve (12.60c), that is, its total turning angle, is also twice the total signed area swept by the unit tangent $\hat{\boldsymbol{t}}_\perp$, (12.55). The turning number is therefore closely related to the area swept by a planar curve's velocity vector. If the curve $\mathcal{C} : t \to \boldsymbol{r}(t)$ is in a plane perpendicular to $\hat{\boldsymbol{a}}$, such that $\boldsymbol{r} = \boldsymbol{r}_\perp$ in the previous discussion, then the signed area swept by the velocity vector $\boldsymbol{v} = \mathrm{d}\boldsymbol{r}/\mathrm{d}t$ is, similarly to (12.53),

$$\tfrac{1}{2}\hat{\boldsymbol{a}} \cdot \int_{\mathcal{C}} \boldsymbol{v} \times \frac{\mathrm{d}\boldsymbol{v}}{\mathrm{d}t}\, \mathrm{d}t = \tfrac{1}{2} \int_{\mathcal{C}} \kappa v^2 \, \mathrm{d}s, \tag{12.61}$$

where $\kappa = \hat{\boldsymbol{a}} \cdot \hat{\boldsymbol{t}} \times \mathrm{d}\hat{\boldsymbol{t}}/\mathrm{d}s$ is the *signed* curvature (12.36), with $v = |\mathrm{d}\boldsymbol{r}/\mathrm{d}t| = \mathrm{d}s/\mathrm{d}t$ the speed, arclength s along the curve, and fixed binormal $\hat{\boldsymbol{a}}$ for this curve in a plane perpendicular to $\hat{\boldsymbol{a}}$. For constant speed, this equals v^2 times the total tangent turning angle $\int_{\mathcal{C}} \mathrm{d}\beta$, a result popularized by Mamikon and Apostol as the *sweeping tangent theorem*, known to Peano in 1887.[8]

[8] Gabriele H. Greco, Sonia Mazzucchi, and Enrico M. Pagani. "Peano on Definition of Surface Area." *Rendiconti Lincei* 27.3 (2016): 251–286.

Arclength

The length of curve $\mathcal{C}$ from point A to point B is determined by the integral

$$\int_{\mathcal{C}} \mathrm{d}s = \int_{\mathcal{C}} |\mathrm{d}\boldsymbol{r}| = \lim_{|\Delta \boldsymbol{r}_n| \to 0} \sum_{n=1}^{N} |\Delta \boldsymbol{r}_n| \tag{12.62}$$

since $\mathrm{d}\boldsymbol{r} = \hat{\boldsymbol{t}}\mathrm{d}s$, with $\boldsymbol{r}_0 = \boldsymbol{r}_A$ and $\boldsymbol{r}_N = \boldsymbol{r}_B$. If a parametrization $\boldsymbol{r}(t)$ is known then

$$\int_{\mathcal{C}} |\mathrm{d}\boldsymbol{r}| = \int_{t_A}^{t_B} \left| \frac{\mathrm{d}\boldsymbol{r}(t)}{\mathrm{d}t} \right| |\mathrm{d}t|, \tag{12.63}$$

where $\boldsymbol{r}_A = \boldsymbol{r}(t_A)$ and $\boldsymbol{r}_B = \boldsymbol{r}(t_B)$. This is *almost* an elementary integral of one real variable, except for the absolute value in $|\mathrm{d}t|$. What does that mean? Again, we can understand that from the limit-of-a-sum definition with $t_0 = t_A$, $t_N = t_B$ and $\Delta t_n = t_n - t_{n-1}$. If $t_A < t_B$ then $\Delta t_n > 0$ and $\mathrm{d}t > 0$, so $|\mathrm{d}t| = \mathrm{d}t$ and all is good. But if $t_B < t_A$ then $\Delta t_n < 0$ and $\mathrm{d}t < 0$, so $|\mathrm{d}t| = -\mathrm{d}t$, and

$$\int_{t_A}^{t_B} (\cdots)|\mathrm{d}t| = \int_{t_B}^{t_A} (\cdots)\mathrm{d}t, \qquad \text{if} \quad t_A > t_B. \tag{12.64}$$

So it is a regular integral but the order of the bounds depend on whether $t_A < t_B$. For the arclength, we do not need to worry and can just take the absolute value after the fact, but for other integrals such as $\int_{\mathcal{C}} f(\boldsymbol{r})\mathrm{d}s$, where $f(\boldsymbol{r})$ might be sign indefinite, we need to handle the bounds more carefully, as just discussed.

For the example of an ellipse, $\mathcal{C} : \boldsymbol{r}(t) = \boldsymbol{a}\cos t + \boldsymbol{b}\sin t$, with $\boldsymbol{a}\cdot\boldsymbol{b} = 0$ and $t = 0 \to 2\pi$, $\mathrm{d}\boldsymbol{r}(t) = (-\boldsymbol{a}\sin t + \boldsymbol{b}\cos t)\,\mathrm{d}t$ and

$$\begin{aligned} \mathcal{L} = \int_{\mathcal{C}} |\mathrm{d}\boldsymbol{r}| &= \int_0^{2\pi} \sqrt{a^2\sin^2 t + b^2\cos^2 t}\,\mathrm{d}t \\ &= 4a\int_0^{\pi/2} \sqrt{1 - k^2\sin^2 t}\,\mathrm{d}t, \end{aligned} \tag{12.65}$$

where $a^2 = \boldsymbol{a}\cdot\boldsymbol{a}$, $b^2 = \boldsymbol{b}\cdot\boldsymbol{b}$ and $0 \le k^2 = (a^2 - b^2)/a^2 \le 1$, assuming (without loss of generality) that $a \ge b$. The last integral is obtained using symmetry properties of the trigonometric functions. It cannot be evaluated in terms of elementary functions, but can be evaluated numerically, such as (12.62) with a sufficiently large, but finite N. It is called the *complete elliptic integral of the second kind.*

Travel time

Another interesting example of line integral is the total travel time T along a curve $\mathcal{C}$ with prescribed speed $v(s)$ at point $P(s)$ of the curve,

$$T = \int_{\mathcal{C}} \mathrm{d}t = \int_{\mathcal{C}} \frac{\mathrm{d}t}{\mathrm{d}s}\mathrm{d}s = \int_{\mathcal{C}} \frac{\mathrm{d}s}{v(s)}, \tag{12.66}$$

where s is arclength from some reference point along the curve $\mathcal{C}$, and $v(s) \ge 0$ is the travel speed along the curve at the point labeled by s. If the curve is parametrized by $\boldsymbol{r}(u)$, with $u = a \to b > a$ (with parameter u since t is taken), then

$$T = \int_{\mathcal{C}} \mathrm{d}t = \int_{\mathcal{C}} \frac{\mathrm{d}s}{v(s)} = \int_a^b \frac{|\mathrm{d}\boldsymbol{r}/\mathrm{d}u|}{v(s(u))}\,\mathrm{d}u. \tag{12.67}$$

For example, for a particle moving without friction under the action of gravity $\boldsymbol{g} = -g\hat{\mathbf{y}}$, along a curve $\mathcal{C} : \boldsymbol{r}(u) = x(u)\hat{\mathbf{x}} + y(u)\hat{\mathbf{y}}$, $u = u_0 \to u_1$, conservation of energy gives

$$\tfrac{1}{2}v^2 + gy = \tfrac{1}{2}v_0^2 + gy_0 \;\Rightarrow v(u) = \sqrt{v_0^2 + 2g(y_0 - y(u))}$$

with v_0 and y_0 the starting speed and height, and the travel time is thus

$$T = \int_{u_0}^{u_1} \frac{\sqrt{(\mathrm{d}x/\mathrm{d}u)^2 + (\mathrm{d}y/\mathrm{d}u)^2}}{\sqrt{v_0^2 + 2g(y_0 - y(u))}}\,|\mathrm{d}u|.$$

12.6 Hanging from curves

Before moving on to surfaces, we consider a mechanical example of curves as a *string*, or *chain*, attached to two points A and B, tugged by forces.

The mathematical model for a string is that it is a curve $\boldsymbol{r}(s)$ that transmits a force called the *tension* $\boldsymbol{t}(s)$ tangent to the curve. That is,

$$\boldsymbol{t}(s) = |\boldsymbol{t}|\,\hat{\boldsymbol{t}} = |\boldsymbol{t}|\,\frac{\mathrm{d}\boldsymbol{r}}{\mathrm{d}s} \tag{12.68}$$

where s is arclength along the curve. A string thus has no resistance to *bending* or *torsion.* This is a mathematical idealization valid when the length is much longer than the cross-sectional diameter, such as a long rope, a long chain, or a strand of hair.

We define $\boldsymbol{t}(s)$ to be the tension at s on section $0 \to s$ of the string. By action–reaction, the force at s acting on section $s \to L$ is $-\boldsymbol{t}(s)$, where L is the total length. Then, considering a section $s \to s + \mathrm{d}s$ of the curve that is pulled by a force per unit length $\boldsymbol{f}(s)$, simple force balance gives (Fig. 12.31)

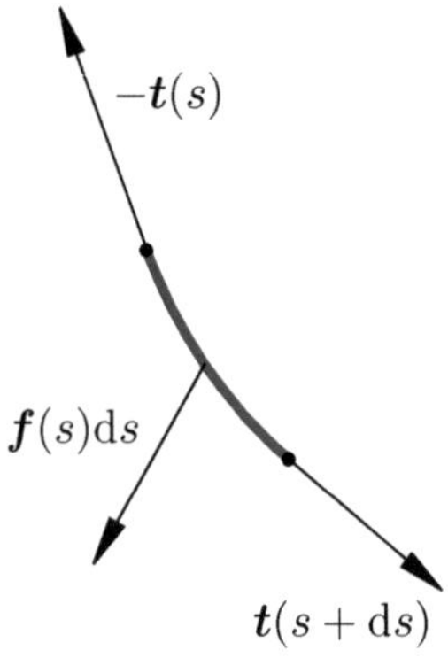

Fig. 12.31 Forces on $(s, s + \mathrm{d}s)$.

$$\boldsymbol{t}(s + \mathrm{d}s) - \boldsymbol{t}(s) + \boldsymbol{f}(s)\,\mathrm{d}s = 0,$$

yielding the differential equation

$$\frac{\mathrm{d}\boldsymbol{t}}{\mathrm{d}s} + \boldsymbol{f}(s) = 0, \tag{12.69}$$

where $\boldsymbol{t}$ is the tension, with $\hat{\boldsymbol{t}} = \mathrm{d}\boldsymbol{r}/\mathrm{d}s$. Integrating from $s = 0$ to L gives

$$\boldsymbol{t}(L) - \boldsymbol{t}(0) = \int_0^L \boldsymbol{f}(s)\,\mathrm{d}s = \boldsymbol{F},$$

where $\boldsymbol{F}$ is the total force on the string.

If there is no force acting on the string, then $\mathrm{d}\boldsymbol{t}/\mathrm{d}s = 0$ and the string is straight under tension, or $\boldsymbol{t}(s) = 0$, and its shape is arbitrary. A less trivial example is that of a uniform chain hanging under its own weight. In that case, $\boldsymbol{f}(s) = \rho(s)\boldsymbol{g}$ where $\rho = M/L$ is the mass density per unit length, assumed constant here, the chain's total mass M divided by its length L. This is the opportune moment to *nondimensionalize*, with the substitutions

$$\boldsymbol{t} \to (\rho g L)\boldsymbol{t}, \qquad s \to Ls, \tag{12.70}$$

that is, $\boldsymbol{t} = (\rho g L)\boldsymbol{t}'$, $s = Ls'$, then dropping primes after the substitutions have been made. The new tension $\boldsymbol{t}$ is dimensionless, and measures the force in units of the total weight $\rho g L$. The new arclength s is nondimensional and measures length in units of L, with $0 \le s \le 1$.

In cartesian directions with $\boldsymbol{g} = -g\hat{\mathbf{y}}$, the force balance equation (12.69) yields

$$\frac{\mathrm{d}\boldsymbol{t}}{\mathrm{d}s} = \hat{\mathbf{y}} \;\Rightarrow\; \boldsymbol{t} = \boldsymbol{t}_0 + s\,\hat{\mathbf{y}}. \tag{12.71a}$$

Let $\boldsymbol{t}_0 = a\,\hat{\mathbf{x}} + b\hat{\mathbf{y}}$, where a and b are constants; then the magnitude and direction of the tension $\boldsymbol{t} = |\boldsymbol{t}|\hat{\boldsymbol{t}}$ are, from (12.71a),

$$|\boldsymbol{t}| = \sqrt{a^2 + (s+b)^2}, \tag{12.71b}$$

$$\hat{\boldsymbol{t}} = \frac{a}{\sqrt{a^2 + (s+b)^2}}\,\hat{\mathbf{x}} + \frac{s+b}{\sqrt{a^2 + (s+b)^2}}\,\hat{\mathbf{y}}. \tag{12.71c}$$

Since $\hat{\boldsymbol{t}} = \mathrm{d}\boldsymbol{r}/\mathrm{d}s = (\mathrm{d}x/\mathrm{d}s)\hat{\mathbf{x}} + (\mathrm{d}y/\mathrm{d}s)\hat{\mathbf{y}}$, (12.71c) gives explicit expressions for $\mathrm{d}x/\mathrm{d}s$ and $\mathrm{d}y/\mathrm{d}s$ that integrate to

$$x(s) = a \operatorname{arcsinh} \frac{s+b}{a}, \quad y(s) = \sqrt{a^2 + (s+b)^2}, \tag{12.72}$$

where the constants of integrations have been set to zero, thus fixing the origin of the axes in some way, and arcsinh is the inverse *hyperbolic sine* function. Eliminating s yields

$$s + b = a \sinh \frac{x}{a} \;\Rightarrow\; y = a \cosh \frac{x}{a}, \tag{12.73}$$

since $\cosh^2 \alpha - \sinh^2 \alpha = 1$.

The shape of a chain hanging under its own weight is a *hyperbolic cosine*, and that curve is called the *catenary* (Fig. 12.32). This is a curve for which we have an explicit $\boldsymbol{r}(s)$ parametrization (12.72) in terms of its arclength s, in addition to lines and circles. There remain two constants, a, b, to determine from the increments $x(1) - x(0)$, $y(1) - y(0)$, between the starting and end points, whose magnitude must be less than the unit length 1.

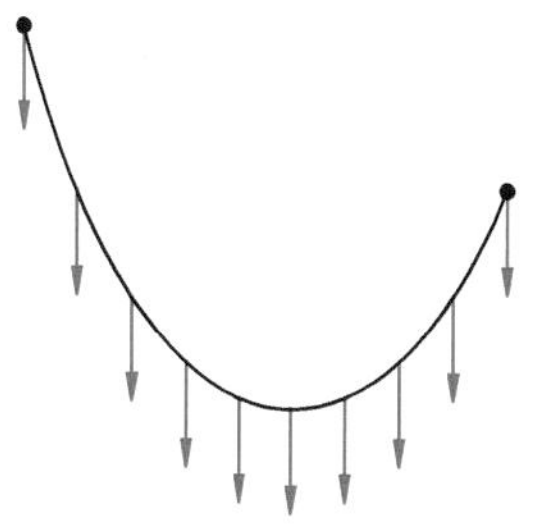

Fig. 12.32 Hanging chain.

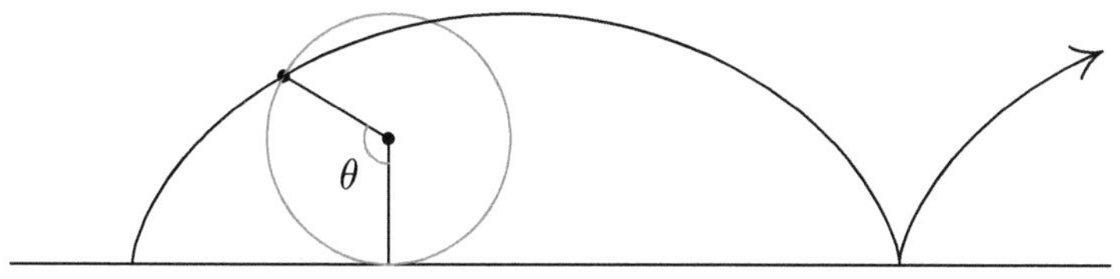

Fig. 12.33 Cycloid (exercise 12.19).

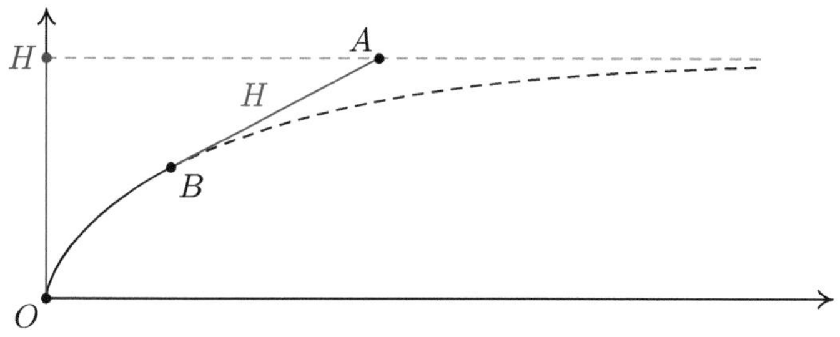

Fig. 12.34 Tractrix (exercise 12.22).

Exercises

(12.1) If (12.1) and (12.2) define the same line, show/explain why

$$\boldsymbol{r}_0 = \frac{h_1 - h_2\cos\theta}{1-\cos^2\theta}\hat{\boldsymbol{n}}_1 + \frac{h_2 - h_1\cos\theta}{1-\cos^2\theta}\hat{\boldsymbol{n}}_2,$$
$$\boldsymbol{v}_0 = \hat{\boldsymbol{n}}_1 \times \hat{\boldsymbol{n}}_2, \tag{12.74}$$

where $\hat{\boldsymbol{n}}_1 \cdot \hat{\boldsymbol{n}}_2 = \cos\theta$. What is special about this $\boldsymbol{r}_0$? Are there other choices for $\boldsymbol{r}_0$ and $\boldsymbol{v}_0$? Explain.

(12.2) Write (12.7a) in cartesian coordinates with the origin at A and the z-axis in direction $\hat{\boldsymbol{\omega}}$, with a suitable orientation of the x- and y-axes. Derive the equivalent of (12.7c).

(12.3) Given two circles in a plane (i) give an algorithm to decide whether one circle is entirely inside the other; (ii) find vector equations for all the lines that are tangent to both circles. Sketch for both intersecting, and nonintersecting circles.

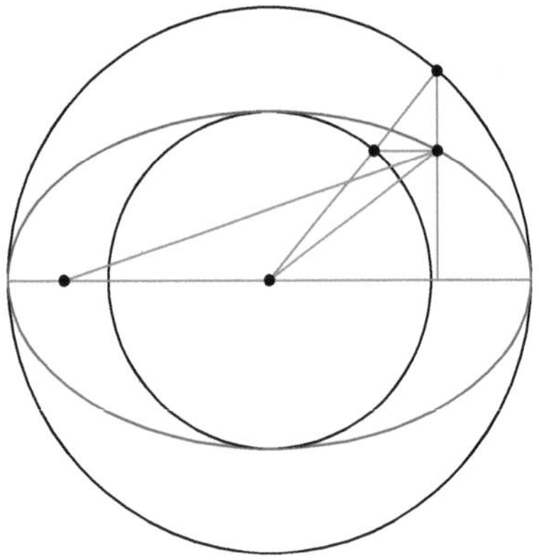

Fig. 12.35 Where is θ?

(12.4) Show that the parameter θ in (12.10) is *not* the angle α between $\boldsymbol{a}$ and $\boldsymbol{r} - \boldsymbol{r}_C$, nor the angle φ between $\boldsymbol{a}$ and $\boldsymbol{r} - \boldsymbol{r}_F$, where C is the center and F a focus of the ellipse. Indicate F, C, P, a, b, θ, α and φ on Fig. 12.35.

(12.5) Show that the points $P \equiv (x, y)$ on the ellipse $x^2/a^2 + y^2/b^2 = 1$ are such that the sum of their distances to the *foci* $F_1 \equiv (-c, 0)$ and $F_2 \equiv (c, 0)$ is equal to $2a$, where $c = \sqrt{a^2 - b^2}$, for $a \geq b$. This is the geometric definition of an ellipse.

(12.6) Show that the points $P \equiv (x, y)$ on the hyperbola $x^2/a^2 - y^2/b^2 = 1$ are such that the *difference* of their distances to the *foci* $F_1 \equiv (-c, 0)$ and $F_2 \equiv (c, 0)$ is equal to $\pm 2a$, where $c = \sqrt{a^2 + b^2}$. This is the geometric definition of a hyperbola.

(12.7) Show that (12.15) can be reduced to the standard equation for a hyperbola, by choosing an appropriate basis and eliminating the parameter.

(12.8) Given two arbitrary fixed points F_1 and F_2 and a constant $2a \geq |F_1F_2|$, consider the curve traced by the point $P(t)$ such that (Fig. 12.36)

$$|\overrightarrow{F_1P}| + |\overrightarrow{F_2P}| = 2a$$

in an arbitrary plane containing F_1 and F_2. Use vector calculus, in particular (10.9), to show that the tangent to that curve at P makes equal angles with the vectors $\overrightarrow{F_1P}$ and $\overrightarrow{F_2P}$. This is a fundamental *optical* property of the ellipse.
(Hint: F_1 and F_2 are fixed but P varies, so $\overrightarrow{F_1P} = \boldsymbol{r}(t) - \boldsymbol{r}_1$ and $\mathrm{d}\overrightarrow{F_1P}/\mathrm{d}t = \mathrm{d}\boldsymbol{r}/\mathrm{d}t$ is tangent to the curve. Recall (10.9) for the derivative of a magnitude.)

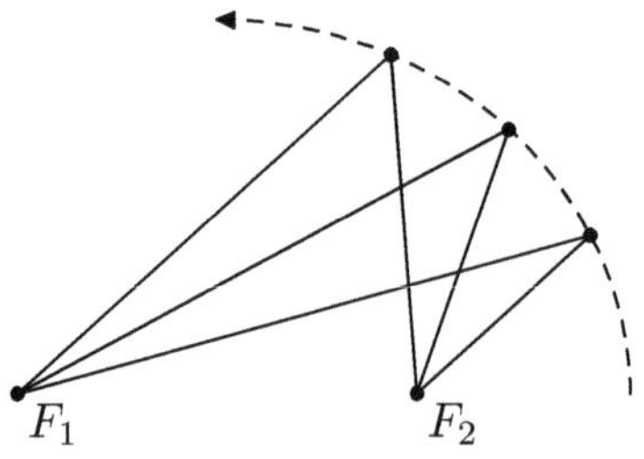

Fig. 12.36 Ellipse.

(12.9) Given two arbitrary fixed points F_1 and F_2 and a constant $2a \leq |F_1F_2|$, consider the curve traced by the point $P(t)$ such that (Fig. 12.37)

$$|\overrightarrow{F_1P}| - |\overrightarrow{F_2P}| = 2a$$

in an arbitrary plane containing F_1 and F_2. Use vector calculus to show that the tangent to that curve at P makes equal angles with the vectors $\overrightarrow{F_1P}$ and $\overrightarrow{F_2P}$. This is a fundamental *optical* property of the hyperbola.

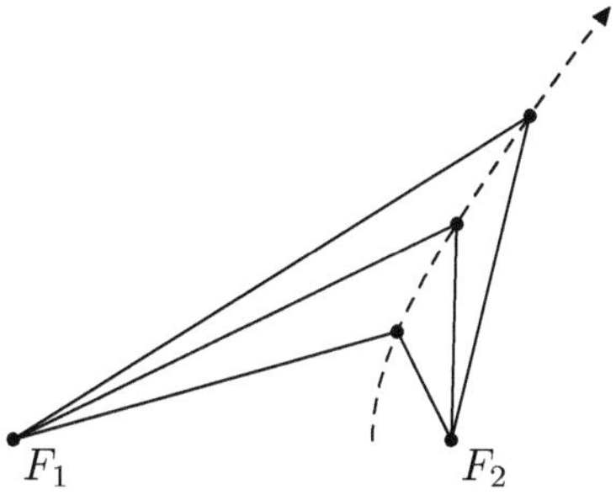

Fig. 12.37 Hyperbola.

(12.10) Given a fixed point F, a direction $\hat{\boldsymbol{a}}$ and a scalar constant $r_* > 0$, show that the locus of points $P(t)$ such that (Fig. 12.38)

$$|\overrightarrow{FP}| = 2r_* + \hat{\boldsymbol{a}} \cdot \overrightarrow{FP}$$

is a parabola.[9] (Hint: pick suitable cartesian coordinates.) Use this vector equation to show that the tangent to the parabola at P makes equal angles with $\hat{\boldsymbol{a}}$ and $\overrightarrow{FP}$. This is the optical property of a parabola. Indicate r_* on the figure.

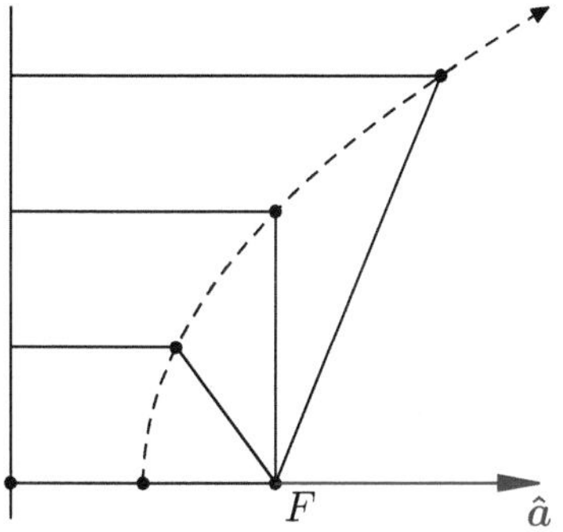

Fig. 12.38 Parabola.

(12.11) Show that (12.13) yields $\mathbf{x}^T\mathbf{A}\mathbf{x} = 1$ that expands out to (12.14), where $\mathbf{A} =$

$$\begin{bmatrix} \cos\alpha & -\sin\alpha \\ \sin\alpha & \cos\alpha \end{bmatrix} \begin{bmatrix} 1/a^2 & 0 \\ 0 & 1/b^2 \end{bmatrix} \begin{bmatrix} \cos\alpha & \sin\alpha \\ -\sin\alpha & \cos\alpha \end{bmatrix}$$

and $\mathbf{x} = (x - x_c, y - y_c)$. Hence $\mathbf{A}$ is a symmetric positive definite matrix with eigenvalue $1/a^2$ with eigenvector $(\cos\alpha, \sin\alpha)$, and $1/b^2$ with $(-\sin\alpha, \cos\alpha)$.

(12.12) For the hyperbolic functions (12.16), show that $\cosh t$ is even, $\sinh t$ is odd, and

$$\begin{aligned} \cosh t \pm \sinh t &= e^{\pm t}, \\ \cosh(a \pm b) &= \cosh a \cosh b \pm \sinh a \sinh b, \\ \sinh(a \pm b) &= \sinh a \cosh b \pm \cosh a \sinh b. \end{aligned} \tag{12.75}$$

Sketch e^t, e^{-t}, $\cosh t$ and $\sinh t$, by hand, on the same graph.

(12.13) Consider $\boldsymbol{r}(\theta) = \boldsymbol{a}\cos\theta + \boldsymbol{b}\sin\theta$ for constant vectors $\boldsymbol{a}$ and $\boldsymbol{b}$, nonorthogonal in general. Find $\boldsymbol{r}$ and $\boldsymbol{v} = \mathrm{d}\boldsymbol{r}/\mathrm{d}\theta$ at $\theta = 0, \pm\pi/2, \pi$. Sketch the curve using those results. Indicate the points where $\boldsymbol{r} \cdot \boldsymbol{v} = 0$. Find θ_* corresponding to one such point. Show that this curve is an ellipse by rewriting $\boldsymbol{r}(\theta)$ in terms of $\boldsymbol{a}' = \boldsymbol{r}(\theta_*)$ and $\boldsymbol{b}' = \boldsymbol{v}(\theta_*)$.

(12.14) Consider $\boldsymbol{r}(t) = \boldsymbol{a}\cosh t + \boldsymbol{b}\sinh t$ for constant vectors $\boldsymbol{a}$ and $\boldsymbol{b}$, nonorthogonal in general. Find the asymptotes as $t \to \pm\infty$. Find $\boldsymbol{r}$ and $\boldsymbol{v} = \mathrm{d}\boldsymbol{r}/\mathrm{d}t$ at $\sinh t = -1, 0, 1$. Sketch the curve using those results. Indicate the point where $\boldsymbol{r} \cdot \boldsymbol{v} = 0$. Find t_* corresponding to that point. Show that this curve is a hyperbola by rewriting $\boldsymbol{r}(t)$ in terms of $\boldsymbol{a}' = \boldsymbol{r}(t_*)$ and $\boldsymbol{b}' = \boldsymbol{v}(t_*)$.

(12.15) Consider the previous two problems using cartesian coordinates and linear algebra. Let $\mathbf{A} = [\mathbf{a}\ \mathbf{b}]$ be the matrix whose columns are the cartesian components of $\boldsymbol{a}$ and $\boldsymbol{b}$. Eliminate the parameter, θ or t, to obtain a quadratic form $\mathbf{x}^T\mathbf{B}\mathbf{x} = 1$. Specify $\mathbf{B}$ in terms of $\mathbf{A}$ when $\boldsymbol{a}$, $\boldsymbol{b}$ are arbitrary 3D vectors. Discuss how one would proceed to reduce the ellipse or hyperbola to standard form.

(12.16) Sketch $\boldsymbol{r}(t) = \boldsymbol{a}\,(1-t^2)/(1+t^2) + \boldsymbol{b}\,2t/(1+t^2) + \boldsymbol{c}$. Identify and show what curve this is.

(12.17) Sketch $\boldsymbol{r}(t) = \boldsymbol{a}\sqrt{1+t^2} + \boldsymbol{b}\,t + \boldsymbol{c}$. Identify and demonstrate what curve this is.

(12.18) Sketch $\boldsymbol{r}(\theta) = \boldsymbol{a}/\cos\theta + \boldsymbol{b}\tan\theta + \boldsymbol{c}$, for constant $\boldsymbol{a}$, $\boldsymbol{b}$, $\boldsymbol{c}$. Identify and demonstrate what curve this is.

(12.19) *Cycloid.* A wheel of radius R rolls without slipping along a straight line (Fig. 12.33). Show that the trajectory of a point on its rim is, in suitable cartesian coordinates,

$$\begin{aligned} x &= R\,(\theta - \sin\theta), \\ y &= R\,(1 - \cos\theta). \end{aligned}$$

(12.20) *Cardioid.* A wheel of radius R rolls without slipping on the outside of a fixed wheel of radius R.

[9] A *paraboloid* surface actually, by rotational symmetry about $\hat{\boldsymbol{a}}$.

Demonstrate that the trajectory of a point P on its rim is (Fig. 12.39)

$$x = R(\sin 2\theta - 2\sin\theta),$$
$$y = R(2\cos\theta - \cos 2\theta),$$

in suitable cartesian coordinates. Let O be the center of the fixed circle, with $\boldsymbol{r}$ the position vector of P from O. If P_0 is the position of P when it touches the fixed circle, show geometrically that $\boldsymbol{r} - \boldsymbol{r}_0$ remains parallel to the line joining the centers and has magnitude $2R(1-\cos\theta)$; thus, a coordinate-free parametrization of the cardioid is

$$\boldsymbol{r} = \boldsymbol{r}_0 + 2(1-\cos\theta)(\cos\theta\,\boldsymbol{r}_0 + \sin\theta\,\hat{\mathbf{z}}\times\boldsymbol{r}_0).$$

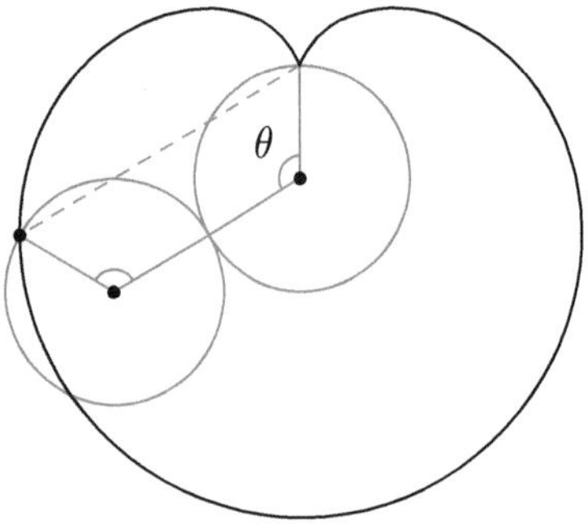

Fig. 12.39 Cardioid.

(12.21) A double pendulum with rigid rods of lengths L_1 and L_2 rotates in a plane such that the rotation rate of L_1 relative to a fixed frame is the same as that of L_2 *relative to* L_1. Show that the trajectory of the end point P of the double pendulum is, in suitable cartesian coordinates, (Fig. 12.40)

$$x = L_2\sin 2\theta - L_1\sin\theta,$$
$$y = L_1\cos\theta - L_2\cos 2\theta,$$

a family of curves known as *limaçons*. Let O be the fixed attach point of the pendulum and $\hat{\boldsymbol{r}}_0$ the direction of rod 1 when the pendulum arms are folded against each other. If P_0 is the point at distance L_2 from O in direction $\hat{\boldsymbol{r}}_0$, show that $\boldsymbol{r} - \boldsymbol{r}_0$ remains parallel to rod L_1 with magnitude $L_1 - 2L_2\cos\theta$; thus, a coordinate-free parametrization of the limaçon is

$$\boldsymbol{r} = L_2\hat{\boldsymbol{r}}_0 + (L_1 - 2L_2\cos\theta)(\cos\theta\,\hat{\boldsymbol{r}}_0 + \sin\theta\,\hat{\mathbf{z}}\times\hat{\boldsymbol{r}}_0).$$

If the rotation rate $\dot{\theta} = \omega$ is constant, predict where the velocity and acceleration have largest magnitudes; then verify by calculation. (Hint: introduce cylindrical coordinate directions with $\hat{\boldsymbol{\rho}}(\theta) = \cos\theta\,\hat{\boldsymbol{r}}_0 + \sin\theta\,\hat{\mathbf{z}}\times\hat{\boldsymbol{r}}_0$ and $\hat{\boldsymbol{\theta}}(\theta) = \hat{\mathbf{z}}\times\hat{\boldsymbol{\rho}}(\theta)$.)

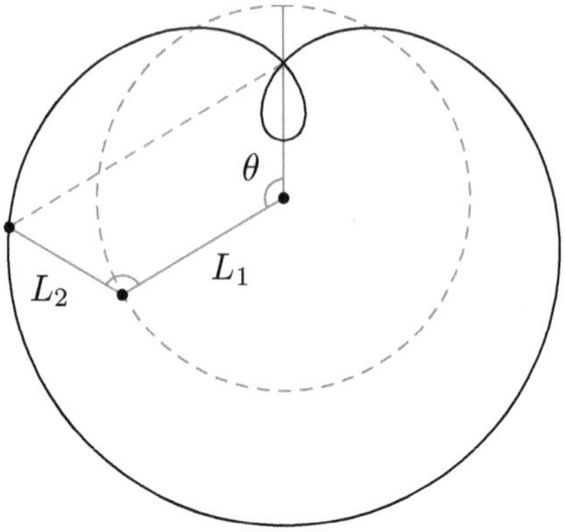

Fig. 12.40 Limacon.

(12.22) *Tractrix.* Bond and his partner are plucked from the sea by an airplane snatching their balloon cable (the *Fulton recovery system*). Assume that the cable remains straight with negligible stretching and that the force balance is between the cable tension and the aerodynamic drag on the couple, such that their velocity remains parallel to the cable. Show that their trajectory is the *tractrix* (Fig. 12.34)

$$x = -H\left(\cos\theta + \ln\tan\frac{\theta}{2}\right), \quad y = H(1-\sin\theta), \tag{12.76}$$

where H is the cable length, equal to the airplane altitude above the pickup point, and θ is the angle between the cable and the airplane velocity.

(12.23) Show that the linear curve $\boldsymbol{r}(t) = (1-t)\boldsymbol{r}_0 + t\boldsymbol{r}_1$ is represented exactly by the cubic Bézier curve (12.26) with control points $\boldsymbol{r}_0$, $\boldsymbol{r}_{1/3}$, $\boldsymbol{r}_{2/3}$, $\boldsymbol{r}_1$, such that

$$\boldsymbol{r}_{1/3} = \tfrac{2}{3}\boldsymbol{r}_0 + \tfrac{1}{3}\boldsymbol{r}_1, \quad \boldsymbol{r}_{2/3} = \tfrac{1}{3}\boldsymbol{r}_0 + \tfrac{2}{3}\boldsymbol{r}_1.$$

(12.24) A quadratic Bézier curve is defined by three controls points $\boldsymbol{r}_0$, $\boldsymbol{r}_1$, $\boldsymbol{r}_2$ as

$$\boldsymbol{r}(t) = (1-t)^2\,\boldsymbol{r}_0 + 2(1-t)t\,\boldsymbol{r}_1 + t^2\,\boldsymbol{r}_2.$$

Show that its velocity $\boldsymbol{v}(t) = (1-t)\boldsymbol{v}_0 + t\boldsymbol{v}_1$, with $\boldsymbol{v}_k = 2(\boldsymbol{r}_{k+1} - \boldsymbol{r}_k)$, and its acceleration

$\boldsymbol{a}_0 = 2(\boldsymbol{r}_2 - 2\boldsymbol{r}_1 + \boldsymbol{r}_0)$. Show that this quadratic is represented exactly by a cubic Bézier curve (12.26) with control points $\boldsymbol{r}_0$, $\boldsymbol{r}_{2/3}$, $\boldsymbol{r}_{4/3}$, $\boldsymbol{r}_2$, such that (Fig. 12.41)

$$\boldsymbol{r}_{2/3} = \tfrac{1}{3}\boldsymbol{r}_0 + \tfrac{2}{3}\boldsymbol{r}_1, \qquad \boldsymbol{r}_{4/3} = \tfrac{2}{3}\boldsymbol{r}_1 + \tfrac{1}{3}\boldsymbol{r}_2.$$

(12.25) Construct quadratic *Lagrange interpolants* $L_j(t)$, that are quadratic polynomials such $L_j(t_i) = \delta_{ij}$, for i and $j = (0, 1, 2)$ with $t_i = i/2$. Use those interpolants to construct a parabola $\boldsymbol{r}(t)$ such that $\boldsymbol{r}(t_i) = \boldsymbol{r}_i$, for given arbitrary $\boldsymbol{r}_0$, $\boldsymbol{r}_1$, $\boldsymbol{r}_2$. Prove that $\dot{\boldsymbol{r}}(t_1) = \boldsymbol{r}_2 - \boldsymbol{r}_0$. Show that the control points for the quadratic Bézier curve passing through those points are $\boldsymbol{r}_0$, $\boldsymbol{r}'_1 = 2\boldsymbol{r}_1 - \frac{1}{2}(\boldsymbol{r}_0 + \boldsymbol{r}_2)$ and $\boldsymbol{r}_2$ (Fig. 12.42).

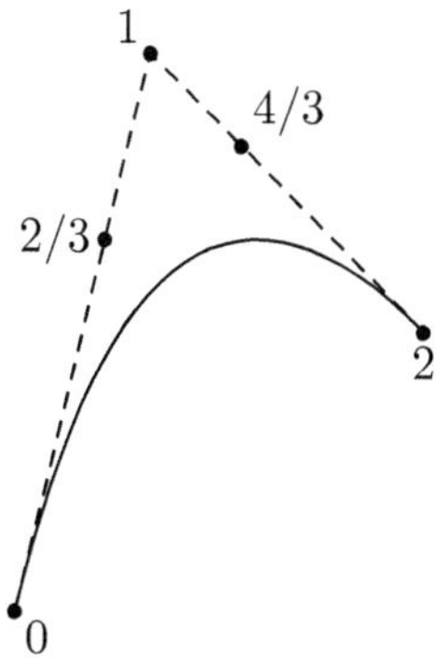

Fig. 12.41 Parabola with cubic Bézier.

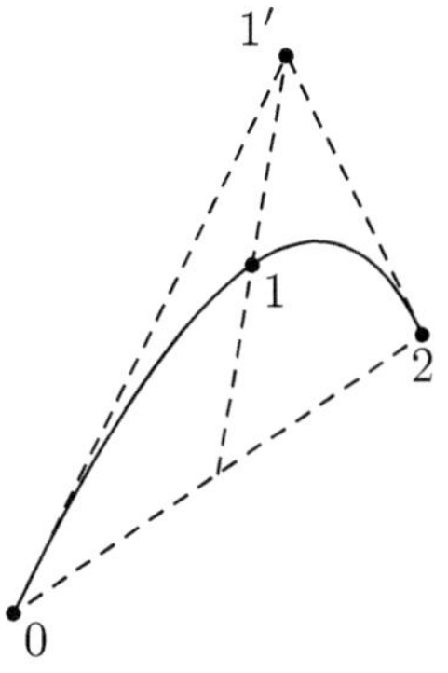

Fig. 12.42 Lagrange interpolants and quadratic Bézier.

(12.26) Generalize the quadratic *Lagrange interpolants* $L_j(t)$ of the previous problem, to a nonuniform *knot sequence* $t_0 < t_1 < t_2$ with $L_j(t_i) = \delta_{ij}$. Construct the parabola such that $\boldsymbol{r}(t_i) = \boldsymbol{r}_i$. What are the quadratic Bézier control points in this case? Sketch an example with $t_0 = 0$, $t_1 = 2/3$, and $t_2 = 1$.

(12.27) Prove the recurrences (12.24) and (12.25).

(12.28) For the cubic Bézier curve (12.26) with control points 0, 1, 2, 3, derive the velocity and acceleration formula (12.28), and show that they expand to

$$\begin{aligned}\dot{\boldsymbol{r}} &= 3(1-t)^2(\boldsymbol{r}_1 - \boldsymbol{r}_0) \\ &\quad + 6(1-t)t(\boldsymbol{r}_2 - \boldsymbol{r}_1) + 3t^2(\boldsymbol{r}_3 - \boldsymbol{r}_2), \\ \ddot{\boldsymbol{r}} &= 6(1-t)(\boldsymbol{r}_0 - 2\boldsymbol{r}_1 + \boldsymbol{r}_2) + 6t(\boldsymbol{r}_1 - 2\boldsymbol{r}_2 + \boldsymbol{r}_3).\end{aligned}$$

So the velocity $\dot{\boldsymbol{r}}$ is the quadratic Bézier representation of the successive piecewise linear velocities $\boldsymbol{v}_k = 3(\boldsymbol{r}_{k+1} - \boldsymbol{r}_k)$,[10] and the acceleration $\ddot{\boldsymbol{r}}$ is the linear Bézier representation of the successive accelerations[11] $\boldsymbol{a}_k = 2(\boldsymbol{v}_{k+1} - \boldsymbol{v}_k)$.

(12.29) What is the most general $\boldsymbol{r}(t)$ for a curve with constant acceleration? What is the most general $\boldsymbol{r}(s)$ for a curve with constant $\mathrm{d}^2\boldsymbol{r}/\mathrm{d}s^2$ where s is arclength along the curve? Are these the same curves?

(12.30) Give an example of a curve with constant curvature. Are such curves necessarily planar?

(12.31) Show/explain the following identities for the curvature of curve $\boldsymbol{r}(t)$,

$$\kappa = \left|\frac{\mathrm{d}\hat{\boldsymbol{t}}}{\mathrm{d}s}\right| = \frac{1}{|\dot{\boldsymbol{r}}|}\left|\frac{\mathrm{d}}{\mathrm{d}t}\left(\frac{\dot{\boldsymbol{r}}}{|\dot{\boldsymbol{r}}|}\right)\right| = \frac{|\ddot{\boldsymbol{r}} - (\ddot{\boldsymbol{r}}\cdot\hat{\boldsymbol{t}})\hat{\boldsymbol{t}}|}{|\dot{\boldsymbol{r}}|^2} = \frac{|\dot{\boldsymbol{r}} \times \ddot{\boldsymbol{r}}|}{|\dot{\boldsymbol{r}}|^3}. \quad (12.77)$$

(12.32) For a planar curve parametrized in cartesian coordinates as $y = f(x)$, show how (12.33a) yields the (signed) curvature as

$$\kappa = \frac{y_{xx}}{\left(1 + y_x^2\right)^{3/2}}, \quad (12.78)$$

where $y_x = \mathrm{d}y/\mathrm{d}x$, $y_{xx} = \mathrm{d}^2y/\mathrm{d}x^2$. Derive the same formula using (12.36).

(12.33) Given $\boldsymbol{r}(t)$, show how to best compute its curvature and its torsion.

[10]The piecewise linear $\boldsymbol{r}(t)$ curve goes from $\boldsymbol{r}_k$ to $\boldsymbol{r}_{k+1}$ in $\Delta t = 1/3$ to go from $\boldsymbol{r}_0$ to $\boldsymbol{r}_3$ in time $\Delta t = 1$.

[11]The piecewise linear $\boldsymbol{v}(t)$ curve goes from $\boldsymbol{v}_k$ to $\boldsymbol{v}_{k+1}$ in $\Delta t = 1/2$ to go from $\boldsymbol{v}_0$ to $\boldsymbol{v}_2$ in time $\Delta t = 1$.

(12.34) A curve $\mathcal{C}$ is specified as $(x(t), y(t), z(t))$ in cartesian coordinates. Show that its curvature and binormal are given by (with $\dot{x} = \mathrm{d}x/\mathrm{d}t$, etc.)

$$\kappa\hat{\boldsymbol{b}} = \frac{(\dot{y}\ddot{z} - \dot{z}\ddot{y})\,\hat{\mathbf{x}} + (\dot{z}\ddot{x} - \dot{x}\ddot{z})\,\hat{\mathbf{y}} + (\dot{x}\ddot{y} - \dot{y}\ddot{x})\,\hat{\mathbf{z}}}{(\dot{x}^2 + \dot{y}^2 + \dot{z}^2)^{3/2}}. \tag{12.79}$$

(12.35) What is a curve with constant jerk? Find a general $\boldsymbol{r}(t)$ for such a curve.

(12.36) Can a curve have curvature but no torsion? Can it have torsion but no curvature? Give examples of such curves if yes; explain if no.

(12.37) Consider $\boldsymbol{r}(t) = a\cos\omega t\,\hat{\mathbf{x}} + a\sin\omega t\,\hat{\mathbf{y}} + bt\,\hat{\mathbf{z}}$, where a, b and ω are constant real numbers and $\hat{\mathbf{x}}, \hat{\mathbf{y}}, \hat{\mathbf{z}}$ is a cartesian basis. Sketch the curve. Calculate the velocity, acceleration and jerk for this curve. Find $\hat{\boldsymbol{t}}$, $\hat{\boldsymbol{n}}$, and $\hat{\boldsymbol{b}}$. What are the curvature and torsion for this curve?

(12.38) Show that $\mathrm{d}\hat{\boldsymbol{b}}/\mathrm{d}s = -\tau\hat{\boldsymbol{n}}$ from (12.33a), (12.35), and $\hat{\boldsymbol{b}} = \hat{\boldsymbol{t}} \times \hat{\boldsymbol{n}}$.

(12.39) Verify that (12.39) with $\boldsymbol{u} = \hat{\boldsymbol{t}}$, $\hat{\boldsymbol{n}}$, and $\hat{\boldsymbol{b}}$, yields (12.38).

(12.40) Consider $\boldsymbol{r}(\theta) = \boldsymbol{a}\cos\theta + \boldsymbol{b}\sin\theta + \boldsymbol{c}$, where $\boldsymbol{a}\cdot\boldsymbol{b} = 0$, and $\theta = 0 \to 2\pi$. Show that this is a plane curve. Derive explicit θ-integrals for the length and area of that curve. One of these integrals is elementary, the other is an *elliptic integral.*

(12.41) For a curve in the (x, y)-plane parametrized as $y = f(x)$, find the position vector $\boldsymbol{r}(x)$ and show that the arclength $\mathrm{d}s = \mathrm{d}x\sqrt{1 + (\mathrm{d}y/\mathrm{d}x)^2}$. Construct x-integral expressions for the length $\mathcal{L}$ of the curve and for the area A swept by the radius vector. Illustrate and explain your derivation.

(12.42) A planar curve is given in polar coordinates as distance from the origin as a function of azimuthal angle, $\rho = \rho(\varphi)$ for $\varphi_0 \le \varphi \le \varphi_f$. What is the position vector $\boldsymbol{r}(\varphi)$ for the curve? Show that $\mathrm{d}s = \mathrm{d}\varphi\,\sqrt{\rho^2 + (\mathrm{d}\rho/\mathrm{d}\varphi)^2}$. Construct φ-integral expressions for the length $\mathcal{L}$ of the curve and for the area A swept by the radius vector. Illustrate and explain your derivation.

(12.43) Calculate the curvature and arclength of the parabola $y = ax^2/2$ as functions of x. Show how $ax = \sinh t$ with (12.75) enables evaluating arclength.

(12.44) Derive explicit θ-integrals for the length and area of the cardioid and the limaçon in Fig. 12.39.

(12.45) Calculate $\int_{\mathcal{C}} \mathrm{d}\boldsymbol{r}$, $\int_{\mathcal{C}} \boldsymbol{r}\cdot\mathrm{d}\boldsymbol{r}$, and $\int_{\mathcal{C}} \boldsymbol{r}\times\mathrm{d}\boldsymbol{r}$ along the curve of exercise 12.40 from $\boldsymbol{r}(0)$ to $\boldsymbol{r}(-3\pi/2)$. Show and justify your work.

(12.46) Calculate $\int_{\mathcal{C}} \mathrm{d}\boldsymbol{r}$, $\int_{\mathcal{C}} \boldsymbol{r}\cdot\mathrm{d}\boldsymbol{r}$, and $\int_{\mathcal{C}} \boldsymbol{r}\times\mathrm{d}\boldsymbol{r}$ for the curve $\mathcal{C}$ defined by

$$\boldsymbol{r}(t) = R\,(\cos(2k\pi t)\,\hat{\mathbf{x}} + \sin(2k\pi t)\,\hat{\mathbf{y}}) + Ht\,\hat{\mathbf{z}},$$

with $t = 0 \to 1$.

(12.47) Express $\boldsymbol{r}$, $\mathrm{d}\boldsymbol{r}$, $\boldsymbol{r}\cdot\mathrm{d}\boldsymbol{r}$ and $\boldsymbol{r}\times\mathrm{d}\boldsymbol{r}$ in cartesian, cylindrical and spherical coordinates.

(12.48) If $\hat{\boldsymbol{a}}$ is any fixed unit vector and β is an azimuthal angle about $\hat{\boldsymbol{a}}$, show that

$$\boldsymbol{r} = \boldsymbol{r}_\perp + h\hat{\boldsymbol{a}},$$
$$\mathrm{d}\boldsymbol{r} = \hat{\boldsymbol{r}}_\perp \mathrm{d}r_\perp + \hat{\boldsymbol{a}} \times \boldsymbol{r}_\perp\,\mathrm{d}\beta + \hat{\boldsymbol{a}}\,\mathrm{d}h.$$

(12.49) Derive/justify (12.55).

(12.50) *Line source and vortex.* If $\mathcal{C}'$ is the infinite line $\boldsymbol{r}' = t'\hat{\mathbf{z}}$, for $t' = -\infty \to \infty$, show that

$$\int_{\mathcal{C}'} \frac{\boldsymbol{r} - \boldsymbol{r}'}{|\boldsymbol{r} - \boldsymbol{r}'|^3}\,|\mathrm{d}\boldsymbol{r}'| = 2\frac{\hat{\boldsymbol{r}}_\perp}{r_\perp}, \tag{12.80}$$

$$\int_{\mathcal{C}'} \frac{\mathrm{d}\boldsymbol{r}' \times (\boldsymbol{r} - \boldsymbol{r}')}{|\boldsymbol{r} - \boldsymbol{r}'|^3} = 2\,\hat{\mathbf{z}} \times \frac{\hat{\boldsymbol{r}}_\perp}{r_\perp}. \tag{12.81}$$

Use θ as the parameter with $\boldsymbol{r} - \boldsymbol{r}' = \boldsymbol{r}_\perp + \hat{\mathbf{z}}\;r_\perp/\tan\theta$ (Fig. 12.43). Integral (12.80)is the electric field induced be a *line charge*, and the velocity field of a *line source.* Integral (12.81) is the magnetic field induced by a *line current*, and the velocity field of a *line vortex.*

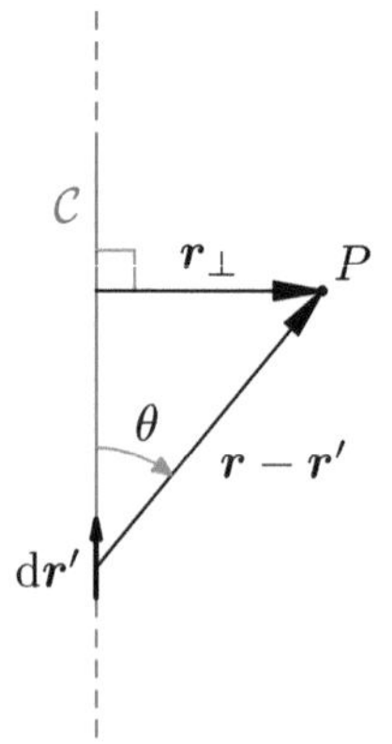

Fig. 12.43 Line charge and vortex.

(12.51) The magnetic field induced by a constant electric current J flowing along the z-axis follows form the Biot–Savart law and (12.81) as

$$\boldsymbol{B}(\boldsymbol{r}) = \frac{\mu_0 J}{2\pi}\frac{\hat{\mathbf{z}} \times \boldsymbol{r}}{|\hat{\mathbf{z}} \times \boldsymbol{r}|^2} = \frac{\mu_0 J}{2\pi}\frac{\hat{\boldsymbol{\varphi}}}{\rho}, \tag{12.82}$$

where μ_0 is the magnetic constant. Calculate $\oint_{\mathcal{C}} \boldsymbol{B} \cdot \mathrm{d}\boldsymbol{r}$ and $\oint_{\mathcal{C}} \boldsymbol{B} \times \mathrm{d}\boldsymbol{r}$ when $\mathcal{C}$ is the circle of radius R in the (x, y)-plane centered at the origin. Sketch $\boldsymbol{B}(\boldsymbol{r})$ and $\mathcal{C}$. How do the integrals depend on R? Analyze these integrals when $\mathcal{C}$ is any closed loop in 3D space, as done for (12.56).

(12.52) *Work-energy theorem.* If, according to Newton, $\boldsymbol{F}(\boldsymbol{r}(t)) = m\,\mathrm{d}\boldsymbol{v}/\mathrm{d}t$ with $\boldsymbol{v}(t) = \mathrm{d}\boldsymbol{r}/\mathrm{d}t$, for an arbitrary vector field $\boldsymbol{F}(\boldsymbol{r})$, show that

$$\int_{\mathcal{C}} \boldsymbol{F}(\boldsymbol{r}) \cdot \mathrm{d}\boldsymbol{r} = \tfrac{1}{2}m\left(v_B^2 - v_A^2\right)$$

for any curve $\mathcal{C}$ from A to B parameterized by $\boldsymbol{r}(t)$, where m is a constant and $v_A = |\boldsymbol{v}|$ at A, $v_B = |\boldsymbol{v}|$ at B. Does this mean that the integral is path independent?

(12.53) Show that $\oint_{\mathcal{C}} \boldsymbol{r} \times \mathrm{d}\boldsymbol{r} = 2A\,\hat{\boldsymbol{n}}$, where $\mathcal{C}$ is a simple closed curve in a plane perpendicular to $\hat{\boldsymbol{n}}$, and A is the area enclosed by the curve, even if O is *not* in the plane of the curve.

(12.54) Let $\boldsymbol{r} = \overrightarrow{OP}$. Show that for any piecewise smooth closed curve $\mathcal{C}$

$$\boldsymbol{A} = \frac{1}{2}\oint_{\mathcal{C}} \boldsymbol{r} \times \mathrm{d}\boldsymbol{r}$$

is independent of the origin O.

(12.55) Calculate the circulation of $\boldsymbol{v}(\boldsymbol{r}) = \boldsymbol{\omega} \times \boldsymbol{r}$ about the curve of exercise 12.40 using an explicit θ-integral from $\theta = 0$ to $\theta = \alpha$, with $\boldsymbol{\omega}$ arbitrary but independent of $\boldsymbol{r}$.

(12.56) A curve $\mathcal{C}$ is parametrized in cartesian coordinates as $(x(t), y(t), z(t))$. Show that the winding and turning numbers about the z axis, (12.56), (12.58), are given by (with $\dot{x} = \mathrm{d}x/\mathrm{d}t$, etc.)

$$2\pi k = \int_{\mathcal{C}} \frac{\hat{\mathbf{z}} \times \boldsymbol{r}}{|\hat{\mathbf{z}} \times \boldsymbol{r}|^2} \cdot \mathrm{d}\boldsymbol{r} = \int_{t_0}^{t_1} \frac{x\dot{y} - y\dot{x}}{x^2 + y^2}\mathrm{d}t,$$

$$2\pi m = \int_{\mathcal{C}} \frac{\hat{\mathbf{z}} \times \hat{\boldsymbol{t}}}{|\hat{\mathbf{z}} \times \hat{\boldsymbol{t}}|^2} \cdot \mathrm{d}\hat{\boldsymbol{t}} = \int_{t_0}^{t_1} \frac{\dot{x}\ddot{y} - \dot{y}\ddot{x}}{\dot{x}^2 + \dot{y}^2}\mathrm{d}t.$$

(12.57) The unit tangent $\hat{\boldsymbol{t}}(t)$ to a planar curve $\boldsymbol{r}(t)$ maps the curve to a point on the unit circle. Indicate that map on the unit circle for the planar curve in Fig. 12.44 and estimate $\int_{\mathcal{C}} \hat{\mathbf{z}} \times \hat{\boldsymbol{t}} \cdot \mathrm{d}\hat{\boldsymbol{t}}$, where $\hat{\mathbf{z}}$ is the unit normal to the plane of the curve.

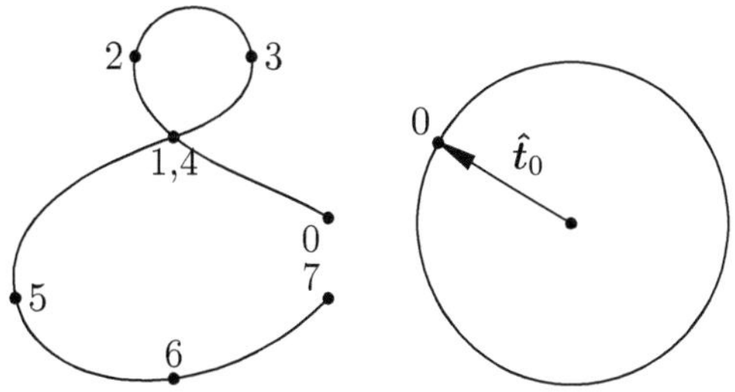

Fig. 12.44 The unit tangent $\hat{\boldsymbol{t}}$ maps the curve to the unit circle.

(12.58) For a particle sliding without friction along a curve $y = f(x)$, under the action of gravity $-g\hat{\mathbf{y}}$, find the explicit integral for the total travel time from $x = a$ to $x = b > a$. Construct and calculate one or more explicit examples.

(12.59) A pendulum with a particle of mass m attached at the end of a string of length ℓ oscillates under the action of gravity $-g\hat{\mathbf{y}}$, so that its trajectory is a circular arc $\boldsymbol{r}(\theta) = \ell\,(\hat{\mathbf{x}}\sin\theta - \hat{\mathbf{y}}\cos\theta)$. If the particle is released from rest at $0 < \theta_0 < \pi$, show that the time to reach $\theta = 0$ is

$$T = \sqrt{\frac{\ell}{2g}}\int_0^{\theta_0} \frac{\mathrm{d}\theta}{\sqrt{\cos\theta - \cos\theta_0}} = \sqrt{\frac{\ell}{g}}\int_0^{\pi/2} \frac{\mathrm{d}\varphi}{\sqrt{1 - k^2\sin^2\varphi}},$$

where $k = \sin(\theta_0/2)$, and $\sin(\theta/2) = k\sin\varphi$. The latter integral is called the *complete elliptic integral of the first kind.* It cannot be calculated in terms of elementary functions.

(12.60) A particle slides without friction along the upside down cycloid $x(\theta) = R(\theta - \sin\theta)$, $y(\theta) = R(\cos\theta - 1)$, under the action of gravity $-g\hat{\mathbf{y}}$. The particle is released from rest at $0 \le \theta_0 < \pi$. Show that the time to reach the bottom at $\theta = \pi$ is $T = \pi\sqrt{R/g}$, independent of θ_0, using the change of variables $\cos(\theta/2) = k\cos\varphi$ with $k = \cos(\theta_0/2)$. This is the *tautochrone* property of a cycloid.

Surfaces

13

13.1 Representations

A surface $\mathcal{S}$ can be specified *implicitly* as the set of points P in 3D euclidean space such that $f(P) = 0$, where f is a scalar function of position P, with radius vector $\boldsymbol{r} = x\hat{\mathbf{x}}+y\hat{\mathbf{y}}+z\hat{\mathbf{z}}$ in cartesian coordinates. For example, a sphere of radius R centered at the origin O has

$$f(P) = |\overrightarrow{OP}|^2 - R^2 = \boldsymbol{r}\cdot\boldsymbol{r} - R^2 = x^2 + y^2 + z^2 - R^2 = 0,$$

and a plane perpendicular to $\boldsymbol{a} = a_1\hat{\mathbf{x}} + a_2\hat{\mathbf{y}} + a_3\hat{\mathbf{z}}$ has (7.8c)

$$f(P) = \boldsymbol{a}\cdot\boldsymbol{r} - b = a_1 x + a_2 y + a_3 z - b = 0.$$

A surface $\mathcal{S}$ is *parametrized explicitly* by specifying the radius vector $\boldsymbol{r}$ as a vector function of *two* real variables (u, v) in a subset $\mathcal{U}$ of $\mathbb{R}^2$

$$(u, v) \in \mathcal{U} \to P \in \mathcal{S},\ \overrightarrow{OP} = \boldsymbol{r}(u, v), \tag{13.1}$$

which provides a *map* from $\mathcal{U}$ in the (u, v) plane to the surface $\mathcal{S}$ in $\mathbb{E}^3$. The parameters (u, v) are coordinates for points P on the surface $\mathcal{S}$. The vector function $\boldsymbol{r}(u, v)$ is the *two-parameter* family of solutions to the implicit scalar equation $f(\boldsymbol{r}) = 0$ defining the surface, that is,

$\forall$ means *for all*, or *for any*.

$$f(\boldsymbol{r}(u, v)) = 0, \qquad \forall (u, v) \in \mathcal{U}.$$

Plane example For the plane $\boldsymbol{a}\cdot\boldsymbol{r} - b = 0$, with given vector $\boldsymbol{a}$ and scalar b, let $\Delta\boldsymbol{r}_1$ and $\Delta\boldsymbol{r}_2$ be any two nonparallel vectors that are both perpendicular to $\boldsymbol{a}$, and thus *tangent* to the plane; then the radius vector of any point on the plane can be specified explicitly by

$$\boldsymbol{r}(u, v) = \boldsymbol{r}_0 + u\,\Delta\boldsymbol{r}_1 + v\,\Delta\boldsymbol{r}_2, \tag{13.2}$$

with $(u, v) \in \mathbb{R}^2$, and $\boldsymbol{r}_0$ is the position vector of any point P_0 in the plane, such that $\boldsymbol{a}\cdot\boldsymbol{r}_0 = b$. For example, $\boldsymbol{r}_0 = \boldsymbol{a}\,b/a^2$, which is actually the point on the plane nearest to the origin O. If the plane is specified by three points P_0, P_1, P_2 with position vectors $\boldsymbol{r}_0$, $\boldsymbol{r}_1$, $\boldsymbol{r}_2$, respectively, then

$$\Delta\boldsymbol{r}_1 = \boldsymbol{r}_1 - \boldsymbol{r}_0, \qquad \Delta\boldsymbol{r}_2 = \boldsymbol{r}_2 - \boldsymbol{r}_1, \qquad \boldsymbol{a} = \Delta\boldsymbol{r}_1 \times \Delta\boldsymbol{r}_2.$$

Representation (13.2) is explicit: any (u, v) provides the position vector $\overrightarrow{OP} = \boldsymbol{r}(u, v)$ for a point P that is guaranteed to be on the plane. Indeed, substituting that $\boldsymbol{r}(u,v)$ (13.2) into the plane equation $\boldsymbol{a}\cdot\boldsymbol{r}=b$ yields an identity

$$\boldsymbol{a}\cdot\boldsymbol{r}(u,v)=\boldsymbol{a}\cdot\boldsymbol{r}_0+u\,\boldsymbol{a}\cdot\Delta\boldsymbol{r}_1+v\,\boldsymbol{a}\cdot\Delta\boldsymbol{r}_2=b,$$

since $\boldsymbol{a}\cdot\boldsymbol{r}_0=b$, and $\boldsymbol{a}\cdot\Delta\boldsymbol{r}_1=0=\boldsymbol{a}\cdot\Delta\boldsymbol{r}_2$. The representation is obviously not unique since any two linearly independent tangent vectors $\Delta\boldsymbol{r}_1$ and $\Delta\boldsymbol{r}_2$ will do. The range of the parameters is the entire real (u,v)-plane, $-\infty<u<\infty$, $-\infty<v<\infty$, to cover the entire plane.

For the triangle with vertices P_0, P_1, P_2, any point P on that triangle surface has a position vector $\boldsymbol{r}=\overrightarrow{OP}$ that can be parametrized as (13.2) with (u,v) in the domain $0\le v\le u\le 1$ (Fig. 13.1).

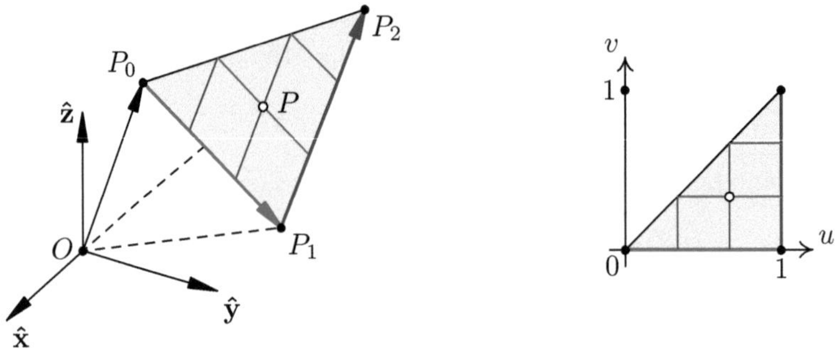

Fig. 13.1 Map of triangle P_0, P_1, P_2 to the unit triangle in the (u,v) plane.

Sphere For the sphere of radius R centered at O, we can solve

$$x^2+y^2+z^2-R^2=0$$

for z as a function of x,y, to obtain the parametrizations

$$\boldsymbol{r}(x,y)=x\,\hat{\mathbf{x}}+y\,\hat{\mathbf{y}}\pm\sqrt{R^2-x^2-y^2}\,\hat{\mathbf{z}}, \tag{13.3}$$

defined for the disk $x^2+y^2\le R^2$. Two vector functions $\boldsymbol{r}(x,y)$ are needed to describe the sphere, one for the upper hemisphere with $z>0$, and one for the lower hemisphere with $z<0$. This is inconvenient if we want the whole sphere, but these coordinates are very good near the poles, where $x^2+y^2\ll R^2$, and[1]

$$z\simeq\pm R\left(1-\frac{x^2+y^2}{2R^2}\right). \tag{13.4}$$

[1] From the (binomial) Taylor series

$$(1+x)^\alpha=1+\alpha\,x+\frac{\alpha(\alpha-1)}{2}\,x^2+\cdots$$

for $|x|<1$.

The same surface $x^2+y^2+z^2=R^2$ can be parametrized using *spherical coordinates* (θ,φ) as

$$\boldsymbol{r}(\theta,\varphi)=R\sin\theta\cos\varphi\,\hat{\mathbf{x}}+R\sin\theta\sin\varphi\,\hat{\mathbf{y}}+R\cos\theta\,\hat{\mathbf{z}}, \tag{13.5}$$

where θ is the polar angle between $\hat{\mathbf{z}}$ and $\boldsymbol{r}$, while φ is the azimuthal (or longitude) angle between the $(\hat{\mathbf{z}},\hat{\mathbf{x}})$ plane, and the $(\hat{\mathbf{z}},\boldsymbol{r})$ plane. This parametrization is better than (13.3), in that it describes the entire sphere uniquely if $0 \le \theta \le \pi$ and $0 \le \varphi < 2\pi$, or $-\pi < \varphi \le \pi$, but the poles at $\theta = 0$ or π are problematic since φ is undefined there. The parametrization (13.5) can be derived geometrically from Fig. 13.2, or somewhat more algebraically through the following Pythagoras reduction

$$\begin{cases} z^2 + \rho^2 = R^2 & \Leftrightarrow \quad z = R\cos\theta,\ \rho = R\sin\theta, \\ x^2 + y^2 = \rho^2 & \Leftrightarrow \quad x = \rho\cos\varphi,\ y = \rho\sin\varphi. \end{cases} \tag{13.6}$$

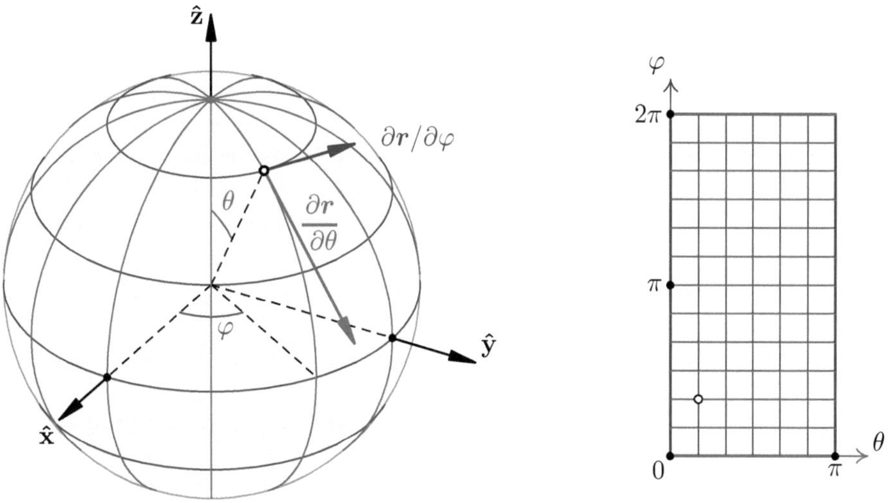

Fig. 13.2 Map of the sphere to (θ, φ) rectangle, and tangent vectors $\partial \boldsymbol{r}/\partial\theta$, $\partial \boldsymbol{r}/\partial\varphi$.

13.2 Coordinate curves and tangent plane

If one of the parameters is held fixed, $v = v_0$ say, we obtain the coordinate curve

$$\mathcal{C}_{v_0} : u \in \mathbb{R} \to \overrightarrow{OP} = \boldsymbol{r}(u, v_0),$$

parametrized by u and labeled by v_0. This is a family of curves, one for each v_0. We typically refer to these curves as the u-coordinate curves, curves along which only u varies, though any one of those curves can be labeled by the fixed value of the other surface parameter, $v = v_0$. For the sphere parametrized as in (13.5), $\boldsymbol{r}(\theta, \varphi_0)$ is the φ_0 *meridian*, the semi-circle from the north and to the south pole, whose plane is an angle φ_0 from the *prime meridian* plane. Likewise,

$$\mathcal{C}_{u_0} : v \to \boldsymbol{r}(u_0, v)$$

describes another family of coordinate curves parametrized by v and labeled by u_0. For the sphere, $\boldsymbol{r}(\theta_0, \varphi)$ is the θ_0 *parallel*, a circle parallel to the equatorial circle. Meridians and parallels are illustrated in Fig. 13.2. The set of all such *coordinate curves* generates (or sweeps) the surface. It may be useful to think of these curves as the fibers of a cloth surface.

The vectors

$$\frac{\partial \boldsymbol{r}}{\partial u}, \quad \frac{\partial \boldsymbol{r}}{\partial v},$$

are tangent to their respective parametric curves and hence to the surface at the corresponding $\boldsymbol{r}(u,v)$ point. The partial derivative notation ∂ automatically specifies which coordinate curve is involved. These two vectors taken at the same point $\boldsymbol{r}(u,v)$ define the *tangent plane* at that point, that is, the set of points P' whose position vectors are

$$\overrightarrow{OP'} = \boldsymbol{r}' = \boldsymbol{r}(u,v) + \alpha\left(\frac{\partial \boldsymbol{r}}{\partial u}\right) + \beta\left(\frac{\partial \boldsymbol{r}}{\partial v}\right), \tag{13.7}$$

where the tangent vectors are evaluated at fixed (u,v). We need to distinguish between the surface vector function $\boldsymbol{r}(u,v)$ and the plane vector function $\boldsymbol{r}'(\alpha,\beta;u,v)$. That tangent plane is the set of all $\boldsymbol{r}'$ such that

$$\boldsymbol{N}(u,v) \cdot (\boldsymbol{r}' - \boldsymbol{r}(u,v)) = 0, \tag{13.8}$$

where

$$\boldsymbol{N}(u,v) = \frac{\partial \boldsymbol{r}}{\partial u} \times \frac{\partial \boldsymbol{r}}{\partial v} \tag{13.9}$$

is normal to the surface at point P with $\overrightarrow{OP} = \boldsymbol{r}(u,v)$.

The normal $\boldsymbol{N}(u,v)$ should be unique, nonzero and differentiable if $\boldsymbol{r}(u,v)$ is a suitable parametrization of the smooth surface $\mathcal{S}$, and $\mathcal{S}$ has a tangent plane at that point. The normal $\boldsymbol{N}$ can vanish or not exist at isolated points, or along isolated curves in $\mathcal{U}$, where the parametrization is then *singular*, as illustrated hereafter. The singularities may be a limitation of the map $\boldsymbol{r}(u,v)$, such as the poles in spherical coordinates, or actual singular points for the surface, such as the vertex of a cone.

Orthogonal coordinates

The (u,v) coordinates are *orthogonal* if the coordinate curves intersect at right angles, that is, if the tangent vectors $\partial \boldsymbol{r}/\partial u$ and $\partial \boldsymbol{r}/\partial v$ are orthogonal to each other at every point $\boldsymbol{r}(u,v)$,

$$\frac{\partial \boldsymbol{r}}{\partial u} \cdot \frac{\partial \boldsymbol{r}}{\partial v} = 0, \tag{13.10}$$

except perhaps at some singular points where one or both tangent vectors vanish.

Coordinate curves and their tangent vectors are illustrated for a triangular patch in Fig. 13.1, where $\partial \boldsymbol{r}/\partial u = \boldsymbol{r}_1 - \boldsymbol{r}_0$ and $\partial \boldsymbol{r}/\partial v = \boldsymbol{r}_2 - \boldsymbol{r}_1$, for all (u,v). These are convenient coordinates for the triangle $P_0P_1P_2$, although they are not orthogonal, in general. The spherical coordinate curves are illustrated in Fig. 13.2. Their tangent vectors can be calculated from (13.5),

$$\boldsymbol{r}(\theta,\varphi) = r\cos\varphi\sin\theta\,\hat{\mathbf{x}} + r\sin\varphi\sin\theta\,\hat{\mathbf{y}} + r\cos\theta\,\hat{\mathbf{z}},$$

with $r = R$ fixed but arbitrary, yielding

$$\begin{aligned}\frac{\partial \boldsymbol{r}}{\partial \theta} &= r\cos\varphi\cos\theta\,\hat{\mathbf{x}} + r\sin\varphi\cos\theta\,\hat{\mathbf{y}} - r\sin\theta\,\hat{\mathbf{z}},\\ \frac{\partial \boldsymbol{r}}{\partial \varphi} &= -\,r\sin\varphi\sin\theta\,\hat{\mathbf{x}} + r\cos\varphi\sin\theta\,\hat{\mathbf{y}}.\end{aligned} \tag{13.11}$$

They are clearly orthogonal,

$$\frac{\partial \boldsymbol{r}}{\partial \theta}\cdot\frac{\partial \boldsymbol{r}}{\partial \varphi} = 0, \quad \forall\,\theta,\varphi.$$

However, $\partial\boldsymbol{r}/\partial\varphi = 0$ at $\theta = 0,\ \pi$, and φ, $\partial\boldsymbol{r}/\partial\theta$ and $\boldsymbol{N}$ are undefined there. These are the two poles, north and south, and the coordinates are *singular* at those two points.

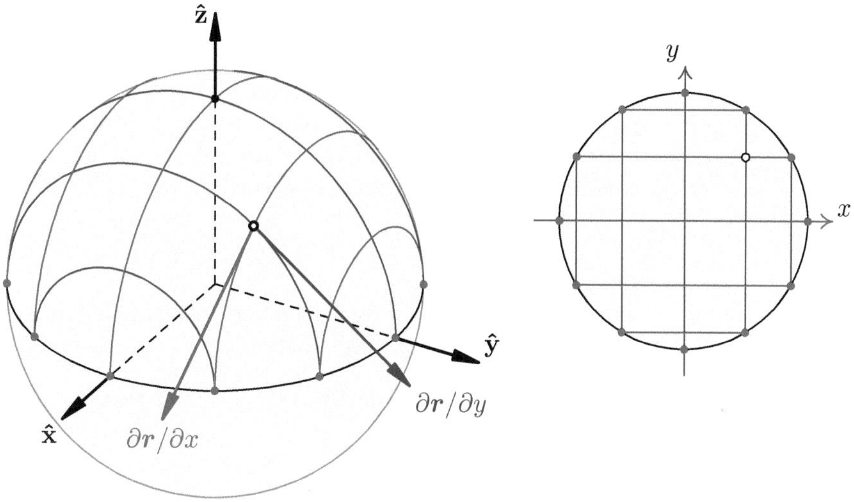

Fig. 13.3 Map of the hemisphere to the (x, y) disk, and tangent vectors $\partial\boldsymbol{r}/\partial x$, $\partial\boldsymbol{r}/\partial y$.

The cartesian coordinate curves for the upper hemisphere are illustrated in Fig. 13.3. Their tangent vectors can be calculated from (13.3). It is left as an exercise to verify that these coordinates are not orthogonal,

$$\frac{\partial \boldsymbol{r}}{\partial x}\cdot\frac{\partial \boldsymbol{r}}{\partial y} \neq 0,$$

except at the pole $(x, y) = (0, 0)$. Worse, the tangent vectors become parallel, and their magnitudes go to infinity, as (x, y) approaches the equator, $x^2 + y^2 = R^2$. These coordinates are *singular* everywhere along the equator. Coordinates are *singular* wherever

$$|\boldsymbol{N}| = \left|\frac{\partial \boldsymbol{r}}{\partial u}\times\frac{\partial \boldsymbol{r}}{\partial v}\right| \to 0 \text{ or } \infty.$$

Conformal coordinates

Coordinates are *conformal* (or *isothermal*) if they are orthogonal (13.10), *and* the coordinate tangents have the same magnitudes

$$\left|\frac{\partial \boldsymbol{r}}{\partial u}\right| = \left|\frac{\partial \boldsymbol{r}}{\partial v}\right|. \tag{13.12}$$

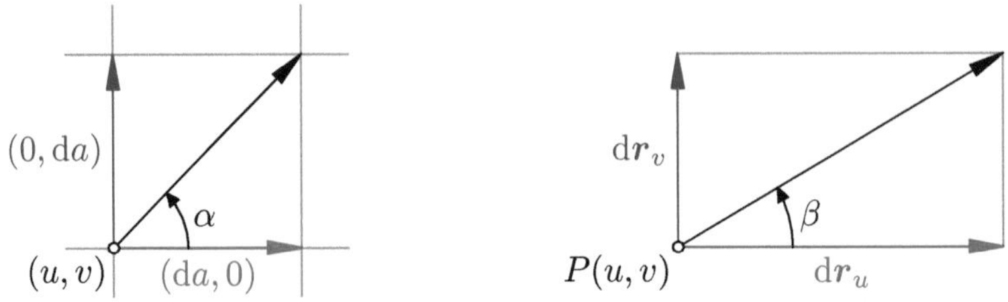

Fig. 13.4 Mapping of $(\mathrm{d}u, \mathrm{d}v)$ at (u, v), to $\mathrm{d}\boldsymbol{r}$ at $\boldsymbol{r}(u, v)$ for orthogonal coordinates.

Conformal coordinates preserve angles between the map in the (u, v) plane, and the surface $\boldsymbol{r}(u, v)$. This is most easily understood by considering the mapping of displacements vectors $(\mathrm{d}u, \mathrm{d}v)$, from point (u, v) to $(u+\mathrm{d}u, v+\mathrm{d}v)$ in the (u, v) plane, to the displacement vectors $\mathrm{d}\boldsymbol{r}(u, v)$, between points $\boldsymbol{r}(u, v)$ and $\boldsymbol{r}(u+\mathrm{d}u, v+\mathrm{d}v)$ on the surface $\mathcal{S}$. The chain rule links these displacements according to

$$\mathrm{d}\boldsymbol{r} = \frac{\partial \boldsymbol{r}}{\partial u}\mathrm{d}u + \frac{\partial \boldsymbol{r}}{\partial v}\mathrm{d}v. \tag{13.13}$$

Thus, two orthogonal displacements, $(\mathrm{d}a, 0)$ and $(0, \mathrm{d}a)$, in the (u, v) plane are mapped to[2] (Fig. 13.4)

$$(\mathrm{d}a, 0) \to \mathrm{d}\boldsymbol{r}_u = (\partial \boldsymbol{r}/\partial u)\,\mathrm{d}a,$$
$$(0, \mathrm{d}a) \to \mathrm{d}\boldsymbol{r}_v = (\partial \boldsymbol{r}/\partial v)\,\mathrm{d}a.$$

[2]We could use a finite Δa, divide everything through by Δa, then take the limit $\Delta a \to 0$. The differential notation $\mathrm{d}a$ keeps all that implicit, and helps to focus on the essence of the current issue.

Clearly, the 90° angle at point (u, v) on the map is preserved at $P(u, v)$ on the surface $\mathcal{S}$ only if the coordinates are orthogonal (13.10). However, the displacement $(\mathrm{d}u, \mathrm{d}v) = (\mathrm{d}a, \mathrm{d}a)$ in the (u, v) plane is mapped to

$$(\mathrm{d}a, \mathrm{d}a) \to \mathrm{d}\boldsymbol{r} = \mathrm{d}\boldsymbol{r}_u + \mathrm{d}\boldsymbol{r}_v,$$

as illustrated in Fig. 13.4, and the angle β between $\mathrm{d}\boldsymbol{r}$ and $\mathrm{d}\boldsymbol{r}_u$ equals the angle $\alpha = 45°$ between the displacements $(\mathrm{d}a, 0)$ and $(\mathrm{d}a, \mathrm{d}a)$ in the (u, v), only if the coordinate partial derivatives have equal magnitudes (13.12), in addition to being orthogonal.

Cartesian coordinates (x, y) for the hemisphere (13.3) are certainly not conformal since they are not orthogonal. Spherical coordinates (θ, φ) are orthogonal; however, it follows from (13.11) that

$$\left|\frac{\partial \boldsymbol{r}}{\partial \theta}\right| = r \neq \left|\frac{\partial \boldsymbol{r}}{\partial \varphi}\right| = r\sin\theta,$$

so spherical coordinates are not conformal, except at the equator where $\sin\theta = 1$.

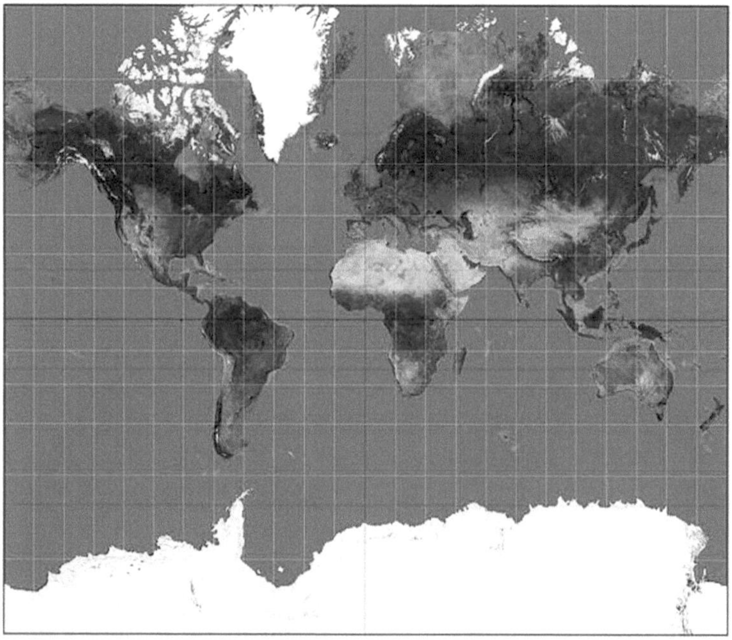

Fig. 13.5 Mercator projection of the world. Reproduced from Strebe (2011) Wikimedia Commons, under a Creative Commons Attribution-ShareAlike 3.0 (CC BY 3.0) `https://commons.wikimedia.org/wiki/File:Mercator_projection_SW.jpg`

Conformal coordinates are very desirable in ship and airplane navigation since angles measured on a conformal map are the same as the angles on the actual surface. *Gerardus Mercator* is credited with creating the first conformal map of the Earth in year 1569. He projected the spherical Earth model onto a cylinder tangential to the sphere at the equator, then stretched the cylinder axis coordinate to obtain conformal coordinates. We can do both steps at once with our calculus. The objective is to use coordinates (φ, v), instead of the spherical coordinates (θ, φ), such that the magnitude of $\partial \boldsymbol{r}/\partial v$ equals that of $\partial \boldsymbol{r}/\partial \varphi$ everywhere. Now, by the chain rule

$$\frac{\partial \boldsymbol{r}}{\partial v} = \frac{\partial \boldsymbol{r}}{\partial \theta}\frac{\mathrm{d}\theta}{\mathrm{d}v}$$

and since $|\partial \boldsymbol{r}/\partial \theta| = r$, $|\partial \boldsymbol{r}/\partial \varphi| = r \sin\theta$, we need $\mathrm{d}\theta/\mathrm{d}v = -\sin\theta$, that is,

$$v(\theta) = -\ln\tan(\theta/2).$$

This yields the conformal Mercator parametrization

$$\boldsymbol{r}(\varphi, v) = \frac{r}{\cosh v}\,\hat{\boldsymbol{\rho}}(\varphi) + r\tanh v\,\hat{\mathbf{z}}, \tag{13.14}$$

where $\hat{\boldsymbol{\rho}} = \cos\varphi\,\hat{\mathbf{x}} + \sin\varphi\,\hat{\mathbf{y}}$, and $-\pi < \varphi \le \pi$, $-\infty < v < \infty$. Obviously, the map has to be truncated in the v direction, but the log helps a lot. For instance, latitude 82° south to 82° north, gives a range $-2.66 < v < 2.66$, while $-\pi < \varphi \le \pi$, so such a truncated map of the sphere is almost square in the (φ, v) plane (Fig. 13.5).

13.3 Surface element

The *line element* $\mathrm{d}\boldsymbol{r}$ arises as the limit of secant vectors, when computing tangents and integrals along a curve $\mathcal{C}$. For a surface $\mathcal{S}$, with an infinite set of tangents at a point but one unit normal $\hat{\boldsymbol{n}}$, the primary differential form is the *surface element*

$$\mathrm{d}\boldsymbol{S} = \hat{\boldsymbol{n}}\,\mathrm{d}A,$$

which represents the *area vector* of a small surface patch shrinking to that point.

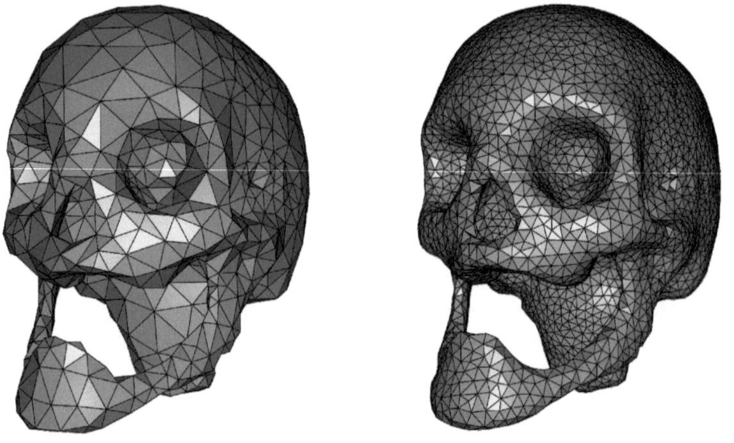

Fig. 13.6 Skull surface mesh refinement $\Delta\boldsymbol{S} \to 0$. Courtesy of Rineau and Yvinec, CGAL, https://www.cgal.org.

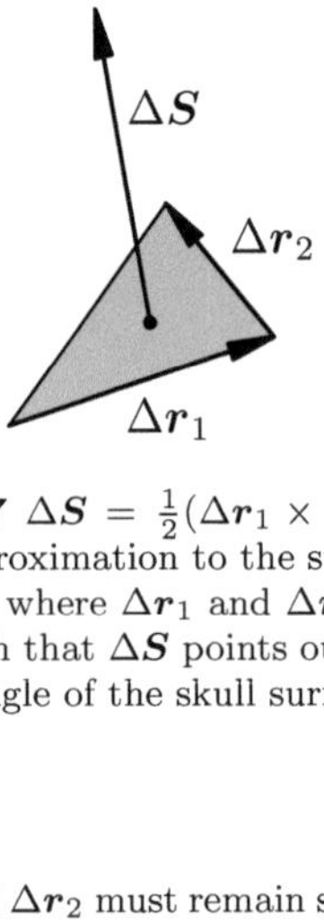

Fig. 13.7 $\Delta\boldsymbol{S} = \frac{1}{2}(\Delta\boldsymbol{r}_1 \times \Delta\boldsymbol{r}_2)$ is a finite approximation to the surface element $\mathrm{d}\boldsymbol{S}$, where $\Delta\boldsymbol{r}_1$ and $\Delta\boldsymbol{r}_2$ are oriented such that $\Delta\boldsymbol{S}$ points outward for each triangle of the skull surface.

Consider, for example, any one of the small triangles that make up the surface of the discretized skull in Fig. 13.6. That small triangle has area vector

$$\Delta\boldsymbol{S} = \tfrac{1}{2}\Delta\boldsymbol{r}_1 \times \Delta\boldsymbol{r}_2 = \hat{\boldsymbol{n}}\,\Delta A,$$

where $\Delta\boldsymbol{r}_1$ and $\Delta\boldsymbol{r}_2$ are two of the edges of the patch (Fig, 13.7). The magnitude of $\Delta\boldsymbol{S}$ is the surface area ΔA of the small triangle, and its direction $\hat{\boldsymbol{n}}$ is orthogonal to the triangle, and thus to the surface in the limit of the triangle shrinking to a point in a suitable manner.[3]

It is assumed that all triangles are oriented such that the unit normal $\hat{\boldsymbol{n}}$ varies smoothly from one triangle to its neighbors, in the limit, thereby *orienting* the surface. For the skull or a sphere, this defines an inside and an outside. For an open surface such as a single triangle, this defines a top and a bottom. Some surfaces are not orientable, the simplest example of a nonorientable surface is the *Möbius strip* investigated in exercise 13.19. Many small triangles are needed to obtain a sufficiently smooth representation of the skull, and $\mathrm{d}\boldsymbol{S}$ is meant to implicitly represent that limit process at a point on the surface (Fig. 13.6). It is useful to think of $\mathrm{d}\boldsymbol{S}$ as a vector of infinitesimal magnitude $\mathrm{d}A$ and direction $\hat{\boldsymbol{n}}$ normal to the surface at that point, but there is no vector field $\boldsymbol{S}(\boldsymbol{r})$ whose differential is $\mathrm{d}\boldsymbol{S}$.

[3] $\Delta\boldsymbol{r}_1$ and $\Delta\boldsymbol{r}_2$ must remain sufficiently independent in the limit; that is,

$$\sin\theta_{12} = \frac{|\Delta\boldsymbol{r}_1 \times \Delta\boldsymbol{r}_2|}{|\Delta\boldsymbol{r}_1||\Delta\boldsymbol{r}_2|}$$

must remain bounded away from 0, strictly, in the limit (exercise 13.27).

$\mathrm{d}\boldsymbol{S}$ is a differential *2-form*.

If a parametric representation $\boldsymbol{r}(u,v)$ for the surface is known then

$$\mathrm{d}\boldsymbol{S} = \mathrm{d}\boldsymbol{r}_u \times \mathrm{d}\boldsymbol{r}_v = \left(\frac{\partial \boldsymbol{r}}{\partial u} \times \frac{\partial \boldsymbol{r}}{\partial v}\right) \mathrm{d}u\mathrm{d}v, \qquad (13.15)$$

since the right-hand side represents the area of the parallelogram formed by the line elements

$$\mathrm{d}\boldsymbol{r}_u = \frac{\partial \boldsymbol{r}}{\partial u}\,\mathrm{d}u, \qquad \mathrm{d}\boldsymbol{r}_v = \frac{\partial \boldsymbol{r}}{\partial v}\,\mathrm{d}v,$$

along the u and v coordinate curves, respectively. A parametrization $\boldsymbol{r}(u,v)$ partitions the surface into *quadrilaterals* whose surface elements $\mathrm{d}\boldsymbol{S}$ are the vector sum of two triangle elements (4.17). The surface orientation is determined by the (u,v) order.

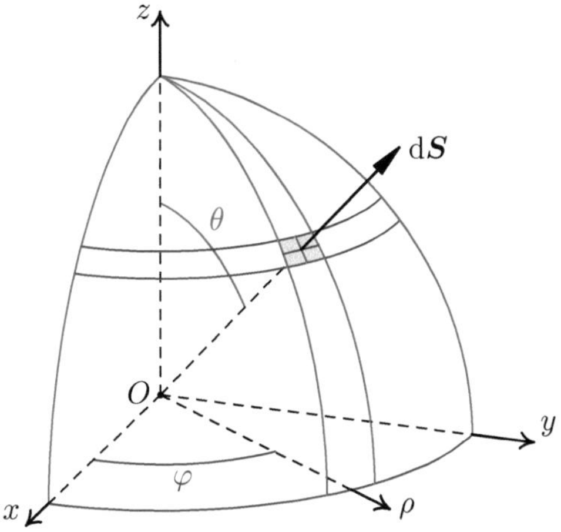

Fig. 13.8 Spherical surface element $\mathrm{d}\boldsymbol{S}(\theta,\varphi) = \mathrm{d}\boldsymbol{r}_\theta \times \mathrm{d}\boldsymbol{r}_\varphi$.

For the example of spherical coordinates (θ,φ) (Fig. 13.8)

$$\mathrm{d}\boldsymbol{S} = \mathrm{d}\boldsymbol{r}_\theta \times \mathrm{d}\boldsymbol{r}_\varphi = \left(\frac{\partial \boldsymbol{r}}{\partial \theta} \times \frac{\partial \boldsymbol{r}}{\partial \varphi}\right) \mathrm{d}\theta\mathrm{d}\varphi.$$

The coordinate tangents have been calculated in (13.11), and we could use those to calculate the cross product. Since $\boldsymbol{r} = r\hat{\boldsymbol{r}}$, and the θ and φ derivatives of $\hat{\boldsymbol{r}}$ have been studied as unit-rate rotations in (10.24a), we could use those results as well. But these spherical tangents come up often, so it is useful to rederive them from the following simple geometric reasoning, illustrated in Fig. 13.9.

The line element in the θ direction is

$\hat{\boldsymbol{\varphi}} = (\hat{\mathbf{z}} \times \hat{\boldsymbol{r}})/\sin\theta, \quad \hat{\boldsymbol{\theta}} = \hat{\boldsymbol{\varphi}} \times \hat{\boldsymbol{r}}.$

$$\mathrm{d}\boldsymbol{r}_\theta = \boldsymbol{r}(r,\theta+\mathrm{d}\theta,\varphi) - \boldsymbol{r}(r,\theta,\varphi) = r\mathrm{d}\theta\,\hat{\boldsymbol{\theta}},$$

since it is in the direction $\hat{\boldsymbol{\theta}}$, and its magnitude is the arc of angle $\mathrm{d}\theta$ and radius r. Likewise,

$$\mathrm{d}\boldsymbol{r}_\varphi = \boldsymbol{r}(r,\theta,\varphi+\mathrm{d}\varphi) - \boldsymbol{r}(r,\theta,\varphi) = \rho\mathrm{d}\varphi\,\hat{\boldsymbol{\varphi}},$$

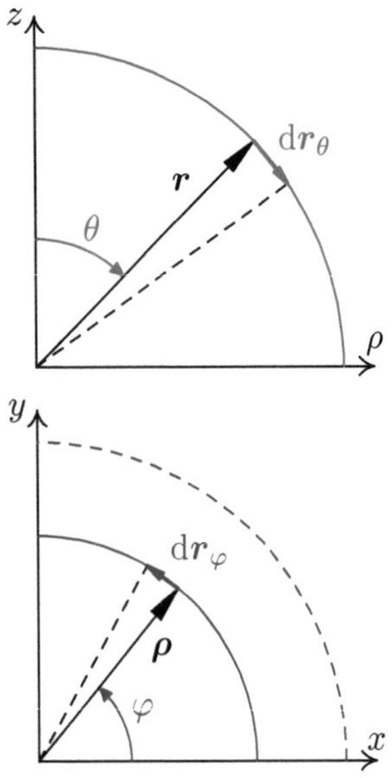

Fig. 13.9 Line elements $\mathrm{d}\boldsymbol{r}_\theta$, $\mathrm{d}\boldsymbol{r}_\varphi$.

since it is in the direction $\hat{\boldsymbol{\varphi}}$, and its magnitude is the arc of angle $\mathrm{d}\varphi$ but radius $\rho = r\sin\theta$ now. Thus,

$$\frac{\partial \boldsymbol{r}}{\partial \theta} = r\,\hat{\boldsymbol{\theta}}, \qquad \frac{\partial \boldsymbol{r}}{\partial \varphi} = r\sin\theta\,\hat{\boldsymbol{\varphi}}. \tag{13.16}$$

These derivatives are orthogonal, and $\hat{\boldsymbol{\theta}} \times \hat{\boldsymbol{\varphi}} = \hat{\boldsymbol{r}}$, so the surface element for a sphere of radius r is

$$\mathrm{d}\boldsymbol{S} = \frac{\partial \boldsymbol{r}}{\partial \theta} \times \frac{\partial \boldsymbol{r}}{\partial \varphi}\,\mathrm{d}\theta\mathrm{d}\varphi = \hat{\boldsymbol{r}}\,r^2\sin\theta\,\mathrm{d}\theta\mathrm{d}\varphi, \tag{13.17}$$

where $\hat{\boldsymbol{r}} = \boldsymbol{r}/r$, $\hat{\boldsymbol{\varphi}} = \hat{\mathbf{z}} \times \hat{\boldsymbol{r}}/\sin\theta$, and $\hat{\boldsymbol{\theta}} = \hat{\boldsymbol{\varphi}} \times \hat{\boldsymbol{r}}$ (Fig. 10.3).

Equal area map

The Mercator conformal projection was a breakthrough for useful navigational maps. Mercator achieved that by appropriate *stretching* of the polar angle coordinate $\theta \to v = \ln\tan\theta/2$, but that distorts areas even more than spherical coordinates (Fig. 13.5).

To preserve proportional areas, we need a map $\boldsymbol{r}(u,v)$ for which the coordinate area amplification factor, $|\partial\boldsymbol{r}/\partial u \times \partial\boldsymbol{r}/\partial v|$, is constant across the map. Since spherical coordinates (θ,φ) are orthogonal

$$\left|\frac{\partial \boldsymbol{r}}{\partial \theta} \times \frac{\partial \boldsymbol{r}}{\partial \varphi}\right| = \left|\frac{\partial \boldsymbol{r}}{\partial \theta}\right| \left|\frac{\partial \boldsymbol{r}}{\partial \varphi}\right| = \left|\frac{\partial \boldsymbol{r}}{\partial \theta}\right| r\sin\theta,$$

and it is simple then to now *squeeze* the polar angle coordinate to preserve areas. We need a change of variable $\theta \to v(\theta)$ such that

$$\left|\frac{\partial \boldsymbol{r}}{\partial v}\right| = \left|\frac{\partial \boldsymbol{r}}{\partial \theta}\right| \left|\frac{\mathrm{d}\theta}{\mathrm{d}v}\right| = \frac{r}{\sin\theta},$$

that is, $|\mathrm{d}v/\mathrm{d}\theta| = \sin\theta$ and thus $v = \cos\theta$. Projection of the world on a $(\varphi, \cos\theta)$ plane is shown in Fig. 13.10. Such a map was introduced by Johann Lambert in 1772. For navigation, this distorts angles even more than spherical coordinates. We can preserve angles, or areas, but not both. In the case of spherical coordinates, these are literally opposite requirements: preserving angles requires stretching by a $1/\sin\theta$ factor, whereas preserving areas requires squeezing by a $\sin\theta$ factor.

13.4 Surface integrals

A standard surface integral is

$$\int_S \boldsymbol{V} \cdot \mathrm{d}\boldsymbol{S}, \tag{13.18}$$

which represents the *flux* of the vector field $\boldsymbol{V}$ through the surface $\mathcal{S}$. If $\boldsymbol{V}(\boldsymbol{r})$ is the velocity of a fluid, water or air, at point $\boldsymbol{r}$, then (13.18) is the net amount of fluid volume flowing through that surface $\mathcal{S}$ per unit

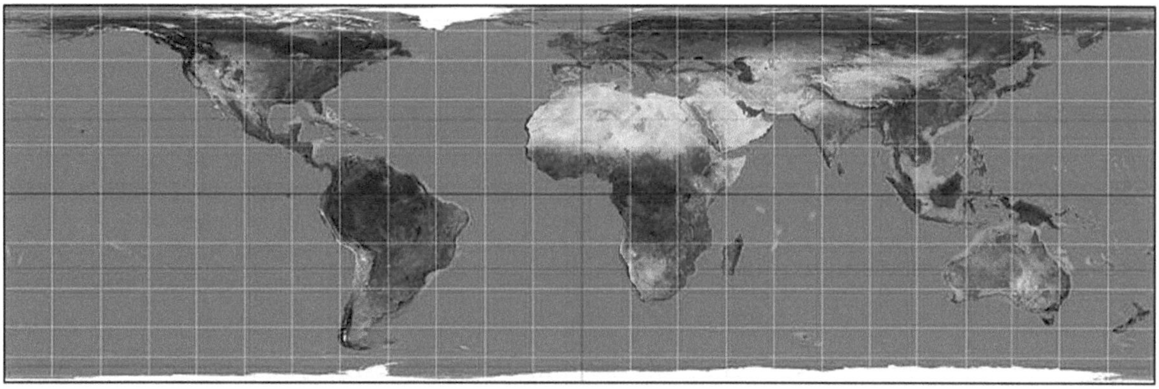

Fig. 13.10 Lambert cylindrical equal area projection of the world. Reproduced from Strebe (2011) Wikimedia Commons under a Creative Commons Attribution ShareAlike 3.0 (CC BY-SA 3.0) `https://commons.wikimedia.org/wiki/File:Lambert_cylindrical_equal-area_projection_SW.jpg`

time. Indeed, $\delta\boldsymbol{\ell} = \boldsymbol{V}\,\delta t$ is the displacement of a fluid particle in time δt and

$$\delta\boldsymbol{\ell} \cdot \mathrm{d}\boldsymbol{S} = (\delta\boldsymbol{\ell} \cdot \hat{\boldsymbol{n}})\,\mathrm{d}A$$

is the volume of fluid passing through the surface element $\mathrm{d}\boldsymbol{S} = \hat{\boldsymbol{n}}\mathrm{d}A$ in time δt. We used δt instead of $\mathrm{d}t$ to help keep our differentials straight; $\mathrm{d}\boldsymbol{S}$ is a surface patch whose magnitude and direction have nothing to do with δt (Fig. 13.11).

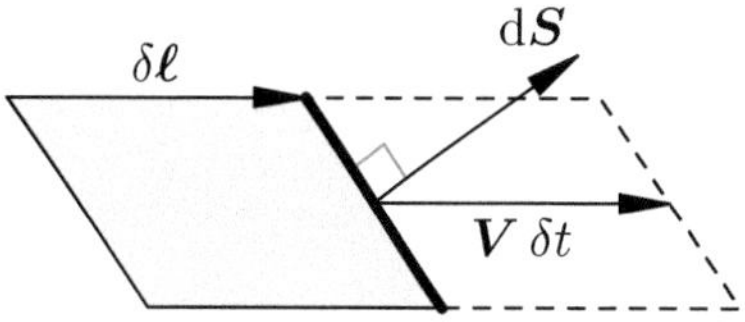

Fig. 13.11 Flux through $\mathrm{d}\boldsymbol{S}$ in δt.

We can make sense of that flux integral as the usual *limit of a sum*,

$$\int_{\mathcal{S}} \boldsymbol{V} \cdot \mathrm{d}\boldsymbol{S} = \lim_{\Delta S_n \to 0} \sum_{n=1}^{N} \boldsymbol{V}_n \cdot \Delta\boldsymbol{S}_n, \tag{13.19}$$

where the sum is over, say, a partition of the surface (such as the skull surface, Fig. 13.6) into N triangles and $\boldsymbol{V}_n$ is, for instance, $\boldsymbol{V}(G_n)$ at the center of area G_n of triangle n whose area vector is $\Delta\boldsymbol{S}_n$. Another good choice is to define $\boldsymbol{V}_n$ as the average of $\boldsymbol{V}$ at the vertices of triangle n. The limit $\Delta S_n \to 0$ means that the surface partition is uniformly refined with $N \to \infty$. That surface mesh refinement limit is intuitive; however, there are pitfalls illustrated by the "pleated venetian lantern" example of H. A. Schwarz and G. Peano (exercise 13.27, Fig. 13.12).

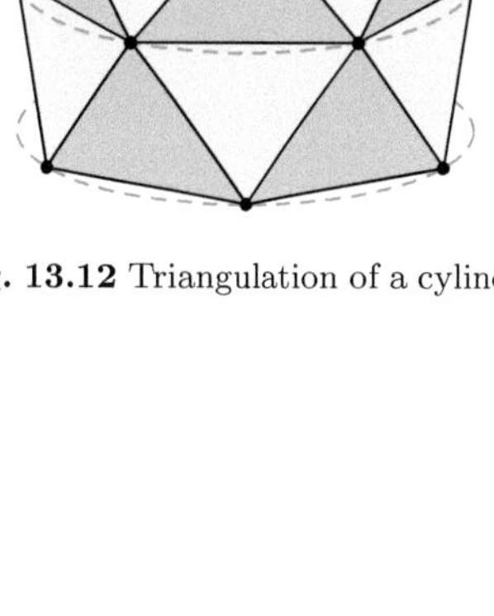

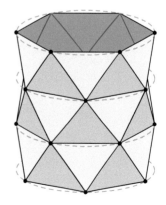

Fig. 13.12 Triangulation of a cylinder.

If a parametric representation $\boldsymbol{r}(u, v)$ is known for the surface then we can partition the surface into quadrilaterals, and the single sum over all quadrilaterals becomes a double integral over u and v in the limit,

$$\int_{\mathcal{S}} \boldsymbol{V} \cdot \mathrm{d}\boldsymbol{S} = \iint_{\mathcal{U}} \boldsymbol{V} \cdot \left(\frac{\partial \boldsymbol{r}}{\partial u} \times \frac{\partial \boldsymbol{r}}{\partial v}\right) \mathrm{d}u\,\mathrm{d}v, \tag{13.20}$$

where $\mathcal{U}$ is the (u, v) domain that corresponds to the surface $\mathcal{S}$. We write the sum over all area patches $\mathrm{d}\boldsymbol{S}$, formed from triangles or quadrilaterals, as a *single sum over all patches* because it is a single sum. The corresponding sums over the surface coordinates u and v are a double sum, a sum over all u's for fixed v, then a sum of the u-sums over all v's, or vice versa.

Other examples of surface integrals that show up in applications are

$$\int_{\mathcal{S}} p\,\mathrm{d}\boldsymbol{S}, \quad \int_{\mathcal{S}} \boldsymbol{V} \times \mathrm{d}\boldsymbol{S}, \quad \int_{\mathcal{S}} f\,|\mathrm{d}\boldsymbol{S}|, \;\ldots.$$

The first one represents the net pressure force on $\mathcal{S}$ if $p(\boldsymbol{r})$ is the pressure at $\boldsymbol{r}$. The second one arises in computing torques from pressure forces, with $\boldsymbol{V} = p\boldsymbol{r}$ for example, and the last one occurs with $f = 1$ when computing the total area of surface $\mathcal{S}$. For such integrals, involving only the area element $\mathrm{d}A = |\mathrm{d}\boldsymbol{S}|$, computing the cross product of the tangent vectors is not necessary; dot products suffice since

$$\left|\frac{\partial \boldsymbol{r}}{\partial u} \times \frac{\partial \boldsymbol{r}}{\partial v}\right|^2 = \left(\frac{\partial \boldsymbol{r}}{\partial u} \cdot \frac{\partial \boldsymbol{r}}{\partial u}\right)\left(\frac{\partial \boldsymbol{r}}{\partial v} \cdot \frac{\partial \boldsymbol{r}}{\partial v}\right) - \left(\frac{\partial \boldsymbol{r}}{\partial u} \cdot \frac{\partial \boldsymbol{r}}{\partial v}\right)^2, \tag{13.21}$$

from the vector identity $(\boldsymbol{a} \times \boldsymbol{b}) \cdot (\boldsymbol{c} \times \boldsymbol{d}) = (\boldsymbol{a} \cdot \boldsymbol{c})(\boldsymbol{b} \cdot \boldsymbol{d}) - (\boldsymbol{a} \cdot \boldsymbol{d})(\boldsymbol{b} \cdot \boldsymbol{c})$.

Example 1 For the triangle T with vertices and orientation A, B, C, we can directly evaluate the surface integral

$$\int_T \mathrm{d}\boldsymbol{S} = \frac{1}{2}\overrightarrow{AB} \times \overrightarrow{BC},$$

since it is the vector sum of all surface elements $\mathrm{d}\boldsymbol{S} = \hat{\boldsymbol{n}}\,\mathrm{d}A$ but $\hat{\boldsymbol{n}}$ is the same at all points of the triangle, so that is the total triangle area times the unit normal to the triangle. We can also use the parametrization $\boldsymbol{r}(u, v) = \boldsymbol{r}_A + u\,\overrightarrow{AB} + v\,\overrightarrow{BC}$, for the sake of example, and

$$\int_T \mathrm{d}\boldsymbol{S} = \int_0^1 \mathrm{d}u \int_0^u \frac{\partial \boldsymbol{r}}{\partial u} \times \frac{\partial \boldsymbol{r}}{\partial v}\,\mathrm{d}v = \overrightarrow{AB} \times \overrightarrow{BC} \int_0^1 \mathrm{d}u \int_0^u \mathrm{d}v = \frac{1}{2}\overrightarrow{AB} \times \overrightarrow{BC}.$$

Example 2 The surface integral $\int_T \mathrm{d}A$ is the sum of the magnitudes of the surface elements, not the *vector sum* as in the previous example. Therefore, it is the total triangle area. It can also be computed as

$$\int_T |\mathrm{d}\boldsymbol{S}| = \int_0^1 \mathrm{d}u \int_0^u \left|\frac{\partial \boldsymbol{r}}{\partial u} \times \frac{\partial \boldsymbol{r}}{\partial v}\right| \mathrm{d}v = \frac{1}{2}\left|\overrightarrow{AB} \times \overrightarrow{BC}\right|.$$

Example 3 The flux of the position vector $\boldsymbol{r}$ through the triangle A, B, C

$$\int_T \boldsymbol{r} \cdot \mathrm{d}\boldsymbol{S} = \frac{1}{2}(\boldsymbol{r}_A \cdot \hat{\boldsymbol{n}}) \int_T \mathrm{d}A = \frac{1}{2}(\boldsymbol{r}_A \cdot \hat{\boldsymbol{n}}) \left|\overrightarrow{AB} \times \overrightarrow{BC}\right|,$$

since $\mathrm{d}\boldsymbol{S} = \hat{\boldsymbol{n}}\,\mathrm{d}A$ but $\hat{\boldsymbol{n}}$ is the same at all points of the triangle, and $\boldsymbol{r} \cdot \hat{\boldsymbol{n}} = \boldsymbol{r}_A \cdot \hat{\boldsymbol{n}}$ for all points in that plane. Thus, it can be pulled out of the sum, leaving the sum as the total area times $(\boldsymbol{r}_A \cdot \hat{\boldsymbol{n}})$, where $(\boldsymbol{r}_A \cdot \hat{\boldsymbol{n}})$ is, in fact, the distance from the origin to the plane of the triangle.

That integral is therefore three times the *signed* volume of the pyramid with O as its vertex and base A, B, C. The volume of a pyramid is base times height/3, the triangle area is 1/2 of the parallelogram area, and $\boldsymbol{r}_A \cdot \hat{\boldsymbol{n}} = h$ is the height, with a sign that depends on which side of A, B, C is the origin O.

Since a surface can be partitioned into triangles, we conclude that the flux of the radius vector $\boldsymbol{r}$ through a surface is three times the *signed* volume swept by the radius vector (Fig. 13.13)

$$\int_{\mathcal{S}} \boldsymbol{r} \cdot \mathrm{d}\boldsymbol{S} = \iint_{\mathcal{U}} \boldsymbol{r} \cdot \left(\frac{\partial \boldsymbol{r}}{\partial u} \times \frac{\partial \boldsymbol{r}}{\partial v} \right) \mathrm{d}u \, \mathrm{d}v = \pm 3V, \tag{13.22}$$

similarly to the signed area swept by the radius vector (12.54).

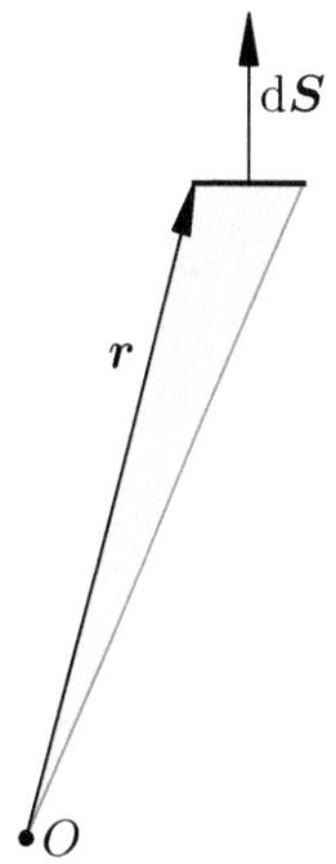

Fig. 13.13 Volume swept by radius vector $\mathrm{d}V = \frac{1}{3}\boldsymbol{r} \cdot \mathrm{d}\boldsymbol{S}$.

Example 4 Using spherical coordinates and the surface element (13.17), the total surface area A of a sphere $\mathcal{S}$ of radius R is

$$A = \int_{\mathcal{S}} |\mathrm{d}\boldsymbol{S}| = \int_0^{2\pi} \int_0^{\pi} R^2 \sin\theta \, \mathrm{d}\theta \, \mathrm{d}\varphi = 4\pi R^2, \tag{13.23}$$

which is also $3/R$ times the volume swept by the radius vector in this special case since $|\mathrm{d}\boldsymbol{S}| = \hat{\boldsymbol{n}} \cdot \mathrm{d}\boldsymbol{S} = \hat{\boldsymbol{r}} \cdot \mathrm{d}\boldsymbol{S} = R^{-1}\boldsymbol{r} \cdot \mathrm{d}\boldsymbol{S}$. Thus, $V = \frac{4}{3}\pi R^3$.

Area of spherical triangle

A more general and beautiful result can be obtained for a geodesic triangle on a sphere (Fig. 6.3). We begin with a proper geodesic triangle where the arcs are the shortest arclengths between the vertices, and all angles are between 0 and π. Picking spherical coordinates (θ, φ) with OA as the polar axis, AB as the prime meridian, and ABC as right-handed, the area of the triangle on a unit sphere is

$$A = \int_0^{\alpha} \int_0^{\theta(\varphi)} \sin\theta \, \mathrm{d}\theta \, \mathrm{d}\varphi = \int_0^{\alpha} (1 - \cos\theta(\varphi)) \, \mathrm{d}\varphi, \tag{13.24}$$

where $\theta(\varphi)$ is the solution of the geodesic equation (7.12) that can be rewritten

$$c_* \cos\theta = -s_* \cos(\varphi - \varphi_*) \sin\theta, \tag{13.25}$$

where $s_* = \sin\theta_*$, $c_* = \cos\theta_*$, and $\sin\theta$ are all positive. We can solve for $\sin\theta$ by squaring both sides and using Pythagoras, $\cos^2\theta = 1 - \sin^2\theta$. Substituting in (13.25) yields

$$\cos\theta = \frac{-s_* \cos(\varphi - \varphi_*)}{\sqrt{1 - s_*^2 \sin^2(\varphi - \varphi_*)}}. \tag{13.26}$$

Using this expression in (13.24) with the substitution $u = s_* \sin(\varphi - \varphi_*)$ yields

$$A = \alpha + \int_{u_0}^{u_\alpha} \frac{du}{\sqrt{1 - u^2}} = \alpha + \arccos u_0 - \arccos u_\alpha, \tag{13.27}$$

$$\frac{\mathrm{d}(\arccos u)}{\mathrm{d}u} = \frac{-1}{\sqrt{1 - u^2}}$$

where

$$\begin{aligned} u_0 &= -s_* \sin\varphi_* && = \cos\beta, \\ u_\alpha &= s_* \sin(\alpha - \varphi_*) && = \sin\theta_*(\sin\alpha\cos\varphi_* - \cos\alpha\sin\varphi_*) \\ & && = -\sin\alpha\sin\beta\cos c + \cos\alpha\cos\beta = -\cos\gamma, \end{aligned}$$

from (7.12) and (6.26) for $\cos\gamma$. Since $\arccos(-\cos\gamma) = \pi - \gamma$, (13.27) yields

$$A = (\alpha + \beta + \gamma - \pi)R^2, \tag{13.28}$$

for a sphere[4] of radius R. The quantity $\alpha + \beta + \gamma - \pi$ is called the *spherical excess*. It vanishes for a flat triangle in the limit $R \to \infty$, with arclength a, b, c fixed. Although we considered a proper geodesic triangle in this derivation, the reader can verify that formula (13.28) applies to other geodesic triangles such as the outer triangle consisting of the full sphere minus the small triangle, or the triangles where one of the arcs is the long arc around the sphere, as long as the angles α, β, γ are the *inner* angles of the triangle.

[4] For a geometric proof of this classic result, see exercise 13.28.

13.5 Green's theorem

Surface integrals such as the flux integral (13.20) thus reduce to double integrals of a function $f(u, v)$ over a domain $\mathcal{U}$ in the coordinate space $(u, v) \in \mathbb{R}^2$. For example, for a surface $\mathcal{S} : (u, v) \in \mathcal{U} \to \boldsymbol{r}(u, v) \in \mathbb{E}^3$, and a vector field $\boldsymbol{V}(\boldsymbol{r}) \in \mathbb{E}^3$,

$$\int_{\mathcal{S}} \boldsymbol{V} \cdot \mathrm{d}\boldsymbol{S} = \iint_{\mathcal{U}} f(u, v)\, \mathrm{d}u\, \mathrm{d}v,$$

where

$$f(u, v) = \boldsymbol{V}(\boldsymbol{r}(u, v)) \cdot \left(\frac{\partial \boldsymbol{r}}{\partial u} \times \frac{\partial \boldsymbol{r}}{\partial v}\right).$$

Those double integrals can be computed as iterated integrals[5]

$$\iint_{\mathcal{U}} f(u, v)\, \mathrm{d}u\, \mathrm{d}v = \int_{v_B}^{v_T} \left(\int_{u_l(v)}^{u_r(v)} f(u, v)\, \mathrm{d}u\right) \mathrm{d}v, \tag{13.29a}$$

$$= \int_{u_L}^{u_R} \left(\int_{v_b(u)}^{v_t(u)} f(u, v)\, \mathrm{d}v\right) \mathrm{d}u, \tag{13.29b}$$

where the bounds depend on the order of integration, as illustrated in Fig. 13.14.

[5] if $f(u, v)$ is continuous on the closure of $\mathcal{U}$, or more generally as long as $\iint_{\mathcal{U}} |f(u, v)| \mathrm{d}u\mathrm{d}v$ exists (Fubini's theorem).

Performing the u-integral first, we treat v as a constant parameter and find an antiderivative of $f(u, v)$ with v-fixed, that is, a function $G(u, v)$ such that

$$\frac{\partial G}{\partial u} = f(u, v).$$

Then the *fundamental theorem of calculus* yields

$$\int_{v_B}^{v_T} \left(\int_{u_l(v)}^{u_r(v)} \frac{\partial G}{\partial u}\, \mathrm{d}u\right) \mathrm{d}v$$

$$= \int_{v_B}^{v_T} \left[G\left(u_r(v), v\right)\right) - G\left(u_\ell(v)\right), v)\right] \mathrm{d}v. \tag{13.30a}$$

The double integral has been reduced to a single integral of a function

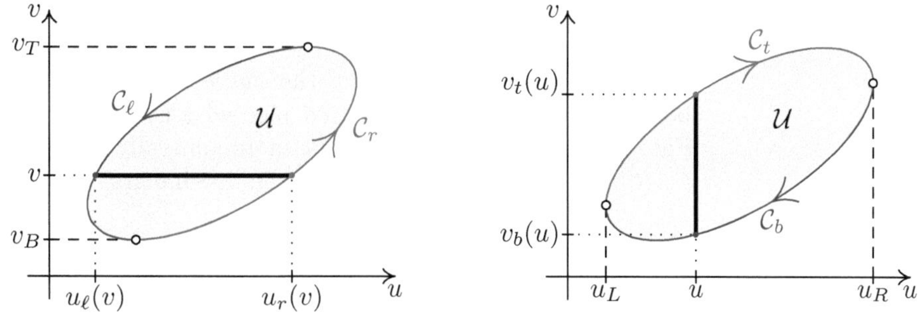

Fig. 13.14 Bounds for iterated integrals (13.29a) (left) and (13.29b) (right).

of v. We might be content with that result, but we can go further by noticing that the final v-integral is, in fact, a line integral of $G(u, v)$ over the boundary curve $\partial\mathcal{U}$ of region $\mathcal{U}$, oriented *counterclockwise*. Indeed, with reference to the curves illustrated in Fig. 13.14,

$$\begin{aligned} \int_{v_B}^{v_T} G\left(u_r(v), v)\right)\mathrm{d}v &= \int_{\mathcal{C}_r} G(u,v)\mathrm{d}v, \\ -\int_{v_B}^{v_T} G\left(u_\ell(v), v)\right)\mathrm{d}v &= \int_{\mathcal{C}_\ell} G(u,v)\mathrm{d}v, \end{aligned} \tag{13.30b}$$

where $\mathcal{C}_r : (u, v) = (u_r(v), v)$ with $v = v_B \to v_T$ is the right boundary curve, bottom to top, while $\mathcal{C}_\ell : (u, v) = (u_\ell(v), v)$ with $v = v_T \to v_B$ is the left boundary curve, *top to bottom*. The two line integrals together yield a line integral over the *entire* boundary $\partial\mathcal{U}$, *counterclockwise*,

$$\iint_{\mathcal{U}} \frac{\partial G}{\partial u}\,\mathrm{d}u\,\mathrm{d}v = \int_{\mathcal{C}_r} G(u,v)\mathrm{d}v + \int_{\mathcal{C}_l} G(u,v)\mathrm{d}v = \oint_{\partial\mathcal{U}} G(u,v)\mathrm{d}v. \tag{13.31a}$$

Proceeding similarly for the reverse order of integration, v first, then u, yields

$$\iint_{\mathcal{U}} \frac{\partial F}{\partial v}\mathrm{d}u\mathrm{d}v = \int_{\mathcal{C}_t} F(u,v)\mathrm{d}u + \int_{\mathcal{C}_b} F(u,v)\mathrm{d}u = \oint_{\partial\mathcal{U}} F(u,v)\,\mathrm{d}u, \tag{13.31b}$$

where the line integral around the closed boundary $\partial\mathcal{U}$ is now *clockwise*, the line integral over the top $\mathcal{C}_t$ from left to right plus that over the bottom $\mathcal{C}_b$ from right to left. Since reversing the orientation changes the sign of the integral, $\oint = -\oint$, and (13.31a), (13.31b) apply to *any* sufficiently regular $G(u, v)$ and $F(u, v)$, the two results can be combined into the useful form of *Green's theorem*,

$$\iint_{\mathcal{U}} \left(\frac{\partial G}{\partial u} - \frac{\partial F}{\partial v}\right)\mathrm{d}u\mathrm{d}v = \oint_{\partial\mathcal{U}} F\mathrm{d}u + G\mathrm{d}v. \tag{13.31c}$$

This fundamental theorem holds for any sufficiently regular functions $F(u, v)$, $G(u, v)$ and region $\mathcal{U}$. Sufficient conditions are that $F(u, v)$, $G(u, v)$ be continuously differentiable (C^1) inside and on $\partial\mathcal{U}$, and that $\partial\mathcal{U}$ be piecewise smooth.

It is important to recognize that the variables u and v are not independent in the line integral on the right-hand side of (13.31c); they are connected as specified by the curve boundary $\partial\mathcal{U}$. That curve boundary could be parametrized piecewise by u or v as illustrated in Fig. 13.14 and eqn (13.30b), but more generally by a parametrization $t \in [t_0, t_1] \to (u(t), v(t))$ such that the line integral becomes

$$\oint_{\partial\mathcal{U}} F\mathrm{d}u + G\mathrm{d}v = \int_{t_0}^{t_1} \left(F(u(t), v(t))\frac{\mathrm{d}u}{\mathrm{d}t} + G(u(t), v(t))\frac{\mathrm{d}v}{\mathrm{d}t} \right) \mathrm{d}t. \tag{13.31d}$$

This is particularly useful since regions $\mathcal{U} \in \mathbb{R}^2$ are usually defined by specifying their boundary $\partial\mathcal{U}$.

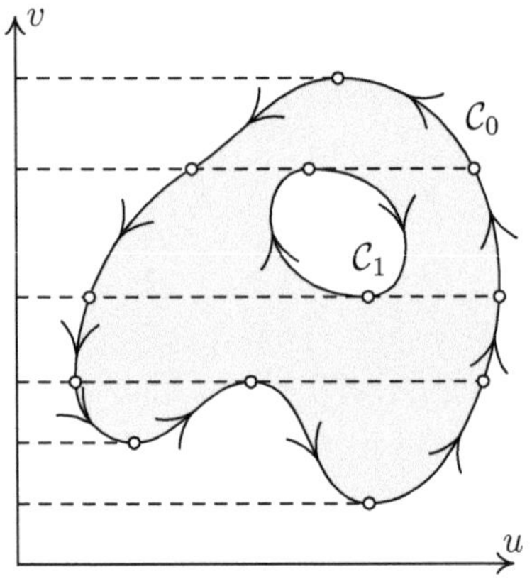

Fig. 13.15 Green's theorem (13.31c) with (13.31e) for a domain whose boundary $\partial\mathcal{U}$ consists of the *counterclockwise* outer boundary curve $\mathcal{C}_0$ and the inner *clockwise* closed curve $\mathcal{C}_1$.

Green's theorem applies to multiconnected domains (domains with holes) and nonconvex boundaries, as can be shown by suitable segmentation of the domain into simple subdomains. This is illustrated in Fig. 13.15 with a segmentation suitable for v-parametrization of the boundary and integration of $\partial G/\partial u$ integrands. A different segmentation would be required for u-parametrization and integration of $\partial F/\partial v$ integrands, as was the case already for the simple domain in Fig. 13.14.

This would be tedious in practice, but it is only needed in theory to show that Green's theorem readily applies to such domains with the outer boundary *counterclockwise* but the inner boundaries oriented *clockwise*; the inside of the domain is to the *left* when traveling along any of the boundaries. Thus, the line integral over the boundary $\partial\mathcal{U}$ in Green's theorem (13.31c) should be interpreted as the sum of the integral over the outer boundary $\mathcal{C}_0$ oriented *counterclockwise* (CCW) and the integrals over however many inner boundaries $\mathcal{C}_1, \ldots, \mathcal{C}_k$ oriented *clockwise.* That is equivalent to orienting all boundaries CCW and *subtracting* the integrals over the inner boundaries

$$\begin{aligned} \oint_{\partial\mathcal{U}} &= \oint_{\mathcal{C}_0} + \oint_{\mathcal{C}_1} + \cdots + \oint_{\mathcal{C}_k} \\ &= \oint_{\mathcal{C}_0} - \oint_{\mathcal{C}_1} - \cdots - \oint_{\mathcal{C}_k}. \end{aligned} \tag{13.31e}$$

Exercises

(13.1) Sketch the (u, v) map of the triangle A, B, C when $\boldsymbol{r} = \boldsymbol{r}_A + u\,\overrightarrow{AB} + v\,\overrightarrow{AC}$.

(13.2) Define spherical coordinates for the 3D sphere $x_1^2 + x_2^2 + x_3^2 + x_4^2 = R^2$ in $\mathbb{R}^4$. What is its area element? What is its total surface area?

(13.3) Derive Newton's binomial series used in (13.4) and valid for $|x| < 1$ if α is not a positive integer,

$$\begin{aligned} (1+x)^\alpha &= 1 + \alpha x + \alpha(\alpha-1)\frac{x^2}{2} + \cdots \\ &= \sum_{n=0}^{\infty} \frac{\alpha\cdots(\alpha-n+1)}{n!}x^n. \end{aligned} \tag{13.32}$$

(13.4) Calculate the coordinate tangent vectors for

the hemisphere $x^2 + y^2 + z^2 = R^2$, $z \geq 0$, parametrized using x and y. Show that these coordinates are not orthogonal except at $(x, y) = (0, 0)$. Show that the tangent vectors become parallel and infinite at the equator.

(13.5) Find $\boldsymbol{r}(x, y)$ for surface specified $z = f(x, y)$ in cartesian coordinates. Compute the tangent vectors and show that they are not orthogonal to each other in general. Determine the class of functions $f(x, y)$ for which (x, y) are orthogonal coordinates on the surface $z = f(x, y)$ and interpret geometrically. Show that the surface and area elements are, with $z_x = \partial z/\partial x$, $z_y = \partial z/\partial y$,

$$\begin{aligned} \mathrm{d}\boldsymbol{S} &= (\hat{\mathbf{z}} - \hat{\mathbf{x}}\, z_x - \hat{\mathbf{y}}\, z_y)\, \mathrm{d}x\mathrm{d}y, \\ \mathrm{d}A &= \sqrt{1 + z_x^2 + z_y^2}\, \mathrm{d}x\mathrm{d}y. \end{aligned} \tag{13.33}$$

(13.6) For a surface $\rho = f(\varphi, z)$ in cylindrical coordinates (ρ, φ, z), show that the surface element $\mathrm{d}\boldsymbol{S} = \hat{\boldsymbol{n}}\mathrm{d}A$, with $\partial_\varphi \rho = \partial \rho/\partial \varphi$ and $\partial_z \rho = \partial \rho/\partial z$, is

$$\begin{aligned} \mathrm{d}\boldsymbol{S} &= (\rho\hat{\boldsymbol{\rho}} - \hat{\boldsymbol{\varphi}}\, \partial_\varphi \rho - \hat{\mathbf{z}}\, \partial_z \rho)\, \mathrm{d}\varphi\mathrm{d}z, \\ \mathrm{d}A &= \sqrt{\rho^2 + (\partial_\varphi \rho)^2 + (\partial_z \rho)^2}\, \mathrm{d}\varphi\mathrm{d}z. \end{aligned} \tag{13.34}$$

(13.7) Verify that Mercator coordinates (13.14) are conformal. Find the surface element $\mathrm{d}\boldsymbol{S}$ in terms of those coordinates.

(13.8) Show the steps to solve the differential equation $\mathrm{d}\theta/\mathrm{d}v = \sin\theta$. Use the classic tangent half-angle substitution $t = \tan\theta/2$, and the double-angle identities.

(13.9) *Stereographic coordinates* map a point (x, y, z) on the sphere $x^2 + y^2 + z^2 = R^2$ to the point $(u, v, 0)$ that is the intersection of the $z = 0$ plane and the line passing through the south pole $(0, 0, -R)$ and the point (x, y, z) (Fig. 13.16). Find (u, v) as functions of (x, y, z), then (x, y, z) as functions of (u, v) to derive

$$\boldsymbol{r}(u, v) = R\left(\frac{2uR}{R^2 + u^2 + v^2}\, \hat{\mathbf{x}} + \frac{2vR}{R^2 + u^2 + v^2}\, \hat{\mathbf{y}} + \frac{R^2 - u^2 - v^2}{R^2 + u^2 + v^2}\, \hat{\mathbf{z}}\right). \tag{13.35}$$

Show that these coordinates are *conformal*.

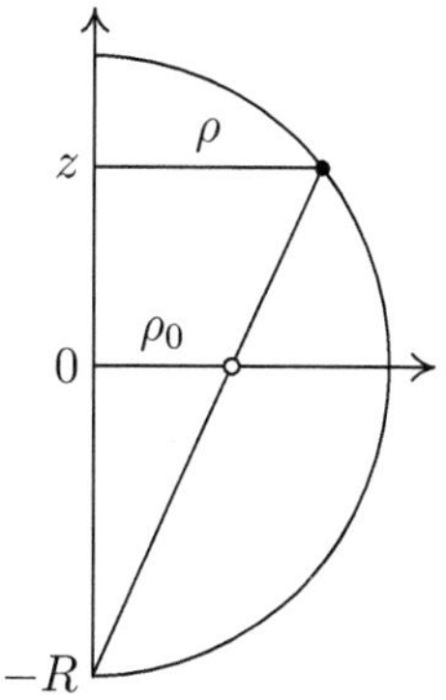

Fig. 13.16 Stereographic coordinates.

(13.10) *Cone.* Show that the cone $\hat{\mathbf{z}} \cdot \boldsymbol{r} = |\boldsymbol{r}| \cos\theta_0$ with apex at the origin and cone angle θ_0, can be parametrized as

$$\boldsymbol{r}(u, v) = v \tan\theta_0\, (\hat{\mathbf{x}} \cos u + \hat{\mathbf{y}} \sin u) + v\, \hat{\mathbf{z}},$$

where $\{\hat{\mathbf{x}}, \hat{\mathbf{y}}, \hat{\mathbf{z}}\}$ is a standard cartesian basis. Are these orthogonal coordinates? Conformal? Find the surface element $\mathrm{d}\boldsymbol{S}(u, v)$. Show that $x^2 + y^2 = z^2 \tan^2\theta_0$.

(13.11) *Developing the cone.* In spherical coordinates (r, θ, φ), the cone is simply $\theta = \theta_0$. Slice the cone along $\varphi = \pi$; then it can be *developed*, that is, it can be mapped isometrically to the cartesian plane $(u, v) = (r\cos\varphi', r\sin\varphi')$ with $r\varphi' = \rho\varphi$, to preserve length along parallels (r, θ_0 fixed). Length along the meridians (θ_0, φ fixed) is r and is preserved. Show that

$$u = r\cos(\varphi \sin\theta_0), \quad v = r\sin(\varphi\sin\theta_0). \tag{13.36}$$

Find $\partial\boldsymbol{r}/\partial u$, $\partial\boldsymbol{r}/\partial v$, and show that these vectors are orthogonal and of equal and constant magnitudes. Find the surface element $\mathrm{d}\boldsymbol{S}(u, v)$.

(13.12) *Elliptic paraboloïd.* Show that the surface specified in cartesian coordinates as $z = x^2/a^2 + y^2/b^2$ can be parametrized as

$$\boldsymbol{r}(u, v) = \hat{\mathbf{x}}\, av\cos u + \hat{\mathbf{y}}\, bv\sin u + \hat{\mathbf{z}}\, v^2. \tag{13.37}$$

Describe the u and v coordinate curves. Are the coordinates orthogonal? Calculate the tangent basis $\partial\boldsymbol{r}/\partial u$, $\partial\boldsymbol{r}/\partial v$ and the surface element $\mathrm{d}\boldsymbol{S}$.

(13.13) *Elliptic hyperboloïd.* Show that the surface specified as $z^2/c^2 = 1 + x^2/a^2 + y^2/b^2$, in cartesian coordinates, can be parametrized as

$$\boldsymbol{r}(u, v) = \hat{\mathbf{x}}\, a\cos u\sinh v + \hat{\mathbf{y}}\, b\sin u\sinh v + \hat{\mathbf{z}}\, c\cosh v. \tag{13.38}$$

Describe the u and v coordinate curves. Are the coordinates orthogonal? Calculate the tangent basis $\partial \boldsymbol{r}/\partial u$, $\partial \boldsymbol{r}/\partial v$ and the surface element $\mathrm{d}\boldsymbol{S}$.

(13.14) *Ellipsoid.* Consider the surface parametrized as

$$\boldsymbol{r}(u,v) = \hat{\mathbf{x}}\, a \cos u \cos v + \hat{\mathbf{y}}\, b \sin u \cos v + \hat{\mathbf{z}}\, c \sin v, \quad (13.39)$$

where a, b, and c are real positive constants. Describe the coordinate curves. Are u and v orthogonal coordinates for that surface? Calculate the tangent basis $\partial \boldsymbol{r}/\partial u$, $\partial \boldsymbol{r}/\partial v$ and the surface element $\mathrm{d}\boldsymbol{S}$. Consider cartesian coordinates (x, y, z) such that $\boldsymbol{r} = x\hat{\mathbf{x}}+y\hat{\mathbf{y}}+z\hat{\mathbf{z}}$ and derive the implicit equation $f(x, y, z) = 0$ for this surface.

(13.15) *Helicoid.* Visualize the surface parametrized by

$$\boldsymbol{r}(u,v) = \hat{\mathbf{x}}\, u \cos v + \hat{\mathbf{y}}\, u \sin v + \hat{\mathbf{z}}\, v. \quad (13.40)$$

Show that the u coordinate curves are straight lines and the v coordinate curves are helices. Find the tangent basis $\partial_u \boldsymbol{r}, \partial_v \boldsymbol{r}$, and the surface element $\mathrm{d}\boldsymbol{S}$.

(13.16) The Lambert equal area parametrization for a sphere of radius R is

$$\boldsymbol{r}(\varphi, v) = R\left(\sqrt{1-v^2}\,\hat{\boldsymbol{\rho}}(\varphi) + v\,\hat{\mathbf{z}}\right)$$

with $\hat{\boldsymbol{\rho}}(\varphi) = \hat{\mathbf{x}} \cos\varphi + \hat{\mathbf{y}} \sin\varphi$. Show that the area element $\mathrm{d}A = R^2 \mathrm{d}\varphi \mathrm{d}v$.

(13.17) *Torus.* Consider a circle of radius a perpendicular to and centered on another circle of radius $R > a$. A torus is generated by rotating the circle of radius a about that of radius R (Fig. 13.17). Show that this yields the parametrization

$$\boldsymbol{r}(u,v) = (R + a\cos v)\cos u\,\hat{\mathbf{x}} + (R + a\cos v)\sin u\,\hat{\mathbf{y}} + a \sin v\,\hat{\mathbf{z}}. \quad (13.41)$$

What are the ranges of u and v needed to cover the entire torus uniquely? Are these orthogonal coordinates? Justify. Calculate the surface element $\mathrm{d}\boldsymbol{S}(u, v)$ and the total surface area of the torus. Show that the implicit equation of that torus is

$$4R^2(a^2 - z^2) = \left(x^2 + y^2 + z^2 - R^2 - a^2\right)^2.$$

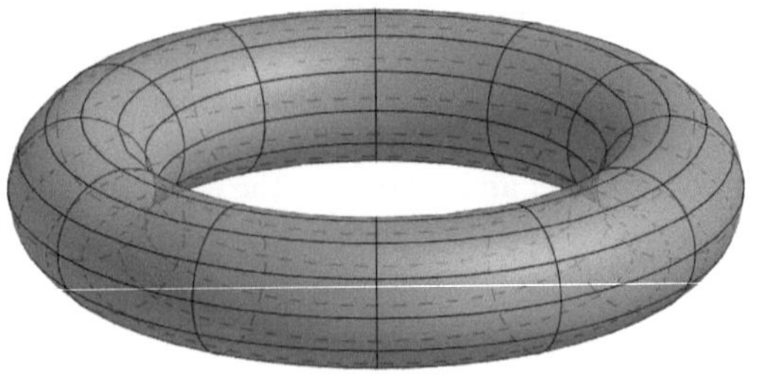

Fig. 13.17 Torus.

(13.18) Describe the surface parametrized by

$$\boldsymbol{r}(u,v) = R\cos 2u\,\hat{\mathbf{x}} + R\sin 2u\,\hat{\mathbf{y}} + v\,\hat{\mathbf{z}}$$

with $0 \le u < \pi$, $-h \le v \le h$, and R and h are positive constants. What is the surface element and what is the total surface area? Show that $\partial \boldsymbol{r}/\partial u$, $\partial \boldsymbol{r}/\partial v$ are continuous and unique at all points on the surface, in particular $\lim_{u\to\pi} \hat{\boldsymbol{n}}(u,v) = \hat{\boldsymbol{n}}(0,v)$.

(13.19) The *Möbius strip* (Fig. 13.18) is parametrized as

$$\boldsymbol{r}(u,v) = (R + v\sin u)\,\cos 2u\,\hat{\mathbf{x}} + (R + v\sin u)\,\sin 2u\,\hat{\mathbf{y}} + v\cos u\,\hat{\mathbf{z}} \quad (13.42)$$

with $u_0 \le u < u_0 + \pi$, $-h \le v \le h$, for any real u_0. Show that

$$\lim_{u\to u_0+\pi} \boldsymbol{r}(u,0) = \boldsymbol{r}(u_0,0)$$

but the normal $\boldsymbol{N} = \partial_u \boldsymbol{r} \times \partial_v \boldsymbol{r}$ flips

$$\lim_{u\to u_0+\pi} \boldsymbol{N}(u,0) = -\boldsymbol{N}(u_0,0).$$

Thus, moving continuously across the surface along $v = 0$, we come back to the same point but with *opposite* normal $\boldsymbol{N}$ to the surface. This is the classic example of a surface that is not orientable. A unique and continuous normal cannot be defined at every point. Up is also down. This is true for every point by using the trajectory $u = u_0 + t$, $v = v_0 \cos t$, which returns to the $t = 0$ starting point $\boldsymbol{r}(u_0, v_0)$ when $t = \pi$.

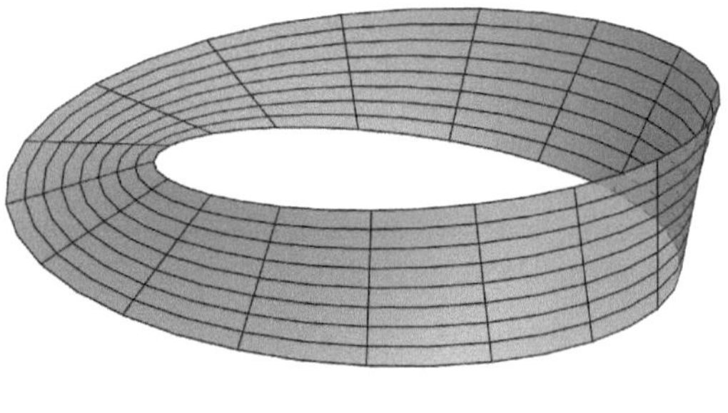

Fig. 13.18 Möbius strip.

(13.20) Compute the percentage of surface area that lies north of the *arctic circle* on the Earth. Show your work; don't just google the answer! The arctic circle is at $\theta = 23.4°$. What is the physical origin of that angle?

(13.21) Characterize $\mathrm{d}\boldsymbol{S}$ and $\boldsymbol{r} \cdot \mathrm{d}\boldsymbol{S}$ geometrically, then use those insights to evaluate $\int_{\mathcal{S}} \boldsymbol{r} \cdot \mathrm{d}\boldsymbol{S}$ efficiently when $\mathcal{S}$ is

(i) the square $0 \le x \le a$, $0 \le y \le a$ at $z = b$,

(ii) the tetrahedron with vertices A, B, C, D in 3D space,

(iii) the sphere of radius R centered at the origin O, and

(iv) the sphere of radius R centered at $\boldsymbol{r} = \boldsymbol{r}_c$.

(13.22) $\int_{\mathcal{S}} (\boldsymbol{r}/r^3) \cdot \mathrm{d}\boldsymbol{S}$ is a fundamental integral—the flux of the inverse square law field through a surface $\mathcal{S}$. Identify $\mathrm{d}\boldsymbol{S}$ and $\boldsymbol{r} \cdot \mathrm{d}\boldsymbol{S}$ geometrically, then use those insights to calculate the integral when $\mathcal{S}$ is the sphere of radius R centered at the origin. How does the result depend on R?

(13.23) Calculate

$$\oint_{\mathcal{S}} \frac{\boldsymbol{r} - \boldsymbol{r}_1}{|\boldsymbol{r} - \boldsymbol{r}_1|^3} \cdot \mathrm{d}\boldsymbol{S}, \tag{13.43}$$

where $\mathcal{S} : |\boldsymbol{r}| = R$ is the sphere of radius R centered at the origin and $\boldsymbol{r}_1$ is a constant vector. Use spherical coordinates with polar axis $\hat{\mathbf{z}}$ in the $\boldsymbol{r}_1$ direction, so that $\boldsymbol{r}_1 = r_1 \hat{\mathbf{z}}$ with $r_1 \ge 0$. The integral involves absolute values and depends on whether $r_1 \gtrless R$.

(13.24) A surface can be specified as height above a plane, $z = h(x, y)$ in cartesian coordinates. Similarly, a surface can be specified as distance from the origin, $r = f(\theta, \varphi)$, in spherical coordinates (r, θ, φ). Calculate the coordinate tangent vectors for such parametrization. Show that the surface element

$$\mathrm{d}\boldsymbol{S}(\theta, \varphi) = \left(\hat{\boldsymbol{r}}\, r^2 \sin\theta - \hat{\boldsymbol{\theta}}\, r \frac{\partial r}{\partial \theta} \sin\theta - \hat{\boldsymbol{\varphi}}\, r \frac{\partial r}{\partial \varphi}\right) \mathrm{d}\theta \mathrm{d}\varphi \tag{13.44}$$

and that

$$\int_{\mathcal{S}} \frac{\boldsymbol{r}}{r^3} \cdot \mathrm{d}\boldsymbol{S}$$

is independent of $f(\theta, \varphi)$ (Fig. 13.19).
Use that insight to calculate that flux integral by evaluating it on an equivalent but more convenient surface $\mathcal{S}' : r = g(\theta, \varphi)$ when $\mathcal{S}$ is

(i) the triangle $x/a + y/b + z/c = 1$, with a, b, c and x, y, z positive,

(ii) a disk of radius R in the $z = H$ plane centered on the z axis,

(iii) a sphere of radius R centered at a distance $H < R$ from the origin, and

(iv) a sphere of radius R centered at a distance $H > R$ from the origin.

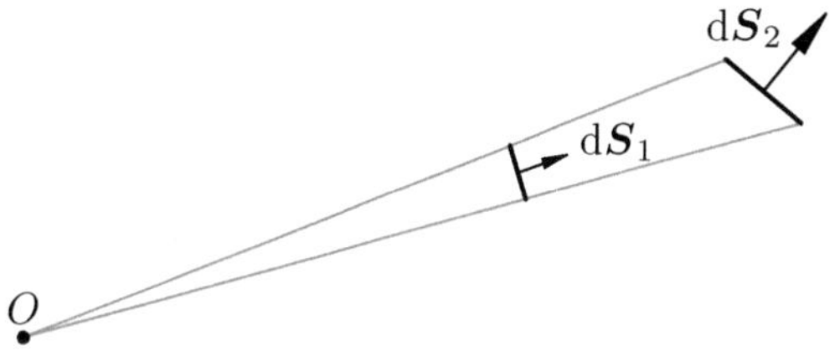

Fig. 13.19 Surface elements $\mathrm{d}\boldsymbol{S}_1$ and $\mathrm{d}\boldsymbol{S}_2$ with the same solid angle $(\hat{\boldsymbol{r}}/r_1^2) \cdot \mathrm{d}\boldsymbol{S}_1 = (\hat{\boldsymbol{r}}/r_2^2) \cdot \mathrm{d}\boldsymbol{S}_2$.

(13.25) For a surface $\mathcal{S}$ parametrized by $\boldsymbol{r}(u, v)$, with $\boldsymbol{r} = r\,\hat{\boldsymbol{r}}$, and $(u, v) \in \mathcal{U}$, show that

$$\int_{\mathcal{S}} \frac{\boldsymbol{r}}{r^3} \cdot \mathrm{d}\boldsymbol{S} = \iint_{\mathcal{U}} \frac{\boldsymbol{r}}{r^3} \cdot \frac{\partial \boldsymbol{r}}{\partial u} \times \frac{\partial \boldsymbol{r}}{\partial v} \mathrm{d}u\mathrm{d}v = \iint_{\mathcal{U}} \hat{\boldsymbol{r}} \cdot \frac{\partial \hat{\boldsymbol{r}}}{\partial u} \times \frac{\partial \hat{\boldsymbol{r}}}{\partial v} \mathrm{d}u\mathrm{d}v. \tag{13.45}$$

Explain why the latter integral is a signed area on the unit sphere.

(13.26) The pressure *outside* a sphere $\mathcal{S}$ of radius R centered at $\boldsymbol{r}_c$ is $p(\boldsymbol{r}) = p_0 + \boldsymbol{a} \cdot \boldsymbol{r}$, where p_0 and $\boldsymbol{a}$ are constants. The force on surface element $\mathrm{d}\boldsymbol{S} = \hat{\boldsymbol{n}} \mathrm{d}A$ is $\mathrm{d}\boldsymbol{f} = -p\,\mathrm{d}\boldsymbol{S} = -p\hat{\boldsymbol{n}} \mathrm{d}A$, where $\hat{\boldsymbol{n}}$ is the unit *outward* normal to the sphere, since pressure is a force per unit area with a direction *locally* perpendicular to the surface. Show that the net force on the sphere is

$$\boldsymbol{F} = \int_{\mathcal{S}} \mathrm{d}\boldsymbol{f} = -\frac{4}{3}\pi R^3\, \boldsymbol{a}. \qquad (13.46)$$

Calculate the torque on the sphere about an arbitrary point with position vector $\boldsymbol{r}_1$,

$$\boldsymbol{T}_1 = \int_{\mathcal{S}} (\boldsymbol{r} - \boldsymbol{r}_1) \times \mathrm{d}\boldsymbol{f}.$$

(Hint: $\boldsymbol{r} - \boldsymbol{r}_1 = (\boldsymbol{r} - \boldsymbol{r}_c) + (\boldsymbol{r}_c - \boldsymbol{r}_1)$.)

(13.27) Consider the cylinder of radius R and height H, $\boldsymbol{r}(\varphi, z) = R(\hat{\mathbf{x}} \cos\varphi + \hat{\mathbf{y}} \sin\varphi) + z\hat{\mathbf{z}}$, $-\pi < \varphi \le \pi$, $0 \le z \le H$. Partition the cylinder into congruent triangles as in Fig. 13.12. The vertices of one such triangle are $\boldsymbol{r}_0 = \boldsymbol{r}(0,0)$, $\boldsymbol{r}_1 = \boldsymbol{r}(\Delta\varphi, \Delta z)$, $\boldsymbol{r}_2 = \boldsymbol{r}(-\Delta\varphi, \Delta z)$, for $\Delta\varphi = \pi/m$, $\Delta z = H/n$, where m and n are positive integers. Let $\Delta\boldsymbol{r}_1 = \boldsymbol{r}_1 - \boldsymbol{r}_0$, $\Delta\boldsymbol{r}_2 = \boldsymbol{r}_2 - \boldsymbol{r}_1$. Show that (i)

$$\begin{aligned}\Delta\boldsymbol{S} &= \frac{1}{2}\Delta\boldsymbol{r}_1 \times \Delta\boldsymbol{r}_2 \\ &= \hat{\mathbf{x}}\, R\Delta z \sin\Delta\varphi + \hat{\mathbf{z}}\, R^2 \sin\Delta\varphi(1 - \cos\Delta\varphi) \\ &\approx \hat{\mathbf{x}}\, R\Delta\varphi\Delta z + \hat{\mathbf{z}}\, \frac{R^2}{2}(\Delta\varphi)^3.\end{aligned}$$

Hence, if $\Delta z \le O(\Delta\varphi^2)$ as $\Delta\varphi \to 0$, then $\Delta\boldsymbol{S}$ does *not* tend to the $\hat{\mathbf{x}}$ direction perpendicular to the cylinder at that limit point $\boldsymbol{r}(0,0) = R\hat{\mathbf{x}}$;
(ii) the sum of the areas of all the triangles is

$$2mn|\Delta\boldsymbol{S}| = 2m \sin\frac{\pi}{m} R \sqrt{H^2 + 4R^2n^2 \sin^4 \frac{\pi}{2m}},$$

and diverges if $n/m^2 \to \infty$ as $m \to \infty$;[6] and
(iii) $\Delta\boldsymbol{S}$ and the total polyhedral area yield the correct results if $m = 2n \to \infty$ (Fig. 13.12 has $m = 2n = 6$), among other possibilities. Quadrilaterals, whose $\Delta\boldsymbol{S}$ is the *vector* sum of two triangle area vectors with a common edge (4.17), also behave properly no matter how m, n tend to infinity.

(13.28) Derive (13.28) geometrically. A, B, C are the vertices of a spherical triangle and A'', B'', C'' are their antipodes (Fig. 13.20). First, argue that the *lune* $ABA''CA$—the spherical surface between the meridians ABA'' and ACA''—has area $2\alpha R^2$, where α is the angle between the meridians at A. Similarly, the area of the lune $BCB''AB$ is $2\beta R^2$, and that of lune $CAC''BC$ is $2\gamma R^2$. Next, argue that the area of the spherical triangle ABC'' equals that of triangle $A''B''C$. Then, the area of those three lunes add up to a full hemisphere area, *plus* two extra copies of the triangle ABC area. Deduce (13.28).[7]

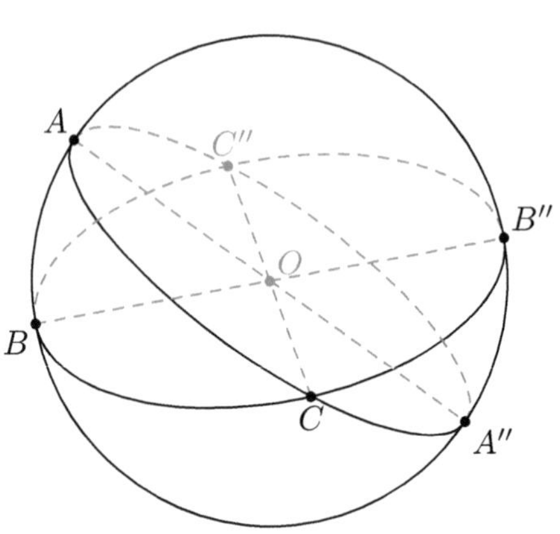

Fig. 13.20 Spherical triangle area.

(13.29) Calculate $\iint_{\mathcal{U}} \mathrm{d}u\mathrm{d}v$ where $\mathcal{U}$ is the region inside

$$\begin{aligned}\partial\mathcal{U} : &(u(t), v(t)) \\ &= (a_1, a_2)\, \cos t + (b_1, b_2)\, \sin t + (c_1, c_2),\end{aligned}$$

where $a_1, a_2, b_1, b_2, c_1, c_2$ are constants and t is real with $t = 0 \to 2\pi$.

(13.30) Parametrize the boundary of the triangle $\mathcal{U}$ with vertices at (u_0, v_0), (u_1, v_1), (u_2, v_2) and calculate $\iint_{\mathcal{U}} u\, \mathrm{d}u\mathrm{d}v$.

(13.31) Show that the area A of a region $\mathcal{U}$ bounded by the simple closed curve $\partial\mathcal{U}$ in the cartesian (x, y) plane is given by the line integrals

$$\begin{aligned}A = \iint_{\mathcal{U}} \mathrm{d}x\mathrm{d}y &= \oint_{\partial\mathcal{U}} x\mathrm{d}y = -\oint_{\partial\mathcal{U}} y\mathrm{d}x \\ &= \frac{1}{2}\oint_{\partial\mathcal{U}} x\mathrm{d}y - y\mathrm{d}x. \qquad (13.47)\end{aligned}$$

Compare with (4.16) and (12.53).

[6] For historical background on this example, see, for example, Gabriele H. Greco, Sonia Mazzucchi, and Enrico M. Pagani. "Peano on Definition of Surface Area." *Rendiconti Lincei* **27**.3 (2016): 251–286. `https://doi.org/10.4171/rlm/734`
[7] §97 in I. Todhunter, *Spherical Trigonometry*, Macmillan, 1886. `https://www.gutenberg.org/ebooks/19770`

Curves on Surfaces

14

A curve on a surface is a composite function $t \in \mathbb{R} \to \boldsymbol{r}(u(t), v(t))$ for the position vector $\boldsymbol{r} = x\hat{\mathbf{x}} + y\hat{\mathbf{y}} + z\hat{\mathbf{z}}$ of a point P in 3D euclidean space $\mathbb{E}^3$, with surface coordinates $(u, v) \in \mathcal{U} \subset \mathbb{R}^2$ (Fig. 14.1).

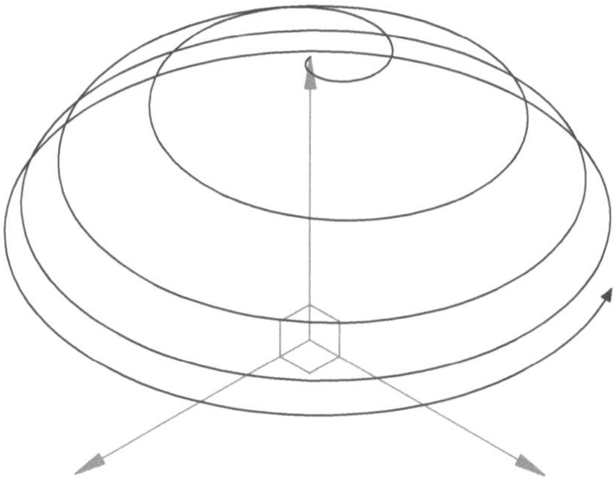

Fig. 14.1 A curve $\boldsymbol{r}(\theta(t), \varphi(t))$ spiraling from the pole to the equator on the unit sphere $\boldsymbol{r}(\theta, \varphi) = \sin\theta\cos\varphi\,\hat{\mathbf{x}} + \sin\theta\sin\varphi\,\hat{\mathbf{y}} + \cos\theta\,\hat{\mathbf{z}}$, with $(\theta, \varphi) = (\arctan at,\ \omega t)$.

14.1 Velocity and arclength

The velocity $\dot{\boldsymbol{r}} = \mathrm{d}\boldsymbol{r}/\mathrm{d}t$ of a particle traveling on a surface $\boldsymbol{r}(u, v)$ follows from the chain rule as

$$\frac{\mathrm{d}\boldsymbol{r}}{\mathrm{d}t} = \frac{\partial \boldsymbol{r}}{\partial u}\frac{\mathrm{d}u}{\mathrm{d}t} + \frac{\partial \boldsymbol{r}}{\partial v}\frac{\mathrm{d}v}{\mathrm{d}t}. \tag{14.1}$$

The dot product of (14.1) with itself gives the square of the speed $\mathrm{d}s/\mathrm{d}t = |\dot{\boldsymbol{r}}|$ as

$$\dot{s}^2 = \frac{\partial \boldsymbol{r}}{\partial u} \cdot \frac{\partial \boldsymbol{r}}{\partial u}\,\dot{u}^2 + 2\frac{\partial \boldsymbol{r}}{\partial u} \cdot \frac{\partial \boldsymbol{r}}{\partial v}\,\dot{u}\dot{v} + \frac{\partial \boldsymbol{r}}{\partial v} \cdot \frac{\partial \boldsymbol{r}}{\partial v}\,\dot{v}^2, \tag{14.2}$$

where $\dot{s} = \mathrm{d}s/\mathrm{d}t$, $\dot{u} = \mathrm{d}u/\mathrm{d}t$ and $\dot{v} = \mathrm{d}v/\mathrm{d}t$. This yields the *first fundamental form*, expressing the arclength element $\mathrm{d}s$ in terms of the coordinate differentials $\mathrm{d}u$, $\mathrm{d}v$ as

$$\mathrm{d}s^2 = \frac{\partial \boldsymbol{r}}{\partial u} \cdot \frac{\partial \boldsymbol{r}}{\partial u}\,\mathrm{d}u^2 + 2\frac{\partial \boldsymbol{r}}{\partial u} \cdot \frac{\partial \boldsymbol{r}}{\partial v}\,\mathrm{d}u\,\mathrm{d}v + \frac{\partial \boldsymbol{r}}{\partial v} \cdot \frac{\partial \boldsymbol{r}}{\partial v}\,\mathrm{d}v^2. \tag{14.3a}$$

This first fundamental form (14.3) is commonly written in one of the following more compact notations,

$$\begin{aligned} ds^2 &= E\,du^2 + 2F\,du\,dv + G\,dv^2, \\ &= \begin{bmatrix} du & dv \end{bmatrix} \begin{bmatrix} E & F \\ F & G \end{bmatrix} \begin{bmatrix} du \\ dv \end{bmatrix} = \begin{bmatrix} du^1 & du^2 \end{bmatrix} \begin{bmatrix} g_{11} & g_{12} \\ g_{12} & g_{22} \end{bmatrix} \begin{bmatrix} du^1 \\ du^2 \end{bmatrix}, \\ &= g_{11}\,du^1du^1 + 2g_{12}\,du^1\,du^2 + g_{22}\,du^2du^2 = g_{ij}du^idu^j, \end{aligned} \tag{14.3b}$$

where $E = g_{11}$, $F = g_{12} = g_{21}$, $G = g_{22}$ are the metric coefficients

$$g_{11} = \frac{\partial \boldsymbol{r}}{\partial u} \cdot \frac{\partial \boldsymbol{r}}{\partial u}, \quad g_{12} = \frac{\partial \boldsymbol{r}}{\partial u} \cdot \frac{\partial \boldsymbol{r}}{\partial v}, \quad g_{22} = \frac{\partial \boldsymbol{r}}{\partial v} \cdot \frac{\partial \boldsymbol{r}}{\partial v}. \tag{14.3c}$$

du^idu^j uses superscripts i, j, not exponents.

The E, F, G notation was introduced by Gauss. The expression $ds^2 = g_{ij}du^idu^j$ uses summation over indices repeated *once lower and once upper*. That compact notation generalizes to higher dimensions and is discussed in §14.4.

The metric coefficients g_{ij} determine lengths, angles, and areas on the surface. The cross product of the velocity basis vectors yields the unit normal to the surface $\hat{\boldsymbol{n}}$ and Jacobian J,

$$\frac{\partial \boldsymbol{r}}{\partial u} \times \frac{\partial \boldsymbol{r}}{\partial v} = J\,\hat{\boldsymbol{n}}, \tag{14.4a}$$

where

$$J = \left| \frac{\partial \boldsymbol{r}}{\partial u} \times \frac{\partial \boldsymbol{r}}{\partial v} \right| = \sqrt{EG - F^2} = \sqrt{g_{11}g_{22} - g_{12}^2} \tag{14.4b}$$

is the square root of the metric determinant (13.21), such that the surface element

$$d\boldsymbol{S} = \frac{\partial \boldsymbol{r}}{\partial u} \times \frac{\partial \boldsymbol{r}}{\partial v}\,dudv = \hat{\boldsymbol{n}}\,J\,dudv. \tag{14.4c}$$

14.2 Acceleration and curvature

The acceleration of a particle traveling on a surface is the derivative of (14.1)

$$\begin{aligned} \ddot{\boldsymbol{r}} &= \frac{\partial^2 \boldsymbol{r}}{\partial u^2}\dot{u}^2 + 2\frac{\partial^2 \boldsymbol{r}}{\partial u \partial v}\dot{u}\dot{v} + \frac{\partial^2 \boldsymbol{r}}{\partial v^2}\dot{v}^2 + \frac{\partial \boldsymbol{r}}{\partial u}\ddot{u} + \frac{\partial \boldsymbol{r}}{\partial v}\ddot{v}, \\ &= \dot{u}^i\dot{u}^j\,\partial_i\partial_j\boldsymbol{r} + \ddot{u}^i\,\partial_i\boldsymbol{r}, \end{aligned} \tag{14.5}$$

where $\ddot{\boldsymbol{r}} = d^2\boldsymbol{r}/dt^2$, $\ddot{u} = d^2u/dt^2$, $\ddot{v} = d^2v/dt^2$. The compact expression $\ddot{\boldsymbol{r}} = \dot{u}^i\dot{u}^j\partial_i\partial_j\boldsymbol{r} + \ddot{u}^i\partial_i\boldsymbol{r}$ uses summation over repeated indices, discussed in §14.4.

The velocity $\dot{\boldsymbol{r}}$ and acceleration $\ddot{\boldsymbol{r}}$ contain geometric information about the trajectory, as derived in (12.30) and (12.32),

$$\dot{\boldsymbol{r}} = \dot{s}\,\hat{\boldsymbol{t}}, \qquad \ddot{\boldsymbol{r}} = \ddot{s}\,\hat{\boldsymbol{t}} + \dot{s}^2\kappa\hat{\boldsymbol{n}}_c, \tag{14.6}$$

where $\dot{s} = |\dot{\boldsymbol{r}}|$ is the speed, $\hat{\boldsymbol{t}}$ is the unit tangent to the curve, $\ddot{s}$ is the acceleration along the curve, and $\hat{\boldsymbol{n}}_c$ is the curve turn direction orthogonal to $\hat{\boldsymbol{t}}$, with curvature κ such that $\mathrm{d}\hat{\boldsymbol{t}}/\mathrm{d}s = \kappa\hat{\boldsymbol{n}}_c$. For a curve on a surface, the curve tangent $\hat{\boldsymbol{t}}$ is also tangent to the surface, and orthogonal to the surface normal $\hat{\boldsymbol{n}}$, such that $\hat{\boldsymbol{n}} \times \hat{\boldsymbol{t}}$ is a unit vector and $\{\hat{\boldsymbol{n}}, \hat{\boldsymbol{t}}, \hat{\boldsymbol{n}} \times \hat{\boldsymbol{t}}\}$ form an orthonormal moving frame, called a *Darboux frame.*

The curve normal $\hat{\boldsymbol{n}}_c$ is not orthogonal to the surface, in general, but is orthogonal to $\hat{\boldsymbol{t}}$ by definition and can be expressed in terms of the surface normal $\hat{\boldsymbol{n}}$ and the surface tangent $\hat{\boldsymbol{n}} \times \hat{\boldsymbol{t}}$ such that

$$\kappa\,\hat{\boldsymbol{n}}_c = \kappa_n\,\hat{\boldsymbol{n}} + \kappa_g\,\hat{\boldsymbol{n}} \times \hat{\boldsymbol{t}}, \tag{14.7a}$$

with $\kappa^2 = \kappa_n^2 + \kappa_g^2$, where κ_n is the *normal curvature*, characteristic of the surface at that point in direction $\hat{\boldsymbol{t}}$, and κ_g is the curve's *geodesic curvature.* Those two curvatures follow from (14.6) and (14.7a) as

$$\kappa_n = \frac{\ddot{\boldsymbol{r}}\cdot\hat{\boldsymbol{n}}}{|\dot{\boldsymbol{r}}|^2}, \qquad \kappa_g = \frac{\ddot{\boldsymbol{r}}\cdot(\hat{\boldsymbol{n}}\times\dot{\boldsymbol{r}})}{|\dot{\boldsymbol{r}}|^3}. \tag{14.7b}$$

Geodesics, discussed in §14.5, are curves such that $\kappa_g = 0$, and the "geodesic curvature" κ_g is thus a measure of *how far from geodesic* a curve is.[1] The curvature of a geodesic is actually κ_n. These κ_n, κ_g are *signed* curvatures, relative to the orientation specified by the surface normal $\hat{\boldsymbol{n}}$ and the curve direction $\dot{\boldsymbol{r}}$. Substituting for $\ddot{\boldsymbol{r}}$ from (14.5) into (14.7b) gives the normal curvature as

$$\kappa_n|\dot{\boldsymbol{r}}|^2 = \hat{\boldsymbol{n}}\cdot\frac{\partial^2\boldsymbol{r}}{\partial u^2}\,\dot{u}^2 + 2\hat{\boldsymbol{n}}\cdot\frac{\partial^2\boldsymbol{r}}{\partial u\partial v}\dot{u}\dot{v} + \hat{\boldsymbol{n}}\cdot\frac{\partial^2\boldsymbol{r}}{\partial v^2}\,\dot{v}^2. \tag{14.7c}$$

Since $|\dot{\boldsymbol{r}}| = \mathrm{d}s/\mathrm{d}t$, this yields the *second fundamental form* for the surface

$$\kappa_n\mathrm{d}s^2 = \hat{\boldsymbol{n}}\cdot\frac{\partial^2\boldsymbol{r}}{\partial u^2}\,\mathrm{d}u^2 + 2\hat{\boldsymbol{n}}\cdot\frac{\partial^2\boldsymbol{r}}{\partial u\partial v}\mathrm{d}u\mathrm{d}v + \hat{\boldsymbol{n}}\cdot\frac{\partial^2\boldsymbol{r}}{\partial v^2}\,\mathrm{d}v^2, \tag{14.8a}$$

commonly written in compact form as

$$\begin{aligned}\kappa_n\mathrm{d}s^2 &= L\,\mathrm{d}u^2 + 2M\mathrm{d}u\mathrm{d}v + N\,\mathrm{d}v^2\\ &= \begin{bmatrix}\mathrm{d}u & \mathrm{d}v\end{bmatrix}\begin{bmatrix}L & M\\ M & N\end{bmatrix}\begin{bmatrix}\mathrm{d}u\\ \mathrm{d}v\end{bmatrix}\\ &= \begin{bmatrix}\mathrm{d}u^1 & \mathrm{d}u^2\end{bmatrix}\begin{bmatrix}b_{11} & b_{12}\\ b_{12} & b_{22}\end{bmatrix}\begin{bmatrix}\mathrm{d}u^1\\ \mathrm{d}u^2\end{bmatrix} = b_{ij}\,\mathrm{d}u^i\mathrm{d}u^j,\end{aligned} \tag{14.8b}$$

where $L = b_{11}$, $M = b_{12} = b_{21}$, $N = b_{22}$ are the *curvature coefficients*[2]

$$b_{11} = \hat{\boldsymbol{n}}\cdot\frac{\partial^2\boldsymbol{r}}{\partial u^2}, \quad b_{12} = \hat{\boldsymbol{n}}\cdot\frac{\partial^2\boldsymbol{r}}{\partial u\partial v}, \quad b_{22} = \hat{\boldsymbol{n}}\cdot\frac{\partial^2\boldsymbol{r}}{\partial v^2}, \tag{14.8c}$$

the dot products of the surface unit normal $\hat{\boldsymbol{n}}$ with the second derivatives of the surface position vector $\boldsymbol{r}(u, v)$. The L, M, N notation is common in introductory differential geometry. The compact index notation expression $\kappa_n\mathrm{d}s^2 = b_{ij}\mathrm{d}u^i\mathrm{d}u^j$ generalizes to higher dimensions and is introduced in §14.4. The normal curvature κ_n and the curvic b_{ij} are studied in §14.7.

[1] "Tangential curvature" would have been a better name, contrasting with "normal curvature." This definition of κ_g dates back to O. Bonnet, *Mémoire sur la théorie générale des surfaces*, 1848.

[2] Or *curvic* b_{ij}.

14.3 Meridians and parallels

The sphere is a familiar, though very special, example of a curved surface. It is suitably parametrized as

$$\boldsymbol{r}(\theta,\varphi) = \hat{\mathbf{x}}\, R\sin\theta\cos\varphi + \hat{\mathbf{y}}\, R\sin\theta\sin\varphi + \hat{\mathbf{z}}\, R\cos\theta \tag{14.9}$$

in terms of the polar (or *co-latitude*) angle θ and azimuthal (or *longitude*) angle φ, with R as the sphere radius and $\{\hat{\mathbf{x}}, \hat{\mathbf{y}}, \hat{\mathbf{z}}\}$ the usual cartesian basis.

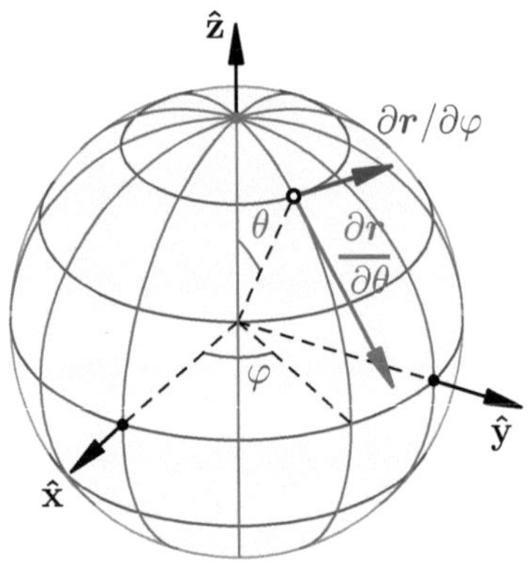

Fig. 14.2 Meridians and parallels.

That parametrization yields two families of curves on the sphere, the *meridians* corresponding to varying θ for fixed φ, and the *parallels* with fixed θ but varying φ. The local tangent vectors to the meridians and parallels are, respectively (Fig. 14.2)

$$\begin{aligned}\frac{\partial \boldsymbol{r}}{\partial\theta} &= \hat{\mathbf{x}}\, R\cos\theta\cos\varphi + \hat{\mathbf{y}}\, R\cos\theta\sin\varphi - \hat{\mathbf{z}}\, R\sin\theta = R\,\hat{\boldsymbol{\theta}},\\ \frac{\partial \boldsymbol{r}}{\partial\varphi} &= -\hat{\mathbf{x}}\, R\sin\theta\sin\varphi + \hat{\mathbf{y}}\, R\sin\theta\cos\varphi = R\,\sin\theta\,\hat{\boldsymbol{\varphi}}.\end{aligned} \tag{14.10a}$$

These are orthogonal coordinates since $\partial\boldsymbol{r}/\partial\theta \cdot \partial\boldsymbol{r}/\partial\varphi = 0$, such that the first fundamental form (14.3)

$$\mathrm{d}s^2 = R^2\,\mathrm{d}\theta^2 + R^2\sin^2\theta\,\mathrm{d}\varphi^2, \tag{14.10b}$$

with metric (14.3c)

$$\begin{bmatrix} E & F\\ F & G\end{bmatrix} \equiv \begin{bmatrix} g_{11} & g_{12}\\ g_{21} & g_{22}\end{bmatrix} = \begin{bmatrix} R^2 & 0\\ 0 & R^2\sin^2\theta\end{bmatrix}. \tag{14.10c}$$

The second derivatives of $\boldsymbol{r}(\theta,\varphi)$ are

$$\begin{aligned}\frac{\partial^2\boldsymbol{r}}{\partial\theta^2} &= R\,\frac{\partial\hat{\boldsymbol{\theta}}}{\partial\theta} = -R\,\hat{\boldsymbol{r}},\\ \frac{\partial^2\boldsymbol{r}}{\partial\varphi\partial\theta} = R\,\frac{\partial\hat{\boldsymbol{\theta}}}{\partial\varphi} &= \frac{\partial^2\boldsymbol{r}}{\partial\theta\partial\varphi} = R\frac{\partial(\sin\theta\,\hat{\boldsymbol{\varphi}})}{\partial\theta} = R\cos\theta\,\hat{\boldsymbol{\varphi}},\\ \frac{\partial^2\boldsymbol{r}}{\partial\varphi^2} &= R\,\sin\theta\,\frac{\partial\hat{\boldsymbol{\varphi}}}{\partial\varphi} = -R\sin\theta\,\hat{\boldsymbol{\rho}},\end{aligned} \tag{14.11a}$$

where

$$\hat{\boldsymbol{\rho}} = \hat{\mathbf{x}}\cos\varphi + \hat{\mathbf{y}}\sin\varphi = \hat{\boldsymbol{\theta}}\cos\theta + \hat{\boldsymbol{r}}\sin\theta$$

is the cylindrical radial direction and $\hat{\boldsymbol{n}} = \hat{\boldsymbol{r}} = \boldsymbol{r}/R$ is the unit radial direction, orthogonal to the sphere, such that $\hat{\boldsymbol{n}}\cdot\hat{\boldsymbol{\rho}} = \sin\theta$. The *second fundamental form* (14.8) thus reads

$$\kappa_n\mathrm{d}s^2 = -R\,\mathrm{d}\theta^2 - R\sin^2\theta\,\mathrm{d}\varphi^2 \tag{14.11b}$$

with *curvic* (14.8c)

$$\begin{bmatrix} L & M\\ M & N\end{bmatrix} \equiv \begin{bmatrix} b_{11} & b_{12}\\ b_{21} & b_{22}\end{bmatrix} = \begin{bmatrix} -R & 0\\ 0 & -R\sin^2\theta\end{bmatrix}. \tag{14.11c}$$

Meridians

A particle moving along meridians, with constant longitude φ but varying polar angle $\theta(t)$ will have velocity and acceleration

$$\begin{aligned} \dot{\boldsymbol{r}} &= \dot{s}\hat{\boldsymbol{t}} = \frac{\partial \boldsymbol{r}}{\partial \theta}\dot{\theta} = R\dot{\theta}\,\hat{\boldsymbol{\theta}}, \\ \ddot{\boldsymbol{r}} &= \ddot{s}\hat{\boldsymbol{t}} + \kappa \dot{s}^2 \hat{\boldsymbol{n}}_c = \frac{\partial \boldsymbol{r}}{\partial \theta}\ddot{\theta} + \frac{\partial^2 \boldsymbol{r}}{\partial \theta^2}\dot{\theta}^2 = R\ddot{\theta}\,\hat{\boldsymbol{\theta}} - R\dot{\theta}^2\,\hat{\boldsymbol{r}} \end{aligned} \tag{14.12a}$$

with *Darboux frame* $\{\hat{\boldsymbol{n}}, \hat{\boldsymbol{t}}, \hat{\boldsymbol{n}} \times \hat{\boldsymbol{t}}\} = \{\hat{\boldsymbol{r}}, \hat{\boldsymbol{\theta}}, \hat{\boldsymbol{\varphi}}\}$, varying with θ. The normal and geodesic curvatures of meridians are (14.7b) are

$$\kappa_n = \frac{\ddot{\boldsymbol{r}} \cdot \hat{\boldsymbol{n}}}{|\dot{\boldsymbol{r}}|^2} = -\frac{1}{R}, \qquad \kappa_g = \frac{\ddot{\boldsymbol{r}} \cdot (\hat{\boldsymbol{n}} \times \dot{\boldsymbol{r}})}{|\dot{\boldsymbol{r}}|^3} = 0. \tag{14.12b}$$

Thus, meridians are geodesics on a sphere, with curvature $\kappa = -\kappa_n = 1/R$. The normal curvature, κ_n also follows from (14.11b) divided by (14.10b) with $\mathrm{d}\varphi = 0$ for meridians. Meridians are great circles, with radius R, such that

$$\frac{\mathrm{d}^2 \boldsymbol{r}}{\mathrm{d}s^2} = -\frac{\hat{\boldsymbol{r}}}{R} = \kappa_n \hat{\boldsymbol{n}}. \tag{14.12c}$$

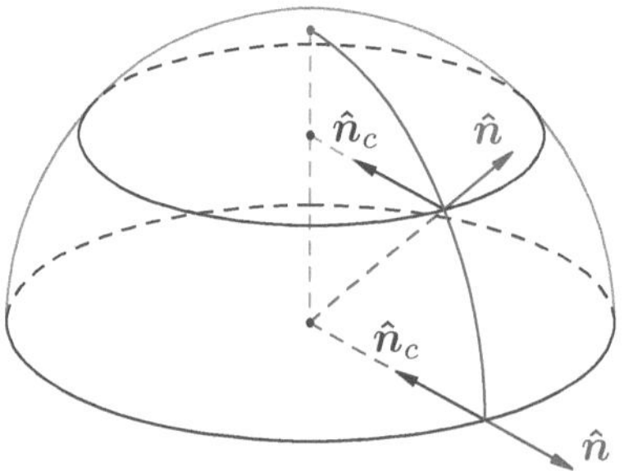

Fig. 14.3 Meridians (red) are geodesics. Parallels (blue) are not, except at the equator.

Parallels

A particle moving along parallels, with constant polar angle θ but varying longitude $\varphi(t)$, has velocity and acceleration

$$\begin{aligned} \dot{\boldsymbol{r}} &= \dot{s}\hat{\boldsymbol{t}} = \frac{\partial \boldsymbol{r}}{\partial \varphi}\dot{\varphi} = R\sin\theta\,\dot{\varphi}\,\hat{\boldsymbol{\varphi}}, \\ \ddot{\boldsymbol{r}} &= \ddot{s}\hat{\boldsymbol{t}} + \kappa \dot{s}^2 \hat{\boldsymbol{n}}_c = \frac{\partial \boldsymbol{r}}{\partial \varphi}\ddot{\varphi} + \frac{\partial^2 \boldsymbol{r}}{\partial \varphi^2}\dot{\varphi}^2 = R\sin\theta\ddot{\varphi}\,\hat{\boldsymbol{\varphi}} - R\sin\theta\,\dot{\varphi}^2\,\hat{\boldsymbol{\rho}} \end{aligned} \tag{14.13a}$$

with *Darboux frame* $\{\hat{\boldsymbol{n}}, \hat{\boldsymbol{t}}, \hat{\boldsymbol{n}} \times \hat{\boldsymbol{t}}\} = \{\hat{\boldsymbol{r}}, \hat{\boldsymbol{\varphi}}, -\hat{\boldsymbol{\theta}}\}$ varying with φ. The normal and geodesic curvatures (14.7b) of parallels are

$$\kappa_n = \frac{\ddot{\boldsymbol{r}} \cdot \hat{\boldsymbol{n}}}{|\dot{\boldsymbol{r}}|^2} = -\frac{1}{R}, \qquad \kappa_g = \frac{\ddot{\boldsymbol{r}} \cdot (\hat{\boldsymbol{n}} \times \dot{\boldsymbol{r}})}{|\dot{\boldsymbol{r}}|^3} = \frac{1}{R}\frac{\cos\theta}{\sin\theta}. \tag{14.13b}$$

Thus, parallels are not geodesics on a sphere, except at the equator where $\theta = \pi/2$ (Fig. 14.3). The normal curvature, κ_n also follows from (14.11b) divided by (14.10b) with $\mathrm{d}\theta = 0$ for parallels. Parallels are circles $\boldsymbol{r} = \boldsymbol{\rho} + z\hat{\boldsymbol{z}}$ with radius $|\boldsymbol{\rho}| = \rho = R\sin\theta$ and height $z = R\cos\theta$ fixed, such that

$$\begin{aligned} \frac{\mathrm{d}^2 \boldsymbol{r}}{\mathrm{d}s^2} = \frac{\mathrm{d}^2 \boldsymbol{\rho}}{\mathrm{d}s^2} = -\frac{\hat{\boldsymbol{\rho}}}{\rho} &= -\frac{\hat{\boldsymbol{r}}\sin\theta + \hat{\boldsymbol{\theta}}\cos\theta}{R\sin\theta} \\ &= \kappa_n \hat{\boldsymbol{r}} + \kappa_g\,(\hat{\boldsymbol{r}} \times \hat{\boldsymbol{\varphi}}). \end{aligned} \tag{14.13c}$$

14.4 General index notation

The surface velocity basis vectors (14.1)

$$\boldsymbol{a}_1 \triangleq \frac{\partial \boldsymbol{r}}{\partial u}, \qquad \boldsymbol{a}_2 \triangleq \frac{\partial \boldsymbol{r}}{\partial v} \tag{14.14}$$

define a basis for the tangent plane at point $\boldsymbol{r}(u,v)$ on a surface $\mathcal{S}$. Simple surface parametrizations, such as the familiar (Monge) parametrization as height above a plane $z = h(x,y)$ with coordinates $(u,v) = (x,y)$, for example, yield *non*orthogonal surface coordinates such that

$$\frac{\partial \boldsymbol{r}}{\partial u} \cdot \frac{\partial \boldsymbol{r}}{\partial v} = \boldsymbol{a}_1 \cdot \boldsymbol{a}_2 = g_{12} = F \neq 0.$$

Handling nonorthogonal coordinates requires using a second set of basis vectors, reciprocal to the velocity basis vectors, and the requisite calculations are facilitated by generalized index notation that uses *lower* indices $\boldsymbol{a}_i$ (subscripts) and *upper* indices $\boldsymbol{a}^i$ (superscripts). The manipulation rules for such dual indices are known as *Ricci calculus.*

The velocity basis $\{\boldsymbol{a}_1, \boldsymbol{a}_2\}$ has a *reciprocal* basis $\{\breve{\boldsymbol{a}}_1, \breve{\boldsymbol{a}}_2\}$ in the same tangent plane, such that $\boldsymbol{a}_i \cdot \breve{\boldsymbol{a}}_j = \delta_{ij}$, previewed in (3.28) and (6.17). In *Ricci calculus*, the reciprocal basis vectors are denoted with *upper indices*, $\breve{\boldsymbol{a}}_i \equiv \boldsymbol{a}^i$, where i is now a superscript, *not* an exponent. The reciprocals $\{\boldsymbol{a}^1, \boldsymbol{a}^2\}$ of $\{\boldsymbol{a}_1, \boldsymbol{a}_2\}$ have the explicit expressions

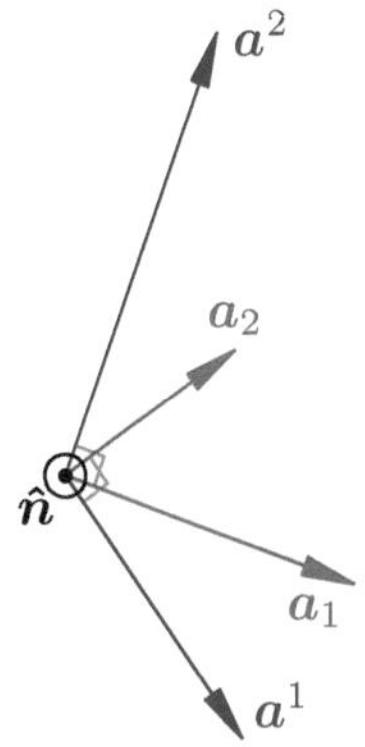

Fig. 14.4 Reciprocal bases.

$$\boldsymbol{a}^1 = \frac{\boldsymbol{a}_2 \times \hat{\boldsymbol{n}}}{\boldsymbol{a}_1 \cdot \boldsymbol{a}_2 \times \hat{\boldsymbol{n}}}, \qquad \boldsymbol{a}^2 = \frac{\hat{\boldsymbol{n}} \times \boldsymbol{a}_1}{\boldsymbol{a}_2 \cdot \hat{\boldsymbol{n}} \times \boldsymbol{a}_1}, \tag{14.15}$$

where $\hat{\boldsymbol{n}} = (\boldsymbol{a}_1 \times \boldsymbol{a}_2)/|\boldsymbol{a}_1 \times \boldsymbol{a}_2|$ is the unit normal to the surface (Fig. 14.4). The denominators in (14.15) both equal the Jacobian $J = \boldsymbol{a}_1 \times \boldsymbol{a}_2 \cdot \hat{\boldsymbol{n}} = |\boldsymbol{a}_1 \times \boldsymbol{a}_2| \neq 0$ (14.4b). The mutual orthonormality of those two bases is written

$$\boldsymbol{a}_i \cdot \boldsymbol{a}^j = \delta_i^j, \tag{14.16}$$

now keeping lower and upper indices at their position in the Kronecker delta δ_i^j, which is still 1 if $i = j$, and 0 otherwise.

Both the coordinate tangents $\{\boldsymbol{a}_1, \boldsymbol{a}_2\}$ and their reciprocals $\{\boldsymbol{a}^1, \boldsymbol{a}^2\}$, can serve as a basis for the tangent plane at $\boldsymbol{r}(u,v)$. The components of a tangent vector $\boldsymbol{p}$ in one basis are obtained by dotting $\boldsymbol{p}$ with its reciprocal basis. That follows directly from the orthonormality relations (14.16). In Ricci notation, these expansions and projections are written

$$\boldsymbol{p} = \begin{cases} p^i \, \boldsymbol{a}_i \\ p_i \, \boldsymbol{a}^i \end{cases} \Leftrightarrow \begin{cases} p^i = \boldsymbol{a}^i \cdot \boldsymbol{p}, \\ p_i = \boldsymbol{a}_i \cdot \boldsymbol{p}, \end{cases} \tag{14.17}$$

with *summation over repeated indices* but now *only when the repeated indices are in different levels, one upper, one lower, in the same term.* Thus, for a 2D surface,

$$p^i q_i = p^1 q_1 + p^2 q_2, \qquad p_i q^i = p_1 q^1 + p_2 q^2,$$

but

$$p_i q_i = (p_1 q_1,\, p_2 q_2), \qquad p^i q^i = (p^1 q^1, p^2 q^2).$$

A consequence of (14.17) is that the components $(\dot u, \dot v)$ of the velocity $\dot{\boldsymbol{r}}$ of a particle traveling on the surface, are

$$\dot{\boldsymbol{r}} = \frac{\partial \boldsymbol{r}}{\partial u}\dot u + \frac{\partial \boldsymbol{r}}{\partial v}\dot v = \dot u\, \boldsymbol{a}_1 + \dot v\, \boldsymbol{a}_2 \Rightarrow \begin{cases} \dot u = \boldsymbol{a}^1 \cdot \dot{\boldsymbol{r}}, \\ \dot v = \boldsymbol{a}^2 \cdot \dot{\boldsymbol{r}}. \end{cases}$$

Thus, in Ricci index notation, the coordinates should be written with *upper* indices

$$(u, v) \equiv (u^1, u^2), \tag{14.18}$$

where the superscripts are indices, *not exponents,* and the velocity of a particle on the surface is

$$\dot{\boldsymbol{r}} = \frac{\partial \boldsymbol{r}}{\partial u}\dot u + \frac{\partial \boldsymbol{r}}{\partial v}\dot v = \frac{\partial \boldsymbol{r}}{\partial u^i}\,\dot u^i = \dot u^i\, \boldsymbol{a}_i, \tag{14.19}$$

with sum over the index $i = 1, 2$, that is, $\dot u^i\, \boldsymbol{a}_i = \dot u^1\, \boldsymbol{a}_1 + \dot u^2\, \boldsymbol{a}_2$. The speed square is (14.2)

$$\dot{\boldsymbol{r}} \cdot \dot{\boldsymbol{r}} = \frac{\partial \boldsymbol{r}}{\partial u^i}\dot u^i \cdot \frac{\partial \boldsymbol{r}}{\partial u^j}\dot u^j = \boldsymbol{a}_i \cdot \boldsymbol{a}_j \dot u^i \dot u^j = g_{ij} \dot u^i \dot u^j, \tag{14.20}$$

with double sums over i and j. If the square of a component is needed, it is written as an explicit product $(u)^2 \equiv u^1 u^1$ and $(v)^2 \equiv u^2 u^2$.

A standard approach for distinguishing surface indices running from 1 to 2, from full space indices running from 1 to 3, is to write the former using Greek letters, thus writing $\dot{\boldsymbol{r}} = \dot u^\alpha \boldsymbol{a}_\alpha$ for velocities tangent to the surface with coordinates (u^1, u^2), but $\dot{\boldsymbol{r}} = \dot u^i \boldsymbol{a}_i$ for general velocity in 3D coordinate space (u^1, u^2, u^3). We will not make that distinction unless absolutely necessary because most of the formulas apply to spaces of any dimension. For a surface, we will remember that the third coordinate u^3 is constant, $u^3 = 0$ say, and $\mathrm{d}u^3 = 0$, $\dot u^3 = 0$ with $\boldsymbol{a}_3 = \boldsymbol{a}^3 = \hat{\boldsymbol{n}}$.

Covariant and contravariant quantities

Our presumably judicious choice of surface parametrization $\boldsymbol{r}(u, v)$ comes with velocity basis vectors, $\boldsymbol{a}_i = \partial \boldsymbol{r}/\partial u^i$, that are a convenient basis for the tangent plane at that point. These basis vectors and their reciprocals are vector functions, $\boldsymbol{a}_i(u, v)$, $\boldsymbol{a}^i(u, v)$, that vary from point to point, in general. If a particle moves at constant speed but the tangent vectors $\boldsymbol{a}_i$ increase in magnitude with u^i, for example, then the components $\dot u^i$ will *decrease* proportionally to maintain constant speed. That is, the velocity components $\dot u^i$ have a geometric variation factor that is *contrary* to the variation of the velocity basis vectors $\boldsymbol{a}_i$.

Likewise, the reciprocal vectors $\boldsymbol{a}^i$ will vary inversely to $\boldsymbol{a}_i$, decreasing in magnitude if $\boldsymbol{a}_i$ increases since $\boldsymbol{a}_i \cdot \boldsymbol{a}^j = \delta_i^j$. Quantities with a geometric amplification factor that is inverse to that of the velocity basis vectors $\boldsymbol{a}_i$ are called *contravariant* and denoted with an *upper index,* v^i,

$\boldsymbol{a}^i$. Quantities with a geometric amplification factor equal to that of the velocity basis $\boldsymbol{a}_i$ are called *covariant* and denoted with a *lower index* such as v_i, $\boldsymbol{a}_i$.

A derivative with respect to a coordinate, $\partial/\partial u^i$, has a geometric variation factor that is inverse to that of the coordinate differentials $\mathrm{d}u^i$. For example, moving along a u^1-curve, with fixed u^2, the change per unit length of a function f is

$$\frac{\mathrm{d}f}{\mathrm{d}s} = \frac{\partial f}{\partial u^1}\frac{\mathrm{d}u^1}{\mathrm{d}s}.$$

If $|\boldsymbol{a}_1| = |\partial \boldsymbol{r}/\partial u^1| = |\mathrm{d}s/\mathrm{d}u^1|$ increases, for example, then $|\mathrm{d}u^1/\mathrm{d}s|$ decreases and $\partial f/\partial u^1$ will have a proportional increase to compensate for the shrinking $|\mathrm{d}u^1/\mathrm{d}s|$. So $\partial/\partial u^i$ has a covarying geometric factor, and will be abbreviated as the covariant operator ∂_i, thus writing

$$\boldsymbol{a}_i = \frac{\partial \boldsymbol{r}}{\partial u^i} = \partial_i \boldsymbol{r}. \tag{14.21}$$

The index i in $\partial/\partial u^i$ is thus considered a lower index, and expressions such as

$$\begin{aligned} \frac{\partial f}{\partial u^i}\,\mathrm{d}u^i &\equiv \frac{\partial f}{\partial u^1}\,\mathrm{d}u^1 + \cdots + \frac{\partial f}{\partial u^n}\,\mathrm{d}u^n, \\ \frac{\partial v^i}{\partial u^i} &\equiv \frac{\partial v^1}{\partial u^1} + \cdots + \frac{\partial v^n}{\partial u^n} \end{aligned} \tag{14.22}$$

are sums over i in the summation convention with upper and lower indices.

We could, of course, orthonormalize the basis vectors $\boldsymbol{a}_i \to \boldsymbol{q}_i$ using Gram–Schmidt, and work with an orthonormal tangent plane basis $\boldsymbol{q}_i(u, v)$. However, that introduces square roots, and turns out to be more cumbersome since these are vector functions, and we need to take multiple derivatives of those functions to investigate geodesics and curvature, for example. If the parametrization $\boldsymbol{r}(u, v)$ is suitable for the geometry, then $\boldsymbol{a}_i = \partial_i \boldsymbol{r}$ is the best basis to use, and the reciprocals $\boldsymbol{a}^i$ provide straightforward access to the components.[3]

Metrics and first fundamental form

Since $\{\boldsymbol{a}_1, \boldsymbol{a}_2\}$ and $\{\boldsymbol{a}^1, \boldsymbol{a}^2\}$ span the same tangent plane and are mutually orthonormal (14.16), the expansion of one basis in terms of the other reads

$$\begin{aligned} \boldsymbol{a}_i &= (\boldsymbol{a}_i \cdot \boldsymbol{a}_j)\,\boldsymbol{a}^j = g_{ij}\,\boldsymbol{a}^j, \\ \boldsymbol{a}^i &= (\boldsymbol{a}^i \cdot \boldsymbol{a}^j)\,\boldsymbol{a}_j = g^{ij}\boldsymbol{a}_j, \end{aligned} \tag{14.23}$$

[3]If the columns of matrix $\mathbf{A} = [\mathbf{a}_1\ \mathbf{a}_2]$ are the cartesian components of the covariant basis, the QR decomposition (§ 9.5) yields $\mathbf{A} = [\mathbf{a}_1\ \mathbf{a}_2] = \mathbf{QR}$, then the columns of $[\mathbf{a}^1\ \mathbf{a}^2] = \mathbf{QR}^{-\mathsf{T}}$ are the cartesian components of the reciprocal basis. The matrix $(\mathbf{A}^\mathsf{T}\mathbf{A})^{-1}\mathbf{A}^\mathsf{T} = \mathbf{R}^{-1}\mathbf{Q}^\mathsf{T} = [\mathbf{a}^1\ \mathbf{a}^2]^\mathsf{T}$ is a left inverse of $\mathbf{A}$ (aka a pseudo-inverse), whose *rows* are the reciprocal basis. For surfaces in 3D, matrix $\mathbf{A} = [\mathbf{a}_1\ \mathbf{a}_2]$ and its reciprocal $[\mathbf{a}^1\ \mathbf{a}^2]$ are 3-by-2 .

showing that the *metric* coefficients g_{ij} of the basis vectors $\boldsymbol{a}_i$ are their components in the reciprocal basis $\boldsymbol{a}^j$, and vice versa. The index positions are kept for the *metrics* also, with

$$g_{ij} = \boldsymbol{a}_i \cdot \boldsymbol{a}_j, \qquad g^{ij} = \boldsymbol{a}^i \cdot \boldsymbol{a}^j.$$

The metrics are clearly symmetric $g_{ij} = g_{ji}$, and $g^{ij} = g^{ji}$, and it follows from (14.16) and (14.23) that they are inverses of each other,

$$g_{ik}\, g^{kj} = \delta_i^j = g^{jk}\, g_{ki}. \tag{14.24}$$

The metrics transform contravariant into covariant components and vice versa, that is, multiplication by the appropriate metric *lowers or raises indices.* For any vector $\boldsymbol{p}$ in the tangent space, $\boldsymbol{p} = p^i \boldsymbol{a}_i = p_i \boldsymbol{a}^i$ and

$$p_i = g_{ij} p^j, \qquad p^i = g^{ij} p_j. \tag{14.25}$$

The dot product between any two vectors $\boldsymbol{p}$ and $\boldsymbol{q}$ in the tangent space reads

$$\boldsymbol{p} \cdot \boldsymbol{q} = g_{ij} p^i q^j = p_i q^j = p^i q_j = g^{ij} p_i q_j, \tag{14.26}$$

in terms of their contravariant, or mixed, or covariant components. In particular, the dot product keeps a simple form

$$\boldsymbol{p} \cdot \boldsymbol{q} = \begin{cases} p_1 q^1 + p_2 q^2 + \cdots + p_n q^n = p_i q^j, \\ p^1 q_1 + p^2 q_2 + \cdots + p^n q_n = p^i q_j, \end{cases}$$

when $\boldsymbol{p}$ is expressed in one basis, and $\boldsymbol{q}$ in the other.

The metrics arise in computing distances, angles and areas. The square of the arclength element is $\mathrm{d}s^2 = \mathrm{d}\boldsymbol{r} \cdot \mathrm{d}\boldsymbol{r}$ with $\mathrm{d}\boldsymbol{r} = \partial_i \boldsymbol{r}\, \mathrm{d}u^i$, yielding the *first fundamental form* for a surface as

$$\mathrm{d}s^2 = \left(\partial_i \boldsymbol{r}\, \mathrm{d}u^i\right) \cdot \left(\partial_j \boldsymbol{r}\, \mathrm{d}u^j\right) = \boldsymbol{a}_i \cdot \boldsymbol{a}_j\, \mathrm{d}u^i \mathrm{d}u^j = g_{ij}\, \mathrm{d}u^i \mathrm{d}u^j, \tag{14.27}$$

where $\mathrm{d}s^2 = \mathrm{d}\boldsymbol{r} \cdot \mathrm{d}\boldsymbol{r} = (\mathrm{d}s)^2$, not $\mathrm{d}(s^2) = 2s\, \mathrm{d}s$, as seen in (14.3). For a 2D surface, the determinant of the 2-by-2 metric is

$$g \triangleq g_{11} g_{22} - g_{12} g_{21} = \epsilon^{ij3} g_{i1} g_{j2}, \tag{14.28a}$$

and using the vector identity $(\boldsymbol{a} \times \boldsymbol{b}) \cdot (\boldsymbol{a} \times \boldsymbol{b}) = (\boldsymbol{a} \cdot \boldsymbol{a})(\boldsymbol{b} \cdot \boldsymbol{b}) - (\boldsymbol{a} \cdot \boldsymbol{b})^2$, this yields the area Jacobian (14.4), (14.15), as

$$J = |\boldsymbol{a}_1 \times \boldsymbol{a}_2| = |\partial_1 \boldsymbol{r} \times \partial_2 \boldsymbol{r}| = \sqrt{g}, \tag{14.28b}$$

and the surface element

$$\mathrm{d}\boldsymbol{S} = \partial_1 \boldsymbol{r} \times \partial_2 \boldsymbol{r}\, \mathrm{d}u^1 \mathrm{d}u^2 = \hat{\boldsymbol{n}}\, \sqrt{g}\, \mathrm{d}u^1 \mathrm{d}u^2. \tag{14.28c}$$

14.5 Geodesics

A geodesic is a curve that turns only because the surface forces it to. On a plane, it is a straight line. On a curved surface, the geodesic only turns in a direction $\hat{\boldsymbol{n}}_c$ normal to the surface. If $\hat{\boldsymbol{n}}$ is the unit normal to the surface, then a geodesic's turn direction $\hat{\boldsymbol{n}}_c = \pm\hat{\boldsymbol{n}}$. From (14.6) and (14.7a) this implies that the acceleration along a geodesic has the form

$$\ddot{\boldsymbol{r}} = \ddot{s}\,\hat{\boldsymbol{t}} + \kappa_n \dot{s}^2\,\hat{\boldsymbol{n}}$$

such that

$$\ddot{\boldsymbol{r}} \cdot \hat{\boldsymbol{n}} \times \dot{\boldsymbol{r}} = 0, \tag{14.29}$$

since $\dot{\boldsymbol{r}} = \dot{s}\,\hat{\boldsymbol{t}}$; thus, the "geodesic curvature" $\kappa_g = 0$ for a geodesic (14.7b). To determine the geodesics, we can further restrict to arbitrary constant speed such that $\ddot{s} = 0$, $\dot{s} = \dot{s}_0$ and $\ddot{\boldsymbol{r}} \cdot \dot{\boldsymbol{r}} = 0$. The acceleration when traveling at constant speed $\dot{s}_0$ along a geodesic is thus purely in the surface normal direction

$$\ddot{\boldsymbol{r}} = \kappa_n \dot{s}_0^2\,\hat{\boldsymbol{n}}. \tag{14.30}$$

Geodesic equations

Since the surface normal $\hat{\boldsymbol{n}}$ is orthogonal to the surface tangents $\partial\boldsymbol{r}/\partial u$, $\partial\boldsymbol{r}/\partial v$, the geodesics equations follow from (14.30) as

$$\ddot{\boldsymbol{r}} \cdot \frac{\partial \boldsymbol{r}}{\partial u} = 0, \quad \ddot{\boldsymbol{r}} \cdot \frac{\partial \boldsymbol{r}}{\partial v} = 0. \tag{14.31}$$

Substituting for $\ddot{\boldsymbol{r}}$ from (14.5) yields two coupled second order differential equations for the surface coordinates $(u(t), v(t))$, namely,

$$\frac{\partial \boldsymbol{r}}{\partial u} \cdot \frac{\partial \boldsymbol{r}}{\partial u}\ddot{u} + \frac{\partial \boldsymbol{r}}{\partial u} \cdot \frac{\partial \boldsymbol{r}}{\partial v}\ddot{v} + \frac{\partial \boldsymbol{r}}{\partial u} \cdot \frac{\partial^2 \boldsymbol{r}}{\partial u^2}\dot{u}^2 + 2\frac{\partial \boldsymbol{r}}{\partial u} \cdot \frac{\partial^2 \boldsymbol{r}}{\partial u \partial v}\dot{u}\dot{v} + \frac{\partial \boldsymbol{r}}{\partial u} \cdot \frac{\partial^2 \boldsymbol{r}}{\partial v^2}\dot{v}^2 = 0, \tag{14.32a}$$

$$\frac{\partial \boldsymbol{r}}{\partial v} \cdot \frac{\partial \boldsymbol{r}}{\partial u}\ddot{u} + \frac{\partial \boldsymbol{r}}{\partial v} \cdot \frac{\partial \boldsymbol{r}}{\partial v}\ddot{v} + \frac{\partial \boldsymbol{r}}{\partial v} \cdot \frac{\partial^2 \boldsymbol{r}}{\partial u^2}\dot{u}^2 + 2\frac{\partial \boldsymbol{r}}{\partial v} \cdot \frac{\partial^2 \boldsymbol{r}}{\partial u \partial v}\dot{u}\dot{v} + \frac{\partial \boldsymbol{r}}{\partial v} \cdot \frac{\partial^2 \boldsymbol{r}}{\partial v^2}\dot{v}^2 = 0. \tag{14.32b}$$

The formulas are more compact and easier to manipulate in index notation with $(u, v) \to u^i \equiv (u^1, u^2)$ and summation over indices appearing in a *lower* and *upper* position in the same term. The indices run only from 1 to 2 here, but the index notation formulas are identical to those in n-dimensional spaces where indices run from 1 to n. The surface is assumed smooth enough that $\boldsymbol{r}(u, v)$ is at least twice continuously differentiable (C^2), so the mixed partials commute $\partial_i\partial_j = \partial_j\partial_i$.

In index notation, eqns (14.31) with the acceleration (14.5) read

$$(\partial_i \boldsymbol{r}) \cdot \ddot{\boldsymbol{r}} = (\partial_i \boldsymbol{r}) \cdot (\ddot{u}^j\,\partial_j \boldsymbol{r} + \dot{u}^j \dot{u}^k\,\partial_j \partial_k \boldsymbol{r}) = 0,$$

yielding the *geodesic equations* (14.32) as

$$g_{ij}\ddot{u}^j + \Gamma_{ijk}\dot{u}^j\dot{u}^k = 0, \tag{14.33}$$

where[4]

$$\Gamma_{ijk} = \partial_i \boldsymbol{r} \cdot \partial_j \partial_k \boldsymbol{r} = \frac{\partial \boldsymbol{r}}{\partial u^i} \cdot \frac{\partial^2 \boldsymbol{r}}{\partial u^j \partial u^k} \tag{14.34}$$

is the *Christoffel symbol of the first kind*, symmetric in its last two indices $\Gamma_{ijk} = \Gamma_{ikj}$, and $g_{ij} = \boldsymbol{a}_i \cdot \boldsymbol{a}_j$ is the covariant metric (14.23).

Multiplying (14.33) by the inverse metric g^{li}, such that $g^{li}g_{ij} = \delta^l_j$, then renaming $l \to i$, yields the diagonalized form of the *geodesic equations*

$$\ddot{u}^i + \Gamma^i_{jk}\dot{u}^j\dot{u}^k = 0, \tag{14.35}$$

where

$$\Gamma^i_{jk} = g^{il}\Gamma_{ljk} \tag{14.36}$$

is the *Christoffel symbol of the second kind.* Equations (14.35) can also be derived by projecting (14.30) on the contravariant basis $\boldsymbol{a}^i$ instead of the covariant basis $\boldsymbol{a}_i = \partial_i \boldsymbol{r}$. This yields an equivalent expression for the Christoffel symbol of the second kind,

$$\Gamma^i_{jk} = \boldsymbol{a}^i \cdot \frac{\partial^2 \boldsymbol{r}}{\partial u^j \partial u^k}, \tag{14.37}$$

with $\boldsymbol{a}^i$ defined in (14.15) for a surface in 3D space, and (14.23) more generally. From (14.24), the inverse of (14.36) is

$$\Gamma_{ijk} = g_{il}\Gamma^l_{jk}. \tag{14.38}$$

The geodesic equations (14.35) are two coupled second order ordinary differential equations for $(u^1(t), u^2(t)) \equiv (u(t), v(t))$. From a mechanical point of view, these are, in fact, Newton's second order vector differential equation $m\ddot{\boldsymbol{r}} = \boldsymbol{F}(\boldsymbol{r}, \dot{\boldsymbol{r}})$, where the force $\boldsymbol{F}(\boldsymbol{r}, \dot{\boldsymbol{r}})$ is the purely normal force that keeps the particle on the surface. We can solve them, in principle or numerically, as an *initial value problem* from a starting point $(u(0), v(0))$ with an initial velocity $(\dot{u}(0), \dot{v}(0))$, or as a *boundary value problem* from a starting point $(u(0), v(0))$ to an end point $(u(1), v(1))$. For the latter, note that the geodesic equations (14.35) are homogeneous in time so we can rescale time arbitrarily and pick the end time as $t = 1$. In other words, the constant travel speed $\dot{s}_0$ is arbitrary, and the total length can only be determined after the geodesic curve has been computed.

Christoffel symbols and Gauss formulas

The Christoffel symbols (14.34) and (14.37) are the components of $\partial_j \partial_k \boldsymbol{r}$ in the contravariant or covariant bases. In general, there is a third component in direction $\hat{\boldsymbol{n}}$ normal to the surface given by b_{ij} (14.8c), and thus

$$\frac{\partial^2 \boldsymbol{r}}{\partial u^i \partial u^j} = \Gamma_{kij}\boldsymbol{a}^k + b_{ij}\hat{\boldsymbol{n}} = \Gamma^k_{ij}\boldsymbol{a}_k + b_{ij}\hat{\boldsymbol{n}}. \tag{14.39}$$

[4] Erwin Kreyszig defines instead

$$\Gamma_{ijk} = \partial_i \partial_j \boldsymbol{r} \cdot \partial_k \boldsymbol{r},$$

eqn (45.6) in his *Differential Geometry*, Dover, 1991.

These are the *Gauss formulas*, defining Γ_{kij}, b_{ij} as the components of the second derivatives $\partial_i\partial_j\boldsymbol{r}$ in the contravariant basis $\{\boldsymbol{a}^1, \boldsymbol{a}^2, \hat{\boldsymbol{n}}\}$, and Γ^k_{ij} and b_{ij} as their components in the covariant basis $\{\partial_1\boldsymbol{r}, \partial_2\boldsymbol{r}, \hat{\boldsymbol{n}}\}$.

The Christoffel symbols of the first kind, Γ_{ijk}, are the dot products of "velocities" with "accelerations." In dynamics, that leads to kinetic energy $\dot{\boldsymbol{r}} \cdot \ddot{\boldsymbol{r}} = \frac{1}{2}\frac{\mathrm{d}}{\mathrm{d}t}(\dot{\boldsymbol{r}} \cdot \dot{\boldsymbol{r}})$. Likewise,

$$\Gamma_{iji} = \frac{\partial \boldsymbol{r}}{\partial u^i} \cdot \frac{\partial^2 \boldsymbol{r}}{\partial u^j \partial u^i} = \frac{1}{2}\frac{\partial}{\partial u^j}\left(\frac{\partial \boldsymbol{r}}{\partial u^i} \cdot \frac{\partial \boldsymbol{r}}{\partial u^i}\right) = \frac{1}{2}\frac{\partial g_{ii}}{\partial u^j}, \tag{14.40}$$

with *no sum* over the repeated i that occurs in the same upper, or lower position in each term.[5] Formula (14.40) applies in any dimension. For a surface of dimension 2, it gives 4+2=6 of the eight symbols, because of the identity $\Gamma_{ijk} = \Gamma_{ikj}$. The missing two symbols are

$$\Gamma_{122} = \frac{\partial g_{12}}{\partial u^2} - \frac{1}{2}\frac{\partial g_{22}}{\partial u^1}, \qquad \Gamma_{211} = \frac{\partial g_{12}}{\partial u^1} - \frac{1}{2}\frac{\partial g_{11}}{\partial u^2}. \tag{14.41}$$

[5] In Ricci calculus with upper and lower indices, summation is only when an index is repeated once lower and once upper in the same term.

A general formula, applicable to higher dimension, can be derived from the product rule applied to $g_{jk} = \partial_j\boldsymbol{r} \cdot \partial_k\boldsymbol{r}$, yielding

$$\partial_i g_{jk} = \Gamma_{kij} + \Gamma_{jki} = g_{kl}\Gamma^l_{ij} + g_{jl}\Gamma^l_{ki}. \tag{14.42}$$

Rewriting this formula twice for $(i, j, k) \to (j, k, i) \to (k, i, j)$ gives two similar equations for the three unknowns Γ_{ijk}, Γ_{jki}, Γ_{kij}, thanks to the $\Gamma_{jik} = \Gamma_{jki}$ symmetry. Adding the last two and subtracting the first yields

$$\Gamma_{ijk} = \tfrac{1}{2}\left(\partial_j g_{ki} + \partial_k g_{ij} - \partial_i g_{jk}\right), \tag{14.43}$$

where $\partial_i = \partial/\partial u^i$. This formula shows that the Christoffel symbols can be evaluated through derivatives of the metric g_{ij} only, they are *intrinsic* in the language of differential geometry, and there is no need to compute and project the second derivatives of $\boldsymbol{r}(u, v)$. Next, we consider an earthy example of this machinery, while exercise 14.15 lays out a shorter path to the same solution.

Geodesics on a sphere

To illustrate these concepts, consider a sphere of radius r parametrized by polar and azimuthal angles $\boldsymbol{r}(\theta, \varphi)$ (13.5). Thus, with $(u^1, u^2) \equiv (\theta, \varphi)$, the coordinate tangents are (13.16),

$$\boldsymbol{a}_1 \equiv \partial_\theta \boldsymbol{r} = r\hat{\boldsymbol{\theta}}, \qquad \boldsymbol{a}_2 \equiv \partial_\varphi \boldsymbol{r} = r\sin\theta\hat{\boldsymbol{\varphi}}. \tag{14.44}$$

The metric $g_{ij} = \boldsymbol{a}_i \cdot \boldsymbol{a}_j = g_{ji}$ and its inverse are

$$\begin{bmatrix} g_{11} & g_{12} \\ g_{21} & g_{22} \end{bmatrix} = \begin{bmatrix} r^2 & 0 \\ 0 & r^2\sin^2\theta \end{bmatrix}, \quad \begin{bmatrix} g^{11} & g^{12} \\ g^{21} & g^{22} \end{bmatrix} = \begin{bmatrix} r^{-2} & 0 \\ 0 & (r\sin\theta)^{-2} \end{bmatrix}. \tag{14.45}$$

The unit normal to the sphere is $\hat{\boldsymbol{n}} = \hat{\boldsymbol{r}}$. The contravariant basis can be obtained from (14.15) with $J = \boldsymbol{a}_1 \times \boldsymbol{a}_2 \cdot \hat{\boldsymbol{r}} = r^2\sin\theta$, or without cross products from (14.23) and (14.45) as

$$\begin{aligned} \boldsymbol{a}^1 &= (\boldsymbol{a}_2 \times \hat{\boldsymbol{r}})/J = g^{11}\boldsymbol{a}_1 + g^{12}\boldsymbol{a}_2 = \hat{\boldsymbol{\theta}}/r, \\ \boldsymbol{a}^2 &= (\hat{\boldsymbol{r}} \times \boldsymbol{a}_1)/J = g^{21}\boldsymbol{a}_1 + g^{22}\boldsymbol{a}_2 = \hat{\boldsymbol{\varphi}}/(r\sin\theta). \end{aligned} \tag{14.46}$$

These spherical coordinates are orthogonal; hence, $\boldsymbol{a}_i$ and $\boldsymbol{a}^i$ are parallel, but have inverse magnitudes.

For geodesics, we can restrict to the unit sphere, $r = 1$, since the surface lengthscale drops out of the geodesic equations. The only non-zero metric derivative is $\partial_1 g_{22} = 2\sin\theta\cos\theta = \sin 2\theta$. The eight Christoffel symbols reduce to six since $\Gamma_{ijk} = \Gamma_{ikj}$; they follow from (14.40) and (14.41) as

$$\begin{bmatrix} \Gamma_{111} & \Gamma_{112}(=\Gamma_{121}) & \Gamma_{122} \\ \Gamma_{211} & \Gamma_{212}(=\Gamma_{221}) & \Gamma_{222} \end{bmatrix} = \begin{bmatrix} 0 & 0 & -\frac{1}{2}\sin 2\theta \\ 0 & \frac{1}{2}\sin 2\theta & 0 \end{bmatrix}.$$

The symbols of the second kind, such that $\Gamma^i_{jk} = \Gamma^i_{kj}$, are obtained from (14.36) by premultiplying by the inverse metric

$$\begin{aligned} \begin{bmatrix} \Gamma^1_{11} & \Gamma^1_{12}(=\Gamma^1_{21}) & \Gamma^1_{22} \\ \Gamma^2_{11} & \Gamma^2_{12}(=\Gamma^2_{21}) & \Gamma^2_{22} \end{bmatrix} &= \begin{bmatrix} g^{11} & g^{12} \\ g^{12} & g^{22} \end{bmatrix} \begin{bmatrix} \Gamma_{111} & \Gamma_{112} & \Gamma_{122} \\ \Gamma_{211} & \Gamma_{212} & \Gamma_{222} \end{bmatrix} \\ &= \begin{bmatrix} 0 & 0 & -\sin\theta\cos\theta \\ 0 & \cot\theta & 0 \end{bmatrix}, \end{aligned} \tag{14.47}$$

where $\cot\theta = \cos\theta/\sin\theta$ is the cotangent (Fig. 14.5). The geodesic equations (14.35) read

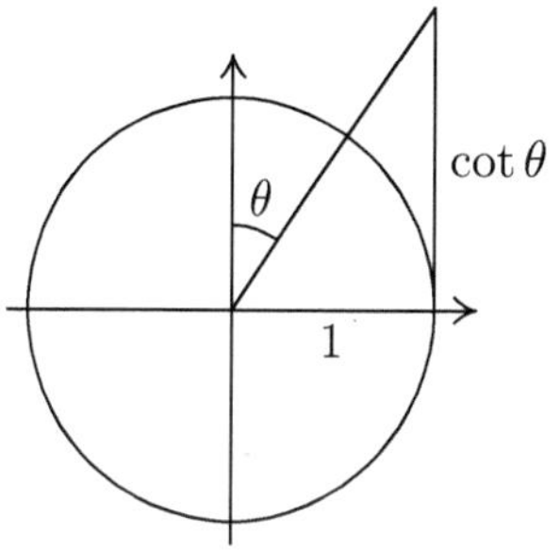

Fig. 14.5 Cotangent.

$$\begin{cases} \ddot{\theta} = \sin\theta\cos\theta\,(\dot{\varphi})^2, \\ \ddot{\varphi} = -2\cot\theta\,(\dot{\theta}\dot{\varphi}). \end{cases} \tag{14.48}$$

Multiplying the first by $\dot{\theta}$, and the second by $\dot{\varphi}$, yields

$$\begin{cases} \frac{1}{2}\dfrac{\mathrm{d}}{\mathrm{d}t}(\dot{\theta})^2 = \dot{\theta}\,\sin\theta\cos\theta\,(\dot{\varphi})^2, \\ \frac{1}{2}\dfrac{\mathrm{d}}{\mathrm{d}t}(\dot{\varphi})^2 = -2\cot\theta\,\dot{\theta}\,(\dot{\varphi})^2. \end{cases} \tag{14.49}$$

The latter is readily separated, and

$$\frac{\mathrm{d}}{\mathrm{d}t}\ln(\dot{\varphi})^2 = -4\frac{\mathrm{d}}{\mathrm{d}t}\ln\sin\theta \;\Rightarrow\; \dot{\varphi} = C_1/\sin^2\theta,$$

where C_1 is a constant of integration. Inserting this $\dot{\varphi}$ in the $\dot{\theta}$ equation gives

$$\frac{\mathrm{d}}{\mathrm{d}t}(\dot{\theta})^2 = 2C_1^2\frac{\cos\theta}{\sin^3\theta}\,\dot{\theta} \;\Rightarrow\; (\dot{\theta})^2 = C_2^2 - \frac{C_1^2}{\sin^2\theta},$$

where C_2^2 is another constant. Eliminating $\mathrm{d}t$ between $\dot{\varphi}$ and $\dot{\theta}$ then yields

$$\mathrm{d}\varphi = \frac{a\,\mathrm{d}\theta}{\sin^2\theta\sqrt{1 - a^2/\sin^2\theta}},$$

where $a = C_1/C_2$. The substitution $u = \cot\theta$, with $\mathrm{d}u = -\mathrm{d}\theta/\sin^2\theta$ and $1+u^2 = 1+\cot^2\theta = 1/\sin^2\theta$, leads to $\mathrm{d}\varphi = \mathrm{d}(\arccos(u/u_*))$, where $u_* = \sqrt{1-a^2}/a$, yielding the geodesics as

$$\cot\theta = \cot\theta_* \cos(\varphi - \varphi_*), \tag{14.50}$$

where $u_* = \cot\theta_*$ and φ_* are constants to be determined from initial or boundary conditions. (θ_*, φ_*) are the spherical coordinates of the point on the particular geodesic that is closest to the pole. The geodesics from $(\theta, \varphi) = (45°, \varphi_0)$ are shown in Fig. 14.6 in the longitude φ and latitude $90° - \theta$ coordinate plane.

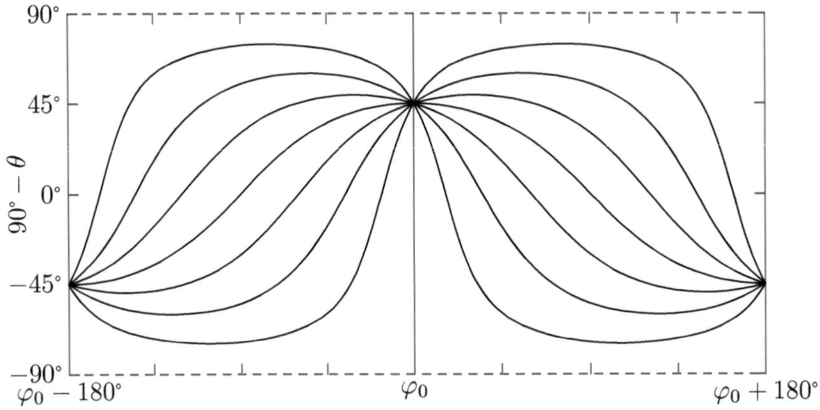

Fig. 14.6 Geodesics for spherical coordinates from $(\theta_0 = 45°, \varphi_0)$.

Geometrically, the geodesics are *arcs of great circles*, that is, the geodesic between two points A and B on a sphere is the (shortest) arc of the circle passing through A and B that is concentric with the sphere (exercise 14.24). The rate of change of the unit tangent $\mathrm{d}\hat{\boldsymbol{t}}/\mathrm{d}s = \mathrm{d}^2\boldsymbol{r}/\mathrm{d}s^2$ along those arcs is parallel to the surface normal $\hat{\boldsymbol{n}} = \hat{\boldsymbol{r}}$, and that is the geodesic condition (14.33).

Note that combining $(\dot{\theta})^2 = C_2^2 - C_1^2/\sin^2\theta$ with $\dot{\varphi} = C_1/\sin^2\theta$ yields

$$(\dot{\theta})^2 + \sin^2\theta(\dot{\varphi})^2 = C_2^2,$$

but arclength in spherical coordinates on a sphere of radius r is

$$\mathrm{d}s^2 = \mathrm{d}\boldsymbol{r}\cdot\mathrm{d}\boldsymbol{r} = r^2\mathrm{d}\theta^2 + r^2\sin^2\theta\,\mathrm{d}\varphi^2,$$

so $rC_2 = \dot{s}_0$ is the constant speed along the geodesic.

14.6 Minimizing arclength

It should be clear that the tight geodesics,[6] which turn only because they really have to, are the shortest paths between two points on a surface. The curve length $\mathcal{L}$ along an arbitrary curve $\mathcal{C} : t \in \mathbb{R} \to \boldsymbol{r}(u(t), v(t))$ on the surface $\boldsymbol{r}(u, v)$ is

$$\mathcal{L} = \int_{\mathcal{C}} |\mathrm{d}\boldsymbol{r}| = \int_{t_0}^{t_f} \sqrt{g_{ij}\, \dot{u}^i\, \dot{u}^j}\, \mathrm{d}t \tag{14.51}$$

from expressions (14.19) for $\mathrm{d}\boldsymbol{r}$, and (14.27) for $\mathrm{d}s = \sqrt{\mathrm{d}\boldsymbol{r} \cdot \mathrm{d}\boldsymbol{r}}$. The metric coefficients g_{ij} are functions of $(u^1, u^2) \equiv (u, v)$, which are functions of the time-like parameter t.

Minimizing curve length between two points thus involves finding a curve, that is, $(u(t), v(t))$ in the coordinate space, which minimizes $\mathcal{L}$ given in (14.51) as an integral over the unknown functions $(u(t), v(t))$. How do we do that?

[6] To find the geodesic between points A and B on a slippery sphere: fix one end of a string at A and pull it from B until it is tight and unable to slip any further. For more general surfaces, we need to imagine two physical copies of the surface on top of one another, and the string slipping and tightening in between those two identical surfaces, unable to "pop out" of the surface.

Euler–Lagrange equations

Let's illustrate the idea on an example that is both more general than (14.51) and simpler by dealing with a single variable u to begin with. The standard formulation is to consider

$$\mathcal{F} = \int_a^b L(t, u, \dot{u})\, \mathrm{d}t \tag{14.52}$$

with $t \in \mathbb{R}$ and a, b fixed. $\mathcal{F}$ is a *functional*, a function of the function $u(t)$. We are interested in the extrema of $\mathcal{F}$ over all functions $u(t)$ with fixed starting and ending values $u(a) = u_0$, $u(b) = u_1$.

Let $u_*(t)$ be the unknown function $u(t)$ that yields an extremum of $\mathcal{F}$. To derive an equation for $u_*(t)$, we consider *variations* of $u(t)$ about $u_*(t)$. In physics and engineering that is usually written

$$u(t) = u_*(t) + \delta u(t),$$

where $\delta u(t)$ is an arbitrary variation of $u_*(t)$ that vanishes at the end points $\delta u(a) = 0 = \delta u(b)$, since $u(a)$ and $u(v)$ are fixed. Substituting in the integral (14.52) for $\mathcal{F}$, and Taylor expanding in powers of δu, the terms linear in δu provide the *first variation* of $\mathcal{F}$ as

$$\delta\mathcal{F} = \int_a^b \left(\frac{\partial L}{\partial u} \delta u + \frac{\partial L}{\partial \dot{u}} \delta \dot{u} \right) \mathrm{d}t \tag{14.53}$$

by the chain rule applied to $L(u, \dot{u})$. The partial derivatives are evaluated at $u(t) = u_*(t)$, and $\delta\mathcal{F} = 0$ for any $\delta u(t)$ that vanishes at the end points, if $u_*(t)$ is an extremum. This is *calculus of variations*.

A somewhat cleaner approach is to define

$$\delta u(t) = \epsilon\, v(t),$$

where $v(t)$ is an arbitrary function that vanishes at the end points $v(a) = 0 = v(b)$, and ϵ is a real variable. Assuming for the moment that $v(t)$ is an arbitrary but fixed function transforms $\mathcal{F}$ into an actual function of ϵ only, $\mathcal{F}(\epsilon)$, and that function should have an extremum at $\epsilon = 0$ if $u_*(t)$ is an extremizer, that is,

$$\left.\frac{\mathrm{d}\mathcal{F}}{\mathrm{d}\epsilon}\right|_{\epsilon=0} = \int_a^b \left(\frac{\partial L}{\partial u}v + \frac{\partial L}{\partial \dot{u}}\dot{v}\right)\,\mathrm{d}t = 0, \tag{14.54}$$

for any $v(t)$ with $v(a) = 0 = v(b)$. Clearly, this is the same result as saying that the first variation $\delta\mathcal{F} = 0$ in (14.53).

The remaining step is that (14.54) must hold for any $v(t)$, but $v(t)$ and $\dot{v}(t)$ are different, yet related functions. That connection is unraveled through integration by parts, aka product rule, on the second term, and (14.54) becomes

$$\int_a^b \left(\frac{\partial L}{\partial u}v + \frac{\mathrm{d}}{\mathrm{d}t}\left(\frac{\partial L}{\partial \dot{u}}v\right) - v\frac{\mathrm{d}}{\mathrm{d}t}\left(\frac{\partial L}{\partial \dot{u}}\right)\right)\mathrm{d}t$$
$$= \left[\frac{\partial L}{\partial \dot{u}}v\right]_a^b - \int_a^b \left(\frac{\mathrm{d}}{\mathrm{d}t}\left(\frac{\partial L}{\partial \dot{u}}\right) - \frac{\partial L}{\partial u}\right)v\,\mathrm{d}t = 0.$$

The boundary term vanishes since $v(a) = 0 = v(b)$. The remaining integral must vanish for any $v(t)$, and that can only happen if[7]

$$\frac{\mathrm{d}}{\mathrm{d}t}\left(\frac{\partial L}{\partial \dot{u}}\right) = \frac{\partial L}{\partial u}. \tag{14.55}$$

These are the *Euler–Lagrange equations* for the functional (14.52). If the functional has several variable functions $(u^1(t), u^2(t), \ldots)$, then we simply have one such equation for every variable,

$$\frac{\mathrm{d}}{\mathrm{d}t}\left(\frac{\partial L}{\partial \dot{u}^i}\right) = \frac{\partial L}{\partial u^i}. \tag{14.56}$$

These are the equations that the extremizers $u_*^i(t)$ must satisfy.

Geodesics as extremizers

The Euler–Lagrange equations for the extrema of the curve length (14.51) are

$$\frac{\mathrm{d}}{\mathrm{d}t}\left(\frac{\partial}{\partial \dot{u}^i}\sqrt{g_{jk}\,\dot{u}^j\,\dot{u}^k}\right) = \frac{\partial}{\partial u^i}\sqrt{g_{jk}\,\dot{u}^j\,\dot{u}^k}$$

where the metric g_{jk} depends on the coordinates u^i but not the coordinate velocities $\dot{u}^i$. Therefore, those Euler–Lagrange equations become

$$\frac{\mathrm{d}}{\mathrm{d}t}\left(\frac{g_{ij}\dot{u}^j}{\sqrt{\cdots}}\right) = \frac{1}{2}\frac{(\partial_i g_{jk})}{\sqrt{\cdots}}\,\dot{u}^j\dot{u}^k, \tag{14.57}$$

[7] If a sum $f_1g^1 + f_2g^2 + \cdots + f_ng^n = 0$ for any g^k, then $f_k = 0$ for all k. Pick $g^k = \delta_1^k, \delta_2^k, \ldots$.

where $\partial_i = \partial/\partial u^i$, and we have used $\partial \dot{u}^j/\partial \dot{u}^i = \delta^j_i$, and $g_{ij}\dot{u}^j = g_{ik}\dot{u}^k$. Now

$$\sqrt{\cdots} = \sqrt{g_{jk}\,\dot{u}^j\,\dot{u}^k} = \frac{\mathrm{d}s}{\mathrm{d}t}$$

is the speed along the curve, and we can assume that it is constant along this curve. Of course, this constant speed along a given curve varies from curve to curve when considering *all the possible curves* from point A to point B in the functional (14.51). For a single curve, we can assume constant speed, undetermined until we actually find the particular length along that curve from A to B.

With constant $\sqrt{\cdots}$, (14.57) simplifies to

$$\frac{\mathrm{d}}{\mathrm{d}t}\left(g_{ij}\dot{u}^j\right) = \tfrac{1}{2}(\partial_i g_{jk})\,\dot{u}^j\dot{u}^k, \tag{14.58}$$

and by the chain rule on the left-hand side,

$$g_{ij}\ddot{u}^j + (\partial_k g_{ij})\dot{u}^j\dot{u}^k = \tfrac{1}{2}(\partial_i g_{jk})\,\dot{u}^j\dot{u}^k, \tag{14.59}$$

which is the geodesic equations (14.33), with the explicit expression (14.43) for Γ_{ijk} since

$$(\partial_k g_{ij})\dot{u}^j\dot{u}^k = (\partial_j g_{ik})\dot{u}^j\dot{u}^k = \tfrac{1}{2}(\partial_k g_{ij} + \partial_j g_{ik})\dot{u}^j\dot{u}^k.$$

Therefore, geodesics are extremizers of curve length. They correspond to local minima or saddles. On the sphere for instance, the shorter path from A to B is the smallest great circle arc between them, but the complementary arc, around the world the other way, is a saddle for curve length. Almost all perturbations of that arc increase length, but a special kind of perturbation, sliding it perpendicular to itself one way or the other, reduces the length.

14.7 Normal curvature

The normal curvature κ_n (14.7) is the minimum curvature of a surface curve passing through a point P on the surface in a given direction $\hat{\boldsymbol{t}}$. It is the curvature of the geodesic passing through that point in that direction. It is a characteristic of the surface and a function of $\hat{\boldsymbol{t}}$. If $\boldsymbol{r}(u(t), v(t))$ is a surface curve with $\dot{\boldsymbol{r}} = \mathrm{d}\boldsymbol{r}/\mathrm{d}t = \dot{u}^i\partial_i\boldsymbol{r}$ and $\hat{\boldsymbol{t}} = \dot{\boldsymbol{r}}/|\dot{\boldsymbol{r}}|$, then, as derived in (14.2) and (14.7c),

$$\begin{aligned} |\dot{\boldsymbol{r}}|^2 &= (\partial_i\boldsymbol{r}\cdot\partial_j\boldsymbol{r})\,\dot{u}^i\dot{u}^j = g_{ij}\,\dot{u}^i\dot{u}^j, \\ \kappa_n\,|\dot{\boldsymbol{r}}|^2 &= (\hat{\boldsymbol{n}}\cdot\partial_i\partial_j\boldsymbol{r})\,\dot{u}^i\dot{u}^j = b_{ij}\,\dot{u}^i\dot{u}^j, \end{aligned} \tag{14.60a}$$

where $g_{ij} = \partial_i\boldsymbol{r}\cdot\partial_j\boldsymbol{r}$ is the metric and $b_{ij} = \hat{\boldsymbol{n}}\cdot\partial_i\partial_j\boldsymbol{r}$ the "curvic" (14.8c), with $i = 1, 2$ and $j = 1, 2$ for a 2D surface, and $\hat{\boldsymbol{n}}$ is the unit normal to the surface. The curve parameter t can be eliminated since $|\dot{\boldsymbol{r}}| = \mathrm{d}s/\mathrm{d}t$, $\dot{u}^i = \mathrm{d}u^i/\mathrm{d}t$, where s is arclength, to obtain the *first* and *second fundamental forms,* respectively,

$$\begin{aligned} \mathrm{d}s^2 &= g_{ij}\,\mathrm{d}u^i\mathrm{d}u^j, \\ \kappa_n\mathrm{d}s^2 &= b_{ij}\,\mathrm{d}u^i\mathrm{d}u^j. \end{aligned} \tag{14.60b}$$

The metric g_{ij} is symmetric and so is the curvic b_{ij} since $\partial_i\partial_j \boldsymbol{r} = \partial_j\partial_i \boldsymbol{r}$,

$$\begin{aligned} g_{ij} &= \partial_i \boldsymbol{r} \cdot \partial_j \boldsymbol{r} = g_{ji}, \\ b_{ij} &= \hat{\boldsymbol{n}} \cdot \partial_i\partial_j \boldsymbol{r} = b_{ji}. \end{aligned} \tag{14.61}$$

The identity $\hat{\boldsymbol{n}} \cdot \partial_j \boldsymbol{r} = 0$ (14.4) implies that $\partial_i(\hat{\boldsymbol{n}} \cdot \partial_j \boldsymbol{r}) = 0$ whose expansion yields

$$b_{ij} = \hat{\boldsymbol{n}} \cdot \partial_i\partial_j \boldsymbol{r} = -\partial_i \hat{\boldsymbol{n}} \cdot \partial_j \boldsymbol{r}. \tag{14.62}$$

This second expression for b_{ij} (14.62) transforms the second fundamental form (14.60b) into

$$\kappa_n \, \mathrm{d}s^2 = -(\mathrm{d}u^i \, \partial_i \hat{\boldsymbol{n}}) \cdot (\mathrm{d}u^j \, \partial_j \boldsymbol{r}) = -\mathrm{d}\hat{\boldsymbol{n}} \cdot \mathrm{d}\boldsymbol{r}. \tag{14.63a}$$

Thus, $\kappa_n \mathrm{d}s^2 = -\mathrm{d}\hat{\boldsymbol{n}} \cdot \mathrm{d}\boldsymbol{r}$ is coordinate invariant, independent of the choice of coordinates (u^1, u^2) for that surface, as is the arclength element $\mathrm{d}s^2 = \mathrm{d}\boldsymbol{r} \cdot \mathrm{d}\boldsymbol{r}$, and the normal curvature in direction $\hat{\boldsymbol{t}} = \mathrm{d}\boldsymbol{r}/\mathrm{d}s$ is

$$\kappa_n = -\frac{\mathrm{d}\hat{\boldsymbol{n}}}{\mathrm{d}s} \cdot \hat{\boldsymbol{t}} = \hat{\boldsymbol{n}} \cdot \frac{\mathrm{d}\hat{\boldsymbol{t}}}{\mathrm{d}s} \tag{14.63b}$$

with $\mathrm{d}\hat{\boldsymbol{n}} = \hat{\boldsymbol{n}}(\boldsymbol{r} + \mathrm{d}\boldsymbol{r}) - \hat{\boldsymbol{n}}(\boldsymbol{r})$. This recovers κ_n in (14.7b) with arclength parametrization $\mathrm{d}s = \mathrm{d}t$.

The normal curvature κ_n has a sign; it is negative if the surface normal $\hat{\boldsymbol{n}}$ *diverges* along the curve (convex), and positive when it *converges* (concave) (Fig. 14.7). The sign of κ_n thus depends on the surface orientation. Expression (14.63b) for κ_n is invariant under a change of surface coordinates (u^1, u^2). It depends only on the unit normal $\hat{\boldsymbol{n}}$ and its rate of change in the tangent direction $\pm\hat{\boldsymbol{t}}$, with $\kappa_n(\boldsymbol{r}, \hat{\boldsymbol{t}}) = \kappa_n(\boldsymbol{r}, -\hat{\boldsymbol{t}})$.

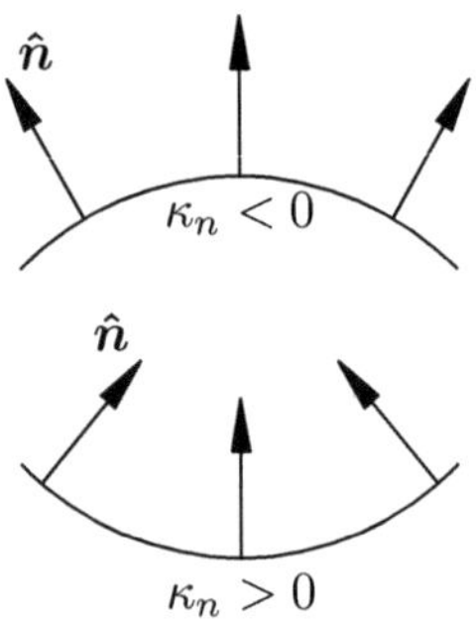

Fig. 14.7 Converging and diverging surface normal $\hat{\boldsymbol{n}}$.

Weingarten formulas

The right-hand side of (14.62) implies that $-b_{ij}$ is the j contravariant component of $\partial_i \hat{\boldsymbol{n}}$, since $\partial_j \boldsymbol{r} = \boldsymbol{a}_j$, that is, from (14.17) and (14.62),

$$\boldsymbol{a}_j \cdot \partial_i \hat{\boldsymbol{n}} = -b_{ij} \quad \Leftrightarrow \quad \partial_i \hat{\boldsymbol{n}} = -b_{ij} \boldsymbol{a}^j + c\,\hat{\boldsymbol{n}},$$

but the component $c = 0$ since $\hat{\boldsymbol{n}} \cdot \hat{\boldsymbol{n}} = 1 \Rightarrow \hat{\boldsymbol{n}} \cdot \partial_i \hat{\boldsymbol{n}} = 0$. Thus, $\partial_i \hat{\boldsymbol{n}}$ is a vector in the tangent plane such that from (14.17) and (14.23),

$$\partial_i \hat{\boldsymbol{n}} = -b_{ij} \boldsymbol{a}^j = -b_{ij} g^{jk} \boldsymbol{a}_k = -b_i^k \boldsymbol{a}_k. \tag{14.64a}$$

These are the *Weingarten formulas.* In matrix notation, they are

$$\begin{aligned} \begin{bmatrix} \partial_1 \hat{\boldsymbol{n}} \\ \partial_2 \hat{\boldsymbol{n}} \end{bmatrix} &= -\begin{bmatrix} b_{11} & b_{12} \\ b_{21} & b_{22} \end{bmatrix} \begin{bmatrix} \boldsymbol{a}^1 \\ \boldsymbol{a}^2 \end{bmatrix} \\ &= -\begin{bmatrix} b_{11} & b_{12} \\ b_{21} & b_{22} \end{bmatrix} \begin{bmatrix} g^{11} & g^{12} \\ g^{21} & g^{22} \end{bmatrix} \begin{bmatrix} \partial_1 \boldsymbol{r} \\ \partial_2 \boldsymbol{r} \end{bmatrix}. \end{aligned} \tag{14.64b}$$

For an arbitrary $\mathrm{d}\boldsymbol{r} = \mathrm{d}u^i \partial_i \boldsymbol{r} \equiv \mathrm{d}u \, \partial_u \boldsymbol{r} + \mathrm{d}v \, \partial_v \boldsymbol{r}$, the corresponding coordinates differentials are $\mathrm{d}u^i = \boldsymbol{a}^i \cdot \mathrm{d}\boldsymbol{r}$; then from (14.64a) and $b_{ij} = b_{ji}$,

$$\mathrm{d}\hat{\boldsymbol{n}} = \mathrm{d}u^i \partial_i \hat{\boldsymbol{n}} = -b_{ij} \boldsymbol{a}^i \boldsymbol{a}^j \cdot \mathrm{d}\boldsymbol{r}. \tag{14.64c}$$

The symmetric tensor $-b_{ij}\boldsymbol{a}^i\boldsymbol{a}^j$ thus maps surface displacements $\mathrm{d}\boldsymbol{r}$ to differentials $\mathrm{d}\hat{\boldsymbol{n}}$, both in the local tangent plane.[8] The matrix of that tensor in cartesian coordinates is 3-by-3, symmetric with rank ≤ 2.

For a sphere with $\hat{\boldsymbol{n}} = \hat{\boldsymbol{r}}$ and $(u^1, u^2) = (\theta, \varphi)$, we can compute the curvic $[b_{ij}]$ directly from (14.64a) using (10.24) and (14.46),

$$\partial_1\hat{\boldsymbol{n}} = \partial_\theta\hat{\boldsymbol{r}} = \hat{\boldsymbol{\theta}} = r\boldsymbol{a}^1,$$
$$\partial_2\hat{\boldsymbol{n}} = \partial_\varphi\hat{\boldsymbol{r}} = \sin\theta\,\hat{\boldsymbol{\varphi}} = r\sin^2\theta\,\boldsymbol{a}^2,$$
$$\Rightarrow \begin{bmatrix} b_{11} & b_{12} \\ b_{21} & b_{22} \end{bmatrix} = \begin{bmatrix} -r & 0 \\ 0 & -r\sin^2\theta \end{bmatrix}. \tag{14.65}$$

14.8 Principal curvatures

A natural question is to ask what directions $\hat{\boldsymbol{t}}$ lead to maximum and minimum $\kappa_n(\hat{\boldsymbol{t}})$ at a fixed point on the surface. That requires optimizing κ_n in (14.60b) over all $(\mathrm{d}u^1, \mathrm{d}u^2)$ with fixed $\mathrm{d}s^2 = g_{ij}\mathrm{d}u^i\mathrm{d}u^j$, that is, to optimize

$$\kappa = \frac{b_{ij}\,\mathrm{d}u^i\mathrm{d}u^j}{g_{kl}\,\mathrm{d}u^k\mathrm{d}u^l} = \frac{b_{ij}\,\dot{u}^i\dot{u}^j}{g_{kl}\,\dot{u}^k\dot{u}^l},$$

where κ is the normal curvature (dropping the subscript n here), and $\dot{u}^i = \mathrm{d}u^i/\mathrm{d}t$, for an arbitrary curve parameter t. The optimization is over all such directions $(\dot{u}^1, \dot{u}^2)$ at a fixed point (u^1, u^2). The extrema thus satisfy

$$\frac{\partial\kappa}{\partial\dot{u}^i} = 0 \Rightarrow (b_{ij} - \kappa\,g_{ij})\,\dot{u}^j = 0, \tag{14.66}$$

where symmetry of b_{ij} and g_{ij} has been used. This is a generalized eigenvalue problem

$$\mathbf{B}\,\dot{\mathbf{u}} = \kappa\,\mathbf{G}\,\dot{\mathbf{u}}, \tag{14.67}$$

in matrix notation, where $\mathbf{G}$ and $\mathbf{B}$ are 2-by-2 real and symmetric matrices, and $\dot{\mathbf{u}} = (\dot{u}^1, \dot{u}^2)$ are the eigenvectors. The eigenvalues κ_1, κ_2 are real,[9] and there are two distinct eigenvectors $\{\dot{\mathbf{u}}_1, \dot{\mathbf{u}}_2\}$ that are $\mathbf{G}$-orthogonal and can be normalized such that

$$\dot{\mathbf{u}}_\alpha^{\mathsf{T}}\mathbf{G}\,\dot{\mathbf{u}}_\beta = \dot{u}_\alpha^k\,g_{kl}\,\dot{u}_\beta^l = \delta_{\alpha\beta}. \tag{14.68}$$

This normalization amounts to enforcing arclength parametrization, that is, $t = s$ and $\dot{u}^i = \mathrm{d}u^i/\mathrm{d}s$, where the Greek subscripts α, β are the eigenmode labels, *not* covariant indices.

$\mathbf{G}$-orthogonality of coordinate directions $(\dot{u}^1, \dot{u}^2)$ is regular orthogonality in the tangent plane of the surface, since $\boldsymbol{p}\cdot\boldsymbol{q} = g_{ij}p^iq^j$ for any vectors $\boldsymbol{p}$ and $\boldsymbol{q}$ in the tangent plane. Therefore, there are two orthogonal principal directions tangent to the surface,

$$\hat{\boldsymbol{t}}_1 = \dot{u}_1^i\,\partial_i\boldsymbol{r}, \quad \hat{\boldsymbol{t}}_2 = \dot{u}_2^j\,\partial_j\boldsymbol{r}, \tag{14.69a}$$

[9] Let $[\mathbf{a}_1, \mathbf{a}_2] = \mathbf{A} = \mathbf{QR}$, where $\mathbf{Q}^{\mathsf{T}}\mathbf{Q} = \mathbf{I}$; then $\mathbf{G} = \mathbf{A}^{\mathsf{T}}\mathbf{A} = \mathbf{R}^{\mathsf{T}}\mathbf{R}$ and $\mathbf{Bv} = \kappa\mathbf{Gv}$ is equivalent to

$$(\mathbf{R}^{-\mathsf{T}}\mathbf{B}\mathbf{R}^{-1})(\mathbf{Rv}) = \kappa(\mathbf{Rv})$$

which is a regular symmetric eigenvalue problem for κ with eigenvectors $\mathbf{Rv}$, and we proved that symmetric matrices have real eigenvalues and orthogonal eigenvectors.

[8] The map from a point P on a surface to the normal $\hat{\boldsymbol{n}}(P)$ on the unit sphere S^2 is the *Gauss map*, and the matrix (14.64a) $\mathbf{S} = -\mathbf{B}\mathbf{G}^{-1} = [-b_i^j]$ is the *shape operator* such that $b_i^jp^iq_j = b_i^jq^ip_j$ for any $\boldsymbol{p}, \boldsymbol{q}$ in the tangent plane, since $b_i^jp^iq_j = b_{ik}g^{kj}p^iq_j = b_{ik}p^iq^k = b_{ik}q^ip^k = b_{ik}g^{kj}q^ip_j = b_i^jq^ip_j$.

with

$$\hat{\boldsymbol{t}}_\alpha \cdot \hat{\boldsymbol{t}}_\beta = \partial_i \boldsymbol{r} \cdot \partial_j \boldsymbol{r}\, \dot{u}^i_\alpha\, \dot{u}^j_\beta = g_{ij}\, \dot{u}^i_\alpha\, \dot{u}^j_\beta = \delta_{\alpha\beta},$$

corresponding to the *principal curvatures* κ_1 and κ_2, which are the maximum and minimum curvatures of geodesics passing through that point $\boldsymbol{r}(u^1, u^2)$ on the surface.

An arbitrary direction $\hat{\boldsymbol{t}} = \mathrm{d}u^i/\mathrm{d}s\, \partial_i \boldsymbol{r}$ can be represented in terms of those orthogonal principal directions $\{\hat{\boldsymbol{t}}_1, \hat{\boldsymbol{t}}_2\}$ as

$$\hat{\boldsymbol{t}} = \hat{\boldsymbol{t}}_1 \cos\varphi + \hat{\boldsymbol{t}}_2 \sin\varphi = (\dot{u}^i_1 \cos\varphi + \dot{u}^i_2 \sin\varphi)\, \partial_i \boldsymbol{r}. \tag{14.69b}$$

The directional derivative of the unit normal $\hat{\boldsymbol{n}}$ with respect to arclength in that $\hat{\boldsymbol{t}}$ direction is then, from (14.64a),

$$\begin{aligned} \frac{\mathrm{d}\hat{\boldsymbol{n}}}{\mathrm{d}s} = \frac{\mathrm{d}u^i}{\mathrm{d}s} \partial_i \hat{\boldsymbol{n}} &= -(\dot{u}^i_1 \cos\varphi + \dot{u}^i_2 \sin\varphi)\, b_{ij} g^{jk}\, \partial_k \boldsymbol{r} \\ &= -\kappa_1 \hat{\boldsymbol{t}}_1 \cos\varphi - \kappa_2 \hat{\boldsymbol{t}}_2 \sin\varphi. \end{aligned} \tag{14.69c}$$

since $\dot{u}^i_\alpha b_{ij} g^{jk} = \kappa_\alpha \dot{u}^k_\alpha$ and $\dot{u}^k_\alpha \partial_k \boldsymbol{r} = \hat{\boldsymbol{t}}_\alpha$. The normal curvature κ_n in that $\hat{\boldsymbol{t}}$ direction then follows from (14.63b) as

$$\kappa_n = -\frac{\mathrm{d}\hat{\boldsymbol{n}}}{\mathrm{d}s} \cdot \hat{\boldsymbol{t}} = \kappa_1 \cos^2\varphi + \kappa_2 \sin^2\varphi. \tag{14.69d}$$

The sum $\kappa_1 + \kappa_2$ and product $\kappa_1 \kappa_2$ of the principal curvatures are the trace and determinant of the 2-by-2 matrix $\mathbf{BG}^{-1}$. The trace, determinant and eigenvalues are invariant under similarity transformations, corresponding to an arbitrary change of basis in the tangent plane, and

$$\begin{aligned} \kappa_1 + \kappa_2 &= \operatorname{tr}\left(\mathbf{BG}^{-1}\right) = b_{ik}\, g^{ki}, \\ \kappa_1 \kappa_2 &= \det\left(\mathbf{BG}^{-1}\right) = \frac{\det \mathbf{B}}{\det \mathbf{G}} = \frac{b}{g}, \end{aligned} \tag{14.70}$$

where, from (14.3) and (14.8),

$$\begin{aligned} b &\triangleq \det \mathbf{B} = \epsilon^{ij3} b_{i1} b_{j2} = LN - M^2, \\ g &\triangleq \det \mathbf{G} = \epsilon^{kl3} g_{k1} g_{l2} = EG - F^2. \end{aligned} \tag{14.71}$$

The product of the principal curvatures, is the *Gaussian curvature* $\kappa_1 \kappa_2 = K$, and their sum is twice the *mean curvature* $H = (\kappa_1 + \kappa_2)/2$. That sum occurs in connection with the *surface tension* of membranes and interfaces, such as soap films, and the mean curvature is zero everywhere if such surfaces *minimize area* for a prescribed boundary as discussed in §14.10. From the Weingarten formulas (14.64a) and (14.70),

$$\kappa_1 + \kappa_2 = b_{ik} g^{ki} = -\boldsymbol{a}^k \cdot \partial_k \hat{\boldsymbol{n}} = -\boldsymbol{\nabla} \cdot \hat{\boldsymbol{n}},. \tag{14.72}$$

where $\boldsymbol{\nabla} \cdot \hat{\boldsymbol{n}}$ is the divergence of $\hat{\boldsymbol{n}}$, discussed in later sections.[10] Mean and Gaussian curvatures are discussed further below.

[10]For 3D space $\boldsymbol{\nabla} \cdot \hat{\boldsymbol{n}} = \boldsymbol{a}^1 \cdot \partial_1 \hat{\boldsymbol{n}} + \boldsymbol{a}^2 \cdot \partial_2 \hat{\boldsymbol{n}} + \boldsymbol{a}^3 \cdot \partial_3 \hat{\boldsymbol{n}}$, but $\boldsymbol{a}^3 = \hat{\boldsymbol{n}}$ here and $\hat{\boldsymbol{n}} \cdot \partial_3 \hat{\boldsymbol{n}} = 0$.

Umbilics

An *umbilical point*, or *umbilic*, is a point where $\kappa_1 = \kappa_2$, and it follows from (14.69d) that κ is the same in all directions at such points, and any direction is a principal direction; thus, at umbilical points, from (14.66) and (14.64a),

$$b_{ij} = \kappa\, g_{ij} \quad \text{and} \quad \partial_i \hat{\boldsymbol{n}} = -\kappa\, \partial_i \boldsymbol{r}. \tag{14.73}$$

For example, every point of a sphere of radius r is umbilic with $\kappa = -1/r$. Indeed from (14.45) and (14.65), $b_{ij} = -g_{ij}/r$; thus, $\kappa = -1/r$, the mean curvature $H = -1/r$ and the Gaussian curvature $K = 1/r^2$. Every point of a plane is umbilic with $\kappa = 0$. The paraboloid $z = a(x^2+y^2)/2$ has an umbilic at $(x, y) = (0, 0)$ with $\kappa = a$, and any surface is locally of that form in the neighborhood of an umbilic. The ellipsoid of revolution $x^2/a^2+y^2/a^2+z^2/c^2 = 1$ has umbilics at $(x, y, z) = (0, 0, \pm c)$.

A surface is locally spherical at an umbilic, and if *every point* of a surface is umbilic, then the surface *must* be a sphere or a plane. Indeed, from (14.73) and commutation of mixed derivatives of $\hat{\boldsymbol{n}}$ and $\boldsymbol{r}$ for sufficiently smooth surfaces,

$$\partial_1\partial_2\hat{\boldsymbol{n}} - \partial_2\partial_1\hat{\boldsymbol{n}} = 0 = \partial_2\kappa\, \partial_1\boldsymbol{r} - \partial_1\kappa\, \partial_2\boldsymbol{r},$$

so $\partial_1\kappa = 0 = \partial_2\kappa$ since $\partial_1\boldsymbol{r}$ and $\partial_2\boldsymbol{r}$ are independent, and κ must be constant.

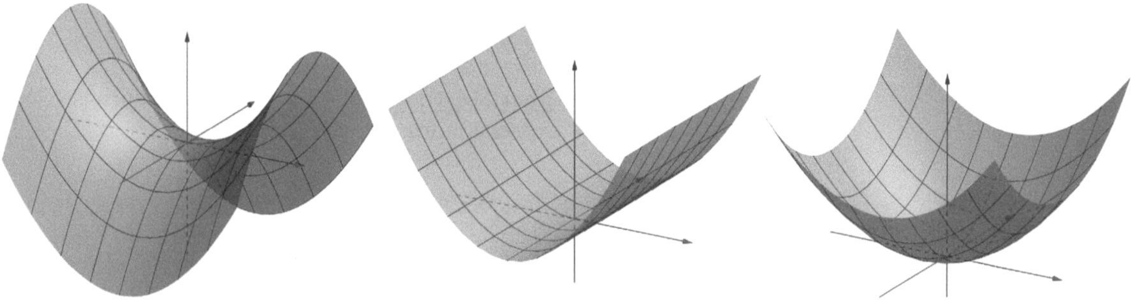

Fig. 14.8 Hyperbolic paraboloid, parabolic cylinder, and elliptic paraboloid, left to right (drawn with *GeoGebra*).

14.9 Gaussian curvature

The product of the principal curvatures κ_1, κ_2, is the *Gaussian curvature* (14.70)

$$K = \kappa_1\kappa_2 \equiv \frac{1}{R_1 R_2},$$

where R_1, R_2 are the principal radii of curvature. The Gaussian "curvature" is a curvature *squared*, with units of inverse length square. The sign of the Gaussian curvature at a point determines the local shape of the surface.

In a neighborhood of one of its points, the surface is (Fig. 14.8, exercise 14.35)

- a *hyperbolic paraboloid* with $z \simeq \frac{1}{2}\left(|\kappa_1|\, x^2 - |\kappa_2|\, y^2\right)$ if $K < 0$,
- a *parabolic cylinder* with $z \simeq \frac{1}{2}|\kappa_1|\, x^2$ if $K = 0$ with $H \neq 0$, and
- an *elliptic paraboloid* with $z \simeq \frac{1}{2}\left(|\kappa_1|\, x^2 + |\kappa_2|\, y^2\right)$ if $K > 0$,

in cartesian coordinates with the origin at the current surface point, $\hat{\mathbf{x}}$, $\hat{\mathbf{y}}$ aligned with the principal directions $\hat{\boldsymbol{t}}_1$ and $\hat{\boldsymbol{t}}_2$, and $\hat{\mathbf{z}}$ parallel to $\hat{\boldsymbol{n}}$. In particular, if the Gaussian curvature is zero *everywhere*, except at singular points, then it can be *developed*; that is, it can be mapped to a plane without distortion. For example, a cone is developable with its apex as a singular point. If we slice the cone open along a curve from its apex to its base, we can lay it flat on a plane without stretching or crumpling, thus preserving all lengths, and angles and areas. If $H = 0 = K$, the surface is locally of higher order $z = O(x^3, x^2y, xy^2, y^3)$ and higher derivatives determine its shape.

Geometric formula for Gaussian curvature

From the definition of b_{ij} (14.62) and the mixed and double cross product identities, the determinant of the curvic b_{ij} is

$$\begin{aligned} b \triangleq \epsilon^{ij3} b_{i1} b_{j2} &= b_{11}b_{22} - b_{12}b_{21} \\ &= (\partial_1\hat{\boldsymbol{n}} \cdot \partial_1\boldsymbol{r})(\partial_2\hat{\boldsymbol{n}} \cdot \partial_2\boldsymbol{r}) - (\partial_1\hat{\boldsymbol{n}} \cdot \partial_2\boldsymbol{r})(\partial_2\hat{\boldsymbol{n}} \cdot \partial_1\boldsymbol{r}) \\ &= \left(\frac{\partial\hat{\boldsymbol{n}}}{\partial u} \times \frac{\partial\hat{\boldsymbol{n}}}{\partial v}\right) \cdot \left(\frac{\partial\boldsymbol{r}}{\partial u} \times \frac{\partial\boldsymbol{r}}{\partial v}\right). \end{aligned}$$

Since $\partial_u\boldsymbol{r} \times \partial_v\boldsymbol{r} = \sqrt{g}\,\hat{\boldsymbol{n}}$ (14.4), and the Gaussian curvature is $K = b/g$ (14.70), the above expression for b yields

$$K\sqrt{g} = \frac{\partial\hat{\boldsymbol{n}}}{\partial u} \times \frac{\partial\hat{\boldsymbol{n}}}{\partial v} \cdot \hat{\boldsymbol{n}}. \tag{14.74}$$

Multiplying both sides by $\mathrm{d}u\,\mathrm{d}v$ gives the *Rodrigues–Gauss formula*

$$K \mathrm{d}A_{\boldsymbol{r}} = \mathrm{d}A_{\hat{\boldsymbol{n}}},$$

with the surface area elements

$$\begin{aligned} \mathrm{d}A_{\boldsymbol{r}} &= \partial_u\boldsymbol{r} \times \partial_v\boldsymbol{r} \cdot \hat{\boldsymbol{n}}\, \mathrm{d}u\mathrm{d}v, \\ \mathrm{d}A_{\hat{\boldsymbol{n}}} &= \partial_u\hat{\boldsymbol{n}} \times \partial_v\hat{\boldsymbol{n}} \cdot \hat{\boldsymbol{n}}\, \mathrm{d}u\mathrm{d}v, \end{aligned}$$

traced by $\boldsymbol{r}(u, v)$ on the surface and by $\hat{\boldsymbol{n}}(u, v)$ on the unit sphere, respectively, when the coordinates (u, v) span a rectangle of sides $\mathrm{d}u$ and $\mathrm{d}v$. This gives a geometric interpretation of the Gaussian curvature K at a point P on a surface as the limit of the ratio of the area spanned by the unit normal $\hat{\boldsymbol{n}}(P)$ on the unit sphere to that spanned by P on the surface when the latter area shrinks to P. However, while $\partial_u\boldsymbol{r}\times\partial_v\boldsymbol{r}\cdot\hat{\boldsymbol{n}} = \sqrt{g} > 0$ by definition, $\partial_u\hat{\boldsymbol{n}} \times \partial_v\hat{\boldsymbol{n}} \cdot \hat{\boldsymbol{n}}$ does not have to be positive and can change sign as P moves along the surface.

For a sphere $\mathcal{S}$ of radius R, with $\boldsymbol{r} = R\hat{\boldsymbol{n}}$ and $\partial_u\boldsymbol{r}\times\partial_v\boldsymbol{r} = R^2\partial_u\hat{\boldsymbol{n}}\times\partial_v\hat{\boldsymbol{n}}$ this recovers $K = 1/R^2$, and $\int_{\mathcal{S}} K\mathrm{d}A = 4\pi$, the surface area of the unit sphere. For an arbitrary ellipsoïd, the Gaussian curvature K is not constant over the surface, but $\hat{\boldsymbol{n}}$ will sweep once around the unit sphere positively and $\oint_{\mathcal{S}} K \mathrm{d}A = 4\pi$ also.

For a torus $\mathcal{S}$, generated by the rotation of a circle of radius a perpendicular to and centered on a circle of radius $R > a$ about the latter,[11] the unit normal $\hat{\boldsymbol{n}}$ will sweep the entire unit sphere, with $\partial_u \hat{\boldsymbol{n}} \times \partial_v \hat{\boldsymbol{n}} \cdot \hat{\boldsymbol{n}} \geq 0$ for the outer part of the torus where $\cos\theta \geq 0$ and $\hat{\boldsymbol{n}}$ diverges in all directions. On the inner part of the torus, where $\cos\theta < 0$ and $\hat{\boldsymbol{n}}$ diverges in the poloidal direction (θ) but converges in the toroidal direction (φ), $\hat{\boldsymbol{n}}$ will again sweep the entire unit sphere but now with $\partial_u \hat{\boldsymbol{n}} \times \partial_v \hat{\boldsymbol{n}} \cdot \hat{\boldsymbol{n}} \leq 0$ such that the integral of K over the entire torus $\oint_{\mathcal{S}} K \mathrm{d}A = 0$.

[11] Torus:

$$\boldsymbol{r}(\varphi, \theta) = (R + a\cos\theta)\hat{\boldsymbol{\rho}}(\varphi) + a\sin\theta\, \hat{\mathbf{z}},$$

with $0 \leq \varphi < 2\pi$, $-\pi/2 \leq \theta < 3\pi/2$, and $\hat{\boldsymbol{\rho}} = \hat{\mathbf{x}}\cos\varphi + \hat{\mathbf{y}}\sin\varphi$.

Theorema Egregium

Gauss's *Theorema Egregium*[12] states that the Gaussian curvature $K = \kappa_1\kappa_2 = b/g$ (14.70) is fully determined by the metric g_{ij}, and its derivatives along the surface without explicit reference to the direction $\hat{\boldsymbol{n}}$ perpendicular to the surface. This is a surprising result in view of (14.74) and the fact that curvature is precisely about characterizing the bending of the surface in the $\hat{\boldsymbol{n}}$ direction.

[12] *"Egregious"* now means "really bad," but used to mean awesome!

A quick proof of the Theorema Egregium is achieved by dotting the Gauss equations (14.39) for $\partial_i\partial_1\boldsymbol{r}$ with $\partial_j\partial_2\boldsymbol{r}$ (or vice versa) to obtain

$$\partial_i\partial_1\boldsymbol{r} \cdot \partial_j\partial_2\boldsymbol{r} = g_{kl}\Gamma^k_{i1}\Gamma^l_{j2} + b_{i1}b_{j2},$$

since $\partial_k\boldsymbol{r} \cdot \partial_j\partial_2\boldsymbol{r} = \Gamma_{kj2} = g_{kl}\Gamma^l_{j2}$, and $\hat{\boldsymbol{n}} \cdot \partial_j\partial_k\boldsymbol{r} = b_{jk}$, from (14.38) and (14.62). Using the resulting expression for $b_{i1}b_{j2}$ gives

$$\begin{aligned} K\,g = b = \epsilon^{ij3} b_{i1} b_{j_2} &= \epsilon^{ij3}\left(\partial_i\partial_1\boldsymbol{r} \cdot \partial_j\partial_2\boldsymbol{r} - g_{kl}\Gamma^k_{i1}\Gamma^l_{j2}\right), \\ &= \epsilon^{ij3}\left(\partial_i\Gamma_{1j2} - g_{kl}\Gamma^k_{i1}\Gamma^l_{j2}\right), \end{aligned} \tag{14.75}$$

because

$$\epsilon^{ij3}\partial_i\partial_1\boldsymbol{r} \cdot \partial_j\partial_2\boldsymbol{r} = \epsilon^{ij3}\left(\partial_i\left(\partial_1\boldsymbol{r} \cdot \partial_j\partial_2\boldsymbol{r}\right) - \partial_1\boldsymbol{r} \cdot \partial_i\partial_j\partial_2\boldsymbol{r}\right) = \epsilon^{ij3}\partial_i\Gamma_{1j2},$$

since $\epsilon^{ij3}\partial_i\partial_j = \partial_1\partial_2 - \partial_2\partial_1 = 0$ when the partials commute, which is guaranteed if the surface parametrization $\boldsymbol{r}(u^1, u^2)$ is C^3, and $\Gamma_{klm} = \partial_k\boldsymbol{r} \cdot \partial_l\partial_m\boldsymbol{r}$ by definition (14.34). This proves the Theorema Egregium since the Christoffel symbols can be expressed in terms of the metric alone (14.36), (14.43).

A consequence of this theorem is that *isometries* preserve Gaussian curvature K. An isometry is a map that preserves all distances, and thus the metric. Therefore, it is not possible to map a sphere to a plane while preserving all distances, and thus all angles *and* areas, since $K = 1/R^2$ for a sphere of radius R, while $K = 0$ for a plane. We can have the angle preserving Mercator map (13.14) or the area preserving Lambert map $(u, v) = (\varphi, \cos\theta)$ (Fig. 13.10), but there is no map of the sphere to a plane that preserves both angles and areas.

Gauss–Bonnet theorem

Using the orthonormal moving frame $\{\hat{\boldsymbol{a}}_1, \hat{\boldsymbol{a}}^2 = \hat{\boldsymbol{n}} \times \hat{\boldsymbol{a}}_1, \hat{\boldsymbol{n}}\}$, we derive the following expression for the Gaussian curvature K (exercise 14.53),

$$K\sqrt{g} = \partial_2 \left(\hat{\boldsymbol{a}}^2 \cdot \partial_1 \hat{\boldsymbol{a}}_1\right) - \partial_1 \left(\hat{\boldsymbol{a}}^2 \cdot \partial_2 \hat{\boldsymbol{a}}_1\right). \tag{14.76}$$

Then by Green's theorem (13.31c)

$$\begin{aligned}\int_{\mathcal{S}} K \, \mathrm{d}A &= \iint_{\mathcal{U}} K\sqrt{g} \, \mathrm{d}u^1 \mathrm{d}u^2 \\ &= -\oint_{\partial\mathcal{U}} \hat{\boldsymbol{a}}^2 \cdot \left(\partial_1 \hat{\boldsymbol{a}}_1 \, \mathrm{d}u^1 + \partial_2 \hat{\boldsymbol{a}}_1 \, \mathrm{d}u^2\right) = -\oint_{\partial\mathcal{S}} \hat{\boldsymbol{a}}^2 \cdot \mathrm{d}\hat{\boldsymbol{a}}_1, \end{aligned} \tag{14.77}$$

where $\mathcal{S}$ is the image of $\mathcal{U}$ mapped by $\boldsymbol{r}(u,v)$. The boundary $\partial\mathcal{S}$ of the surface patch $\mathcal{S}$ is a (piecewise smooth) curve with unit tangent $\hat{\boldsymbol{t}}$. The right-handed moving frame $\{\hat{\boldsymbol{t}}, \hat{\boldsymbol{u}} = \hat{\boldsymbol{n}} \times \hat{\boldsymbol{t}}, \hat{\boldsymbol{n}}\}$ for $\partial\mathcal{S}$ is a rotation by angle ψ about $\hat{\boldsymbol{n}}$ of the surface frame $\{\hat{\boldsymbol{a}}_1, \hat{\boldsymbol{a}}^2, \hat{\boldsymbol{n}}\}$,[13] with

[13]For a sphere, $\{\hat{\boldsymbol{a}}_1, \hat{\boldsymbol{a}}^2, \hat{\boldsymbol{n}}\} = \{\hat{\boldsymbol{\theta}}, \hat{\boldsymbol{\varphi}}, \hat{\boldsymbol{r}}\}$ with $\{\hat{\boldsymbol{t}}, \hat{\boldsymbol{u}}, \hat{\boldsymbol{n}}\} \equiv \{\mathbf{e}_1', \mathbf{e}_2', \mathbf{e}_3'\}$ in Fig. 8.7.

$$\begin{bmatrix}\hat{\boldsymbol{t}} \\ \hat{\boldsymbol{u}}\end{bmatrix} = \begin{bmatrix}\cos\psi & \sin\psi \\ -\sin\psi & \cos\psi\end{bmatrix} \begin{bmatrix}\hat{\boldsymbol{a}}_1 \\ \hat{\boldsymbol{a}}^2\end{bmatrix} \Leftrightarrow \begin{bmatrix}\hat{\boldsymbol{a}}_1 \\ \hat{\boldsymbol{a}}^2\end{bmatrix} = \begin{bmatrix}\cos\psi & -\sin\psi \\ \sin\psi & \cos\psi\end{bmatrix} \begin{bmatrix}\hat{\boldsymbol{t}} \\ \hat{\boldsymbol{u}}\end{bmatrix}.$$

Angle ψ is a function of arclength s along the curve, $\psi(s)$. The differential of $\hat{\boldsymbol{a}}_1 = \cos\psi \, \hat{\boldsymbol{t}} - \sin\psi \, \hat{\boldsymbol{u}}$ along the curve is thus

$$\mathrm{d}\hat{\boldsymbol{a}}_1 = (-\sin\psi \, \hat{\boldsymbol{t}} - \cos\psi \, \hat{\boldsymbol{u}}) \, \mathrm{d}\psi + (\cos\psi \, \mathrm{d}\hat{\boldsymbol{t}} - \sin\psi \, \mathrm{d}\hat{\boldsymbol{u}}),$$

while $\hat{\boldsymbol{a}}^2 = \sin\psi \, \hat{\boldsymbol{t}} + \cos\psi \, \hat{\boldsymbol{u}}$. Then, using $\hat{\boldsymbol{t}} \cdot \hat{\boldsymbol{u}} = 0$ so $\hat{\boldsymbol{u}} \cdot \mathrm{d}\hat{\boldsymbol{t}} = -\hat{\boldsymbol{t}} \cdot \mathrm{d}\hat{\boldsymbol{u}}$ gives

$$-\hat{\boldsymbol{a}}^2 \cdot \mathrm{d}\hat{\boldsymbol{a}}_1 = \mathrm{d}\psi - \hat{\boldsymbol{u}} \cdot \mathrm{d}\hat{\boldsymbol{t}} = \mathrm{d}\psi - \kappa_g \mathrm{d}s,$$

because $\hat{\boldsymbol{u}} \cdot \mathrm{d}\hat{\boldsymbol{t}} = (\hat{\boldsymbol{n}} \times \hat{\boldsymbol{t}}) \cdot \mathrm{d}\hat{\boldsymbol{t}} = \kappa_g \, \mathrm{d}s$, where κ_g is the geodesic curvature (14.7b) since with arclength parametrization $\dot{\boldsymbol{r}} = \mathrm{d}\boldsymbol{r}/\mathrm{d}s = \hat{\boldsymbol{t}}$ and $\ddot{\boldsymbol{r}} = \mathrm{d}^2\boldsymbol{r}/\mathrm{d}s^2 = \mathrm{d}\hat{\boldsymbol{t}}/\mathrm{d}s$. Substituting that expression for $\hat{\boldsymbol{a}}^2 \cdot \mathrm{d}\hat{\boldsymbol{a}}_1$ in (14.77) yields the *Gauss–Bonnet theorem*

$$\int_{\mathcal{S}} K \, \mathrm{d}A = \oint_{\partial\mathcal{S}} \mathrm{d}\psi - \oint_{\partial\mathcal{S}} \kappa_g \, \mathrm{d}s. \tag{14.78a}$$

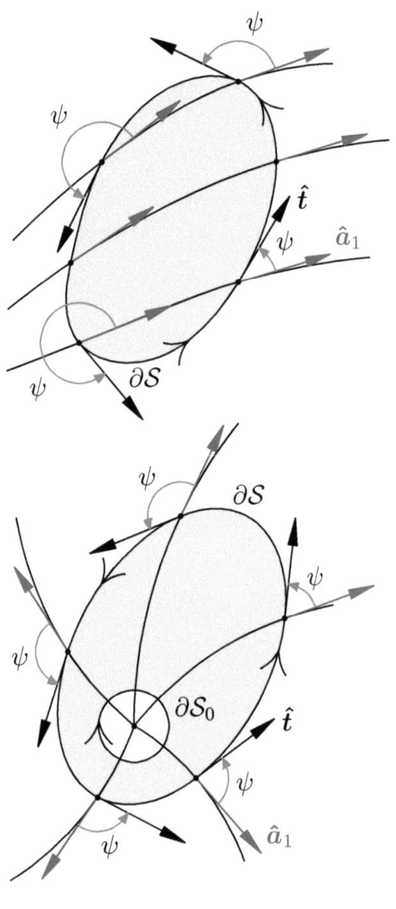

Fig. 14.9 Coordinate curves traversing or radiating through a surface patch $\mathcal{S}$.

The parametrization $\boldsymbol{r}(u,v)$ should be sufficiently smooth in $\mathcal{U}$ in order to apply Green's theorem (13.31c) to $G(u,v) = \hat{\boldsymbol{a}}^2 \cdot \partial_v \hat{\boldsymbol{a}}_1$ and $F(u,v) = \hat{\boldsymbol{a}}^2 \cdot \partial_u \hat{\boldsymbol{a}}_1$. There should not be poles such as for polar or spherical coordinates inside $\mathcal{U}$. If the surface is smooth, we may use other coordinates, that have no poles in the surface patch $\mathcal{S}$ under consideration. Using spherical coordinates on a sphere or an ellipsoïd, for example, we pick polar axes that do not go through $\mathcal{S}$ (Fig. 14.9, top), or we isolate those singular points with additional boundaries $\partial\mathcal{S}_0$ (Fig. 14.9, bottom, exercise 14.55). Green's theorem (§13.5) indeed applies to domains with holes (Fig. 13.15).

In the first case, for coordinate curves that traverse a smooth $\partial\mathcal{S}$ (Fig. 14.9, top), as the meridians crossing a closed contour that does not include either pole on a sphere, the integral $\oint_{\partial\mathcal{S}} \mathrm{d}\psi = 2\pi$, since the unit tangent $\hat{\boldsymbol{t}}$ will do a full positive turn around $\hat{\boldsymbol{n}}$ with respect to $\hat{\boldsymbol{a}}_1$ as we travel around the curve $\partial\mathcal{S}$ back to the starting point.

In the second case (Fig. 14.9, bottom), for coordinate singularities, such as the poles for spherical coordinates on a sphere or ellipsoid, we exclude those points using a small cutout such as the intersection of the surface with a cylinder of small radius ϵ around the pole. The integral sum of $\mathrm{d}\psi$ around that additional boundary, $\partial\mathcal{S}_0$ say, and the sum around the outer $\partial\mathcal{S}$ vanish since the tangent $\hat{\boldsymbol{t}}$ and the coordinate direction $\hat{\boldsymbol{a}}_1$ both rotate a full turn around the pole, so there is no net rotation of $\hat{\boldsymbol{t}}$ relative to $\hat{\boldsymbol{a}}_1$. However, the integral of the geodesic curvature $\oint_{\partial\mathcal{S}_0} \kappa_g \, \mathrm{d}s \to -2\pi$ as $\epsilon \to 0$, because in that limit, the inner curve tends to a flat circle of radius ϵ and geodesic curvature $\kappa_g = -1/\epsilon$ since $\hat{\boldsymbol{n}} \times \hat{\boldsymbol{t}}$ points toward $\mathcal{S}$. The net result from this inner boundary is again a 2π term.

In either case, for a smooth simply connected surface patch $\mathcal{S}$ with smooth boundary $\partial\mathcal{S}$, the *Gauss–Bonnet theorem* reads

$$\int_{\mathcal{S}} K \, \mathrm{d}A = 2\pi - \oint_{\partial\mathcal{S}} \kappa_g \, \mathrm{d}s, \tag{14.78b}$$

where the 2π arises from $\oint_{\partial\mathcal{S}} \mathrm{d}\psi$, or $-\oint_{\partial\mathcal{S}_0} \kappa_g \mathrm{d}s$ in the limit of $\partial\mathcal{S}_0$ shrinking to the coordinate pole.

For a piecewise smooth boundary $\partial\mathcal{S}$, such as the geodesic triangles shown in Fig. 6.3, where the boundary consists of three smooth arcs $\partial\mathcal{S} = \mathcal{C}_1 + \mathcal{C}_2 + \mathcal{C}_3$, there are two ways to handle the corners. The simplest way is to smooth out the corners with small arcs whose contributions to $\oint \kappa_g \mathrm{d}s$ is the radian lengths of those arcs, equal to the tangent turning angles at those corners, then $\oint \mathrm{d}\psi = 2\pi$ still and, in the limit of vanishing rounding arcs,

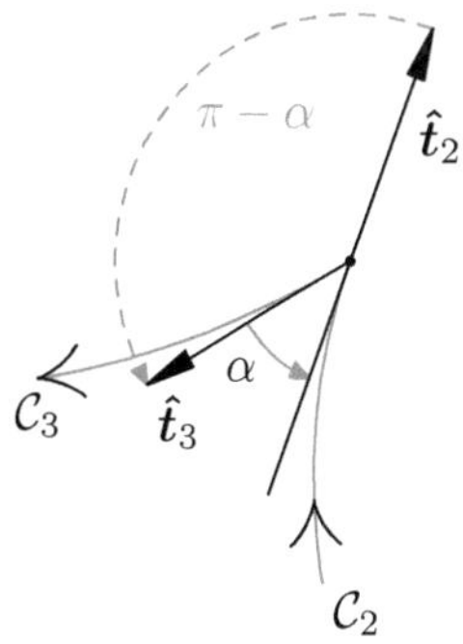

Fig. 14.10 Abrupt tangent turn at a corner.

$$\oint_{\partial\mathcal{S}} \kappa_g \mathrm{d}s \to (\pi - \alpha) + (\pi - \beta) + (\pi - \gamma) + \sum_{k=1}^{3} \int_{\mathcal{C}_k} \kappa_g \mathrm{d}s,$$

where α, β, γ are the *inner* angles at the corners, and $\pi - \alpha, \pi - \beta, \pi - \gamma$, the corresponding external angles by which the tangent abruptly turns at each corner (Fig. 14.10).

The other way is to directly handle the corners by going back to (14.78a) with the integral written as a sum over each piecewise smooth arc of the contour. Then the net tangent turning angle $\oint \mathrm{d}\psi$ equals 2π *minus* the sum of the tangent turning angles at the corners; that is,

$$\oint_{\partial\mathcal{S}} \mathrm{d}\psi = \sum_{k=1}^{3} \int_{\mathcal{C}_k} \mathrm{d}\psi = 2\pi - (\pi - \alpha) - (\pi - \beta) - (\pi - \gamma).$$

Either way, the result is

$$\int_{\mathcal{S}} K \, \mathrm{d}A = \alpha + \beta + \gamma - \pi - \sum_{k=1}^{3} \int_{\mathcal{C}_k} \kappa_g \mathrm{d}s. \tag{14.78c}$$

For a geodesic triangle, this reads

$$\int_{\mathcal{S}} K\,\mathrm{d}A = \alpha + \beta + \gamma - \pi, \tag{14.78d}$$

since $\kappa_g = 0$ on each geodesic arc, recovering the result (13.28) for a sphere.

For a more general surface $\mathcal{S}$, we partition $\mathcal{S}$ into F triangular surface patches $j = 1, \ldots, F$, with smooth parametrizations, possibly varying from one triangle to the next if necessary to avoid coordinate singularities. Assuming that the surface boundary $\partial\mathcal{S}$ is smooth, and using (14.78c) for each triangle, we obtain[14]

$$\begin{aligned}\int_{\mathcal{S}} K\,\mathrm{d}A &= \sum_{j=1}^{F}(\alpha_j + \beta_j + \gamma_j - \pi) - \sum_{j=1}^{F}\sum_{k=1}^{3}\int_{\mathcal{C}_{j,k}} \kappa_g \mathrm{d}s, \\ &= 2\pi V_i + \pi V_b - \pi F + \oint_{\partial\mathcal{S}} \kappa_g \mathrm{d}s,\end{aligned} \tag{14.78e}$$

[14] Following Manfredo P. do Carmo, *Differential Geometry of Curves and Surfaces*, Dover, 2016, p. 277, though we assume smooth boundaries here to simplify the presentation.

where V_i is the number of internal vertices, V_b is the number of vertices on the boundary $\partial\mathcal{S}$, and F the number of triangles or *faces*. This is because at each internal vertex, the sum of all the internal angles at that vertex add up to 2π. That sum adds to π at a boundary vertex, since we assume the boundary is smooth, that is, locally straight. The κ_g integrals over each internal edge cancel out since they are computed twice for each adjoining triangle, but in opposite directions; hence, they cancel out. The boundary $\partial\mathcal{S}$ can consist of several disconnected closed curves $\partial\mathcal{S}_m$, oriented such that $\hat{\boldsymbol{n}} \times \hat{\boldsymbol{t}}_m$ points towards $\mathcal{S}$.

To further reduce (14.78e), let E_i be the number of internal edges, and E_b the number of edges on the boundary $\partial\mathcal{S}$. On the closed boundary $\partial\mathcal{S}$, $E_b = V_b$. For a single triangle $F = 1$, $E_i = 0$, $E_b = 3$. For two triangles, joined along one edge, $F = 2$, $E_i = 1$, and $E_b = 4$. Adding another vertex and thus one edge on the boundary together with an internal edge to that new vertex adds one triangle, so $(F, E_i, E_b) \to (F+1, E_i+1, E_b+1)$. Adding one vertex in the middle of one of those triangles adds three faces and three internal edges, etc. (Fig. 14.11). Thus, we have

$$3F = 2E_i + E_b.$$

Then with $V = V_i + V_b$, $E = E_i + E_b$ and $V_b = E_b$, we obtain

$$2V_i + V_b - F = 2(V + F - E) = 2\chi,$$

where $\chi = V + F - E$ is the Euler characteristic of the surface. It is an invariant, independent of the number of triangles. For a sphere, four vertices with four triangles and six edges provide a nice partition (the vertices of a regular tetrahedron, for example); hence, $V + F - E = 2$. Adding one vertex inside a triangle adds three edges and three triangles, minus onr that was partitioned; hence, $V + F - E = 2$ still. For a sphere with a hole, one triangle is missing; thus, $V + F - E = 1$. For a cylinder, or a torus, two triangles are missing; hence, $V + F - E = 0$. Since

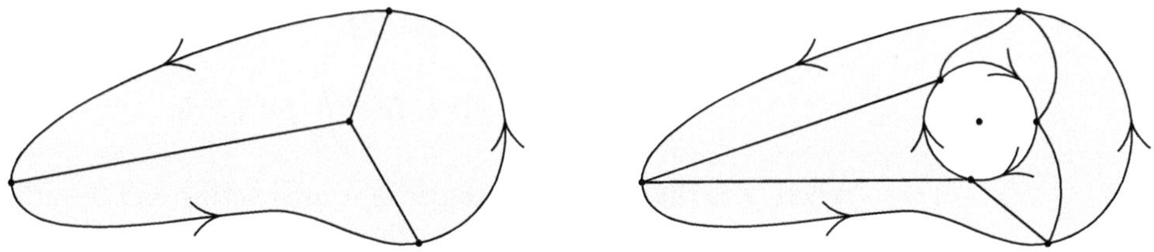

$V_i = 1, V_b = 3 = E_b, E_i = 3, F = 3$ $\qquad$ $V_i = 0, V_b = 6 = E_b, E_i = 6, F = 6$

Fig. 14.11 Vertices, Edges and Faces examples.

$\chi = V + F - E$ is also invariant when we remove one edge, $E \to E - 1$, and thus one face, $F \to F - 1$, this implies that we can mix triangles with quadrilaterals and any surface polygonal section in our partitioning of $\mathcal{S}$ to determine the value of $\chi = V + F - E$. The Gauss–Bonnet theorem (14.78e) in its more general form thus reads

$$\int_{\mathcal{S}} K \,\mathrm{d}A = 2\pi\chi(\mathcal{S}) - \oint_{\partial\mathcal{S}} \kappa_g \mathrm{d}s, \tag{14.78f}$$

where

$$\chi(\mathcal{S}) = V + F - E \tag{14.79}$$

is the Euler characteristic of the surface. The boundary, or boundaries, should be oriented such that $\hat{\boldsymbol{n}} \times \hat{\boldsymbol{t}}$ points toward $\mathcal{S}$, where $\hat{\boldsymbol{t}}$ is the unit tangent to the boundary, and boundary corners can be handled as for a triangle (14.78c) through the κ_g integral in the limit of vanishing arcs rounding out the corners.

One remarkable consequence of (14.78f) is that for a surface without boundary, such as a whole sphere or a whole torus,

$$\int_{\mathcal{S}} K \,\mathrm{d}A = 2\pi\chi(\mathcal{S}),$$

so that integral is a topological invariant, it equals 4π for any closed surface that can be smoothly deformed to a sphere, and 0 for any surface that can be smoothly deformed to a torus, for example.

14.10 Surface tension and mean curvature

Interfaces between air and water, and soap films, are physical examples of surfaces. Those surfaces are the 2D equivalent of strings; they do not resist bending, twisting, or shearing, but can transmit tensile forces. Surface tension $\gamma \geq 0$ is a scalar quantity with units of force per unit length. It is the 2D equivalent of pressure p, but while pressure provides a compressive force, surface tension provides an extensional force, tangent to the surface.

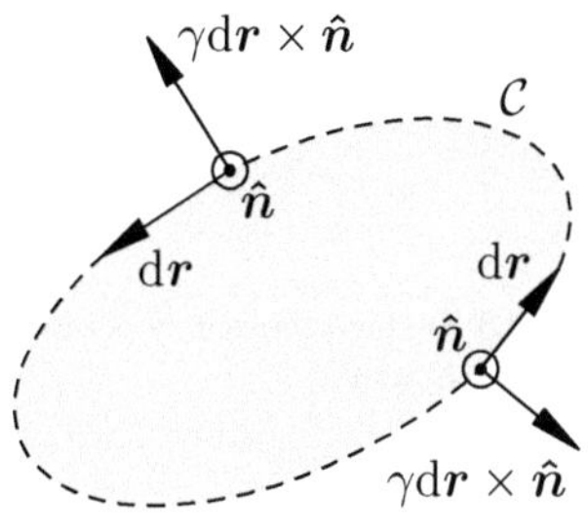

Fig. 14.12 Surface tension force around a patch with boundary $\mathcal{C}$.

In static equilibrium, the net surface tension force on the closed curve boundary $\mathcal{C}$ of an arbitrary surface patch $\mathcal{S}$ must balance the net external force on the patch; thus (Fig. 14.12),

$$\oint_{\mathcal{C}} \gamma \mathrm{d}\boldsymbol{r} \times \hat{\boldsymbol{n}} + \int_{\mathcal{S}} \boldsymbol{f} \, \mathrm{d}A = 0, \tag{14.80a}$$

where $\boldsymbol{f}$ is the external force per unit area acting on the surface element of area $\mathrm{d}A$, a pressure force for example, and the curve boundary $\mathcal{C} = \partial\mathcal{S}$ is oriented counterclockwise around the unit surface normal $\hat{\boldsymbol{n}}$. The line element along the curve boundary is $\mathrm{d}\boldsymbol{r}$ and γ and $\hat{\boldsymbol{n}}$ vary along the boundary (that is, $\hat{\boldsymbol{n}}$ is "inside" the integral in (14.80a)).

The line integral in (14.80a) can be turned into an integral over the patch after parametrization of $\mathcal{S} : \boldsymbol{r}(u, v)$ with $(u, v) \in \mathcal{U}$, such that $\mathrm{d}\boldsymbol{r} = \partial_u \boldsymbol{r} \, \mathrm{d}u + \partial_v \boldsymbol{r} \, \mathrm{d}v$, $\sqrt{g}\, \hat{\boldsymbol{n}} = \partial_u \boldsymbol{r} \times \partial_v \boldsymbol{r}$ and $\mathrm{d}A = \sqrt{g}\, \mathrm{d}u\mathrm{d}v$ (14.28c). Using Green's theorem (13.31c), the line integral in (14.80a) becomes

$$\begin{aligned}\oint_{\mathcal{C}} \gamma \mathrm{d}\boldsymbol{r} \times \hat{\boldsymbol{n}} &= \oint_{\partial\mathcal{U}} \gamma \left(\partial_u \boldsymbol{r} \, \mathrm{d}u + \partial_v \boldsymbol{r} \, \mathrm{d}v \right) \times \hat{\boldsymbol{n}} \\ &= \iint_{\mathcal{U}} \left(\partial_u \left(\gamma \partial_v \boldsymbol{r} \times \hat{\boldsymbol{n}} \right) + \partial_v \left(\gamma \hat{\boldsymbol{n}} \times \partial_u \boldsymbol{r} \right) \right) \mathrm{d}u\mathrm{d}v,\end{aligned}$$

and (14.80a) reads

$$\iint_{\mathcal{U}} \left(\partial_u \left(\gamma \partial_v \boldsymbol{r} \times \hat{\boldsymbol{n}} \right) + \partial_v \left(\gamma \hat{\boldsymbol{n}} \times \partial_u \boldsymbol{r} \right) + \boldsymbol{f} \sqrt{g} \right) \mathrm{d}u\mathrm{d}v = 0. \tag{14.80b}$$

Since this holds for any subpatch $\mathcal{U}$ of the entire surface, the integrand must vanish everywhere, yielding the partial differential equation

$$\partial_u \left(\partial_v \boldsymbol{r} \times \gamma \hat{\boldsymbol{n}} \right) + \partial_v \left(\gamma \hat{\boldsymbol{n}} \times \partial_u \boldsymbol{r} \right) + \sqrt{g}\, \boldsymbol{f} = 0, \tag{14.81a}$$

which reduces to

$$\partial_v \boldsymbol{r} \times \partial_u \left(\gamma \hat{\boldsymbol{n}} \right) + \partial_v \left(\gamma \hat{\boldsymbol{n}} \right) \times \partial_u \boldsymbol{r} + \sqrt{g}\, \boldsymbol{f} = 0, \tag{14.81b}$$

since $\partial_u \partial_v \boldsymbol{r} = \partial_v \partial_u \boldsymbol{r}$.

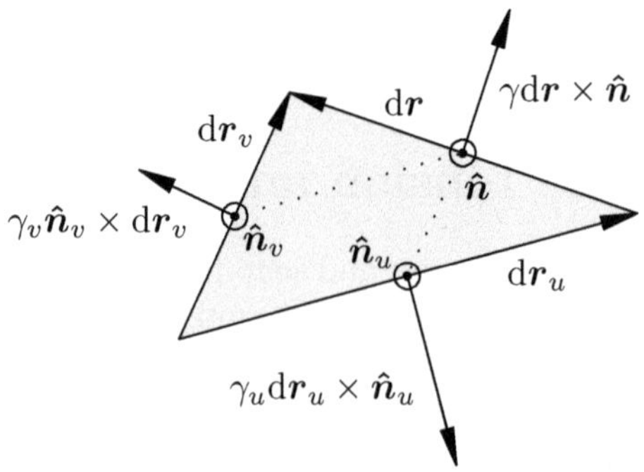

Fig. 14.13 Surface tension γ around a triangular element.

This equation can also be derived without explicit reference to Green's theorem by considering a triangular coordinate element with vertices at $\boldsymbol{r}(u, v)$, $\boldsymbol{r}(u + \mathrm{d}u, v)$, $\boldsymbol{r}(u, v + \mathrm{d}v)$ (Fig. 14.13). Balance of tension and surface forces for that element of area $\mathrm{d}A = \frac{1}{2}|\partial_u \boldsymbol{r} \times \partial_v \boldsymbol{r}|\,\mathrm{d}u\mathrm{d}v = \frac{1}{2}\sqrt{g}\,\mathrm{d}u\mathrm{d}v$ reads

$$\gamma_u \mathrm{d}\boldsymbol{r}_u \times \hat{\boldsymbol{n}}_u + \gamma \mathrm{d}\boldsymbol{r} \times \hat{\boldsymbol{n}} - \gamma_v \mathrm{d}\boldsymbol{r}_v \times \hat{\boldsymbol{n}}_v + \tfrac{1}{2}\boldsymbol{f}\sqrt{g}\,\mathrm{d}u\mathrm{d}v = 0,$$

where the $\gamma\hat{\boldsymbol{n}}$'s are evaluated at the *midpoints* of each corresponding segment (by the midpoint or trapezoidal rules).[15] Substituting $\mathrm{d}\boldsymbol{r} = \mathrm{d}\boldsymbol{r}_v - \mathrm{d}\boldsymbol{r}_u$ and rearranging gives

$$\mathrm{d}\boldsymbol{r}_v \times (\gamma\hat{\boldsymbol{n}} - \gamma_v\hat{\boldsymbol{n}}_v) + (\gamma\hat{\boldsymbol{n}} - \gamma_u\hat{\boldsymbol{n}}_u) \times \mathrm{d}\boldsymbol{r}_u + \tfrac{1}{2}\boldsymbol{f}\sqrt{g}\,\mathrm{d}u\mathrm{d}v = 0,$$

yielding (14.81b) since $\gamma\hat{\boldsymbol{n}} - \gamma_v\hat{\boldsymbol{n}}_v = \frac{1}{2}\mathrm{d}u\,\partial_u(\gamma\hat{\boldsymbol{n}})$, $\gamma\hat{\boldsymbol{n}} - \gamma_u\hat{\boldsymbol{n}}_u = \frac{1}{2}\mathrm{d}v\,\partial_v(\gamma\hat{\boldsymbol{n}})$, and $\mathrm{d}\boldsymbol{r}_u = \mathrm{d}u\,\partial_u\boldsymbol{r}$, $\mathrm{d}\boldsymbol{r}_v = \mathrm{d}v\,\partial_v\boldsymbol{r}$.

[15] For example, with $\boldsymbol{N} \triangleq \gamma\hat{\boldsymbol{n}}$,

$$\boldsymbol{N}_u = \boldsymbol{N}(u + \mathrm{d}u/2, v),$$

or

$$\boldsymbol{N}_u = \frac{1}{2}\left(\boldsymbol{N}(u,v) + \boldsymbol{N}(u + \mathrm{d}u, v)\right).$$

Balancing pressure

If the force $\boldsymbol{f}$ is purely normal to the surface, as the net pressure force on the surface of a bubble, for example, with $\boldsymbol{f} = \hat{\boldsymbol{n}}\,\Delta p$, where $\Delta p = p_- - p_+$ is the pressure difference across the surface in the direction of $\hat{\boldsymbol{n}}$, then dotting (14.81b) with $\partial_u\boldsymbol{r}$ and $\partial_v\boldsymbol{r}$, using $\boldsymbol{a} \times \boldsymbol{b} \cdot \boldsymbol{c} = \boldsymbol{b} \times \boldsymbol{c} \cdot \boldsymbol{a}$ and $\partial_u\boldsymbol{r} \times \partial_v\boldsymbol{r} = \sqrt{g}\,\hat{\boldsymbol{n}}$, yields

$$\partial_u\boldsymbol{r} \cdot \partial_v\boldsymbol{r} \times \partial_u(\gamma\hat{\boldsymbol{n}}) = \sqrt{g}\,\hat{\boldsymbol{n}} \cdot \partial_u(\gamma\hat{\boldsymbol{n}}) = \sqrt{g}\,\partial_u\gamma = 0, \tag{14.82a}$$
$$\partial_v\boldsymbol{r} \cdot \partial_v(\gamma\hat{\boldsymbol{n}}) \times \partial_u\boldsymbol{r} = \sqrt{g}\,\hat{\boldsymbol{n}} \cdot \partial_v(\gamma\hat{\boldsymbol{n}}) = \sqrt{g}\,\partial_v\gamma = 0, \tag{14.82b}$$

since $\hat{\boldsymbol{n}} \cdot \partial\hat{\boldsymbol{n}} = 0$. Thus, the surface tension γ must be constant along the surface. Dotting (14.81b) with $\hat{\boldsymbol{n}}$ now, using the contravariant basis (14.15) such that $\sqrt{g}\,\boldsymbol{a}^u = \partial_v\boldsymbol{r} \times \hat{\boldsymbol{n}}$ and $\sqrt{g}\,\boldsymbol{a}^v = \hat{\boldsymbol{n}} \times \partial_u\boldsymbol{r}$, yields the *Young–Laplace equation*

$$p_+ - p_- = -\gamma\left(\boldsymbol{a}^u \cdot \partial_u\hat{\boldsymbol{n}} + \boldsymbol{a}^v \cdot \partial_v\hat{\boldsymbol{n}}\right) = -\gamma\boldsymbol{\nabla} \cdot \hat{\boldsymbol{n}} = 2H\,\gamma, \tag{14.82c}$$

where $2H = \kappa_1 + \kappa_2$ is the sum of the principal curvatures (14.72). The mean curvature H is negative if $\hat{\boldsymbol{n}}$ diverges, in which case the pressure p_+ on the $\hat{\boldsymbol{n}}$ side of the surface is $p_+ < p_-$, and vice versa. For a spherical bubble of radius R, the principal curvatures are $\kappa_1 = \kappa_2 = -1/R$; then, (14.82c) yields $p_+ - p_- = -2\gamma/R < 0$, as the pressure jump from inside to outside the bubble.

Minimizing area

If the force acting on the surface is negligible, then the force balance equation (14.81b) reduces to zero mean curvature, (14.82c)

$$-\,\boldsymbol{\nabla} \cdot \hat{\boldsymbol{n}} = 2H = 0. \tag{14.83}$$

This equation implies that the surface minimizes surface area. Indeed, the total area of surface $\mathcal{S} : (u,v) \in \mathcal{U} \to \boldsymbol{r}(u,v) \in \mathbf{E}^3$ is

$$A = \iint_{\mathcal{U}} |\partial_u\boldsymbol{r} \times \partial_v\boldsymbol{r}|\,\mathrm{d}u\mathrm{d}v.$$

A neighboring surface with the same boundary $\partial\mathcal{S}$ has the parametrization

$$\mathcal{S}' : (u,v) \in \mathcal{U} \to \boldsymbol{r}'(u,v) = \boldsymbol{r}(u,v) + \epsilon\, a(u,v)\hat{\boldsymbol{n}}(u,v), \tag{14.84}$$

where $a(u,v)$ is an arbitrary smooth function with $\max|a(u,v)| = 1$ that vanishes at the boundary, ϵ is an arbitrary small real parameter, and $\hat{\boldsymbol{n}}(u,v)$ is the unit normal to the surface $\mathcal{S}$. Tangent perturbations of $\boldsymbol{r}(u,v)$ do not change the surface, only the parametrization, to first order in ϵ. We are interested in the variation of the surface area to first order in ϵ. We have

$$\begin{aligned}\partial_u\boldsymbol{r}' \times \partial_v\boldsymbol{r}' &= (\partial_u\boldsymbol{r} + \epsilon\partial_u(a\hat{\boldsymbol{n}})) \times (\partial_v\boldsymbol{r} + \epsilon\partial_v(a\hat{\boldsymbol{n}}))\\ &= (\partial_u\boldsymbol{r} \times \partial_v\boldsymbol{r}) + \epsilon\big(\partial_u\boldsymbol{r} \times \partial_v(a\hat{\boldsymbol{n}}) + \partial_u(a\hat{\boldsymbol{n}}) \times \partial_v\boldsymbol{r}\big) + O(\epsilon^2),\end{aligned}$$

where $O(\epsilon^2)$ are terms of order ϵ^2 or higher. Then the dot product of that vector with itself yields

$$\begin{aligned}|\partial_u\boldsymbol{r}' \times \partial_v\boldsymbol{r}'|^2 &= |\partial_u\boldsymbol{r} \times \partial_v\boldsymbol{r}|^2\\ &\quad + 2\epsilon a\sqrt{g}\,(\hat{\boldsymbol{n}}\cdot\partial_u\boldsymbol{r}\times\partial_v\hat{\boldsymbol{n}} + \hat{\boldsymbol{n}}\cdot\partial_u\hat{\boldsymbol{n}}\times\partial_v\boldsymbol{r}) + O(\epsilon^2)\\ &= g + 2\epsilon\, ag\,(\boldsymbol{a}^u\cdot\partial_u\hat{\boldsymbol{n}} + \boldsymbol{a}^v\cdot\partial_v\hat{\boldsymbol{n}}) + O(\epsilon^2)\\ &= g + 2\epsilon\, ag\,\boldsymbol{\nabla}\cdot\hat{\boldsymbol{n}} + O(\epsilon^2),\end{aligned}$$

because $\partial_u\boldsymbol{r} \times \partial_v\boldsymbol{r} = \sqrt{g}\,\hat{\boldsymbol{n}}$ so the terms proportional to $\partial_u a$ and $\partial_v a$ drop out in the dot product.Then $\sqrt{g}\,\boldsymbol{a}^u = \partial_v\boldsymbol{r}\times\hat{\boldsymbol{n}}$, $\sqrt{g}\,\boldsymbol{a}^v = \hat{\boldsymbol{n}}\times\partial_u\boldsymbol{r}$ and (14.72) have been used. Now since $(1+2\epsilon\alpha)^{1/2} = 1 + \epsilon\alpha + O(\epsilon^2)$, the surface area as a function of ϵ is then

$$A(\epsilon) = A(0) + \epsilon\iint_{\mathcal{U}} a(u,v)\,\boldsymbol{\nabla}\cdot\hat{\boldsymbol{n}}\,\sqrt{g}\,\mathrm{d}u\mathrm{d}v + O(\epsilon^2) \tag{14.85}$$

and the surface $\mathcal{S}$ minimizes area if $\mathrm{d}A(\epsilon)/\mathrm{d}\epsilon = 0$ at $\epsilon = 0$. Since this must be true for any choice of perturbation $a(u,v)$, the minimal surface must have zero mean curvature (14.72),

$$\boldsymbol{\nabla}\cdot\hat{\boldsymbol{n}} = \boldsymbol{a}^u\cdot\partial_u\hat{\boldsymbol{n}} + \boldsymbol{a}^v\cdot\partial_v\hat{\boldsymbol{n}} = -2H = 0. \tag{14.86}$$

Thus, $\boldsymbol{\nabla}\cdot\hat{\boldsymbol{n}} = 0$ for a surface of zero mean curvature. If the surface is represented implicitly as a level set of a scalar field, $f(\boldsymbol{r}) = 0$ say, then its unit normal $\hat{\boldsymbol{n}}$ is the normalized gradient

$$\hat{\boldsymbol{n}} = \frac{\boldsymbol{\nabla}f}{|\boldsymbol{\nabla}f|}. \tag{14.87}$$

Finding a surface of zero mean curvature then involves finding a scalar field $f(\boldsymbol{r})$ that solves the nonlinear partial differential equation

$$\boldsymbol{\nabla}\cdot\left(\frac{\boldsymbol{\nabla}f}{|\boldsymbol{\nabla}f|}\right) = 0, \tag{14.88}$$

with $f(\boldsymbol{r}(t)) = 0$ on the prescribed boundary curve $\boldsymbol{r}(t)$.

The original example of a minimal surface is the *catenoid*, a surface of revolution (Fig. 14.14),

$$\boldsymbol{r}(\varphi, z) = \rho(z)\, \hat{\boldsymbol{\rho}}(\varphi) + z\, \hat{\mathbf{z}}, \tag{14.89}$$

where $\rho = a\cosh(z/a)$ is a *catenary* curve (12.73), with $\hat{\boldsymbol{\rho}}(\varphi) = \cos\varphi\,\hat{\mathbf{x}} + \sin\varphi\,\hat{\mathbf{y}}$. Indeed, for such surface of revolution (14.109),

$$\boldsymbol{\nabla}\cdot\hat{\boldsymbol{n}} = -2H = \frac{1}{\rho\,(\mathrm{d}\rho/\mathrm{d}z)}\frac{\mathrm{d}}{\mathrm{d}z}\left(\frac{\rho}{\sqrt{1+(\mathrm{d}\rho/\mathrm{d}z)^2}}\right),$$

and this is a minimal surface if $\boldsymbol{\nabla}\cdot\hat{\boldsymbol{n}} = 0$, that is, if $\rho^2 = a^2(1+(\mathrm{d}\rho/\mathrm{d}z)^2)$, where a is an arbitrary positive constant. This yields a separable first-order differential equation that is readily integrated to obtain

$$\frac{\mathrm{d}\rho}{\mathrm{d}z} = \pm\sqrt{\frac{\rho^2}{a^2} - 1} \Rightarrow \rho = a\cosh\frac{z}{a}, \tag{14.90}$$

and the catenoid is the only smooth surface of revolution with zero mean curvature.

The problem of minimal surfaces, named *Plateau's problem* after Joseph Plateau (1801–1883), who formulated the laws governing the structures of soap films and their intersections, has been an active research area in mathematics, beautifully illustrated and motivated by soap films and bubbles.[16] Figure 14.15 shows the helicoid (13.40), another example of a minimal surface.

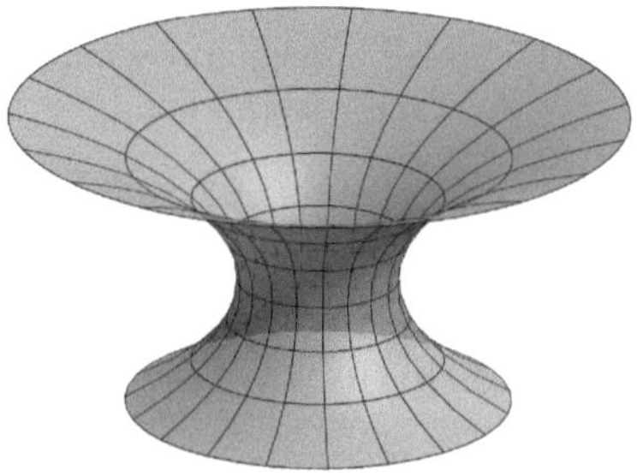

Fig. 14.14 The catenoid, the only smooth minimal surface of revolution.

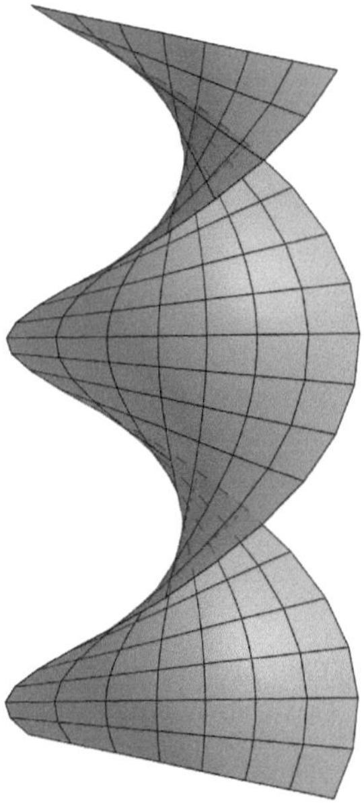

Fig. 14.15 The helicoid (13.40) is also a minimal surface.

[16] F. J. Almgren and J. E. Taylor, *The Geometry of Soap Films and Soap Bubbles*, Scientific American, **235**, (1976), 82–93.

Exercises

(14.1) Given two nonparallel unit vectors $\hat{\boldsymbol{a}}_1$, $\hat{\boldsymbol{a}}_2$, define $\hat{\boldsymbol{a}}_1 \times \hat{\boldsymbol{a}}_2 = \hat{\boldsymbol{n}} \sin\theta$ and

$$\hat{\boldsymbol{a}}^1 = \hat{\boldsymbol{a}}_2 \times \hat{\boldsymbol{n}}, \quad \hat{\boldsymbol{a}}^2 = \hat{\boldsymbol{n}} \times \hat{\boldsymbol{a}}_1.$$

Let (α_1, α_2) be the *parallel* projections of an arbitrary vector $\boldsymbol{v}$ in the $(\hat{\boldsymbol{a}}_1, \hat{\boldsymbol{a}}_2)$ subspace, and (β_1, β_2) its *perpendicular* projections onto $\hat{\boldsymbol{a}}_1$, $\hat{\boldsymbol{a}}_2$, respectively (Fig. 14.16).

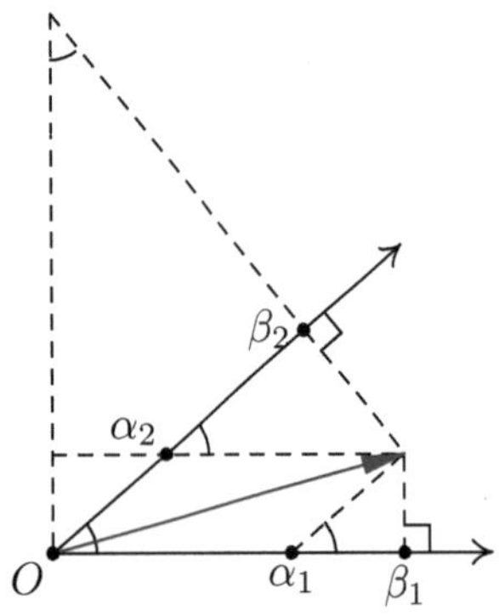

Fig. 14.16 Geometric interpretation of covariant and contravariant components (exercise 14.1).

Show geometrically that

$$\boldsymbol{v} = \alpha_1 \hat{\boldsymbol{a}}_1 + \alpha_2 \hat{\boldsymbol{a}}_2 = \frac{\beta_1}{\sin\theta} \hat{\boldsymbol{a}}^1 + \frac{\beta_2}{\sin\theta} \hat{\boldsymbol{a}}^2.$$

In terms of covariant and contravariant bases (14.15), with $\boldsymbol{a}_1 = \sqrt{g_{11}}\, \hat{\boldsymbol{a}}_1$ and $\boldsymbol{a}_2 = \sqrt{g_{22}}\, \hat{\boldsymbol{a}}_2$, show that these read

$$\begin{aligned} \boldsymbol{v} &= \frac{\alpha_1}{\sqrt{g_{11}}} \boldsymbol{a}_1 + \frac{\alpha_2}{\sqrt{g_{22}}} \boldsymbol{a}_2 \\ &= \beta_1 \sqrt{g_{11}}\, \boldsymbol{a}^1 + \beta_2 \sqrt{g_{22}}\, \boldsymbol{a}^2. \end{aligned} \tag{14.91}$$

This provides geometric interpretations for contravariant and covariant components of a vector

$$\boldsymbol{v} = v^1 \boldsymbol{a}_1 + v^2 \boldsymbol{a}_2 = v_1 \boldsymbol{a}^1 + v_2 \boldsymbol{a}^2.$$

The contravariant components $(v^1, v^2) = (\alpha_1/\sqrt{g_{11}}, \alpha_2/\sqrt{g_{22}})$ are inversely proportional to the magnitudes $|\boldsymbol{a}_1| = \sqrt{g_{11}}$, $|\boldsymbol{a}_2| = \sqrt{g_{22}}$, while the covariant components $(v_1, v_2) = (\beta_1\sqrt{g_{11}}, \beta_2\sqrt{g_{22}})$ are proportional to those magnitudes.

(14.2) Show how (14.24) follows from (14.16) and (14.23); then, with $g \triangleq g_{11}g_{22} - g_{12}g_{21}$, obtain

$$\begin{bmatrix} g^{11} & g^{12} \\ g^{21} & g^{22} \end{bmatrix} = \frac{1}{g} \begin{bmatrix} g_{22} & -g_{12} \\ -g_{21} & g_{11} \end{bmatrix}. \tag{14.92}$$

(14.3) Let $\boldsymbol{a}_i \triangleq \partial_i \boldsymbol{r}$ and $\sqrt{g}\, \hat{\boldsymbol{n}} = \boldsymbol{a}_1 \times \boldsymbol{a}_2$, as in (14.28b). Show that

$$\hat{\boldsymbol{n}} \times \partial_i \boldsymbol{r} = \sqrt{g}\, \epsilon_{ij3}\, \boldsymbol{a}^j. \tag{14.93}$$

That is $\hat{\boldsymbol{n}} \times \partial_1 \boldsymbol{r} = \sqrt{g}\, \boldsymbol{a}^2$ and $\hat{\boldsymbol{n}} \times \partial_2 \boldsymbol{r} = -\sqrt{g}\, \boldsymbol{a}^1$.

(14.4) For a surface parametrized as $z = f(x, y)$ in cartesian coordinates x, y, z, show that the covariant and contravariant bases are

$$\boldsymbol{a}_1 = \hat{\mathbf{x}} + \hat{\mathbf{z}} z_x, \quad \boldsymbol{a}_2 = \hat{\mathbf{y}} + \hat{\mathbf{z}} z_y,$$

$$\boldsymbol{a}^1 = \frac{\hat{\mathbf{x}}(1 + z_y^2) - z_x z_y \hat{\mathbf{y}} + \hat{\mathbf{z}} z_x}{1 + z_x^2 + z_y^2},$$

$$\boldsymbol{a}^2 = \frac{-\hat{\mathbf{x}} z_x z_y + \hat{\mathbf{y}}(1 + z_x^2) + \hat{\mathbf{z}} z_y}{1 + z_x^2 + z_y^2},$$

where $z_x = \partial z/\partial x$, $z_y = \partial z/\partial y$. Verify that $\boldsymbol{a}_i \cdot \boldsymbol{a}^j = \delta_i^j$.

(14.5) For a surface $z = f(x, y)$ in cartesian coordinates x, y, z, show that the first fundamental form and the metrics are

$$\mathrm{d}s^2 = (1 + z_x^2)\,\mathrm{d}x^2 + 2 z_x z_y\,\mathrm{d}x\mathrm{d}y + (1 + z_y^2)\,\mathrm{d}y^2,$$

$$\mathsf{G} = \begin{bmatrix} 1 + z_x^2 & z_x z_y \\ z_x z_y & 1 + z_y^2 \end{bmatrix},$$

$$\mathsf{G}^{-1} = \frac{1}{1 + z_x^2 + z_y^2} \begin{bmatrix} 1 + z_y^2 & -z_x z_y \\ -z_x z_y & 1 + z_x^2 \end{bmatrix}.$$

(14.6) For a surface $z = f(x, y)$ in cartesian coordinates x, y, z, use the previous two problems to verify formula (14.23).

(14.7) A surface of revolution is given as $\rho(t)$, $z(t)$ in cylindrical coordinates (ρ, φ, z) (10.14). Find the position vector $\boldsymbol{r}(\varphi, t)$; then show that the covariant and contravariant basis vectors and the unit normal read, with $\dot{\rho} = \mathrm{d}\rho/\mathrm{d}t$, $\dot{z} = \mathrm{d}z/\mathrm{d}t$,

$$\begin{aligned} \boldsymbol{a}_\varphi &= \rho\, \hat{\boldsymbol{\varphi}}, \quad \boldsymbol{a}_t = \dot{\rho} \hat{\boldsymbol{\rho}} + \dot{z} \hat{\mathbf{z}}, \\ \boldsymbol{a}^\varphi &= \frac{\hat{\boldsymbol{\varphi}}}{\rho}, \quad \boldsymbol{a}^t = \frac{\dot{\rho} \hat{\boldsymbol{\rho}} + \dot{z} \hat{\mathbf{z}}}{\dot{\rho}^2 + \dot{z}^2}, \end{aligned} \tag{14.94}$$

$$\hat{\boldsymbol{n}} = \frac{\dot{z} \hat{\boldsymbol{\rho}} - \dot{\rho} \hat{\mathbf{z}}}{\sqrt{\dot{\rho}^2 + \dot{z}^2}}.$$

Find the metrics and show that the first fundamental form

$$\mathrm{d}s^2 = \rho^2 \mathrm{d}\varphi^2 + (\dot\rho^2 + \dot z^2)\,\mathrm{d}t^2. \tag{14.95}$$

Show that meridians (φ fixed) are geodesics, but parallels (t fixed) are not, in general.

(14.8) Expand the formula (14.35), for each index and all sums explicitly.

(14.9) Show that u^1-coordinate curves, along which u^1 varies but u^2 is constant, are geodesics if

$$\partial_1\partial_1 \boldsymbol{r} = \Gamma^1_{11}\partial_1\boldsymbol{r} + b_{11}\hat{\boldsymbol{n}} \quad \Leftrightarrow \quad \Gamma^2_{11} = 0.$$

Similarly, u^2-coordinate curves are geodesics if $\Gamma^1_{22} = 0$.

(14.10) Show that the acceleration $\ddot{\boldsymbol{r}}$ of a particle moving on a surface $\boldsymbol{r}(u^1, u^2)$ is

$$\ddot{\boldsymbol{r}} = \left(\ddot u^i + \dot u^j \dot u^k \Gamma^i_{jk}\right)\partial_i \boldsymbol{r} + \dot u^i \dot u^j\, b_{ij}\hat{\boldsymbol{n}}. \tag{14.96}$$

Thus, a particle of mass m moving freely along the surface follows geodesics and experiences a force $\boldsymbol{F} = m^{-1}\dot u^i \dot u^k\, b_{ij}\hat{\boldsymbol{n}}$ that keeps it on the surface.

(14.11) Verify formula (14.41). Derive general formula (14.43) for Γ_{ijk}.

(14.12) Calculate the metric and show that the Christoffel symbols are all 0 for the cylinder

$$\boldsymbol{r}(\varphi, z) = R(\hat{\mathbf{x}}\cos\varphi + \hat{\mathbf{y}}\sin\varphi) + z\,\hat{\mathbf{z}} = R\hat{\boldsymbol{\rho}}(\varphi) + z\hat{\mathbf{z}}.$$

(14.13) Show that helices

$$\boldsymbol{r}(t) = R(\hat{\mathbf{x}}\cos\omega t + \hat{\mathbf{y}}\sin\omega t) + at\,\hat{\mathbf{z}}$$

are geodesics for the cylinder

$$\boldsymbol{r}(u, v) = R(\hat{\mathbf{x}}\cos u + \hat{\mathbf{y}}\sin u) + v\,\hat{\mathbf{z}},$$

for any constant a and ω.

(14.14) Compute the three second derivatives $\partial_i\partial_j\boldsymbol{r}$ for spherical coordinates (θ, φ) from (10.23) or (13.5); then project (14.5) onto the tangent vectors (13.16) to verify the geodesic equations (14.48).

(14.15) Derive the geodesic equations (14.48) for the sphere in spherical coordinates from the Euler–Lagrange equations (14.58).

(14.16) Show that

$$\tfrac{1}{2}\frac{\mathrm{d}}{\mathrm{d}t}(g_{ij}\dot u^i \dot u^j) = g_{ij}\ddot u^j \dot u^i + \tfrac{1}{2}(\partial_k g_{ij})\dot u^i \dot u^j \dot u^k,$$

where $g_{ij} = g_{ji} = g_{ij}(u^1, \ldots, u^n)$, and conclude that (14.59) conserves $g_{ij}\dot u^i \dot u^j$ (Hint: rename dummy indices.) From a mechanical point of view, this is conservation of kinetic energy for a particle of constant mass m moving under the sole action of a normal force keeping the particle on the surface.

(14.17) Show that (14.57), with $K = g_{lm}\dot u^l \dot u^m$, yields

$$g_{ij}\ddot u^j + (\partial_k g_{ij})\dot u^j \dot u^k - \frac{1}{2}(\partial_i g_{jk})\,\dot u^j \dot u^k = \frac{g_{ij}\dot u^j}{2K}\frac{\mathrm{d}K}{\mathrm{d}t}.$$

Since solutions of (14.59) conserve K, as shown in the previous problem, conclude that those are solutions of (14.57) also. The converse is not true since there are solutions of (14.57) for which K is not constant. For geodesics on a sphere, for example, $\varphi(t) = \varphi_0$, $\theta(t) = \omega_0 t$ is a solution of the geodesic equations (14.48), with constant speed, but $\varphi(t) = \varphi_0$, $\theta(t) = at^2$ is not a solution of the geodesic equations. It is the same curve, but the speed is not constant.

(14.18) Show that another way to justify passing from (14.57) to (14.58) is to divide both sides of (14.57) by $\dot s(t) = \sqrt{g_{lm}\dot u^l \dot u^m} > 0$ and change variable $t \to s$ such that $\mathrm{d}s = \sqrt{g_{lm}\dot u^l \dot u^m}\,\mathrm{d}t$; then (14.57) becomes (14.58) with s in place of t, and $\mathrm{d}u/\mathrm{d}s$ in place of $\dot u \equiv \mathrm{d}u/\mathrm{d}t$.

(14.19) For orthogonal coordinates, the metric is diagonal $g_{ij} = \partial_i\boldsymbol{r}\cdot\partial_j\boldsymbol{r} = 0$ if $i \neq j$. Revert to classic notation $(u^1, u^2) \equiv (u, v)$ and define the metric (or *Lamé*) coefficients $h_u = |\partial\boldsymbol{r}/\partial u|$, $h_v = |\partial\boldsymbol{r}/\partial v|$ such that the first fundamental form (14.27)

$$\mathrm{d}s^2 = h_u^2\,\mathrm{d}u^2 + h_v^2\,\mathrm{d}v^2.$$

Show that the geodesic equations reduce to

$$\begin{cases} h_u \dfrac{\mathrm{d}}{\mathrm{d}t}(h_u \dot u) = -\dot v\left(\dot u\,\partial_v h_u^2 - \dot v\,\partial_u h_v^2\right)/2, \\ h_v \dfrac{\mathrm{d}}{\mathrm{d}t}(h_v \dot v) = \dot u\left(\dot u\,\partial_v h_u^2 - \dot v\,\partial_u h_v^2\right)/2, \end{cases} \tag{14.97}$$

and that $(h_u\dot u)^2 + (h_v\dot v)^2$ is conserved by these equations. Apply these to the sphere $\boldsymbol{r}(\theta, \varphi)$ (13.5) to obtain (14.48).

(14.20) Investigate geodesics for a torus (13.41).

[17] Historically called *isothermal* coordinates.

(14.21) Conformal coordinates[17] have $\partial_i \boldsymbol{r} \cdot \partial_j \boldsymbol{r} = h^2 \delta_{ij}$. Revert to classic notation $(u^1, u^2) \equiv (u, v)$ so that the first fundamental form (14.27)

$$ds^2 = h^2 \left(du^2 + dv^2\right),$$

where $h = |\partial_u \boldsymbol{r}| = |\partial_v \boldsymbol{r}| = f(u,v)$, in general, with $\partial_u \boldsymbol{r} \cdot \partial_v \boldsymbol{r} = 0$. Show that the geodesic equations (14.58) reduce to

$$\begin{cases} d(h\dot{u})/dt = -\dot{v}\left(\dot{u}\,\partial_v h - \dot{v}\,\partial_u h\right), \\ d(h\dot{v})/dt = \dot{u}\left(\dot{u}\,\partial_v h - \dot{v}\,\partial_u h\right). \end{cases} \tag{14.98}$$

(14.22) Show that for a sphere of radius r in Mercator coordinates (φ, v) (13.14)

$$ds^2 = \frac{r^2}{\cosh^2 v}\left(d\varphi^2 + dv^2\right),$$

and that the geodesic equations are

$$\begin{aligned} \ddot{\varphi} &= 2\dot{\varphi}\dot{v}\,\tanh v, \\ \frac{d}{dt}\left(\frac{\dot{v}}{\cosh v}\right) &= -\dot{\varphi}^2\,\frac{\sinh v}{\cosh^2 v}. \end{aligned} \tag{14.99}$$

Multiplying the first by $\dot{\varphi}$ yields $(\dot{\varphi})^2 = C\cosh^4 v$.

(14.23) Express the constants b and c in (14.50) in terms of (i) initial conditions (θ_0, φ_0) with $(d\theta/d\varphi)_0 = \alpha_0$, and (ii) boundary conditions (θ_0, φ_0) and (θ_1, φ_1).

(14.24) Let (θ_A, φ_A) be the spherical coordinates of a point A on a sphere centered at O. The geodesics through A are simply the *meridians* of the spherical coordinates with (O, A) as the polar axis. Let (u, v) be the spherical coordinates with respect to those axes. Use your knowledge of orthogonal transformations (Chapter 8) to find parametric equations for the geodesics through A in the (θ, φ) plane. Compare graphically with (14.50).

(14.25) Show that the Euler–Lagrange equation (14.55) remains unchanged if the integrand $L = L(u, \dot{u}, t)$ is an explicit function of t in the functional $\mathcal{F}$ (14.52).

(14.26) Show that if $L = L(u, \dot{u})$ is *not* an explicit function of t; then the Euler–Lagrange equation (14.55) can be written as

$$\frac{d}{dt}\left(\dot{u}\frac{\partial L}{\partial \dot{u}} - L\right) = 0, \tag{14.100}$$

and thus $\dot{u}\,\partial L/\partial \dot{u} - L =$ *constant.*

(14.27) Show that if $L = L(u^1, u^2, \ldots, \dot{u}^1, \dot{u}^2, \ldots)$ is *not* an explicit function of t; then the Euler–Lagrange equations (14.56) yield (with sum over i)

$$\frac{d}{dt}\left(\dot{u}^i\frac{\partial L}{\partial \dot{u}^i} - L\right) = 0. \tag{14.101}$$

Show that when $L = a_{ij}\dot{u}^i\dot{u}^j - V$ where $a_{ij} = a_{ji}$, with a_{ij} and V functions of the u^k but not $\dot{u}^k$, this conserved quantity is $E = a_{ij}\dot{u}^i\dot{u}^j + V$. In Lagrangian mechanics, this is the total energy of the system. If $L = f(x)$ where $x = g_{ij}\dot{u}^i\dot{u}^j$ with $g_{ij} = g_{ji}$ functions of u^k but not $\dot{u}^k$, show that this is conservation of $2x\,df/dx - f$. For $f(x) = x^p$ this is conservation of $(2p-1)x^p$.

(14.28) *Brachistochrone.* (Fig. 14.18) A particle starting from rest slides without friction under the action of gravity g, dropping by height H in horizontal distance L. Find the slide shape that minimizes the travel time T (12.67). In cartesian coordinates (Hx, Hy), so x and y are nondimensional, with y pointing *downward* in the direction of gravity and $\dot{x} = dx/dy$, show that

$$\sqrt{\frac{2g}{H}}\,T = \int_0^1 \frac{\sqrt{\dot{x}^2+1}}{\sqrt{y}}\,dy \equiv \int_0^1 L(y, x, \dot{x})\,dy.$$

This is (14.52) with $(t, u, \dot{u}) \to (y, x, \dot{x})$.
Show that the Euler–Lagrange equation yields

$$\frac{dx}{dy} = f(ky),$$

for some function f to be determined, and constant k. Use the substitution $ky = \sin^2(\theta/2) = (1-\cos\theta)/2$ to obtain

$$x = R\,(\theta - \sin\theta), \quad y = R\,(1-\cos\theta), \tag{14.102}$$

that is, the equation of a *cycloid*, corresponding to the position of a point on a circle of radius $R = (2k)^{-1}$ rolling along the x-axis. The end point $x = L/H$ at $y = 1$, yields

$$R = \frac{1}{1-\cos\theta_1}, \quad \frac{L}{H} = \frac{\theta_1 - \sin\theta_1}{1-\cos\theta_1}, \tag{14.103}$$

where the latter is an increasing function from $0 \to \infty$ in $\theta_1 = 0 \to 2\pi$, so there is a unique θ_1 for any $0 \le L/H < \infty$. Note that this solution assumed $x = x(y)$, yet for $\theta_1 > \pi$, that is, $L/H > \pi/2$, it yields a *multivalued* $x(y)$.

(14.29) *Brachistochrone 2.* (Fig. 14.18) Repeat the previous problem with x as the independent variable instead of y. Show that the travel time is now given by

$$\sqrt{\frac{2g}{H}}\,T = \int_0^{L/H} \frac{\sqrt{1+\dot{y}^2}}{\sqrt{y}}\,\mathrm{d}x \equiv \int_0^{L/H} L(x,y,\dot{y})\,\mathrm{d}x$$

with $\dot{y} = \mathrm{d}y/dx$. This is (14.52) with $(t,u,\dot{u}) \to (x,y,\dot{y})$. Show that (14.55) yields $2y\ddot{y}+\dot{y}^2 = -1$. Since $L(y,\dot{y})$ does not depend on x explicitly, we can use (14.100). Show that $\dot{y} = F(ky)$; then integrate that separable ODE using $ky = \sin^2(\theta/2)$.

(14.30) *Brachistochrone 3.* Find the curve of the previous problems using a parametrization $(x(t), y(t))$, where $t = 0 \to 1$ is a nondimensional time parameter, so the actual time is $t_* = t\,T$, where T is the total travel time. Show that the travel time is now given by

$$\sqrt{\frac{2g}{H}}\,T = \int_0^1 \sqrt{\frac{\dot{x}^2+\dot{y}^2}{y}}\,\mathrm{d}t,$$

and this is in the form of (14.51) with $(u^1,u^2) = (x,y)$ and $g_{ij} = y^{-1}\delta_{ij}$. Use $(\dot{x}^2+\dot{y}^2)/y = C$ constant along the shortest time curve, as in (14.57), then use (14.58) to show that $\dot{x}/y = D$ is constant; thus, $\dot{y}^2 = Cy - D^2y^2$. This yields a separable ODE for $y(t)$ that can be integrated by completing the square, using the substitution $u = 2D^2y/C - 1$, and the well-known $\mathrm{d}u/\sqrt{1-u^2} = \mathrm{d}(\arccos u)$. Find $x(t)$, $y(t)$, and the constants of integration.

(14.31) Show that the planar, simple (=non-self intersecting) closed curve of length $\mathcal{L}$ that maximizes the enclosed area A is a circle. Assuming counterclockwise orientation, recall why the area A and length $\mathcal{L}$ are given by the integrals

$$A = \int_0^1 (x\dot{y} - y\dot{x})\,\mathrm{d}t, \qquad \mathcal{L} = \int_0^1 \sqrt{\dot{x}^2+\dot{y}^2}\,\mathrm{d}t.$$

In deriving the Euler–Lagrange equations (14.56), a boundary term arises but vanishes when the end points are fixed. Explain what happens to the boundary term in this problem. Solve the problem using a *Lagrange multiplier* λ by optimizing the functional

$$\mathcal{F} = A - \lambda\mathcal{L} = \int_0^1 \left(x\dot{y} - y\dot{x} - \lambda\sqrt{\dot{x}^2+\dot{y}^2}\right)\mathrm{d}t.$$

Since $\partial_x L = \dot{y}$ and $\partial_y L = -\dot{x}$, both equations (14.56) are readily integrated.

(14.32) Show the steps in the derivation of the eigenvalue problem (14.66), (i) using explicit expansions of the sums in the definition of κ that consist of only three terms in 2D, and (ii) using index notation that applies to any dimension.

(14.33) Fill in and justify the steps in the derivation of (14.69d) and (14.69c).

(14.34) Use (14.73) to prove that if every point of a surface is an umbilic, then κ is a constant, and the surface must be a plane or a sphere.

(14.35) A smooth surface can be locally parametrized using cartesian coordinates with $\hat{\mathbf{z}}$ in the surface normal direction, $\hat{\mathbf{z}} = \hat{\boldsymbol{n}}$, as

$$z = ax^2 + 2b\,xy + cy^2 + (\cdots) = \begin{bmatrix} x & y \end{bmatrix}\begin{bmatrix} a & b \\ b & c \end{bmatrix}\begin{bmatrix} x \\ y \end{bmatrix} + (\cdots),$$

where $(\cdots)$ are terms that go to zero with x and y faster than quadratically (by Taylor series expansion). Calculate the metric g_{ij}, the curvic b_{ij}, then the mean, Gaussian, and principal curvatures κ_1, κ_2 for that surface at $(x,y) = (0,0)$, where a, b, c are arbitrary real constants.

(14.36) If the mean and Gaussian curvatures are H and K, respectively, show that the principal curvatures are the solutions of $\kappa^2 - 2H\kappa + K = 0$, that is,

$$\kappa_1 = H + \sqrt{H^2-K}, \quad \kappa_2 = H - \sqrt{H^2-K},$$

and that umbilics are points where $H^2 = K$.

(14.37) For a surface parametrized by $z = f(x,y)$ where x, y, z are cartesian coordinates, show that the metric $\mathbf{G} = [g_{ij}]$ and the curvic $\mathbf{B} = [b_{ij}]$ are

$$\mathbf{G} = \begin{bmatrix} 1+z_x^2 & z_xz_y \\ z_xz_y & 1+z_y^2 \end{bmatrix}, \quad \mathbf{B} = \frac{1}{\sqrt{1+z_x^2+z_y^2}}\begin{bmatrix} z_{xx} & z_{xy} \\ z_{xy} & z_{yy} \end{bmatrix}, \tag{14.104}$$

with $z_x = \partial_x z$, etc, and that the mean H and Gaussian K curvatures are given by

$$2H = \frac{(1+z_y^2)z_{xx} - 2z_xz_yz_{xy} + (1+z_x^2)z_{yy}}{\left(1+z_x^2+z_y^2\right)^{3/2}}, \quad K = \frac{z_{xx}z_{yy} - z_{xy}^2}{\left(1+z_x^2+z_y^2\right)^2}. \tag{14.105}$$

(14.38) A surface is parametrized as $\boldsymbol{r}(u,v) = x(u,v)\hat{\mathbf{x}} + y(u,v)\hat{\mathbf{y}} + z(u,v)\hat{\mathbf{z}} \in \mathbf{E}^3$, where $\hat{\mathbf{x}}$, $\hat{\mathbf{y}}$, $\hat{\mathbf{z}}$ are the usual fixed cartesian basis vectors, with $(u, v) \in \mathbb{R}^2$. Following Gauss, define

$$\boldsymbol{n} = A\hat{\mathbf{x}} + B\hat{\mathbf{y}} + C\hat{\mathbf{z}} = \partial_u \boldsymbol{r} \times \partial_v \boldsymbol{r},$$
$$D = \boldsymbol{n} \cdot \partial_u \partial_u \boldsymbol{r},\ D' = \boldsymbol{n} \cdot \partial_u \partial_v \boldsymbol{r},\ D'' = \boldsymbol{n} \cdot \partial_v \partial_v \boldsymbol{r},$$
$$E = \partial_u \boldsymbol{r} \cdot \partial_u \boldsymbol{r},\ F = \partial_u \boldsymbol{r} \cdot \partial_v \boldsymbol{r},\ G = \partial_v \boldsymbol{r} \cdot \partial_v \boldsymbol{r}.$$

Show that $A^2 + B^2 + C^2 = EG - F^2$ and that the mean H and Gaussian K curvatures are given by

$$2H = \frac{GD + ED'' - 2FD'}{(EG - F^2)^{3/2}},\qquad K = \frac{DD'' - (D')^2}{(A^2 + B^2 + C^2)^2}. \tag{14.106}$$

(14.39) A surface is parametrized as $\boldsymbol{r}(u, v) \in \mathbf{E}^3$ with $(u, v) \in \mathbb{R}^2$. Let

$$\boldsymbol{n} = (\partial_u \boldsymbol{r} \times \partial_v \boldsymbol{r}) = \sqrt{g}\, \hat{\boldsymbol{n}},$$
$$L = \hat{\boldsymbol{n}} \cdot \partial_u \partial_u \boldsymbol{r},\ M = \hat{\boldsymbol{n}} \cdot \partial_u \partial_v \boldsymbol{r},\ N = \hat{\boldsymbol{n}} \cdot \partial_v \partial_v \boldsymbol{r},$$
$$E = \partial_u \boldsymbol{r} \cdot \partial_u \boldsymbol{r},\ F = \partial_u \boldsymbol{r} \cdot \partial_v \boldsymbol{r},\ G = \partial_v \boldsymbol{r} \cdot \partial_v \boldsymbol{r}.$$

Show that $g = EG - F^2$ and that the mean H and Gaussian K curvatures are given by

$$2H = \frac{GL + EN - 2FM}{EG - F^2},\qquad K = \frac{LN - M^2}{EG - F^2}. \tag{14.107}$$

(14.40) A surface of revolution is parametrized as $\boldsymbol{r}(\varphi, t) = \rho(t)\hat{\boldsymbol{\rho}}(\varphi) + z(t)\hat{\mathbf{z}}$ in cylindrical coordinates (ρ, φ, z) (10.14). Show that the metric $\mathsf{G} = [g_{ij}]$ and the curvic $\mathsf{B} = [b_{ij}]$, where $g_{ij} = \partial_i \boldsymbol{r} \cdot \partial_j \boldsymbol{r}$ and $b_{ij} = \hat{\boldsymbol{n}} \cdot \partial_i \partial_j \boldsymbol{r}$, are, with $\dot{\rho} = \mathrm{d}\rho/\mathrm{d}t$, $\dot{z} = \mathrm{d}z/\mathrm{d}t$,

$$\mathsf{G} = \begin{bmatrix} \rho^2 & 0 \\ 0 & \dot{\rho}^2 + \dot{z}^2 \end{bmatrix},$$
$$\mathsf{B} = \frac{1}{\sqrt{\dot{\rho}^2 + \dot{z}^2}} \begin{bmatrix} -\rho\dot{z} & 0 \\ 0 & \ddot{\rho}\dot{z} - \dot{\rho}\ddot{z} \end{bmatrix},$$

so that the mean H and Gaussian K curvatures are given by

$$2H = \frac{\rho(\ddot{\rho}\dot{z} - \dot{\rho}\ddot{z}) - \dot{z}(\dot{\rho}^2 + \dot{z}^2)}{\rho(\dot{\rho}^2 + \dot{z}^2)^{3/2}},\qquad K = \frac{\dot{z}(\dot{\rho}\ddot{z} - \dot{z}\ddot{\rho})}{\rho(\dot{\rho}^2 + \dot{z}^2)^2}. \tag{14.108}$$

Find the principal curvatures and show that the coordinate curves are lines of curvature (that is, everywhere tangent to the principal curvature directions). Show that the expression for the mean curvature in (14.108) can be reduced to

$$H\frac{\mathrm{d}\rho^2}{\mathrm{d}t} = \frac{\mathrm{d}}{\mathrm{d}t}\left(\frac{-\rho\dot{z}}{\sqrt{\dot{\rho}^2 + \dot{z}^2}}\right). \tag{14.109}$$

(14.41) Find the principal curvatures for a circular cylinder $\boldsymbol{r}(\varphi, z) = R\hat{\boldsymbol{\rho}}(\varphi) + z\hat{\mathbf{z}}$.

(14.42) Find the principal curvatures for a cone $\boldsymbol{r}(\varphi, z) = a\,z\hat{\boldsymbol{\rho}}(\varphi) + z\hat{\mathbf{z}}$ (a constant).

(14.43) Calculate the metric g_{ij} and the curvic b_{ij} for the torus parametrization (13.41). Find the mean, Gaussian, and principal curvatures. Where does the Gaussian curvature vanish on that surface?

(14.44) Consider the ellipsoid

$$\boldsymbol{r}(u, v) = a\hat{\mathbf{x}} \cos u \cos v + b\hat{\mathbf{y}} \sin u \cos v + c\hat{\mathbf{z}} \sin v,$$

with $\hat{\mathbf{x}}, \hat{\mathbf{y}}, \hat{\mathbf{z}}$ a cartesian basis, and a, b, c real and positive. Find the mean and Gaussian curvatures. Find the equations that locate the umbilics.

(14.45) Consider the hyperboloid

$$\boldsymbol{r}(u, v) = a\hat{\mathbf{x}} \cos u \cosh v + b\hat{\mathbf{y}} \sin u \cosh v + c\hat{\mathbf{z}} \sinh v,$$

with $\hat{\mathbf{x}}, \hat{\mathbf{y}}, \hat{\mathbf{z}}$ a cartesian basis, and a, b, c real and positive. Find the mean and Gaussian curvatures.

(14.46) Fill in and justify the steps to derive (14.82) from (14.81b).

(14.47) From the Gauss (14.39) and Weingarten (14.64a) formulas

$$\partial_i \partial_j \boldsymbol{r} = \partial_i \boldsymbol{a}_j = \partial_j \boldsymbol{a}_i = \Gamma^k_{ij} \boldsymbol{a}_k + b_{ij} \hat{\boldsymbol{n}}, \tag{14.110a}$$
$$\partial_i \hat{\boldsymbol{n}} = -b_{ij} \boldsymbol{a}^j = -b_{ij} g^{jk} \boldsymbol{a}_k, \tag{14.110b}$$

together with $\boldsymbol{a}_i \cdot \boldsymbol{a}^j = \delta^j_i$ and $\hat{\boldsymbol{n}} \cdot \boldsymbol{a}^j = 0$, show that

$$\partial_i \boldsymbol{a}^j = -\Gamma^j_{ik} \boldsymbol{a}^k + b_{ik} g^{kj} \hat{\boldsymbol{n}}. \tag{14.110c}$$

(14.48) For $\sqrt{g} = \boldsymbol{a}_1 \times \boldsymbol{a}_2 \cdot \hat{\boldsymbol{n}} = J$, as defined in (14.15), (14.28b), show that

$$\partial_i \sqrt{g} = \Gamma^k_{ik} \sqrt{g} = \left(\Gamma^1_{i1} + \Gamma^2_{i2}\right) \sqrt{g},$$
$$\partial_i g = 2g\, \Gamma^k_{ik}. \tag{14.111}$$

(14.49) An arbitrary vector field $\boldsymbol{v}(u^1, u^2)$ *tangent to the surface* can be expanded in terms of the local covariant $\{\boldsymbol{a}_1, \boldsymbol{a}_2\} = \{\partial_1 \boldsymbol{r}, \partial_2 \boldsymbol{r}\}$ or contravariant bases $\{\boldsymbol{a}^1, \boldsymbol{a}^2\}$ as

$$\boldsymbol{v} = v^k \boldsymbol{a}_k = v_k \boldsymbol{a}^k.$$

The *covariant derivatives* of $\boldsymbol{v}$ are its surface coordinate derivatives $\partial_j \boldsymbol{v}$ projected onto the surface bases. Show that

$$\begin{aligned} \nabla_j v^i &\triangleq \boldsymbol{a}^i \cdot \partial_j \boldsymbol{v} &&= \partial_j v^i + \Gamma^i_{jk} v^k, \\ \nabla_j v_i &\triangleq \boldsymbol{a}_i \cdot \partial_j \boldsymbol{v} &&= \partial_j v_i - \Gamma^k_{ij} v_k. \end{aligned} \tag{14.112}$$

In particular, if the rate of change of a vector $\boldsymbol{v}$ is only perpendicular to the surface (that is, *parallel transport* of $\boldsymbol{v}$), such as if $\boldsymbol{v}$ is the unit tangent to a geodesic, then $\partial_j v^i \neq 0 \neq \partial_j v_i$ in general, but $\nabla_j v^i = 0 = \nabla_j v_i$.

For a tensor $\boldsymbol{\mathcal{T}} = T_{ij} \boldsymbol{a}^i \boldsymbol{a}^j = T^{ij} \boldsymbol{a}_i \boldsymbol{a}_j$, we have similarly

$$\begin{aligned} \nabla_k T^{ij} &\triangleq \boldsymbol{a}^i \cdot (\partial_k \boldsymbol{\mathcal{T}}) \cdot \boldsymbol{a}^j \\ &= \partial_k T^{ij} + \Gamma^i_{kl} T^{lj} + \Gamma^j_{kl} T^{il}, \\ \nabla_k T_{ij} &\triangleq \boldsymbol{a}_i \cdot (\partial_k \boldsymbol{\mathcal{T}}) \cdot \boldsymbol{a}_j \\ &= \partial_k T_{ij} - \Gamma^l_{ki} T_{lj} - \Gamma^l_{kj} T_{il}. \end{aligned} \tag{14.113}$$

Use this and (14.42) to show that the covariant derivatives of the metrics vanish

$$\nabla_k g_{ij} = 0 = \nabla_k g^{ij}. \tag{14.114}$$

This also follows from

$$g_{ij} \boldsymbol{a}^i \boldsymbol{a}^j = \boldsymbol{a}_i \boldsymbol{a}^i = \boldsymbol{\mathcal{P}} = \boldsymbol{a}^i \boldsymbol{a}_i = g^{ij} \boldsymbol{a}_i \boldsymbol{a}_j,$$

where

$$\boldsymbol{\mathcal{P}} = \boldsymbol{a}_1 \boldsymbol{a}^1 + \boldsymbol{a}_2 \boldsymbol{a}^2 = \boldsymbol{a}^1 \boldsymbol{a}_1 + \boldsymbol{a}^2 \boldsymbol{a}_2 = \boldsymbol{\mathcal{I}} - \hat{\boldsymbol{n}}\hat{\boldsymbol{n}}$$

is the projector onto the tangent plane, perpendicular to $\hat{\boldsymbol{n}}$, and $\boldsymbol{\mathcal{I}}$ is the identity tensor, such that $\boldsymbol{\mathcal{P}} \cdot \boldsymbol{v} = \boldsymbol{v}$, $\forall \boldsymbol{v}$ tangent to the surface; thus, $\boldsymbol{a}_i \cdot \partial_k \boldsymbol{\mathcal{P}} \cdot \boldsymbol{a}_j = 0 = \boldsymbol{a}^i \cdot \partial_k \boldsymbol{\mathcal{P}} \cdot \boldsymbol{a}^j$.

(14.50) In spherical coordinates $(u^1, u^2) = (\theta, \varphi)$, the covariant and contravariant bases are, respectively, (13.16)

$$\begin{aligned} (\boldsymbol{a}_\theta, \boldsymbol{a}_\varphi) &= (\partial_\theta \boldsymbol{r}, \partial_\varphi \boldsymbol{r}) = \left(r\,\hat{\boldsymbol{\theta}},\, r \sin\theta\, \hat{\boldsymbol{\varphi}}\right), \\ (\boldsymbol{a}^\theta, \boldsymbol{a}^\varphi) &= (\boldsymbol{a}_\varphi \times \hat{\boldsymbol{n}}, \hat{\boldsymbol{n}} \times \boldsymbol{a}_\theta)\,/J \\ &= \left(\hat{\boldsymbol{\theta}}/r,\, \hat{\boldsymbol{\varphi}}/(r\sin\theta)\right), \end{aligned}$$

with unit outward normal $\hat{\boldsymbol{n}} = \hat{\boldsymbol{r}}$ and $J = \boldsymbol{a}_1 \times \boldsymbol{a}_2 \cdot \hat{\boldsymbol{n}} = r^2 \sin\theta \equiv \sqrt{g}$ (14.15). An arbitrary tangent vector field $\boldsymbol{v}(P)$ can be expanded with either bases as

$$\boldsymbol{v} = v^\theta\, \partial_\theta \boldsymbol{r} + v^\varphi\, \partial_\varphi \boldsymbol{r} = v_\theta\, \boldsymbol{a}^\theta + v_\varphi\, \boldsymbol{a}^\varphi.$$

Verify *ab initio* using (10.24) that

$$\begin{aligned} \partial_\theta \boldsymbol{v} &= -v^\theta r \hat{\boldsymbol{r}} + \left(\partial_\theta v^\theta\right) \partial_\theta \boldsymbol{r} \\ &\quad + \left(\partial_\theta v^\varphi + v^\varphi \cot\theta\right) \partial_\varphi \boldsymbol{r}, \\ \partial_\varphi \boldsymbol{v} &= -v^\varphi r \sin^2\theta\, \hat{\boldsymbol{r}} \\ &\quad + \left(\partial_\varphi v^\theta - v^\varphi \sin\theta\cos\theta\right) \partial_\theta \boldsymbol{r} \\ &\quad + \left(v^\theta \cot\theta + \partial_\varphi v^\varphi\right) \partial_\varphi \boldsymbol{r}. \end{aligned} \tag{14.115}$$

Compare with (14.112) using (14.47). Show that if the tangent vector $\boldsymbol{v}$ is parallel transported along meridians ($\partial_\theta \boldsymbol{v} \parallel \hat{\boldsymbol{r}}$), then $\boldsymbol{v}$ maintains constant angle with $\hat{\boldsymbol{\theta}}$ (and $\hat{\boldsymbol{\varphi}}$), but when parallel transported along parallels ($\partial_\varphi \boldsymbol{v} \parallel \hat{\boldsymbol{r}}$) then

$$\partial_\varphi \begin{bmatrix} v^\theta \\ v^\varphi \sin\theta \end{bmatrix} = \begin{bmatrix} 0 & \cos\theta \\ -\cos\theta & 0 \end{bmatrix} \begin{bmatrix} v^\theta \\ v^\varphi \sin\theta \end{bmatrix},$$

such that $\boldsymbol{v}$ rotates about $\hat{\boldsymbol{r}}$ at rate $-\cos\theta$ with respect to $\hat{\boldsymbol{\theta}}, \hat{\boldsymbol{\varphi}}$ in the tangent plane. In this parallel transport case, show that

$$\partial_\varphi \boldsymbol{v} = \hat{\mathbf{z}} \times \boldsymbol{v} - \cos\theta\, \hat{\boldsymbol{r}} \times \boldsymbol{v} = -\sin\theta\, \hat{\boldsymbol{\theta}} \times \boldsymbol{v}.$$

(14.51) Show that

$$\epsilon^{ij3} \left(\partial_i \partial_1 \boldsymbol{r}\right) \cdot \left(\partial_j \partial_2 \boldsymbol{r}\right) = \epsilon^{ij3} \partial_i \Gamma_{1j2} = \epsilon^{ij3} \partial_j \Gamma_{2i1}$$

and deduce another expression for $g\,K$ in (14.75).

(14.52) Show that for any differentiable vector $\boldsymbol{a}$, $\boldsymbol{b}$ and scalar β functions of some variables, with ∂_k indicating derivative with respect to one of those variables,

$$\begin{aligned} \boldsymbol{a} \cdot \boldsymbol{a} = 1 &\Rightarrow\ \boldsymbol{a} \cdot \partial_k \boldsymbol{a} = 0, \\ \boldsymbol{a} \cdot \boldsymbol{b} = 0 &\Rightarrow\ \boldsymbol{a} \cdot \partial_k (\beta \boldsymbol{b}) = \beta\, \boldsymbol{a} \cdot \partial_k \boldsymbol{b}. \end{aligned} \tag{14.116}$$

(14.53) Let $\boldsymbol{a}_i = \partial_i \boldsymbol{r}$ and $\hat{\boldsymbol{a}}_i = \boldsymbol{a}_i / |\boldsymbol{a}_i|$, where $\boldsymbol{r}(u^1, u^2) \equiv \boldsymbol{r}(u, v)$ parametrizes a surface, and $\partial_i = \partial/\partial u^i$. Use (14.116) and (14.23), or (14.110c), together with (14.62), to show that for the orthonormal frame $\{\hat{\boldsymbol{a}}_1, \hat{\boldsymbol{a}}^2 = \hat{\boldsymbol{n}} \times \hat{\boldsymbol{a}}_1, \hat{\boldsymbol{n}}\}$ (Fig. 14.17)

$$\partial_i \hat{\boldsymbol{a}}_1 = (\hat{\boldsymbol{a}}^2 \cdot \partial_i \hat{\boldsymbol{a}}_1)\, \hat{\boldsymbol{a}}^2 + (\hat{\boldsymbol{n}} \cdot \partial_i \hat{\boldsymbol{a}}_1)\, \hat{\boldsymbol{n}},$$

then

$$\begin{aligned} \partial_i \hat{\boldsymbol{a}}_1 \cdot \partial_j \hat{\boldsymbol{a}}^2 &= (\hat{\boldsymbol{n}} \cdot \partial_i \hat{\boldsymbol{a}}_1)(\hat{\boldsymbol{n}} \cdot \partial_j \hat{\boldsymbol{a}}^2), \\ &= \frac{\hat{\boldsymbol{n}} \cdot \partial_i \boldsymbol{a}_1}{\sqrt{g_{11}}} \frac{\hat{\boldsymbol{n}} \cdot \partial_j \boldsymbol{a}^2}{\sqrt{g^{22}}} \\ &= \frac{b_{i1}}{\sqrt{g_{11}}} \frac{b_{jk} g^{k2}}{\sqrt{g^{22}}}. \end{aligned}$$

Next, using $\epsilon^{ij3} b_{i1} b_{j1} = 0$ and $g^{22} = g_{11}/g$ (14.92), show that

$$\begin{aligned}\epsilon^{ij3}\partial_i \hat{\boldsymbol{a}}_1 \cdot \partial_j \hat{\boldsymbol{a}}^2 &= \epsilon^{ij3}\frac{b_{i1}}{\sqrt{g_{11}}}\frac{b_{jk}\, g^{k2}}{\sqrt{g^{22}}} \\ &= \epsilon^{ij3}\frac{b_{i1}}{\sqrt{g_{11}}}\frac{b_{j2}\, g^{22}}{\sqrt{g^{22}}} \\ &= \epsilon^{ij3}\frac{b_{i1} b_{j2}}{\sqrt{g}} = K\sqrt{g}.\end{aligned}$$

Then, using $\epsilon^{ij3}\partial_i\partial_j = 0$, deduce equation (14.76). Repeating this calculation with the orthonormal frame $\{\hat{\boldsymbol{a}}^1, \hat{\boldsymbol{a}}_2, \hat{\boldsymbol{n}}\}$ yields the alternate formula

$$K\sqrt{g} = \partial_1\left(\hat{\boldsymbol{a}}^1 \cdot \partial_2 \hat{\boldsymbol{a}}_2\right) - \partial_2\left(\hat{\boldsymbol{a}}^1 \cdot \partial_1 \hat{\boldsymbol{a}}_2\right).$$

(14.54) Verify (14.76) for a sphere of radius R with $(u^1, u^2) = (\theta, \varphi)$ and $K = 1/R^2$, $\hat{\boldsymbol{a}}_1 = \hat{\boldsymbol{\theta}}$, $\hat{\boldsymbol{a}}^2 = \hat{\boldsymbol{\varphi}}$.

(14.55) Show that (14.77) is *incorrect* for a spherical cap bounded by the parallel at $\theta = \theta_1$ on a sphere of radius R for which $K = 1/R^2$, $\hat{\boldsymbol{a}}_1 = \hat{\boldsymbol{\theta}}$, $\hat{\boldsymbol{a}}^2 = \hat{\boldsymbol{\varphi}}$. This is because of the spherical coordinate singularity at the poles. Show that (14.77) is *correct* for a spherical annulus with polar angle $0 < \theta_0 \le \theta \le \theta_1 < \pi$, and two boundary curves $\partial\mathcal{S}_0$, $\partial\mathcal{S}_1$. Show that the limit $\theta_0 \to 0$ gives the correct result for the annulus area.

(14.56) Show that (14.78b) gives the correct result for a spherical cap $0 \le \varphi < 2\pi$, $0 \le \theta \le \theta_0$. Discuss the κ_g integral for $\theta_0 \to 0^+$, $\theta_0 = \pi/2$ and $\theta_0 \to \pi^-$.

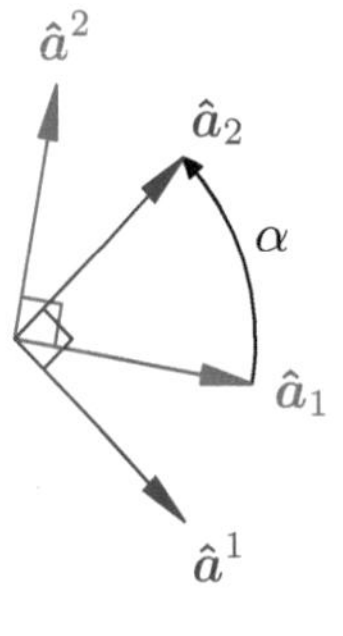

Fig. 14.17 Orthonormal Darboux frames $\{\hat{\boldsymbol{a}}_1, \hat{\boldsymbol{a}}^2, \hat{\boldsymbol{n}}\}$ and $\{\hat{\boldsymbol{a}}^1, \hat{\boldsymbol{a}}_2, \hat{\boldsymbol{n}}\}$, based on coordinate tangents and their reciprocals.

(14.57) Let $\boldsymbol{a}_i = \partial_i \boldsymbol{r}$ and $\hat{\boldsymbol{a}}_i = \boldsymbol{a}_i/|\boldsymbol{a}_i|$, where $\boldsymbol{r}(u^1, u^2) \equiv \boldsymbol{r}(u, v)$ parametrizes a surface, and $\partial_i = \partial/\partial u^i$. Consider the right-handed frames $\{\hat{\boldsymbol{a}}_1, \hat{\boldsymbol{a}}^2 = \hat{\boldsymbol{n}} \times \hat{\boldsymbol{a}}_1, \hat{\boldsymbol{n}}\}$ and $\{\hat{\boldsymbol{a}}^1 = \hat{\boldsymbol{a}}_2 \times \hat{\boldsymbol{n}}, \hat{\boldsymbol{a}}_2, \hat{\boldsymbol{n}}\}$, with α the angle between the coordinate curves, from $\hat{\boldsymbol{a}}_1$ to $\hat{\boldsymbol{a}}_2$ (Fig. 14.17). Show that the geodesic curvatures (14.7b) of the coordinate curves are

$$\begin{aligned}\kappa_{g1} &= \frac{\hat{\boldsymbol{a}}^2 \cdot \partial_1 \hat{\boldsymbol{a}}_1}{\sqrt{g_{11}}}, \\ \kappa_{g2} &= -\frac{\hat{\boldsymbol{a}}^1 \cdot \partial_2 \hat{\boldsymbol{a}}_2}{\sqrt{g_{22}}}.\end{aligned} \tag{14.117a}$$

Then, writing $\{\hat{\boldsymbol{a}}_1, \hat{\boldsymbol{a}}^2\}$ in terms of $\{\hat{\boldsymbol{a}}^1, \hat{\boldsymbol{a}}_2\}$, show that (14.76) yields Liouville's Gaussian curvature formula

$$\begin{aligned}K\sqrt{g} &= \partial_1\partial_2\alpha \\ &\quad + \partial_2(\kappa_{g1}\sqrt{g_{11}}) - \partial_1(\kappa_{g2}\sqrt{g_{22}})\end{aligned} \tag{14.117b}$$

(14.58) Use (14.116), together with Γ^k_{ij} (14.37) and g^{ij} (14.92), to show that the geodesic curvatures of the coordinate curves (14.117a) and the Gaussian curvature K satisfy the equations

$$\begin{aligned}\kappa_{g1} &= \sqrt{\frac{g}{g_{11}}}\frac{\Gamma^2_{11}}{g_{11}}, \\ \kappa_{g2} &= -\sqrt{\frac{g}{g_{22}}}\frac{\Gamma^1_{22}}{g_{22}},\end{aligned} \tag{14.117c}$$

and

$$\begin{aligned}K\sqrt{g} &= \partial_1\partial_2\alpha \\ &\quad + \partial_2\left(\frac{\sqrt{g}}{g_{11}}\Gamma^2_{11}\right) + \partial_1\left(\frac{\sqrt{g}}{g_{22}}\Gamma^1_{22}\right).\end{aligned} \tag{14.117d}$$

(14.59) From problem 14.53 and using (14.116), show that

$$\begin{aligned}\epsilon^{ij3}\partial_i \hat{\boldsymbol{a}}_1 \cdot \partial_j \hat{\boldsymbol{a}}^2 &= \epsilon^{ij3}\partial_j\left(\frac{\sqrt{g}}{g_{11}}\boldsymbol{a}^2 \cdot \partial_i \boldsymbol{a}_1\right) \\ &= \epsilon^{ij3}\partial_j\left(\frac{\sqrt{g}}{g_{11}}\Gamma^2_{i1}\right),\end{aligned}$$

and derive the following compact formula for the Gaussian curvature K,

$$\begin{aligned}&K\sqrt{g} \\ &= \partial_2\left(\frac{\sqrt{g}}{g_{11}}\Gamma^2_{11}\right) - \partial_1\left(\frac{\sqrt{g}}{g_{11}}\Gamma^2_{12}\right).\end{aligned} \tag{14.118a}$$

Repeat the calculation for the basis $\{\boldsymbol{a}^1, \boldsymbol{a}_2, \hat{\boldsymbol{n}}\}$, to obtain

$$\begin{aligned}&K\sqrt{g} \\ &= \partial_1\left(\frac{\sqrt{g}}{g_{22}}\Gamma^1_{22}\right) - \partial_2\left(\frac{\sqrt{g}}{g_{22}}\Gamma^1_{12}\right).\end{aligned} \tag{14.118b}$$

(14.60) For orthogonal coordinates $g_{ij} = g_{ii}\delta_{ij}$ and $g^{ij} = \delta^{ij}/g_{ii}$ (no sums over i), then (no sum over k)

$$\Gamma^k_{ij} = g^{km}\Gamma_{mij} = \frac{1}{g_{kk}}\Gamma_{kij}. \qquad (14.119)$$

Use (14.42) to show that the Gaussian curvature formula (14.117d) and (14.118a) reduce to

$$K = -\frac{1}{2\sqrt{g}}\left(\partial_2\left(\frac{\partial_2 g_{11}}{\sqrt{g}}\right) + \partial_1\left(\frac{\partial_1 g_{22}}{\sqrt{g}}\right)\right). \qquad (14.120)$$

Check the $K = 1/R^2$ result for spherical coordinates (θ, φ).

(14.61) For conformal coordinates $g_{ij} = h^2\delta_{ij}$, show that the Gaussian curvature K in (14.120) reduces to Liouville's formula (Liouville 1850, p. 590)

$$K = -\frac{1}{h^2}(\partial_1\partial_1 + \partial_2\partial_2)\ln h. \qquad (14.121)$$

Check the $K = 1/R^2$ result for Mercator coordinates (13.14).

(14.62) From the Gauss (14.39) and Weingarten (14.64a) formulas, show that $\partial_1\partial_2\hat{\boldsymbol{n}} = \partial_2\partial_1\hat{\boldsymbol{n}}$ implies the compatibility equations

$$\partial_1 b_2^k + b_2^m\Gamma^k_{1m} = \partial_2 b_1^k + b_1^m\Gamma^k_{2m}. \qquad (14.122)$$

(14.63) From the Gauss (14.39) and Weingarten formulas (14.64a), show that

$$\partial_k(\partial_i\partial_j\boldsymbol{r}) = \left(\partial_k\Gamma^l_{ij} + \Gamma^m_{ij}\Gamma^l_{km} - b_{ij}b^l_k\right)\partial_l\boldsymbol{r} + \left(\Gamma^l_{ij}b_{kl} + \partial_k b_{ij}\right)\hat{\boldsymbol{n}}. \qquad (14.123)$$

Then, from $\partial_j\partial_k\partial_i\boldsymbol{r} = \partial_k\partial_j\partial_i\boldsymbol{r}$, conclude that

$$\partial_j\Gamma^l_{ik} - \partial_k\Gamma^l_{ij} + \Gamma^m_{ik}\Gamma^l_{jm} - \Gamma^m_{ij}\Gamma^l_{km} = b_{ik}b^l_j - b_{ij}b^l_k, \qquad (14.124)$$

$$\partial_j b_{ik} - \partial_k b_{ij} + \Gamma^l_{ik}b_{jl} - \Gamma^l_{ij}b_{kl} = 0. \qquad (14.125)$$

The second of these compatibility conditions are the *Mainardi–Codazzi* equations. The first one brings out the *Riemann curvature tensor*

$$R^l_{ijk} \triangleq \partial_j\Gamma^l_{ik} - \partial_k\Gamma^l_{ij} + \Gamma^m_{ik}\Gamma^l_{jm} - \Gamma^m_{ij}\Gamma^l_{km}. \qquad (14.126)$$

Lowering the upper index l through multiplication by the metric g_{ml}, yields the *covariant Riemann curvature tensor*

$$R_{mijk} = g_{ml}R^l_{ijk},$$

in terms of which (14.124) reads

$$R_{mijk} = b_{ik}b_{jm} - b_{ij}b_{km}. \qquad (14.127)$$

In particular, $R_{2121} = b_{11}b_{22} - b_{12}b_{21} = b = \det \mathbf{B}$, proving that Gaussian curvature $K = b/g$ depends only on the metric, since R^l_{ijk} depends only on the Christoffel symbols that depend only on the metric (14.43).

(14.64) Since $b_{ij} = b_{ji}$, deduce from (14.127) that

$$R_{mijk} = R_{jkmi} = -R_{imjk} = -R_{mikj}. \qquad (14.128)$$

(14.65) Show that (14.125) is trivially satisfied when $j = k$, and is equivalent to (14.122) when $j \neq k$, in which case $(j,k) = (1,2)$ or $(2,1)$ for a 2D surface. Start from (14.125) and use $b_{ik} = b^m_i g_{mk}$.

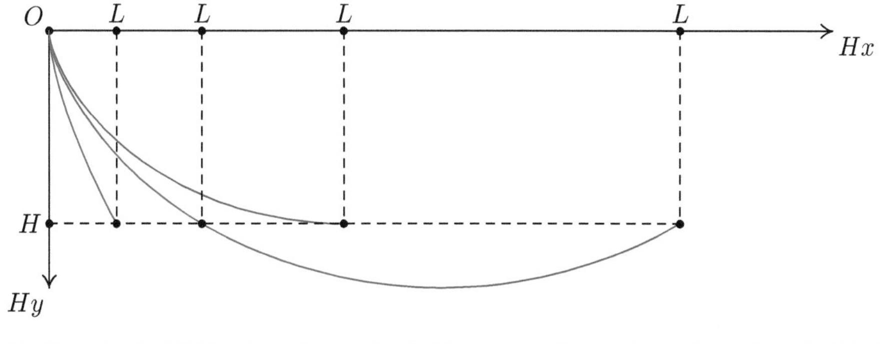

Fig. 14.18 (Exercise 14.28) Brachistochrones for L/H corresponding to $\theta_1 = \pi/3, 2\pi/3, \pi, 4\pi/3$ in (14.103).

Curvilinear Coordinates

15

15.1 Coordinates and dimensions

In 3D space, a curve $\mathcal{C}$ can be defined by *two* implicit equations, or one explicit vector function $\boldsymbol{r}(t)$ that maps a point t in an interval $I = (t_0, t_f) \subset \mathbb{R}$ to a point P on $\mathcal{C} \subset \mathbb{E}^3$,

$$t \in I \to P \in \mathcal{C},\ \overrightarrow{OP} = \boldsymbol{r}(t).$$

The variable t is the *coordinate* of point P on that curve $\mathcal{C}$ in that representation. A curve is a 1-dimensional space, its *codimension* is $3 - 1 = 2$, in 3D space. The codimension is the number of implicit equations needed to define the subspace, and is the dimension of the ambient space minus the number of coordinates of the subspace.

A surface $\mathcal{S}$ can be defined by *one* implicit equation, or explicitly by a vector function $\boldsymbol{r}(u, v)$ that maps a point (u, v) in a subset $\mathcal{U}$ of $\mathbb{R}^2$ to a point P on the surface $\mathcal{S} \subset \mathbb{E}^3$,

$$(u, v) \in \mathcal{U} \to P \in \mathcal{S},\ \overrightarrow{OP} = \boldsymbol{r}(u, v),$$

and (u, v) are the coordinates of point P on the surface $\mathcal{S}$ in that representation. A surface is a 2-dimensional space; its codimension is $3-2 = 1$ in 3D space.

A volume $\mathcal{V}$, can be defined by *inequalities*, with bounding surfaces that define a subset of $\mathbb{E}^3$, for example $|\boldsymbol{r}| < R$ is the sphere of radius R centered at the origin O. A volume can be represented by an explicit vector function $\boldsymbol{r}(u, v, w)$ of three variables (u, v, w) in a subset $\mathcal{U}$ of $\mathbb{R}^3$,

$$(u, v, w) \in \mathcal{U} \to P \in \mathcal{V},\ \overrightarrow{OP} = \boldsymbol{r}(u, v, w).$$

The coordinates (u, v, w) are *curvilinear* in the sense that coordinate curves are not straight lines, in general. A *volume* $\mathcal{V}$ is a 3-dimensional subspace of 3D space; its codimension is 0. The word 'volume' is also used for the measure V of that set; thus, the volume V of the set $|\boldsymbol{r}| < R$ is $\frac{4}{3}\pi R^3$, and a volume *set* as a nonzero volume *measure*.

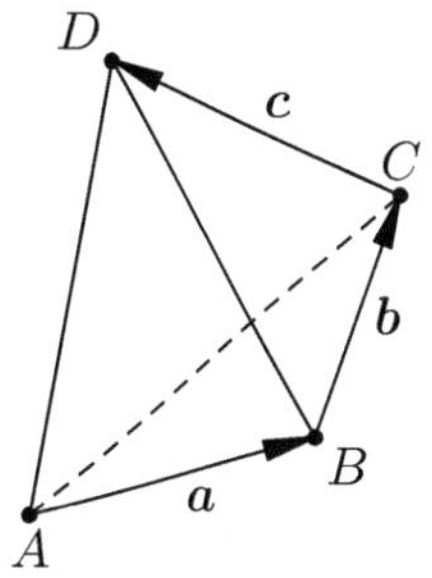

Fig. 15.1 Tetrahedron.

Example 1 A *tetrahedron* is the space between four nonparallel planes. Any point P inside the tetrahedron with vertices A, B, C, D can be parametrized as (Fig. 15.1)

$$\boldsymbol{r} = \boldsymbol{r}_A + u\,\boldsymbol{a} + v\,\boldsymbol{b} + w\,\boldsymbol{c}, \qquad (15.1)$$

with $\boldsymbol{r}_A = \overrightarrow{OA}$, $\boldsymbol{a} = \overrightarrow{AB}$, $\boldsymbol{b} = \overrightarrow{BC}$, $\boldsymbol{c} = \overrightarrow{CD}$, and $0 \le w \le v \le u \le 1$.

Example 2 A *sphere* of radius R centered at $\boldsymbol{r}_c$ is bounded by one closed surface, and has the implicit equation $|\boldsymbol{r}-\boldsymbol{r}_c| \leq R$. In cartesian coordinates this translates into the implicit *inequality*

$$(x-x_c)^2+(y-y_c)^2+(z-z_c)^2 \leq R^2. \tag{15.2}$$

That sphere can be parametrized as

$$\boldsymbol{r} = x\hat{\mathbf{x}} + y\hat{\mathbf{y}} + z\hat{\mathbf{z}} \tag{15.3}$$

with x, y, z constrained by (15.2). A convenient parametrization for that sphere is

$$\boldsymbol{r} = \boldsymbol{r}_c + s\sin\theta\,(\hat{\mathbf{x}}\cos\varphi + \hat{\mathbf{y}}\sin\varphi) + \hat{\mathbf{z}}\,s\cos\theta, \tag{15.4}$$

where $s = |\boldsymbol{r}-\boldsymbol{r}_c|$ is the distance to the center C of the sphere not to the origin, θ is the polar angle, φ is the azimuthal (or longitude) angle, and the coordinate domain $\mathcal{U}$ is $0 \leq s \leq R$, $0 \leq \theta \leq \pi$, $0 \leq \varphi < 2\pi$.

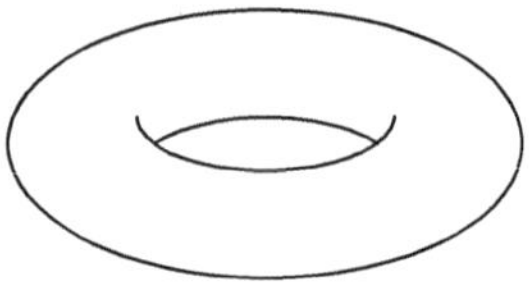

Fig. 15.2 Torus.

Example 3 A *torus* is the set of points within a distance a of a circle $\boldsymbol{r} = \boldsymbol{r}_C(\varphi)$

$$\min_{\varphi} |\boldsymbol{r} - \boldsymbol{r}_C(\varphi)| \leq a.$$

For a circle of radius R centered at O in the xy plane, the torus is parametrized in terms of (s, θ, φ) as (Fig. 15.2)

$$\boldsymbol{r} = (R + s\cos\theta)(\hat{\mathbf{x}}\cos\varphi + \hat{\mathbf{y}}\sin\varphi) + \hat{\mathbf{z}}\,s\sin\theta, \tag{15.5}$$

with $0 \leq s \leq a \leq R$, $0 \leq \theta < 2\pi$, $0 \leq \varphi < 2\pi$.

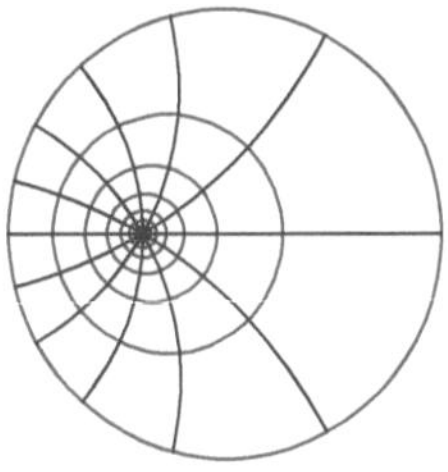

Fig. 15.3 Toroidal coordinates.

Example 4 A torus can also be described by *toroidal coordinates* that are *bipolar coordinates* (u, w) (7.14) in the (x, z) plane rotated by angle v about the z-axis (Fig. 15.3) as

$$\begin{aligned}\boldsymbol{r} &= x\,\hat{\mathbf{x}} + y\,\hat{\mathbf{y}} + z\,\hat{\mathbf{z}} \\ &= \frac{R_*}{\cosh w - \cos u}\Big(\sinh w\,(\hat{\mathbf{x}}\cos v + \hat{\mathbf{y}}\sin v) + \hat{\mathbf{z}}\sin u\Big)\end{aligned} \tag{15.6}$$

with $R_* = \sqrt{R^2 - a^2}$, and

$$-\pi < u \leq \pi, \quad -\pi < v \leq \pi, \quad \tfrac{1}{2}\ln\frac{R+R_*}{R-R_*} \leq w.$$

If the torus is centered at C, with its equator in a plane perpendicular to $\hat{\boldsymbol{n}}$, it suffices to add $\boldsymbol{r}_c$ to (15.5) or (15.6), and replace $\hat{\mathbf{x}}, \hat{\mathbf{y}}, \hat{\mathbf{z}}$ by an orthonormal basis $\hat{\boldsymbol{a}}, \hat{\boldsymbol{b}}, \hat{\boldsymbol{n}}$.

Example 5 An ellipsoid can be parametrized using non-orthogonal coordinates (15.63), akin to spherical coordinates, or with orthogonal ellipsoidal coordinates corresponding to the three real roots $t = (u, v, w)$ of the equation

$$\frac{x^2}{a^2-t} + \frac{y^2}{b^2-t} + \frac{z^2}{c^2-t} = 1.$$

15.2 Coordinate curves and surfaces

A volume parametrization $\boldsymbol{r}(u, v, w)$ includes a triple set of *coordinate curves*, one curve for every fixed (u, v) whose tangent is $\partial \boldsymbol{r}/\partial w$, for example. The same $\boldsymbol{r}(u, v, w)$ defines a triple set of surfaces, one surface for every fixed w with tangent planes defined by $\partial \boldsymbol{r}/\partial u$ and $\partial \boldsymbol{r}/\partial v$, and normal $\boldsymbol{N}_{uv} = \partial \boldsymbol{r}/\partial u \times \partial \boldsymbol{r}/\partial v$, for example (Fig. 15.4). The parametrization $\boldsymbol{r}(u, v, w)$ is assumed sufficiently smooth, at least twice continuously differentiable (C^2) so the second partial derivatives commute.

At any point $\boldsymbol{r}(u, v, w)$, there are three coordinate curve tangents,

$$\frac{\partial \boldsymbol{r}}{\partial u}, \quad \frac{\partial \boldsymbol{r}}{\partial v}, \quad \frac{\partial \boldsymbol{r}}{\partial w} \tag{15.7}$$

and three coordinate surface normals

$$\boldsymbol{N}_{uv} = \frac{\partial \boldsymbol{r}}{\partial u} \times \frac{\partial \boldsymbol{r}}{\partial v}, \quad \boldsymbol{N}_{vw} = \frac{\partial \boldsymbol{r}}{\partial v} \times \frac{\partial \boldsymbol{r}}{\partial w}, \quad \boldsymbol{N}_{wu} = \frac{\partial \boldsymbol{r}}{\partial w} \times \frac{\partial \boldsymbol{r}}{\partial u}. \tag{15.8}$$

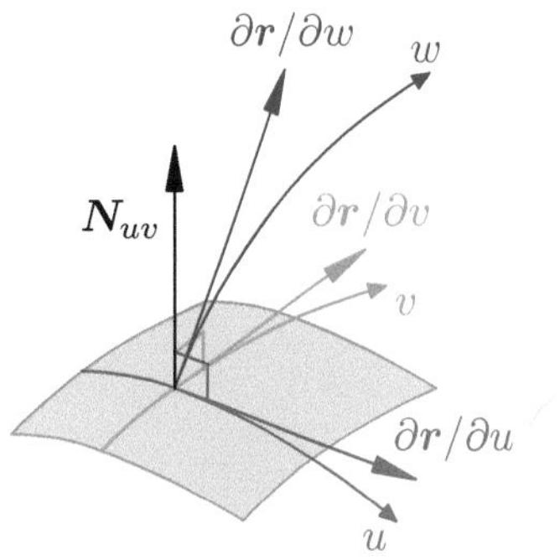

Fig. 15.4 w isosurface with tangents $\partial \boldsymbol{r}/\partial u$, $\partial \boldsymbol{r}/\partial v$ and normal $\boldsymbol{N}_{uv}$.

The *Jacobian determinant*, or simply *Jacobian*,

$$J = \frac{\partial \boldsymbol{r}}{\partial u} \times \frac{\partial \boldsymbol{r}}{\partial v} \boldsymbol{\cdot} \frac{\partial \boldsymbol{r}}{\partial w}, \tag{15.9}$$

determines whether the coordinates are a locally valid parametrization of 3D space. The Jacobian is the dot product of a tangent vector with its corresponding surface normal

$$J = \boldsymbol{N}_{uv} \boldsymbol{\cdot} \frac{\partial \boldsymbol{r}}{\partial w} = \boldsymbol{N}_{vw} \boldsymbol{\cdot} \frac{\partial \boldsymbol{r}}{\partial u} = \boldsymbol{N}_{wu} \boldsymbol{\cdot} \frac{\partial \boldsymbol{r}}{\partial v},$$

as follows from the cyclic property of the mixed product $\boldsymbol{a} \times \boldsymbol{b} \boldsymbol{\cdot} \boldsymbol{c} = \boldsymbol{b} \times \boldsymbol{c} \boldsymbol{\cdot} \boldsymbol{a}$.

If $J = 0$, then the tangent vectors are locally in the same plane, and the coordinates are *singular* at that point. That can happen at some limit points, or limit curves, but J certainly cannot be zero everywhere; otherwise, we do not have a volume parametrization, and J should not change sign, since that would correspond to the volume being turned inside out. The coordinate ordering (u, v, w) defines the sign of the Jacobian and right-handed ordering has $J > 0$. Some examples are discussed later in this section.

A vector function $\boldsymbol{F}$ of position $\boldsymbol{r}$ (a vector *field*, $\boldsymbol{F}(\boldsymbol{r})$) can be expanded in terms of either the coordinate tangents or surface normals as

$$\boldsymbol{F} = \begin{cases} \dfrac{1}{J}\left(\boldsymbol{F} \boldsymbol{\cdot} \dfrac{\partial \boldsymbol{r}}{\partial u}\right) \dfrac{\partial \boldsymbol{r}}{\partial v} \times \dfrac{\partial \boldsymbol{r}}{\partial w} + \cdots \\[2ex] \dfrac{1}{J}\left(\boldsymbol{F} \boldsymbol{\cdot} \dfrac{\partial \boldsymbol{r}}{\partial v} \times \dfrac{\partial \boldsymbol{r}}{\partial w}\right) \dfrac{\partial \boldsymbol{r}}{\partial u} + \cdots, \end{cases} \tag{15.10}$$

where the dots $(\cdots)$ stand for two similar terms obtained by cyclic permutation of (u, v, w). The components in one basis are obtained by projecting onto the other and dividing by the Jacobian.

Tangents and normals bases

In Ricci index notation with $(u, v, w) \equiv (u^1, u^2, u^3)$ and $\partial_i \boldsymbol{r} = \partial \boldsymbol{r}/\partial u^i$, the coordinate tangent vectors define the *covariant* basis

$$\boldsymbol{a}_i = \partial_i \boldsymbol{r}. \tag{15.11}$$

The coordinate surface normals $\boldsymbol{N}_{jk} = \partial_j \boldsymbol{r} \times \partial_k \boldsymbol{r} = \boldsymbol{a}_j \times \boldsymbol{a}_k$, such that $\boldsymbol{a}_i \cdot \boldsymbol{N}_{jk} = \epsilon_{ijk} J$, yield the *contravariant* basis when normalized by $J = \boldsymbol{a}_1 \times \boldsymbol{a}_2 \cdot \boldsymbol{a}_3$,

$$\boldsymbol{a}^1 = \frac{\boldsymbol{N}_{23}}{J}, \quad \boldsymbol{a}^2 = \frac{\boldsymbol{N}_{31}}{J}, \quad \boldsymbol{a}^3 = \frac{\boldsymbol{N}_{12}}{J}, \tag{15.12a}$$

with $\boldsymbol{a}^1 \times \boldsymbol{a}^2 \cdot \boldsymbol{a}^3 = J^{-1}$. In Ricci index notation

$$\boldsymbol{a}^i = \epsilon^{ijk} \frac{\boldsymbol{N}_{jk}}{2J} \longleftrightarrow \boldsymbol{N}_{ij} = J \,\epsilon_{ijk}\, \boldsymbol{a}^k, \tag{15.12b}$$

where the factor of 2 arises from the sums over j, k that contribute two equal terms.

These bases are *biorthogonal*, that is,

$$\delta_i^j = \begin{cases} 1 \text{ if } i = j, \\ 0 \text{ if } i \neq j. \end{cases}$$

$$\boldsymbol{a}_i \cdot \boldsymbol{a}^j = \delta_i^j, \tag{15.13a}$$

but not orthonormal themselves, and

$$g_{ij} = \boldsymbol{a}_i \cdot \boldsymbol{a}_j, \qquad \boldsymbol{a}^i \cdot \boldsymbol{a}^j = g^{ij} \tag{15.13b}$$

are the *metrics* of both bases such that

$$\boldsymbol{a}_i = g_{ij} \boldsymbol{a}^j, \qquad \boldsymbol{a}^i = g^{ij} \boldsymbol{a}_j. \tag{15.13c}$$

The metrics are symmetric, positive definite matrices and inverses of each other (14.23),

$$g_{ik} g^{kj} = \delta_i^j = g^{jk} g_{ki}. \tag{15.13d}$$

Orthogonal coordinates

The coordinates are *orthogonal* if their tangent vectors are orthogonal at every point

$$\partial_i \boldsymbol{r} \cdot \partial_j \boldsymbol{r} = 0 \ \text{ if } i \neq j. \tag{15.14}$$

The surface normals $\boldsymbol{N}_{jk}$ are then also orthogonal to each other and parallel to their respective tangent vector. Thus, the covariant and contravariant vectors have the same direction when the coordinates are orthogonal, and it is convenient then to work with the *scale factors*[1] $h_i = |\partial_i \boldsymbol{r}|$ and the local orthonormal basis $\hat{\boldsymbol{u}}_i$ such that (*no sum* over i)

$$\partial_i \boldsymbol{r} = h_i \hat{\boldsymbol{u}}_i. \tag{15.15}$$

The Jacobian for orthogonal coordinates is simply the product of those scale factors,

$$J = |\partial_1 \boldsymbol{r}|\, |\partial_2 \boldsymbol{r}|\, |\partial_3 \boldsymbol{r}| = h_1 h_2 h_3. \tag{15.16}$$

[1]aka *Lamé coefficients*, after Gabriel Lamé, "Mémoire sur les coordonnées curvilignes," *Journal de mathématiques pures et appliquées,* **5**, (1840), 313–347, although what Lamé called h is here $1/h$.

The basis vectors $\hat{\boldsymbol{u}}_i$ and the scale factors h_i vary with the coordinates, in general (Fig. 15.5). However, just as Christoffel symbols Γ_{ijk} can be expressed in terms of derivatives of the metric g_{ij} (14.43), the derivatives of the orthonormal basis vectors $\hat{\boldsymbol{u}}_i$ can be expressed in terms of derivatives of the scale factors h_i. It is left as an exercise (15.11 or 15.12), to verify that (*no sums*)

$$(\partial_i \hat{\boldsymbol{u}}_j) \cdot \hat{\boldsymbol{u}}_k = \delta_{ik} \frac{\partial_j h_i}{h_j} - \delta_{ij} \frac{\partial_k h_i}{h_k}, \tag{15.17a}$$

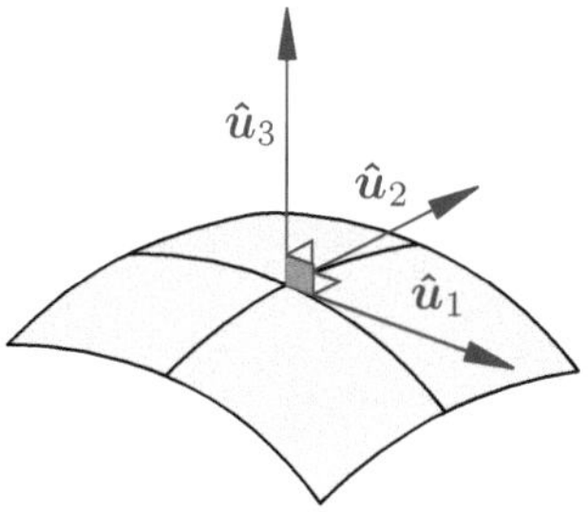

Fig. 15.5 Orthogonal directions.

yielding

$$\partial_i \hat{\boldsymbol{u}}_j = \hat{\boldsymbol{u}}_i \frac{\partial_j h_i}{h_j} - \delta_{ij} \left(\hat{\boldsymbol{u}}_1 \frac{\partial_1 h_i}{h_1} + \hat{\boldsymbol{u}}_2 \frac{\partial_2 h_i}{h_2} + \hat{\boldsymbol{u}}_3 \frac{\partial_3 h_i}{h_3} \right). \tag{15.17b}$$

As for surfaces, coordinates are *conformal* if they are orthogonal, *and* the scale factors are equal at every point; that is, $h_1 = h_2 = h_3 = h(u, v, w)$. Thus, conformal coordinates have

$$\partial_i \boldsymbol{r} \cdot \partial_j \boldsymbol{r} = h^2 \delta_{ij}. \tag{15.18}$$

However, in 3D space, the class of conformal maps is very restricted by Liouville's theorem, as explored in exercise 15.17.

Cartesian coordinates $\boldsymbol{r} = x\hat{\mathbf{x}} + y\hat{\mathbf{y}} + z\hat{\mathbf{z}}$ are orthogonal and conformal

$$\frac{\partial \boldsymbol{r}}{\partial x} = \hat{\mathbf{x}}, \quad \frac{\partial \boldsymbol{r}}{\partial y} = \hat{\mathbf{y}}, \quad \frac{\partial \boldsymbol{r}}{\partial z} = \hat{\mathbf{z}}. \tag{15.19}$$

The basis vectors have magnitude 1 everywhere and the Jacobian $J = 1$.

Cylindrical coordinates $\boldsymbol{r} = \rho\hat{\boldsymbol{\rho}}(\varphi) + z\hat{\mathbf{z}}$ have

$\hat{\boldsymbol{\rho}}(\varphi) = \cos\varphi\,\hat{\mathbf{x}} + \sin\varphi\,\hat{\mathbf{y}}$

$$\frac{\partial \boldsymbol{r}}{\partial \rho} = \hat{\boldsymbol{\rho}}, \quad \frac{\partial \boldsymbol{r}}{\partial \varphi} = \rho\hat{\boldsymbol{\varphi}}, \quad \frac{\partial \boldsymbol{r}}{\partial z} = \hat{\mathbf{z}}, \tag{15.20a}$$

since $\mathrm{d}\hat{\boldsymbol{\rho}}/\mathrm{d}\varphi = \hat{\boldsymbol{\varphi}}$ (10.18). They are orthogonal with scale factors

$$h_\rho = 1, \quad h_\varphi = \rho, \quad h_z = 1, \tag{15.20b}$$

and Jacobian $J = h_\rho h_\varphi h_z = \rho$. The basis vector derivatives have been computed in (10.18), and now using (15.17b),

$$\partial_\varphi \hat{\boldsymbol{\rho}} = \hat{\boldsymbol{\varphi}} \frac{\partial_\rho h_\varphi}{h_\rho} = \hat{\boldsymbol{\varphi}}, \qquad \partial_\varphi \hat{\boldsymbol{\varphi}} = -\hat{\boldsymbol{\rho}} \frac{\partial_\rho h_\varphi}{h_\rho} - \hat{\mathbf{z}} \frac{\partial_z h_\varphi}{h_z} = -\hat{\boldsymbol{\rho}}. \tag{15.20c}$$

Spherical coordinates $\boldsymbol{r} = r\hat{\boldsymbol{r}}(\theta, \varphi)$ have

$\hat{\boldsymbol{r}}(\theta, \varphi) = \sin\theta\,\hat{\boldsymbol{\rho}}(\varphi) + \cos\theta\,\hat{\mathbf{z}}$

$$\frac{\partial \boldsymbol{r}}{\partial r} = \hat{\boldsymbol{r}}, \quad \frac{\partial \boldsymbol{r}}{\partial \theta} = r\,\hat{\boldsymbol{\theta}}, \quad \frac{\partial \boldsymbol{r}}{\partial \varphi} = r\sin\theta\,\hat{\boldsymbol{\varphi}}, \tag{15.21a}$$

since $\partial\hat{\boldsymbol{r}}/\partial\theta = \hat{\boldsymbol{\theta}}$ and $\partial\hat{\boldsymbol{r}}/\partial\varphi = \sin\theta\,\hat{\boldsymbol{\varphi}}$ from (10.23), or (13.16). They are orthogonal with scale factors

$$h_r = 1, \quad h_\theta = r, \quad h_\varphi = r\sin\theta, \tag{15.21b}$$

and Jacobian $J = h_r h_\theta h_\varphi = r^2 \sin\theta$. The derivatives of the direction vectors have been computed in (10.23), and now as a check of (15.17b),

$$\begin{aligned}
\partial_\theta \hat{\boldsymbol{\theta}} &= -\hat{\boldsymbol{r}}\,\frac{\partial_r h_\theta}{h_r} - \hat{\boldsymbol{\varphi}}\,\frac{\partial_\varphi h_\theta}{h_\varphi} = -\hat{\boldsymbol{r}}, \\
\partial_\varphi \hat{\boldsymbol{\theta}} &= \hat{\boldsymbol{\varphi}}\frac{\partial_\theta h_\varphi}{h_\theta} = \hat{\boldsymbol{\varphi}}\cos\theta, \\
\partial_\theta \hat{\boldsymbol{\varphi}} &= \hat{\boldsymbol{\theta}}\frac{\partial_\varphi h_\theta}{h_\varphi} = 0, \\
\partial_\varphi \hat{\boldsymbol{\varphi}} &= -\hat{\boldsymbol{r}}\,\frac{\partial_r h_\varphi}{h_r} - \hat{\boldsymbol{\theta}}\,\frac{\partial_\theta h_\varphi}{h_\theta} = -\hat{\boldsymbol{r}}\sin\theta - \hat{\boldsymbol{\theta}}\cos\theta.
\end{aligned} \tag{15.21c}$$

Tetrahedral coordinates $\boldsymbol{r} = \boldsymbol{r}_A + u\boldsymbol{a} + v\boldsymbol{b} + w\boldsymbol{c}$, (15.1), have

$$\begin{gathered}
\frac{\partial \boldsymbol{r}}{\partial u} = \boldsymbol{a}, \quad \frac{\partial \boldsymbol{r}}{\partial v} = \boldsymbol{b}, \quad \frac{\partial \boldsymbol{r}}{\partial w} = \boldsymbol{c}, \\
\frac{\boldsymbol{N}_{vw}}{J} = \frac{\boldsymbol{b}\times\boldsymbol{c}}{\boldsymbol{b}\times\boldsymbol{c}\cdot\boldsymbol{a}}, \quad \frac{\boldsymbol{N}_{wu}}{J} = \frac{\boldsymbol{c}\times\boldsymbol{a}}{\boldsymbol{c}\times\boldsymbol{a}\cdot\boldsymbol{b}}, \quad \frac{\boldsymbol{N}_{uv}}{J} = \frac{\boldsymbol{a}\times\boldsymbol{b}}{\boldsymbol{a}\times\boldsymbol{b}\cdot\boldsymbol{c}}.
\end{gathered} \tag{15.22}$$

They are orthogonal only if $\boldsymbol{a}$, $\boldsymbol{b}$, $\boldsymbol{c}$ are mutually orthogonal.

15.3 Line, surface, and volume elements

Line elements

To compute line integrals and measure *distances* in terms of our (u, v, w) coordinates, we need the coordinate line elements

$$\mathrm{d}\boldsymbol{r}_u = \frac{\partial \boldsymbol{r}}{\partial u}\mathrm{d}u, \quad \mathrm{d}\boldsymbol{r}_v = \frac{\partial \boldsymbol{r}}{\partial v}\mathrm{d}v, \quad \mathrm{d}\boldsymbol{r}_w = \frac{\partial \boldsymbol{r}}{\partial w}\mathrm{d}w.$$

The line element for an arbitrary coordinate increment $(\mathrm{d}u, \mathrm{d}v, \mathrm{d}w)$ is

$$\begin{aligned}
\mathrm{d}\boldsymbol{r} &= \boldsymbol{r}(u + \mathrm{d}u, v + \mathrm{d}v, w + \mathrm{d}w) - \boldsymbol{r}(u, v, w), \\
&= \frac{\partial \boldsymbol{r}}{\partial u}\mathrm{d}u + \frac{\partial \boldsymbol{r}}{\partial v}dv + \frac{\partial \boldsymbol{r}}{\partial w}\mathrm{d}w = \mathrm{d}u^i \partial_i \boldsymbol{r},
\end{aligned} \tag{15.23}$$

which is the chain rule in differential form. The arclength element $\mathrm{d}s$ follows from

$$\mathrm{d}s^2 = \mathrm{d}\boldsymbol{r}\cdot\mathrm{d}\boldsymbol{r} = g_{ij}\,\mathrm{d}u^i \mathrm{d}u^j, \tag{15.24}$$

where $g_{ij} = \partial_i \boldsymbol{r}\cdot\partial_j \boldsymbol{r}$ is the metric of the displacement basis, and $\mathrm{d}s^2 = (\mathrm{d}s)^2$ not $\mathrm{d}(s^2) = 2s\,\mathrm{d}s$.

Surface elements

To compute surface integrals and measure *areas* in terms of coordinates (u, v, w), we need the coordinate *surface elements.* The surface element on a w-isosurface, for example, is

$$\mathrm{d}\boldsymbol{S}_{uv} = \mathrm{d}\boldsymbol{r}_u \times \mathrm{d}\boldsymbol{r}_v = \boldsymbol{N}_{uv}\,\mathrm{d}u\mathrm{d}v, \tag{15.25a}$$

while the surface elements on a u-isosurface and a v-isosurface are, respectively,

$$\mathrm{d}\boldsymbol{S}_{vw} = \boldsymbol{N}_{vw}\,\mathrm{d}v\mathrm{d}w, \quad \mathrm{d}\boldsymbol{S}_{wu} = \boldsymbol{N}_{wu}\,\mathrm{d}w\mathrm{d}u. \tag{15.25b}$$

The surface element for an arbitrary quadrilateral surface patch perpendicular to $\hat{\boldsymbol{n}}$ at point $\boldsymbol{r}(u, v, w)$ is

$$\begin{aligned}\mathrm{d}\boldsymbol{S} &= \mathrm{d}\boldsymbol{r}_1 \times \mathrm{d}\boldsymbol{r}_2 = \partial_i \boldsymbol{r}\,\mathrm{d}u_1^i \times \partial_j \boldsymbol{r}\,\mathrm{d}u_2^j = \boldsymbol{N}_{ij}\mathrm{d}u_1^i \mathrm{d}u_2^j \\ &= \boldsymbol{N}_{uv}\begin{vmatrix}\mathrm{d}u_1 & \mathrm{d}u_2 \\ \mathrm{d}v_1 & \mathrm{d}v_2\end{vmatrix} + \boldsymbol{N}_{vw}\begin{vmatrix}\mathrm{d}v_1 & \mathrm{d}v_2 \\ \mathrm{d}w_1 & \mathrm{d}w_2\end{vmatrix} + \boldsymbol{N}_{wu}\begin{vmatrix}\mathrm{d}w_1 & \mathrm{d}w_2 \\ \mathrm{d}u_1 & \mathrm{d}u_2\end{vmatrix},\end{aligned} \tag{15.25c}$$

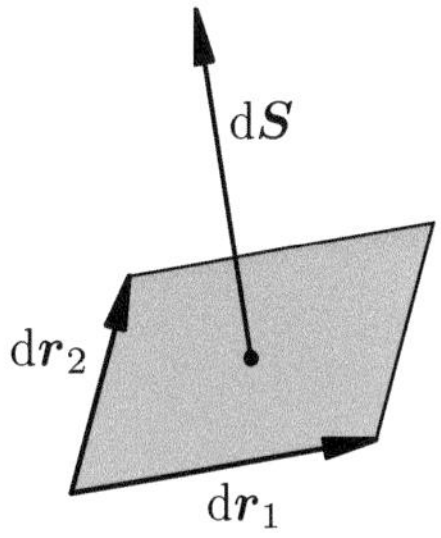

Fig. 15.6 Quadrilateral surface element.

where the differentials $\mathrm{d}\boldsymbol{r}_1$, $\mathrm{d}\boldsymbol{r}_2$, are the edges of the surface patch, perpendicular to $\mathrm{d}\boldsymbol{S} = \hat{\boldsymbol{n}}\,\mathrm{d}A$, and nonparallel to each other, but otherwise arbitrary (Fig. 15.6). Indeed, a surface $\mathcal{S}$ is specified in u^i coordinates as a composite function

$$\boldsymbol{s}(t^1, t^2) = \boldsymbol{r}(u^1(t^1, t^2), u^2(t^1, t^2), u^3(t^1, t^2)).$$

The surface element for $\mathcal{S}$ is then

$$\begin{aligned}\mathrm{d}\boldsymbol{S} = \frac{\partial \boldsymbol{s}}{\partial t^1} \times \frac{\partial \boldsymbol{s}}{\partial t^2}\,\mathrm{d}t^1\mathrm{d}t^2 &= \frac{\partial \boldsymbol{r}}{\partial u^i}\frac{\partial u^i}{\partial t^1} \times \frac{\partial \boldsymbol{r}}{\partial u^j}\frac{\partial u^j}{\partial t^2}\mathrm{d}t^1\mathrm{d}t^2 \\ &= \boldsymbol{N}_{ij}\frac{\partial u^i}{\partial t^1}\frac{\partial u^j}{\partial t^2}\mathrm{d}t^1\mathrm{d}t^2,\end{aligned}$$

and that is (15.25c) with $\mathrm{d}u_1^i = \partial u^i/\partial t^1\,\mathrm{d}t^1$ and $\mathrm{d}u_2^i = \partial u^i/\partial t^2\,\mathrm{d}t^2$, for $i = 1, 2, 3$.

Volume element

Finally, to compute volume integrals and measure *volumes* in terms of coordinates (u, v, w), we need the volume element $\mathrm{d}V$, that is the dot product of a coordinate line element with the corresponding surface element,

$$\begin{aligned}\mathrm{d}V &= \mathrm{d}\boldsymbol{r}_u \cdot \mathrm{d}\boldsymbol{S}_{vw} = \mathrm{d}\boldsymbol{r}_v \cdot \mathrm{d}\boldsymbol{S}_{wu} = \mathrm{d}\boldsymbol{r}_w \cdot \mathrm{d}\boldsymbol{S}_{uv} \\ &= \left(\frac{\partial \boldsymbol{r}}{\partial u} \times \frac{\partial \boldsymbol{r}}{\partial v} \cdot \frac{\partial \boldsymbol{r}}{\partial w}\right)\mathrm{d}u\mathrm{d}v\mathrm{d}w = J\,\mathrm{d}u\mathrm{d}v\mathrm{d}w.\end{aligned} \tag{15.26}$$

This is the determinant of the coordinate line elements by the cyclic properties of mixed products, $\boldsymbol{a} \cdot \boldsymbol{b} \times \boldsymbol{c} = \boldsymbol{b} \cdot \boldsymbol{c} \times \boldsymbol{a}$. We assume that the coordinates (u, v, w) are ordered such that $J \geq 0$. That is why spherical coordinates are listed as (r, θ, φ), and not (r, φ, θ), for instance, where θ is the polar angle. The latter ordering gives $J \leq 0$.

General domain element

The differential scalar measures, $\mathrm{d}s$ for arclength, $\mathrm{d}A$ for surface area, and $\mathrm{d}V$ for volume, given by

$$\mathrm{d}s = \left|\frac{\mathrm{d}\boldsymbol{r}}{\mathrm{d}u}\right| \mathrm{d}u, \quad \mathrm{d}A = \left|\frac{\partial \boldsymbol{r}}{\partial u^1} \times \frac{\partial \boldsymbol{r}}{\partial u^2}\right| \mathrm{d}u^1 \mathrm{d}u^2,$$

$$\mathrm{d}V = \left|\frac{\partial \boldsymbol{r}}{\partial u^1} \times \frac{\partial \boldsymbol{r}}{\partial u^2} \cdot \frac{\partial \boldsymbol{r}}{\partial u^3}\right| \mathrm{d}u^1 \mathrm{d}u^2 \mathrm{d}u^3,$$

can be expressed in a unified form in terms of the metric $g_{ij} = \partial_i \boldsymbol{r} \cdot \partial_j \boldsymbol{r}$, as

$$\mathrm{d}\mu = \sqrt{g}\, \mathrm{d}u^1 \mathrm{d}u^2 \ldots \mathrm{d}u^n, \tag{15.27}$$

where $g = \det[g_{ij}]$ is the determinant of the metric. Expression (15.27) readily applies to higher dimensional subspaces, such as an n-dimensional "surface" (or *manifold*) embedded in an m-dimensional space, with $0 < n \leq m$.

For $n = 1$, (15.27) is immediate since $g = g_{11} = (\mathrm{d}\boldsymbol{r}/\mathrm{d}u) \cdot (\mathrm{d}\boldsymbol{r}/\mathrm{d}u)$. For $n = 2$, it follows from (14.28b). For $n = 3$, using the contravariant expansion of the covariant basis, $\boldsymbol{a}_i = g_{ij}\boldsymbol{a}^j$ (14.23), and $\boldsymbol{a}^i \times \boldsymbol{a}^j \cdot \boldsymbol{a}^k = \epsilon^{ijk}(\boldsymbol{a}^1 \times \boldsymbol{a}^2 \cdot \boldsymbol{a}^3)$,

$$\begin{aligned}(\boldsymbol{a}_1 \times \boldsymbol{a}_2 \cdot \boldsymbol{a}_3)&(\boldsymbol{a}_1 \times \boldsymbol{a}_2 \cdot \boldsymbol{a}_3) \\ &= (\boldsymbol{a}_1 \times \boldsymbol{a}_2 \cdot \boldsymbol{a}_3)(\epsilon^{ijk} g_{1i} g_{2j} g_{3k})(\boldsymbol{a}^1 \times \boldsymbol{a}^2 \cdot \boldsymbol{a}^3) = J g J^{-1} = g.\end{aligned}$$

Another derivation that generalizes to higher dimension $n \leq m$ in m-dimensional euclidean space $\mathbb{R}^m$, is to orthogonalize the basis vectors $\partial \boldsymbol{r}/\partial u^i = \boldsymbol{a}_i$ using Gram–Schmidt. In matrix notation, with the basis vectors as the columns of matrix $\mathsf{A} = [\mathsf{a}_1 \ldots \mathsf{a}_n]$, this is the QR decomposition $\mathsf{A} = \mathsf{QR}$ where $\mathsf{Q}^\mathsf{T}\mathsf{Q} = \mathbf{I}$ and R is upper triangular n-by-n. The diagonal elements of R are the length of the edges of the rectified parallelotope and the volume V is thus the product of the diagonal elements, that is, $V = \det \mathsf{R}$. This is

$$V = \sqrt{\det(\mathsf{A}^\mathsf{T}\mathsf{A})} \tag{15.28}$$

because $\det(\mathsf{A}^\mathsf{T}\mathsf{A}) = \det(\mathsf{R}^\mathsf{T}\mathsf{Q}^\mathsf{T}\mathsf{QR}) = \det(\mathsf{R}^\mathsf{T}\mathsf{R}) = (\det \mathsf{R})^2$. Since the metric $g_{ij} = \boldsymbol{a}_i \cdot \boldsymbol{a}_j$, that is, $\mathsf{G} = \mathsf{A}^\mathsf{T}\mathsf{A}$ in matrix form, this yields formula (15.27).

15.4 Inverse map

The map $\boldsymbol{r}(u, v, w) \equiv \boldsymbol{r}(u^1, u^2, u^3)$ specifies (the radius vector of) a point P in 3D euclidean space as a function of the coordinates $(u, v, w) \equiv (u^1, u^2, u^3)$. Expression (15.23)

$$\mathrm{d}\boldsymbol{r} = \frac{\partial \boldsymbol{r}}{\partial u}\mathrm{d}u + \frac{\partial \boldsymbol{r}}{\partial v}dv + \frac{\partial \boldsymbol{r}}{\partial w}\mathrm{d}w = \mathrm{d}u^i \partial_i \boldsymbol{r}. \tag{15.29}$$

for the line element $\mathrm{d}\boldsymbol{r}$ is the starting point for constructing the *inverse map* $u^i(\boldsymbol{r})$.

For a given arbitrary $\mathrm{d}\boldsymbol{r}$ from a starting point $\boldsymbol{r}$, eqn (15.29) yields the corresponding coordinate increments $\mathrm{d}u^i$ provided the tangent vectors $\partial_i \boldsymbol{r}$ form a basis, that is provided $J = \partial_1 \boldsymbol{r} \times \partial_2 \boldsymbol{r} \cdot \partial_3 \boldsymbol{r} \neq 0$. The solution $\mathrm{d}u^i$ of (15.29) for given $\mathrm{d}\boldsymbol{r}$ is obtained by dotting (15.29) with the contravariant basis $\boldsymbol{a}^i$ (15.12), yielding

$$\mathrm{d}u^i = \boldsymbol{a}^i \cdot \mathrm{d}\boldsymbol{r} = \frac{\epsilon^{ijk}}{2J} \boldsymbol{N}_{jk} \cdot \mathrm{d}\boldsymbol{r}, \tag{15.30}$$

where $\boldsymbol{N}_{jk} = \partial_j \boldsymbol{r} \times \partial_k \boldsymbol{r}$ are the surface normals, functions of $(u^1, u^2, u^3) \equiv (u, v, w)$.

For example, picking $\mathrm{d}\boldsymbol{r} = \mathrm{d}s\, \hat{\boldsymbol{N}}_{12} \equiv \mathrm{d}s\, \hat{\boldsymbol{N}}_{uv}$ in the direction of the local normal $\boldsymbol{N}_{12}$ to the u^3-isosurface, yields

$$\frac{\mathrm{d}}{\mathrm{d}s} \begin{bmatrix} u \\ v \\ w \end{bmatrix} = \frac{1}{J} \begin{bmatrix} \boldsymbol{N}_{vw} \cdot \hat{\boldsymbol{N}}_{uv} \\ \boldsymbol{N}_{wu} \cdot \hat{\boldsymbol{N}}_{uv} \\ \boldsymbol{N}_{uv} \cdot \hat{\boldsymbol{N}}_{uv} \end{bmatrix} = \begin{bmatrix} f(u, v, w) \\ g(u, v, w) \\ h(u, v, w) \end{bmatrix}, \tag{15.31}$$

These are coupled differential equations yielding $(u(s), v(s), w(s))$, and the corresponding curve $\boldsymbol{r}(u(s), v(s), w(s))$ is perpendicular to $w \equiv u^3$-isosurfaces from an arbitrary starting point, with s the arclength along that curve. These curves are the $w \equiv u^3$ gradient lines.

More generally, to determine $u(\boldsymbol{r}_1), v(\boldsymbol{r}_1), w(\boldsymbol{r}_1)$ for any point $\boldsymbol{r}_1$ we can start from any (u_0, v_0, w_0) at $\boldsymbol{r}_0 = \boldsymbol{r}(u_0, v_0, w_0)$ and integrate along the straight line $\boldsymbol{r}(t) = \boldsymbol{r}_0 + t(\boldsymbol{r}_1 - \boldsymbol{r}_0)$ with $t = 0 \to 1$. The coordinates $u^i(t)$ follow from (15.30) through the differential equations

$$\frac{\mathrm{d}u^i}{\mathrm{d}t} = \frac{\epsilon^{ijk}}{2J} \boldsymbol{N}_{jk} \cdot \frac{\mathrm{d}\boldsymbol{r}}{\mathrm{d}t} = \frac{\epsilon^{ijk}}{2J} \boldsymbol{N}_{jk} \cdot (\boldsymbol{r}_1 - \boldsymbol{r}_0). \tag{15.32}$$

The normals $\boldsymbol{N}_{jk}$ and Jacobian J are continuous functions of the coordinates (u^1, u^2, u^3) such that (15.32) is again a system of three coupled differential equations for $(u^1(t), u^2(t), u^3(t)) \equiv (u(t), v(t), w(t))$.

Thus, given $\boldsymbol{r}(u^i)$, we can use the ODEs (15.31) or (15.32) to construct the inverse map $u^i(\boldsymbol{r})$. Clearly, this inverse map construction is possible only if $J \neq 0$, which is seen as the condition for a *locally* one-to-one invertible map $\mathrm{d}u^i \longleftrightarrow \mathrm{d}\boldsymbol{r}$.

In dimension $n > 1$, $J \neq 0$, does not guarantee *globally* one-to-one mapping. For example, cylindrical coordinates

$$\boldsymbol{r}(\rho, \varphi, z) = \rho \hat{\boldsymbol{\rho}}(\varphi) + z\, \hat{\mathbf{z}} = \rho \cos\varphi\, \hat{\mathbf{x}} + \rho \sin\varphi\, \hat{\mathbf{y}} + z\, \hat{\mathbf{z}}, \tag{15.33}$$

with tangent vectors (15.20a), have Jacobian $J = \rho \geq 0$, that is nonzero if $\rho \neq 0$, so the map is locally one-to-one for $\rho \neq 0$, yet it is not globally one to one since $\boldsymbol{r}(\rho, \varphi, z) = \boldsymbol{r}(\rho, \varphi + 2\pi, z)$. On the z-axis, where $\rho = 0$, the coordinates are singular and not locally one to one, since φ is arbitrary.

Spherical coordinates

$$\boldsymbol{r}(r, \theta, \varphi) = r\hat{\boldsymbol{r}}(\theta, \varphi) = r \sin\theta \cos\varphi\, \hat{\mathbf{x}} + r \sin\theta \sin\varphi\, \hat{\mathbf{y}} + r \cos\theta\, \hat{\mathbf{z}}, \tag{15.34}$$

with tangent vectors (15.21a) have $J = r^2 \sin\theta > 0$, except along the polar axis where $\theta = 0$ or π, φ is undefined, and $J = 0$. The singularity is even stronger at $r = 0$, where both θ and φ are undefined. The spherical-to-cartesian map $(r, \theta, \varphi) \to (x, y, z)$ can be inverted analytically, as done in (7.4). In the ODE inversion procedure, one computes the coordinate tangents $\partial_i \boldsymbol{r}$ and the surface normals $\boldsymbol{N}_{jk}$, obtaining for example

$$\boldsymbol{N}_{31}(r, \theta, \varphi) = \frac{\partial \boldsymbol{r}}{\partial \varphi} \times \frac{\partial \boldsymbol{r}}{\partial r} = r \sin\theta\, \hat{\boldsymbol{\theta}};$$

then one integrates (15.30) in direction $\hat{\boldsymbol{\theta}} = \hat{\boldsymbol{N}}_{31}$, yielding

$$\frac{\mathrm{d}r}{\mathrm{d}s} = 0, \quad \frac{\mathrm{d}\varphi}{\mathrm{d}s} = 0, \quad \frac{\mathrm{d}\theta}{\mathrm{d}s} = \frac{1}{r}, \tag{15.35}$$

to construct $\theta(\boldsymbol{r})$ from arbitrary starting points, and similarly for $r(\boldsymbol{r})$, $\varphi(\boldsymbol{r})$.

15.5 Volume integrals

The definition of a volume integral as the limit of a sum is straightforward. We imagine the volume $\mathcal{V}$ being partitioned into N *tetrahedra* or *hexahedra*, then

$$\int_{\mathcal{V}} f(\boldsymbol{r})\, \mathrm{d}V = \lim_{\Delta V_n \to 0} \sum_{n=1}^{N} \bar{f}_n\, \Delta V_n \tag{15.36}$$

where ΔV_n is the volume of the nth polyhedron, and $\bar{f}_n$ is an estimate of the average of $f(\boldsymbol{r})$ in the nth polyhedron, for instance, the average of the values at the vertices of that polyhedron. As for surfaces, the limit $\Delta V_n \to 0$ should correspond to an acceptable mesh refinement where $\Delta V_n \to 0$ as the cube of its diameter. Flattening polyhedra, or shrinking them to a line, is not an acceptable limit.

If an explicit parametrization $\boldsymbol{r}(u, v, w)$ for the volume is known, we can use the volume element (15.26) and write the volume integral in $\boldsymbol{r}$ space as an iterated triple integral over u, v, w,

$$\int_{\mathcal{V}} f(\boldsymbol{r})\, \mathrm{d}V = \iiint_{\mathcal{U}} f(\boldsymbol{r}(u, v, w))\, J(u, v, w)\, \mathrm{d}u \mathrm{d}v \mathrm{d}w, \tag{15.37}$$

where $J = \partial_u \boldsymbol{r} \times \partial_v \boldsymbol{r} \cdot \partial_w \boldsymbol{r} \geq 0$ is the Jacobian of the coordinates, and the u, v, w limits of integration should be chosen such that the volume of the domain

$$V = \int_{\mathcal{V}} \mathrm{d}V = \iiint_{\mathcal{U}} J(u, v, w)\, \mathrm{d}u \mathrm{d}v \mathrm{d}w > 0.$$

We write the sum over volume elements $\mathrm{d}V$ as a single sum, since it is the limit of a single sum over all polyhedra, but the iterated integrals over the scalar coordinates u, v, w are a triple sum.

For example, the volume element for orthogonal spherical coordinates $\boldsymbol{r} = r\hat{\boldsymbol{r}}(\theta, \varphi)$ is, from (15.21a),

$$\mathrm{d}V = r^2 \sin\theta \, \mathrm{d}r\mathrm{d}\theta\mathrm{d}\varphi,$$

and the volume of a sphere of radius R is

$$V = \int_0^{2\pi}\int_0^{\pi}\int_0^{R} r^2 \sin\theta \, \mathrm{d}r\mathrm{d}\theta\mathrm{d}\varphi = \frac{4}{3}\pi R^3. \tag{15.38}$$

For the torus, parametrized as in (15.5) that are orthogonal coordinates with

$$\left|\frac{\partial \boldsymbol{r}}{\partial s}\right| = 1, \quad \left|\frac{\partial \boldsymbol{r}}{\partial \theta}\right| = s, \quad \left|\frac{\partial \boldsymbol{r}}{\partial \varphi}\right| = R + s\cos\theta$$

and thus $\mathrm{d}V = s(R + s\cos\theta)\,\mathrm{d}s\mathrm{d}\theta\mathrm{d}\varphi$, so the torus volume is

$$V = \int_0^{2\pi}\int_0^{2\pi}\int_0^{a} s(R + s\cos\theta)\,\mathrm{d}s\mathrm{d}\theta\mathrm{d}\varphi = 2\pi^2 Ra^2 = (2\pi R)(\pi a^2). \tag{15.39}$$

For the tetrahedron (15.1), $\boldsymbol{r}(u, v, w) = \boldsymbol{r}_A + u\,\boldsymbol{a} + v\,\boldsymbol{b} + w\,\boldsymbol{c}$, the volume element

$$\mathrm{d}V = (\boldsymbol{a} \times \boldsymbol{b} \cdot \boldsymbol{c})\,\mathrm{d}u\mathrm{d}v\mathrm{d}w$$

and the volume

$$V = \int_0^{1}\int_0^{u}\int_0^{v} (\boldsymbol{a} \times \boldsymbol{b} \cdot \boldsymbol{c})\,\mathrm{d}w\mathrm{d}v\mathrm{d}u = \tfrac{1}{6}\,\boldsymbol{a} \times \boldsymbol{b} \cdot \boldsymbol{c}. \tag{15.40}$$

15.6 Change of coordinates

A change of coordinates from $(u^1, \ldots, u^n)$ to $(\bar{u}^1, \ldots, \bar{u}^n)$ corresponds to a change of tangent space basis $\partial\boldsymbol{r}/\partial u^i \to \partial\boldsymbol{r}/\partial\bar{u}^i$. The transformations of one basis into the other follow from the chain rule as, with sums over j,

$$\frac{\partial \boldsymbol{r}}{\partial \bar{u}^i} = \frac{\partial \boldsymbol{r}}{\partial u^j}\frac{\partial u^j}{\partial \bar{u}^i}, \tag{15.41a}$$

$$\frac{\partial \boldsymbol{r}}{\partial u^i} = \frac{\partial \boldsymbol{r}}{\partial \bar{u}^j}\frac{\partial \bar{u}^j}{\partial u^i}. \tag{15.41b}$$

In matrix notation, the forward transformation (15.41a) reads

$$\begin{bmatrix} \dfrac{\partial \boldsymbol{r}}{\partial \bar{u}^1} \\ \dfrac{\partial \boldsymbol{r}}{\partial \bar{u}^2} \\ \dfrac{\partial \boldsymbol{r}}{\partial \bar{u}^3} \end{bmatrix} = \begin{bmatrix} \dfrac{\partial u^1}{\partial \bar{u}^1} & \dfrac{\partial u^2}{\partial \bar{u}^1} & \dfrac{\partial u^3}{\partial \bar{u}^1} \\ \dfrac{\partial u^1}{\partial \bar{u}^2} & \dfrac{\partial u^2}{\partial \bar{u}^2} & \dfrac{\partial u^3}{\partial \bar{u}^2} \\ \dfrac{\partial u^1}{\partial \bar{u}^3} & \dfrac{\partial u^2}{\partial \bar{u}^3} & \dfrac{\partial u^3}{\partial \bar{u}^3} \end{bmatrix} \begin{bmatrix} \dfrac{\partial \boldsymbol{r}}{\partial u^1} \\ \dfrac{\partial \boldsymbol{r}}{\partial u^2} \\ \dfrac{\partial \boldsymbol{r}}{\partial u^3} \end{bmatrix}, \tag{15.42}$$

where the *Jacobian matrices*

$$\mathbf{J} = \mathbf{J}^{u}_{\bar{u}} = \left[\frac{\partial u^j}{\partial \bar{u}^i}\right], \qquad \mathbf{J}^{\bar{u}}_{u} = \left[\frac{\partial \bar{u}^j}{\partial u^i}\right] = \mathbf{J}^{-1},$$

have i as the row index and j as the column index.[2] The backward transformation (15.41b) involves the Jacobian $\mathbf{J}_u^{\bar{u}} = \mathbf{J}^{-1}$. The Jacobians are inverse of each other as implied by (15.41), and verified directly by the chain rule since

$$\frac{\partial u^k}{\partial \bar{u}^i}\frac{\partial \bar{u}^j}{\partial u^k} = \frac{\partial \bar{u}^j}{\partial \bar{u}^i} = \delta_i^j, \tag{15.43a}$$

$$\frac{\partial \bar{u}^k}{\partial u^i}\frac{\partial u^j}{\partial \bar{u}^k} = \frac{\partial \bar{u}^j}{\partial \bar{u}^i} = \delta_i^j. \tag{15.43b}$$

[2] Some writers switch rows and columns in the definition of the Jacobian matrix **J**.

Cartesian to curvilinear example

In 3D, with cartesian coordinates $x^j \equiv (x, y, z)$ and curvilinear coordinates $u^i \equiv (u, v, w)$, then $\partial \boldsymbol{r}/\partial x^j = \mathbf{e}_j \equiv (\hat{\mathbf{x}}, \hat{\mathbf{y}}, \hat{\mathbf{z}})$ and (15.41) yields

$$\frac{\partial \boldsymbol{r}}{\partial u^i} = \frac{\partial \boldsymbol{r}}{\partial x^j}\frac{\partial x^j}{\partial u^i} = \frac{\partial x}{\partial u^i}\hat{\mathbf{x}} + \frac{\partial y}{\partial u^i}\hat{\mathbf{y}} + \frac{\partial z}{\partial u^i}\hat{\mathbf{z}},$$

with

$$\mathbf{J}_u^x = \begin{bmatrix} \frac{\partial x^1}{\partial u^1} & \frac{\partial x^2}{\partial u^1} & \frac{\partial x^3}{\partial u^1} \\ \frac{\partial x^1}{\partial u^2} & \frac{\partial x^2}{\partial u^2} & \frac{\partial x^3}{\partial u^2} \\ \frac{\partial x^1}{\partial u^3} & \frac{\partial x^2}{\partial u^3} & \frac{\partial x^3}{\partial u^3} \end{bmatrix} \equiv \begin{bmatrix} \frac{\partial x}{\partial u} & \frac{\partial y}{\partial u} & \frac{\partial z}{\partial u} \\ \frac{\partial x}{\partial v} & \frac{\partial y}{\partial v} & \frac{\partial z}{\partial v} \\ \frac{\partial x}{\partial w} & \frac{\partial y}{\partial w} & \frac{\partial z}{\partial w} \end{bmatrix}, \tag{15.44}$$

and the cartesian components of the tangent vectors $\partial \boldsymbol{r}/\partial u^i$ along rows.

Covariant transformations

Transformations of covariant components and metric follow from their definition and (15.41) as

$$\begin{aligned} \bar{p}_i &\triangleq \boldsymbol{p}\cdot\frac{\partial \boldsymbol{r}}{\partial \bar{u}^i} = \boldsymbol{p}\cdot\frac{\partial \boldsymbol{r}}{\partial u^j}\frac{\partial u^j}{\partial \bar{u}^i} = p_j\frac{\partial u^j}{\partial \bar{u}^i}, \\ \bar{g}_{ij} &\triangleq \frac{\partial \boldsymbol{r}}{\partial \bar{u}^i}\cdot\frac{\partial \boldsymbol{r}}{\partial \bar{u}^j} = \frac{\partial \boldsymbol{r}}{\partial u^k}\frac{\partial u^k}{\partial \bar{u}^i}\cdot\frac{\partial \boldsymbol{r}}{\partial u^l}\frac{\partial u^l}{\partial \bar{u}^j} = g_{kl}\frac{\partial u^k}{\partial \bar{u}^i}\frac{\partial u^l}{\partial \bar{u}^j}. \end{aligned} \tag{15.45}$$

In matrix notation with $\mathsf{p} = [p_i]$ as a column vector of covariant components, and $\mathbf{J} = [\partial u^j/\partial \bar{u}^i]$, that is

$$\bar{\mathsf{p}} = \mathbf{J}\,\mathsf{p}, \quad \bar{\mathsf{G}} = \mathbf{J}\mathsf{G}\mathbf{J}^{\mathsf{T}}.$$

The covariant components thus transform like the basis vectors (15.41a), (15.42), and that is the definition of "covariance," in general. The inverse transformations follow from (15.43) or directly from (15.41),

$$\begin{aligned} p_i &= \boldsymbol{p}\cdot\frac{\partial \boldsymbol{r}}{\partial u^i} = \boldsymbol{p}\cdot\frac{\partial \boldsymbol{r}}{\partial \bar{u}^j}\frac{\partial \bar{u}^j}{\partial u^i} = \bar{p}_j\frac{\partial \bar{u}^j}{\partial u^i}, \\ g_{ij} &= \frac{\partial \boldsymbol{r}}{\partial u^i}\cdot\frac{\partial \boldsymbol{r}}{\partial u^j} = \cdots = \bar{g}_{kl}\frac{\partial \bar{u}^k}{\partial u^i}\frac{\partial \bar{u}^l}{\partial u^j}, \end{aligned} \tag{15.46}$$

that is $\mathsf{p} = \mathbf{J}^{-1}\,\bar{\mathsf{p}}$ and $\mathsf{G} = \mathbf{J}^{-1}\bar{\mathsf{G}}\,\mathbf{J}^{-\mathsf{T}}$.

Contravariant transformations

The contravariant metric g^{ij} is the inverse of the covariant metric g_{ij}; hence, its transformation rule is the inverse of (15.45)

$$\bar{g}^{ij} = g^{kl} \frac{\partial \bar{u}^i}{\partial u^k} \frac{\partial \bar{u}^j}{\partial u^l}. \tag{15.47}$$

The transformation of contravariant vector components can then be obtained from the metric and covariant components, $\bar{p}^i = \bar{g}^{ij}\bar{p}_j$, or directly from the identity

$$\boldsymbol{p} = \frac{\partial \boldsymbol{r}}{\partial \bar{u}^i}\, \bar{p}^i = \frac{\partial \boldsymbol{r}}{\partial u^j}\, p^j, \tag{15.48}$$

where $\boldsymbol{p}$ is an arbitrary tangent vector. Using (15.41) to transform the last expression yields

$$\frac{\partial \boldsymbol{r}}{\partial \bar{u}^i}\, \bar{p}^i = \frac{\partial \boldsymbol{r}}{\partial \bar{u}^i} \frac{\partial \bar{u}^i}{\partial u^j}\, p^j,$$

and since the basis vectors $\partial \boldsymbol{r}/\partial \bar{u}^i$ are linearly independent, this gives the contravariant components transformation as

$$\bar{p}^i = \frac{\partial \bar{u}^i}{\partial u^j}\, p^j. \tag{15.49}$$

In matrix notation, writing $\mathsf{p}' = \mathsf{G}^{-1}\mathsf{p} = [p^i]$ for the column vector of contravariant components, with $\mathsf{p} = [p_i]$ and $\mathsf{J} = [\partial u^j/\partial \bar{u}^i]$ as before, this is

$$\bar{\mathsf{G}}^{-1} = \mathsf{J}^{-\mathsf{T}}\mathsf{G}^{-1}\mathsf{J}^{-1}, \qquad \bar{\mathsf{p}}' = \mathsf{J}^{-\mathsf{T}}\,\mathsf{p}'$$

The inverse transformations likewise follow from (15.48) and (15.41), or directly from (15.43),

$$p^i = \frac{\partial u^i}{\partial \bar{u}^j}\, \bar{p}^j. \tag{15.50}$$

That is, $\mathsf{p}' = \mathsf{J}^{\mathsf{T}}\,\bar{\mathsf{p}}'$.

Intrinsic definitions

The transformation rules (15.45) and (15.49) can be obtained, and indeed serve as the definitions of covariant and contravariant vectors and tensors, without reference to the bases $\boldsymbol{a}_i = \partial_i \boldsymbol{r}$ and $\boldsymbol{a}^i = g^{ij}\boldsymbol{a}_j$.

§ 16.4 identifies the contravariant basis vectors as the gradients of the coordinates: $\boldsymbol{a}^i = \boldsymbol{\nabla} u^i$.

The starting point is the first fundamental form that measures distance and is a geometric invariant, independent of the choice of coordinates, such that

$$\mathrm{d}s^2 = \bar{g}_{ij}\mathrm{d}\bar{u}^i\mathrm{d}\bar{u}^j = g_{kl}\mathrm{d}u^k\mathrm{d}u^l, \tag{15.51}$$

for two sets of coordinates $(u^1, \ldots, u^n)$ and $(\bar{u}^1, \ldots, \bar{u}^n)$. From the chain rule,

$$\mathrm{d}u^k = \frac{\partial u^k}{\partial \bar{u}^i}\,\mathrm{d}\bar{u}^i, \quad \mathrm{d}u^l = \frac{\partial u^l}{\partial \bar{u}^j}\,\mathrm{d}\bar{u}^j, \tag{15.52}$$

substituting these differentials in (15.51) that holds for any $(\mathrm{d}\bar{u}^1, \ldots, \mathrm{d}\bar{u}^n)$ yields

$$\bar{g}_{ij} = g_{kl} \frac{\partial u^k}{\partial \bar{u}^i} \frac{\partial u^l}{\partial \bar{u}^j}. \tag{15.53}$$

The inverse metric g^{ij} then transforms as in (15.47). Next, we can define the covariant components p_k of an arbitrary vector $\boldsymbol{p}$ as those components that transform similarly to (15.53) as

$$\bar{p}_i = \frac{\partial u^k}{\partial \bar{u}^i}\, p_k, \tag{15.54}$$

and its contravariant components $p^i \triangleq g^{ij} p_j$ that transform, from (15.47) and (15.54), as

$$\bar{p}^i = \frac{\partial \bar{u}^i}{\partial u^k}\, p^k, \tag{15.55}$$

such that the dot product of two arbitrary vectors $\boldsymbol{p}$ and $\boldsymbol{q}$ defined as

$$\boldsymbol{p} \cdot \boldsymbol{q} \triangleq p_i q^i = g_{ij} p^i q^j = p^i q_i = g^{ij} p_i q_j, \tag{15.56}$$

is invariant under a change of coordinates $(u^1, \ldots, u^n) \to (\bar{u}^1, \ldots, \bar{u}^n)$.

Surface element transformations

The transformations of surface elements readily follow from (15.41). For a surface parametrization $\boldsymbol{r}(u^1, u^2)$ and a change of coordinates (u^1, u^2) $\to (\bar{u}^1, \bar{u}^2)$, the surface element becomes

$$\begin{aligned} \mathrm{d}\bar{\boldsymbol{S}} = \frac{\partial \boldsymbol{r}}{\partial \bar{u}^1} \times \frac{\partial \boldsymbol{r}}{\partial \bar{u}^2} \mathrm{d}\bar{u}^1 \mathrm{d}\bar{u}^2 &= \frac{\partial \boldsymbol{r}}{\partial u^i} \times \frac{\partial \boldsymbol{r}}{\partial u^j} \frac{\partial u^i}{\partial \bar{u}^1} \frac{\partial u^j}{\partial \bar{u}^2} \mathrm{d}\bar{u}^1 \mathrm{d}\bar{u}^2 \\ &= \frac{\partial \boldsymbol{r}}{\partial u^1} \times \frac{\partial \boldsymbol{r}}{\partial u^2}\, \epsilon_{ij3} \frac{\partial u^i}{\partial \bar{u}^1} \frac{\partial u^j}{\partial \bar{u}^2} \mathrm{d}\bar{u}^1 \mathrm{d}\bar{u}^2 \\ &= \frac{\partial \boldsymbol{r}}{\partial u^1} \times \frac{\partial \boldsymbol{r}}{\partial u^2}\, J \,\mathrm{d}\bar{u}^1 \mathrm{d}\bar{u}^2, \end{aligned} \tag{15.57}$$

using the identity $\boldsymbol{a}_i \times \boldsymbol{a}_j = \epsilon_{ij3}\, \boldsymbol{a}_1 \times \boldsymbol{a}_2$, where

$$J = \epsilon_{ij3} \frac{\partial u^i}{\partial \bar{u}^1} \frac{\partial u^j}{\partial \bar{u}^2} = \begin{vmatrix} \dfrac{\partial u^1}{\partial \bar{u}^1} & \dfrac{\partial u^2}{\partial \bar{u}^1} \\ \dfrac{\partial u^1}{\partial \bar{u}^2} & \dfrac{\partial u^2}{\partial \bar{u}^2} \end{vmatrix} = \det \mathbf{J} \equiv \frac{\partial(u^1, u^2)}{\partial(\bar{u}^1, \bar{u}^2)}$$

is the determinant of the 2-by-2 Jacobian matrix $\mathbf{J}$.

Since the surface element $\mathrm{d}\boldsymbol{S}$ in terms of the (u^1, u^2) coordinates is

$$\mathrm{d}\boldsymbol{S} = \frac{\partial \boldsymbol{r}}{\partial u^1} \times \frac{\partial \boldsymbol{r}}{\partial u^2}\, \mathrm{d}u^1 \mathrm{d}u^2,$$

(15.57) yields the *change of variable formula* for iterated double integrals as,

$$\mathrm{d}u^1 \mathrm{d}u^2 \to J \,\mathrm{d}\bar{u}^1 \mathrm{d}\bar{u}^2, \tag{15.58}$$

writing "$\to$" instead of "$=$" because the coordinates correspond to different partitions of the same domain. Thus, $\mathrm{d}u^1 \mathrm{d}u^2$ is the area of a rectangle in the (u^1, u^2) domain, but $J \mathrm{d}\bar{u}^1 \mathrm{d}\bar{u}^2$ is the area of the quadrangle in the (u^1, u^2) domain, which is the image of a rectangle of sides $(\mathrm{d}\bar{u}^1, \mathrm{d}\bar{u}^2)$ in the $(\bar{u}^1, \bar{u}^2)$ domain. The equality is (15.57), not $\mathrm{d}\bar{\boldsymbol{S}} = \mathrm{d}\boldsymbol{S}$.

For example, a domain $\mathcal{A}$ in an euclidean plane is naturally parametrized in cartesian coordinates $(u^1, u^2) \equiv (x, y)$ with surface element

$$\mathrm{d}\boldsymbol{S} = \frac{\partial \boldsymbol{r}}{\partial x} \times \frac{\partial \boldsymbol{r}}{\partial y} \mathrm{d}x\mathrm{d}y = \hat{\mathbf{z}}\, \mathrm{d}x\mathrm{d}y, \tag{15.59}$$

but the particular domain $\mathcal{A}$ might be better parametrized using other coordinates. For example, polar coordinates (ρ, φ) with $x = \rho\cos\varphi$, $y = \rho\sin\varphi$, are better suited for the annular sector

$$\mathcal{A} : (x, y) \in \mathbb{R}^2 \quad \text{s.t.} \begin{cases} \rho_1 < \sqrt{x^2 + y^2} < \rho_2, \\ \varphi_1 < \operatorname{atan2}(y, x) < \varphi_2. \end{cases}$$

The surface element in those coordinates $(\bar{u}^1, \bar{u}^2) \equiv (\rho, \varphi)$ is

$$\mathrm{d}\bar{\boldsymbol{S}} = \frac{\partial \boldsymbol{r}}{\partial \rho} \times \frac{\partial \boldsymbol{r}}{\partial \varphi} \mathrm{d}\rho\mathrm{d}\varphi = \frac{\partial \boldsymbol{r}}{\partial x} \times \frac{\partial \boldsymbol{r}}{\partial y} \begin{vmatrix} \frac{\partial x}{\partial \rho} & \frac{\partial y}{\partial \rho} \\ \frac{\partial x}{\partial \varphi} & \frac{\partial y}{\partial \varphi} \end{vmatrix} \mathrm{d}\rho\mathrm{d}\varphi = \hat{\mathbf{z}}\, \rho\, \mathrm{d}\rho\mathrm{d}\varphi. \tag{15.60}$$

Volume element transformations

Likewise, the volume element in terms of the $(\bar{u}^1, \bar{u}^2, \bar{u}^3)$ coordinates is

$$\begin{aligned} \mathrm{d}\bar{V} &= \frac{\partial \boldsymbol{r}}{\partial \bar{u}^1} \times \frac{\partial \boldsymbol{r}}{\partial \bar{u}^2} \cdot \frac{\partial \boldsymbol{r}}{\partial \bar{u}^3}\, \mathrm{d}\bar{u}^1\mathrm{d}\bar{u}^2\mathrm{d}\bar{u}^3 \\ &= \frac{\partial \boldsymbol{r}}{\partial u^i} \times \frac{\partial \boldsymbol{r}}{\partial u^j} \cdot \frac{\partial \boldsymbol{r}}{\partial u^k} \frac{\partial u^i}{\partial \bar{u}^1} \frac{\partial u^j}{\partial \bar{u}^2} \frac{\partial u^k}{\partial \bar{u}^3}\, \mathrm{d}\bar{u}^1\mathrm{d}\bar{u}^2\mathrm{d}\bar{u}^3 \\ &= \frac{\partial \boldsymbol{r}}{\partial u^1} \times \frac{\partial \boldsymbol{r}}{\partial u^2} \cdot \frac{\partial \boldsymbol{r}}{\partial u^3}\, J\, \mathrm{d}\bar{u}^1\mathrm{d}\bar{u}^2\mathrm{d}\bar{u}^3 \end{aligned} \tag{15.61}$$

where

$$J = \epsilon_{ijk} \frac{\partial u^i}{\partial \bar{u}^1} \frac{\partial u^j}{\partial \bar{u}^2} \frac{\partial u^k}{\partial \bar{u}^3} = \det \mathbf{J} \equiv \frac{\partial(u^1, u^2, u^3)}{\partial(\bar{u}^1, \bar{u}^2, \bar{u}^3)}$$

is the determinant of the 3-by-3 Jacobian matrix $\mathbf{J}$, and the identity $\boldsymbol{a}_i \times \boldsymbol{a}_j \cdot \boldsymbol{a}_k = \epsilon_{ijk}\, \boldsymbol{a}_1 \times \boldsymbol{a}_2 \cdot \boldsymbol{a}_3$ has been used. The volume element for the (u^1, u^2, u^3) coordinates is

$$\mathrm{d}V = \frac{\partial \boldsymbol{r}}{\partial u^1} \times \frac{\partial \boldsymbol{r}}{\partial u^2} \cdot \frac{\partial \boldsymbol{r}}{\partial u^3}\, \mathrm{d}u^1\mathrm{d}u^2\mathrm{d}u^3,$$

and (15.61) yields the *change of variable formula* for iterated triple integrals as

$$\mathrm{d}u^1\mathrm{d}u^2\mathrm{d}u^3 \to J\, \mathrm{d}\bar{u}^1\mathrm{d}\bar{u}^2\mathrm{d}\bar{u}^3, \tag{15.62}$$

where J is the Jacobian determinant in (15.61), again writing "$\to$" instead of "=" since these correspond to different partitions of the same domain. The equality is (15.61), not $\mathrm{d}\bar{V} = \mathrm{d}V$.

It is clear from (15.57) and (15.61) that the Jacobian determinant

$$J = \frac{\partial(u^1, \dots, u^n)}{\partial(\bar{u}^1, \dots, \bar{u}^n)} \geq 0$$

of the change of variables $(u^1, \dots, u^n) \to (\bar{u}^1, \dots, \bar{u}^n)$ should be *sign definite*, that is, everywhere positive (or negative, but then reordering the coordinates changes its sign). The Jacobian may vanish at singular points, or singular curves, but certainly not everywhere otherwise $(\bar{u}^1, \dots, \bar{u}^n)$ are not valid coordinates.

Volume of an ellipsoid by successive transformations

An ellipsoid is nicely parametrized by

$$\boldsymbol{r}(s,\theta,\varphi) = s\,(\boldsymbol{a}\sin\theta\cos\varphi + \boldsymbol{b}\sin\theta\sin\varphi + \boldsymbol{c}\cos\theta) \tag{15.63}$$

with $0 \leq s \leq 1$, $0 \leq \theta \leq \pi$, $-\pi < \varphi \leq \pi$. The vectors $\boldsymbol{a}$, $\boldsymbol{b}$, $\boldsymbol{c}$ are fixed and linearly independent. If they are mutually orthogonal, they are the principal radii of the ellipsoid. To compute the volume, we would need to calculate the coordinate tangent vectors, then the Jacobian to evaluate the volume as

$$V = \int_0^{2\pi}\int_0^{\pi}\int_0^1 \frac{\partial \boldsymbol{r}}{\partial s} \times \frac{\partial \boldsymbol{r}}{\partial \theta} \cdot \frac{\partial \boldsymbol{r}}{\partial \varphi}\, \mathrm{d}s\mathrm{d}\theta\mathrm{d}\varphi.$$

That is easy enough, but we can do the calculation even more simply using successive coordinate transformations starting with the *tetrahedral coordinates* (15.40)

$$\boldsymbol{r}(u,v,w) = u\,\boldsymbol{a} + v\,\boldsymbol{b} + w\,\boldsymbol{c}$$

with Jacobian

$$J_1 = \frac{\partial \boldsymbol{r}}{\partial u} \times \frac{\partial \boldsymbol{r}}{\partial v} \cdot \frac{\partial \boldsymbol{r}}{\partial w} = \boldsymbol{a} \times \boldsymbol{b} \cdot \boldsymbol{c},$$

and coordinates in the unit sphere $u^2+v^2+w^2 \leq 1$, followed by *spherical coordinates* (s,θ,φ) for that unit sphere with Jacobian

$$J_2 = \frac{\partial(u,v,w)}{\partial(s,\theta,\varphi)} = s^2\sin\theta,$$

and volume $4\pi/3$ (15.38), that is,

$$V = \int_0^{2\pi}\int_0^{\pi}\int_0^1 \left(\frac{\partial \boldsymbol{r}}{\partial u} \times \frac{\partial \boldsymbol{r}}{\partial v} \cdot \frac{\partial \boldsymbol{r}}{\partial w}\right) \frac{\partial(u,v,w)}{\partial(s,\theta,\varphi)}\, \mathrm{d}s\mathrm{d}\theta\mathrm{d}\varphi = \frac{4\pi}{3}\,\boldsymbol{a} \times \boldsymbol{b} \cdot \boldsymbol{c}.$$

Carnot cycle

As another example of coordinate transformations, consider a Carnot cycle for a perfect gas of volume V and pressure P. The Carnot cycle is an idealized thermodynamic cycle that consists of (Fig. 15.7)

(1) an isothermal expansion from $0 \to 1$ with $PV = T_1$,

(2) an adiabatic expansion from $1 \to 2$ with $PV^\gamma = S_1$,

(3) an isothermal compression from $2 \to 3$ with $PV = T_0$, and

(4) an adiabatic compression from $3 \to 0$ with $PV^\gamma = S_0$,

where S_0, S_1, T_0, T_1 are positive constants, and $\gamma = c_P/c_V > 1$ is the ratio of heat capacities.

The work W done by the gas when its volume changes from V_a to V_b is $\int_{V_a}^{V_b} P\mathrm{d}V$. This is most easily understood for a gas in a cylinder with a piston, since work = force × displacement, P = force/area, and

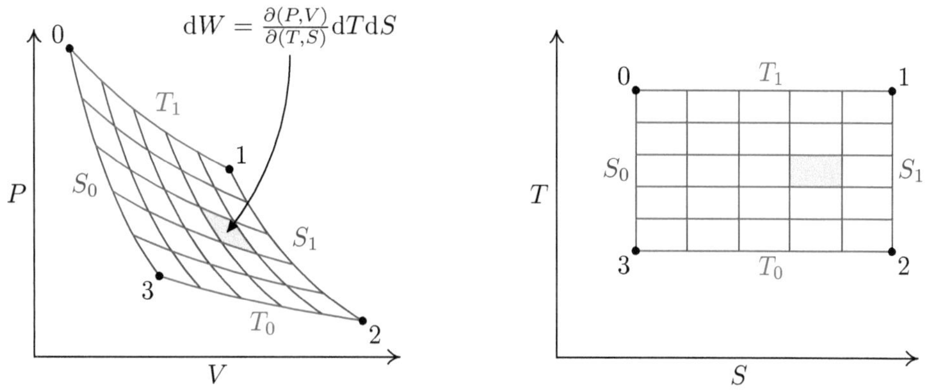

Fig. 15.7 Carnot cycle. Left: (P, V) diagram. Right: (T, S) diagram. Schematic.

dV=area × displacement. Thus, the area inside the cycle in the (P, V) plane is the net work performed by the gas during one cycle.

The area W of the (P, V) domain

$$C : \begin{cases} T_0 \le PV \le T_1, \\ S_0 \le PV^\gamma \le S_1, \end{cases}$$

corresponding to the Carnot cycle, can be calculated by elementary methods, but the calculation is most straightforward using the change of variables

$$T = PV, \quad S = PV^\gamma \tag{15.64}$$

with

$$\begin{aligned} J = \frac{\partial(T,S)}{\partial(P,V)} &= \frac{\partial T}{\partial P}\frac{\partial S}{\partial V} - \frac{\partial T}{\partial V}\frac{\partial S}{\partial P} \\ &= VP\gamma V^{\gamma-1} - PV^\gamma = (\gamma-1)S. \end{aligned} \tag{15.65}$$

The Jacobian of the inverse transformation $(T, S) \to (P, V)$ is simply the inverse of (15.65), that is,

$$\frac{\partial(P,V)}{\partial(T,S)} = J^{-1} = \frac{1}{(\gamma-1)S}. \tag{15.66}$$

This change of variables maps the curved Carnot domain C in the (P, V) plane to a simple rectangle in the (T, S) plane. The area in the (P, V) plane is then

$$\begin{aligned} W = \iint_C \mathrm{d}P\mathrm{d}V &= \int_{T_0}^{T_1}\int_{S_0}^{S_1} \frac{\partial(P,V)}{\partial(T,S)}\,\mathrm{d}S\mathrm{d}T \\ &= \int_{T_0}^{T_1}\int_{S_0}^{S_1} \frac{1}{(\gamma-1)S}\,\mathrm{d}S\mathrm{d}T = \frac{T_1-T_0}{\gamma-1}\ln\frac{S_1}{S_0}, \end{aligned} \tag{15.67}$$

from the change of variable formula (15.58). In thermodynamics, our T is nR times the *temperature* in kelvins, where R is the gas constant and n the number of moles, and $c_V \ln(S_1/S_0)$ is the *entropy* jump from state 0 to 1 (and thus 3 to 2 also).

Exercises

(15.1) Show that bipolar coordinates (u, w), in (15.6) with $v = 0$, are conformal. Show that the identities $\cos^2 u + \sin^2 u = 1$ and $\cosh^2 w - \sinh^2 w = 1$ yield

$$\left(x - \frac{R_*}{\tanh w}\right)^2 + z^2 = \frac{R_*^2}{\sinh^2 w},$$
$$x^2 + \left(z - \frac{R_*}{\tan u}\right)^2 = \frac{R_*^2}{\sin^2 u}.$$

What curves are these? Find the xz plane area element $\mathrm{d}A$ in terms of u and w.

(15.2) Show that toroidal coordinates (15.6) are orthogonal but not conformal. Find the volume element $\mathrm{d}V$ in terms of u, v, w.

(15.3) *Elliptic coordinates* (u, v) are defined by

$$x = \cosh u \cos v, \quad y = \sinh u \sin v,$$

where x, y are standard cartesian coordinates in a 2D euclidean plane. Show that the u-coordinate curves are ellipses, and the v-coordinate curves are hyperbolas. Sketch the curves. Show that these are conformal coordinates in the plane and find the surface element $\mathrm{d}A$ in terms of u and v.

(15.4) A *spheroid* is an *ellipsoid* of revolution with the standard cartesian equation

$$\frac{x^2}{a^2} + \frac{y^2}{a^2} + \frac{z^2}{c^2} = 1.$$

The Earth is a slightly oblate ($a > c$) spheroid as a result of centrifugal acceleration. Parametrize the spheroid in two different ways: (i) as a modification of spherical coordinates, and (ii) by rotation of elliptic coordinates. Is either system of coordinates orthogonal? conformal?

(15.5) The determinant of the coordinate tangents is J in (15.9). Show that the determinant of the surface normals (15.8) and the determinant of the metric (15.13b) both equal J^2,

$$\boldsymbol{N}_{uv} \times \boldsymbol{N}_{vw} \cdot \boldsymbol{N}_{wu} = J^2 = g \triangleq \det[g_{ij}]. \quad (15.68)$$

(15.6) Show that the covariant (15.11) and contravariant bases (15.12) satisfy

$$\begin{aligned} \boldsymbol{a}_i \times \boldsymbol{a}_j &= J\, \epsilon_{ijk}\, \boldsymbol{a}^k, \\ \boldsymbol{a}^i \times \boldsymbol{a}^j &= J^{-1} \epsilon^{ijk} \boldsymbol{a}_k. \end{aligned} \quad (15.69)$$

(15.7) The Christoffel symbols (14.34), (14.37), are the components of the derivatives of covariant basis vectors $\partial_i \boldsymbol{a}_j = \partial_i \partial_j \boldsymbol{r} = \partial_j \boldsymbol{a}_i$, such that

$$\partial_i \boldsymbol{a}_j \triangleq \Gamma_{kij}\, \boldsymbol{a}^k = \Gamma_{ij}^k\, \boldsymbol{a}_k, \quad (15.70a)$$

here for 3D euclidean space with indices in $\{1, 2, 3\}$, and sums over k. There are $3^2 - 3 = 6$ vectors $\partial_i \partial_j \boldsymbol{r} = \partial_j \partial_i \boldsymbol{r}$ and thus $3 * 6 = 18$ components Γ_{kij} and Γ_{ij}^k. The Christoffel symbols are given in terms of derivatives of the metric, as derived in (14.43).
Show that

$$\partial_i \boldsymbol{a}^j = -\Gamma_{ik}^j \boldsymbol{a}^k = -g^{jm} \Gamma_{mik}\, g^{kl} \boldsymbol{a}_l, \quad (15.70b)$$

and

$$\begin{aligned} \partial_i J &= \Gamma_{ik}^k J, \\ \partial_i g_{jk} &= \Gamma_{ij}^l\, g_{lk} + \Gamma_{ik}^l\, g_{lj}, \\ \partial_i g^{jk} &= -\Gamma_{il}^j\, g^{lk} - \Gamma_{il}^k\, g^{lj}, \end{aligned} \quad (15.70c)$$

then, using (15.69),

$$\begin{aligned} \boldsymbol{a}^i \times \partial_i \boldsymbol{a}^j &= 0, \\ \boldsymbol{a}^i \cdot \partial_i \left(\boldsymbol{a}^j \times \boldsymbol{a}^k\right) &= 0, \end{aligned} \quad (15.70d)$$

where $J = \boldsymbol{a}_1 \times \boldsymbol{a}_2 \cdot \boldsymbol{a}_3 \triangleq \sqrt{g}$ is the Jacobian determinant, $\boldsymbol{a}^i$ are the contravariant vectors such that $\boldsymbol{a}^i \cdot \boldsymbol{a}_j = \delta_j^i$, with $g_{ij} = \boldsymbol{a}_i \cdot \boldsymbol{a}_j$ the covariant metric, and $g^{ij} = \boldsymbol{a}^i \cdot \boldsymbol{a}^j$ the contravariant metric. These are the two fundamental vector calculus identities: the curl of a gradient, $\boldsymbol{\nabla} \times \boldsymbol{\nabla}(\cdot)$, and the divergence of a curl, $\boldsymbol{\nabla} \cdot (\boldsymbol{\nabla} \times (\cdot))$, vanish, as discussed in Chapter 16.
The identities (15.70d) are for 3D euclidean space with sums over $i = 1, 2, 3$. Those identities are *not* true in general for a 2D surface as illustrated in the next problem. Compare (15.70a), (15.70b) to (14.39).

(15.8) For spherical coordinates (θ, φ, r) where $\{\hat{\boldsymbol{a}}^1, \hat{\boldsymbol{a}}^2, \hat{\boldsymbol{a}}^3\} = \{\hat{\boldsymbol{\theta}}/r, \hat{\boldsymbol{\varphi}}/(r \sin\theta), \hat{\boldsymbol{r}}\}$, show that

$$\boldsymbol{a}^1 \times \partial_1 \boldsymbol{a}^1 + \boldsymbol{a}^2 \times \partial_2 \boldsymbol{a}^1 = \frac{\hat{\boldsymbol{\varphi}}}{r^2}$$

while $\boldsymbol{a}^i \partial_i \times \boldsymbol{a}^1 = \boldsymbol{a}^i \times \partial_i \boldsymbol{a}^1$

$$= \boldsymbol{a}^1 \times \partial_1 \boldsymbol{a}^1 + \boldsymbol{a}^2 \times \partial_2 \boldsymbol{a}^1 + \boldsymbol{a}^3 \times \partial_3 \boldsymbol{a}^1 = 0.$$

Similarly, show that

$$\begin{aligned}&\boldsymbol{a}^1\cdot\partial_1(\boldsymbol{a}^1\times\boldsymbol{a}^2)+\boldsymbol{a}^2\cdot\partial_2(\boldsymbol{a}^1\times\boldsymbol{a}^2)\\&=\frac{\hat{\boldsymbol{\theta}}}{r}\cdot\frac{\partial}{\partial\theta}\left(\frac{\hat{\boldsymbol{r}}}{r^2\sin\theta}\right)+\frac{\hat{\boldsymbol{\varphi}}}{r\sin\theta}\cdot\frac{\partial}{\partial\varphi}\left(\frac{\hat{\boldsymbol{r}}}{r^2\sin\theta}\right)\\&=\frac{2}{r^3\sin\theta},\end{aligned}$$

while $\boldsymbol{a}^i\partial_i\cdot\left(\boldsymbol{a}^1\times\boldsymbol{a}^2\right)=\boldsymbol{a}^i\cdot\partial_i\left(\boldsymbol{a}^1\times\boldsymbol{a}^2\right)$

$$=\boldsymbol{a}^1\cdot\partial_1(\boldsymbol{a}^1\times\boldsymbol{a}^2)+\boldsymbol{a}^2\cdot\partial_2(\boldsymbol{a}^1\times\boldsymbol{a}^2)+\boldsymbol{a}^3\cdot\partial_3(\boldsymbol{a}^1\times\boldsymbol{a}^2)=0.$$

(15.9) From (14.43) and (15.70c), show that

$$\Gamma^k_{ik}=\tfrac{1}{2}g^{kl}\,\partial_i g_{kl}=J^{-1}\partial_i J. \qquad (15.71)$$

(15.10) Use (15.70b) to show that the contraction of the Riemann curvature tensor R^l_{ijk} (14.126) over the first and last indices vanishes, $R^i_{ijk}=0$ (sum over i).

(15.11) For orthogonal coordinates with $\partial_i\boldsymbol{r}=h_i\hat{\boldsymbol{u}}_i$ (no sum) and $\hat{\boldsymbol{u}}_i\cdot\hat{\boldsymbol{u}}_j=\delta_{ij}$, show that the Christoffel symbols (14.34), (14.43), here with indices in $\{1,2,3\}$, become (*no sums*)

$$\begin{aligned}\Gamma_{kij}&=\tfrac{1}{2}\left(\delta_{jk}\,\partial_i g_{jj}+\delta_{ki}\,\partial_j g_{ii}-\delta_{ij}\,\partial_k g_{ii}\right)\\&=\tfrac{1}{2}\left(\delta_{jk}\,\partial_i h_j^2+\delta_{ki}\,\partial_j h_k^2-\delta_{ij}\,\partial_k h_i^2\right).\end{aligned} \qquad (15.72)$$

In particular, $\Gamma_{kij}=0$ when all indices are distinct (i.e. when $\epsilon_{ijk}\neq 0$). Then, using

$$\Gamma_{kij}=\partial_k\boldsymbol{r}\cdot(\partial_i\partial_j\boldsymbol{r})=h_k\hat{\boldsymbol{u}}_k\cdot\partial_i(h_j\hat{\boldsymbol{u}}_j),$$

derive (15.17). For orthogonal coordinates, upper and lower indices become equivalent since $\boldsymbol{a}_i=h_i\hat{\boldsymbol{u}}_i$ and $\boldsymbol{a}^i=\hat{\boldsymbol{u}}_i/h_i=H_i\hat{\boldsymbol{u}}_i$. In the above formula, indices are repeated on the right hand side but there is no sum since they are all in lower positions.

(15.12) For orthogonal coordinates with $\partial_i\boldsymbol{r}=h_i\hat{\boldsymbol{u}}_i$ (no sum) and $\hat{\boldsymbol{u}}_i\cdot\hat{\boldsymbol{u}}_j=\delta_{ij}$, thus $\partial_i(\partial_j\boldsymbol{r}\cdot\partial_k\boldsymbol{r})=\partial_i(h_jh_k)\delta_{jk}$. Use $\partial_i\partial_k\boldsymbol{r}=\partial_k\partial_i\boldsymbol{r}$ to show that

$$\hat{\boldsymbol{u}}_k\cdot\frac{\partial_i\hat{\boldsymbol{u}}_j}{h_i}+\hat{\boldsymbol{u}}_j\cdot\frac{\partial_k\hat{\boldsymbol{u}}_i}{h_k}=\delta_{jk}\frac{\partial_i h_k}{h_ih_k}-\delta_{ij}\frac{\partial_k h_i}{h_kh_i}$$

Rewriting this equation for $(i,j,k)\to(j,k,i)\to(k,i,j)$ then adding two of those equations and subtracting the third, deduce (15.17)

$$\hat{\boldsymbol{u}}_k\cdot\partial_i\hat{\boldsymbol{u}}_j=\delta_{ki}\frac{\partial_j h_i}{h_j}-\delta_{ij}\frac{\partial_k h_i}{h_k}. \qquad (15.73)$$

For $i=1$ that is

$$\partial_1\begin{bmatrix}\hat{\boldsymbol{u}}_1\\\hat{\boldsymbol{u}}_2\\\hat{\boldsymbol{u}}_3\end{bmatrix}=\begin{bmatrix}0&-\dfrac{\partial_2h_1}{h_2}&-\dfrac{\partial_3h_1}{h_3}\\\dfrac{\partial_2h_1}{h_2}&0&0\\\dfrac{\partial_3h_1}{h_3}&0&0\end{bmatrix}\begin{bmatrix}\hat{\boldsymbol{u}}_1\\\hat{\boldsymbol{u}}_2\\\hat{\boldsymbol{u}}_3\end{bmatrix}$$

that can be written

$$\partial_1\hat{\boldsymbol{u}}_j=\boldsymbol{\omega}_1\times\hat{\boldsymbol{u}}_j=\left(\hat{\boldsymbol{u}}_2\frac{\partial_3h_1}{h_3}-\hat{\boldsymbol{u}}_3\frac{\partial_2h_1}{h_2}\right)\times\hat{\boldsymbol{u}}_j,$$

and similarly for $\partial_2\hat{\boldsymbol{u}}_j$ and $\partial_3\hat{\boldsymbol{u}}_j$ by cyclic permutation of $(1,2,3)$. This derivation of (15.73) is similar to that for (14.43) but avoids using Γ_{ijk} explicitly.

(15.13) For orthogonal coordinates with $\partial_i\boldsymbol{r}=h_i\hat{\boldsymbol{u}}_i$ (no sum), and $\hat{\boldsymbol{u}}_i\cdot\hat{\boldsymbol{u}}_j=\delta_{ij}$, use $\partial_i\partial_j\hat{\boldsymbol{u}}_i=\partial_j\partial_i\hat{\boldsymbol{u}}_i$ for $i\neq j$ together with (15.17) to deduce the following Lamé (1840) relations (with i,j,k distinct here),

$$\partial_i\left(\frac{\partial_i h_j}{h_i}\right)+\partial_j\left(\frac{\partial_j h_i}{h_j}\right)=-\frac{\partial_k h_i}{h_k}\frac{\partial_k h_j}{h_k}, \qquad (15.74)$$

and

$$\partial_j\left(\frac{\partial_k h_i}{h_k}\right)=\frac{\partial_j h_i}{h_j}\frac{\partial_k h_j}{h_k}. \qquad (15.75)$$

The latter can be written

$$\partial_j\partial_k h_i=\frac{1}{h_j}\partial_j h_i\,\partial_k h_j+\frac{1}{h_k}\partial_k h_i\,\partial_j h_k. \qquad (15.76)$$

(15.14) Orthogonal coordinates have $\partial_i\boldsymbol{r}\cdot\partial_j\boldsymbol{r}=0$ when $i\neq j$. Derive an identity from $\partial_k(\partial_i\boldsymbol{r}\cdot\partial_j\boldsymbol{r})=0$. Rewrite that identity for $(i,j,k)\to(j,k,i)\to(k,i,j)$, for distinct indices ($\epsilon_{ijk}\neq 0$), add two of those and subtract the third to show that

$$\Gamma_{ijk}=\partial_i\boldsymbol{r}\cdot\partial_j\partial_k\boldsymbol{r}=0, \qquad (\epsilon_{ijk}\neq 0)$$

which can also be obtained from (15.72). Conclude that coordinate tangents are principal directions for the coordinate isosurfaces; that is, *orthogonal coordinate curves are lines of curvature* for the coordinate surfaces.[3] For example, $\partial_1\boldsymbol{r}$ and $\partial_2\boldsymbol{r}$ are principal directions at any point on a u^3-isosurface whose unit normal is $\hat{\boldsymbol{n}}=\hat{\boldsymbol{u}}_3=H_3\partial_3\boldsymbol{r}$, since

$$b_{12}=\hat{\boldsymbol{u}}_3\cdot\partial_1\partial_2\boldsymbol{r}=0,\quad g_{12}=\partial_1\boldsymbol{r}\cdot\partial_2\boldsymbol{r}=0,$$

so $[b_{ij}]$ and $[g_{ij}]$ are diagonal for u^3-isosurfaces parametrized by (u^1,u^2).

[3] *Lines of curvature* are curves that are everywhere tangent to the principal curvature directions.

(15.15) For *conformal* coordinates with $\partial_i \boldsymbol{r} \cdot \partial_j \boldsymbol{r} = h^2 \delta_{ij}$, use $\partial_i(\partial_j \boldsymbol{r} \cdot \partial_j \boldsymbol{r}) = \partial_i(h^2)$ (*no sum*), and $\partial_j(\partial_i \boldsymbol{r} \cdot \partial_j \boldsymbol{r}) = 0$ for $i \neq j$, to show that

$$\Gamma_{ijj} = \partial_i \boldsymbol{r} \cdot \partial_j \partial_j \boldsymbol{r} = -h \partial_i h, \qquad (i \neq j)$$

which can also be obtained from (15.72). Conclude that $b_{11} = b_{22} = -\partial_3 h$ on u^3-isosurfaces, together with $b_{12} = 0 = g_{12}$ as shown in the previous problem, and $g_{11} = g_{22} = h^2$ by definition of conformal. So

$$\kappa_1 = b_{11}/g_{11} = \partial_3(h^{-1}) = b_{22}/g_{22} = \kappa_2$$

and $\kappa = \partial_3(h^{-1})$ in all directions at all points on u^3-isosurfaces, and similarly for u^1 and u^2 isosurfaces. Thus, all points are *umbilics* (14.73), and *conformal coordinate surfaces must be planes or spheres.* This is Liouville's theorem, and a major limitation of conformal coordinates in dimension $n \geq 3$. For example, there cannot be 3D conformal coordinates for an oblate spheroid, or for a torus.

(15.16) Consider the inversion map

$$\boldsymbol{r}(u,v,w) = \frac{u\,\hat{\mathbf{x}} + v\,\hat{\mathbf{y}} + w\,\hat{\mathbf{z}}}{u^2+v^2+w^2}. \qquad (15.77a)$$

As a map from $\mathbf{u} = (u,v,w) \in \mathbb{R}^3$ to $\mathbf{x} = (x,y,z) \in \mathbb{R}^3$, this is

$$\mathbf{x} = \frac{\mathbf{u}}{\mathbf{u}^\mathsf{T}\mathbf{u}} \Leftrightarrow \mathbf{u} = \frac{\mathbf{x}}{\mathbf{x}^\mathsf{T}\mathbf{x}},$$

and $x^2+y^2+z^2 = (u^2+v^2+w^2)^{-1}$. Show that

$$\begin{cases} \partial_u \boldsymbol{r} \cdot \partial_v \boldsymbol{r} = \partial_v \boldsymbol{r} \cdot \partial_w \boldsymbol{r} = \partial_w \boldsymbol{r} \cdot \partial_u \boldsymbol{r} = 0, \\ \partial_u \boldsymbol{r} \cdot \partial_u \boldsymbol{r} = \partial_v \boldsymbol{r} \cdot \partial_v \boldsymbol{r} = \partial_w \boldsymbol{r} \cdot \partial_w \boldsymbol{r} \\ \qquad = \dfrac{1}{(u^2+v^2+w^2)^2}. \end{cases} \qquad (15.77b)$$

Hence, this map is conformal. Show that the coordinate isosurfaces are spheres and that the inside of the unit sphere is mapped to the outside of the unit sphere, and vice versa. Show that spheres through the origin are mapped to planes, for example the sphere

$$x^2+y^2+(z-a)^2 = a^2 \longleftrightarrow w = (2a)^{-1},$$

in which case (u,v) are *stereographic coordinates* for those spheres (Fig. 15.8).

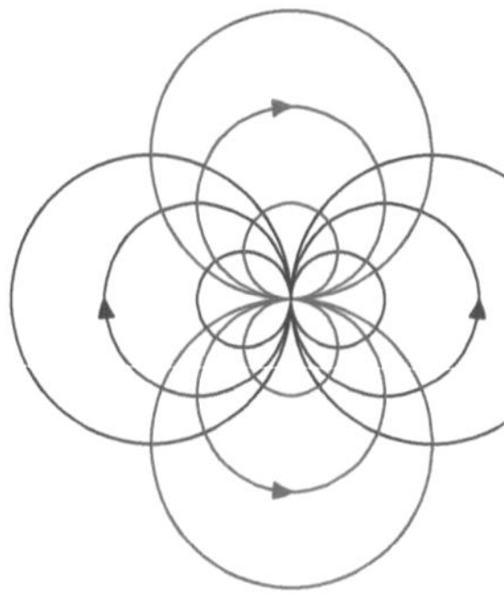

Fig. 15.8 (15.77a) $\boldsymbol{r}(u,v_0,0)$ (red), $\boldsymbol{r}(u_0,v,0)$ (blue).

(15.17) *Liouville's theorem on conformal coordinates.* For conformal coordinates with $\partial_i \boldsymbol{r} \cdot \partial_j \boldsymbol{r} = h^2 \delta_{ij}$, (15.17b) becomes, with $L = \ln h = -\ln H$,

$$\begin{aligned} \partial_i \hat{\mathbf{u}}_j &= \hat{\mathbf{u}}_i \partial_j L \\ &\quad - \delta_{ij}\left(\hat{\mathbf{u}}_1 \partial_1 L + \hat{\mathbf{u}}_2 \partial_2 L + \hat{\mathbf{u}}_3 \partial_3 L\right), \end{aligned} \qquad (15.78a)$$

that is,

$$\begin{aligned} \partial_i \hat{\mathbf{u}}_i &= -\hat{\mathbf{u}}_j \partial_j L - \hat{\mathbf{u}}_k \partial_k L, \\ \partial_i \hat{\mathbf{u}}_j &= \hat{\mathbf{u}}_i \partial_j L, \end{aligned} \qquad (15.78b)$$

for distinct i,j,k. Show that $\partial_j \partial_i \hat{\mathbf{u}}_j = \partial_i \partial_j \hat{\mathbf{u}}_j$ implies that (cf. exercise 15.13)

$$\begin{aligned} (\partial_i\partial_i + \partial_j\partial_j)L + (\partial_k L)(\partial_k L) &= 0, \\ \partial_i \partial_k L - (\partial_i L)(\partial_k L) &= 0, \end{aligned} \qquad (15.79)$$

and in terms of H these equations read (*no sums*)

$$\begin{aligned} (\partial_i\partial_i + \partial_j\partial_j)H &= \frac{(\partial_i H)^2 + (\partial_j H)^2 + (\partial_k H)^2}{H}, \\ \partial_i \partial_k H &= 0. \end{aligned}$$

The second, $\partial_i \partial_k H = 0$, was obtained in a somewhat different manner in exercise 15.15. It implies that $H = f_1(u_1) + f_2(u_2) + f_3(u_3)$, where f_k is a function of conformal coordinate u_k only. Substituting this form in the first H equation yields, for $(i,j) = (1,2)$,

$$f_1'' + f_2'' = \frac{(f_1')^2 + (f_2')^2 + (f_3')^2}{f_1 + f_2 + f_3}$$

and two similar equations for $(i,j) = (2,3)$, and $(3,1)$, where the prime $()'$ here denotes derivative with respect to the function argument. The right-hand side (RHS) is identical for all three equations. The first equation for $f_1'' + f_2''$ that is independent of u_3 implies that the RHS is independent of u_3. The other two equations, for $f_2'' + f_3''$ and $f_3'' + f_1''$, in turn show that the RHS is independent of u_1 and u_2, so it is constant.

That implies that $f_1'' = f_2'' = f_3'' = C$ constant. Integrating and ensuring that the RHS $= 2C$ yields

$$\frac{1}{h} = H = \tfrac{1}{2}C(u_1^2 + u_2^2 + u_3^2) \quad \text{or} \quad H = D,$$

where D is another constant, and $C = 0$ when $H = D$. So the only nontrivial conformal map in 3D is the inversion map (15.77b).

This is Liouville's original proof published in his annotated edition of Monge's book, accelerated here with index notation.[4]

(15.18) The equations for $L = \ln h$ in the Liouville proof can be obtained more generally from the vanishing of the Ricci curvature tensor in euclidean space. The Ricci tensor R_{ij} is the contraction of the Riemann curvature tensor R^l_{ijk} (14.126) over the second or third index (equal up to a sign change)

$$\begin{aligned} R_{ij} &\equiv R^k_{ijk} \\ &= \partial_j \Gamma^k_{ik} - \partial_k \Gamma^k_{ij} + \Gamma^m_{ik}\Gamma^k_{jm} - \Gamma^m_{ij}\Gamma^k_{km}. \end{aligned}$$

For conformal coordinates, $g_{ij} = h^2 \delta_{ij}$ and

$$\begin{aligned} \Gamma^k_{ij} = g^{kl}\Gamma_{lij} &= \frac{\delta^{kl}}{h^2}\Gamma_{lij} \\ &= \delta^k_i \partial_j L + \delta^k_j \partial_i L - \delta_{ij}\delta^{kl}\partial_l L. \end{aligned}$$

Then in $\mathbb{R}^N$ with $\delta^k_k = N$, show that $\Gamma^k_{ik} = N\partial_i L$ and

$$\begin{aligned} R_{ij} = (N-2)&\left(\partial_i \partial_j L - \partial_i L \partial_j L\right) \\ &+ \delta_{ij}\delta^{kl}\left(\partial_k \partial_l L + (N-2)\partial_k L \partial_l L\right). \end{aligned}$$

For euclidean space $R_{ij} = 0$. Show that these reduce to (15.79) for $N = 3$.

(15.19) Integrate the nonlinear coupled ODEs (15.35) starting from $(r_0, \theta_0, \varphi_0)$.

(15.20) A curve $\boldsymbol{r}(t)$ is given in terms of spherical coordinates as $(r(t), \theta(t), \varphi(t))$ for $t = t_0 \to t_f$. Derive an explicit expression for the length of that curve as a t-integral in terms of the functions $r(t)$, $\theta(t)$, $\varphi(t)$.

(15.21) A curve $\boldsymbol{r}(t)$ is given in terms of (u, v, w) coordinates as $(u(t), v(t), w(t))$ for $t = t_0 \to t_f$. Derive an explicit expression for the length of that curve as a t-integral in terms of $(u(t), v(t), w(t))$.

(15.22) Show that $\bar{g}^{ik}\bar{g}_{kj} = \delta^i_j$ for the transformed metrics defined in (15.45) and (15.47).

(15.23) Let $\bar{p}^i = \bar{g}^{ij}\bar{p}_j$ and $p^k = g^{kl}p_l$. Show (15.55) from (15.47) and (15.54).

(15.24) Calculate the area of the ellipse $x^2/a^2 + y^2/b^2 = 1$ and the volume of the ellipsoid $x^2/a^2 + y^2/b^2 + z^2/c^2 = 1$ by mapping them to a disk and a sphere, respectively.

(15.25) Consider the coordinates $(r, \varphi) \to (x, y) = (r\cos\varphi, r\sin\varphi)$. Calculate the Jacobian matrix and determinant of that transformation, and of the inverse transformation $(x, y) \to (r, \varphi)$. Express the latter in terms of (x, y). Briefly justify your work.

(15.26) Consider the coordinates $(x, y, z) = (r\sin\theta\cos\varphi, r\sin\theta\sin\varphi, r\cos\theta)$. Calculate the Jacobian matrix and determinant of that transformation, and of the inverse transformation $(x, y, z) \to (r, \theta, \varphi)$. Express the latter in terms of (x, y, z).

(15.27) Verify (15.67) using a direct calculation in the (P, V) domain. Show your work.

(15.28) Calculate the (x, y) area such that $0 < \alpha_1 \le xy \le \alpha_2$ and $0 < \beta_1 x \le y \le \beta_2 x$. Sketch the domain and select suitable coordinates.

(15.29) Calculate the (x, y) area such that $0 < \alpha_1 \le xy \le \alpha_2$ and $0 < \beta_1 \le xy^2 \le \beta_2$. Sketch the domain and select suitable coordinates.

(15.30) Calculate the (x, y) area such that

$$\begin{aligned} &0 < \alpha_1 \le xy \le \alpha_2, \\ &0 < \beta_1 \le x^2 - y^2 \le \beta_2. \end{aligned}$$

Sketch the domain and select suitable coordinates.

(15.31) Calculate the area between the curves with

$$x^2 + y^2 = 2\alpha_1 x,\ 2\alpha_2 x,\ 2\beta_1 y,\ 2\beta_2 y,$$

for positive α_1, α_2, β_1, β_2. Sketch.

(15.32) Calculate the integral

$$\int_{-\infty}^{\infty}\int_{-\infty}^{\infty} e^{-(x^2+y^2)} \mathrm{d}x\mathrm{d}y$$

using polar coordinates. Deduce the value of the gaussian integral $\int_{-\infty}^{\infty} e^{-x^2}\mathrm{d}x = \sqrt{\pi}$.

(15.33) Calculate

$$\int_{-\infty}^{\infty}\int_{-\infty}^{\infty} (a^2 + x^2 + y^2)^{\alpha} \mathrm{d}x\mathrm{d}y,$$

for $a \neq 0$ and α real, using suitable coordinates. For what values of α does the integral not exist?

[4] Joseph Liouville, "Extension au cas des trois dimensions de la question du tracé géographique," Note VI, in Gaspard Monge, *Application de l'analyse à la géométrie*, 5eme Edition, Revue, Corrigée et Annotée par M. Liouville, 1850.

Fields

16

In physics, a *field* is a function of position P in physical space. A field can be a *scalar* field, a *vector* field, or a *tensor* field.

16.1 Scalar fields

A *scalar field* is a scalar function of position P in physical space such as *temperature* $T(P)$ or *pressure* $p(P)$ at some point P in a gas. A scalar field is denoted

$$f(P), f(\boldsymbol{r}), f(x, y, z), f(r, \theta, \varphi), \ldots,$$

where the meaning of the function $f(\cdot)$ varies with how position P is specified. For example, the distance from P to the origin O is the scalar field

$$|\overrightarrow{OP}| = |\boldsymbol{r}| = r = \sqrt{\rho^2 + z^2} = \sqrt{x^2 + y^2 + z^2},$$

in spherical, cylindrical, and cartesian coordinates, respectively, where $\boldsymbol{r} = \overrightarrow{OP}$ is the position vector of P with respect to a chosen origin O (§7.1).

A standard way to visualize scalar fields is through plots of its level sets, or *contours*. A *level set* of $f(P)$ is the set of all points P for which the field has a given value, f_0 say,

$$f(P) = f_0.$$

In 3D, the level set $f(P) = f_0$ is the implicit equation of a surface, an *isosurface* of $f(P)$. For example, $|\boldsymbol{r}| = r_0$ is a sphere of radius r_0 centered at the origin O. If P is restricted to a surface $\mathcal{S}$ with implicit equation $g(P) = 0$, then the level set $f(P) = f_0$ with $g(P) = 0$ is the implicit equation of a curve on that surface, an *isocurve* of $f(P)$ on $\mathcal{S}$.

Examples of scalar fields

(1) The scalar field

$$f(P) = \boldsymbol{a} \cdot \boldsymbol{r} = |\boldsymbol{a}| s = a_x x + a_y y + a_z z,$$

where $s = \hat{\boldsymbol{a}} \cdot \boldsymbol{r}$ is the distance from O in direction $\hat{\boldsymbol{a}}$, has level sets $\boldsymbol{a} \cdot \boldsymbol{r} = f_0$, which are *planes* perpendicular to $\boldsymbol{a}$ in 3D, and *lines* perpendicular to $\boldsymbol{a}$ in 2D (Fig. 16.1);

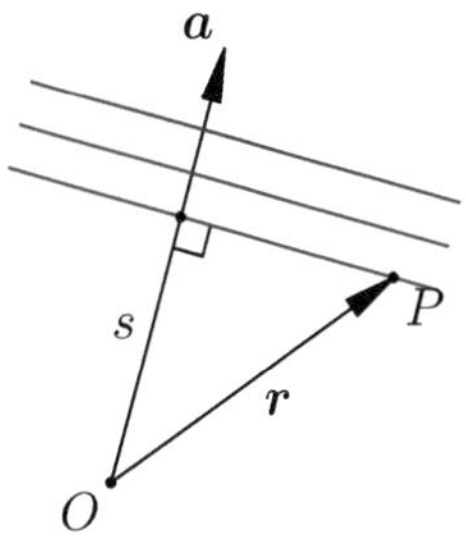

Fig. 16.1 Contours of $f(P) = \boldsymbol{a} \cdot \boldsymbol{r}$.

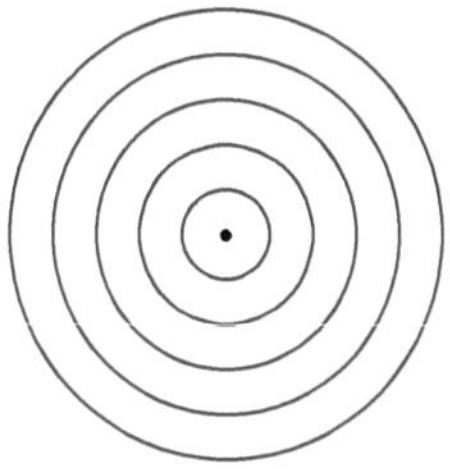

Fig. 16.2 Level sets of $|AP| = |\boldsymbol{r} - \boldsymbol{r}_A|$.

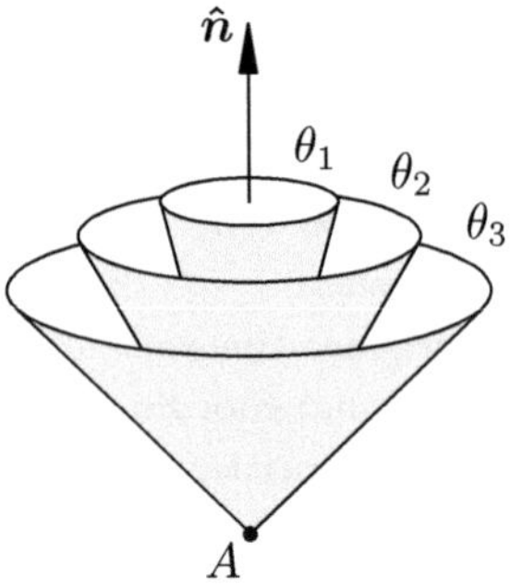

Fig. 16.3 θ isosurfaces.

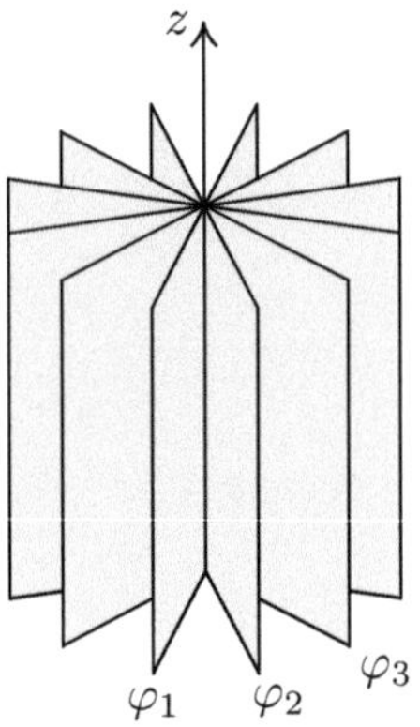

Fig. 16.4 φ isosurfaces.

(2) The level-sets of distance to point A,

$$|AP| = |\boldsymbol{r} - \boldsymbol{r}_A| = \sqrt{(x - x_A)^2 + (y - y_A)^2 + (z - z_A)^2},$$

are *spheres* centered at A in 3D, and *circles* centered at A in 2D (Fig. 16.2);

(3) The level sets of $\theta(P) = \arccos(\hat{\boldsymbol{n}} \cdot \overrightarrow{AP})$, the angle between $\hat{\boldsymbol{n}}$ and $\overrightarrow{AP}$, are *cones* with apex at A and axis $\hat{\boldsymbol{n}}$ in 3D, and *radials* from A in 2D (Fig. 16.3);

(4) The level sets of $\varphi(P)$, the angle between the planes $(\hat{\mathbf{z}}, \hat{\mathbf{x}})$ and $(\hat{\mathbf{z}}, \boldsymbol{r})$, are *half-planes* starting at the z-axis in 3D, and *radials* from the origin in 2D (Fig. 16.4);

(5) The level sets of $f(P) = \min(d(P, A), d(P, B))$, the distance from P to the nearest of two points A or B, are spheres or partial spheres (Fig. 16.5); and

(6) The level sets of $f(P) = d(P, AB)$, the distance to a line segment AB, are cylinders with axis AB capped by hemispheres in 3D, and parallel lines equidistant from AB capped by semi-circles in 2D, known as the *stadium shape*, Fig. 16.5.

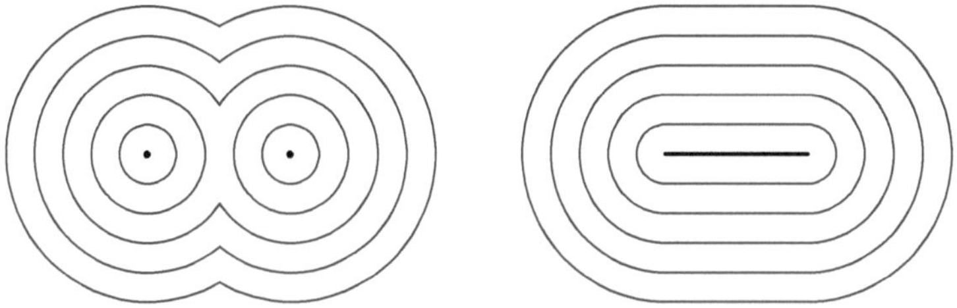

Fig. 16.5 Contours of distance to nearest of two points, and to a line segment.

Daily weather reports provide visualizations of the temperature and pressure fields at the Earth's surface. Contours of equal pressure are called *isobars*. Contours of equal temperature are called *isotherms*. Figure 16.6) shows atmospheric pressure contours over the North Atlantic on October 30, 2012. The tight almost circular contours on the center left, with a box indicating "STORM," is hurricane Sandy hitting the New York City area. The letter **L** indicates the approximate center of a low pressure area. There is another low pressure area, a *cyclone*, on the top right between Scotland and Iceland, and another lower center right over the Azores islands, west of Portugal and Morocco. The bold letter **H** indicate the approximate center of high pressure areas, *anticyclones*. Norway is on the top right of the figure, and Wisconsin is center left, just left of Lake Michigan and below Lake Superior.

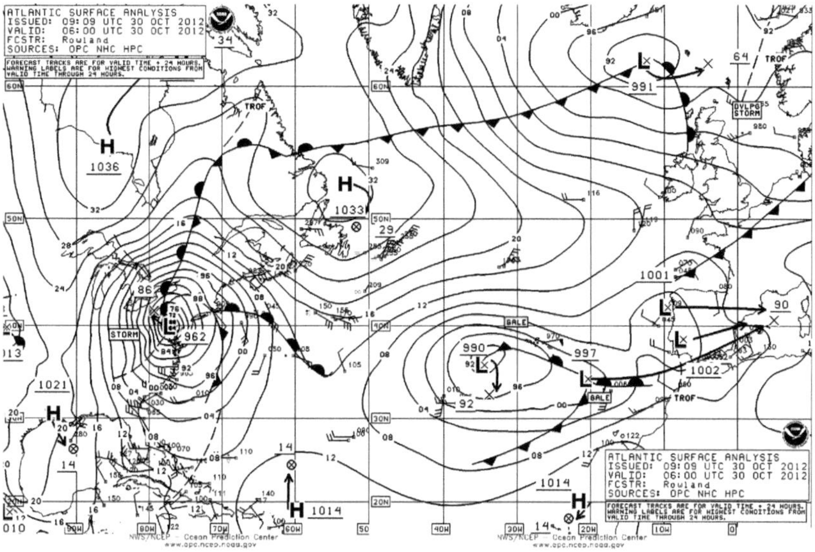

Fig. 16.6 North Atlantic surface pressure contours. Hurricane Sandy (left), Oct 30, 2012.

16.2 Vector fields

A *vector field* $\boldsymbol{F}(P)$ is a vector function of position P such as the wind velocity $\boldsymbol{V}(P)$, an electric field $\boldsymbol{E}(P)$, or a magnetic field $\boldsymbol{B}(P)$ at point P in physical space. A vector field is denoted

$$\boldsymbol{F}(P), \boldsymbol{F}(\boldsymbol{r}), \boldsymbol{F}(x, y, z), \boldsymbol{F}(r, \theta, \varphi), \ldots,$$

where the meaning of the vector function $\boldsymbol{F}(\cdot)$ varies with how point P is specified. For example,

(1) The vector field

$$\boldsymbol{F}(P) = \overrightarrow{OP} = \boldsymbol{r} = r\hat{\boldsymbol{r}} = \rho\hat{\boldsymbol{\rho}} + z\hat{\mathbf{z}} = x\hat{\mathbf{x}} + y\hat{\mathbf{y}} + z\hat{\mathbf{z}},$$

is the distance and direction from the origin, in vector, spherical, cylindrical and cartesian representations. This simple $\boldsymbol{F}(\boldsymbol{r}) = \boldsymbol{r}$ is the vector field equivalent of the function $f(x) = x$;

(2) The vector field

$$\boldsymbol{F}(P) = \frac{\overrightarrow{AP}}{|AP|} = \frac{\boldsymbol{r} - \boldsymbol{r}_A}{|\boldsymbol{r} - \boldsymbol{r}_A|} \tag{16.1}$$

is the (unit) direction vector pointing away from A (Fig. 16.7);

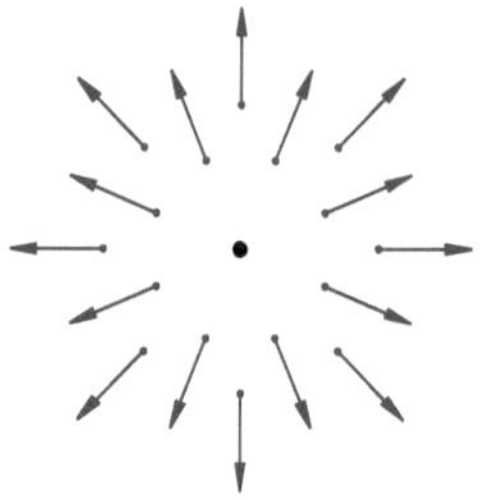

Fig. 16.7 Unit radial vector field (16.1).

(3) The *inverse square law* is the field

$$\boldsymbol{F}(P) = \frac{\hat{\boldsymbol{r}}}{r^2} = \frac{\rho\hat{\boldsymbol{\rho}} + z\hat{\mathbf{z}}}{(\rho^2 + z^2)^{3/2}} = \frac{x\hat{\mathbf{x}} + y\hat{\mathbf{y}} + z\hat{\mathbf{z}}}{(x^2 + y^2 + z^2)^{3/2}}, \tag{16.2}$$

in spherical, cylindrical, and cartesian coordinates. This is the electric field induced by a point charge, or the gravity field of a point mass, or the velocity field of a point source of fluid, up to physical unit factors (Fig. 16.8);

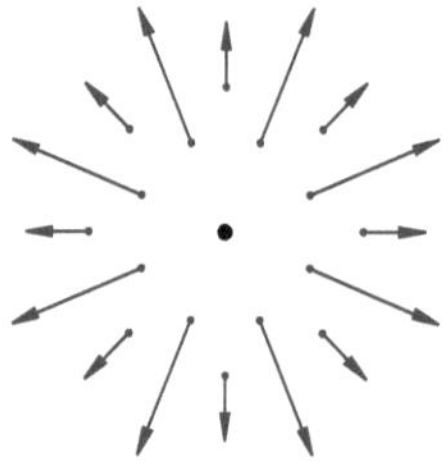

Fig. 16.8 Inverse square law (16.2).

(4) The *line vortex* field

$$\boldsymbol{F}(P) = \frac{\hat{\mathbf{z}} \times \boldsymbol{r}}{|\hat{\mathbf{z}} \times \boldsymbol{r}|^2} = \frac{\hat{\boldsymbol{\varphi}}}{\rho} = \frac{x\,\hat{\mathbf{y}} - y\,\hat{\mathbf{x}}}{x^2 + y^2} \tag{16.3}$$

is the velocity field induced by a line vortex in fluid dynamics, or the magnetic field induced by a line current in electromagnetism, up to physical unit factors (Fig. 16.9); and

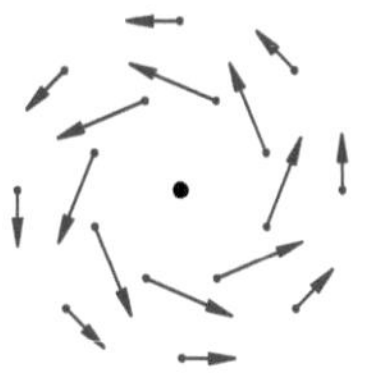

Fig. 16.9 Line vortex $\hat{\boldsymbol{\varphi}}/\rho$ (16.3).

(5) The vector field

$$\boldsymbol{F}(P) = \frac{\hat{\mathbf{z}} \times \boldsymbol{s}_1}{|\hat{\mathbf{z}} \times \boldsymbol{s}_1|^2} - \frac{\hat{\mathbf{z}} \times \boldsymbol{s}_2}{|\hat{\mathbf{z}} \times \boldsymbol{s}_2|^2}, \tag{16.4}$$

with $\boldsymbol{s}_1 = \overrightarrow{F_1P} = \boldsymbol{r} - \boldsymbol{r}_1$, and $\boldsymbol{s}_2 = \overrightarrow{F_2P} = \boldsymbol{r} - \boldsymbol{r}_2$, is a vortex dipole in fluids, or the magnetic field induced by two parallel and opposite line currents in electromagnetism. Its quiver plot is shown in Fig. 16.10.

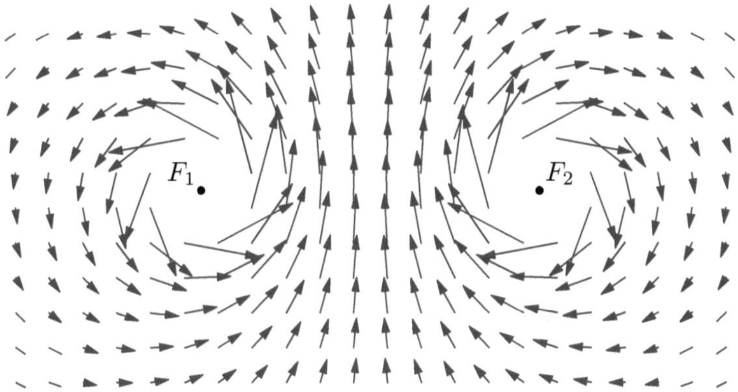

Fig. 16.10 Vortex dipole (16.4) quiver plot.

Field lines

One way to visualize vector fields is to display the actual vector $\boldsymbol{F}(\boldsymbol{r})$ at point $\boldsymbol{r}$ as in Fig. 16.10, but such *quiver plots* quickly become cluttered with arrows overlapping each other. Another way is to plot field lines. A *field line* is a curve $\mathcal{C}$ that is everywhere tangent to the vector field $\boldsymbol{F}(\boldsymbol{r})$. Thus, if $\boldsymbol{r}(t)$ is a parametrization of a field line, then

$$\frac{\mathrm{d}\boldsymbol{r}}{\mathrm{d}t} = \boldsymbol{F}(\boldsymbol{r}(t)), \tag{16.5}$$

since $\mathrm{d}\boldsymbol{r}/\mathrm{d}t$ is a vector tangent to the curve at $\boldsymbol{r}(t)$. For a given vector field $\boldsymbol{F}(\boldsymbol{r})$, this is a vector differential equation for $\boldsymbol{r}(t)$ with t an arbitrary time-like parameter. There is one such field line for every starting point $\boldsymbol{r}_0$. For example,

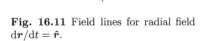

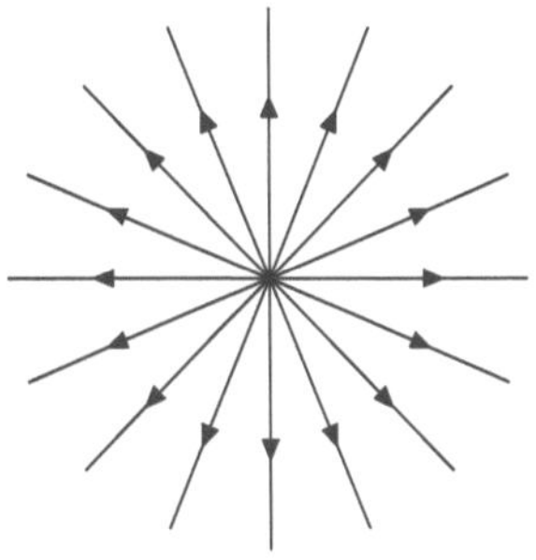

Fig. 16.11 Field lines for radial field $\mathrm{d}\boldsymbol{r}/\mathrm{d}t = \hat{\boldsymbol{r}}$.

(1) For $\boldsymbol{F}(\boldsymbol{r}) = \hat{\boldsymbol{r}}$, the field lines are the solutions of

$$\frac{\mathrm{d}\boldsymbol{r}}{\mathrm{d}t} = \frac{\mathrm{d}r}{\mathrm{d}t}\hat{\boldsymbol{r}} + r\frac{\mathrm{d}\hat{\boldsymbol{r}}}{\mathrm{d}t} = \hat{\boldsymbol{r}} \;\Rightarrow\; \frac{\mathrm{d}r}{\mathrm{d}t} = 1,\; \frac{\mathrm{d}\hat{\boldsymbol{r}}}{\mathrm{d}t} = 0,$$

since $\hat{\boldsymbol{r}} \cdot \mathrm{d}\hat{\boldsymbol{r}}/\mathrm{d}t = 0$. Thus, $\hat{\boldsymbol{r}} = \hat{\boldsymbol{r}}_0$ is constant and $\boldsymbol{r}(t) = \boldsymbol{r}_0 + t\,\hat{\boldsymbol{r}}_0$, corresponding to lines radiating away from the origin (Fig. 16.11);

(2) For the vector field $\boldsymbol{F}(\boldsymbol{r}) = \boldsymbol{\omega} \times \boldsymbol{r}$ with $\boldsymbol{\omega} = \omega\hat{\boldsymbol{\omega}}$ independent of $\boldsymbol{r}$, the field lines obey

$$\frac{\mathrm{d}\boldsymbol{r}}{\mathrm{d}t} = \boldsymbol{\omega} \times \boldsymbol{r}$$

whose solutions are (10.38)

$$\boldsymbol{r}(t) = \boldsymbol{r}_0^{\|} + \cos(\omega t)\,((\hat{\boldsymbol{\omega}} \times \boldsymbol{r}_0) \times \hat{\boldsymbol{\omega}}) + \sin(\omega t)\,(\hat{\boldsymbol{\omega}} \times \boldsymbol{r}_0)$$

and correspond to circles perpendicular to the axis $(O, \boldsymbol{\omega})$ and centered on that axis. This vector field corresponds to *solid body rotation* (Fig. 16.12); and

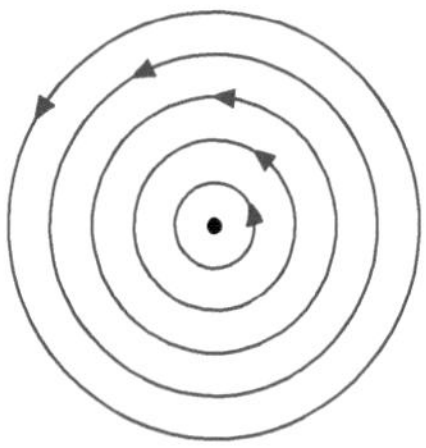

Fig. 16.12 Field lines for solid body rotation $\mathrm{d}\boldsymbol{r}/\mathrm{d}t = \boldsymbol{\omega} \times \boldsymbol{r}$.

(3) The field lines of the vortex dipole field (16.4) are shown in Fig. 16.13, for $\hat{\mathbf{z}}$ pointing out of the page.

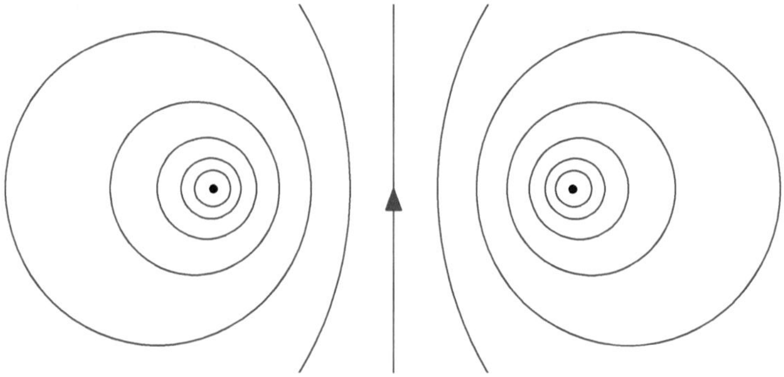

Fig. 16.13 Vortex dipole (16.4) field lines.

16.3 Gradient

A scalar field $f(P)$ is differentiable at point P in euclidean space if it has a linear approximation in a neighborhood at that point, that is, if there exists a vector field $\boldsymbol{V}(P)$ such that

$$f(Q) = f(P) + \boldsymbol{V}(P) \cdot \overrightarrow{PQ} + o(|PQ|), \tag{16.6a}$$

for any point Q close to P. The *little oh* notation $o(|PQ|)$ means an expression that goes to zero faster than the distance $|PQ|$, such that (16.6a) is equivalent to the limit statement

$$\lim_{|PQ| \to 0} \frac{f(Q) - f(P) - \boldsymbol{V}(P) \cdot \overrightarrow{PQ}}{|PQ|} = 0. \tag{16.6b}$$

The vector field $\boldsymbol{V}(P)$, if it exists, is the *gradient* of $f(P)$ at P, written

$$\boldsymbol{V}(P) = \boldsymbol{\nabla} f(P),$$

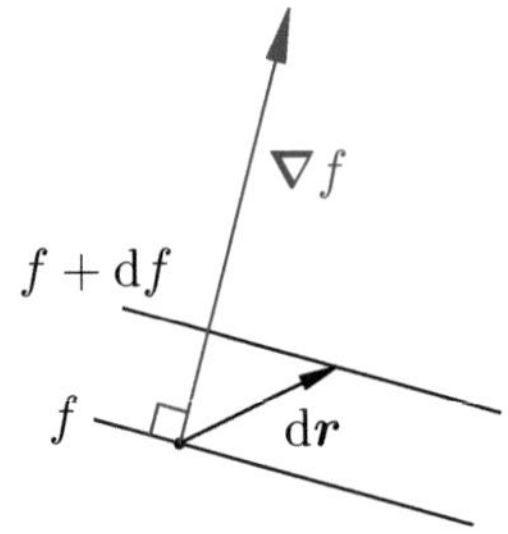

Fig. 16.14 $\mathrm{d}f = \mathrm{d}\boldsymbol{r} \cdot \boldsymbol{\nabla} f$.

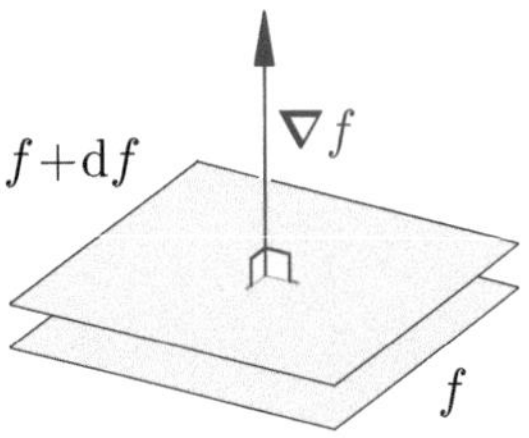

Fig. 16.15 $\boldsymbol{\nabla} f$ is $\perp$ to level sets.

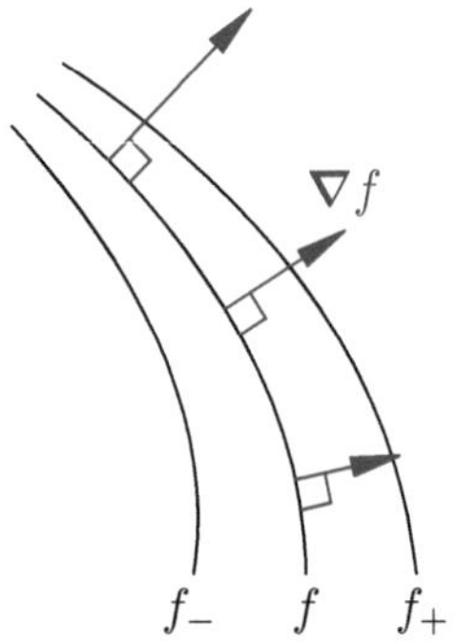

Fig. 16.16 Level sets and $\boldsymbol{\nabla} f$.

and read as "grad f" or "del f." The symbol $\boldsymbol{\nabla}$ is called the *del operator*. Definition (16.6a) can be written in differential form as (Fig. 16.14)

$$\mathrm{d}f = \mathrm{d}\boldsymbol{r} \cdot \boldsymbol{\nabla} f, \tag{16.6c}$$

where $\mathrm{d}f = f(P + \mathrm{d}\boldsymbol{r}) - f(P)$ is the change in f resulting from the arbitrary differential step $\mathrm{d}\boldsymbol{r}$ from P and $\boldsymbol{\nabla} f(P)$ is independent of $\mathrm{d}\boldsymbol{r}$. The gradient $\boldsymbol{\nabla} f(P)$ determines the rate of change of f in *any* direction from point P.

If the limit $Q \to P$ is taken for a fixed but arbitrary direction $\hat{\boldsymbol{s}}$, then (16.6b) with $\overrightarrow{PQ} = \hat{\boldsymbol{s}}\,\Delta s$, or equivalently (16.6c) with $\mathrm{d}\boldsymbol{r} = \hat{\boldsymbol{s}}\,\mathrm{d}s$, gives the *directional derivative* of $f(\boldsymbol{r})$ as

$$\frac{\partial f}{\partial s} = \hat{\boldsymbol{s}} \cdot \boldsymbol{\nabla} f, \tag{16.7}$$

where

$$\frac{\partial f}{\partial s} = \lim_{\Delta s \to 0} \frac{f(P + \hat{\boldsymbol{s}}\,\Delta s) - f(P)}{\Delta s} \equiv \frac{f(P + \hat{\boldsymbol{s}}\,\mathrm{d}s) - f(P)}{\mathrm{d}s}$$

with direction $\hat{\boldsymbol{s}}$ fixed, hence the partial derivative notation $\partial f/\partial s$.

More generally, for $f(\boldsymbol{r}(t))$, (16.6c) yields

$$\frac{\mathrm{d}f}{\mathrm{d}t} = \frac{\mathrm{d}\boldsymbol{r}}{\mathrm{d}t} \cdot \boldsymbol{\nabla} f, \tag{16.8}$$

and, for $f(\boldsymbol{r}(u, v, w))$, (16.6c) or (16.7) gives

$$\frac{\partial f}{\partial u} = \frac{\partial \boldsymbol{r}}{\partial u} \cdot \boldsymbol{\nabla} f, \tag{16.9}$$

where

$$\frac{\partial f}{\partial u} \equiv \frac{f(\boldsymbol{r}(u + \mathrm{d}u, v, \ldots)) - f(\boldsymbol{r}(u, v, \ldots))}{\mathrm{d}u},$$

and similarly for $\partial f/\partial v$ and $\partial f/\partial w$.

Geometric characterization

It follows from (16.6) that $\boldsymbol{\nabla} f$ is perpendicular to the level set of f at $\boldsymbol{r}$, since $\mathrm{d}f = 0$ for any $\mathrm{d}\boldsymbol{r}$ such that $\mathrm{d}\boldsymbol{r} \cdot \boldsymbol{\nabla} f = 0$ (Fig. 16.15), and the magnitude of $\boldsymbol{\nabla} f$ is the rate of change of f with respect to distance in that $\boldsymbol{\nabla} f$ direction, which is the direction of greatest increase of f at $\boldsymbol{r}$ (Fig. 16.16). This is confirmed by the directional derivative (16.7) that shows that $\partial f/\partial s$ is largest when $\hat{\boldsymbol{s}}$ is in the direction of $\boldsymbol{\nabla} f$.

To encapsulate that geometric characterization of $\boldsymbol{\nabla} f$, let $\hat{\boldsymbol{S}}$ be the direction of *steepest* (or greatest) increase of f at $\boldsymbol{r}$, and s the distance in that direction, then

$$\boldsymbol{\nabla} f = \hat{\boldsymbol{S}}\,\frac{\partial f}{\partial s}, \tag{16.10}$$

where this $\partial f/\partial s$ is the rate of change in the steepest increase direction $\hat{\boldsymbol{S}}$, not to be confused with the rate of change in an arbitrary direction $\hat{\boldsymbol{s}}$ in (16.7).

In summary, ∇f, the gradient of a scalar field $f(P)$, is a *vector* pointing in the direction of *greatest increase of* f, whose magnitude is the rate of change of f per distance in that direction. It is perpendicular to level sets of f and its magnitude is inversely proportional to the distance between level sets. The gradient is a vector field $\boldsymbol{F}(P) = \nabla f$ derived from the scalar field $f(P)$.

Fundamental gradients

(1) The gradient of $f(\boldsymbol{r}) = \boldsymbol{a} \cdot \boldsymbol{r}$, with $\boldsymbol{a}$ constant, is (Fig. 16.17)

$$\nabla(\boldsymbol{a} \cdot \boldsymbol{r}) = \boldsymbol{a}. \tag{16.11}$$

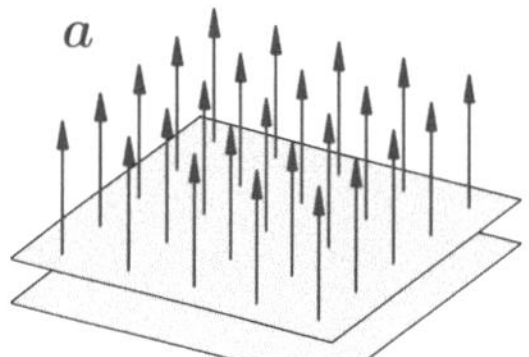

Fig. 16.17 $\nabla(\boldsymbol{a} \cdot \boldsymbol{r}) = \boldsymbol{a}$.

This is immediate from (16.6a) since $\boldsymbol{a} \cdot \boldsymbol{r}' = \boldsymbol{a} \cdot \boldsymbol{r} + \boldsymbol{a} \cdot (\boldsymbol{r}' - \boldsymbol{r})$. Equation (16.6a) defines a local linear approximation, but this $f(\boldsymbol{r}) = \boldsymbol{a} \cdot \boldsymbol{r}$ is already linear. Geometrically, $\boldsymbol{a} \cdot \boldsymbol{r} = |\boldsymbol{a}| s$ where $s = \hat{\boldsymbol{a}} \cdot \boldsymbol{r}$ is the (signed) distance from the origin in the direction $\hat{\boldsymbol{a}}$, and that is the direction of steepest increase. Thus, (16.10) gives

$$\nabla(\boldsymbol{a} \cdot \boldsymbol{r}) = \hat{\boldsymbol{a}} \frac{\partial(|\boldsymbol{a}|s)}{\partial s} = \hat{\boldsymbol{a}}|\boldsymbol{a}| = \boldsymbol{a}.$$

In particular,

$$\begin{aligned} \nabla(\hat{\mathbf{x}} \cdot \boldsymbol{r}) &= \nabla x = \hat{\mathbf{x}}, \\ \nabla(\hat{\mathbf{y}} \cdot \boldsymbol{r}) &= \nabla y = \hat{\mathbf{y}}, \\ \nabla(\hat{\mathbf{z}} \cdot \boldsymbol{r}) &= \nabla z = \hat{\mathbf{z}}. \end{aligned} \tag{16.12}$$

(2) The gradient of $f(\boldsymbol{r}) = |\boldsymbol{r} - \boldsymbol{r}_A| = |AP|$, the distance from point A to P, is (Fig. 16.18)

$$\nabla|\boldsymbol{r} - \boldsymbol{r}_A| = \frac{\boldsymbol{r} - \boldsymbol{r}_A}{|\boldsymbol{r} - \boldsymbol{r}_A|} \tag{16.13}$$

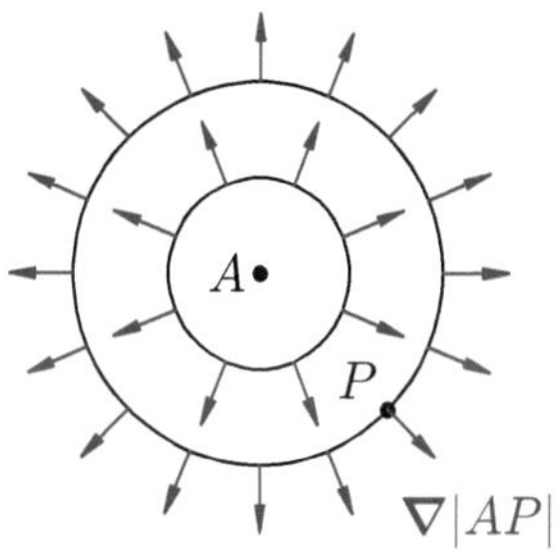

Fig. 16.18 $\nabla|\boldsymbol{r} - \boldsymbol{r}_A|$, (16.13).

since the direction of greatest increase of $|\boldsymbol{r} - \boldsymbol{r}_A|$ is in the direction of $\boldsymbol{r} - \boldsymbol{r}_A$, that is, the unit vector $(\boldsymbol{r} - \boldsymbol{r}_A)/|\boldsymbol{r} - \boldsymbol{r}_A|$, and the rate of change of that distance $|\boldsymbol{r} - \boldsymbol{r}_A|$ in that direction is obviously 1. In particular, the gradient of distance to the origin $|\boldsymbol{r}| = r$, is

$$\nabla r = \hat{\boldsymbol{r}}. \tag{16.14a}$$

Similarly, the gradient of distance to the z-axis is $\nabla \rho = \hat{\boldsymbol{\rho}}$.

(3) The gradient of $f(\boldsymbol{r}) = \theta = \arccos(\hat{\mathbf{z}} \cdot \boldsymbol{r})$, the angle between $\hat{\mathbf{z}}$ and $\boldsymbol{r}$, is (Fig. 16.19)

$$\nabla \theta = \hat{\boldsymbol{\theta}} \frac{\partial \theta}{\partial s} = \frac{\hat{\boldsymbol{\theta}}}{r}, \tag{16.14b}$$

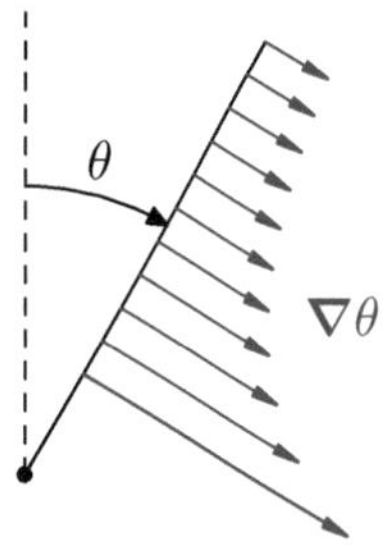

Fig. 16.19 $\nabla \theta$, (16.14b).

since the direction of greatest increase of θ is $\hat{\boldsymbol{\theta}}$ (south), and an infinitesimal step ds in that direction yields an increment dθ such that d$s = r$dθ, and thus $\partial\theta/\partial s = 1/r$, as illustrated in Fig. 16.20. This gradient is not defined at $\theta = 0$ or π and diverges as $r \to 0$.

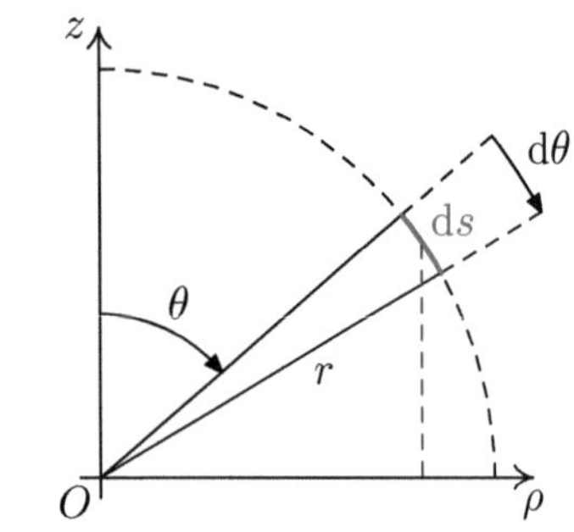

Fig. 16.20 $ds = r d\theta \Rightarrow \partial\theta/\partial s = 1/r$ (16.14b).

(4) The gradient of $f(\boldsymbol{r}) = \varphi = \operatorname{atan2}(\hat{\mathbf{y}} \cdot \boldsymbol{r}, \hat{\mathbf{x}} \cdot \boldsymbol{r})$, the angle between the planes $(\hat{\mathbf{z}}, \hat{\mathbf{x}})$ and $(\hat{\mathbf{z}}, \boldsymbol{r})$, is

$$\nabla\varphi = \hat{\boldsymbol{\varphi}} \frac{\partial \varphi}{\partial s} = \frac{\hat{\boldsymbol{\varphi}}}{\rho} = \frac{\hat{\boldsymbol{\varphi}}}{r \sin\theta}, \tag{16.14c}$$

since $\hat{\boldsymbol{\varphi}}$ is the direction (east) of greatest increase of φ (longitude), and an infinitesimal step ds in that direction yields an increment $d\varphi$ such that $ds = \rho\, d\varphi = r \sin\theta\, d\varphi$, where $\rho = r\sin\theta$ is the distance to the $\hat{\mathbf{z}}$ axis, as illustrated in Fig. 16.21.

Note that $\operatorname{atan2}(y, x)$ jumps by 2π across the negative x axis thus its gradient does not exist there, but $\hat{\boldsymbol{\varphi}}/\rho$ is continuous across that line. It is often useful to think of φ as the continuous *multivalued* function

$$\varphi = \operatorname{atan2}(y, x) \mod 2\pi, \tag{16.15}$$

with $\nabla\varphi = \hat{\boldsymbol{\varphi}}/\rho = -\hat{\mathbf{y}}/\rho$ along $x < 0, y = 0$. The singularity of $\nabla\varphi$ at $\rho = 0$ cannot be avoided, however.

Figures 16.20 and 16.21, to construct $\nabla\theta = \hat{\boldsymbol{\theta}}/r$ and $\nabla\varphi = \hat{\boldsymbol{\varphi}}/\rho$ geometrically, should be compared with Fig. 13.9 that serves the same purpose to derive $\partial\boldsymbol{r}/\partial\theta = r\hat{\boldsymbol{\theta}}$ and $\partial\boldsymbol{r}/\partial\varphi = \rho\hat{\boldsymbol{\varphi}}$ in (13.16). These vector derivatives are inverse of one another. For example, $\nabla\theta$ is the rate of change of θ with respect to $d\boldsymbol{r}$ in the direction of greatest increase of θ, while $\partial\boldsymbol{r}/\partial\theta$ is the rate of change of $\boldsymbol{r}$ with respect to θ with r and φ fixed. Furthermore,

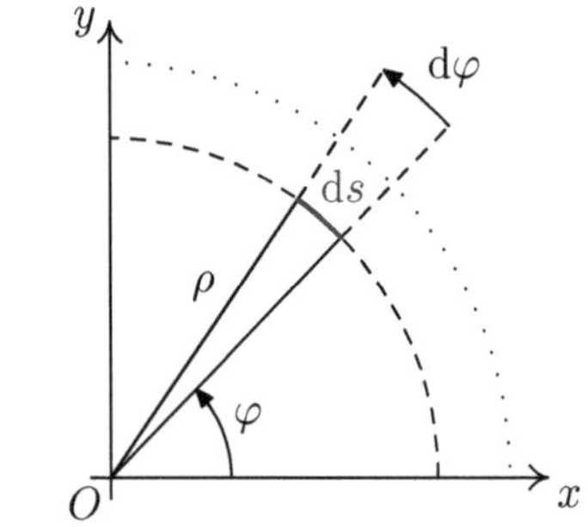

Fig. 16.21 $ds = \rho d\varphi \Rightarrow \partial\varphi/\partial s = 1/\rho$ (16.14c).

$$\begin{aligned} \frac{\partial \boldsymbol{r}}{\partial \theta} \cdot \nabla\theta &= 1, & \frac{\partial \boldsymbol{r}}{\partial \theta} \cdot \nabla\varphi &= 0, \\ \frac{\partial \boldsymbol{r}}{\partial \varphi} \cdot \nabla\theta &= 0, & \frac{\partial \boldsymbol{r}}{\partial \varphi} \cdot \nabla\varphi &= 1. \end{aligned} \tag{16.16}$$

This is a special case of a general result derived in (16.27).

Gradient rules

The summation and product rules hold for gradients,

$$\nabla(f + g) = \nabla f + \nabla g, \tag{16.17}$$

$$\nabla(fg) = (\nabla f)g + f(\nabla g). \tag{16.18}$$

These can be derived from the differentiation rules

$$d(f + g) = df + dg, \qquad d(fg) = (df)g + f(dg),$$

together with (16.6c), or as special cases of the chain rule (16.19).

For a scalar function $f(u, v, \ldots)$ of multiple scalar fields $u(\boldsymbol{r}), v(\boldsymbol{r}), \ldots$, the *chain rule* in differential form reads

$$df = \frac{\partial f}{\partial u} du + \frac{\partial f}{\partial v} dv + \cdots$$

then, applying (16.6c) to the differential of each scalar field yields

$$\mathrm{d}\boldsymbol{r} \cdot \boldsymbol{\nabla} f = \frac{\partial f}{\partial u}\mathrm{d}\boldsymbol{r} \cdot \boldsymbol{\nabla} u + \frac{\partial f}{\partial v}\mathrm{d}\boldsymbol{r} \cdot \boldsymbol{\nabla} v + \cdots .$$

Since this equality holds for any $\mathrm{d}\boldsymbol{r}$, it implies the *chain rule for gradients*

$$\boldsymbol{\nabla} f(u, v, \ldots) = \frac{\partial f}{\partial u}\boldsymbol{\nabla} u + \frac{\partial f}{\partial v}\boldsymbol{\nabla} v + \cdots . \tag{16.19}$$

Examples

If (x, y, z) are the usual cartesian coordinates, then (16.19) with (16.12) gives

$$\boldsymbol{\nabla} f(x, y, z) = \frac{\partial f}{\partial x}\,\hat{\mathbf{x}} + \frac{\partial f}{\partial y}\,\hat{\mathbf{y}} + \frac{\partial f}{\partial z}\,\hat{\mathbf{z}}. \tag{16.20}$$

That is the standard cartesian formula for the gradient, but the combination of the differentiation rules with the fundamental gradients can often bypass cumbersome cartesian calculations. For example,

$$\boldsymbol{\nabla} r^{-1} = \boldsymbol{\nabla}\left(\sqrt{x^2 + y^2 + z^2}\right)^{-1} = -\frac{x}{(x^2 + y^2 + z^2)^{3/2}}\,\hat{\mathbf{x}} + \cdots ,$$

but the chain rule (16.19) with (16.14a) gives [1]

$$\boldsymbol{\nabla} r^{-1} = \frac{\mathrm{d}(r^{-1})}{\mathrm{d}r}\,\boldsymbol{\nabla} r = -\frac{\hat{\boldsymbol{r}}}{r^2}. \tag{16.22}$$

The latter derivation is simpler and yields a clearer result.

[1] For a general $f(r)$, that is

$$\boldsymbol{\nabla} f(r) = \frac{\mathrm{d}f}{\mathrm{d}r}\,\boldsymbol{\nabla} r = \frac{\mathrm{d}f}{\mathrm{d}r}\,\hat{\boldsymbol{r}}. \tag{16.21}$$

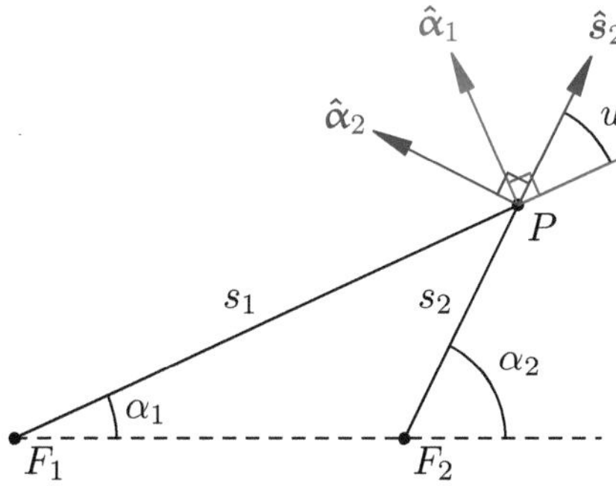

Fig. 16.22 Bipolar coordinates $u = \alpha_2 - \alpha_1,\ v = \ln(s_1/s_2)$.

As another example, consider bipolar coordinates $u = \alpha_2 - \alpha_1$ and $v = \ln(s_2/s_1)$, illustrated in Fig. 7.12, where s_1 is the distance to F_1 and α_1 the angle from (F_1, F_2) to (F_1, P), and similarly for s_2 and α_2, as illustrated in Fig. 16.22. The chain rule (16.19) with (16.13) and (16.14b) gives

$$\begin{aligned} \boldsymbol{\nabla} u &= \boldsymbol{\nabla}\alpha_2 - \boldsymbol{\nabla}\alpha_1 = \frac{\hat{\boldsymbol{\alpha}}_2}{s_2} - \frac{\hat{\boldsymbol{\alpha}}_1}{s_1}, \\ \boldsymbol{\nabla} v &= \boldsymbol{\nabla}\ln\frac{s_2}{s_1} = \frac{1}{s_2}\boldsymbol{\nabla} s_2 - \frac{1}{s_1}\boldsymbol{\nabla} s_1 = \frac{\hat{\boldsymbol{s}}_2}{s_2} - \frac{\hat{\boldsymbol{s}}_1}{s_1}. \end{aligned} \tag{16.23}$$

Since $\hat{\boldsymbol{\alpha}}_1$ and $\hat{\boldsymbol{\alpha}}_2$ are the 90° rotation of $\hat{\boldsymbol{s}}_1$ and $\hat{\boldsymbol{s}}_2$, respectively, then likewise $\boldsymbol{\nabla} u$ is the 90° rotation of $\boldsymbol{\nabla} v$; therefore,

$$\boldsymbol{\nabla} u \cdot \boldsymbol{\nabla} v = 0, \qquad |\boldsymbol{\nabla} u| = |\boldsymbol{\nabla} v|,$$

showing that bipolar coordinates are orthogonal and conformal. Deriving these results in cartesian coordinates is considerably more cumbersome and opaque.

16.4 Curvilinear coordinates

The chain rule (16.19) readily yields expressions for $\boldsymbol{\nabla} f$ in other coordinate systems. For instance, in spherical coordinates (r, θ, φ),

$$\begin{aligned} \boldsymbol{\nabla} f(r,\theta,\varphi) &= \frac{\partial f}{\partial r}\boldsymbol{\nabla} r + \frac{\partial f}{\partial \theta}\boldsymbol{\nabla}\theta + \frac{\partial f}{\partial \varphi}\boldsymbol{\nabla}\varphi \\ &= \frac{\partial f}{\partial r}\hat{\boldsymbol{r}} + \frac{\partial f}{\partial \theta}\frac{\hat{\boldsymbol{\theta}}}{r} + \frac{\partial f}{\partial \varphi}\frac{\hat{\boldsymbol{\varphi}}}{r\sin\theta}, \end{aligned} \tag{16.24}$$

since $\boldsymbol{\nabla} r = \hat{\boldsymbol{r}}$, $\boldsymbol{\nabla}\theta = \hat{\boldsymbol{\theta}}/r$, $\boldsymbol{\nabla}\varphi = \hat{\boldsymbol{\varphi}}/(r\sin\theta)$, from (16.14). In cylindrical coordinates (ρ, φ, z), with $\boldsymbol{\nabla}\rho = \hat{\boldsymbol{\rho}}$, $\boldsymbol{\nabla}\varphi = \hat{\boldsymbol{\varphi}}/\rho$, and $\boldsymbol{\nabla} z = \hat{\mathbf{z}}$, (16.14), we obtain

$$\begin{aligned} \boldsymbol{\nabla} f(\rho,\varphi,z) &= \frac{\partial f}{\partial \rho}\boldsymbol{\nabla}\rho + \frac{\partial f}{\partial \varphi}\boldsymbol{\nabla}\varphi + \frac{\partial f}{\partial z}\boldsymbol{\nabla} z \\ &= \frac{\partial f}{\partial \rho}\hat{\boldsymbol{\rho}} + \frac{\partial f}{\partial \varphi}\frac{\hat{\boldsymbol{\varphi}}}{\rho} + \frac{\partial f}{\partial z}\hat{\mathbf{z}}. \end{aligned} \tag{16.25}$$

For general curvilinear coordinates (u, v, w),

$$\boldsymbol{\nabla} f(u,v,w) = \frac{\partial f}{\partial u}\boldsymbol{\nabla} u + \frac{\partial f}{\partial v}\boldsymbol{\nabla} v + \frac{\partial f}{\partial w}\boldsymbol{\nabla} w. \tag{16.26}$$

Curvilinear coordinates specify the position vector $\boldsymbol{r}$ in terms of (u, v, w) coordinates with Jacobian (15.9)

$$J = \frac{\partial \boldsymbol{r}}{\partial u} \times \frac{\partial \boldsymbol{r}}{\partial v} \cdot \frac{\partial \boldsymbol{r}}{\partial w} > 0,$$

such that the map $(u, v, w) \longleftrightarrow \boldsymbol{r}$ is one-to-one, and a scalar field $f(\boldsymbol{r})$ can be viewed as a function of $\boldsymbol{r}$ or of the coordinates (u, v, w). The inverse map specifies the scalars (u, v, w) in terms of $\boldsymbol{r}$, and these are the scalar fields that appear in (16.26). Equation (16.9) gives

$$\frac{\partial f}{\partial u} = \frac{\partial \boldsymbol{r}}{\partial u}\cdot\boldsymbol{\nabla} f, \quad \frac{\partial f}{\partial v} = \frac{\partial \boldsymbol{r}}{\partial v}\cdot\boldsymbol{\nabla} f, \quad \frac{\partial f}{\partial w} = \frac{\partial \boldsymbol{r}}{\partial w}\cdot\boldsymbol{\nabla} f.$$

These hold for any field f and in particular for $f = u$, v, or w, yielding (Fig. 16.23)

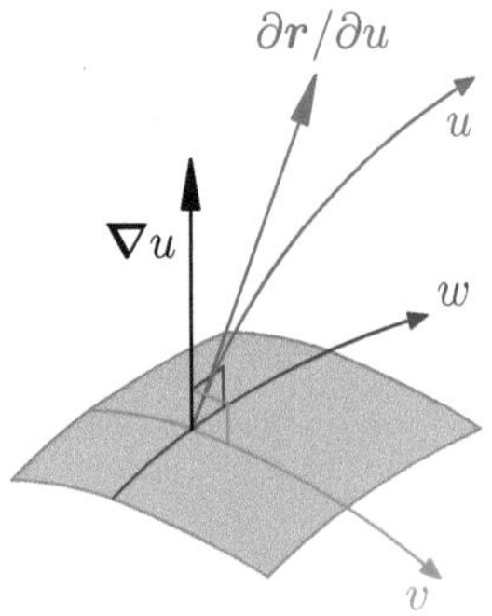

Fig. 16.23 $\boldsymbol{\nabla} u$ is orthogonal to u-isosurface, and to $\partial \boldsymbol{r}/\partial v$ and $\partial \boldsymbol{r}/\partial w$.

$$\begin{bmatrix} \boldsymbol{\nabla} u \\ \boldsymbol{\nabla} v \\ \boldsymbol{\nabla} w \end{bmatrix} \cdot \begin{bmatrix} \dfrac{\partial \boldsymbol{r}}{\partial u} & \dfrac{\partial \boldsymbol{r}}{\partial v} & \dfrac{\partial \boldsymbol{r}}{\partial w} \end{bmatrix} = \begin{bmatrix} 1 & 0 & 0 \\ 0 & 1 & 0 \\ 0 & 0 & 1 \end{bmatrix}. \tag{16.27}$$

Hence, the gradient basis $\{\nabla u, \nabla v, \nabla w\}$ is the *contravariant basis* (15.12), the reciprocal of the velocity basis $\{\partial \boldsymbol{r}/\partial u, \partial \boldsymbol{r}/\partial v, \partial \boldsymbol{r}/\partial w\}$. The contravariant basis vectors are the surface normals (15.8) divided by the Jacobian (15.12); thus,

$$\nabla u = \frac{1}{J}\frac{\partial \boldsymbol{r}}{\partial v} \times \frac{\partial \boldsymbol{r}}{\partial w}, \tag{16.28a}$$

$$\frac{\partial \boldsymbol{r}}{\partial u} = J\, \nabla v \times \nabla w, \tag{16.28b}$$

and similarly for ∇v, ∇w, $\partial \boldsymbol{r}/\partial v$, $\partial \boldsymbol{r}/\partial w$ by cyclic rotation of (u, v, w), with

$$\nabla u \times \nabla v \cdot \nabla w = \frac{1}{J}. \tag{16.29}$$

The velocity basis $\{\partial \boldsymbol{r}/\partial u, \partial \boldsymbol{r}/\partial v, \partial \boldsymbol{r}/\partial w\}$ consists of tangents to the coordinate curves along which two coordinates are fixed. The gradient basis $\{\nabla u, \nabla v, \nabla w\}$ consists of normals to the coordinate surfaces across which one coordinate is fixed.

Velocity and gradient bases

In index notation with $(u, v, w) \equiv (u^1, u^2, u^3)$, the coordinate tangents yield the covariant basis

$$\boldsymbol{a}_i = \frac{\partial \boldsymbol{r}}{\partial u^i} \equiv \partial_i \boldsymbol{r}, \tag{16.30}$$

while the coordinate gradients make up the contravariant basis

$$\boldsymbol{a}^i = \nabla u^i. \tag{16.31}$$

These bases are biorthogonal (16.27) (Fig. 16.24),

$$\boldsymbol{a}_i \cdot \boldsymbol{a}^j = \frac{\partial \boldsymbol{r}}{\partial u^i} \cdot \nabla u^j = \delta_i^j, \tag{16.32}$$

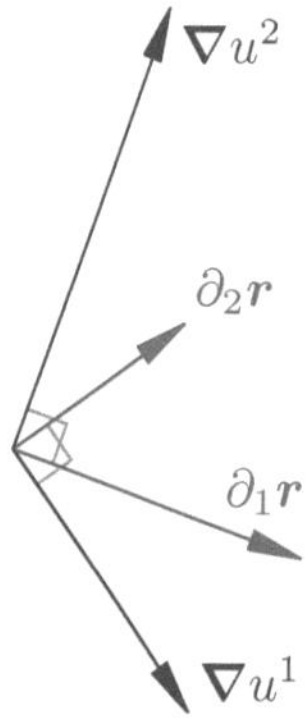

Fig. 16.24 Coordinate tangents and gradients are reciprocal bases.

with metrics

$$\begin{aligned} g_{ij} &= \boldsymbol{a}_i \cdot \boldsymbol{a}_j = \partial_i \boldsymbol{r} \cdot \partial_j \boldsymbol{r}, \\ g^{ij} &= \boldsymbol{a}^i \cdot \boldsymbol{a}^j = \nabla u^i \cdot \nabla u^j. \end{aligned} \tag{16.33}$$

Equations (16.28) read

$$\begin{aligned} \nabla u^i &= \frac{\epsilon^{ijk}}{2J}\, \partial_j \boldsymbol{r} \times \partial_k \boldsymbol{r} \;\longleftrightarrow\; \partial_i \boldsymbol{r} \times \partial_j \boldsymbol{r} = J \epsilon_{ijk} \nabla u^k, \\ \partial_i \boldsymbol{r} &= J \frac{\epsilon_{ijk}}{2}\, \nabla u^j \times \nabla u^k \;\longleftrightarrow\; \nabla u^i \times \nabla u^j = \frac{\epsilon^{ijk}}{J} \partial_k \boldsymbol{r}, \end{aligned} \tag{16.34}$$

with summation over lower indices repeated as upper indices. The factors of 2 arise from the sums over j and k, which contribute two equal terms.[2]

[2] Or, equivalently, from $\epsilon^{ijk}\epsilon_{jkl} = 2\delta_l^i$.

The two bases are identical in cartesian coordinates such that

$$\boldsymbol{r} = x\hat{\mathbf{x}} + y\hat{\mathbf{y}} + z\hat{\mathbf{z}} \equiv x^i \mathbf{e}_i = x_i \mathbf{e}^i \tag{16.35}$$

with $\partial \boldsymbol{r}/\partial x^i = \mathbf{e}_i = \boldsymbol{\nabla} x_i$, and $\mathbf{e}_i \cdot \mathbf{e}_j = \delta_{ij}$, $x_i = x^i$, $\mathbf{e}_i = \mathbf{e}^i$, $J = 1$. There is no need to distinguish between upper and lower indices for cartesian coordinates, except perhaps for the summation convention.

The coordinate tangents are the natural basis for velocities since

$$\frac{\mathrm{d}\boldsymbol{r}}{\mathrm{d}t} = \dot{u}\frac{\partial \boldsymbol{r}}{\partial u} + \dot{v}\frac{\partial \boldsymbol{r}}{\partial v} + \dot{w}\frac{\partial \boldsymbol{r}}{\partial w} \equiv \dot{u}^i\, \partial_i \boldsymbol{r} = \dot{u}^i\, \boldsymbol{a}_i, \tag{16.36}$$

while the coordinate gradients are the natural basis for gradients since

$$\boldsymbol{\nabla} f = \frac{\partial f}{\partial u}\boldsymbol{\nabla} u + \frac{\partial f}{\partial v}\boldsymbol{\nabla} v + \frac{\partial f}{\partial w}\boldsymbol{\nabla} w = \frac{\partial f}{\partial u^i}\boldsymbol{\nabla} u^i = \boldsymbol{a}^i \partial_i f, \tag{16.37}$$

but the coordinates in one basis are obtained by projecting onto the other,

$$\dot{u}^i = \dot{\boldsymbol{r}} \cdot \boldsymbol{\nabla} u^i, \qquad \partial_i f = \partial_i \boldsymbol{r} \cdot \boldsymbol{\nabla} f. \tag{16.38}$$

An arbitrary vector field $\boldsymbol{F}(\boldsymbol{r})$ can be expanded in terms of either basis as

$$\boldsymbol{F} = F^i\, \partial_i \boldsymbol{r} = F_i\, \boldsymbol{\nabla} u^i \quad \Leftrightarrow \quad \begin{cases} F^i = \boldsymbol{F} \cdot \boldsymbol{\nabla} u^i, \\ F_i = \boldsymbol{F} \cdot \partial_i \boldsymbol{r}. \end{cases} \tag{16.39}$$

Switching from one expansion to the other is done by *raising or lowering indices* using the metrics (16.33)

$$F_i = g_{ij} F^j, \quad F^i = g^{ij} F_j, \tag{16.40}$$

with sum over the lower index j repeated as an upper index.

Gradient field

A vector field $\boldsymbol{F}(P)$ is a gradient if there exists a scalar field $f(P)$ such that $\boldsymbol{F} = \boldsymbol{\nabla} f$ at any point P. That scalar field $f(P)$ is the *scalar potential* of $\boldsymbol{F}(P)$. The scalar potential $f(P)$ is defined up to an arbitrary constant. In the coordinate gradient basis,

$$\boldsymbol{F} = F_i \boldsymbol{\nabla} u^i = \boldsymbol{\nabla} f = (\partial_i f)\, \boldsymbol{\nabla} u^i, \tag{16.41a}$$

implies that $F_i = \partial_i f$, where $\partial_i \equiv \partial/\partial u^i$ and $F_i = \boldsymbol{F} \cdot \partial_i \boldsymbol{r}$ are the covariant components (16.39). If the F_i's are continuously differentiable (C^1), then f, if it exists, would be twice continuously differentiable (C^2), and its partial derivatives would commute $\partial_i \partial_j f = \partial_j \partial_i f$. Therefore, existence of a scalar field $f(P)$ such that $\boldsymbol{F} = \boldsymbol{\nabla} f$ requires the *integrability conditions* $\partial_i F_j = \partial_j F_i$, that is,

$$\frac{\partial}{\partial u^i}\left(\boldsymbol{F} \cdot \frac{\partial \boldsymbol{r}}{\partial u^j}\right) = \frac{\partial}{\partial u^j}\left(\boldsymbol{F} \cdot \frac{\partial \boldsymbol{r}}{\partial u^i}\right). \tag{16.41b}$$

In cartesian coordinates the integrability conditions (16.41b) for

$$\boldsymbol{F} = F_x \hat{\mathbf{x}} + F_y \hat{\mathbf{y}} + F_z \hat{\mathbf{z}} = F_x \boldsymbol{\nabla} x + F_y \boldsymbol{\nabla} y + F_z \boldsymbol{\nabla} z$$

are

$$\frac{\partial F_y}{\partial x} = \frac{\partial F_x}{\partial y}, \quad \frac{\partial F_z}{\partial y} = \frac{\partial F_y}{\partial z}, \quad \frac{\partial F_x}{\partial z} = \frac{\partial F_z}{\partial x}. \tag{16.42}$$

Thus, solid body rotation $\boldsymbol{F} = \hat{\mathbf{z}} \times \boldsymbol{r} = x\hat{\mathbf{y}} - y\hat{\mathbf{x}}$ is *not* a gradient field, since

$$\frac{\partial F_y}{\partial x} = \frac{\partial x}{\partial x} = 1 \neq \frac{\partial F_x}{\partial y} = \frac{\partial(-y)}{\partial y} = -1,$$

but centrifugal acceleration $\boldsymbol{F} = (\hat{\mathbf{z}} \times \boldsymbol{r}) \times \hat{\mathbf{z}} = x\hat{\mathbf{x}} + y\hat{\mathbf{y}}$ is a gradient since $\partial F_i/\partial x^j = 0$ for all $i \neq j$,

$$\frac{\partial F_y}{\partial x} = \frac{\partial y}{\partial x} = 0 = \frac{\partial F_x}{\partial y} = \frac{\partial x}{\partial y} = 0.$$

Indeed, this $\boldsymbol{F} = x\hat{\mathbf{x}} + y\hat{\mathbf{y}} = \boldsymbol{\nabla}(x^2 + y^2)/2$.

In spherical coordinates, the integrability conditions (16.41b) for

$$\begin{aligned} \boldsymbol{F} &= F_r\,\hat{\boldsymbol{r}} + F_\theta\,\hat{\boldsymbol{\theta}} + F_\varphi\,\hat{\boldsymbol{\varphi}} \\ &= F_r\boldsymbol{\nabla} r + (rF_\theta)\boldsymbol{\nabla}\theta + (r\sin\theta F_\varphi)\boldsymbol{\nabla}\varphi, \end{aligned} \tag{16.43a}$$

are

$$\frac{\partial(rF_\theta)}{\partial r} = \frac{\partial F_r}{\partial \theta}, \quad \frac{\partial(r\sin\theta F_\varphi)}{\partial \theta} = \frac{\partial(rF_\theta)}{\partial \varphi}, \quad \frac{\partial F_r}{\partial \varphi} = \frac{\partial(r\sin\theta F_\varphi)}{\partial r}. \tag{16.43b}$$

This is (16.41b) for $F_i = \boldsymbol{F} \cdot \partial_i \boldsymbol{r}$ with the spherical coordinate tangents (15.21a),

$$\boldsymbol{F} \cdot \frac{\partial \boldsymbol{r}}{\partial r} = F_r, \quad \boldsymbol{F} \cdot \frac{\partial \boldsymbol{r}}{\partial \theta} = rF_\theta, \quad \boldsymbol{F} \cdot \frac{\partial \boldsymbol{r}}{\partial \varphi} = r\sin\theta F_\varphi, \tag{16.43c}$$

but we bypassed these explicit projections on the covariant basis by directly rewriting $\boldsymbol{F}$ in terms of the gradient basis $(\boldsymbol{\nabla} r, \boldsymbol{\nabla}\theta, \boldsymbol{\nabla}\varphi)$ (16.14) in (16.43a). Thus $\boldsymbol{F} = \hat{\mathbf{z}} \times \boldsymbol{r} = r\sin\theta\,\hat{\boldsymbol{\varphi}} = x\hat{\mathbf{y}} - y\hat{\mathbf{x}}$ considered above, is still *not* a gradient field, of course, since

$$\frac{\partial F_r}{\partial \varphi} = 0 \neq \frac{\partial(r\sin\theta F_\varphi)}{\partial r} = 2r\sin^2\theta,$$

but $\boldsymbol{F} = (\hat{\mathbf{z}} \times \boldsymbol{r}) \times \hat{\mathbf{z}} = x\hat{\mathbf{x}} + y\hat{\mathbf{y}} = \rho\hat{\boldsymbol{\rho}} = \boldsymbol{\nabla}\rho^2/2$ still is a gradient.

These examples illustrate that whether a vector field is a gradient should not depend on the choice of coordinates. The geometric invariant that vanishes when a vector field $\boldsymbol{F}$ is a gradient is the *curl* of $\boldsymbol{F}$.

16.5 Curl and del

In 3D, the integrability conditions (16.41b) for $\boldsymbol{F}$ to be a gradient can be written

$$\epsilon^{ijk}\partial_i\left(\boldsymbol{F} \cdot \partial_j \boldsymbol{r}\right) = \epsilon^{ijk}\partial_i F_j = 0, \tag{16.44}$$

with sums over i and j. This form suggests that $\epsilon^{ijk}\partial_i F_j$ are the three components of a vector, but a vector is a geometric object invariant under a change of coordinates. The vector associated with those components is the *curl* of $\boldsymbol{F}$ defined as

$$\operatorname{curl}\boldsymbol{F} = \epsilon^{ijk}\left(\partial_i F_j\right)\, J^{-1}\partial_k \boldsymbol{r}, \tag{16.45}$$

where $\partial_i = \partial/\partial u^i$, $F_j = \boldsymbol{F}\cdot\partial_j \boldsymbol{r}$, and $J = \partial_1 \boldsymbol{r}\times\partial_2 \boldsymbol{r}\cdot\partial_3 \boldsymbol{r}$ is the coordinate Jacobian equal to the square root of the metric determinant $J = \sqrt{g}$ (15.68).

Curl $\boldsymbol{F}$ defined in (16.45) is $\boldsymbol{\nabla}\times\boldsymbol{F}$ where $\boldsymbol{\nabla}$ is the *del operator* defined in arbitrary coordinates $(u, v, w) = (u^1, u^2, u^3)$ as

$$\boldsymbol{\nabla} = \boldsymbol{\nabla} u \,\frac{\partial}{\partial u} + \boldsymbol{\nabla} v \,\frac{\partial}{\partial v} + \boldsymbol{\nabla} w \,\frac{\partial}{\partial w} = \boldsymbol{\nabla} u^i \,\partial_i, \tag{16.46}$$

such that $\boldsymbol{\nabla} f = \boldsymbol{\nabla} u^i \,\partial_i f$ is the gradient of f as in (16.37). Indeed,

$$\boldsymbol{\nabla}\times\boldsymbol{F} = \boldsymbol{\nabla} u^i \,\partial_i \times \boldsymbol{F} = \boldsymbol{\nabla} u^i \times \partial_i \boldsymbol{F}, \tag{16.47a}$$

$$\boldsymbol{\nabla} u^i \times \boldsymbol{\nabla} u^j = \frac{\epsilon^{ijk}}{J}\partial_k \boldsymbol{r}$$

then, expanding $\partial_i \boldsymbol{F} = (\partial_i \boldsymbol{F}\cdot\partial_j \boldsymbol{r})\,\boldsymbol{\nabla} u^j$ as in (16.39) and using (16.34) yields

$$\begin{aligned}\boldsymbol{\nabla}\times\boldsymbol{F} &= \boldsymbol{\nabla} u^i \times \boldsymbol{\nabla} u^j \,(\partial_i \boldsymbol{F}\cdot\partial_j \boldsymbol{r}) \\ &= \boldsymbol{\nabla} u^i \times \boldsymbol{\nabla} u^j \,(\partial_i(\boldsymbol{F}\cdot\partial_j \boldsymbol{r}) - \boldsymbol{F}\cdot\partial_i\partial_j \boldsymbol{r}) \\ &= J^{-1}\epsilon^{ijk}\partial_k \boldsymbol{r}\,\left(\partial_i F_j\right),\end{aligned} \tag{16.47b}$$

because $\boldsymbol{\nabla} u^i \times \boldsymbol{\nabla} u^j \,\partial_i\partial_j \boldsymbol{r} = 0$ by symmetry of $\partial_i\partial_j = \partial_j\partial_i$ and anti-symmetry of the cross product $\boldsymbol{\nabla} u^i\times\boldsymbol{\nabla} u^j = -\boldsymbol{\nabla} u^j\times\boldsymbol{\nabla} u^i$. Thus, $\boldsymbol{\nabla}\times\boldsymbol{F}$ in (16.47b) equals curl $\boldsymbol{F}$ in (16.45). With $(u, v, w) \equiv (u^1, u^2, u^3)$ and Jacobian J, the curl formula (16.45), (16.47b) expands to

$$J = \frac{\partial \boldsymbol{r}}{\partial u}\times\frac{\partial \boldsymbol{r}}{\partial v}\cdot\frac{\partial \boldsymbol{r}}{\partial w}$$

$$\begin{aligned}\boldsymbol{\nabla}\times\boldsymbol{F} = &\frac{1}{J}\frac{\partial \boldsymbol{r}}{\partial u}\left(\frac{\partial}{\partial v}\left(\boldsymbol{F}\cdot\frac{\partial \boldsymbol{r}}{\partial w}\right) - \frac{\partial}{\partial w}\left(\boldsymbol{F}\cdot\frac{\partial \boldsymbol{r}}{\partial v}\right)\right) \\ &+ \frac{1}{J}\frac{\partial \boldsymbol{r}}{\partial v}\left(\frac{\partial}{\partial w}\left(\boldsymbol{F}\cdot\frac{\partial \boldsymbol{r}}{\partial u}\right) - \frac{\partial}{\partial u}\left(\boldsymbol{F}\cdot\frac{\partial \boldsymbol{r}}{\partial w}\right)\right) \\ &+ \frac{1}{J}\frac{\partial \boldsymbol{r}}{\partial w}\left(\frac{\partial}{\partial u}\left(\boldsymbol{F}\cdot\frac{\partial \boldsymbol{r}}{\partial v}\right) - \frac{\partial}{\partial v}\left(\boldsymbol{F}\cdot\frac{\partial \boldsymbol{r}}{\partial u}\right)\right).\end{aligned} \tag{16.47c}$$

Alternative derivations of this general formula are sketched in exercises 16.24, 16.25 and 16.28. It is essentially Stokes' theorem (16.60).

The del operator $\boldsymbol{\nabla}$ is invariant under a change of coordinates from $(u^1, u^2, u^3) \to (v^1, v^2, v^3)$ since by the chain rule

$$\boldsymbol{\nabla} u^i \,\frac{\partial}{\partial u^i} = \left(\boldsymbol{\nabla} v^k \,\frac{\partial u^i}{\partial v^k}\right)\left(\frac{\partial v^l}{\partial u_i}\frac{\partial}{\partial v^l}\right) = \boldsymbol{\nabla} v^k \,\frac{\partial}{\partial v^k}, \tag{16.48}$$

because

$$\frac{\partial u^i}{\partial v^k}\frac{\partial v^l}{\partial u_i} = \frac{\partial v^l}{\partial v^k} = \delta^l_k.$$

Likewise,

$$\boldsymbol{\nabla}\times\boldsymbol{F} = \boldsymbol{\nabla} u^i \times \frac{\partial \boldsymbol{F}}{\partial u^i} = \boldsymbol{\nabla} v^k \times \frac{\partial \boldsymbol{F}}{\partial v^k},$$

and the curl $\boldsymbol{\nabla}\times\boldsymbol{F}$ is invariant under a change of coordinates.

Del and curl in standard coordinates

Expressions (16.46) and (16.47) hold for general coordinates $(u^1, u^2, u^3) \equiv (u, v, w)$. In particular, the del operator is

$$\begin{aligned}\boldsymbol{\nabla} &= \hat{\mathbf{x}}\,\frac{\partial}{\partial x} + \hat{\mathbf{y}}\,\frac{\partial}{\partial y} + \hat{\mathbf{z}}\,\frac{\partial}{\partial z} \\ &= \hat{\boldsymbol{\rho}}\,\frac{\partial}{\partial \rho} + \frac{\hat{\boldsymbol{\varphi}}}{\rho}\frac{\partial}{\partial \varphi} + \hat{\mathbf{z}}\,\frac{\partial}{\partial z} \\ &= \hat{\boldsymbol{r}}\,\frac{\partial}{\partial r} + \frac{\hat{\boldsymbol{\theta}}}{r}\frac{\partial}{\partial \theta} + \frac{\hat{\boldsymbol{\varphi}}}{r\sin\theta}\frac{\partial}{\partial \varphi}\end{aligned} \tag{16.49}$$

in cartesian, cylindrical, and spherical coordinates, respectively, from (16.12) and (16.14). A vector field $\boldsymbol{F}(\boldsymbol{r})$ is expanded as

$$\begin{aligned}\boldsymbol{F} &= F_x\,\hat{\mathbf{x}} + F_y\,\hat{\mathbf{y}} + F_z\,\hat{\mathbf{z}}, \\ &= F_\rho\,\hat{\boldsymbol{\rho}} + F_\varphi\,\hat{\boldsymbol{\varphi}} + F_z\,\hat{\mathbf{z}}, \\ &= F_r\,\hat{\boldsymbol{r}} + F_\theta\,\hat{\boldsymbol{\theta}} + F_\varphi\,\hat{\boldsymbol{\varphi}},\end{aligned} \tag{16.50}$$

and the corresponding expressions for its curl are then, from (16.47c),

$$\begin{aligned}&\boldsymbol{\nabla}\times\boldsymbol{F} \\ &= \hat{\mathbf{x}}\left(\partial_y F_z - \partial_z F_y\right) + \hat{\mathbf{y}}\left(\partial_z F_x - \partial_x F_z\right) + \hat{\mathbf{z}}\left(\partial_x F_y - \partial_y F_x\right),\end{aligned} \tag{16.51a}$$

$$\begin{aligned}&= \frac{\hat{\boldsymbol{\rho}}}{\rho}\left(\frac{\partial F_z}{\partial\varphi} - \frac{\partial(\rho F_\varphi)}{\partial z}\right) + \hat{\boldsymbol{\varphi}}\left(\frac{\partial F_\rho}{\partial z} - \frac{\partial F_z}{\partial \rho}\right) \\ &\qquad + \frac{\hat{\mathbf{z}}}{\rho}\left(\frac{\partial(\rho F_\varphi)}{\partial\rho} - \frac{\partial F_\rho}{\partial\varphi}\right),\end{aligned} \tag{16.51b}$$

$$\begin{aligned}&= \frac{\hat{\boldsymbol{r}}}{r\sin\theta}\left(\frac{\partial(\sin\theta\,F_\varphi)}{\partial\theta} - \frac{\partial F_\theta}{\partial\varphi}\right) + \frac{\hat{\boldsymbol{\theta}}}{r\sin\theta}\left(\frac{\partial F_r}{\partial\varphi} - \frac{\partial(r\sin\theta\,F_\varphi)}{\partial r}\right) \\ &\qquad + \frac{\hat{\boldsymbol{\varphi}}}{r}\left(\frac{\partial(rF_\theta)}{\partial r} - \frac{\partial F_r}{\partial\theta}\right).\end{aligned} \tag{16.51c}$$

In the spherical coordinates case, for example, $(u^1, u^2, u^3) \equiv (r, \theta, \varphi)$,

$$\partial_1\boldsymbol{r} \equiv \frac{\partial\boldsymbol{r}}{\partial r} = \hat{\boldsymbol{r}}, \quad \partial_2\boldsymbol{r} \equiv \frac{\partial\boldsymbol{r}}{\partial\theta} = r\hat{\boldsymbol{\theta}}, \quad \partial_3\boldsymbol{r} \equiv \frac{\partial\boldsymbol{r}}{\partial\varphi} = r\sin\theta\hat{\boldsymbol{\varphi}},$$

from (15.21a) with Jacobian $J = r^2\sin\theta$, so the covariant components of $\boldsymbol{F}$ are

$$F_1 \equiv \boldsymbol{F}\cdot\partial_r\boldsymbol{r} = F_r, \quad F_2 \equiv \boldsymbol{F}\cdot\partial_\theta\boldsymbol{r} = rF_\theta, \quad F_3 \equiv \boldsymbol{F}\cdot\partial_\varphi\boldsymbol{r} = r\sin\theta\,F_\varphi.$$

Substituting in (16.45) or (16.47c) yields the spherical form of $\boldsymbol{\nabla}\times\boldsymbol{F}$ in (16.51). The cartesian and cylindrical expressions are derived similarly. Alternative direct derivations of formulas (16.51) are provided in the examples below and in (16.93) for general orthogonal coordinates.

Curl properties

The defining property is, of course, that the curl of a gradient vanishes (16.44)

$$\boldsymbol{\nabla} \times \boldsymbol{\nabla} f = 0 \tag{16.52a}$$

C^2 means twice continuously differentiable.

for any C^2 scalar field $f(\boldsymbol{r})$. This is easy to remember symbolically from the vanishing of the cross product of parallel vectors $\boldsymbol{a} \times \alpha \boldsymbol{a} = 0$. However, this analogy must be used with care since $\boldsymbol{\nabla}$ is a vector differential operator; thus, for example

$$\boldsymbol{\nabla} \times x \boldsymbol{\nabla} y = \boldsymbol{\nabla} x \times \boldsymbol{\nabla} y = \hat{\mathbf{z}} \neq 0.$$

The curl is a differential operator that obeys the summation and product rules

$$\boldsymbol{\nabla} \times (\boldsymbol{F} + \boldsymbol{G}) = \boldsymbol{\nabla} \times \boldsymbol{F} + \boldsymbol{\nabla} \times \boldsymbol{G}, \tag{16.52b}$$

$$\boldsymbol{\nabla} \times (f \boldsymbol{G}) = \boldsymbol{\nabla} f \times \boldsymbol{G} + f \boldsymbol{\nabla} \times \boldsymbol{G}. \tag{16.52c}$$

These are straightforward to prove and remember from the regular summation and product rules for derivatives applied to (16.47), with the understanding that $(\boldsymbol{\nabla} \times f)\boldsymbol{G}$ does not make sense for a scalar field f, but $\boldsymbol{\nabla} f \times \boldsymbol{G}$ does.

Curl examples

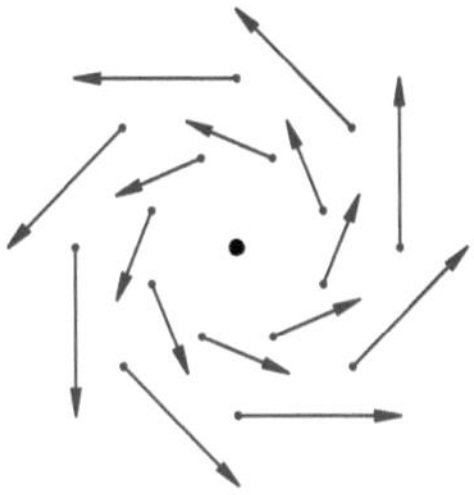

Fig. 16.25 $\boldsymbol{\nabla} \times F(\rho)\hat{\boldsymbol{\varphi}} = \frac{1}{\rho}\frac{\mathrm{d}(\rho F)}{\mathrm{d}\rho}\hat{\mathbf{z}}$.

The curl of solid body rotation $\boldsymbol{F} = \hat{\mathbf{z}} \times \boldsymbol{r} = x\hat{\mathbf{y}} - y\hat{\mathbf{x}} = \rho\hat{\boldsymbol{\varphi}}$ is straightforward to calculate in cartesian or cylindrical coordinates (16.51)

$$\boldsymbol{\nabla} \times (\hat{\mathbf{z}} \times \boldsymbol{r}) = \hat{\mathbf{z}} \left(\frac{\partial x}{\partial x} - \frac{\partial(-y)}{\partial y} \right) = 2\hat{\mathbf{z}} = \frac{\hat{\mathbf{z}}}{\rho} \frac{\partial(\rho^2)}{\partial \rho}. \tag{16.53}$$

Thus, the curl of solid body rotation is *twice the rotation rate* (here $\hat{\mathbf{z}}$).

The curl of $\boldsymbol{F} = F(\rho)\hat{\boldsymbol{\varphi}}$ can be calculated directly using fundamental gradients (16.14) and curl properties,

$$\begin{aligned} \boldsymbol{\nabla} \times F(\rho)\hat{\boldsymbol{\varphi}} &= \boldsymbol{\nabla} \times (\rho F(\rho))\boldsymbol{\nabla}\varphi \\ &= \boldsymbol{\nabla}(\rho F(\rho)) \times \boldsymbol{\nabla}\varphi = \frac{\mathrm{d}(\rho F)}{\mathrm{d}\rho}\hat{\boldsymbol{\rho}} \times \frac{\hat{\boldsymbol{\varphi}}}{\rho} = \frac{1}{\rho}\frac{\mathrm{d}(\rho F)}{\mathrm{d}\rho}\hat{\mathbf{z}}, \end{aligned} \tag{16.54}$$

essentially re-deriving (16.51b) on the fly (Fig. 16.25).

The curl of the inverse square law field (16.2) $\boldsymbol{F} = \hat{\boldsymbol{r}}/r^2$ is tedious to calculate in cartesian or cylindrical coordinates. It is straightforward using the formidable spherical coordinates formula (16.51), but more simply from (16.22) and (16.52a), for $r \neq 0$,

$$\boldsymbol{\nabla} \times \frac{\hat{\boldsymbol{r}}}{r^2} = \boldsymbol{\nabla} \times \boldsymbol{\nabla}(-r^{-1}) = 0. \tag{16.55}$$

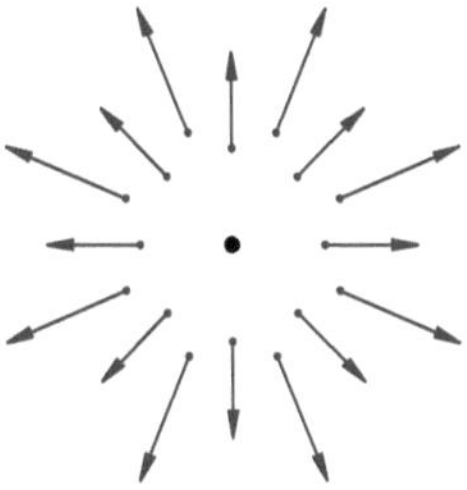

Fig. 16.26 $\boldsymbol{\nabla} \times F(r)\hat{\boldsymbol{r}} = 0$.

In fact, from (16.14a), (16.19), and (16.52c), the curl of any radial field vanishes (Fig. 16.26),

$$\boldsymbol{\nabla} \times F(r)\hat{\boldsymbol{r}} = \boldsymbol{\nabla} F(r) \times \boldsymbol{\nabla} r = \frac{\mathrm{d}F}{\mathrm{d}r}\hat{\boldsymbol{r}} \times \hat{\boldsymbol{r}} = 0. \tag{16.56}$$

The curl of the line vortex field (16.3) is also best obtained using fundamental properties, for $\rho \neq 0$,

$$\nabla \times \left(\frac{\hat{\mathbf{z}} \times \boldsymbol{r}}{|\hat{\mathbf{z}} \times \boldsymbol{r}|^2} \right) = \nabla \times \frac{\hat{\boldsymbol{\varphi}}}{\rho} = \nabla \times \nabla \varphi = 0, \tag{16.57}$$

but the curl of any other azimuthal field does not vanish identically as shown by (16.54) that only vanishes for $F(\rho) = 1/\rho$.

The curl properties (16.52) can be used together with the proper form for del (16.49) to derive the expressions for $\nabla \times \boldsymbol{F}$ in (16.51). In cartesian coordinates, with $\hat{\mathbf{x}}, \hat{\mathbf{y}}, \hat{\mathbf{z}}$ fixed, we find

$$\begin{aligned} \nabla \times \boldsymbol{F} &= \nabla \times (F_x \hat{\mathbf{x}} + F_y \hat{\mathbf{y}} + F_z \hat{\mathbf{z}}) \\ &= (\nabla F_x) \times \hat{\mathbf{x}} + (\nabla F_y) \times \hat{\mathbf{y}} + (\nabla F_z) \times \hat{\mathbf{z}} \\ &= \hat{\mathbf{x}} \, (\partial_y F_z - \partial_z F_y) + \hat{\mathbf{y}} \, (\partial_z F_x - \partial_x F_z) + \hat{\mathbf{z}} \, (\partial_x F_y - \partial_y F_x), \end{aligned} \tag{16.58}$$

since $\nabla = \hat{\mathbf{x}} \partial_x + \hat{\mathbf{y}} \partial_z + \hat{\mathbf{z}} \partial_z$, thus recovering (16.51a).

For cylindrical and spherical coordinates, the direction vectors $\hat{\boldsymbol{\rho}}, \hat{\boldsymbol{\varphi}}, \hat{\boldsymbol{r}}, \hat{\boldsymbol{\theta}}$ are functions of the coordinates. One approach to computing $\nabla \times \boldsymbol{F}$ is to apply ∇ to the direction vectors using their known derivatives (10.19), (10.24a), or (15.17b). For example, consider a meridional field $\boldsymbol{F} = F_\theta \hat{\boldsymbol{\theta}}$ with F_θ a differentiable function of (r, θ, φ) and $\hat{\boldsymbol{\theta}}$ varying with (θ, φ) but not r. From the curl properties (16.52) and del in spherical coordinates (16.49),

$$\begin{aligned} \nabla \times (F_\theta \hat{\boldsymbol{\theta}}) &= (\nabla F_\theta) \times \hat{\boldsymbol{\theta}} + F_\theta \nabla \times \hat{\boldsymbol{\theta}} \\ &= \frac{\partial F_\theta}{\partial r} \hat{\boldsymbol{r}} \times \hat{\boldsymbol{\theta}} + \frac{\partial F_\theta}{\partial \varphi} \frac{\hat{\boldsymbol{\varphi}} \times \hat{\boldsymbol{\theta}}}{r \sin\theta} + F_\theta \left(\frac{\hat{\boldsymbol{\theta}}}{r} \times \frac{\partial \hat{\boldsymbol{\theta}}}{\partial \theta} + \frac{\hat{\boldsymbol{\varphi}}}{r \sin\theta} \times \frac{\partial \hat{\boldsymbol{\theta}}}{\partial \varphi} \right) \\ &= \frac{\partial F_\theta}{\partial r} \hat{\boldsymbol{\varphi}} - \frac{\partial F_\theta}{\partial \varphi} \frac{\hat{\boldsymbol{r}}}{r \sin\theta} + F_\theta \left(\frac{\hat{\boldsymbol{\theta}}}{r} \times (-\hat{\boldsymbol{r}}) + \frac{\hat{\boldsymbol{\varphi}}}{r \sin\theta} \times \cos\theta \, \hat{\boldsymbol{\varphi}} \right) \\ &= \left(\frac{\partial F_\theta}{\partial r} + \frac{F_\theta}{r} \right) \hat{\boldsymbol{\varphi}} - \frac{\partial F_\theta}{\partial \varphi} \frac{\hat{\boldsymbol{r}}}{r \sin\theta} = \frac{\hat{\boldsymbol{\varphi}}}{r} \frac{\partial (r F_\theta)}{\partial r} - \frac{\partial F_\theta}{\partial \varphi} \frac{\hat{\boldsymbol{r}}}{r \sin\theta}, \end{aligned}$$

which matches (16.51c) for $F_r = F_\varphi = 0$. The full (16.51c) can be obtained by proceeding similarly for the $\hat{\boldsymbol{r}}$ and $\hat{\boldsymbol{\varphi}}$ terms.

A better way to handle the varying direction vectors $\hat{\boldsymbol{\rho}}, \hat{\boldsymbol{\varphi}}, \hat{\boldsymbol{r}}, \hat{\boldsymbol{\theta}}$ is to use the identity $\nabla \times \nabla f = 0$ and fundamental gradients (16.14). For the same $\boldsymbol{F} = F_\theta \hat{\boldsymbol{\theta}}$, we begin with $\hat{\boldsymbol{\theta}} = r \nabla \theta$ (16.14b) to rewrite the direction vectors as multiples of gradients. Then using (16.52),

$$\begin{aligned} \nabla \times (F_\theta \hat{\boldsymbol{\theta}}) &= \nabla \times (r F_\theta \nabla \theta) = \nabla (r F_\theta) \times \nabla \theta = \nabla (r F_\theta) \times \hat{\boldsymbol{\theta}} / r \\ &= \frac{\partial (r F_\theta)}{\partial r} \frac{\hat{\boldsymbol{r}} \times \hat{\boldsymbol{\theta}}}{r} + \frac{\partial (r F_\theta)}{\partial \varphi} \frac{\hat{\boldsymbol{\varphi}} \times \hat{\boldsymbol{\theta}}}{r^2 \sin\theta} \\ &= \frac{\partial (r F_\theta)}{\partial r} \frac{\hat{\boldsymbol{\varphi}}}{r} - \frac{\partial F_\theta}{\partial \varphi} \frac{\hat{\boldsymbol{r}}}{r \sin\theta}, \end{aligned} \tag{16.59}$$

which again matches (16.51c) for $F_r = F_\varphi = 0$. This second approach is more efficient, and is applied to general orthogonal coordinates in §16.9.

16.6 Stokes' theorem

It follows directly from the expression for the curl $\boldsymbol{\nabla} \times \boldsymbol{F}$ in general coordinates (16.47c) that (Fig. 16.27)

$$\boldsymbol{\nabla} \times \boldsymbol{F} \cdot \left(\frac{\partial \boldsymbol{r}}{\partial u} \times \frac{\partial \boldsymbol{r}}{\partial v} \right) = \frac{\partial}{\partial u} \left(\boldsymbol{F} \cdot \frac{\partial \boldsymbol{r}}{\partial v} \right) - \frac{\partial}{\partial v} \left(\boldsymbol{F} \cdot \frac{\partial \boldsymbol{r}}{\partial u} \right), \tag{16.60}$$

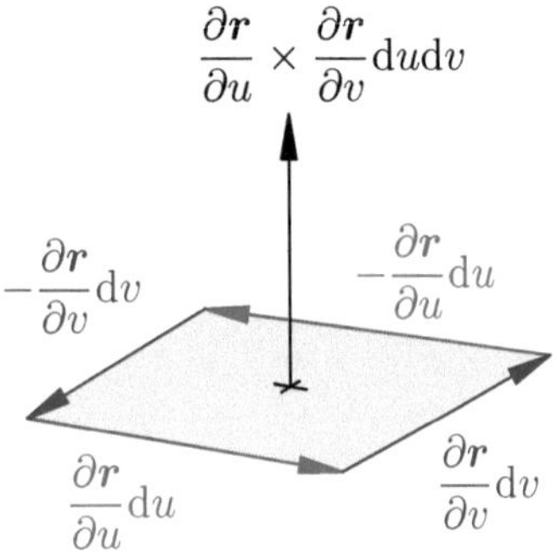

Fig. 16.27 Line elements $\mathrm{d}\boldsymbol{r}$ around a (u, v) coordinate surface element. Net circulation of a vector field $\boldsymbol{F}$ around that element is the curl dotted with the surface element eqns (16.60), (16.61), $(\boldsymbol{\nabla} \times \boldsymbol{F}) \cdot (\partial_u \boldsymbol{r} \times \partial_v \boldsymbol{r})\, \mathrm{d}u\, \mathrm{d}v$.

with two similar equations by cyclic rotations of (u, v, w). Multiplying both sides of (16.60) by $\mathrm{d}u\, \mathrm{d}v$, integrating over a region $(u, v) \in \mathcal{U} \in \mathbb{R}^2$ for fixed w, and using Green's theorem (13.31c) yields *Stokes' theorem*

$$\int_{\mathcal{S}} \boldsymbol{\nabla} \times \boldsymbol{F} \cdot \mathrm{d}\boldsymbol{S} = \oint_{\partial \mathcal{S}} \boldsymbol{F} \cdot \mathrm{d}\boldsymbol{r}, \tag{16.61}$$

where

$$\mathrm{d}\boldsymbol{S} = \frac{\partial \boldsymbol{r}}{\partial u} \times \frac{\partial \boldsymbol{r}}{\partial v} \mathrm{d}u\mathrm{d}v, \qquad \mathrm{d}\boldsymbol{r} = \frac{\partial \boldsymbol{r}}{\partial u}\mathrm{d}u + \frac{\partial \boldsymbol{r}}{\partial v}\mathrm{d}v$$

are the surface and line elements of a surface $\mathcal{S}$ with boundary curve $\partial\mathcal{S}$. The surface $\mathcal{S}$ is the image of $\mathcal{U} \in \mathbb{R}^2$ through the map $\boldsymbol{r}(u, v, w)$ with w fixed, while $\partial\mathcal{S}$ is the boundary of $\mathcal{S}$ that is the image of the boundary of $\mathcal{U}$ through the same map. Since the coordinates (u, v, w) are arbitrary, Stokes' theorem (16.61) is a general result valid for any continuously differentiable vector field $\boldsymbol{F}(P)$ and orientable surface $\mathcal{S}$ with boundary $\partial\mathcal{S}$. The orientation of $\mathcal{S}$ and $\partial\mathcal{S}$ must match according to the right-hand rule, that is, if $\hat{\boldsymbol{n}}(P)$ is the normal to the surface and $\hat{\boldsymbol{t}}(P)$ the tangent to the boundary curve $\partial\mathcal{S}$ at point P, then the tangent $\hat{\boldsymbol{n}} \times \hat{\boldsymbol{t}}$ must point toward the surface. The surface $\mathcal{S}$ can have holes with multiple properly oriented boundaries $\partial\mathcal{S}_k$, as follows from Green's theorem (13.31e).

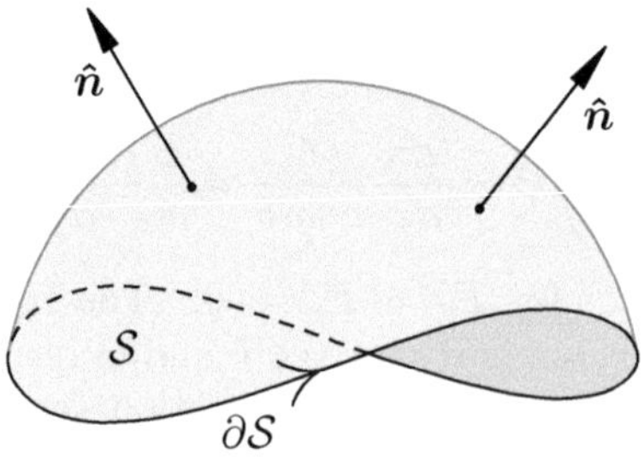

Fig. 16.28 Surface $\mathcal{S}$ with boundary $\partial\mathcal{S}$ and right hand rule orientation for Stokes' theorem.

Stokes' theorem states that the flux of $\boldsymbol{\nabla} \times \boldsymbol{F}$ through a surface $\mathcal{S}$ equals the circulation of $\boldsymbol{F}$ around the curve boundary of $\mathcal{S}$ (Fig. 16.28). It is a fundamental theorem of vector calculus and provides a geometric interpretation for the curl as circulation per area in the limit of that area shrinking to a point P,

$$\hat{\boldsymbol{n}} \cdot \boldsymbol{\nabla} \times \boldsymbol{F} = \lim_{A \to 0} \frac{\oint_{\partial A} \boldsymbol{F} \cdot \mathrm{d}\boldsymbol{r}}{A}, \tag{16.62}$$

where $\hat{\boldsymbol{n}}$ is the unit normal to the surface patch of vanishing area A. This patch can be a disk or a triangle or any shape suitable to the geometry, as illustrated in Fig. 16.27 and obtained in eqn (16.60).

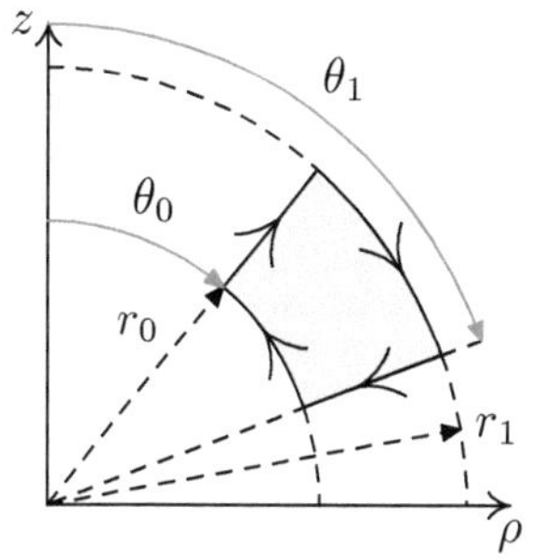

Fig. 16.29 (r, θ) sector.

Consider, for example, the circulation of vector field $\boldsymbol{F} = F_r\hat{\boldsymbol{r}} + F_\theta\hat{\boldsymbol{\theta}} + F_\varphi\hat{\boldsymbol{\varphi}}$ about the meridional sector sketched in Fig. 16.29 with fixed φ and

$$(r_0, \theta_0,) \to (r_1, \theta_0) \to (r_1, \theta_1) \to (r_0, \theta_1) \to (r_0, \theta_0),$$

where the components F_r, F_θ, F_φ are arbitrary continuously differentiable functions of spherical coordinates (r, θ, φ), and $\{\hat{\boldsymbol{r}}, \hat{\boldsymbol{\theta}}, \hat{\boldsymbol{\varphi}}\}$ are the usual spherical direction vectors. That contour consists of two radial segments along which $\boldsymbol{F}\cdot \mathrm{d}\boldsymbol{r} = \boldsymbol{F}\cdot\hat{\boldsymbol{r}}\,\mathrm{d}r = F_r dr$ and two meridional segments along which $\boldsymbol{F}\cdot\mathrm{d}\boldsymbol{r} = \boldsymbol{F}\cdot r\hat{\boldsymbol{\theta}}\mathrm{d}\theta = rF_\theta\mathrm{d}\theta$. The circulation around that closed curve $\mathcal{C}$ is then, keeping φ implicit since it is fixed,

$$\oint_{\mathcal{C}} \boldsymbol{F}\cdot\mathrm{d}\boldsymbol{r} = \int_{r_0}^{r_1} \left(F_r(r,\theta_0) - F_r(r,\theta_1)\right)\mathrm{d}r + \int_{\theta_0}^{\theta_1} \left(r_1F_\theta(r_1,\theta) - r_0F_\theta(r_0,\theta)\right)\mathrm{d}\theta.$$

Both of these integrals can be written as double integrals using the fundamental theorem of calculus

$$f(b) - f(a) = \int_a^b \frac{\mathrm{d}f}{\mathrm{d}x}\,\mathrm{d}x, \tag{16.63}$$

yielding

$$\oint_{\mathcal{C}} \boldsymbol{F}\cdot\mathrm{d}\boldsymbol{r} = \int_{r_0}^{r_1}\int_{\theta_0}^{\theta_1}\left(-\frac{\partial F_r}{\partial\theta} + \frac{\partial\,(rF_\theta)}{\partial r}\right)\mathrm{d}\theta\mathrm{d}r. \tag{16.64a}$$

Now the flux of $\boldsymbol{\nabla}\times\boldsymbol{F}$ through the surface patch $\mathcal{S}$ enclosed by $\mathcal{C} = \partial\mathcal{S}$ with $\mathrm{d}\boldsymbol{S} = (\partial_r\boldsymbol{r}\times\partial_\theta\boldsymbol{r})\mathrm{d}r\mathrm{d}\theta = (\hat{\boldsymbol{r}}\times r\hat{\boldsymbol{\theta}})\mathrm{d}r\mathrm{d}\theta = \hat{\boldsymbol{\varphi}}r\mathrm{d}r\mathrm{d}\theta$ is

$$\int_{\mathcal{S}} \boldsymbol{\nabla}\times\boldsymbol{F}\cdot\mathrm{d}\boldsymbol{S} = \int_{r_0}^{r_1}\int_{\theta_0}^{\theta_1} \boldsymbol{\nabla}\times\boldsymbol{F}\cdot\hat{\boldsymbol{\varphi}}\,r\,\mathrm{d}r\mathrm{d}\theta. \tag{16.64b}$$

Stokes' theorem states that (16.64a) equals (16.64b) for any r_0, r_1, θ_0, θ_1. Since we assume continuity of the integrands that requires

$$\boldsymbol{\nabla}\times\boldsymbol{F}\cdot\hat{\boldsymbol{\varphi}}\,r = -\frac{\partial F_r}{\partial\theta} + \frac{\partial\,(rF_\theta)}{\partial r},$$

verifying Stokes' theorem, and illustrating another way to derive (16.51c) from it.

The curl came to light in magnetostatics as the differential version of Ampère's law. Ampère's law states that a steady current $\boldsymbol{J}$ induces a magnetic field $\boldsymbol{B}$ such that the circulation of $\boldsymbol{B}$ around a closed curve $\mathcal{C}$ is proportional to the net current through any surface orientable $\mathcal{S}$ with boundary $\partial\mathcal{S} = \mathcal{C}$,

$$\oint_{\partial\mathcal{S}} \boldsymbol{B}\cdot\mathrm{d}\boldsymbol{r} = \mu_0\int_{\mathcal{S}} \boldsymbol{j}\cdot\mathrm{d}\boldsymbol{S}, \tag{16.65}$$

where $\boldsymbol{j}$ is the current density. Stokes' theorem leads to the local version of Ampère's law, $\boldsymbol{\nabla}\times\boldsymbol{B} = \mu_0\boldsymbol{j}$, since the integral version holds for any surface $\mathcal{S}$.

16.7 Divergence

We have shown that a vector field is a gradient when it has no curl (§16.5). Here, we ask the next question: when is a vector field a curl? That is, given a vector field $\boldsymbol{F}(P)$, is there another vector field $\boldsymbol{A}(P)$ such that $\boldsymbol{F} = \boldsymbol{\nabla} \times \boldsymbol{A}$?

If it exists, that vector field $\boldsymbol{A}$ is the *vector potential* of $\boldsymbol{F}$, and $\boldsymbol{F}$ is a *solenoidal* field.[3] The vector potential $\boldsymbol{A}$ is defined up to an arbitrary gradient field since the curl of a gradient vanishes (16.52a); thus, for any smooth scalar field f,

[3] A "curl" field, $\boldsymbol{F} = \boldsymbol{\nabla} \times \boldsymbol{A}$, is called a *solenoidal field.*

$$\boldsymbol{\nabla} \times (\boldsymbol{A} + \boldsymbol{\nabla} f) = \boldsymbol{\nabla} \times \boldsymbol{A}.$$

From (16.47), $\boldsymbol{\nabla} \times \boldsymbol{A} = J^{-1}\epsilon^{ijk}\left(\partial_j A_k\right)\partial_i \boldsymbol{r}$. Expanding $\boldsymbol{F} = F^i \partial_i \boldsymbol{r}$ in the velocity basis, yields

$$\boldsymbol{F} = \boldsymbol{\nabla} \times \boldsymbol{A} \Leftrightarrow F^i = J^{-1}\epsilon^{ijk}\partial_j A_k.$$

The integrability condition for $\boldsymbol{F}$ to be a curl is thus

$$J = \frac{\partial \boldsymbol{r}}{\partial u} \times \frac{\partial \boldsymbol{r}}{\partial v} \cdot \frac{\partial \boldsymbol{r}}{\partial w}.$$

$$\partial_i \left(J F^i\right) = 0, \tag{16.66}$$

because $\epsilon^{ijk}\partial_i\partial_j A_k = 0$ since $\epsilon^{ijk} = -\epsilon^{jik}$ but $\partial_i\partial_j = \partial_j\partial_i$.

The left-hand side of eqn (16.66) corresponds to a scalar invariant called the *divergence* of $\boldsymbol{F}$ given by

$$\operatorname{div} \boldsymbol{F} = J^{-1}\partial_i\left(J F^i\right) = \frac{1}{J}\frac{\partial}{\partial u^i}\left(J\,\boldsymbol{F} \cdot \boldsymbol{\nabla} u^i\right), \tag{16.67a}$$

in general coordinates (u^1, u^2, u^3). This divergence $\operatorname{div} \boldsymbol{F} = \boldsymbol{\nabla} \cdot \boldsymbol{F}$. Indeed, from (16.34), (16.46), and $\epsilon^{ijk}\partial_i\partial_j\boldsymbol{r} = 0 = \epsilon^{ijk}\partial_i\partial_k\boldsymbol{r}$,

$$\boldsymbol{\nabla} u^i = \frac{\epsilon^{ijk}}{2J}\partial_j \boldsymbol{r} \times \partial_k \boldsymbol{r}.$$

$$\begin{aligned}
\boldsymbol{\nabla} \cdot \boldsymbol{F} &= \boldsymbol{\nabla} u^i \cdot \partial_i \boldsymbol{F} \\
&= J^{-1}\tfrac{1}{2}\epsilon^{ijk}\partial_j \boldsymbol{r} \times \partial_k \boldsymbol{r} \cdot \partial_i \boldsymbol{F} \\
&= J^{-1}\partial_i\left(\tfrac{1}{2}\epsilon^{ijk}\partial_j \boldsymbol{r} \times \partial_k \boldsymbol{r} \cdot \boldsymbol{F}\right) \\
&= J^{-1}\partial_i\left(J\boldsymbol{\nabla} u^i \cdot \boldsymbol{F}\right) = J^{-1}\partial_i\left(J F^i\right).
\end{aligned} \tag{16.67b}$$

For $(u, v, w) = (u^1, u^2, u^3)$, (16.67b) reads[4]

[4] Note that we have multiplied through by the Jacobian J for display purposes. The divergence theorem is emerging.

$$J\boldsymbol{\nabla} \cdot \boldsymbol{F} = \frac{\partial}{\partial u}(J\boldsymbol{F} \cdot \boldsymbol{\nabla} u) + \frac{\partial}{\partial v}(J\boldsymbol{F} \cdot \boldsymbol{\nabla} v) + \frac{\partial}{\partial w}(J\boldsymbol{F} \cdot \boldsymbol{\nabla} w) \tag{16.68}$$

$$= \frac{\partial}{\partial u}\left(\boldsymbol{F} \cdot \frac{\partial \boldsymbol{r}}{\partial v} \times \frac{\partial \boldsymbol{r}}{\partial w}\right) + \frac{\partial}{\partial v}\left(\boldsymbol{F} \cdot \frac{\partial \boldsymbol{r}}{\partial w} \times \frac{\partial \boldsymbol{r}}{\partial u}\right) + \frac{\partial}{\partial w}\left(\boldsymbol{F} \cdot \frac{\partial \boldsymbol{r}}{\partial u} \times \frac{\partial \boldsymbol{r}}{\partial v}\right).$$

The latter expression follows from $J\boldsymbol{\nabla} u = \partial \boldsymbol{r}/\partial v \times \partial \boldsymbol{r}/\partial w$ (16.28), and similarly for $J\boldsymbol{\nabla} v$ and $J\boldsymbol{\nabla} w$. Invariance of $\boldsymbol{\nabla} \cdot \boldsymbol{F}$ follows from invariance of $\boldsymbol{\nabla}$ (16.48).

Divergence in standard coordinates

For a vector field $\boldsymbol{F}(P)$ expressed in cartesian, cylindrical, or spherical coordinates as in (16.50), the contravariant components for $(u^1, u^2, u^3) = (x, y, z)$, (ρ, φ, z), or (r, θ, φ), with corresponding Jacobians $J = 1$, $J = \rho$, or $J = r^2 \sin\theta$, are obtained from (16.14), respectively,

$$
\begin{aligned}
&F^1 \equiv \boldsymbol{F} \cdot \boldsymbol{\nabla} x = F_x, \quad F^2 \equiv \boldsymbol{F} \cdot \boldsymbol{\nabla} y = F_y, \quad F^3 \equiv \boldsymbol{F} \cdot \boldsymbol{\nabla} z = F_z, \\
&F^1 \equiv \boldsymbol{F} \cdot \boldsymbol{\nabla} \rho = F_\rho, \quad F^2 \equiv \boldsymbol{F} \cdot \boldsymbol{\nabla} \varphi = \frac{F_\varphi}{\rho}, \quad F^3 \equiv \boldsymbol{F} \cdot \boldsymbol{\nabla} z = F_z, \\
&F^1 \equiv \boldsymbol{F} \cdot \boldsymbol{\nabla} r = F_r, \quad F^2 \equiv \boldsymbol{F} \cdot \boldsymbol{\nabla} \theta = \frac{F_\theta}{r}, \quad F^3 \equiv \boldsymbol{F} \cdot \boldsymbol{\nabla} \varphi = \frac{F_\varphi}{r \sin\theta}.
\end{aligned}
$$

Substituting in the general coordinate formula (16.68), we obtain

$$
\begin{aligned}
\boldsymbol{\nabla} \cdot \boldsymbol{F} &= \frac{\partial F_x}{\partial x} + \frac{\partial F_y}{\partial y} + \frac{\partial F_z}{\partial z} \\
&= \frac{1}{\rho} \frac{\partial (\rho F_\rho)}{\partial \rho} + \frac{1}{\rho} \frac{\partial F_\varphi}{\partial \varphi} + \frac{\partial F_z}{\partial z} \\
&= \frac{1}{r^2} \frac{\partial (r^2 F_r)}{\partial r} + \frac{1}{r \sin\theta} \frac{\partial (\sin\theta \, F_\theta)}{\partial \theta} + \frac{1}{r \sin\theta} \frac{\partial F_\varphi}{\partial \varphi}.
\end{aligned}
\tag{16.69}
$$

Divergence properties

The defining property is of course that the divergence of a curl vanishes (16.66),

$$\boldsymbol{\nabla} \cdot \boldsymbol{\nabla} \times \boldsymbol{A} = 0, \tag{16.70a}$$

for any C^2 vector field $\boldsymbol{A}$. This is easy to remember symbolically from the vector identity $\boldsymbol{a} \cdot \boldsymbol{a} \times \boldsymbol{b} = 0$, though one must be careful with this analogy because $\boldsymbol{\nabla}$ is a vector *differential operator*; thus, for example,

$$\boldsymbol{\nabla} \cdot x \boldsymbol{\nabla} \times y \hat{\mathbf{z}} = \boldsymbol{\nabla} x \cdot \boldsymbol{\nabla} y \times \hat{\mathbf{z}} = \hat{\mathbf{x}} \cdot \hat{\mathbf{y}} \times \hat{\mathbf{z}} = 1 \neq 0.$$

The divergence is a scalar differential operator that obeys the summation and product rules

$$
\begin{aligned}
\boldsymbol{\nabla} \cdot (\boldsymbol{F} + \boldsymbol{G}) &= \boldsymbol{\nabla} \cdot \boldsymbol{F} + \boldsymbol{\nabla} \cdot \boldsymbol{G}, \\
\boldsymbol{\nabla} \cdot (f \boldsymbol{G}) &= (\boldsymbol{\nabla} f) \cdot \boldsymbol{G} + f \boldsymbol{\nabla} \cdot \boldsymbol{G}.
\end{aligned}
\tag{16.70b}
$$

As for the curl, these follow from (16.68) and the standard differentiation rules, with the extra understanding that the divergence of a scalar is not defined; thus, $(\boldsymbol{\nabla} \cdot f) \boldsymbol{G}$ is nonsense, and $(\boldsymbol{\nabla} f) \cdot \boldsymbol{G}$ is the correct expression. The dot and cross products, $\boldsymbol{a} \cdot \boldsymbol{b}$ and $\boldsymbol{a} \times \boldsymbol{b}$, are defined between vectors $\boldsymbol{a}$ and $\boldsymbol{b}$, not vector and scalar.

Divergence examples

The divergence of $\boldsymbol{r} = x\hat{\mathbf{x}} + y\hat{\mathbf{y}} + z\hat{\mathbf{z}}$ and that of $\hat{\mathbf{z}} \times \boldsymbol{r} = x\hat{\mathbf{y}} - y\hat{\mathbf{x}}$ are easy to calculate in cartesian coordinates

$$\begin{aligned} \nabla \cdot \boldsymbol{r} &= \frac{\partial x}{\partial x} + \frac{\partial y}{\partial y} + \frac{\partial z}{\partial z} = 3, \\ \nabla \cdot (\hat{\mathbf{z}} \times \boldsymbol{r}) &= \frac{\partial(-y)}{\partial x} + \frac{\partial x}{\partial y} = 0. \end{aligned} \tag{16.71}$$

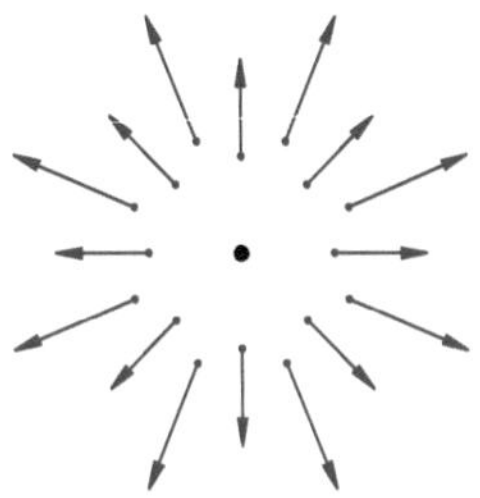

Fig. 16.30 $\nabla \cdot (F(r)\hat{\boldsymbol{r}}) = \frac{1}{r^2}\frac{\mathrm{d}(r^2 F)}{\mathrm{d}r}$.

The divergence of the inverse square law (16.2) is straightforward in spherical coordinates if (16.69) is available, or, directly, using (16.14a) and the product rule (16.70b) with $\boldsymbol{r} = r\hat{\boldsymbol{r}}$, we find for $r \neq 0$,

$$\nabla \cdot \frac{\boldsymbol{r}}{r^3} = \left(\nabla \frac{1}{r^3}\right) \cdot \boldsymbol{r} + \frac{1}{r^3} \nabla \cdot \boldsymbol{r} = -\frac{3\hat{\boldsymbol{r}}}{r^4} \cdot \boldsymbol{r} + \frac{3}{r^3} = 0. \tag{16.72}$$

This approach applies to all radial fields $\boldsymbol{F} = F(r)\hat{\boldsymbol{r}}$ that *diverge* in general, except for the inverse square law $F(r) = 1/r^2$ (Fig. 16.30).

Similarly, the divergence of the line vortex field (16.3) is a straightforward application of the cylindrical coordinate formula (16.69) since $\hat{\mathbf{z}} \times \boldsymbol{r} = \rho\hat{\boldsymbol{\varphi}}$, but is easily calculated as

$$\nabla \cdot \frac{\hat{\mathbf{z}} \times \boldsymbol{r}}{|\hat{\mathbf{z}} \times \boldsymbol{r}|^2} = \left(\nabla \frac{1}{\rho^2}\right) \cdot \rho\hat{\boldsymbol{\varphi}} + \frac{1}{\rho^2} \nabla \cdot (\hat{\mathbf{z}} \times \boldsymbol{r}) = 0, \tag{16.73}$$

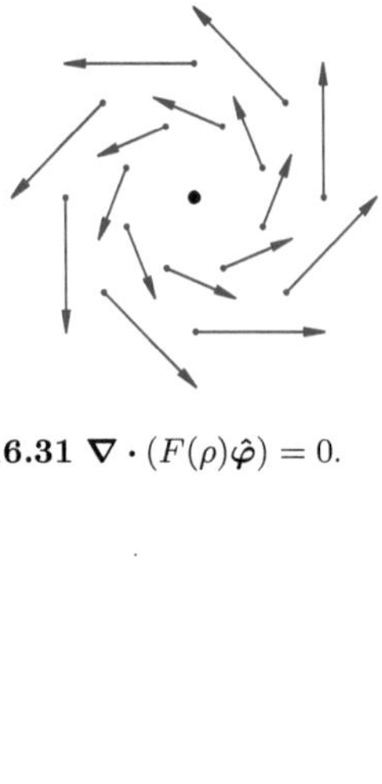

Fig. 16.31 $\nabla \cdot (F(\rho)\hat{\boldsymbol{\varphi}}) = 0$.

for $\rho \neq 0$, since $\nabla \rho^{-2} = -2\hat{\boldsymbol{\rho}}/\rho^3$ is orthogonal to $\hat{\boldsymbol{\varphi}}$. This method applies to any azimuthal field $\boldsymbol{F} = F(\rho)\hat{\boldsymbol{\varphi}}$ whose divergence vanishes for any differentiable $F(\rho)$ (Fig. 16.31).

The divergence properties (16.70) can also be used to derive the cartesian, cylindrical, and spherical expressions (16.69), either by differentiating the direction vectors as illustrated for the curl or, in a better way, by expressing them as cross products of gradients and taking advantage of the identity $\nabla \cdot \nabla \times \boldsymbol{F} = 0$ combined with $\nabla \times f\nabla g = \nabla f \times \nabla g$ from (16.52) that yield the identity

$$\nabla \cdot (\nabla f \times \nabla g) = 0. \tag{16.74}$$

For example,

$$\begin{aligned} \nabla \cdot (F_\theta \hat{\boldsymbol{\theta}}) &= \nabla \cdot (F_\theta\, \hat{\boldsymbol{\varphi}} \times \hat{\boldsymbol{r}}) = \nabla \cdot (F_\theta\, r \sin\theta\, \nabla\varphi \times \nabla r) \\ &= \nabla(F_\theta r \sin\theta) \cdot (\nabla\varphi \times \nabla r) = \frac{1}{r\sin\theta} \frac{\partial(r\sin\theta\, F_\theta)}{\partial\theta}, \end{aligned} \tag{16.75}$$

since $\nabla r = \hat{\boldsymbol{r}}$, $\nabla\varphi = \hat{\boldsymbol{\varphi}}/(r\sin\theta)$, $\nabla \cdot (\nabla\varphi \times \nabla r) = \nabla \cdot (\nabla \times \varphi\nabla r) = 0$ from (16.74), and using the spherical form of ∇ (16.49) in the last step. This approach can be used to compute $\nabla \cdot \boldsymbol{F}$ in arbitrary orthogonal coordinates (§16.9).

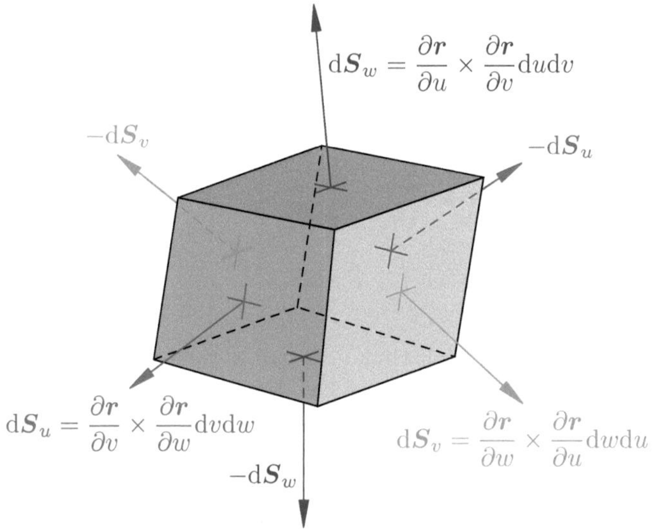

Fig. 16.32 Surface elements $d\boldsymbol{S}$ for a (u, v, w) curvilinear volume element. The net outward flux of a vector field $\boldsymbol{F}$ through that box surface boundary is the divergence times the volume of the box $(\boldsymbol{\nabla} \cdot \boldsymbol{F})\, J\, du\, dv\, dw$ (16.76).

16.8 Divergence theorem

The divergence theorem, and a geometric interpretation for the divergence, readily follows from the general formula (16.68)

$$J\,\boldsymbol{\nabla} \cdot \boldsymbol{F} = \frac{\partial}{\partial u}(JF^u) + \frac{\partial}{\partial v}(JF^v) + \frac{\partial}{\partial w}(JF^w) \tag{16.76}$$

$$= \frac{\partial}{\partial u}\left(\boldsymbol{F} \cdot \frac{\partial \boldsymbol{r}}{\partial v} \times \frac{\partial \boldsymbol{r}}{\partial w}\right) + \frac{\partial}{\partial v}\left(\boldsymbol{F} \cdot \frac{\partial \boldsymbol{r}}{\partial w} \times \frac{\partial \boldsymbol{r}}{\partial u}\right) + \frac{\partial}{\partial w}\left(\boldsymbol{F} \cdot \frac{\partial \boldsymbol{r}}{\partial u} \times \frac{\partial \boldsymbol{r}}{\partial v}\right)$$

in arbitrary coordinates (u, v, w), with Jacobian J. Multiplying both sides of (16.76) by the coordinate differentials $du\, dv\, dw$ (Fig. 16.32), and integrating over a coordinate "brick," that is, a rectangular domain in (u, v, w) coordinate space,

$J = \frac{\partial \boldsymbol{r}}{\partial u} \times \frac{\partial \boldsymbol{r}}{\partial v} \cdot \frac{\partial \boldsymbol{r}}{\partial w}.$

$$\mathcal{U} : u_0 \le u \le u_1, \quad v_0 \le v \le v_1, \quad w_0 \le w \le w_1, \tag{16.77}$$

where $u_0, u_1, v_0, v_1, w_0, w_1$ are constants, yields

$$\begin{aligned} \int_{\mathcal{V}} \boldsymbol{\nabla} \cdot \boldsymbol{F}\, dV &= \int_{\mathcal{S}_{u_1}} \boldsymbol{F} \cdot d\boldsymbol{S}_u - \int_{\mathcal{S}_{u_0}} \boldsymbol{F} \cdot d\boldsymbol{S}_u \\ &+ \int_{\mathcal{S}_{v_1}} \boldsymbol{F} \cdot d\boldsymbol{S}_v - \int_{\mathcal{S}_{v_0}} \boldsymbol{F} \cdot d\boldsymbol{S}_v + \int_{\mathcal{S}_{w_1}} \boldsymbol{F} \cdot d\boldsymbol{S}_w - \int_{\mathcal{S}_{w_0}} \boldsymbol{F} \cdot d\boldsymbol{S}_w, \end{aligned} \tag{16.78}$$

where $dV = J\, du\, dv\, dw$ is the measure of the volume element, and[5]

$$d\boldsymbol{S}_u = \frac{\partial \boldsymbol{r}}{\partial v} \times \frac{\partial \boldsymbol{r}}{\partial w}\, dv\, dw = J \boldsymbol{\nabla} u\, dv dw$$

is the surface element on a u-isosurface, and similarly for $d\boldsymbol{S}_v$, $d\boldsymbol{S}_w$. The 3D domain $\mathcal{V}$ is the image of $\mathcal{U}$ under the map $\boldsymbol{r}(u, v, w)$, and the six boundary surfaces are the u_0, u_1, v_0, v_1, w_0 and w_1-isosurfaces

[5] The u-isosurface element $d\boldsymbol{S}_u$ was written $d\boldsymbol{S}_{vw}$ in (15.25).

corresponding to the same $\boldsymbol{r}(u,v,w)$. The surface integrals arise from the fundamental theorem of calculus. For example, (Fig. 16.33)

$$\int_{u_0}^{u_1}\int_{v_0}^{v_1}\int_{w_0}^{w_1}\frac{\partial}{\partial w}\left(\boldsymbol{F}\cdot\frac{\partial \boldsymbol{r}}{\partial u}\times\frac{\partial \boldsymbol{r}}{\partial v}\right)\mathrm{d}w\mathrm{d}v\mathrm{d}u$$
$$=\int_{u_0}^{u_1}\int_{v_0}^{v_1}\left[\left(\boldsymbol{F}\cdot\frac{\partial \boldsymbol{r}}{\partial u}\times\frac{\partial \boldsymbol{r}}{\partial v}\right)_{w_1}-\left(\boldsymbol{F}\cdot\frac{\partial \boldsymbol{r}}{\partial u}\times\frac{\partial \boldsymbol{r}}{\partial v}\right)_{w_0}\right]\mathrm{d}v\mathrm{d}u$$
$$=\int_{\mathcal{S}_{w_1}}\boldsymbol{F}\cdot\mathrm{d}\boldsymbol{S}_w-\int_{\mathcal{S}_{w_0}}\boldsymbol{F}\cdot\mathrm{d}\boldsymbol{S}_w.$$

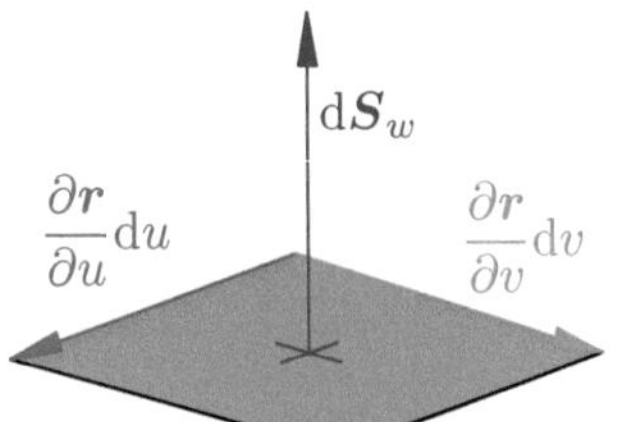

Fig. 16.33 $\mathrm{d}\boldsymbol{S}_w=\frac{\partial \boldsymbol{r}}{\partial u}\times\frac{\partial \boldsymbol{r}}{\partial v}\mathrm{d}u\mathrm{d}v$.

With (u,v,w) ordering such that $J=\partial_u\boldsymbol{r}\times\partial_v\boldsymbol{r}\cdot\partial_w\boldsymbol{r}>0$ the surface normal $\partial_u\boldsymbol{r}\times\partial_v\boldsymbol{r}$ points outside of the domain at the upper bound w_1-isosurface but *inside* on the lower bound w_0 (Fig. 16.32), and similarly for the other four isosurfaces. Thus with the minus signs in front of the lower surfaces, the six surface integrals in (16.78) yield the net *outward* flux of $\boldsymbol{F}$ through the entire surface boundary $\mathcal{S}=\partial\mathcal{V}$ of the volume $\mathcal{V}$, yielding the *divergence theorem*, also known as *Gauss's theorem*,

$$\int_{\mathcal{V}}\boldsymbol{\nabla}\cdot\boldsymbol{F}\,\mathrm{d}V=\oint_{\partial\mathcal{V}}\boldsymbol{F}\cdot\mathrm{d}\boldsymbol{S},\tag{16.79}$$

where $\partial\mathcal{V}$ is the closed surface boundary of volume $\mathcal{V}$ with surface element $\mathrm{d}\boldsymbol{S}=\hat{\boldsymbol{n}}\,\mathrm{d}A$ pointing *outward*. This is a general result that applies to any sufficiently smooth vector field $\boldsymbol{F}(P)$ and volume $\mathcal{V}$ with boundary surface $\mathcal{S}=\partial\mathcal{V}$ with unit normal $\hat{\boldsymbol{n}}$ pointing outward, and it provides a geometric interpretation for the divergence of $\boldsymbol{F}$ as the limit of the ratio of the flux of $\boldsymbol{F}$ through a closed surface over the volume enclosed by that surface

$$\boldsymbol{\nabla}\cdot\boldsymbol{F}=\lim_{V\to 0}\frac{\oint_{\partial V}\boldsymbol{F}\cdot\mathrm{d}\boldsymbol{S}}{V}.\tag{16.80}$$

The divergence theorem (16.79) can be applied to a spherical coordinate brick (Fig. 16.34)

$$r_0\le r\le r_1,\quad \theta_0\le\theta\le\theta_1,\quad \varphi_0\le\varphi\le\varphi_1,$$

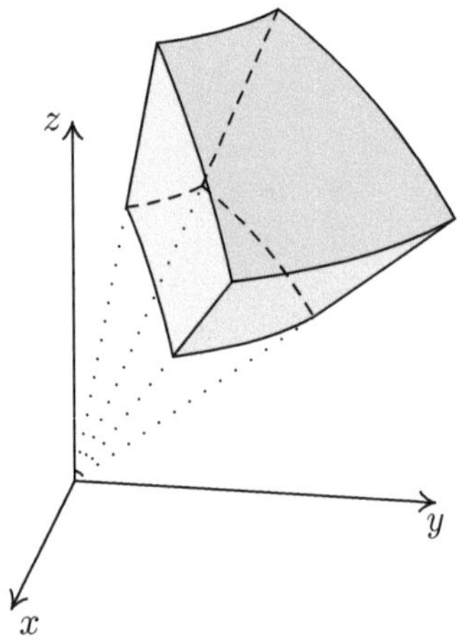

Fig. 16.34 (r,θ,φ) brick.

for constant $r_0,r_1,\theta_0,\theta_1,\varphi_0,\varphi_1$ to derive $\boldsymbol{\nabla}\cdot\boldsymbol{F}$ in spherical coordinates (16.69), for example, similarly to the Stokes' theorem example for the curl in (16.64). This is left as exercise 16.36.

The above derivation is quite general since the coordinates are arbitrary, but the assumption that the boundary consists of coordinate isosurfaces requires further discussion. Coordinate singularities also necessitate a closer look. For example, for a sphere of radius R with the usual spherical coordinates (r,θ,φ) and

$$0\le r\le R,\quad 0\le\theta\le\pi,\quad 0\le\varphi<2\pi,$$

the $r=0$ isosurface reduces to 1 point, the $\theta=0$ and π isosurfaces reduce to line segments, and the flux outward of the $\varphi=0$ and 2π isosurfaces are equal and opposite so they cancel out such that the entire flux comes from the $r = R$ isosurface. A cleaner argument considers the limit $\epsilon \to 0$ for the brick

$$\epsilon R \le r \le R, \quad \epsilon \le \theta \le \pi - \epsilon, \quad 0 \le \varphi \le 2\pi - \epsilon.$$

For a more general derivation, we prove the divergence theorem for a tetrahedral domain, then tile an arbitrary domain with such tetrahedra. Consider a coordinate tetrahedron where three of the edges correspond to u, v, and w coordinate curves, and the three faces with those edges are coordinate isosurfaces (Fig. 16.35). Let P_0, P_1, P_2, P_3 be the vertices of that tetrahedron with coordinates $(u, v, w) = (0,0,0)$, $(1,0,0)$, $(0,1,0)$, $(0,0,1)$, respectively.[6] Let $\mathcal{S}_k$ be the face opposite to vertex P_k, such that $\mathcal{S}_1$, $\mathcal{S}_2$, $\mathcal{S}_3$ are coordinate isosurfaces with $u = 0$, $v = 0$, $w = 0$, respectively. The face $\mathcal{S}_0$ with vertices P_1, P_2, P_3 is assumed to be such that at any point P on $\mathcal{S}_0$ its unit outward normal $\hat{\boldsymbol{n}}(P)$ satisfies

$$\hat{\boldsymbol{n}} \cdot \partial_u \boldsymbol{r} > 0, \quad \hat{\boldsymbol{n}} \cdot \partial_v \boldsymbol{r} > 0, \quad \hat{\boldsymbol{n}} \cdot \partial_w \boldsymbol{r} > 0. \tag{16.81}$$

[6]For a tetrahedron with flat faces, $\boldsymbol{r}(u,v,w) = \boldsymbol{r}_0 + u\,\boldsymbol{a}_1 + v\,\boldsymbol{a}_2 + w\,\boldsymbol{a}_3$, where $\boldsymbol{a}_k$ is the position vector of vertex P_k from vertex P_0.

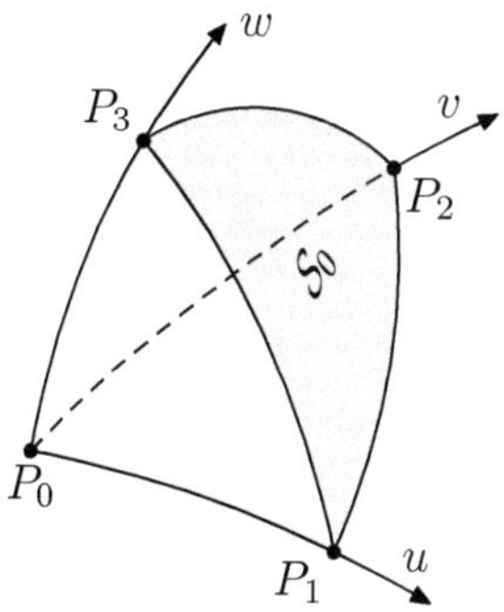

Fig. 16.35 (u, v, w) tetrahedron.

This condition guarantees that all coordinate curves flow outward of the patch $\mathcal{S}_0$, not in and out or tangentially, and $\mathcal{S}_0$ can be equally well parametrized as

$$\begin{aligned}
&(1): 0 \le u \le 1, \quad 0 \le v \le v_{12}(u), \quad w = w_0(u,v),\\
&(2): 0 \le v \le 1, \quad 0 \le w \le w_{23}(u), \quad u = u_0(v,w),\\
&(3): 0 \le w \le 1, \quad 0 \le u \le u_{31}(u), \quad v = v_0(w,u).
\end{aligned}$$

We use parametrization (1) and the fundamental theorem of calculus to integrate the $\partial/\partial w$ term on the right-hand side of (16.76), yielding

$$\int_0^1 \int_0^{v_{12}} \int_0^{w_0} \frac{\partial}{\partial w}(JF^w)\, \mathrm{d}w\mathrm{d}v\mathrm{d}u = \int_0^1 \int_0^{v_{12}} \left[(JF^w)_{\mathcal{S}_0} - (JF^w)_{\mathcal{S}_3}\right] \mathrm{d}v\mathrm{d}u,$$

but the integral on the surface $\mathcal{S}_0$ where $w = w_0(u,v)$ is not the flux of $\boldsymbol{F}$ through that surface anymore because $(\partial_u \boldsymbol{r} \times \partial_v \boldsymbol{r})$ is not normal to that surface. The parametrization for that face $\mathcal{S}_0$ is $\boldsymbol{s}(u,v) = \boldsymbol{r}(u, v, w_0(u,v))$ with surface normal

$$\frac{\partial \boldsymbol{s}}{\partial u} \times \frac{\partial \boldsymbol{s}}{\partial v} = \left(\frac{\partial \boldsymbol{r}}{\partial u} \times \frac{\partial \boldsymbol{r}}{\partial v}\right) + \frac{\partial w_0}{\partial u}\left(\frac{\partial \boldsymbol{r}}{\partial w} \times \frac{\partial \boldsymbol{r}}{\partial v}\right) + \frac{\partial w_0}{\partial v}\left(\frac{\partial \boldsymbol{r}}{\partial u} \times \frac{\partial \boldsymbol{r}}{\partial w}\right),$$

such that its surface element $\mathrm{d}\boldsymbol{S}_0 = (\partial \boldsymbol{s}/\partial u \times \partial \boldsymbol{s}/\partial v)\, \mathrm{d}u\mathrm{d}v$ satisfies[7]

$$\frac{\partial \boldsymbol{r}}{\partial w} \cdot \mathrm{d}\boldsymbol{S}_0 = \frac{\partial \boldsymbol{r}}{\partial w} \cdot \mathrm{d}\boldsymbol{S}_w = J\, \mathrm{d}u\mathrm{d}v \tag{16.82}$$

[7]Similarly for the other parametrizations

$$\mathrm{d}\boldsymbol{S}_0 \cdot \partial_u \boldsymbol{r} = J\mathrm{d}v\mathrm{d}w,$$

$$\mathrm{d}\boldsymbol{S}_0 \cdot \partial_v \boldsymbol{r} = J\mathrm{d}w\mathrm{d}u.$$

since the dot product with $\partial \boldsymbol{r}/\partial w$ cancels out the two extra terms in $\partial_u \boldsymbol{s} \times \partial_v \boldsymbol{s}$ that involve cross products with $\partial \boldsymbol{r}/\partial w$. Now the full $\boldsymbol{F}$ field has the tangent basis expansion

$$\boldsymbol{F} = F^u \partial_u \boldsymbol{r} + F^v \partial_v \boldsymbol{r} + F^w \partial_w \boldsymbol{r},$$

with $(F^u, F^v, F^w) = \boldsymbol{F} \cdot (\boldsymbol{\nabla} u, \boldsymbol{\nabla} v, \boldsymbol{\nabla} w)$, then on surfaces $\mathcal{S}_0$ and $\mathcal{S}_3$

$$JF^w \mathrm{d}u\mathrm{d}v = \boldsymbol{F} \cdot \frac{\partial \boldsymbol{r}}{\partial u} \times \frac{\partial \boldsymbol{r}}{\partial v} \mathrm{d}u\mathrm{d}v = F^w \partial_w \boldsymbol{r} \cdot \mathrm{d}\boldsymbol{S}.$$

Thus, the integral of the $\partial(JF^w)/\partial w$ term in (16.76) gives the flux of the $F^w \partial_w \boldsymbol{r}$ component of $\boldsymbol{F}$ out of face $\mathcal{S}_0$ minus that into face $\mathcal{S}_3$. That component is tangent to faces $\mathcal{S}_1$ and $\mathcal{S}_2$ so it has zero flux through those faces; therefore, the flux of $F^w \partial_w \boldsymbol{r}$ out of face $\mathcal{S}_0$ minus that into face $\mathcal{S}_3$ equals the flux of that component out of all four faces,

$$\int_0^1 \int_0^{v_{12}} \int_0^{w_0} \frac{\partial}{\partial w}(JF^w)\, \mathrm{d}w\mathrm{d}v\mathrm{d}u = \oint_{\mathcal{S}} (F^w \partial_w \boldsymbol{r}) \cdot \mathrm{d}\boldsymbol{S}, \tag{16.83}$$

where $\mathcal{S} = \mathcal{S}_0 + \mathcal{S}_1 + \mathcal{S}_2 + \mathcal{S}_3$ is the total surface boundary of the tetrahedron, with the surface element $\mathrm{d}\boldsymbol{S} = \hat{\boldsymbol{n}}\, \mathrm{d}A$ pointing outward.

The same argument using parametrization (2) for $\mathcal{S}_0$, shows that the integral of the first term in (16.76) yields the outward flux of component $F^u \partial \boldsymbol{r}/\partial u$ through the entire surface $\mathcal{S}$, and likewise with parametrization (3), the integral of the second term in (16.76) gives the outward flux of $F^v \partial \boldsymbol{r}/\partial v$ through $\mathcal{S}$, such that the sum of all three net fluxes is the outward flux of the total field $\boldsymbol{F}$ through $\mathcal{S}$, yielding the divergence theorem (16.79) for the tetrahedron.

An arbitrary volume domain $\mathcal{V}$ can be tiled using a collection of such tetrahedra, with cancellations of the outward fluxes through common faces, yielding a proof of the divergence theorem (16.79) for any volume $\mathcal{V}$ with piecewise smooth surface boundary $\mathcal{S} = \partial\mathcal{V}$. That volume $\mathcal{V}$ can be sphere-like or torus-like or any other topology, including inner holes, that is, "Swiss-cheese"-like, with multiple disconnected surface boundary components $\mathcal{S}_k$ oriented outward of $\mathcal{V}$. This is particularly useful to handle fluxes of vector fields with isolated singularities, such as the gravitational and Coulomb field $\boldsymbol{F}(\boldsymbol{r}) = (\boldsymbol{r} - \boldsymbol{r}_1)/|\boldsymbol{r} - \boldsymbol{r}_1|^3$ with an inverse square law singularity at $\boldsymbol{r} = \boldsymbol{r}_1$ (Fig. 17.3).

The divergence arises in electrostatics through *Gauss's law* stating that the outward flux of the electric field $\boldsymbol{E}(P)$ through a closed surface $\mathcal{S}$ is proportional to the net electric charge inside the volume $\mathcal{V}$ enclosed by that surface. If $\rho(P)$ is the electric charge density (charge per volume) at point P, Gauss's law states that

$$\oint_{\partial\mathcal{V}} \boldsymbol{E} \cdot \mathrm{d}\boldsymbol{S} = \frac{1}{\epsilon_0} \int_{\mathcal{V}} \rho\, \mathrm{d}V, \tag{16.84}$$

where ϵ_0 is the electric constant. The differential version of Gauss's law, $\boldsymbol{\nabla} \cdot \boldsymbol{E} = \rho/\epsilon_0$, arises from rewriting the surface flux integral as a volume integral using the divergence theorem since Gauss's law applies to any volume $\mathcal{V}$.

Gradient and curl theorems

Applying the divergence theorem (16.79) to the vector field $\boldsymbol{F}(P) = \boldsymbol{a}\, f(P)$ where $\boldsymbol{a}$ is any *constant* vector yields, using (16.70b),

$$\begin{aligned}\int_{\mathcal{V}} \boldsymbol{\nabla} \cdot (f\boldsymbol{a})\, \mathrm{d}V &= \boldsymbol{a} \cdot \int_{\mathcal{V}} \boldsymbol{\nabla} f\, \mathrm{d}V \\ &= \oint_{\partial\mathcal{V}} f\, \boldsymbol{a} \cdot \mathrm{d}\boldsymbol{S} = \boldsymbol{a} \cdot \oint_{\partial\mathcal{V}} f\, \mathrm{d}\boldsymbol{S}.\end{aligned}$$

Since this holds for any $\boldsymbol{a}$, we obtain the *gradient theorem*

$$\int_{\mathcal{V}} \boldsymbol{\nabla} f\, \mathrm{d}V = \oint_{\partial\mathcal{V}} f\, \mathrm{d}\boldsymbol{S}. \tag{16.85}$$

Thus, in fluid mechanics, the net pressure force on a fluid volume is

$$-\oint_{\partial\mathcal{V}} p\, \hat{\boldsymbol{n}}\, \mathrm{d}A = -\int_{\mathcal{V}} \boldsymbol{\nabla} p\, \mathrm{d}V$$

and the net pressure force on a fluid element of volume $\mathrm{d}V$ is $-\boldsymbol{\nabla} p\, \mathrm{d}V$. The minus sign arises because the scalar pressure field is compressive, but $\hat{\boldsymbol{n}}$ points outward in the divergence theorem with $\mathrm{d}\boldsymbol{S} = \hat{\boldsymbol{n}}\mathrm{d}A$.

Similarly, applying the divergence theorem (16.79) to the vector field $\boldsymbol{F}(P) \times \boldsymbol{a}$ instead of $\boldsymbol{F}(P)$ for any constant vector $\boldsymbol{a}$ yields, using identity (16.101),

$$\begin{aligned}\int_{\mathcal{V}} \boldsymbol{\nabla} \cdot (\boldsymbol{F} \times \boldsymbol{a})\, \mathrm{d}V &= \left(\int_{\mathcal{V}} \boldsymbol{\nabla} \times \boldsymbol{F}\, \mathrm{d}V\right) \cdot \boldsymbol{a} \\ &= \oint_{\partial\mathcal{V}} \boldsymbol{F} \times \boldsymbol{a} \cdot \mathrm{d}\boldsymbol{S} = \left(\oint_{\partial\mathcal{V}} \hat{\boldsymbol{n}} \times \boldsymbol{F}\, \mathrm{d}A\right) \cdot \boldsymbol{a}\end{aligned}$$

where $\mathrm{d}\boldsymbol{S} = \hat{\boldsymbol{n}}\mathrm{d}A$ is the outward pointing surface element. Since this holds for any constant $\boldsymbol{a}$, we obtain the *curl theorem*

$$\int_{\mathcal{V}} \boldsymbol{\nabla} \times \boldsymbol{F}\, \mathrm{d}V = \oint_{\partial\mathcal{V}} \hat{\boldsymbol{n}} \times \boldsymbol{F}\, \mathrm{d}A. \tag{16.86}$$

The divergence and gradient theorems can also be applied in euclidean 2D space by considering $f(x, y)$ and $\boldsymbol{F}(x, y)$ as 3D fields independent of the third coordinate z. We then apply the divergence and gradient theorem to cylindrical domains parallel to the z axis whose intersection with the (x, y)-plane is a closed curve $\mathcal{C}$ with line element $\mathrm{d}\boldsymbol{r} = \hat{\boldsymbol{t}}\, \mathrm{d}s$, enclosing an area A in the (x, y)-plane. Then the surface element on the side of the cylinder is (Fig. 16.36)

$$\mathrm{d}\boldsymbol{S} = \mathrm{d}\boldsymbol{r} \times \mathrm{d}z\, \hat{\mathbf{z}} = \hat{\boldsymbol{t}} \times \hat{\mathbf{z}}\, \mathrm{d}s\mathrm{d}z = \hat{\boldsymbol{n}}\, \mathrm{d}s\mathrm{d}z,$$

where $\hat{\boldsymbol{n}} = \hat{\boldsymbol{t}} \times \hat{\mathbf{z}}$ is the unit outward normal to the cylinder and the curve $\mathcal{C}$, with $\mathrm{d}s = |\mathrm{d}\boldsymbol{r}|$ as the arclength element along the curve. The $\mathrm{d}z$ (or z integrals over arbitrary cylinder heights) drop out and we obtain the

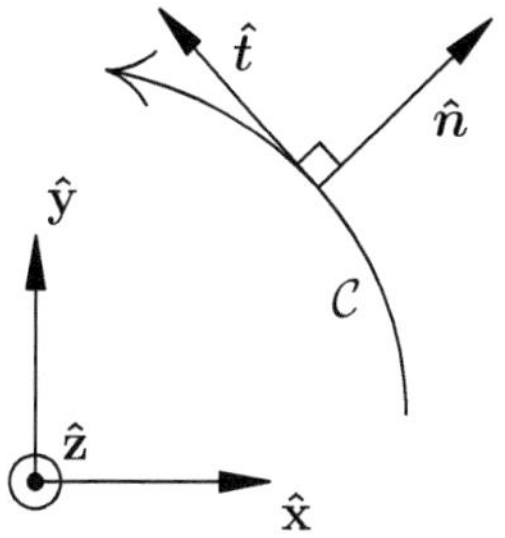

Fig. 16.36 Tangent and unit outward normal.

planar version of the gradient and divergence theorems

$$\int_A \boldsymbol{\nabla} f \, \mathrm{d}A = \oint_C f \, \hat{\boldsymbol{n}} \, \mathrm{d}s, \qquad \int_A \boldsymbol{\nabla} \cdot \boldsymbol{F} \, \mathrm{d}A = \oint_C \boldsymbol{F} \cdot \hat{\boldsymbol{n}} \, \mathrm{d}s. \tag{16.87}$$

These 2D versions also follow directly from Green's theorem (13.31c) by rewriting the latter in divergence instead of curl form.

16.9 Orthogonal coordinates

For general *orthogonal* coordinates (u, v, w), the coordinate tangents $\partial \boldsymbol{r}/\partial u^i$ are orthogonal to each other, and thus the coordinate gradients $\boldsymbol{\nabla} u^i$ are also orthogonal to each other and parallel to their corresponding tangent vectors. It is convenient therefore to define the scale factors h_u, h_v, h_w as the magnitudes of the tangent vectors such that

$$\frac{\partial \boldsymbol{r}}{\partial u} = h_u \hat{\boldsymbol{u}}, \quad \frac{\partial \boldsymbol{r}}{\partial v} = h_v \hat{\boldsymbol{v}}, \quad \frac{\partial \boldsymbol{r}}{\partial w} = h_w \hat{\boldsymbol{w}}, \qquad \boldsymbol{\nabla} u = \frac{\hat{\boldsymbol{u}}}{h_u}, \quad \boldsymbol{\nabla} v = \frac{\hat{\boldsymbol{v}}}{h_v}, \quad \boldsymbol{\nabla} w = \frac{\hat{\boldsymbol{w}}}{h_w}, \tag{16.88}$$

with orthogonal direction vectors ordered such that $\hat{\boldsymbol{u}} \times \hat{\boldsymbol{v}} = \hat{\boldsymbol{w}}$, so the Jacobian is simply

$$J = \frac{\partial \boldsymbol{r}}{\partial u} \times \frac{\partial \boldsymbol{r}}{\partial v} \cdot \frac{\partial \boldsymbol{r}}{\partial w} = h_u h_v h_w > 0, \tag{16.89}$$

and the del operator (16.26), (16.46) reads

$$\boldsymbol{\nabla} = \boldsymbol{\nabla} u^i \, \partial_i = \frac{\hat{\boldsymbol{u}}}{h_u} \frac{\partial}{\partial u} + \frac{\hat{\boldsymbol{v}}}{h_v} \frac{\partial}{\partial v} + \frac{\hat{\boldsymbol{w}}}{h_w} \frac{\partial}{\partial w}. \tag{16.90}$$

An arbitrary vector field $\boldsymbol{F}(P)$ is expanded as

$$\boldsymbol{F} = F_u \hat{\boldsymbol{u}} + F_v \hat{\boldsymbol{v}} + F_w \hat{\boldsymbol{w}}, \tag{16.91}$$

Watch out for meaning of F_u, F_v, F_w.

where these orthogonal components $\boldsymbol{F} \cdot \hat{\boldsymbol{u}}_i = (F_u, F_v, F_w)$ should not be confused with the covariant components $\boldsymbol{F} \cdot \partial_i \boldsymbol{r} = (h_u F_u, h_v F_v, h_w F_w)$.

Curl

To compute $\boldsymbol{\nabla} \times \boldsymbol{F}$, we write $\boldsymbol{F}$ in terms of the coordinate gradients

$$\boldsymbol{F} = (h_u F_u) \boldsymbol{\nabla} u + (h_v F_v) \boldsymbol{\nabla} v + (h_w F_w) \boldsymbol{\nabla} w, \tag{16.92}$$

then use $\boldsymbol{\nabla} \times \boldsymbol{\nabla} f = 0$ (16.52) as for $\boldsymbol{\nabla} \times (F_\theta \hat{\boldsymbol{\theta}})$ in (16.59), to obtain

$$\begin{aligned} \boldsymbol{\nabla} \times \boldsymbol{F} = \frac{\hat{\boldsymbol{u}}}{h_v h_w} & \left(\frac{\partial (h_w F_w)}{\partial v} - \frac{\partial (h_v F_v)}{\partial w} \right) \\ + \frac{\hat{\boldsymbol{v}}}{h_u h_w} & \left(\frac{\partial (h_u F_u)}{\partial w} - \frac{\partial (h_w F_w)}{\partial u} \right) \\ + \frac{\hat{\boldsymbol{w}}}{h_u h_v} & \left(\frac{\partial (h_v F_v)}{\partial u} - \frac{\partial (h_u F_u)}{\partial v} \right). \end{aligned} \tag{16.93}$$

This formula also follows directly from the general one (16.45), of course.

Divergence

For $\nabla \cdot \boldsymbol{F}$, we write $\boldsymbol{F}$ in terms of coordinate gradient cross products

$$\boldsymbol{F} = \frac{JF_u}{h_u} \nabla v \times \nabla w + \frac{JF_v}{h_v} \nabla w \times \nabla u + \frac{JF_w}{h_w} \nabla u \times \nabla v, \qquad (16.94)$$

with $J = h_u h_v h_w$ (16.89), then use $\nabla \cdot (\nabla f \times \nabla g) = 0$ as done for $\nabla \cdot (F_\theta \hat{\boldsymbol{\theta}})$ in (16.75) to calculate

$$\nabla \cdot \left(\frac{JF_u}{h_u} \nabla v \times \nabla w \right) = \nabla \left(\frac{JF_u}{h_u} \right) \cdot \nabla v \times \nabla w = \frac{1}{J} \frac{\partial}{\partial u} \left(\frac{JF_u}{h_u} \right)$$

with ∇ as in (16.90) and $\nabla u \cdot \nabla v \times \nabla w = 1/J$ (16.29). Similar expressions for $F_v \hat{\boldsymbol{v}}$ and $F_w \hat{\boldsymbol{w}}$ follow by cyclic rotation of (u, v, w). Thus, the divergence of $\boldsymbol{F} = F_u \hat{\mathbf{u}} + F_v \hat{\boldsymbol{v}} + F_w \hat{\boldsymbol{w}}$ in orthogonal coordinates (u, v, w) is

$$\nabla \cdot \boldsymbol{F} = \frac{1}{J} \left(\frac{\partial}{\partial u} \left(\frac{JF_u}{h_u} \right) + \frac{\partial}{\partial v} \left(\frac{JF_v}{h_v} \right) + \frac{\partial}{\partial w} \left(\frac{JF_w}{h_w} \right) \right), \qquad (16.95)$$

$J = h_u h_v h_w.$

that also follows directly from the general formula (16.68).

Standard coordinates

For cartesian coordinates (x, y, z) with tangent vectors

$$\left(\frac{\partial \boldsymbol{r}}{\partial x}, \frac{\partial \boldsymbol{r}}{\partial y}, \frac{\partial \boldsymbol{r}}{\partial z} \right) = (\hat{\mathbf{x}}, \hat{\mathbf{y}}, \hat{\mathbf{z}}),$$

the scale factors are $(h_x, h_y, h_z) = (1, 1, 1)$. For cylindrical coordinates $(u, v, w) = (\rho, \varphi, z)$, the tangent vectors are

$$\left(\frac{\partial \boldsymbol{r}}{\partial r}, \frac{\partial \boldsymbol{r}}{\partial \varphi}, \frac{\partial \boldsymbol{r}}{\partial z} \right) = (\hat{\boldsymbol{\rho}}, \rho \hat{\boldsymbol{\varphi}}, \hat{\mathbf{z}}).$$

Thus, $\{\hat{\mathbf{u}}, \hat{\boldsymbol{v}}, \hat{\boldsymbol{w}}\} = \{\hat{\boldsymbol{\rho}}, \hat{\boldsymbol{\varphi}}, \hat{\mathbf{z}}\}$ with scale factors $h_\rho = 1$, $h_\varphi = \rho$, $h_z = 1$. In spherical coordinates, $(u, v, w) = (r, \theta, \varphi)$, the tangent vectors are

$$\left(\frac{\partial \boldsymbol{r}}{\partial r}, \frac{\partial \boldsymbol{r}}{\partial \theta}, \frac{\partial \boldsymbol{r}}{\partial \varphi} \right) = (\hat{\boldsymbol{r}}, r \hat{\boldsymbol{\theta}}, r \sin\theta \, \hat{\boldsymbol{\varphi}}).$$

Thus, $\{\hat{\mathbf{u}}, \hat{\boldsymbol{v}}, \hat{\boldsymbol{w}}\} = \{\hat{\boldsymbol{r}}, \hat{\boldsymbol{\theta}}, \hat{\boldsymbol{\varphi}}\}$ with scale factors $h_r = 1$, $h_\theta = r$, $h_\varphi = r \sin\theta$. Substituting these scale factors in the curl formula (16.93) and the divergence formula (16.95) yields (16.51) and (16.69), respectively.

16.10 Vector identities

Two fundamental identities are that the curl of a gradient (16.52a) and the divergence of a curl (16.70a) vanish for any fields whose second derivatives are continuous (C^2),

$$\begin{aligned} \nabla \times \nabla f &= 0, \\ \nabla \cdot \nabla \times \boldsymbol{F} &= 0. \end{aligned} \qquad (16.96)$$

Three other vector identities are

$$\boldsymbol{\nabla} \times (\boldsymbol{\nabla} \times \boldsymbol{F}) = \boldsymbol{\nabla}(\boldsymbol{\nabla} \cdot \boldsymbol{F}) - (\boldsymbol{\nabla} \cdot \boldsymbol{\nabla})\boldsymbol{F}, \tag{16.97}$$

$$\begin{aligned}\boldsymbol{\nabla} \times (\boldsymbol{F} \times \boldsymbol{G}) = (\boldsymbol{G} \cdot \boldsymbol{\nabla})\boldsymbol{F} - (\boldsymbol{\nabla} \cdot \boldsymbol{F})\boldsymbol{G} \\ + (\boldsymbol{\nabla} \cdot \boldsymbol{G})\boldsymbol{F} - (\boldsymbol{F} \cdot \boldsymbol{\nabla})\boldsymbol{G},\end{aligned} \tag{16.98}$$

$$\boldsymbol{F} \times (\boldsymbol{\nabla} \times \boldsymbol{G}) = (\boldsymbol{\nabla}\boldsymbol{G}) \cdot \boldsymbol{F} - (\boldsymbol{F} \cdot \boldsymbol{\nabla})\boldsymbol{G}, \tag{16.99}$$

where $\boldsymbol{\nabla} \cdot \boldsymbol{\nabla} = \nabla^2$ is the *Laplacian* operator discussed in section 16.11. These can be verified using index notation with $\boldsymbol{\nabla} = \boldsymbol{\nabla} u^i\, \partial_i$ and the double cross product identity (4.9)

$$\boldsymbol{a} \times (\boldsymbol{b} \times \boldsymbol{c}) = \boldsymbol{b}(\boldsymbol{a} \cdot \boldsymbol{c}) - \boldsymbol{c}(\boldsymbol{a} \cdot \boldsymbol{b}). \tag{16.100}$$

These identities can be reconstructed on the fly from that double cross product identity combined with the product rules and a little extra care in handling the vector differential operator $\boldsymbol{\nabla}$. Order does not matter in $\boldsymbol{b}(\boldsymbol{a} \cdot \boldsymbol{c}) = (\boldsymbol{a} \cdot \boldsymbol{c})\boldsymbol{b} = (\boldsymbol{c} \cdot \boldsymbol{a})\boldsymbol{b}$ for regular vectors, but the order matters for the vector differential operator $\boldsymbol{\nabla}$,

$$\boldsymbol{\nabla}(\boldsymbol{\nabla} \cdot \boldsymbol{F}) \neq (\boldsymbol{\nabla} \cdot \boldsymbol{F})\boldsymbol{\nabla} \neq (\boldsymbol{F} \cdot \boldsymbol{\nabla})\boldsymbol{\nabla},$$

so we need to decide which of those expressions to select.

The first identity (16.97) is fundamental in fluid mechanics and electromagnetism. The left-hand side of (16.97) is a field, the curl of the curl of $\boldsymbol{F}$, not an operator. Each term on the right of (16.97) should consist of double derivatives of $\boldsymbol{F}$; hence, the $\boldsymbol{b}(\boldsymbol{a} \cdot \boldsymbol{c})$ term of the double cross product must be $\boldsymbol{\nabla}(\boldsymbol{\nabla} \cdot \boldsymbol{F})$, the gradient of the divergence of $\boldsymbol{F}$. The other valid expressions for regular vectors, $(\boldsymbol{a} \cdot \boldsymbol{c})\boldsymbol{b}$ and $(\boldsymbol{c} \cdot \boldsymbol{a})\boldsymbol{b}$, would yield $(\boldsymbol{\nabla} \cdot \boldsymbol{F})\boldsymbol{\nabla}$ and $(\boldsymbol{F} \cdot \boldsymbol{\nabla})\boldsymbol{\nabla}$; but those are differential operators, not fields. Likewise, the right choice for the $-\boldsymbol{c}(\boldsymbol{a} \cdot \boldsymbol{b})$ term of the double cross product is $-\boldsymbol{\nabla} \cdot \boldsymbol{\nabla}\boldsymbol{F} = -\nabla^2\boldsymbol{F}$, the Laplacian of $\boldsymbol{F}$, not $\boldsymbol{F}(\boldsymbol{\nabla} \cdot \boldsymbol{\nabla})$ which would be the Laplacian operator scaled by $\boldsymbol{F}$, not applied to $\boldsymbol{F}$. The explicit verification in general coordinates has one subtlety given in exercise 16.43.

The second identity (16.98) can also be reconstructed from (16.100), together with the product rule since $\boldsymbol{\nabla}$ is a vector *differential operator*; hence, $\boldsymbol{\nabla} \times (\boldsymbol{F} \times \boldsymbol{G})$ represents derivatives of a product and this doubles the number of terms of the resulting expression. The first two terms on the right-hand side of (16.98) are the double cross product (16.100) for derivatives of $\boldsymbol{F}$ while the last two terms are the double cross product for derivatives of $\boldsymbol{G}$. This is straightfoward to demonstrate in general coordinates with $\boldsymbol{\nabla} = \boldsymbol{\nabla} u^i\, \partial_i$, and the summation convention,

$$\begin{aligned}\boldsymbol{\nabla} \times (\boldsymbol{F} \times \boldsymbol{G}) &= \boldsymbol{\nabla} u^i \times \partial_i(\boldsymbol{F} \times \boldsymbol{G}) = \boldsymbol{\nabla} u^i \times (\partial_i\boldsymbol{F} \times \boldsymbol{G} + \boldsymbol{F} \times \partial_i\boldsymbol{G}) \\ &= (\boldsymbol{G} \cdot \boldsymbol{\nabla} u^i)\partial_i\boldsymbol{F} - (\boldsymbol{\nabla} u^i \cdot \partial_i\boldsymbol{F})\boldsymbol{G} \\ &\quad + \boldsymbol{F}(\boldsymbol{\nabla} u^i \cdot \partial_i\boldsymbol{G}) - (\boldsymbol{F} \cdot \boldsymbol{\nabla} u^i)\partial_i\boldsymbol{G} \\ &= (\boldsymbol{G} \cdot \boldsymbol{\nabla})\boldsymbol{F} - (\boldsymbol{\nabla} \cdot \boldsymbol{F})\boldsymbol{G} + (\boldsymbol{\nabla} \cdot \boldsymbol{G})\boldsymbol{F} - (\boldsymbol{F} \cdot \boldsymbol{\nabla})\boldsymbol{G},\end{aligned}$$

which is indeed (16.98).

In the third identity (16.99), the del operator $\boldsymbol{\nabla}$ only applies to $\boldsymbol{G}$, so we must write the first term on the right hand side as $(\boldsymbol{\nabla}\boldsymbol{G})\cdot\boldsymbol{F}$, not $\boldsymbol{\nabla}(\boldsymbol{G}\cdot\boldsymbol{F})$ that would include derivatives of $\boldsymbol{F}$. The second term also involves the tensor $\boldsymbol{\nabla}\boldsymbol{G}$, but $\boldsymbol{F}\cdot\boldsymbol{\nabla}\boldsymbol{G} = (\boldsymbol{F}\cdot\boldsymbol{\nabla})\boldsymbol{G}$ can be interpreted as the scalar operator $\boldsymbol{F}\cdot\boldsymbol{\nabla}$ applied to $\boldsymbol{G}$. The verification of this identity is straightforward, from (16.47a) and (16.100),

$$\begin{aligned}\boldsymbol{F}\times(\boldsymbol{\nabla}\times\boldsymbol{G}) &= \boldsymbol{F}\times\left(\boldsymbol{\nabla}u^i\times\partial_i\boldsymbol{G}\right)\\ &= \boldsymbol{\nabla}u^i\,\partial_i\boldsymbol{G}\cdot\boldsymbol{F} - \boldsymbol{F}\cdot\boldsymbol{\nabla}u^i\,\partial_i\boldsymbol{G}\\ &= \boldsymbol{\nabla}\boldsymbol{G}\cdot\boldsymbol{F} - \boldsymbol{F}\cdot\boldsymbol{\nabla}\boldsymbol{G},\end{aligned}$$

and reveals that

$$\boldsymbol{\nabla}\boldsymbol{G}\cdot\boldsymbol{F} = \boldsymbol{\nabla}u^1(\partial_1\boldsymbol{G}\cdot\boldsymbol{F}) + \boldsymbol{\nabla}u^2(\partial_2\boldsymbol{G}\cdot\boldsymbol{F}) + \boldsymbol{\nabla}u^3(\partial_3\boldsymbol{G}\cdot\boldsymbol{F}).$$

Another useful vector identity is

$$\boldsymbol{\nabla}\cdot(\boldsymbol{F}\times\boldsymbol{G}) = \boldsymbol{G}\cdot(\boldsymbol{\nabla}\times\boldsymbol{F}) - \boldsymbol{F}\cdot(\boldsymbol{\nabla}\times\boldsymbol{G}), \tag{16.101}$$

which follows from the mixed product property

$$\boldsymbol{a}\cdot\boldsymbol{b}\times\boldsymbol{c} = \boldsymbol{c}\cdot\boldsymbol{a}\times\boldsymbol{b} \tag{16.102}$$

together with the product rule since $\boldsymbol{\nabla}$ applies to both $\boldsymbol{F}$ and $\boldsymbol{G}$, so there should be two terms, and $\boldsymbol{F}\times\boldsymbol{G} = -\boldsymbol{G}\times\boldsymbol{F}$ to reconstruct the second term. The verification in general coordinates is straightforward

$$\begin{aligned}\boldsymbol{\nabla}\cdot(\boldsymbol{F}\times\boldsymbol{G}) &= \boldsymbol{\nabla}u^i\cdot\partial_i(\boldsymbol{F}\times\boldsymbol{G})\\ &= \boldsymbol{\nabla}u^i\cdot(\partial_i\boldsymbol{F})\times\boldsymbol{G} + \boldsymbol{\nabla}u^i\cdot\boldsymbol{F}\times(\partial_i\boldsymbol{G})\\ &= \boldsymbol{G}\cdot\boldsymbol{\nabla}u^i\times\partial_i\boldsymbol{F} - \boldsymbol{F}\cdot\boldsymbol{\nabla}u^i\times\partial_i\boldsymbol{G}\\ &= \boldsymbol{G}\cdot\boldsymbol{\nabla}\times\boldsymbol{F} - \boldsymbol{F}\cdot\boldsymbol{\nabla}\times\boldsymbol{G}.\end{aligned}$$

A consequence of (16.101) and (16.52a) is

$$\boldsymbol{\nabla}\cdot(\boldsymbol{\nabla}f\times\boldsymbol{\nabla}g) = 0, \tag{16.103}$$

that also follows from $\boldsymbol{\nabla}f\times\boldsymbol{\nabla}g = \boldsymbol{\nabla}\times(f\boldsymbol{\nabla}g)$, and (16.52a). Some other identities are explored in exercises (16.141) and (16.142).

16.11 Laplacian

The *Laplacian operator* is the divergence of the del operator $\nabla^2 = \boldsymbol{\nabla}\cdot\boldsymbol{\nabla}$. The Laplacian arises in many contexts, a canonical example being the conduction of heat through a material. If $\boldsymbol{q}$ is the heat flux, with units of heat (energy) per unit time and unit area, such that the flux of heat through a surface $\mathcal{S}$ is $\int_{\mathcal{S}}\boldsymbol{q}\cdot\hat{\boldsymbol{n}}\mathrm{d}A$, then the net flux outward of a volume element $\mathrm{d}V$ is

$$\boldsymbol{\nabla}\cdot\boldsymbol{q}\,\mathrm{d}V$$

from the divergence theorem (16.79). Fourier's law of conduction then prescribes the heat flux as proportional to the gradient of temperature T,

$$\boldsymbol{q} = -\lambda\, \boldsymbol{\nabla} T,$$

where $\lambda > 0$ is the material's thermal conductivity, and the negative sign indicates that heat flows from hot to cold. The divergence of the heat flux is thus

$$\boldsymbol{\nabla} \cdot \boldsymbol{q} = -\boldsymbol{\nabla} \cdot (\lambda \boldsymbol{\nabla} T) = -\lambda\, \nabla^2 T,$$

proportional to the Laplacian of temperature if λ is uniform throughout the material. If C is the volumetric heat capacity and Q the volumetric heating rate, then the rate of change of the amount of heat in volume element $\mathrm{d}V$ equals to the rate of heat input $Q\,\mathrm{d}V$ minus the net heat flux out of the volume element, $\boldsymbol{\nabla} \cdot \boldsymbol{q}\,\mathrm{d}V$, yielding the *heat equation*

$$\frac{\partial}{\partial t}(CT) = Q - \boldsymbol{\nabla} \cdot \boldsymbol{q} = Q + \boldsymbol{\nabla} \cdot (\lambda \boldsymbol{\nabla} T) \tag{16.104}$$

for the temperature $T(\boldsymbol{r}, t)$ at point $\boldsymbol{r}$ and time t. In equilibrium, $\partial/\partial t = 0$, and assuming constant conductivity λ, the temperature $T(\boldsymbol{r})$ obeys *Poisson's equation*

$$-\nabla^2 T = Q/\lambda, \tag{16.105}$$

which reduces to *Laplace's equation*, $\nabla^2 T = 0$, for no internal heating ($Q = 0$).

Laplacian in arbitrary coordinates

The Laplacian is a scalar operator whose expression in cartesian coordinates is simply

$$\nabla^2 = \boldsymbol{\nabla} \cdot \boldsymbol{\nabla} = \frac{\partial^2}{\partial x^2} + \frac{\partial^2}{\partial y^2} + \frac{\partial^2}{\partial z^2}. \tag{16.106}$$

The Laplacian in cylindrical and spherical coordinates can be obtained by applying the expressions for the divergence (16.69) to the corresponding expression for $\boldsymbol{\nabla}$ (16.49). They are given below as examples of the general coordinate formula (16.110), (16.113). Likewise, for any orthogonal coordinates (16.88), the divergence (16.95) of $\boldsymbol{\nabla}$ (16.90) yields, with Jacobian $J = h_u h_v h_w$,

$$\begin{aligned} \nabla^2 &= \boldsymbol{\nabla} \cdot \boldsymbol{\nabla} \\ &= \frac{1}{J}\left(\frac{\partial}{\partial u}\left(\frac{J}{h_u^2}\frac{\partial}{\partial u}\right) + \frac{\partial}{\partial v}\left(\frac{J}{h_v^2}\frac{\partial}{\partial v}\right) + \frac{\partial}{\partial w}\left(\frac{J}{h_w^2}\frac{\partial}{\partial w}\right)\right). \end{aligned} \tag{16.107}$$

In general coordinates (u^1, u^2, u^3), the formula for $\boldsymbol{\nabla} \cdot \boldsymbol{F}$ (16.68) involves the contravariant components F^i of $\boldsymbol{F} = F^i \partial_i \boldsymbol{r}$ in the tangent basis, while a gradient is naturally expressed in terms of the gradient basis $\boldsymbol{\nabla} f = \boldsymbol{\nabla} u^i\, \partial_i f$ (16.46). Switching from one basis to the other follows from biorthogonality $\boldsymbol{\nabla} u^i \cdot \partial_j \boldsymbol{r} = \delta^i_j$ (16.32), and involves the metrics $g^{ij} = \boldsymbol{\nabla} u^i \cdot \boldsymbol{\nabla} u^j$ and $g_{ij} = \partial_i \boldsymbol{r} \cdot \partial_j \boldsymbol{r}$, with

$$\boldsymbol{\nabla} u^i = g^{ij} \partial_j \boldsymbol{r}, \qquad \partial_i \boldsymbol{r} = g_{ij} \boldsymbol{\nabla} u^j. \tag{16.108}$$

Thus

$$\boldsymbol{\nabla} f = \boldsymbol{\nabla} u^j \partial_j f = \left(g^{ij} \partial_j f\right) \partial_i \boldsymbol{r},$$

and the Laplacian in arbitrary coordinates follows from (16.67) as

$$\nabla^2 f = \boldsymbol{\nabla} \cdot \boldsymbol{\nabla} f = J^{-1} \partial_i \left(J \, g^{ij} \partial_j f\right). \tag{16.109}$$

The cartesian expression (16.106) corresponds to $J = 1$ and $g^{ij} = \delta^{ij}$, the Kronecker delta. In cylindrical coordinates (ρ, φ, z), (16.107) and (16.109) yield

$$\nabla^2 = \frac{1}{\rho} \frac{\partial}{\partial \rho} \left(\rho \frac{\partial}{\partial \rho}\right) + \frac{1}{\rho^2} \frac{\partial^2}{\partial \varphi^2} + \frac{\partial^2}{\partial z^2}. \tag{16.110}$$

since $\boldsymbol{\nabla} \rho = \hat{\boldsymbol{\rho}}$, $\boldsymbol{\nabla} \varphi = \hat{\boldsymbol{\varphi}}/\rho$, $\boldsymbol{\nabla} z = \hat{\mathbf{z}}$, $J = \rho$, and

$$[g^{ij}] \equiv \begin{bmatrix} 1 & 0 & 0 \\ 0 & 1/\rho^2 & 0 \\ 0 & 0 & 1 \end{bmatrix}. \tag{16.111}$$

In spherical coordinates (r, θ, φ), with $\boldsymbol{\nabla} r = \hat{\boldsymbol{r}}$, $\boldsymbol{\nabla} \theta = \hat{\boldsymbol{\theta}}/r$, $\boldsymbol{\nabla} \varphi = \hat{\boldsymbol{\varphi}}/(r \sin\theta))$, $J = r^2 \sin\theta$,

$$[g^{ij}] \equiv \begin{bmatrix} 1 & 0 & 0 \\ 0 & 1/r^2 & 0 \\ 0 & 0 & 1/(r^2 \sin^2\theta) \end{bmatrix}, \tag{16.112}$$

and (16.109) or (16.107) yields

$$\nabla^2 = \frac{1}{r^2} \frac{\partial}{\partial r} \left(r^2 \frac{\partial}{\partial r}\right) + \frac{1}{r^2 \sin\theta} \frac{\partial}{\partial \theta} \left(\sin\theta \frac{\partial}{\partial \theta}\right) + \frac{1}{r^2 \sin^2\theta} \frac{\partial^2}{\partial \varphi^2}. \tag{16.113}$$

The Laplacian can also be applied to a vector field. The Laplacian of the velocity field $\boldsymbol{V}(P)$ arises in the Navier–Stokes equations of fluid mechanics, and the Laplacian of the electric and magnetic fields, $\boldsymbol{E}(P)$ and $\boldsymbol{B}(P)$, arise in electrodynamics. For a cartesian basis where $\boldsymbol{F} = F_x \hat{\mathbf{x}} + F_y \hat{\mathbf{y}} + F_z \hat{\mathbf{z}}$ and the basis vectors $\{\hat{\mathbf{x}}, \hat{\mathbf{y}}, \hat{\mathbf{z}}\}$ are constants, the Laplacian is simply

$$\nabla^2 \boldsymbol{F} = \hat{\mathbf{x}} \, \nabla^2 F_x + \hat{\mathbf{y}} \, \nabla^2 F_y + \hat{\mathbf{z}} \, \nabla^2 F_z. \tag{16.114}$$

For other coordinate systems, the Laplacian of a vector involves extra terms from the variation of the basis vectors and is computed below with the help of a vector identity.

Laplacian of a vector

For general curvilinear coordinate systems, the basis vectors are not constant and the Laplacian of a vector field involves the derivatives of the basis vectors. The Laplacian of a vector arises in the $\boldsymbol{\nabla}\times(\boldsymbol{\nabla}\times\boldsymbol{F}) = \boldsymbol{\nabla}(\boldsymbol{\nabla}\cdot\boldsymbol{F}) - \nabla^2\boldsymbol{F}$ and that identity can serve to derive expressions for $\nabla^2\boldsymbol{F}$ in various coordinate systems. It is used hereafter to evaluate the Laplacian of the coordinate gradients as $\nabla^2\boldsymbol{\nabla}u^k = \boldsymbol{\nabla}(\nabla^2 u^k)$.

General coordinates

Using $\nabla^2 = \boldsymbol{\nabla}\cdot\boldsymbol{\nabla}$ and $\boldsymbol{F} = F_k\boldsymbol{a}^k$ where $\boldsymbol{a}^k = \boldsymbol{\nabla}u^k$ are the contravariant/gradient basis vectors and $F_k = \boldsymbol{F}\cdot\partial_k\boldsymbol{r} = \boldsymbol{F}\cdot\boldsymbol{a}_k$ are the covariant components, the product rule leads to

$$\begin{aligned}\nabla^2\boldsymbol{F} &= \boldsymbol{\nabla}\cdot\boldsymbol{\nabla}\left(F_k\boldsymbol{a}^k\right) = \boldsymbol{\nabla}\cdot\left(\boldsymbol{\nabla}F_k\,\boldsymbol{a}^k + F_k\boldsymbol{\nabla}\boldsymbol{a}^k\right)\\ &= \nabla^2F_k\,\boldsymbol{a}^k + 2\boldsymbol{\nabla}F_k\cdot\boldsymbol{\nabla}\boldsymbol{a}^k + F_k\nabla^2\boldsymbol{a}^k\end{aligned}$$

with, from $\boldsymbol{\nabla} = \boldsymbol{a}^i\partial_i$ and (15.70b),

$$\boldsymbol{\nabla}F_k\cdot\boldsymbol{\nabla}\boldsymbol{a}^k = \boldsymbol{a}^i\partial_iF_k\cdot\boldsymbol{a}^j\partial_j\boldsymbol{a}^k = -(\partial_iF_k)g^{ij}\Gamma^k_{jl}\,\boldsymbol{a}^l.$$

Then, from identity (16.97),

$$\begin{aligned}\nabla^2\boldsymbol{a}^k &= \boldsymbol{\nabla}\left(\boldsymbol{\nabla}\cdot\boldsymbol{a}^k\right) - \boldsymbol{\nabla}\times\left(\boldsymbol{\nabla}\times\boldsymbol{a}^k\right)\\ &= \boldsymbol{\nabla}\nabla^2u^k = \boldsymbol{a}^l\partial_l\left(J^{-1}\partial_i\left(Jg^{ij}\partial_ju^k\right)\right)\\ &= \boldsymbol{a}^l\partial_l\left(J^{-1}\partial_i\left(Jg^{ik}\right)\right),\end{aligned}$$

because $\boldsymbol{\nabla}\times\boldsymbol{a}^k = \boldsymbol{\nabla}\times\boldsymbol{\nabla}u^k = 0$, and (16.109) was used for ∇^2u^k with $\partial_ju^k = \partial u^k/\partial u^j = \delta^k_j$. Therefore, putting everything together with some renaming of summation indices yields the general formula

$$\nabla^2\boldsymbol{F} = \boldsymbol{a}^k\Big(\nabla^2F_k - 2g^{il}\Gamma^j_{lk}\partial_iF_j + F_j\,\partial_k\big(J^{-1}\partial_i\left(Jg^{ij}\right)\big)\Big), \qquad (16.115)$$

where, from (14.43),

$$2g^{il}\Gamma^j_{lk} = 2g^{il}g^{jm}\Gamma_{mlk} = g^{il}g^{jm}\left(\partial_lg_{km} + \partial_kg_{ml} - \partial_mg_{lk}\right).$$

This expression for the Laplacian of a vector is often called the "vector Laplacian." This is the same ∇^2 operator as in (16.109), but the derivatives of the basis vectors introduce extra terms besides the Laplacian of the components.

For orthogonal coordinates, the metrics g_{ij} and g^{ij} are diagonal, and thus we can reduce (16.115); however, that expression is in terms of the covariant components $F_i = \boldsymbol{F}\cdot\partial_i\boldsymbol{r}$, instead of the orthogonal components $\boldsymbol{F}\cdot\hat{\boldsymbol{u}}^i$, which are used in the case of orthogonal coordinates. Thus, a secondary reduction is necessary to obtain an expression in terms of those orthogonal components. A direct derivation is given below.

Orthogonal coordinates

For general orthogonal coordinates as in §16.9, vector fields $\boldsymbol{F}(P)$ are represented as

$$\boldsymbol{F} = F_u \hat{\boldsymbol{u}} + F_v \hat{\boldsymbol{v}} + F_w \hat{\boldsymbol{w}}, \tag{16.116}$$

in terms of a variable but orthogonal basis $\{\hat{\boldsymbol{u}}, \hat{\boldsymbol{v}}, \hat{\boldsymbol{w}}\}$. The most compact expressions for $\nabla^2 \boldsymbol{F}$ arise from using identity (16.97)

$$\nabla^2 \boldsymbol{F} = \boldsymbol{\nabla}(\boldsymbol{\nabla} \cdot \boldsymbol{F}) - \boldsymbol{\nabla} \times (\boldsymbol{\nabla} \times \boldsymbol{F}),$$

with the expressions for $\boldsymbol{\nabla}$, $\boldsymbol{\nabla} \cdot \boldsymbol{F}$ and $\boldsymbol{\nabla} \times \boldsymbol{F}$ in orthogonal coordinates §16.9; however, the vector components and the scale factors are entangled by that approach. To obtain the orthogonal formula corresponding to the general formula (16.115), we proceed similarly to that derivation. Focusing on the $F_u \hat{\boldsymbol{u}}$ component, the Laplacian $\nabla^2 \boldsymbol{F} = \boldsymbol{\nabla} \cdot \boldsymbol{\nabla} \boldsymbol{F}$ expands to

$$\boldsymbol{\nabla} \cdot \boldsymbol{\nabla}(F_u \hat{\boldsymbol{u}}) = \hat{\boldsymbol{u}}\, \nabla^2 F_u + 2 \boldsymbol{\nabla} F_u \cdot \boldsymbol{\nabla} \hat{\boldsymbol{u}} + F_u \nabla^2 \hat{\boldsymbol{u}} \tag{16.117}$$

with the orthogonal expression for $\boldsymbol{\nabla}$ (16.90).

The $\boldsymbol{\nabla} F_u \cdot \boldsymbol{\nabla} \hat{\boldsymbol{u}}$ can then be evaluated using the Lamé relationships (15.17),

$$\begin{aligned} \partial_u \hat{\boldsymbol{u}} &= -\hat{\boldsymbol{v}} \frac{\partial_v h_u}{h_v} - \hat{\boldsymbol{w}} \frac{\partial_w h_u}{h_w}, \\ \partial_v \hat{\boldsymbol{u}} &= \hat{\boldsymbol{v}} \frac{\partial_u h_v}{h_u}, \quad \partial_w \hat{\boldsymbol{u}} = \hat{\boldsymbol{w}} \frac{\partial_u h_w}{h_u}, \end{aligned} \tag{16.118}$$

leading to

$$\begin{aligned} \boldsymbol{\nabla} F_u \cdot \boldsymbol{\nabla} \hat{\boldsymbol{u}} &= \frac{\partial_u F_u}{h_u} \frac{\partial_u \hat{\boldsymbol{u}}}{h_u} + \frac{\partial_v F_u}{h_v} \frac{\partial_v \hat{\boldsymbol{u}}}{h_v} + \frac{\partial_w F_u}{h_w} \frac{\partial_w \hat{\boldsymbol{u}}}{h_w} \\ &= \frac{\hat{\boldsymbol{v}}}{h_u h_v} \left((\partial_v F_u) \frac{\partial_u h_v}{h_v} - (\partial_u F_u) \frac{\partial_v h_u}{h_u} \right) \\ &\qquad + \frac{\hat{\boldsymbol{w}}}{h_u h_w} \left((\partial_w F_u) \frac{\partial_u h_w}{h_w} - (\partial_u F_u) \frac{\partial_w h_u}{h_u} \right). \end{aligned}$$

Expressions for $\boldsymbol{\nabla} F_v \cdot \boldsymbol{\nabla} \hat{\boldsymbol{v}}$ and $\boldsymbol{\nabla} F_w \cdot \boldsymbol{\nabla} \hat{\boldsymbol{w}}$ can be obtained by cyclic permutation of (u, v, w). For the $\nabla^2 \hat{\boldsymbol{u}}$ term in (16.117) we use the vector identity (16.97)

$$\nabla^2 \hat{\boldsymbol{u}} = \boldsymbol{\nabla}(\boldsymbol{\nabla} \cdot \hat{\boldsymbol{u}}) - \boldsymbol{\nabla} \times (\boldsymbol{\nabla} \times \hat{\boldsymbol{u}}),$$

together with the orthogonal expressions for $\boldsymbol{\nabla}$, $\boldsymbol{\nabla} \cdot \boldsymbol{F}$ and $\boldsymbol{\nabla} \times \boldsymbol{F}$ in §16.9 for $\boldsymbol{F} = \hat{\boldsymbol{u}}$, that is, $(F_u, F_v, F_w) = (1, 0, 0)$, to obtain

$$\boldsymbol{\nabla} \cdot \hat{\boldsymbol{u}} = \frac{1}{J} \frac{\partial}{\partial u} \left(\frac{J}{h_u} \right), \quad \boldsymbol{\nabla} \times \hat{\boldsymbol{u}} = \frac{\hat{\boldsymbol{v}}}{h_u h_w} \frac{\partial h_u}{\partial w} - \frac{\hat{\boldsymbol{w}}}{h_u h_v} \frac{\partial h_u}{\partial v}, \tag{16.119}$$

$$\begin{aligned} \boldsymbol{\nabla} \times (\boldsymbol{\nabla} \times \hat{\boldsymbol{u}}) = &- \frac{\hat{\boldsymbol{u}}}{h_v h_w} \left(\frac{\partial}{\partial v} \left(\frac{h_w}{h_u h_v} \frac{\partial h_u}{\partial v} \right) + \frac{\partial}{\partial w} \left(\frac{h_v}{h_u h_w} \frac{\partial h_u}{\partial w} \right) \right) \\ &+ \frac{\hat{\boldsymbol{v}}}{h_u h_w} \frac{\partial}{\partial u} \left(\frac{h_w}{h_u h_v} \frac{\partial h_u}{\partial v} \right) + \frac{\hat{\boldsymbol{w}}}{h_u h_v} \frac{\partial}{\partial u} \left(\frac{h_v}{h_u h_w} \frac{\partial h_u}{\partial w} \right), \end{aligned}$$

where $J = h_u h_v h_w$. Subtracting from $\boldsymbol{\nabla}(\boldsymbol{\nabla} \cdot \hat{\boldsymbol{u}})$ yields the expression for $\nabla^2 \hat{\boldsymbol{u}}$. Formulas for $\nabla^2 \hat{\boldsymbol{v}}$ and $\nabla^2 \hat{\boldsymbol{w}}$ follow by cyclic permutations of the variables. Collecting all the $\hat{\boldsymbol{u}}$ components of $\nabla^2 \boldsymbol{F}$ we find that

$$\begin{aligned}\hat{\boldsymbol{u}} \cdot \nabla^2 \boldsymbol{F} = \nabla^2 F_u & \\ + 2\left(\partial_u F_v \frac{\partial_v h_u}{h_u^2 h_v} + \partial_u F_w \frac{\partial_w h_u}{h_u^2 h_w} - \partial_v F_v \frac{\partial_u h_v}{h_u h_v^2} - \partial_w F_w \frac{\partial_u h_w}{h_u h_w^2}\right) & \\ + F_u \left(\frac{\partial_u}{h_u}\left(\frac{1}{J}\partial_u \frac{J}{h_u}\right) + \frac{h_u}{J}\partial_v\left(\frac{h_w^2}{J}\partial_v h_u\right) + \frac{h_u}{J}\partial_w\left(\frac{h_v^2}{J}\partial_w h_u\right)\right) & \\ + F_v \left(\frac{\partial_u}{h_u}\left(\frac{1}{J}\partial_v \frac{J}{h_v}\right) - \frac{h_u}{J}\partial_v\left(\frac{h_w^2}{J}\partial_u h_v\right)\right) & \\ + F_w \left(\frac{\partial_u}{h_u}\left(\frac{1}{J}\partial_w \frac{J}{h_w}\right) - \frac{h_u}{J}\partial_w\left(\frac{h_v^2}{J}\partial_u h_w\right)\right), &\end{aligned}$$

and the $\hat{\boldsymbol{v}}$ and $\hat{\boldsymbol{w}}$ components can be obtained by cyclic permutation of (u, v, w).

For example, in cylindrical coordinates, writing

$$\boldsymbol{F} = F_\rho \, \hat{\boldsymbol{\rho}} + F_\varphi \, \hat{\boldsymbol{\varphi}} + F_z \, \hat{\mathbf{z}} = \sum_{k=1}^{3} F_k \, \hat{\boldsymbol{u}}_k,$$

with $\boldsymbol{\nabla}$ as in (16.25) and $\hat{\boldsymbol{\rho}}, \hat{\boldsymbol{\varphi}}$ functions of φ only, a direct calculation yields

$$\sum_{k=1}^{3} \boldsymbol{\nabla} F_k \cdot \boldsymbol{\nabla} \hat{\boldsymbol{u}}_k \equiv \frac{1}{\rho^2}\frac{\partial F_\rho}{\partial \varphi}\frac{\partial \hat{\boldsymbol{\rho}}}{\partial \varphi} + \frac{1}{\rho^2}\frac{\partial F_\varphi}{\partial \varphi}\frac{\partial \hat{\boldsymbol{\varphi}}}{\partial \varphi},$$

with $\partial_\varphi \hat{\boldsymbol{\rho}} = \hat{\boldsymbol{\varphi}}$, $\partial_\varphi \hat{\boldsymbol{\varphi}} = -\hat{\boldsymbol{\rho}}$. Direct application of ∇^2 to $\hat{\boldsymbol{\rho}}$ and $\hat{\boldsymbol{\varphi}}$ yields

$$\nabla^2 \hat{\boldsymbol{\rho}} = -\frac{\hat{\boldsymbol{\rho}}}{\rho^2}, \qquad \nabla^2 \hat{\boldsymbol{\varphi}} = -\frac{\hat{\boldsymbol{\varphi}}}{\rho^2},$$

from (10.18) and (16.110). Expression (16.117) then gives the vector Laplacian in cylindrical coordinates as

$$\begin{aligned}\nabla^2 \boldsymbol{F} = \hat{\boldsymbol{\rho}} \left(\nabla^2 F_\rho - \frac{2}{\rho^2}\frac{\partial F_\varphi}{\partial \varphi} - \frac{F_\rho}{\rho^2}\right) & \\ + \hat{\boldsymbol{\varphi}} \left(\nabla^2 F_\varphi + \frac{2}{\rho^2}\frac{\partial F_\rho}{\partial \varphi} - \frac{F_\varphi}{\rho^2}\right) + \hat{\mathbf{z}} \, \nabla^2 F_z. &\end{aligned} \tag{16.120}$$

Similarly, in spherical coordinates,

$$\boldsymbol{F} = F_r \, \hat{\boldsymbol{r}} + F_\theta \, \hat{\boldsymbol{\theta}} + F_\varphi \, \hat{\boldsymbol{\varphi}} \equiv \sum_{k=1}^{3} F_k \, \hat{\boldsymbol{u}}_k,$$

with $\boldsymbol{\nabla}$ as in (16.24), and $\hat{\boldsymbol{r}}, \hat{\boldsymbol{\theta}}, \hat{\boldsymbol{\varphi}}$, functions of θ and φ only; thus,

$$\boldsymbol{\nabla} F_k \cdot \boldsymbol{\nabla} \hat{\boldsymbol{u}}_k = \frac{1}{r^2}\frac{\partial F_k}{\partial \theta}\frac{\partial \hat{\boldsymbol{u}}_k}{\partial \theta} + \frac{1}{r^2 \sin^2\theta}\frac{\partial F_k}{\partial \varphi}\frac{\partial \hat{\boldsymbol{u}}_k}{\partial \varphi}.$$

Using (10.24) and (16.113) (see exercise 16.70), expression (16.117) yields

$$\begin{aligned}\nabla^2 \boldsymbol{F} \\ &= \hat{\boldsymbol{r}}\left(\nabla^2 F_r - \frac{2}{r^2}\frac{\partial F_\theta}{\partial \theta} - \frac{2}{r^2 \sin\theta}\frac{\partial F_\varphi}{\partial \varphi} - \frac{2F_r}{r^2} - \frac{2F_\theta \cos\theta}{r^2 \sin\theta}\right) \\ &+ \hat{\boldsymbol{\theta}}\left(\nabla^2 F_\theta + \frac{2}{r^2}\frac{\partial F_r}{\partial \theta} - \frac{2\cos\theta}{r^2 \sin^2\theta}\frac{\partial F_\varphi}{\partial \varphi} - \frac{F_\theta}{r^2 \sin^2\theta}\right) \\ &+ \hat{\boldsymbol{\varphi}}\left(\nabla^2 F_\varphi + \frac{2}{r^2 \sin\theta}\frac{\partial F_r}{\partial \varphi} + \frac{2\cos\theta}{r^2 \sin^2\theta}\frac{\partial F_\theta}{\partial \varphi} - \frac{F_\varphi}{r^2 \sin^2\theta}\right).\end{aligned} \tag{16.121}$$

16.12 The heat equation

For a homogeneous material with constant heat capacity C and conductivity λ, and no internal heating ($Q = 0$), the heat equation (16.104) reduces to

$$\frac{\partial T}{\partial t} = \kappa \nabla^2 T, \tag{16.122}$$

where $T(t, \boldsymbol{r})$ is the temperature, and $\kappa = \lambda/C$ is the *thermal diffusivity.* Since $\partial/\partial t$ has units of 1/time, and ∇^2 has units of $1/\text{length}^2$, the thermal diffusivity must have units of $\text{length}^2/\text{time}$, that is, $\text{meters}^2/\text{second}$ in SI units.

It is easy to verify that (16.122) has simple decaying sinusoidal solutions of the form

$$T(t, \boldsymbol{r}) = e^{-\kappa k^2 t} \cos kx,$$

for any real constant k, where x is the usual cartesian coordinate with $\boldsymbol{r} = x\hat{\mathbf{x}} + y\hat{\mathbf{y}} + z\hat{\mathbf{z}}$. Since the equation is linear, we can construct other solutions by linear combinations of those simple exponential solutions.

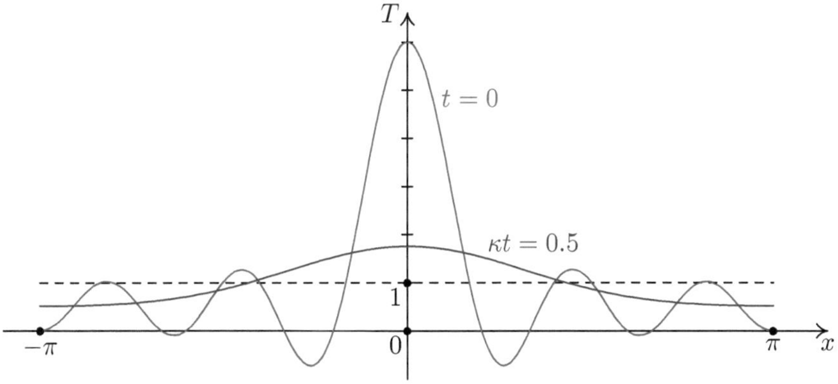

Fig. 16.37 $T(t, x)$ in (16.123) at $t = 0$ and $\kappa t = 0.5$, illustrates smoothing and diffusion.

A fundamental example is the solution corresponding to a sum of $\cos kx$ at $t = 0$, that is,

$$T(t, x) = \sum_{k=0}^{N} e^{-\kappa k^2 t} \cos kx, \tag{16.123}$$

shown in Fig. 16.37 for $N = 5$. There is constructive interference of the $\cos kx$ modes at $x = 0$ that add up to an initial value of $T(0,0) = N + 1 = 6$ at $x = 0$, but destructive interference for all other x's such that the sum tends to $1/2 + \pi$ times a *delta function* as $N \to \infty$, $t \to 0$, for $-\pi \le x \le \pi$ (exercise 19.22),[8]

[8]Although $T(0,x)$ oscillates about $1/2$ for all $x \neq 0$, it doesn't converge to $1/2$ anywhere! $T(0,x)$ is essentially the *Dirichlet kernel* $D_N(x) = 2T(0,x) - 1$.

$$T(0,x) = \sum_{k=0}^{N} \cos kx \to \frac{1}{2} + \pi\,\delta(x). \tag{16.124}$$

The integral

$$\int_{-\pi}^{\pi} T(t,x)\mathrm{d}x = 2\pi,$$

for all times t. This corresponds to the total heat (divided by the heat capacity C) in the domain. The total heat is conserved for all times since there is no flux through the boundaries at $x = \pm\pi$, but the heat is concentrated near $x = 0$ at $t = 0$ and rapidly *diffuses* over the domain. Thus, the heat equation rapidly damps high wavenumbers k, and spreads local hot and cold spots. It *smoothes* and *diffuses.*

General 3D solution

More generally, the heat equation has simple *Fourier mode* solutions of the form

$$T(t,\boldsymbol{r}) = e^{-\kappa|\boldsymbol{k}|^2 t}\, e^{i\boldsymbol{k}\cdot\boldsymbol{r}}, \tag{16.125}$$

where $\boldsymbol{k} = k_x\hat{\mathbf{x}} + k_y\hat{\mathbf{y}} + k_z\hat{\mathbf{z}}$ is any real wavevector, independent of t and $\boldsymbol{r}$, the complex exponential $e^{i\phi} = \cos\phi + i\sin\phi$ by Euler's formula, with i being the imaginary unit such that $i^2 = -1$. This is easy to verify from (16.19) and (16.11),

$$\begin{aligned} \nabla^2 e^{i\boldsymbol{k}\cdot\boldsymbol{r}} &= \boldsymbol{\nabla}\cdot\boldsymbol{\nabla} e^{i\boldsymbol{k}\cdot\boldsymbol{r}} = \boldsymbol{\nabla}\cdot\left(i\boldsymbol{k}e^{i\boldsymbol{k}\cdot\boldsymbol{r}}\right) = -|\boldsymbol{k}|^2 e^{i\boldsymbol{k}\cdot\boldsymbol{r}}, \\ \frac{\partial}{\partial t} e^{-\kappa|\boldsymbol{k}|^2 t} &= -\kappa|\boldsymbol{k}|^2 e^{-\kappa|\boldsymbol{k}|^2 t}, \end{aligned} \tag{16.126}$$

such that (16.125) is indeed a solution of (16.122). Since that equation is linear, the general solution in an unbounded domain is a superposition of all possible Fourier mode solutions, that is,

$$T(t,\boldsymbol{r}) = \iiint_{\mathbb{R}^3} \widetilde{T}(\boldsymbol{k})\, e^{-\kappa|\boldsymbol{k}|^2 t} e^{i\boldsymbol{k}\cdot\boldsymbol{r}} \mathrm{d}k_x \mathrm{d}k_y \mathrm{d}k_z, \tag{16.127}$$

with $\widetilde{T}(-\boldsymbol{k}) = \widetilde{T}^*(\boldsymbol{k})$ for reality, where $()^*$ denotes complex conjugate. The complex amplitudes $\widetilde{T}(\boldsymbol{k})$ can be determined from the initial condition

$$T(0,\boldsymbol{r}) = \iiint_{\mathbb{R}^3} \widetilde{T}(\boldsymbol{k}) e^{i\boldsymbol{k}\cdot\boldsymbol{r}} \mathrm{d}k_x \mathrm{d}k_y \mathrm{d}k_z. \tag{16.128}$$

It is useful here to think of $T(0,\boldsymbol{r})$ as an infinite "vector"[9] with $\boldsymbol{r}$ as the row index, $e^{i\boldsymbol{k}\cdot\boldsymbol{r}}$ as an infinite "matrix" with $\boldsymbol{r}$ as row index and $\boldsymbol{k}$ as column index, and $\widetilde{T}(\boldsymbol{k})$ as an infinite vector with $\boldsymbol{k}$ as row index.

[9]In a Hilbert space of functions.

Then the triple integral on the right-hand side of (16.128) is a matrix–vector multiply. That "matrix" $e^{i\boldsymbol{k}\cdot\boldsymbol{r}}$ is complex orthogonal, up to a $(2\pi)^3$ normalization, so that its inverse is it transpose conjugate divided by $(2\pi)^3$ and

$$\widetilde{T}(\boldsymbol{k}) = \frac{1}{(2\pi)^3} \iiint_{\mathbb{R}^3} T(0, \boldsymbol{r}) e^{-i\boldsymbol{k}\cdot\boldsymbol{r}} \mathrm{d}x \mathrm{d}y \mathrm{d}z, \tag{16.129}$$

in cartesian coordinates with $\boldsymbol{r} = x\hat{\mathbf{x}} + y\hat{\mathbf{y}} + z\hat{\mathbf{z}}$. The integral (16.129) is the *Fourier transform* of $T(0, \boldsymbol{r})$, from real space $\boldsymbol{r}$ to Fourier space $\boldsymbol{k}$, and (16.128) is the inverse Fourier transform from $\boldsymbol{k}$ space back to $\boldsymbol{r}$ space.

Diffusion in a disk

For diffusion of heat from a localized hot spot in a plane or diffusion in a disk or cylinder, polar coordinates (ρ, φ) are more useful than cartesian coordinates $(x, y) = \rho(\cos\varphi, \sin\varphi)$. The heat equation in polar coordinates follows from (16.110) with $\partial/\partial z = 0$

$$\frac{1}{\kappa}\frac{\partial T}{\partial t} = \frac{1}{\rho}\frac{\partial}{\partial \rho}\left(\rho \frac{\partial T}{\partial \rho}\right) + \frac{1}{\rho^2}\frac{\partial^2 T}{\partial \varphi^2}. \tag{16.130}$$

This linear partial differential equation has constant coefficients in t and φ but not ρ; thus, it admits exponential solutions in t and φ of the form

$$T_{k,n}(t, \rho, \varphi) = e^{-\kappa k^2 t}\, e^{in\varphi}\, f_n(\rho), \tag{16.131}$$

where the radial structure function $f_n(\rho)$ must solve the second order differential equation

$$\frac{1}{\rho}\frac{\mathrm{d}}{\mathrm{d}\rho}\left(\rho \frac{\mathrm{d}f_n}{\mathrm{d}\rho}\right) + \left(k^2 - \frac{n^2}{\rho^2}\right) f_n = 0, \tag{16.132a}$$

obtained by substituting $T_{k,n}(t, \rho, \varphi)$ into (16.130). The azimuthal wavenumber n must be an integer for 2π periodicity in φ, but k can be any positive real wavenumber if we consider the infinite plane.

We can eliminate k by introducing a nondimensional radial variable $\chi = k\rho$ such that $\mathrm{d}/\mathrm{d}\rho = k\,\mathrm{d}/\mathrm{d}\chi$ and the differential equation for $f_n(\rho) = B_n(k\rho) = B_n(\chi)$ is *Bessel's equation* of order n

$$\frac{1}{\chi}\frac{\mathrm{d}}{\mathrm{d}\chi}\left(\chi \frac{\mathrm{d}B_n}{\mathrm{d}\chi}\right) + \left(1 - \frac{n^2}{\chi^2}\right) B_n = 0. \tag{16.132b}$$

Bessel's equation has two independent solutions: Bessel functions of the first kind $B_n(\chi) = J_n(\chi)$ (Fig. 16.38) and of the second kind $B_n(\chi) = Y_n(\chi)$. These *special functions* can be obtained by the *Fuchs–Frobenius method* yielding series expansions of the form

Bessel functions, see: NIST Digital Library of Mathematical Functions. Chap. 10. `https://dlmf.nist.gov/`

$$\begin{aligned} J_n(\chi) &= \chi^n \sum_{m=0}^{\infty} c_m \chi^m, \\ Y_n(\chi) &= J_n(\chi) \ln\chi + \sum_{m=0}^{\infty} d_n \chi^{m-n}, \end{aligned} \tag{16.133}$$

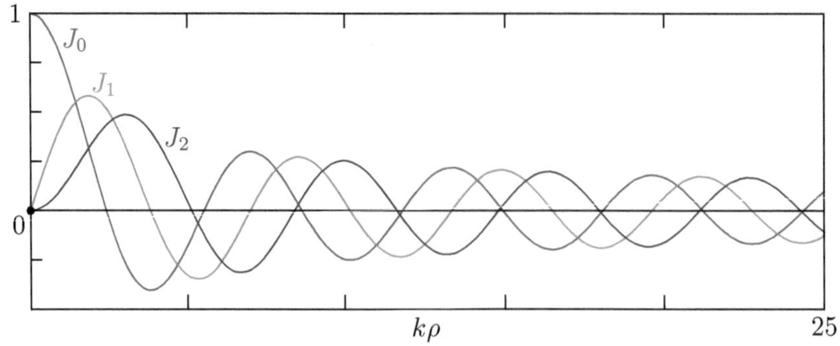

Fig. 16.38 Bessel functions $J_0(k\rho)$, $J_1(k\rho)$, $J_2(k\rho)$.

with $J_n(\chi) \sim \chi^n$ and $Y_n(\chi) \sim \chi^{-n}$, $Y_0(\chi) \sim \ln \chi$ as $\chi \to 0$. Thus ,$J_n(\chi)$ is regular but $Y_n(\chi)$ is singular at $\chi = 0$. This second $Y_n(\chi)$ solution is needed for domains that exclude $\chi = 0$ such as an annulus, but must be rejected for regular solutions that include $\chi = 0$.

Eigenfunctions of the Laplacian

Solutions (16.125) and (16.131) are *eigensolutions* of the heat equation. They keep the same shape for all times but uniformly decay in amplitude at rate $\exp(-\kappa k^2 t)$, in contrast to superpositions of such modes such as (16.123) that change shape and show diffusion. The eigensolutions of the heat equation are *eigenfunctions* of the Laplacian. Indeed,

$$\begin{aligned} \nabla^2 e^{i\boldsymbol{k}\cdot\boldsymbol{r}} &= -|\boldsymbol{k}|^2 e^{i\boldsymbol{k}\cdot\boldsymbol{r}}, \\ \nabla^2\left(e^{in\varphi}J_n(k\rho)\right) &= -k^2\left(e^{in\varphi}J_n(k\rho)\right), \end{aligned} \tag{16.134}$$

where $f_n(\rho) = J_n(k\rho)$ is the Bessel function of the first kind of order n. These two sets of modes, the Fourier modes $\exp(i\boldsymbol{k}\cdot\boldsymbol{r})$ for all real wavevectors $\boldsymbol{k}$, and the Fourier Bessel modes $e^{in\varphi}J_n(k\rho)$ for all integers n and positive real k, are complete sets of complex orthogonal functions that can serve as bases to represent any sufficiently well-behaved function in the plane $(x, y) = \rho(\cos\varphi, \sin\varphi)$.

In particular, Bessel functions can be expressed as superpositions of Fourier modes, and vice versa. For example, when $n = 0$, the function $J_0(k\rho)$ is an axisymmetric eigenmode of the Laplacian. Its wavenumber is k, so it is a sum of *all* Fourier modes with $|\boldsymbol{k}| = k$ fixed, $\boldsymbol{k} = (k_x, k_y) = k(\cos\alpha, \sin\alpha)$ and $\boldsymbol{r} = (x, y) = \rho(\cos\varphi, \sin\varphi)$, such that

$$\boldsymbol{k}\cdot\boldsymbol{r} = k\rho\,(\cos\alpha\cos\varphi + \sin\alpha\sin\varphi) = k\rho\cos(\alpha - \varphi),$$

summed over all α, with equal amplitude $\mathrm{d}\alpha/(2\pi)$ because of axisymmetry, that is,

$$J_0(k\rho) = \frac{1}{2\pi}\int_0^{2\pi} e^{ik\rho\cos(\alpha-\varphi)}\mathrm{d}\alpha = \frac{1}{2\pi}\int_0^{2\pi} e^{ik\rho\cos\beta}\mathrm{d}\beta. \tag{16.135}$$

Similarly, the other Fourier Bessel modes, $e^{in\varphi}J_n(k\rho)$, also have wavenumber k but sinusoidal variation in the azimuthal direction. They are sums of Fourier modes over all angles α with fixed wavenumber k and complex amplitudes $e^{in\alpha}\mathrm{d}\alpha/(2\pi)$,

$$\begin{aligned} e^{in\varphi}J_n(k\rho) &= \frac{(-i)^n}{2\pi}\int_0^{2\pi} e^{in\alpha}e^{ik\rho\cos(\alpha-\varphi)}\mathrm{d}\alpha \\ &= e^{in\varphi}\frac{(-i)^n}{2\pi}\int_0^{2\pi} e^{in\beta}e^{ik\rho\cos\beta}\mathrm{d}\beta. \end{aligned} \tag{16.136a}$$

The $(-i)^n$ is a normalization so that $J_n(k\rho)$ is real with

$$J_n(k\rho) \sim \frac{(k\rho)^n}{2^n n!}, \quad k\rho \to 0, \tag{16.136b}$$

as can be verified from a Taylor series expansion of $\exp(ik\rho\cos\beta)$ in powers of $(k\rho)$ and calculation of the resulting trigonometric integrals. The integral representation can also be used to determine the asymptotic behavior as $k\rho \to \infty$ using complex integration and the method of steepest descent. That yields

$$J_n(k\rho) \sim \sqrt{\frac{2}{\pi k\rho}}\cos\left(k\rho - n\frac{\pi}{2} - \frac{\pi}{4}\right), \quad k\rho \to \infty, \tag{16.136c}$$

which is a very good approximation (Fig. 16.38) long before ∞!

Exercises

(16.1) Find and sketch the field lines of $\boldsymbol{F}(x,y) = x\hat{\mathbf{x}} - y\hat{\mathbf{y}}$.

(16.2) Show that $f(x,y) = \sqrt{|xy|}$ has $\partial f/\partial x = 0 = \partial f/\partial y$ at $(x,y) = (0,0)$, yet $\boldsymbol{\nabla} f$ does not exist, and $f(x,y)$ is not differentiable at $(0,0)$ according to (16.6a).

(16.3) It is tempting to write (16.6c) as $\boldsymbol{\nabla} f = \dfrac{\mathrm{d}f}{\mathrm{d}\boldsymbol{r}}$. Why is that incorrect?

(16.4) Show how to swiftly calculate $\boldsymbol{\nabla} f$ for $f(P) =$ (i) r^{-3}, (ii) $r\sin\theta$, and (iii) z/r^3, where $r = |\boldsymbol{r}|$, $z = \hat{\mathbf{z}}\cdot\boldsymbol{r} = r\cos\theta$.

(16.5) Show how to swiftly calculate $\boldsymbol{\nabla} f$ for

(i) $f(P) = |F_1P| + |F_2P|$, the sum of distances to fixed points F_1 and F_2;

(ii) $f(P) = |F_1P| - |F_2P|$, the difference of those distances; and

(iii) $f(P) = |F_1P|/|F_2P|$, the ratio of those distances.

For the first two, prove that the angle between $\boldsymbol{\nabla} f$ and $\overrightarrow{F_1P}$ equals that between $\boldsymbol{\nabla} f$ and $\overrightarrow{F_2P}$, modulo π, for any P. What are the isosurfaces of $f(P)$?

(16.6) Calculate $\boldsymbol{\nabla}(A/|\boldsymbol{r}-\boldsymbol{r}_1| + B/|\boldsymbol{r}-\boldsymbol{r}_2|)$ where $A, B, \boldsymbol{r}_1, \boldsymbol{r}_2$ are constants, (i) using fundamental gradient properties, and (ii) using cartesian coordinates.

(16.7) Let θ be the polar angle. Calculate $\partial\theta/\partial x$ where $\partial/\partial x$ means (y,z) fixed in cartesian coordinates (x,y,z). Express your result in spherical coordinates (r,θ,φ). Compare cartesian coordinate approach to one using directional derivative and fundamental gradients.

(16.8) For the region $\mathcal{R}$ in a plane defined in cartesian coordinates (x, y) by

$$0 < u_1 \le xy \le u_2, \qquad 0 < v_1 \le y/x \le v_2,$$

consider the coordinates $u = xy$ and $v = y/x$. (i) Sketch the u and v coordinate curves. Highlight $\mathcal{R}$. (ii) Find and sketch $\partial \boldsymbol{r}/\partial u$, $\partial \boldsymbol{r}/\partial v$, $\boldsymbol{\nabla} u$, and $\boldsymbol{\nabla} v$. Are these orthogonal coordinates? (iii) Calculate

$$\frac{\partial \boldsymbol{r}}{\partial u} \cdot \boldsymbol{\nabla} u, \quad \frac{\partial \boldsymbol{r}}{\partial u} \cdot \boldsymbol{\nabla} v, \quad \frac{\partial \boldsymbol{r}}{\partial v} \cdot \boldsymbol{\nabla} u, \quad \frac{\partial \boldsymbol{r}}{\partial v} \cdot \boldsymbol{\nabla} v.$$

(16.9) Show that the vortex dipole field (16.4) is the gradient of the angle $F_1 P F_2$.

(16.10) Show that $\mathrm{d}f/\mathrm{d}t = 0$ for $f(\boldsymbol{r}) = \ln(s_1/s_2)$ and $\mathrm{d}\boldsymbol{r}/\mathrm{d}t = \boldsymbol{F}(\boldsymbol{r})$ given in (16.4).

(16.11) Show that the vortex dipole (16.4) field lines are the contours of $f(P) = \ln(|F_1P|/|F_2P|)$.

(16.12) Show that $\epsilon^{ijk}\partial_i\partial_j \boldsymbol{r} = 0$ for any C^2 function $\boldsymbol{r}(u^1, u^2, u^3)$.

(16.13) Show that $\boldsymbol{\nabla} u^i \times \boldsymbol{\nabla} u^j \cdot \partial_i\partial_j \boldsymbol{r} = 0$ for curvilinear coordinates (u^1, u^2, u^3).

(16.14) Show that $\boldsymbol{\nabla} u^i \times \boldsymbol{\nabla} u^j \, \partial_i(\boldsymbol{F} \cdot \partial_j \boldsymbol{r})$
$= \boldsymbol{\nabla} u^i \times \boldsymbol{\nabla} u^j \, (\partial_i \boldsymbol{F}) \cdot (\partial_j \boldsymbol{r})$.

(16.15) Review, explain and expand (16.32) and (16.34). Use in the following problems.

(16.16) Show that $\partial_k \boldsymbol{r} \cdot \partial_i \boldsymbol{\nabla} u^j = -\partial_i\partial_k \boldsymbol{r} \cdot \boldsymbol{\nabla} u^j$. Use that and (16.34) to verify

$$\boldsymbol{\nabla} \times \boldsymbol{\nabla} u^j = \boldsymbol{\nabla} u^i \times \partial_i \boldsymbol{\nabla} u^j = 0, \qquad (16.137\text{a})$$

$$\boldsymbol{\nabla} \cdot \left(\boldsymbol{\nabla} u^i \times \boldsymbol{\nabla} u^j\right) = \boldsymbol{\nabla} u^k \cdot \partial_k \left(\boldsymbol{\nabla} u^i \times \boldsymbol{\nabla} u^j\right) = 0. \qquad (16.137\text{b})$$

(16.17) For curvilinear coordinates (u^1, u^2, u^3) with $\partial_i \boldsymbol{r} = \partial \boldsymbol{r}/\partial u^i$ and Jacobian $J = \partial_1 \boldsymbol{r} \times \partial_2 \boldsymbol{r} \cdot \partial_3 \boldsymbol{r} > 0$, show that

$$\begin{aligned} \boldsymbol{\nabla} \cdot \left(J^{-1}\partial_i \boldsymbol{r}\right) &= 0, \\ \boldsymbol{\nabla} \times \left(J^{-1}\partial_i \boldsymbol{r} \times \partial_j \boldsymbol{r}\right) &= 0. \end{aligned} \qquad (16.137\text{c})$$

(16.18) Show that

$$\boldsymbol{\nabla} \cdot \partial_i \boldsymbol{r} = \boldsymbol{\nabla} \cdot \boldsymbol{a}_i = \Gamma^j_{ij} = J^{-1}\partial_i J. \qquad (16.137\text{d})$$

(16.19) Show that

$$\boldsymbol{\nabla} \cdot \boldsymbol{\nabla} u^i = \boldsymbol{\nabla} \cdot \boldsymbol{a}^i = J^{-1}\partial_j \left(J g^{ij}\right). \qquad (16.137\text{e})$$

(16.20) Verify that $\boldsymbol{\nabla} \times \boldsymbol{\nabla} f = 0$ for curvilinear coordinates (u^1, u^2, u^3).

(16.21) Verify that $\boldsymbol{\nabla} \cdot \boldsymbol{\nabla} \times \boldsymbol{F} = 0$ for curvilinear coordinates (u^1, u^2, u^3).

(16.22) Derive/explain

$$\boldsymbol{\nabla} \times \boldsymbol{F} = \boldsymbol{a}^i \times \partial_i \left(F_j \boldsymbol{a}^j\right) = J^{-1}\epsilon^{ijk} \boldsymbol{a}_k \, \partial_i F_j, \qquad (16.137\text{f})$$

where $\boldsymbol{a}^i = \boldsymbol{\nabla} u^i$, $\boldsymbol{a}_k = \partial_k \boldsymbol{r}$, and $J = \boldsymbol{a}_1 \times \boldsymbol{a}_2 \cdot \boldsymbol{a}_3$.

(16.23) Derive/explain

$$\boldsymbol{\nabla} \cdot \boldsymbol{F} = \boldsymbol{a}^i \cdot \partial_i \left(F^j \boldsymbol{a}_j\right) = J^{-1}\partial_i (J F^i), \qquad (16.137\text{g})$$

where $\boldsymbol{a}^i = \boldsymbol{\nabla} u^i$, $\boldsymbol{a}_j = \partial_j \boldsymbol{r}$, and $J = \boldsymbol{a}_1 \times \boldsymbol{a}_2 \cdot \boldsymbol{a}_3$.

(16.24) Derive (16.47c) using (u, v, w) instead of (u^1, u^2, u^3). Starting from (16.47a) written

$$\boldsymbol{\nabla} \times \boldsymbol{F} = \boldsymbol{\nabla} u \times \partial_u \boldsymbol{F} + \boldsymbol{\nabla} v \times \partial_v \boldsymbol{F} + \boldsymbol{\nabla} w \times \partial_w \boldsymbol{F},$$

express $\boldsymbol{\nabla} u, \boldsymbol{\nabla} v, \boldsymbol{\nabla} w$ in terms of $\partial_u \boldsymbol{r}, \partial_v \boldsymbol{r}, \partial_w \boldsymbol{r}$ from (16.28), use the identity $(\boldsymbol{a} \times \boldsymbol{b}) \times \boldsymbol{c} = \boldsymbol{b}(\boldsymbol{a} \cdot \boldsymbol{c}) - \boldsymbol{a}(\boldsymbol{b} \cdot \boldsymbol{c})$, and "integrate by parts" $\boldsymbol{a} \cdot \partial_i \boldsymbol{b} = \partial_i(\boldsymbol{a} \cdot \boldsymbol{b}) - \boldsymbol{b} \cdot \partial_i \boldsymbol{a}$. This is an alternate derivation to that in (16.47).

(16.25) Derive (16.60) using (u, v, w) instead of (u^1, u^2, u^3). Starting from (16.47a) written in terms of (u, v, w) as in the previous problem, dot with $\boldsymbol{\nabla} w = J^{-1}\partial_u \boldsymbol{r} \times \partial_v \boldsymbol{r}$ from (16.28), use the vector identity $\boldsymbol{a} \times \boldsymbol{b} \cdot \boldsymbol{c} = \boldsymbol{a} \cdot \boldsymbol{b} \times \boldsymbol{c}$, and "integrate by parts" $\boldsymbol{a} \cdot \partial_i \boldsymbol{b} = \partial_i(\boldsymbol{a} \cdot \boldsymbol{b}) - \boldsymbol{b} \cdot \partial_i \boldsymbol{a}$.

(16.26) Use circulation around cylindrical coordinate patches to calculate $\boldsymbol{\nabla} \times (F(\rho)\hat{\boldsymbol{\varphi}})$ through Stokes' theorem as in Fig. 16.29. Sketch the three patches and explain your derivation.

(16.27) Use circulation around an appropriate surface patch to calculate $\hat{\boldsymbol{\theta}} \cdot \boldsymbol{\nabla} \times \boldsymbol{F}$ using Stokes theorem as in (16.64a) and (16.64b).

(16.28) For curvilinear coordinates (u, v, w) with Jacobian $J > 0$, consider a coordinate surface patch $\mathcal{S} : u_0 \le u \le u_1,\ v_0 \le v \le v_1,\ w = w_0$. Use the fundamental theorem of calculus *backward*, $f(x_1) - f(x_0) = \int_{x_0}^{x_1} (\mathrm{d}f/\mathrm{d}x)\, \mathrm{d}x$, to show that

$$\begin{aligned} &\oint_{\partial \mathcal{S}} \boldsymbol{F} \cdot \mathrm{d}\boldsymbol{r} \\ &= \iint \left\{ \frac{\partial}{\partial u}\left(\boldsymbol{F} \cdot \frac{\partial \boldsymbol{r}}{\partial v}\right) - \frac{\partial}{\partial v}\left(\boldsymbol{F} \cdot \frac{\partial \boldsymbol{r}}{\partial u}\right) \right\} \mathrm{d}u\mathrm{d}v \\ &= \iint \boldsymbol{X} \cdot \left(\frac{\partial \boldsymbol{r}}{\partial u} \times \frac{\partial \boldsymbol{r}}{\partial v}\right) \mathrm{d}u\mathrm{d}v = \int_{\mathcal{S}} \boldsymbol{X} \cdot \mathrm{d}\boldsymbol{S}. \end{aligned}$$

Deduce (16.60) and (16.47c) for $\boldsymbol{X} = \boldsymbol{\nabla} \times \boldsymbol{F}$.

(16.29) Demonstrate the cancellation occurring in (16.67b), that is,

$$\frac{\epsilon^{ijk}}{2}\partial_i(\partial_j \boldsymbol{r} \times \partial_k \boldsymbol{r}) = \partial_u(\partial_v \boldsymbol{r} \times \partial_w \boldsymbol{r}) + \partial_v(\partial_w \boldsymbol{r} \times \partial_u \boldsymbol{r}) + \partial_w(\partial_u \boldsymbol{r} \times \partial_v \boldsymbol{r}) = 0.$$

Use that to derive the general formula for $\nabla \cdot \boldsymbol{F}$ (16.68) without index notation, starting from

$$\nabla \cdot \boldsymbol{F} = \nabla u \cdot \partial_u \boldsymbol{F} + \nabla v \cdot \partial_v \boldsymbol{F} + \nabla w \cdot \partial_w \boldsymbol{F},$$

and using (16.28).

(16.30) Use (16.60) to derive the formula for (i) the $\hat{\mathbf{z}}$ component of $\nabla \times \boldsymbol{F}$ in cylindrical coordinates (ρ, φ, z), and (ii) its $\hat{\boldsymbol{r}}$ component in spherical coordinates (r, θ, φ).

(16.31) Sketch the vector field specified in spherical coordinates (r, θ, φ) as

$$\boldsymbol{F} = 2\frac{\cos\theta}{r^3}\hat{\boldsymbol{r}} + \frac{\sin\theta}{r^3}\hat{\boldsymbol{\theta}}$$

and determine whether it is a gradient field.

(16.32) Use (16.14) and the product rules to calculate $\nabla \times (\rho^a \hat{\boldsymbol{\varphi}})$ and $\nabla \cdot (\rho^a \hat{\boldsymbol{\rho}})$ for constant but arbitrary real a. Sketch the fields.

(16.33) For curvilinear coordinates (u, v, w) with Jacobian $J > 0$, consider a coordinate brick $\mathcal{V} : u_0 \le u \le u_1,\ v_0 \le v \le v_1,\ w_0 \le w \le w_1$. Use the fundamental theorem of calculus *backward*, $f(x_1) - f(x_0) = \int_{x_0}^{x_1} (\mathrm{d}f/\mathrm{d}x)\,\mathrm{d}x$, to show that the *outward flux* of $\boldsymbol{F}$ through the closed surface boundary $\mathcal{S} = \partial\mathcal{V}$ of $\mathcal{V}$ is

$$\oint_{\partial\mathcal{V}} \boldsymbol{F} \cdot \mathrm{d}\boldsymbol{S} = \iiint f \,\mathrm{d}u\mathrm{d}v\mathrm{d}w = \int_{\mathcal{V}} \frac{f}{J}\, dV$$

with

$$f = \frac{\partial}{\partial u}\left(\boldsymbol{F} \cdot \frac{\partial \boldsymbol{r}}{\partial v} \times \frac{\partial \boldsymbol{r}}{\partial w}\right) + \frac{\partial}{\partial v}\left(\boldsymbol{F} \cdot \frac{\partial \boldsymbol{r}}{\partial w} \times \frac{\partial \boldsymbol{r}}{\partial u}\right) + \frac{\partial}{\partial w}\left(\boldsymbol{F} \cdot \frac{\partial \boldsymbol{r}}{\partial u} \times \frac{\partial \boldsymbol{r}}{\partial v}\right).$$

Deduce formula (16.76) for $\nabla \cdot \boldsymbol{F} = f/J$. What are the units of f and of f/J?

(16.34) Use the fundamental theorem of calculus to write the net flux of a vector field $\boldsymbol{F}$ out of a cartesian coordinate "brick,"

$$x_0 \le x \le x_1, \quad y_0 \le y \le y_1, \quad z_0 \le z \le z_1,$$

as a volume integral and deduce the divergence theorem and the form of $\nabla \cdot \boldsymbol{F}$ in cartesian coordinates.

(16.35) Use the fundamental theorem of calculus to write the net flux of a vector field $\boldsymbol{F}$ out of a cylindrical coordinate "brick,"

$$\rho_0 \le \rho \le \rho_1, \quad \varphi_0 \le \varphi \le \varphi_1, \quad z_0 \le z \le z_1,$$

as a volume integral and deduce the divergence theorem and the form of $\nabla \cdot \boldsymbol{F}$ in cylindrical coordinates.

(16.36) Use the fundamental theorem of calculus to write the net flux of a vector field $\boldsymbol{F}$ out of a spherical coordinate "brick," 16.34

$$r_0 \le r \le r_1, \quad \theta_0 \le \theta \le \theta_1, \quad \varphi_0 \le \varphi \le \varphi_1,$$

as a volume integral and deduce the divergence theorem and the form of $\nabla \cdot \boldsymbol{F}$ in spherical coordinates.

(16.37) Use the product rule and $\nabla \cdot \boldsymbol{r} = 3$ to show that

$$\nabla \cdot (F(r)\hat{\boldsymbol{r}}) = r\frac{\mathrm{d}}{\mathrm{d}r}\left(\frac{F}{r}\right) + 3\frac{F}{r}.$$

(16.38) Compute $\nabla \cdot \boldsymbol{F}$ and $\nabla \times \boldsymbol{F}$, with $\boldsymbol{r} = \overrightarrow{OP} = x\hat{\mathbf{x}} + y\hat{\mathbf{y}} + z\hat{\mathbf{z}}$, for the following:

(i) $\boldsymbol{F} = \alpha\boldsymbol{r} + \boldsymbol{\omega} \times \boldsymbol{r}$, with α and $\boldsymbol{\omega}$ constants. Sketch $\boldsymbol{F}(P)$.

(ii) $\boldsymbol{F} = Sy\,\hat{\mathbf{x}}$, where S is constant (shear flow). Sketch $\boldsymbol{F}(P)$.

(iii) $\boldsymbol{F} = \alpha(x\hat{\mathbf{x}} - y\hat{\mathbf{y}})$ with α constant (stagnation point flow). Sketch $\boldsymbol{F}(P)$.

(iv) $\boldsymbol{F} = \boldsymbol{\mathcal{A}} \cdot \boldsymbol{r}$, where $\boldsymbol{\mathcal{A}}$ is a constant tensor so $F_i = A_{ij}x_j$ in cartesian coordinates. This is the linear term in a local Taylor series expansion.

(v) $\boldsymbol{F} = \boldsymbol{\mathcal{S}} \cdot \boldsymbol{r} + \boldsymbol{\omega} \times \boldsymbol{r}$, where $\boldsymbol{\mathcal{S}} = \boldsymbol{\mathcal{S}}^T$ is a constant symmetric tensor and $\boldsymbol{\omega}$ is a fixed vector.

(16.39) Use cartesian coordinates $x^i \equiv (x, y, z)$ to show that $\nabla \cdot (\nabla \times \boldsymbol{V}) = 0$ but $\nabla f \cdot (\nabla \times \boldsymbol{V}) \neq 0$, in general, for any sufficiently differentiable scalar $f(P)$ and vector $\boldsymbol{V}(P)$ fields.

(16.40) Prove in general or disprove with an example

$$\boldsymbol{F} \cdot (\nabla \times \boldsymbol{F}) = 0.$$

(16.41) Use cartesian coordinates $x^i \equiv (x, y, z)$ to show that $\nabla \times \nabla f = 0$ but $\nabla g \times \nabla f \neq 0$ in general, for any sufficiently differentiable $f(P)$ and $g(P)$.

(16.42) Verify (16.97) and (16.98) using index notation for cartesian coordinates $x^i \equiv (x, y, z)$ such that $\nabla x^i = \partial \boldsymbol{r}/\partial x^i \equiv (\hat{\mathbf{x}}, \hat{\mathbf{y}}, \hat{\mathbf{z}})$ are constant.

(16.43) Explain why $(\nabla u^i \times \partial_i \nabla u^j) \times \partial_j \boldsymbol{F} = 0$, then deduce

$$\partial_i \nabla u^j \left(\nabla u^i \cdot \partial_j \boldsymbol{F}\right) = \nabla u^i \left(\partial_i \nabla u^j \cdot \partial_j \boldsymbol{F}\right).$$

Use that result to verify (16.97) in general coordinates.

(16.44) Derive the identity

$$(\nabla \times \boldsymbol{F}) \times \boldsymbol{F} = \boldsymbol{F} \cdot \nabla \boldsymbol{F} - \tfrac{1}{2}\nabla(\boldsymbol{F} \cdot \boldsymbol{F}). \quad (16.138)$$

(16.45) For a vector field in euclidean space $\boldsymbol{F} = F^i \boldsymbol{a}_i = F_i \boldsymbol{a}^i$ with $\boldsymbol{a}_i = \partial_i \boldsymbol{r}$ and $\boldsymbol{a}^i = \nabla u^i$, (16.39). Use (15.70a) and (15.70b) to show that

$$\begin{aligned} \partial_i \boldsymbol{F} &= \left(\partial_i F^j + \Gamma^j_{ik} F^k\right) \boldsymbol{a}_j \\ &= \left(\partial_i F_j - \Gamma^k_{ij} F_k\right) \boldsymbol{a}^j, \quad (16.139) \end{aligned}$$

$$\begin{aligned} \nabla \boldsymbol{F} &= \left(\partial_i F^j + \Gamma^j_{ik} F^k\right) \boldsymbol{a}^i \boldsymbol{a}_j \\ &= \left(\partial_i F_j - \Gamma^k_{ij} F_k\right) \boldsymbol{a}^i \boldsymbol{a}^j. \quad (16.140) \end{aligned}$$

$\nabla \boldsymbol{F}$ is a tensor, the derivative of the vector field $\boldsymbol{F}(\boldsymbol{r})$ such that $\mathrm{d}\boldsymbol{F} = \mathrm{d}\boldsymbol{r} \cdot \nabla \boldsymbol{F}$, for any $\mathrm{d}\boldsymbol{r}$, as in (16.6c) for a scalar field.

(16.46) Show that

$$\begin{aligned} \nabla(\boldsymbol{F} \cdot \boldsymbol{G}) &= (\nabla \boldsymbol{F}) \cdot \boldsymbol{G} + (\nabla \boldsymbol{G}) \cdot \boldsymbol{F} \\ &= \boldsymbol{G} \times (\nabla \times \boldsymbol{F}) + \boldsymbol{G} \cdot \nabla \boldsymbol{F} \\ &\quad + \boldsymbol{F} \times (\nabla \times \boldsymbol{G}) + \boldsymbol{F} \cdot \nabla \boldsymbol{G}. \quad (16.141) \end{aligned}$$

(16.47) Show that

$$\begin{aligned} &(\nabla \times \boldsymbol{F}) \cdot (\nabla \times \boldsymbol{G}) \\ &= (\nabla \boldsymbol{F}) : (\nabla \boldsymbol{G})^{\mathsf{T}} - (\nabla \boldsymbol{F}) : (\nabla \boldsymbol{G}), \quad (16.142) \end{aligned}$$

with $\boldsymbol{ab} : \boldsymbol{cd} = (\boldsymbol{a} \cdot \boldsymbol{d})(\boldsymbol{b} \cdot \boldsymbol{c})$ for regular vectors, and $\boldsymbol{\mathcal{A}} : \boldsymbol{\mathcal{B}} = A_{ij} B_{ji}$ in cartesian coordinates.

(16.48) Using (15.70b) and (16.46), show that

$$\begin{aligned} \nabla \cdot \nabla &= \left(\boldsymbol{a}^i \partial_i\right) \cdot \left(\boldsymbol{a}^j \partial_j\right) \\ &= g^{ij} \left(\partial_i \partial_j - \Gamma^k_{ij} \partial_k\right), \quad (16.143) \end{aligned}$$

and that this equals the Laplacian operator in (16.109).

(16.49) For cylindrical coordinates $(\rho, \varphi, z) \equiv (u^1, u^2, u^3)$, show that

$$[\partial_i \partial_j \boldsymbol{r}] \equiv \begin{bmatrix} 0 & \hat{\boldsymbol{\varphi}} & 0 \\ \hat{\boldsymbol{\varphi}} & -\rho \hat{\boldsymbol{\rho}} & 0 \\ 0 & 0 & 0 \end{bmatrix},$$

and, thus, the Christoffel symbols of the second kind, $\Gamma^k_{ij} = \nabla u^k \cdot \partial_i \partial_j \boldsymbol{r}$, consist of the three matrices $\left[\Gamma^k_{ij}\right]$ for $k = 1, 2, 3$, respectively,

$$\begin{bmatrix} 0 & 0 & 0 \\ 0 & -\rho & 0 \\ 0 & 0 & 0 \end{bmatrix}, \begin{bmatrix} 0 & 1/\rho & 0 \\ 1/\rho & 0 & 0 \\ 0 & 0 & 0 \end{bmatrix}, \begin{bmatrix} 0 & 0 & 0 \\ 0 & 0 & 0 \\ 0 & 0 & 0 \end{bmatrix}.$$

(16.50) For spherical coordinates $(r, \theta, \varphi) \equiv (u^1, u^2, u^3)$, show that

$$[\partial_i \partial_j \boldsymbol{r}] \equiv \begin{bmatrix} 0 & \hat{\boldsymbol{\theta}} & \hat{\boldsymbol{\varphi}} \sin\theta \\ \hat{\boldsymbol{\theta}} & -r\hat{\boldsymbol{r}} & \hat{\boldsymbol{\varphi}}\, r\cos\theta \\ \hat{\boldsymbol{\varphi}} \sin\theta & \hat{\boldsymbol{\varphi}}\, r\cos\theta & -\hat{\boldsymbol{\rho}}\, r\sin\theta \end{bmatrix},$$

where $\hat{\boldsymbol{\rho}} = \hat{\boldsymbol{r}} \sin\theta + \hat{\boldsymbol{\theta}} \cos\theta$, and, thus, the Christoffel symbols of the second kind,

$$\Gamma^k_{ij} = \nabla u^k \cdot \partial_i \partial_j \boldsymbol{r},$$

consist of the three matrices $\left[\Gamma^k_{ij}\right]$ for $k = 1, 2, 3$, respectively,

$$\begin{bmatrix} 0 & 0 & 0 \\ 0 & -r & 0 \\ 0 & 0 & -r\sin^2\theta \end{bmatrix},$$
$$\begin{bmatrix} 0 & 1/r & 0 \\ 1/r & 0 & 0 \\ 0 & 0 & -\frac{1}{2}\sin 2\theta \end{bmatrix}, \begin{bmatrix} 0 & 0 & 1/r \\ 0 & 0 & \cot\theta \\ 1/r & \cot\theta & 0 \end{bmatrix}.$$

(16.51) Consider *elliptical cylindrical coordinates* (u, v, z) where

$$x = a\cosh u \cos v, \quad y = a \sinh u \sin v, \quad (16.144)$$

for cartesian coordinates (x, y, z), with $a > 0$ is a real constant and $u \geq 0$, $-\pi < v \leq \pi$. Find the coordinate tangent vectors $\partial \boldsymbol{r}/\partial u$, $\partial \boldsymbol{r}/\partial v$, $\partial \boldsymbol{r}/\partial z$, the Jacobian J, the covariant metric $[g_{ij}]$, and its inverse $[g^{ij}]$. Show that the Laplacian in these coordinates reads

$$\nabla^2 = \frac{1}{a^2(\sinh^2 u + \sin^2 v)} \left(\frac{\partial^2}{\partial u^2} + \frac{\partial^2}{\partial v^2}\right) + \frac{\partial^2}{\partial z^2},$$

and that $\cosh^2 u - \cos^2 v = \sinh^2 u + \sin^2 v$.

(16.52) Consider *paraboloidal coordinates* (u, v, w) such that the standard cartesian coordinates (x, y, z) are given as

$$\begin{cases} x = uv\cos w, \\ y = uv \sin w, \\ z = \frac{1}{2}(u^2 - v^2) \end{cases} \quad (16.145)$$

with $u \geq 0$, $v \geq 0$, $-\pi < w \leq \pi$. Show that these are orthogonal coordinates. Find the scale factors h_u, h_v, h_w and the Jacobian J. Show that the Laplacian in these coordinates reads

$$\nabla^2 = \frac{1}{u^2+v^2}\left(\frac{1}{u}\frac{\partial}{\partial u}\left(u\frac{\partial}{\partial u}\right) + \frac{1}{v}\frac{\partial}{\partial v}\left(v\frac{\partial}{\partial v}\right)\right) + \frac{1}{u^2v^2}\frac{\partial^2}{\partial w^2}.$$

(16.53) Explain your favorite simplest ways to show that $\nabla\cdot\boldsymbol{r} = 3$ and $\nabla\times\boldsymbol{r} = 0$.

(16.54) Calculate $\nabla\cdot\boldsymbol{\rho}$ and $\nabla\times\boldsymbol{\rho}$, where $\boldsymbol{\rho} = x\hat{\mathbf{x}} + y\hat{\mathbf{y}}$.

(16.55) Consider $\Phi = |\boldsymbol{r}-\boldsymbol{r}_A|^{-1}$ where $\boldsymbol{r}_A$ is constant. Calculate $\nabla\Phi$ and show that $\nabla^2\Phi = 0$. This is the potential of an electric charge in electrostatics, or of a *source* in aerodynamics.

(16.56) Consider $\Phi = \boldsymbol{a}\cdot\boldsymbol{r}/r^3$ with $\boldsymbol{a}$ constant. Calculate $\nabla\Phi$ and show that $\nabla^2\Phi = 0$ for $r \neq 0$. This is the potential of an electric *dipole* in electrostatics, or of a *doublet* in aerodynamics.

(16.57) Consider $\boldsymbol{v} = \boldsymbol{a}\times\boldsymbol{r}$ with $\boldsymbol{a}$ constant. Show that $\nabla\cdot\boldsymbol{v} = 0$ and $\nabla\times\boldsymbol{v} = 2\boldsymbol{a}$ using (i) vector identities, and (ii) cartesian coordinates.

(16.58) Consider $\boldsymbol{A} = \boldsymbol{a}\times\boldsymbol{r}/r^3$ with $\boldsymbol{a}$ constant. Calculate $\nabla\cdot\boldsymbol{A}$ and $\nabla\times\boldsymbol{A}$ and show that $\nabla^2\boldsymbol{A} = 0$ for $r \neq 0$. This is the vector potential of a magnetic *dipole* in magnetostatics.

(16.59) Some students write $\hat{\boldsymbol{r}} = \hat{\mathbf{x}} + \hat{\mathbf{y}} + \hat{\mathbf{z}}$. How is that wrong?

(16.60) Express $\hat{\boldsymbol{\rho}}$, $\hat{\boldsymbol{\varphi}}$, $\hat{\boldsymbol{r}}$, $\hat{\boldsymbol{\theta}}$ in terms of cartesian coordinates x, y, z and directions $\hat{\mathbf{x}}$, $\hat{\mathbf{y}}$, $\hat{\mathbf{z}}$.

(16.61) Show that

$$\hat{\mathbf{z}}\times\boldsymbol{r} = \hat{\mathbf{z}}\times\boldsymbol{\rho} = -y\hat{\mathbf{x}} + x\hat{\mathbf{y}} = \rho\hat{\boldsymbol{\varphi}} = \rho^2\nabla\varphi.$$

Comment on φ and $\nabla\varphi$.

(16.62) Show that

$$\frac{\hat{\mathbf{z}}\times\boldsymbol{r}}{|\hat{\mathbf{z}}\times\boldsymbol{r}|^2} = \frac{-y\hat{\mathbf{x}}+x\hat{\mathbf{y}}}{x^2+y^2} = \frac{\hat{\boldsymbol{\varphi}}}{\rho} = \nabla\varphi = \nabla\times\ln\left(\rho^{-1}\right)\hat{\mathbf{z}}.$$

Discuss the range of validity of these expressions.

(16.63) Show that

$$\frac{\hat{\boldsymbol{r}}}{r^2} = \frac{x\hat{\mathbf{x}}+y\hat{\mathbf{y}}+z\hat{\mathbf{z}}}{(x^2+y^2+z^2)^{3/2}} = \nabla(r^{-1}) = \nabla\times(\varphi\nabla\cos\theta).$$

Discuss the range of validity of these expressions.

(16.64) Show that $\nabla\cdot\hat{\boldsymbol{r}} = 2/r$ and $\nabla\times\hat{\boldsymbol{r}} = 0$ for all $r \neq 0$, using (i) vector identities, and (ii) cartesian coordinates. Sketch $\hat{\boldsymbol{r}}(P)$.

(16.65) Show that $\nabla\cdot\hat{\boldsymbol{\rho}} = 1/\rho$ and $\nabla\times\hat{\boldsymbol{\rho}} = 0$ for all $\rho \neq 0$, using (i) vector identities and fundamental gradients, and (ii) cartesian coordinates. Sketch $\hat{\boldsymbol{\rho}}(P)$.

(16.66) Show that $\nabla\cdot\hat{\boldsymbol{\varphi}} = 0$ and $\nabla\times\hat{\boldsymbol{\varphi}} = \hat{\mathbf{z}}/\rho$ for all $\rho \neq 0$, using (i) vector identities and fundamental gradients, and (ii) cartesian coordinates. Sketch $\hat{\boldsymbol{\varphi}}(P)$.

(16.67) Show that $\nabla\cdot\hat{\boldsymbol{\theta}} = 1/(r\tan\theta)$ and $\nabla\times\hat{\boldsymbol{\theta}} = \hat{\boldsymbol{\varphi}}/r$ for all $\rho \neq 0$, using (i) vector identities and fundamental gradients, and (ii) cartesian coordinates. Sketch $\hat{\boldsymbol{\theta}}(P)$.

(16.68) Calculate $\nabla\cdot(f\hat{\boldsymbol{\varphi}})$ and $\nabla\times(f\hat{\boldsymbol{\varphi}})$ in cylindrical and spherical coordinates using vector identities.

(16.69) Consider $\boldsymbol{E} = \hat{\boldsymbol{r}}/r^2$. Show that $\nabla\cdot\boldsymbol{E} = 0$ and $\nabla\times\boldsymbol{E} = 0$ for all $\boldsymbol{r} \neq 0$, using (i) vector identities and fundamental gradients, (ii) cartesian coordinates, and (iii) spherical coordinates. Sketch $\boldsymbol{E}$.

(16.70) Use vector identities such as $r\nabla\theta = \hat{\boldsymbol{\theta}} = \hat{\boldsymbol{\varphi}}\times\hat{\boldsymbol{r}} = \rho\nabla\varphi\times\nabla r$, together with (16.96), (16.97) and (16.103), for example, to show that

$$\nabla^2\hat{\boldsymbol{r}} = -\frac{2\,\hat{\boldsymbol{r}}}{r^2},$$

$$\nabla^2\hat{\boldsymbol{\theta}} = -\frac{2\cos\theta\,\hat{\boldsymbol{r}}}{r^2\sin\theta} - \frac{\hat{\boldsymbol{\theta}}}{r^2\sin^2\theta},$$

$$\nabla^2\hat{\boldsymbol{\varphi}} = -\frac{\hat{\boldsymbol{\varphi}}}{r^2\sin^2\theta}.$$

(16.71) Consider $\boldsymbol{B} = (\hat{\mathbf{z}}\times\boldsymbol{r})/|\hat{\mathbf{z}}\times\boldsymbol{r}|^2$. Show that $\nabla\cdot\boldsymbol{B} = 0$ and $\nabla\times\boldsymbol{B} = 0$ everywhere, *except* on the z-axis where $\rho = 0$, (i) using vector identities and fundamental gradients, (i) cartesian coordinates, and (iii) cylindrical coordinates. Sketch $\boldsymbol{B}$.

(16.72) Calculate $\nabla\cdot\boldsymbol{v}$ and $\nabla\times\boldsymbol{v}$ for

$$\boldsymbol{v} = A\,\frac{(\boldsymbol{r}-\boldsymbol{a})}{|\boldsymbol{r}-\boldsymbol{a}|^3} + B\,\frac{(\boldsymbol{r}-\boldsymbol{b})}{|\boldsymbol{r}-\boldsymbol{b}|^3},$$

where $A, B, \boldsymbol{a}, \boldsymbol{b}$ are constants using vector identities and fundamental gradients.

(16.73) Calculate ∇f and $\nabla^2 f$ for $f = \boldsymbol{r}\cdot\boldsymbol{A}\cdot\boldsymbol{r} = A_{ij}x^ix^j$, where A_{ij} are constants and x^i are cartesian coordinates.

(16.74) Calculate $\oint_{\mathcal{C}} y\,\mathrm{d}x$ for $\mathcal{C} : \boldsymbol{r}(\theta) = \boldsymbol{a}\cos\theta + \boldsymbol{b}\sin\theta$ with $\boldsymbol{a}, \boldsymbol{b}$ constants, directly and by Stokes.

(16.75) Let A be the area of the triangle with vertices $P_1 \equiv (x_1, y_1)$, $P_2 \equiv (x_2, y_2)$, $P_3 \equiv (x_3, y_3)$ in the cartesian (x, y)-plane and $\mathcal{C} \equiv \partial A$ denotes the boundary of that area, oriented counterclockwise. (i) Calculate $\int_A x^2\, \mathrm{d}A$. Show/explain your work. (ii) Calculate $\oint_{\partial A} \boldsymbol{v} \cdot \mathrm{d}\boldsymbol{r}$ for (a) $\boldsymbol{v} = x\hat{\mathbf{x}} + y\hat{\mathbf{y}}$, and (b) $\boldsymbol{v} = y\hat{\mathbf{x}} - x\hat{\mathbf{y}}$. Sketch $\boldsymbol{v}(P)$.

(16.76) If $\mathcal{C}$ is any simple closed curve in 3D space calculate (i) $\oint_{\mathcal{C}} \boldsymbol{r} \cdot \mathrm{d}\boldsymbol{r}$ and (ii) $\oint_{\mathcal{C}} \boldsymbol{\nabla} f \cdot \mathrm{d}\boldsymbol{r}$ by direct calculation using a parametrization $\boldsymbol{r}(t)$ and by Stokes' theorem.

(16.77) If $\mathcal{C}$ is any closed curve in 3D space not passing through the origin calculate $\oint_{\mathcal{C}} (\boldsymbol{r}/r^3) \cdot \mathrm{d}\boldsymbol{r}$ by direct calculation and by Stokes theorem.

(16.78) Calculate the circulation of the vector field $\boldsymbol{B} = (\hat{\mathbf{z}} \times \boldsymbol{r})/|\hat{\mathbf{z}} \times \boldsymbol{r}|^2$ (i) about a circle of radius R centered at the origin in a plane perpendicular to $\hat{\mathbf{z}}$; (ii) about any closed curve $\mathcal{C}$ in 3D that does not go around the z-axis; and (iii) about any closed curve $\mathcal{C}_0$ that does go around the z-axis. What is wrong with the z-axis anyway? Discuss.

(16.79) Consider $\boldsymbol{v} = \boldsymbol{\omega} \times \boldsymbol{r}$ where $\boldsymbol{\omega}$ is a constant vector, independent of $\boldsymbol{r}$. (i) Evaluate the circulation of $\boldsymbol{v}$ about the circle of radius R centered at the origin in the plane perpendicular to $\boldsymbol{\omega}$ by direct calculation of the line integral; (ii) Calculate the curl of $\boldsymbol{v}$ using vector identities; (iii) calculate the circulation of $\boldsymbol{v}$ about a circle of radius R centered at $\boldsymbol{r}_0$ in the plane perpendicular to $\boldsymbol{n}$.

(16.80) Use Stokes' theorem and suitably chosen vector fields $\boldsymbol{F}$ to show that for any smooth oriented surface $\mathcal{S}$ with boundary curve $\mathcal{C}$ and surface element $\mathrm{d}\boldsymbol{S}$ (cf. (4.17)),

$$\int_{\mathcal{S}} \mathrm{d}\boldsymbol{S} = \frac{1}{2} \oint_{\mathcal{C}} \boldsymbol{r} \times \mathrm{d}\boldsymbol{r}.$$

(16.81) If $\mathcal{C}$ is any closed curve in the (x, y)-plane, calculate $\oint_{\mathcal{C}} \hat{\boldsymbol{n}}\, \mathrm{d}s$ where $\hat{\boldsymbol{n}}(\boldsymbol{r})$ is the unit outside normal to $\mathcal{C}$ at the point $\boldsymbol{r}$ of $\mathcal{C}$ and $\mathrm{d}s = |\mathrm{d}\boldsymbol{r}|$. Show and explain your work.

(16.82) If $\mathcal{S}$ is any closed surface in 3D euclidean space, calculate $\oint_{\mathcal{S}} \hat{\boldsymbol{n}} \mathrm{d}A$ where $\hat{\boldsymbol{n}}(\boldsymbol{r})$ is the unit outside normal to $\mathcal{S}$ at point $\boldsymbol{r}$ on $\mathcal{S}$ and $\mathrm{d}\boldsymbol{S} = \hat{\boldsymbol{n}} \mathrm{d}A$ is the surface area element. Show and explain your work.

(16.83) If $\mathcal{S}$ is any closed surface in 3D, calculate $\oint_{\mathcal{S}} p\, \hat{\boldsymbol{n}}\, \mathrm{d}A$ where $\hat{\boldsymbol{n}}$ is the unit outside normal to $\mathcal{S}$ and $p(\boldsymbol{r}) = (p_0 - \rho g \hat{\mathbf{z}} \cdot \boldsymbol{r})$ where p_0, ρ, and g are constants (This is *Archimedes' principle* with ρ as fluid density and g as the acceleration of gravity.) Calculate the torque, $\oint_{\mathcal{S}} \boldsymbol{r} \times (-p\hat{\boldsymbol{n}})\, dS$ for the same pressure field p.

(16.84) Using both direct surface integral calculation and the divergence theorem, calculate the flux of $\boldsymbol{r}$ through (i) the surface of a sphere of radius R centered at the origin, (ii) the surface of the sphere of radius R centered at $\boldsymbol{r}_0$, and (iii) the surface of a cube of side L with one corner at the origin.

(16.85) Use the divergence theorem to calculate the flux of $\boldsymbol{v} = \boldsymbol{r}/r^3$ through (i) the surface of a sphere of radius ϵ centered at the origin, (ii) any closed surface that does *not* contain the origin, and (iii) an arbitrary closed surface that encloses the origin. Explain carefully. What is wrong with the origin anyway?

(16.86) Calculate $\boldsymbol{\nabla}|\boldsymbol{r} - \boldsymbol{r}_0|^{-1}$ and $\boldsymbol{\nabla} \cdot \big((\boldsymbol{r} - \boldsymbol{r}_0)/|\boldsymbol{r} - \boldsymbol{r}_0|^3\big)$, where $\boldsymbol{r}_0$ is a constant vector, using vector identities and fundamental gradients.

(16.87) What are all the possible values of

$$\oint_{\mathcal{S}} \left(A\, \frac{\boldsymbol{r} - \boldsymbol{a}}{|\boldsymbol{r} - \boldsymbol{a}|^3} + B\, \frac{\boldsymbol{r} - \boldsymbol{b}}{|\boldsymbol{r} - \boldsymbol{b}|^3} \right) \cdot \mathrm{d}\boldsymbol{S},$$

where $A, B, \boldsymbol{a}, \boldsymbol{b}$ are constants and $\mathcal{S}$ is any simple closed surface? Explain/justify carefully.

(16.88) Calculate the flux of

$$\boldsymbol{F}(\boldsymbol{r}) = m_1\, \frac{\boldsymbol{r} - \boldsymbol{r}_1}{|\boldsymbol{r} - \boldsymbol{r}_1|^3} + m_2\, \frac{\boldsymbol{r} - \boldsymbol{r}_2}{|\boldsymbol{r} - \boldsymbol{r}_2|^3}$$

through the surface of a sphere of radius R centered at the origin, where m_1 and m_2 are scalar constants and (i) $|\boldsymbol{r}_1|$ and $|\boldsymbol{r}_2|$ are both less than R; (ii) $|\boldsymbol{r}_1| < R < |\boldsymbol{r}_2|$. Generalize to $\boldsymbol{F}(\boldsymbol{r}) = \sum_{i=1}^{N} m_i(\boldsymbol{r} - \boldsymbol{r}_i)/|\boldsymbol{r} - \boldsymbol{r}_i|^3$.

(16.89) Calculate

$$\boldsymbol{F}(\boldsymbol{r}) = \int_{V'} \frac{\boldsymbol{r} - \boldsymbol{r}'}{|\boldsymbol{r} - \boldsymbol{r}'|^3}\, \mathrm{d}V(\boldsymbol{r}'),$$

where $\boldsymbol{r}$ is any fixed position vector and the integral is over all $\boldsymbol{r}'$ in $V' : |\boldsymbol{r}'| < R$ in 3D euclidean space. This is the gravity field at $\boldsymbol{r}$ due to a sphere of uniform mass density. The integral can be calculated directly using symmetry. The simpler solution is to realize that the integral over $\boldsymbol{r}'$ is essentially a sum over $\boldsymbol{r}_i$ as in the previous exercise so we can figure out the flux of $\boldsymbol{F}(\boldsymbol{r})$ through any closed surface enclosing all of V'. Now by symmetry $\boldsymbol{F}(\boldsymbol{r}) = F(r)\hat{\boldsymbol{r}}$, so knowing the flux is enough to figure out $F(r)$.

(16.90) Show how to obtain (16.132b) from (16.130).

(16.91) Verify that (16.135) is a solution of (16.132a) for $n = 0$.

(16.92) Derive (16.136b).

(16.93) Use computational software to investigate how quickly the Bessel function $J_n(k\rho)$ approaches its asymptotic behavior (16.136c). Compare with Fig. 16.38.

(16.94) Suppose you have all the data to plot the skull surface (Fig. 16.39). How do you compute its volume? Provide an explicit formula or algorithm and specify what data are needed and in what form.

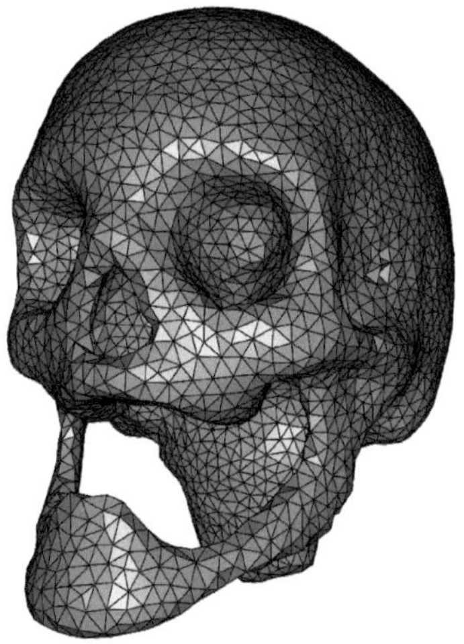

Fig. 16.39 Skull surface. Courtesy of Rineau and Yvinec, *The Computational Geometry Algorithms Library*, `https://www.cgal.org`.

Laplacian and divergence theorem

(16.95) For any smooth real scalar field $\phi(\boldsymbol{r})$ in a domain $\mathcal{V}$ with surface boundary $\mathcal{S} = \partial\mathcal{V}$, show that

$$\int_{\mathcal{V}} (\phi\nabla^2\phi)\,\mathrm{d}V = \oint_{\mathcal{S}} \phi\boldsymbol{\nabla}\phi \cdot \mathrm{d}\boldsymbol{S} - \int_{\mathcal{V}} |\boldsymbol{\nabla}\phi|^2\,\mathrm{d}V. \quad (16.146)$$

Conclude that if ϕ satisfies *Laplace's equation* $\nabla^2\phi = 0$ in $\mathcal{V}$ with $\phi = 0$ or $\hat{\boldsymbol{n}} \cdot \boldsymbol{\nabla}\phi = 0$ on $\mathcal{S}$, then $\phi(\boldsymbol{r}) = 0$ everywhere inside $\mathcal{V}$, or $\phi(\boldsymbol{r})$ is constant if the boundary condition is $\hat{\boldsymbol{n}} \cdot \boldsymbol{\nabla}\phi = 0$ over the entire boundary.

(16.96) *Uniqueness.* Show that if two scalar fields $\phi(\boldsymbol{r})$ and $\psi(\boldsymbol{r})$ satisfy Laplace's $\nabla^2\phi = 0$, $\nabla^2\psi = 0$ in a domain $\mathcal{V}$ with surface boundary $\mathcal{S} = \partial\mathcal{V}$ with the same *Dirichlet* $\phi(\boldsymbol{r}) = \psi(\boldsymbol{r})$, or *Neumann* $\hat{\boldsymbol{n}} \cdot \boldsymbol{\nabla}\phi = \hat{\boldsymbol{n}} \cdot \boldsymbol{\nabla}\psi$ boundary conditions on $\mathcal{S} = \partial\mathcal{V}$, then $\phi(\boldsymbol{r}) = \psi(\boldsymbol{r})$ everywhere inside $\mathcal{V}$. (Hint: consider the difference function $f(\boldsymbol{r}) = \phi(\boldsymbol{r}) - \psi(\boldsymbol{r})$. The functions may differ by an arbitrary constant in the case of Neumann boundary conditions over the entire boundary).

(16.97) If $\boldsymbol{\nabla} \cdot \boldsymbol{\nabla}\phi = \nabla^2\phi = 0$ in a domain $\mathcal{V}$ then the divergence theorem gives

$$\oint_{\mathcal{S}} \boldsymbol{\nabla}\phi \cdot \mathrm{d}\boldsymbol{S} = 0 \quad (16.147)$$

for *any* closed surface $\mathcal{S}$ inside $\mathcal{V}$. Show that

(i) Neumann boundary conditions $\hat{\boldsymbol{n}} \cdot \boldsymbol{\nabla}\phi = f(\boldsymbol{r})$ for all $\boldsymbol{r}$ on the surface boundary $\partial\mathcal{V}$ require the solvability condition $\oint_{\partial\mathcal{V}} f(\boldsymbol{r})\mathrm{d}S = 0$;

(ii) $\phi(\boldsymbol{r})$ cannot have a maximum or a minimum inside $\mathcal{V}$. (Hint: consider an isosurface $\phi = \phi_0$ surrounding the assumed max or min.)

(16.98) For any two smooth scalar fields $\phi(\boldsymbol{r})$ and $\psi(\boldsymbol{r})$ in a domain $\mathcal{V}$ with surface boundary $\mathcal{S} = \partial\mathcal{V}$, derive *Green's identity*

$$\int_{\mathcal{V}} (\phi\nabla^2\psi - \psi\nabla^2\phi)\,\mathrm{d}V = \oint_{\mathcal{S}} (\phi\boldsymbol{\nabla}\psi - \psi\boldsymbol{\nabla}\phi) \cdot \mathrm{d}\boldsymbol{S}. \quad (16.148)$$

(16.99) *Mean value theorem.* Consider (16.148) with $\psi = 1/(4\pi r)$ in a spherical annulus $0 < r_1 \leq r \leq r_2$ around the origin. Show that if ϕ satisfies Laplace's equation $\nabla^2\phi = 0$ then

$$\frac{1}{4\pi r_1^2}\oint_{\mathcal{S}_1} \phi\,\mathrm{d}S_1 = \frac{1}{4\pi r_2^2}\oint_{\mathcal{S}_2} \phi\,\mathrm{d}S_2, \quad (16.149)$$

where $\mathcal{S}_1$ and $\mathcal{S}_2$ are the spherical surfaces of radius r_1 and r_2, respectively. Thus, the average of $\phi(\boldsymbol{r})$ over spheres is independent of radius and in the limit $r_1 \to 0$, the value of ϕ at the origin is equal to its average over any sphere around the origin, as long as the latter is inside the domain where $\nabla^2\phi = 0$. Since the origin is arbitrary, this is true for any *interior* point where $\nabla^2\phi = 0$. (Hint: use (16.147) and the fact that $\psi = 1/r$ is constant on the spheres $\mathcal{S}_1$ and $\mathcal{S}_2$.)

(16.100) Consider scalar eigenfunctions $\phi(\boldsymbol{r})$ of the Laplacian $\nabla^2\phi = \lambda\phi$ in a domain $\mathcal{V}$ with *Fourier-Robin* boundary conditions

$$\alpha\phi + \beta\,\hat{\boldsymbol{n}} \cdot \boldsymbol{\nabla}\phi = 0 \quad (16.150)$$

on $\mathcal{S} = \partial\mathcal{V}$ for some real functions α, β not both zero at the same location. Show that λ must be real and negative. (Hint: consider (16.148) for $\psi = \phi^*$, the complex conjugate of ϕ.)

(16.101) If $\phi(\boldsymbol{r})$ and $\psi(\boldsymbol{r})$ are eigenfunctions of the Laplacian with distinct eigenvalues, that is, $\nabla^2\phi = \lambda\phi$ and $\nabla^2\psi = \mu\psi$ with $\lambda \neq \mu$ both with the same boundary condition (16.150), show that $\phi(\boldsymbol{r})$ and $\psi(\boldsymbol{r})$ are orthogonal in the sense

$$\int_{\mathcal{V}} \phi(\boldsymbol{r})\psi(\boldsymbol{r})\,\mathrm{d}V = 0.$$

(16.102) We have seen that $\boldsymbol{F} = \boldsymbol{\nabla}\phi$ if $\boldsymbol{\nabla} \times \boldsymbol{F} = 0$, and $\boldsymbol{F} = \boldsymbol{\nabla} \times \boldsymbol{A}$ if $\boldsymbol{\nabla} \cdot \boldsymbol{F} = 0$. In general, a vector field is the sum of a gradient and a curl,

$$\boldsymbol{F} = \boldsymbol{\nabla}\phi + \boldsymbol{\nabla} \times \boldsymbol{A}. \tag{16.151}$$

This is the *Helmholtz decomposition*. The scalar potential ϕ is defined up to a constant. The vector potential $\boldsymbol{A}$ is defined up to a gradient field. (i) Show that

$$\nabla^2\phi = \boldsymbol{\nabla} \cdot \boldsymbol{F}, \tag{16.152}$$

and

$$\begin{cases} \nabla^2\boldsymbol{A} = -\boldsymbol{\nabla} \times \boldsymbol{F}, \\ \boldsymbol{\nabla} \cdot \boldsymbol{A} = 0. \end{cases} \tag{16.153}$$

Thus, the potentials are determined by *Poisson equations*, given $\boldsymbol{\nabla} \cdot \boldsymbol{F}$ and $\boldsymbol{\nabla} \times \boldsymbol{F}$, and proper boundary conditions. (ii) Show that this decomposition is orthogonal in the sense that

$$\int_{\mathcal{V}} (\boldsymbol{\nabla}\phi) \cdot (\boldsymbol{\nabla} \times \boldsymbol{A})\,\mathrm{d}V = 0,$$

provided the boundary integral

$$\oint_{\partial\mathcal{V}} \phi\,\hat{\boldsymbol{n}} \cdot \boldsymbol{\nabla} \times \boldsymbol{A}\,\mathrm{d}S = 0,$$

where $\mathrm{d}\boldsymbol{S} = \hat{\boldsymbol{n}}\mathrm{d}S$ is the surface element for the boundary of the domain $\mathcal{V}$. Explicit formulas for ϕ and $\boldsymbol{A}$ exist given $\boldsymbol{\nabla} \cdot \boldsymbol{F}$ and $\boldsymbol{\nabla} \times \boldsymbol{F}$ in $\mathcal{V}$, and $\boldsymbol{F}$ on $\mathcal{S} = \partial\mathcal{V}$.[10]

[10] E.g. Appendix B in David J. Griffiths, *Introduction to Electrodynamics*, 5th Ed., Cambridge University Press, 2023; Appendix B in Kurt E. Oughstun, *Electromagnetic and Optical Pulse Propagation,* 2nd Ed., Springer, 2019.

Electromagnetism

17

Electromagnetism was a major impetus for the development of vector analysis by Josiah Willard Gibbs, Edwin B. Wilson, and Oliver Heaviside. Electromagnetism is all about vector fields circulating along curves and flowing through surfaces. This chapter highlights some fundamental applications of vector calculus.

17.1 Electro and magneto statics

First came *Coulomb's inverse square law* that an electric charge Q_1 located at $\boldsymbol{r}_1$ induces an electric field

$$\boldsymbol{E}(\boldsymbol{r}) = \frac{Q_1}{4\pi\epsilon_0}\frac{\boldsymbol{r}-\boldsymbol{r}_1}{|\boldsymbol{r}-\boldsymbol{r}_1|^3} \tag{17.1}$$

at point $\boldsymbol{r}$, where ϵ_0 is the electric constant. This was generalized to *Gauss's law* that the flux of the electric field $\boldsymbol{E}$ through a closed surface $\mathcal{S}$ is proportional to the sum of charges *inside* the surface

$$\oint_{\mathcal{S}} \boldsymbol{E}\cdot \mathrm{d}\boldsymbol{S} = \frac{Q}{\epsilon_0}, \tag{17.2}$$

where $Q = \sum_n Q_n$ is the total charge inside $\mathcal{S}$. Charges outside the surface do not contribute to the net flux.

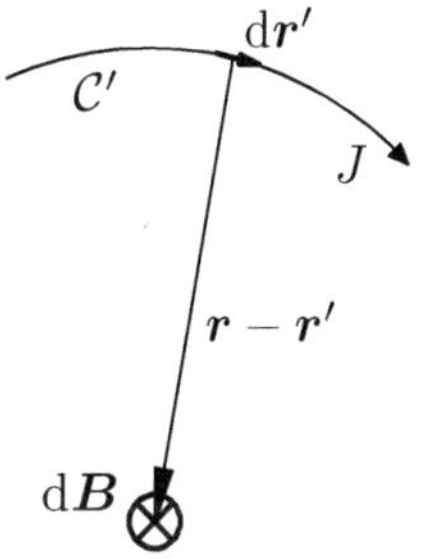

Fig. 17.1 Biot–Savart law.

In magnetostatics (steady currents), there is the *Biot–Savart law* that a current J flowing along a wire $\mathcal{C}'$ induces a magnetic field (Fig. 17.1)

$$\boldsymbol{B}(\boldsymbol{r}) = \frac{\mu_0 J}{4\pi}\int_{\mathcal{C}'}\frac{\mathrm{d}\boldsymbol{r}'\times(\boldsymbol{r}-\boldsymbol{r}')}{|\boldsymbol{r}-\boldsymbol{r}'|^3}, \tag{17.3}$$

where $\boldsymbol{r}'$ is the position vector of a point along the wire, with $\mathrm{d}\boldsymbol{r}'$ the line element along that wire, that is, $\mathcal{C}' : t' \to \boldsymbol{r}'(t')$ for some real parameter t', and μ_0 is the magnetic constant. That curve $\mathcal{C}'$ should be closed or extend to ∞ since a steady current is flowing through it (12.81), (12.82).

Magnetostatics also has *Ampère's law* stating that the circulation of a magnetic field $\boldsymbol{B}(\boldsymbol{r})$ about a closed curve $\mathcal{C}$ is proportional to the total current flowing through any surface $\mathcal{S}$ spanning $\mathcal{C}$ (Fig. 17.2),

$$\oint_{\mathcal{C}} \boldsymbol{B}\cdot \mathrm{d}\boldsymbol{r} = \mu_0 J, \tag{17.4}$$

where J is the net current flowing through a surface $\mathcal{S}$ spanning $\mathcal{C}$ in the direction specified by the right-hand rule and the orientation of $\mathcal{C}$.

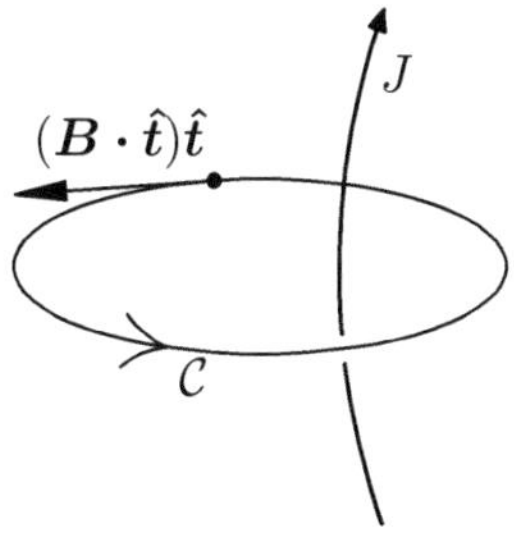

Fig. 17.2 Ampère's law.

Counting and linking numbers

Those laws are not independent. If we substitute Coulomb's law (17.1) into Gauss's law (16.84), all the constants cancel out and we obtain that for any simple closed orientable smooth surface $\mathcal{S}$ (Fig. 17.3)

$$\frac{1}{4\pi}\oint_{\mathcal{S}} \frac{\boldsymbol{r}-\boldsymbol{r}_1}{|\boldsymbol{r}-\boldsymbol{r}_1|^3}\cdot \mathrm{d}\boldsymbol{S} = \begin{cases} 1 & \text{if } \boldsymbol{r}_1 \text{ is inside } \mathcal{S}, \\ 0 & \text{if } \boldsymbol{r}_1 \text{ is outside } \mathcal{S}. \end{cases} \tag{17.5}$$

This is a purely mathematical statement that better be true if both Coulomb's and Gauss's laws are correct. We have several ways to verify this result.

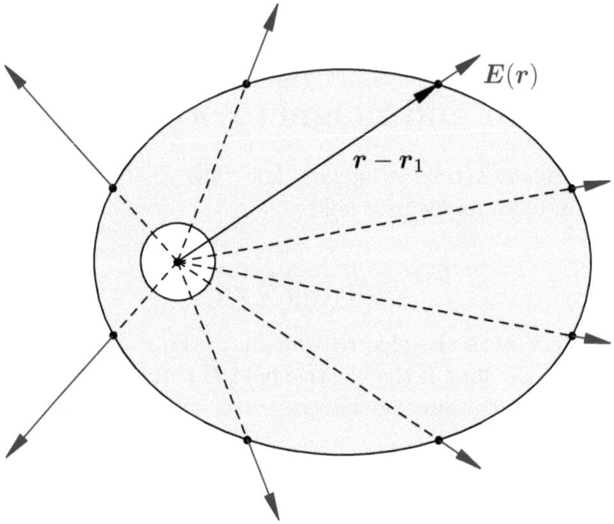

Fig. 17.3 Flux of $\boldsymbol{E}$ through a surface, sketched here for the 2D version of flux through a curve with the field of a line charge $\boldsymbol{E} = (\boldsymbol{r}-\boldsymbol{r}_1)/|\boldsymbol{r}-\boldsymbol{r}_1|^2$.

Using the divergence theorem, we transform the surface integral into a volume integral of the divergence that conveniently vanishes almost everywhere,

$$\boldsymbol{\nabla}\cdot\frac{\boldsymbol{r}-\boldsymbol{r}_1}{|\boldsymbol{r}-\boldsymbol{r}_1|^3} = 0, \quad \boldsymbol{r}\neq\boldsymbol{r}_1.$$

To deal with the singularity at $\boldsymbol{r}_1$, we cut out a little sphere about $\boldsymbol{r}_1$ and easily evaluate the flux through that sphere as 4π. The divergence theorem applied to the volume outside the inner sphere and inside the outer surface, then says that the flux through the outer surface equals that through the inner sphere. This is actually the result derived using spherical coordinates centered at $\boldsymbol{r}_1$, (13.44) (Fig. 13.19) without explicit use of the divergence theorem! In any case, (17.5) is indeed correct and Coulomb's and Gauss's laws are consistent. One consequence of this mathematical result (17.5) is that the integral

$$\frac{1}{4\pi}\oint_{\mathcal{S}} \sum_{n=1}^{N} \frac{\boldsymbol{r}-\boldsymbol{r}_n}{|\boldsymbol{r}-\boldsymbol{r}_n|^3}\cdot \mathrm{d}\boldsymbol{S} = k \tag{17.6}$$

counts the number of points $\boldsymbol{r}_n$ that are *inside* the closed surface $\mathcal{S}$.

Similarly, combining the Biot–Savart (17.3) and Ampère's (17.4) laws leads to (Fig. 17.4)

$$\frac{1}{4\pi}\oint_{\mathcal{C}}\int_{\mathcal{C}'}\frac{\mathrm{d}\boldsymbol{r}'\times(\boldsymbol{r}-\boldsymbol{r}')}{|\boldsymbol{r}-\boldsymbol{r}'|^3}\cdot\mathrm{d}\boldsymbol{r}=l, \tag{17.7}$$

where l is the number of times that the current J flows through the surface $\mathcal{S}$ spanning $\mathcal{C}$ in the direction of the right-hand rule about $\mathcal{C}$. That remarkable double integral gives the *linking number* l of the curves $\mathcal{C}$ and $\mathcal{C}'$. It is named after Gauss for reasons that will become clear as we proceed to demonstrate it. The curve $\mathcal{C}$ is simple and closed, and the curve $\mathcal{C}'$ is also closed or closed at infinity. Again, this is a mathematical statement about the value of an integral, and it better be true if Biot, Savart, and Ampère are all correct.

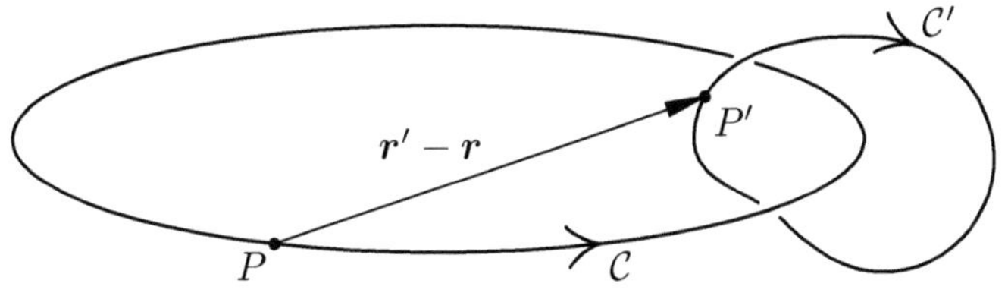

Fig. 17.4 Ampère and Biot–Savart laws yield the linking number (17.7).

The result is easily verified when the wire $\mathcal{C}'$ is the z-axis and $\mathcal{C}$ is a circle centered at O in the (x,y)-plane; then Biot–Savart gives $\boldsymbol{B}(\boldsymbol{r})=\mu_0 J\hat{\boldsymbol{\varphi}}/(2\pi\rho)$ (12.82) with $\mathrm{d}\boldsymbol{r}=\rho\hat{\boldsymbol{\varphi}}\,\mathrm{d}\varphi$ on $\mathcal{C}$ such that $\boldsymbol{B}\cdot\mathrm{d}\boldsymbol{r}=\mu_0 J/(2\pi)\mathrm{d}\varphi$, yielding (17.4) indeed. More generally, the mixed product identity gives

$$\frac{\mathrm{d}\boldsymbol{r}'\times(\boldsymbol{r}-\boldsymbol{r}')}{|\boldsymbol{r}-\boldsymbol{r}'|^3}\cdot\mathrm{d}\boldsymbol{r}=\frac{\boldsymbol{r}-\boldsymbol{r}'}{|\boldsymbol{r}-\boldsymbol{r}'|^3}\cdot\mathrm{d}\boldsymbol{r}\times\mathrm{d}\boldsymbol{r}',$$

that is, the flux of a Coulomb field (17.1) through a surface element $\mathrm{d}\boldsymbol{S}=\mathrm{d}\boldsymbol{r}\times\mathrm{d}\boldsymbol{r}'$. But what surface? Let $\boldsymbol{s}(t,t')=\boldsymbol{r}'(t')-\boldsymbol{r}(t)$ such that $\partial\boldsymbol{s}/\partial t=-\mathrm{d}\boldsymbol{r}/\mathrm{d}t$ and $\partial\boldsymbol{s}/\partial t'=\mathrm{d}\boldsymbol{r}'/\mathrm{d}t'$; then

$$\oint_{\mathcal{C}}\int_{\mathcal{C}'}\frac{\mathrm{d}\boldsymbol{r}'\times(\boldsymbol{r}-\boldsymbol{r}')}{|\boldsymbol{r}-\boldsymbol{r}'|^3}\cdot\mathrm{d}\boldsymbol{r}=\int_{t_0}^{t_1}\int_{t_0'}^{t_1'}\frac{\boldsymbol{s}}{|\boldsymbol{s}|^3}\cdot\frac{\partial\boldsymbol{s}}{\partial t}\times\frac{\partial\boldsymbol{s}}{\partial t'}\,\mathrm{d}t'\,\mathrm{d}t, \tag{17.8}$$

where $\boldsymbol{r}(t_1)=\boldsymbol{r}(t_0)$ and $\boldsymbol{r}'(t_1')=\boldsymbol{r}'(t_0')$ since the curves are closed, and the vector $\boldsymbol{s}(t,t')$ spans a surface, the surface swept by $\mathcal{C}'$ as seen from P. This feels like a torus, but it is trickier; this surface self-intersects and turns inside out!

For example, when $\mathcal{C}$ is a circle of radius R in the (x,y)-plane, and $\mathcal{C}'$ a circle of radius $a<R$ centered at $x=R$ in the (x,z)-plane (Fig. 17.5),

$$\begin{aligned}\mathcal{C}:t\in(0,2\pi)\to\quad&\boldsymbol{r}(t)=R\sin t\,\hat{\mathbf{x}}-R\cos t\,\hat{\mathbf{y}},\\ \mathcal{C}':t'\in(0,2\pi)\to\quad&\boldsymbol{r}'(t')=(R+a\cos t')\,\hat{\mathbf{x}}-a\sin t'\,\hat{\mathbf{z}},\end{aligned}$$

$$\boldsymbol{s}=\boldsymbol{r}'-\boldsymbol{r}=(R-R\sin t+a\cos t')\,\hat{\mathbf{x}}+R\cos t\,\hat{\mathbf{y}}-a\sin t'\,\hat{\mathbf{z}}. \tag{17.9}$$

This is similar to a torus but it is *not* a torus because $\mathcal{C}'$ is translated, not rotated, around $\mathcal{C}$, and it is not a simple surface as a result; it intersects

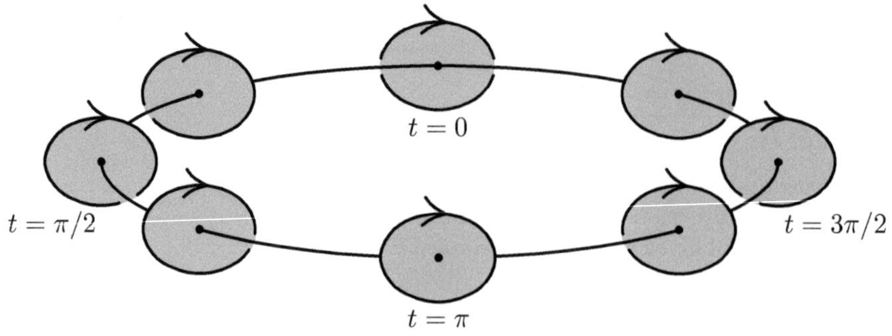

Fig. 17.5 Circles $\mathcal{C}$ and $\mathcal{C}'$ example, (17.9).

itself and turns inside out at $t = 0$ and π! But for the double integral (17.8), we can split it into integrals over two simple surfaces (Fig. 17.6)

$$\mathcal{S}_1 : \boldsymbol{s}(t, t'), \quad 0 \leq t < \pi \qquad 0 \leq t' < 2\pi,$$
$$\mathcal{S}_2 : \boldsymbol{s}(t, t'), \quad \pi \leq t < 2\pi, \quad 0 \leq t' < 2\pi.$$

These are cylindrical surfaces whose axes are the left and right half of circle $\mathcal{C}$ shifted by $R\hat{\mathbf{x}}$. The cylinder endcaps at $t = 0$ and π are not included.

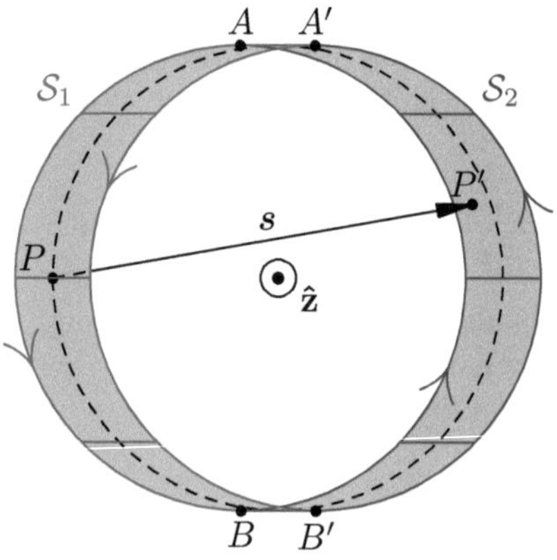

Fig. 17.6 Circles $\mathcal{C}$ and $\mathcal{C}'$ example, (17.9), top view.

Now the $\boldsymbol{s}(t, t')$ integral (17.8) can be evaluated from the counting integral (17.5). Let Φ_A be the outward flux of $\boldsymbol{s}/|\boldsymbol{s}|^3$ through the AA' disk at $t = 0$ and Φ_B the outward flux through BB' at $t = \pi$. They are actually equal by symmetry. The surface normal $\partial \boldsymbol{s}/\partial t \times \partial \boldsymbol{s}/\partial t'$ points outward at all points of $\mathcal{S}_1$. The new origin $|\boldsymbol{s}| = 0$ (point P in the $\boldsymbol{r}$ space) is inside $\mathcal{S}_1$ that needs both endcaps to form a closed surface through which the total flux is 4π. Thus, from (17.5) we find,

$$\int_{\mathcal{S}_1} \frac{\boldsymbol{s}}{|\boldsymbol{s}|^3} \cdot \mathrm{d}\boldsymbol{S}_1 = \int_0^{\pi} \int_0^{2\pi} \frac{\boldsymbol{s}}{|\boldsymbol{s}|^3} \cdot \frac{\partial \boldsymbol{s}}{\partial t} \times \frac{\partial \boldsymbol{s}}{\partial t'} \, \mathrm{d}t' \, \mathrm{d}t, = 4\pi - \Phi_A - \Phi_B.$$

For $\mathcal{S}_2$, the new origin $|\boldsymbol{s}| = 0$ is outside and $\partial \boldsymbol{s}/\partial t \times \partial \boldsymbol{s}/\partial t'$ points inward at all points of $\mathcal{S}_2$. Again the endcaps AA' and BB' are needed to close the surface. So the net flux of $\boldsymbol{s}/|\boldsymbol{s}|^3$ through $\mathcal{S}_2$ closed by the endcaps AB and $A'B'$ oriented *inward* is zero from (17.5); therefore,

$$\int_{\mathcal{S}_2} \frac{\boldsymbol{s}}{|\boldsymbol{s}|^3} \cdot \mathrm{d}\boldsymbol{S}_2 = \int_{\pi}^{2\pi} \int_0^{2\pi} \frac{\boldsymbol{s}}{|\boldsymbol{s}|^3} \cdot \frac{\partial \boldsymbol{s}}{\partial t} \times \frac{\partial \boldsymbol{s}}{\partial t'} \, \mathrm{d}t' \, \mathrm{d}t = \Phi_A + \Phi_B,$$

since the flux inward through the endcaps is minus the outward flux. Thus, the sum of the two integrals over $\mathcal{S}_1$ and $\mathcal{S}_2$ is 4π, verifying (17.7) for $l = 1$. We know from integral (17.5) that the shape of the surface does not matter, so $\mathcal{C}$ and $\mathcal{C}'$ do not have to be circles. If $\mathcal{C}'$ loops around $\mathcal{C}$ several times, we would need to slice and splice the surface swept by $\mathcal{C}'$ into multiple pieces.

Differential versions of Gauss' and Ampère's laws

Defining the charge density $q(\boldsymbol{r})$ as the electric charge per unit volume at point $\boldsymbol{r}$, Gauss's law (16.84) reads

$$\oint_{\mathcal{S}} \boldsymbol{E} \cdot \mathrm{d}\boldsymbol{S} = \frac{Q}{\epsilon_0} = \int_{\mathcal{V}} \frac{q}{\epsilon_0} \, \mathrm{d}V = \int_{\mathcal{V}} \boldsymbol{\nabla} \cdot \boldsymbol{E} \, \mathrm{d}V,$$

where $\mathcal{V}$ is the volume surrounded by the closed surface $\mathcal{S}$ and the divergence theorem has been used to turn the surface flux of $\boldsymbol{E}$ into a volume integral of its divergence. Since this applies to any volume $\mathcal{V}$ no matter how small, and assuming continuity, this yields the differential form of Gauss's law

$$\boldsymbol{\nabla} \cdot \boldsymbol{E} = \frac{q}{\epsilon_0}. \tag{17.10}$$

Similarly for Ampère's law, we define the current density $\boldsymbol{j}(\boldsymbol{r})$ as the the number of electric charges flowing in direction $\boldsymbol{j}$ per unit area such that Ampère's law becomes

$$\oint_{\mathcal{C}} \boldsymbol{B} \cdot \mathrm{d}\boldsymbol{r} = \mu_0 J = \int_{\mathcal{S}} \mu_0 \boldsymbol{j} \cdot \mathrm{d}\boldsymbol{S} = \int_{\mathcal{S}} \boldsymbol{\nabla} \times \boldsymbol{B} \cdot \mathrm{d}\boldsymbol{S},$$

where $\mathcal{S}$ is any surface spanning the closed curve $\mathcal{C}$, oriented following the right-hand rule and the orientation of $\mathcal{C}$, and Stokes' theorem has been used to turn the circulation of $\boldsymbol{B}$ about the closed curve $\mathcal{C}$ into the flux of its curl through the surface $\mathcal{S}$. Since this applies to any surface no matter how small and we assume continuity, this yields the differential form of Ampère's law as

$$\boldsymbol{\nabla} \times \boldsymbol{B} = \mu_0 \boldsymbol{j}. \tag{17.11}$$

17.2 Maxwell's equations

When fluxes are not steady a lot of new physics comes into play. First and foremost came *Faraday's law* that a varying magnetic field flux through a surface spanning a wire circuit generates an electromotive force around that circuit, whether that varying flux arises from a varying magnetic field or circuit. In its modern form, that is

$$\frac{\mathrm{d}}{\mathrm{d}t}\int_{\mathcal{S}} \boldsymbol{B}\cdot \mathrm{d}\boldsymbol{S} = -\oint_{\mathcal{C}} (\boldsymbol{E} + \boldsymbol{v}\times \boldsymbol{B})\cdot \mathrm{d}\boldsymbol{r}, \tag{17.12}$$

where t is time, $\mathcal{C}$ is a closed material curve (a circuit), $\boldsymbol{v}$ is the velocity of material point $\boldsymbol{r}$ on that wire, and $\mathcal{S}$ is a simple surface spanning $\mathcal{C}$. If $\mathcal{S}$ is parametrized as $(u,v)\in\mathcal{U}\to \boldsymbol{r}(t,u,v)$ at any fixed time t, $\mathcal{C}$ is parametrized as $\boldsymbol{r}(t,u(t'),v(t'))$, with t' as the curve coordinate of a material point on the wire, then $\boldsymbol{v} = \partial \boldsymbol{r}(t,t')/\partial t$ is the velocity of that material point. As usual, the orientation of the surface and of its boundary are linked by the right-hand rule.

For $\mathcal{S}$ parametrized as $\boldsymbol{r}(t,u,v)$, with u,v as material labels independent of time t,[1] the surface integral

[1]For example,

$$\boldsymbol{r}(t,u,v) = \hat{\mathbf{y}}Ru\sin v + Ru\cos v(\hat{\mathbf{x}}\cos\omega t - \hat{\mathbf{z}}\sin\omega t)$$

for a disk of radius R rotating about $\hat{\mathbf{y}}$ at angular velocity ω, with $0\le u\le 1$, $0\le v<2\pi$.

$$\begin{aligned}
\frac{\mathrm{d}}{\mathrm{d}t}\int_{\mathcal{S}} \boldsymbol{B}\cdot\mathrm{d}\boldsymbol{S} &= \frac{\mathrm{d}}{\mathrm{d}t}\iint_{\mathcal{U}} \boldsymbol{B}\cdot\frac{\partial \boldsymbol{r}}{\partial u}\times\frac{\partial \boldsymbol{r}}{\partial v}\mathrm{d}u\mathrm{d}v\\
&= \iint_{\mathcal{U}} \frac{\partial \boldsymbol{B}}{\partial t}\cdot\frac{\partial \boldsymbol{r}}{\partial u}\times\frac{\partial \boldsymbol{r}}{\partial v}\mathrm{d}u\mathrm{d}v + \iint_{\mathcal{U}} \boldsymbol{B}\cdot\frac{\partial}{\partial t}\left(\frac{\partial \boldsymbol{r}}{\partial u}\times\frac{\partial \boldsymbol{r}}{\partial v}\right)\mathrm{d}u\mathrm{d}v\\
&= \int_{\mathcal{S}} \frac{\partial \boldsymbol{B}}{\partial t}\cdot\mathrm{d}\boldsymbol{S} + \iint_{\mathcal{U}} \boldsymbol{B}\cdot\left(\frac{\partial \boldsymbol{v}}{\partial u}\times\frac{\partial \boldsymbol{r}}{\partial v}+\frac{\partial \boldsymbol{r}}{\partial u}\times\frac{\partial \boldsymbol{v}}{\partial v}\right)\mathrm{d}u\mathrm{d}v,
\end{aligned}$$

where $\boldsymbol{v} = \partial\boldsymbol{r}(t,u,v)/\partial t$ and we assumed sufficient smoothness that all mixed partials commute. That last integral can be transformed using the mixed product identity $\boldsymbol{a}\cdot\boldsymbol{b}\times\boldsymbol{c} = \boldsymbol{b}\cdot\boldsymbol{c}\times\boldsymbol{a}$, the product rule, and Green's theorem (13.31c),

$$\begin{aligned}
&\iint_{\mathcal{U}} \boldsymbol{B}\cdot\left(\frac{\partial \boldsymbol{v}}{\partial u}\times\frac{\partial \boldsymbol{r}}{\partial v}+\frac{\partial \boldsymbol{r}}{\partial u}\times\frac{\partial \boldsymbol{v}}{\partial v}\right)\mathrm{d}u\mathrm{d}v\\
&\quad= \iint_{\mathcal{U}}\left\{\frac{\partial \boldsymbol{v}}{\partial u}\cdot\left(\frac{\partial \boldsymbol{r}}{\partial v}\times\boldsymbol{B}\right) - \frac{\partial \boldsymbol{v}}{\partial v}\cdot\left(\frac{\partial \boldsymbol{r}}{\partial u}\times\boldsymbol{B}\right)\right\}\mathrm{d}u\mathrm{d}v\\
&\quad= \iint_{\mathcal{U}} \frac{\partial}{\partial u}\left\{\boldsymbol{v}\cdot\left(\frac{\partial \boldsymbol{r}}{\partial v}\times\boldsymbol{B}\right)\right\} - \frac{\partial}{\partial v}\left\{\boldsymbol{v}\cdot\left(\frac{\partial \boldsymbol{r}}{\partial u}\times\boldsymbol{B}\right)\right\}\mathrm{d}u\mathrm{d}v\\
&\quad= \int_{\partial\mathcal{U}} \boldsymbol{v}\cdot\left(\frac{\partial \boldsymbol{r}}{\partial u}\times\boldsymbol{B}\right)\mathrm{d}u + \boldsymbol{v}\cdot\left(\frac{\partial \boldsymbol{r}}{\partial v}\times\boldsymbol{B}\right)\mathrm{d}v\\
&\quad= \int_{\partial\mathcal{U}} (\boldsymbol{B}\times\boldsymbol{v})\cdot\left(\frac{\partial \boldsymbol{r}}{\partial u}\mathrm{d}u + \frac{\partial \boldsymbol{r}}{\partial v}\mathrm{d}v\right) = -\oint_{\partial\mathcal{S}} \boldsymbol{v}\times\boldsymbol{B}\cdot\mathrm{d}\boldsymbol{r},
\end{aligned}$$

where $\partial\boldsymbol{B}/\partial u = (\partial\boldsymbol{r}/\partial u)\cdot\boldsymbol{\nabla}\boldsymbol{B}$ and $\partial\boldsymbol{B}/\partial v = (\partial\boldsymbol{r}/\partial v)\cdot\boldsymbol{\nabla}\boldsymbol{B}$ have been used. Thus, $\boldsymbol{v}\times\boldsymbol{B}$ cancels out from both sides of (17.12) that reduces to

$$\int_{\mathcal{S}} \frac{\partial \boldsymbol{B}}{\partial t}\cdot\mathrm{d}\boldsymbol{S} = -\oint_{\partial\mathcal{S}} \boldsymbol{E}\cdot\mathrm{d}\boldsymbol{r} = -\int_{\mathcal{S}} \boldsymbol{\nabla}\times\boldsymbol{E}\cdot\mathrm{d}\boldsymbol{S},$$

thanks to Stokes' theorem. Since this holds for any surface $\mathcal{S}$ no matter how small, and we assume continuity, this yields the remarkably simple differential form of *Faraday's law*

$$\frac{\partial \boldsymbol{B}}{\partial t} = -\boldsymbol{\nabla}\times\boldsymbol{E}. \tag{17.13}$$

But that equation, together with the vector identity $\boldsymbol{\nabla} \cdot \boldsymbol{\nabla} \times \boldsymbol{E} = 0$, require,

$$\frac{\partial}{\partial t} \boldsymbol{\nabla} \cdot \boldsymbol{B} = 0,$$

and conservation of electric charges implies that

$$\frac{\partial q}{\partial t} = -\boldsymbol{\nabla} \cdot \boldsymbol{j}, \tag{17.14}$$

with $q = \epsilon_0 \boldsymbol{\nabla} \cdot \boldsymbol{E}$, according to Gauss' law (17.10). Thus, Maxwell reasoned that Ampère's law (17.11) had to be generalized to

$$\begin{aligned} \boldsymbol{\nabla} \times \boldsymbol{B} &= \mu_0 \boldsymbol{j} + \mu_0 \epsilon_0 \frac{\partial \boldsymbol{E}}{\partial t}, \\ \boldsymbol{\nabla} \cdot \boldsymbol{B} &= 0, \end{aligned}$$

in order to be consistent with (17.10), (17.13), and (17.14).

Collecting the $\boldsymbol{E}$ and $\boldsymbol{B}$ equations yields *Maxwell's equations*

$$\boldsymbol{\nabla} \times \boldsymbol{E} = -\frac{\partial \boldsymbol{B}}{\partial t}, \tag{17.15a}$$

$$\boldsymbol{\nabla} \cdot \boldsymbol{E} = \frac{q}{\epsilon_0}, \tag{17.15b}$$

$$\boldsymbol{\nabla} \times \boldsymbol{B} = \mu_0 \boldsymbol{j} + \mu_0 \epsilon_0 \frac{\partial \boldsymbol{E}}{\partial t}, \tag{17.15c}$$

$$\boldsymbol{\nabla} \cdot \boldsymbol{B} = 0. \tag{17.15d}$$

... and there was light

One most remarkable consequence of Maxwell's equations, and the modification of Ampère's law to include the $\partial \boldsymbol{E}/\partial t$ term, is that it predicts *electromagnetic waves.* In the absence of charge and current density, that is in vacuum with $q = 0$ and $\boldsymbol{j} = 0$, the equations can be reduced to a single remarkable equation for $\boldsymbol{E}(t, \boldsymbol{r})$ or $\boldsymbol{B}(t, \boldsymbol{r})$.

Taking the curl of (17.15a), using identity (16.97), and substituting for $\boldsymbol{\nabla} \cdot \boldsymbol{E}$ and $\boldsymbol{\nabla} \times \boldsymbol{B}$ from (17.15c), we derive

$$\boldsymbol{\nabla} \times (\boldsymbol{\nabla} \times \boldsymbol{E}) = -\nabla^2 \boldsymbol{E} = -\frac{\partial}{\partial t} \boldsymbol{\nabla} \times \boldsymbol{B} = -\mu_0 \epsilon_0 \frac{\partial^2 \boldsymbol{E}}{\partial t^2},$$

yielding the *wave equation*

$$\frac{\partial^2 \boldsymbol{E}}{\partial t^2} = c^2 \nabla^2 \boldsymbol{E}, \tag{17.16}$$

where $c^2 = 1/(\mu_0 \epsilon_0)$. Taking the curl of (17.15c) and eliminating $\boldsymbol{E}$ yields the same wave equation for $\boldsymbol{B}(t, \boldsymbol{r})$.

These variables have physical units, t is time, so $\partial^2/\partial t^2$ has units of $(1/\text{time}^2)$, $\boldsymbol{\nabla}$ is inverse length and ∇^2 is inverse length square, $\boldsymbol{E}$ is volts/length but its units cancel out from both sides of the linear wave equation (17.16). Thus, the equation tells us that c must have units of $(\text{length}/\text{time})^2$, that is, c is a speed, *the speed of light,*[2]

$$c = \frac{1}{\sqrt{\mu_0 \epsilon_0}} \approx 300\,000 \text{ km/s}. \tag{17.17}$$

[2] $c = 299\,792\,458$ meters/second.

17.3 Electromagnetic waves

Plane waves

It is straightforward to verify that the wave equation (17.16) with $\nabla \cdot \boldsymbol{E} = 0$ has solutions of the form

$$\boldsymbol{E}(t, \boldsymbol{r}) = (a\hat{\mathbf{y}} + b\hat{\mathbf{z}}) \cos k(x - ct),$$

for any constant a, b, k, with $x = \hat{\mathbf{x}} \cdot \boldsymbol{r}$. This is a sinusoidal wave traveling in the $\hat{\mathbf{x}}$ direction at speed c. The wavelength is $2\pi/|k|$ and $|k|$ is the wavenumber, the number of peaks per unit of length. Faraday's law (17.15a) gives the accompanying magnetic field as

$$\boldsymbol{B}(t, \boldsymbol{r}) = \frac{\hat{\mathbf{x}}}{c} \times (a\hat{\mathbf{y}} + b\hat{\mathbf{z}}) \cos k(x - ct).$$

These are *transverse* waves because of $\nabla \cdot \boldsymbol{E} = 0 = \nabla \cdot \boldsymbol{B}$; the fields $\boldsymbol{E}$ and $\boldsymbol{B}$ point in directions $\hat{\mathbf{y}}$ and $\hat{\mathbf{z}}$ perpendicular to the direction of propagation, which is $\hat{\mathbf{x}}$ for this simple solution.

More generally, the wave equation admits plane wave solutions in the form of complex exponential functions

$$\boldsymbol{E}(t, \boldsymbol{r}) = \widetilde{\boldsymbol{E}}\, e^{i(\boldsymbol{k} \cdot \boldsymbol{r} - \omega t)} + c.c., \tag{17.18a}$$

where $\omega = \pm c|\boldsymbol{k}|$ is the signed frequency; $|\boldsymbol{k}|$ is the *wavenumber*, the magnitude of the real *wavevector*

$$\boldsymbol{k} = k_x\hat{\mathbf{x}} + k_y\hat{\mathbf{y}} + k_z\hat{\mathbf{z}},$$

independent of t and $\boldsymbol{r}$ but otherwise arbitrary; $\widetilde{\boldsymbol{E}}$ is an arbitrary complex vector such that $\boldsymbol{k} \cdot \widetilde{\boldsymbol{E}} = 0$; i is the imaginary unit with $i^2 = -1$; $e^{i\phi} = \cos\phi + i \sin\phi$ according to Euler's formula; and *c.c.* denotes *complex conjugate*, that is,

$$\widetilde{\boldsymbol{E}} e^{i\phi} + c.c. = \widetilde{\boldsymbol{E}} e^{i\phi} + \widetilde{\boldsymbol{E}}^* e^{-i\phi^*},$$

where the $(\ )^*$ denotes complex conjugate, so $\boldsymbol{E}(t, \boldsymbol{r})$ is real.

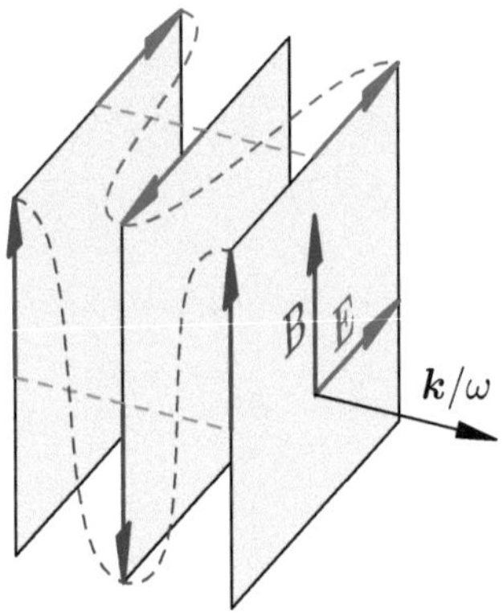

Fig. 17.7 Plane E&M wave.

These are plane waves since they have the same value all along planes perpendicular to $\boldsymbol{k}$ where $\boldsymbol{k} \cdot \boldsymbol{r}$ is constant, at any fixed time t. They travel in the direction of $\boldsymbol{k}/\omega$, that is, $\boldsymbol{k}$ for $\omega = c|\boldsymbol{k}| > 0$ and $-\boldsymbol{k}$ for $\omega = -c|\boldsymbol{k}| < 0$ (Fig. 17.7). They are indeed solutions of the wave equation (17.16) for any real $\boldsymbol{k}$ as long as $\omega = \pm c|\boldsymbol{k}|$ since

$$\begin{aligned}
\nabla e^{i(\boldsymbol{k} \cdot \boldsymbol{r} - \omega t)} &= i\boldsymbol{k}\, e^{i(\boldsymbol{k} \cdot \boldsymbol{r} - \omega t)}, \\
\nabla^2 e^{i(\boldsymbol{k} \cdot \boldsymbol{r} - \omega t)} &= -|\boldsymbol{k}|^2 e^{i(\boldsymbol{k} \cdot \boldsymbol{r} - \omega t)}, \\
\frac{\partial^2}{\partial t^2} e^{i(\boldsymbol{k} \cdot \boldsymbol{r} - \omega t)} &= -\omega^2\, e^{i(\boldsymbol{k} \cdot \boldsymbol{r} - \omega t)}.
\end{aligned}$$

The constraint arising from $\nabla \cdot \boldsymbol{E} = 0$ is that $\boldsymbol{k} \cdot \widetilde{\boldsymbol{E}} = 0$. The associated magnetic field $\boldsymbol{B}(t, \boldsymbol{r})$ follows from Faraday's law (17.15a) as

$$\boldsymbol{B}(t, \boldsymbol{r}) = \frac{\boldsymbol{k}}{\omega} \times \widetilde{\boldsymbol{E}}\, e^{i(\boldsymbol{k} \cdot \boldsymbol{r} - \omega t)} + c.c. \tag{17.18b}$$

Since the wave equation is linear, its general solution is a superposition of all possible plane wave solutions, that is,

$$\boldsymbol{E}(t,\boldsymbol{r}) = \iiint_{\mathbb{R}^3} \left(\widetilde{\boldsymbol{E}}_+(\boldsymbol{k})\, e^{-i|\omega|t} + \widetilde{\boldsymbol{E}}_-(\boldsymbol{k})\, e^{i|\omega|t)}\right) e^{i\boldsymbol{k}\cdot\boldsymbol{r}}\, \mathrm{d}k_x\mathrm{d}k_y\mathrm{d}k_z + c.c. \tag{17.19a}$$

The associated magnetic field $\boldsymbol{B}(t,\boldsymbol{r})$ follows from Faraday's law (17.15a),

$$\boldsymbol{B}(t,\boldsymbol{r}) = \iiint_{\mathbb{R}^3} \left(\widetilde{\boldsymbol{B}}_+(\boldsymbol{k})\, e^{-i|\omega|t} + \widetilde{\boldsymbol{B}}_-(\boldsymbol{k})\, e^{i|\omega|t)}\right) e^{i\boldsymbol{k}\cdot\boldsymbol{r}}\, \mathrm{d}k_x\mathrm{d}k_y\mathrm{d}k_z + c.c. \tag{17.19b}$$

with

$$\widetilde{\boldsymbol{B}}_+(\boldsymbol{k}) = \frac{\boldsymbol{k}}{|\omega|} \times \widetilde{\boldsymbol{E}}_+, \qquad \widetilde{\boldsymbol{B}}_-(\boldsymbol{k}) = \frac{-\boldsymbol{k}}{|\omega|} \times \widetilde{\boldsymbol{E}}_-.$$

In an infinite domain, the amplitudes $\widetilde{\boldsymbol{E}}_+(\boldsymbol{k})$ and $\widetilde{\boldsymbol{E}}_-(\boldsymbol{k})$ can be determined from the *Fourier transforms* of the initial electric and magnetic fields, $\boldsymbol{E}(0,\boldsymbol{r})$ and $\boldsymbol{B}(0,\boldsymbol{r})$.

Wave reflection

Physical boundaries introduce *boundary conditions.* For example, the electric field $\boldsymbol{E}(t,\boldsymbol{r})$ must be perpendicular to a perfect conductor. If an electromagnetic wave travels in a semi-infinite domain bounded by a perfect plane conductor perpendicular to $\hat{\boldsymbol{n}}$, the electric field must have $\hat{\boldsymbol{n}} \times \boldsymbol{E} = 0$ along that plane. This requires another wave to satisfy that boundary condition. Let the total electric field consist of two waves

$$\boldsymbol{E}(t,\boldsymbol{r}) = \widetilde{\boldsymbol{E}}_1 e^{i(\boldsymbol{k}_1\cdot\boldsymbol{r}-\omega_1 t)} + \widetilde{\boldsymbol{E}}_2 e^{i(\boldsymbol{k}_2\cdot\boldsymbol{r}-\omega_2 t)} + c.c.,$$

where $\omega_1 = \pm c|\boldsymbol{k}_1|$, $\omega_2 = \pm c|\boldsymbol{k}_2|$. The boundary condition $\hat{\boldsymbol{n}} \times \boldsymbol{E} = 0$ requires that

$$\hat{\boldsymbol{n}} \times \left(\widetilde{\boldsymbol{E}}_1 e^{i(\boldsymbol{k}_1\cdot\boldsymbol{r}-\omega_1 t)} + \widetilde{\boldsymbol{E}}_2 e^{i(\boldsymbol{k}_2\cdot\boldsymbol{r}-\omega_2 t)} + c.c.\right) = 0$$

for all t and all $\boldsymbol{r}$ along the perfect conductor located at $\hat{\boldsymbol{n}} \cdot \boldsymbol{r} = \lambda$.

Thus, $\boldsymbol{r} = \lambda\hat{\boldsymbol{n}} + \boldsymbol{r}_\perp$ with fixed $\lambda\hat{\boldsymbol{n}}$ on the conducting plane but the perpendicular component $\boldsymbol{r}_\perp$ is free. Since $e^{\pm i\boldsymbol{k}_1\cdot\boldsymbol{r}}$ and $e^{\pm i\boldsymbol{k}_2\cdot\boldsymbol{r}}$ are linearly independent functions of $\boldsymbol{r}_\perp$, and $e^{\pm i\omega_1 t}$ and $e^{\pm i\omega_2 t}$ are linearly independent functions of t, nonzero $\boldsymbol{E}$ solutions require equality of the $\boldsymbol{k}$ components perpendicular to $\hat{\boldsymbol{n}}$, and equality of the frequencies,

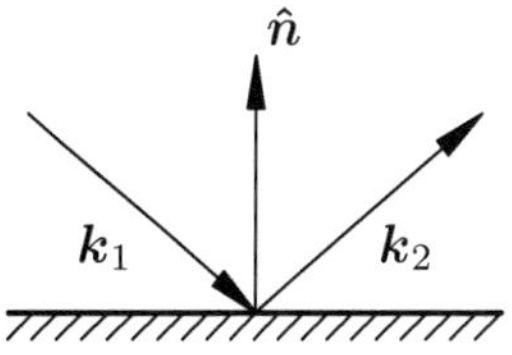

Fig. 17.8 Wave reflection.

$$\boldsymbol{k}_1^\perp = \boldsymbol{k}_2^\perp,$$
$$\omega_1 = \pm c|\boldsymbol{k}_1| = \omega_2 = \pm c|\boldsymbol{k}_2|.$$

Thus, ω_1 and ω_2 must have the same sign, and the wavenumbers must be equal $|\boldsymbol{k}_1| = |\boldsymbol{k}_2|$. Hence, the $\boldsymbol{k}$ components parallel to $\hat{\boldsymbol{n}}$ must have opposite signs[3] (Fig. 17.8),

$$\boldsymbol{k}_1 = \boldsymbol{k}_1^\perp + k_1^\parallel\hat{\boldsymbol{n}}, \qquad \boldsymbol{k}_2 = \boldsymbol{k}_1^\perp - k_1^\parallel\hat{\boldsymbol{n}}. \tag{17.20}$$

[3]Equal signs would mean that the waves are identical and $\widetilde{\boldsymbol{E}}_1 = -\widetilde{\boldsymbol{E}}_2$ to satisfy the boundary condition, yielding a trivial $\boldsymbol{E}(t,\boldsymbol{r}) = 0$.

This is the well-known equal angle reflection. The boundary condition now reduces to

$$\hat{\boldsymbol{n}} \times \widetilde{\boldsymbol{E}}_1\, e^{i\lambda k_1^{\parallel}} = -\hat{\boldsymbol{n}} \times \widetilde{\boldsymbol{E}}_2\, e^{-i\lambda k_1^{\parallel}},$$

so the components parallel to the plane (perpendicular to $\hat{\boldsymbol{n}}$) annihilate each other and the net field is perpendicular to the conducting plane. Since the electric fields are also perpendicular to their wavevectors they can be decomposed into two components orthogonal to each other and to their wavevector,

$$\begin{aligned}\widetilde{\boldsymbol{E}}_1 &= \alpha_1\, \hat{\boldsymbol{k}}_1 \times \hat{\boldsymbol{n}} + \beta_1\, \hat{\boldsymbol{k}}_1 \times (\hat{\boldsymbol{k}}_1 \times \hat{\boldsymbol{n}}),\\ \widetilde{\boldsymbol{E}}_2 &= \alpha_2\, \hat{\boldsymbol{k}}_2 \times \hat{\boldsymbol{n}} + \beta_2\, \hat{\boldsymbol{k}}_2 \times (\hat{\boldsymbol{k}}_2 \times \hat{\boldsymbol{n}}),\end{aligned} \tag{17.21}$$

such that the boundary condition leads to

$$\alpha_2 = -\alpha_1 e^{2i\lambda k_1^{\parallel}}, \qquad \beta_2 = \beta_1 e^{2i\lambda k_1^{\parallel}}.$$

In other words, the boundary condition fully determines $\boldsymbol{k}_2$ and $\widetilde{\boldsymbol{E}}_2$ in terms of $\boldsymbol{k}_1$ and $\widetilde{\boldsymbol{E}}_1$, and vice versa.

If the perfectly conducting boundary is not a simple plane, such that the boundary condition is still $\hat{\boldsymbol{n}} \times \boldsymbol{E} = 0$ but $\hat{\boldsymbol{n}}$ now varies, a single reflected wave will not be enough and the incoming wave is *diffracted*.

Spherical waves

For waves that emanate from a localized source (an antenna), solutions of the wave equation in spherical coordinates are more useful and relevant. A complete expansion in *vector spherical harmonics*[4] similar to the plane wave expansion (17.19) can be made, but we will investigate just the simplest nontrivial examples of such solutions here.

[4] G.M. Vasil et al., "Tensor Calculus in Spherical Coordinates Using Jacobi Polynomials." *J. Comp. Phys.: X,* 3, (2019), 100013. `https://doi.org/10.1016/j.jcpx.2019.100013`

In spherical coordinates, the electric field is $\boldsymbol{E} = \hat{\boldsymbol{r}} E_r + \hat{\boldsymbol{\theta}} E_\theta + \hat{\boldsymbol{\varphi}} E_\varphi$ and the divergence-free constraint reads (16.69)

$$\boldsymbol{\nabla} \cdot \boldsymbol{E} = \frac{1}{r^2}\frac{\partial (r^2 E_r)}{\partial r} + \frac{1}{r \sin\theta}\frac{\partial(\sin\theta\, E_\theta)}{\partial \theta} + \frac{1}{r\sin\theta}\frac{\partial E_\varphi}{\partial \varphi} = 0. \tag{17.22}$$

We begin by looking for a spherically symmetric solution $\boldsymbol{E}(t, \boldsymbol{r}) = \hat{\boldsymbol{r}} E_r(t, r)$. Zero divergence imposes (17.22),

$$\frac{1}{r^2}\frac{\partial\left(r^2 E_r\right)}{\partial r} = 0 \;\Rightarrow E_r(t, r) = \frac{f(t)}{r^2},$$

where $f(t)$ is any function of time t only. The wave equation (17.16) then reduces to

$$\frac{\hat{\boldsymbol{r}}}{r^2}\frac{\mathrm{d}^2 f}{\mathrm{d}t^2} = c^2 f(t) \nabla^2 \left(\frac{\hat{\boldsymbol{r}}}{r^2}\right) = 0,$$

from the vector Laplacian in spherical coordinates (16.121). In other words, no such simple radial waves exists. The only spherically symmetric solution is the static Coulomb field $\hat{\boldsymbol{r}}/r^2$ with constant $f(t)$.

Since the wave equation is linear with coefficients independent of t and φ, it has exponential solutions in those two variables, that is,

$$\boldsymbol{E}(t, \boldsymbol{r}) = e^{-i\omega t} e^{im\varphi} \boldsymbol{F}(r, \theta).$$

For our next attempt, we search for *axisymmetric* waves, independent of φ, that is $m = 0$, and

$$\boldsymbol{E} = e^{-i\omega t} \left(\hat{\boldsymbol{r}} E_r(r, \theta) + \hat{\boldsymbol{\theta}} E_\theta(r, \theta) + \hat{\boldsymbol{\varphi}} E_\varphi(r, \theta) \right).$$

The E_θ and E_φ components must go to 0 on the polar axis; otherwise, the field is grossly singular there. The simplest solution, thus, might have those components proportional to $\sin\theta$ that vanishes at $\theta \to 0$ and $\theta \to \pi$. Inspection of (17.22) shows that this will require the radial component $E_r \propto \cos\theta$ to balance the $E_\theta \propto \sin\theta$ component but the $E_\varphi \propto \sin\theta$ drops out of the divergence because of our axisymmetry assumption. This suggests that there may be *two* classes of axisymmetric solutions: one class with $E_\varphi = 0$ but $E_r \neq 0$, $E_\theta \neq 0$, and another class with $E_\varphi \neq 0$ but $E_r = E_\theta = 0$.

Thus, we look for $\boldsymbol{E}$ wave solutions in either of the forms

$$\boldsymbol{E}_P = e^{-i\omega t} \left(\hat{\boldsymbol{r}}\, f(r) \cos\theta + \hat{\boldsymbol{\theta}}\, g(r) \sin\theta \right), \tag{17.23}$$

or

$$\boldsymbol{E}_T = e^{-i\omega t}\ \hat{\boldsymbol{\varphi}}\, h(r) \sin\theta. \tag{17.24}$$

Fields of the form (17.23) are *poloidal* with components in the radial $\hat{\boldsymbol{r}}$ and meridional $\hat{\boldsymbol{\theta}}$ directions only. Fields of the form (17.24) are *toroidal* with component in the azimuthal $\hat{\boldsymbol{\varphi}}$ direction only.[5] The divergence-free constraint $\boldsymbol{\nabla} \cdot \boldsymbol{E} = 0$ is automatically satisfied for the axisymmetric toroidal field (17.24) but for the poloidal field (17.23) $\boldsymbol{\nabla} \cdot \boldsymbol{E} = 0$ requires (17.22)

$$\frac{1}{r^2} \frac{\mathrm{d}\left(r^2 f(r)\right)}{\mathrm{d}r} + 2\frac{g(r)}{r} = 0. \tag{17.25}$$

[5] Called *zonal* in atmospheric science.

Toroidal $\boldsymbol{E}$ waves

The toroidal field (17.24) automatically satisfies $\boldsymbol{\nabla} \cdot \boldsymbol{E} = 0$, (17.22). The wave equation (17.16) for solutions of the form (17.24) reduces to

$$-\,k^2 \hat{\boldsymbol{\varphi}}\, h(r) \sin\theta = \nabla^2 \left(\hat{\boldsymbol{\varphi}}\, h(r) \sin\theta \right), \tag{17.26}$$

where $k^2 = \omega^2/c^2$ and, although we assumed axisymmetry, $\hat{\boldsymbol{\varphi}}$ depends on φ. The vector Laplacian in spherical coordinates (16.121) gives

$$\nabla^2 \left(\hat{\boldsymbol{\varphi}}\, E_\varphi(r, \theta) \right) = \hat{\boldsymbol{\varphi}} \left(\nabla^2 E_\varphi(r, \theta)) - \frac{E_\varphi(r, \theta)}{r^2 \sin^2\theta} \right). \tag{17.27}$$

Both sides of the wave equation only have $\hat{\boldsymbol{\varphi}}$ components for toroidal fields. Here we assumed further that $E_\varphi(r, \theta) = h(r) \sin\theta$, so the wave[6] equation (17.26) reduces to

$$-k^2\, h(r) \sin\theta = \nabla^2 \big(h(r) \sin\theta \big) - \frac{h(r)}{r^2 \sin\theta}.$$

[6] *Helmholtz* equation actually, since we restricted to $e^{i\omega t}$, so $\partial^2/\partial t^2 \to -\omega^2$.

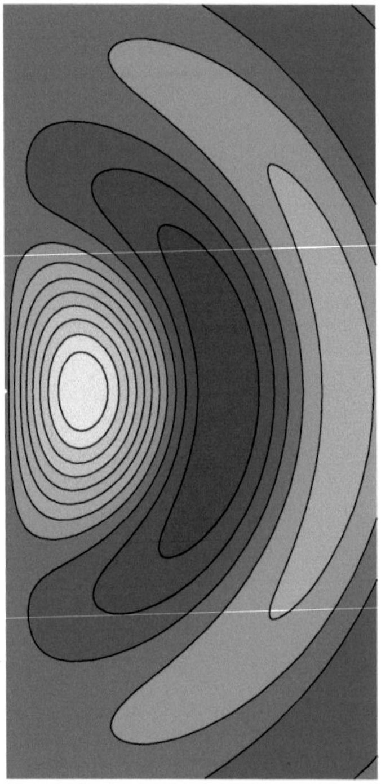

Fig. 17.9 $j_1(kr)\sin\theta$ in $k(\rho, z)$ plane.

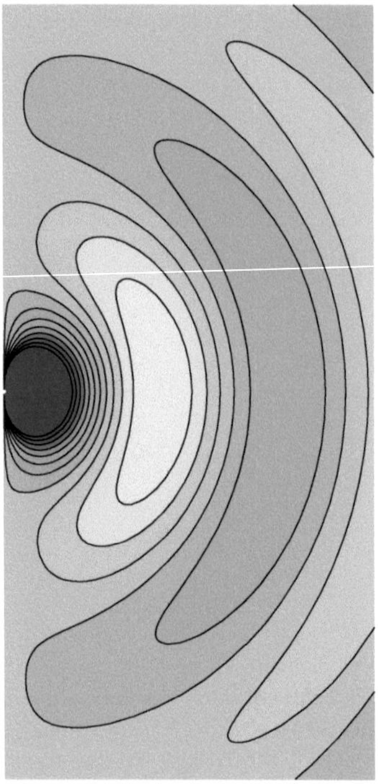

Fig. 17.10 $y_1(kr)\sin\theta$ in $k(\rho, z)$ plane.

Then, using ∇^2 in spherical coordinates (16.113), the $\sin\theta$'s nicely cancel out to yield

$$\frac{\mathrm{d}^2 h}{\mathrm{d}r^2} + \frac{2}{r}\frac{\mathrm{d}h}{\mathrm{d}r} + \left(k^2 - \frac{2}{r^2}\right) h = 0. \tag{17.28}$$

This is the *spherical Bessel equation* of order 1, and the substitution $h(r) = B(kr)/\sqrt{kr}$ transforms it into a Bessel equation (16.132) of order $n = 3/2$ for $B(\chi)$ with $\chi = kr$. One solution of eqn (17.28) that is regular at $\chi = 0$ is the *spherical Bessel function of the first kind of order 1* and it has a simple exact formula[7] (Fig. 17.9),

$$h(r) = j_1(kr) = \frac{\sin(kr)}{(kr)^2} - \frac{\cos(kr)}{(kr)}. \tag{17.29a}$$

For radiation from a dipole antenna—a dipole singularity at $r = 0$ aligned with the z-axis—we also need the spherical Bessel function of the *second kind* of order 1 (Fig. 17.10)

$$h(r) = y_1(kr) = -\frac{\cos(kr)}{(kr)^2} - \frac{\sin(kr)}{(kr)}. \tag{17.29b}$$

To obtain waves radiating from a source at $r = 0$ we need to properly combine those two modes. Since the equation is linear, the general solution is a linear combination of both modes, that is,

$$\boldsymbol{E}_T(t, \boldsymbol{r}) = \hat{\boldsymbol{\varphi}} \sin\theta\, e^{-i\omega t}\big(A\, j_1(kr) + B\, y_1(kr)\big) + c.c.$$

for some arbitrary complex amplitude A and B. Substituting the formula for the real $j_1(kr)$ and $y_1(kr)$ (17.29) and expanding the complex conjugate yields

$$\begin{aligned}\frac{\hat{\boldsymbol{\varphi}}\cdot\boldsymbol{E}_T(t, \boldsymbol{r})}{\sin\theta} &= \left(Ae^{-i\omega t} + A^* e^{i\omega t}\right)\frac{\sin kr}{(kr)^2} - \left(Be^{-i\omega t} + B^* e^{i\omega t}\right)\frac{\cos kr}{(kr)^2} \\ &\quad - \left(Ae^{-i\omega t} + A^* e^{i\omega t}\right)\frac{\cos kr}{kr} - \left(Be^{-i\omega t} + B^* e^{i\omega t}\right)\frac{\sin kr}{kr}.\end{aligned}$$

We pick $k\omega > 0$, since both signs of ω appear, such that $(kr - \omega t)$ will correspond to a wave radiating from $r = 0$. Recalling the trig formulas

$$\begin{aligned}\cos(kr - \omega t) &= \cos kr\, \cos\omega t + \sin kr\, \sin\omega t, \\ \sin(kr - \omega t) &= \sin kr\, \cos\omega t - \cos kr\, \sin\omega t,\end{aligned}$$

we see that A real with $B = iA$ yields the real radiating wave

$$\boldsymbol{E}_T(t, \boldsymbol{r}) = 2A\,\hat{\boldsymbol{\varphi}}\, \sin\theta \left(\frac{\sin(kr - \omega t)}{(kr)^2} - \frac{\cos(kr - \omega t)}{kr}\right). \tag{17.30}$$

This wave is the *far-field* of an oscillating *magnetic dipole*— little loop of oscillating current centered at the origin in the (x, y)-plane. Note that the wave decays like $1/r$, in contrast to the Coulomb field that decays faster like $1/r^2$. One learns in E&M that the electromagnetic energy flux is given by the Poynting vector $\boldsymbol{E} \times \boldsymbol{B}/\mu_0$, and that scales like $1/r^2$, so there is conservation of energy flux.

[7] DLMF: §10.4 Spherical Bessel functions `https://dlmf.nist.gov/10.49`

Poloidal $\boldsymbol{E}$ waves

Since Maxwell's equations show that the magnetic field $\boldsymbol{B}$ and $\partial\boldsymbol{B}/\partial t$ also satisfy the wave equation, and Faraday's law (17.15a) gives $-\partial\boldsymbol{B}/\partial t = \boldsymbol{\nabla}\times\boldsymbol{E}$, it follows that $\boldsymbol{\nabla}\times\boldsymbol{E}$ solves the wave equation whenever $\boldsymbol{E}$ does. For a general axisymmetric toroidal field $\boldsymbol{E}_T = \hat{\boldsymbol{\varphi}}\, E_\varphi(t, r, \theta)$, we obtain a divergence-free poloidal field from (16.51c)

$$k\boldsymbol{E}_P = \boldsymbol{\nabla}\times\boldsymbol{E}_T = \frac{\hat{\boldsymbol{r}}}{r\sin\theta}\frac{\partial(\sin\theta\, E_\varphi)}{\partial\theta} - \frac{\hat{\boldsymbol{\theta}}}{r}\frac{\partial(r\, E_\varphi)}{\partial r}. \tag{17.31}$$

The factor of k keeps the same units for $\boldsymbol{E}_P$ as those of $\boldsymbol{E}_T$. It is left as an exercise to verify that the converse is also true: the curl of an axisymmetric poloidal field is an axisymmetric toroidal field. A consequence of these facts is that a toroidal $\boldsymbol{E}$ field goes with a poloidal $\boldsymbol{B}$ field, and vice versa.

A divergence-free poloidal $\boldsymbol{E}$ wave follows from taking the curl of the toroidal wave $\boldsymbol{E}_T$ derived in (17.30) yielding

$$\boldsymbol{E}_P = 2A\left(2\,\hat{\boldsymbol{r}}\,\cos\theta\,\frac{F(r,t)}{kr} - \hat{\boldsymbol{\theta}}\,\sin\theta\frac{1}{kr}\frac{\partial}{\partial r}\Big(r\,F(r,t)\Big)\right), \tag{17.32}$$

with

$$F(r,t) = \frac{\sin(kr-\omega t)}{(kr)^2} - \frac{\cos(kr-\omega t)}{kr}.$$

This is the far-field of an oscillating *electric dipole*—two equal and opposite electric charges periodically switching sides about the origin along the z-axis.

17.4 Poisson's equation

Going back to electrostatics, we saw that Coulomb's and Gauss's laws are consistent with each other, and Gauss's law is actually a consequence of Coulomb's law. In an infinite domain, Coulomb prescribes the electric vector field knowing the charge density distribution $q(\boldsymbol{r}')$ as

$$\boldsymbol{E}(\boldsymbol{r}) = \frac{1}{4\pi\epsilon_0}\int_{\mathbb{E}^3}\frac{\boldsymbol{r}-\boldsymbol{r}'}{|\boldsymbol{r}-\boldsymbol{r}'|^3}\,q(\boldsymbol{r}')\mathrm{d}V', \tag{17.33}$$

where $\mathrm{d}V'$ is the volume element at $\boldsymbol{r}'$ in 3D euclidean space $\mathbb{E}^3$ and the integral is over all position vectors $\boldsymbol{r}'$ in that space. Gauss' law (17.10) only specifies the divergence of $\boldsymbol{E}$ knowing the charge distribution, but now Maxwell and Faraday add that the curl of $\boldsymbol{E}$ vanishes when $\partial\boldsymbol{B}/\partial t = 0$, and together these two equations

$$\boldsymbol{\nabla}\cdot\boldsymbol{E} = q/\epsilon_0,$$
$$\boldsymbol{\nabla}\times\boldsymbol{E} = 0,$$

fully specify the electric field $\boldsymbol{E}$. The vanishing curl tells us that $\boldsymbol{E}$ is the gradient of a potential ϕ, defined with a minus sign in physics,

$$\boldsymbol{E} = -\boldsymbol{\nabla}\phi. \tag{17.34}$$

Then, Gauss's law leads to *Poisson's equation* for the electric potential

$$\nabla \cdot \boldsymbol{E} = -\nabla \cdot \nabla\phi = -\nabla^2\phi = \frac{q}{\epsilon_0}. \tag{17.35}$$

The solution to this equation in an infinite domain is, of course, that corresponding to (17.33), that is,

$$\phi(\boldsymbol{r}) = \frac{1}{4\pi\epsilon_0}\int_{\mathbf{E}^3}\frac{q(\boldsymbol{r}')}{|\boldsymbol{r}-\boldsymbol{r}'|}\,\mathrm{d}V', \tag{17.36}$$

where

$$G(\boldsymbol{r},\boldsymbol{r}') = \frac{1}{4\pi\epsilon_0|\boldsymbol{r}-\boldsymbol{r}'|} \tag{17.37}$$

is the *fundamental solution,* also called the unbounded space *Green's function* of Poisson's equation corresponding to a unit point charge at $\boldsymbol{r}'$, that is,

$$-\nabla^2 G(\boldsymbol{r},\boldsymbol{r}') = \epsilon_0^{-1}\delta(\boldsymbol{r}-\boldsymbol{r}'), \tag{17.38}$$

where $\delta(\boldsymbol{r}-\boldsymbol{r}')$ is a *Dirac delta density distribution* that is zero everywhere but infinite at $\boldsymbol{r}'$ in such a way that the net charge is 1. Two examples of spherically symmetric functions leading to that 3D delta distribution are

$$\begin{aligned}
&(a)\quad \delta(\boldsymbol{r}-\boldsymbol{r}') = \lim_{R\to 0}\begin{cases}3/(4\pi R^3), & 0 \le |\boldsymbol{r}-\boldsymbol{r}'| \le R\\ 0, & |\boldsymbol{r}-\boldsymbol{r}'| > R,\end{cases}\\
&(b)\quad \delta(\boldsymbol{r}-\boldsymbol{r}') = \lim_{R\to 0}\frac{e^{-|\boldsymbol{r}-\boldsymbol{r}'|^2/R^2}}{(\sqrt{\pi}R)^3}.
\end{aligned} \tag{17.39}$$

If there are boundaries, then Poisson's equation comes with *boundary conditions.* For example, the electric field must be perpendicular to a conductor in electrostatics, that is

$$\hat{\boldsymbol{n}} \times \boldsymbol{E} = 0 = \hat{\boldsymbol{n}} \times \nabla\phi,$$

on the surface of the conductor. Since $\nabla\phi$ is always perpendicular to the isosurfaces of $\phi(\boldsymbol{r})$, this boundary condition says that the conductor is an isosurface, and $\phi(\boldsymbol{r})$ must be constant on the conductor.

Point charge and conducting sphere

A classic example is to determine the electric field induced by a point charge in the presence of a spherical conductor. Let the sphere of radius R be centered at the origin O with the point charge Q_1 located at $\boldsymbol{r}_1$. The normal to the sphere $\hat{\boldsymbol{n}}$ is then the radial direction $\hat{\boldsymbol{r}}$ from the origin, $\hat{\boldsymbol{n}} = \hat{\boldsymbol{r}}$. We can satisfy the boundary condition on the sphere using the *method of images* and vector algebra. We need an image charge Q_2 to negate the tangential component of $\boldsymbol{E}$ induced by Q_1; that is, we need Q_2 and $\boldsymbol{r}_2$ such that when $|\boldsymbol{r}| = R$ (Fig. 17.11)

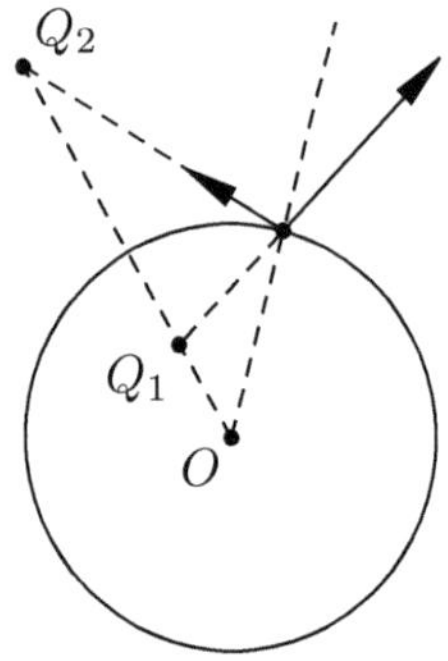

Fig. 17.11 Image charge for a sphere.

$$\left(\frac{Q_1(\boldsymbol{r}-\boldsymbol{r}_1)}{|\boldsymbol{r}-\boldsymbol{r}_1|^3} + \frac{Q_2(\boldsymbol{r}-\boldsymbol{r}_2)}{|\boldsymbol{r}-\boldsymbol{r}_2|^3}\right)\times\boldsymbol{r} = 0 = \frac{Q_1(\boldsymbol{r}\times\boldsymbol{r}_1)}{|\boldsymbol{r}-\boldsymbol{r}_1|^3} + \frac{Q_2(\boldsymbol{r}\times\boldsymbol{r}_2)}{|\boldsymbol{r}-\boldsymbol{r}_2|^3}.$$

Thus, $\boldsymbol{r}_1$ and $\boldsymbol{r}_2$ have to be parallel since $\boldsymbol{r}\times\boldsymbol{r}_2$ must vanish with $\boldsymbol{r}\times\boldsymbol{r}_1$ when $\boldsymbol{r} = R\hat{\boldsymbol{r}}_1$, so the boundary condition becomes

$$\frac{Q_1 r_1}{|\boldsymbol{r}-\boldsymbol{r}_1|^3} = -\frac{Q_2 r_2}{|\boldsymbol{r}-\boldsymbol{r}_2|^3} \tag{17.40}$$

where $r_1 = |\boldsymbol{r}_1|$, $r_2 = |\boldsymbol{r}_2|$ and $|\boldsymbol{r}| = R$. This shows that Q_1 and Q_2 must have opposite signs. Evaluating the magnitudes[8] with $\hat{\boldsymbol{r}}\cdot\hat{\boldsymbol{r}}_1 = \cos\alpha = \hat{\boldsymbol{r}}\cdot\hat{\boldsymbol{r}}_2$, the boundary condition reduces to

[8] $|\boldsymbol{r}-\boldsymbol{r}_1|^2 = (\boldsymbol{r}-\boldsymbol{r}_1)\cdot(\boldsymbol{r}-\boldsymbol{r}_1) = R^2 + r_1^2 - 2Rr_1\cos\alpha$.

$$(Q_1^2 r_1^2)^{1/3}(R^2 + r_2^2 - 2Rr_2\cos\alpha) = (Q_2^2 r_2^2)^{1/3}(R^2 + r_1^2 - 2Rr_1\cos\alpha).$$

Everything is constant except for the variable $0 \le \alpha \le \pi$, so we need $\cos\alpha$ to cancel out, giving $Q_1^2/r_1 = Q_2^2/r_2$, such that the boundary condition becomes

$$r_1(R^2 + r_2^2) = r_2(R^2 + r_1^2).$$

That is a quadratic equation for r_2 with solutions $r_2 = R^2/r_1$, and the trivial solution $r_2 = r_1$. In summary, the image is

$$Q_2 = -\frac{R}{r_1}Q_1 \quad \text{at} \quad \boldsymbol{r}_2 = \frac{R^2}{r_1^2}\boldsymbol{r}_1. \tag{17.41}$$

We could have solved this problem by seeking Q_2 and $\boldsymbol{r}_2$ such that the scalar potential field

$$\phi(\boldsymbol{r}) = \frac{1}{4\pi\epsilon_0}\left(\frac{Q_1}{|\boldsymbol{r}-\boldsymbol{r}_1|} + \frac{Q_2}{|\boldsymbol{r}-\boldsymbol{r}_2|}\right) \tag{17.42}$$

is constant on $|\boldsymbol{r}| = R$; then the α derivative should vanish, yielding (17.40). Potential lines and field lines are displayed in Fig. 17.12 in (ρ, z)-planes for a charge inside a sphere.

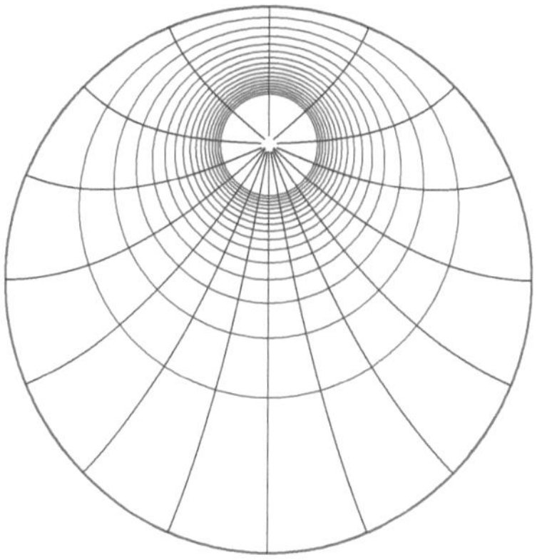

Fig. 17.12 Potential and field lines (17.42).

Poisson integral formula

The potential function (17.42) with $Q_1 = 1$ and $\boldsymbol{r}_2 = (R/r_1)^2\,\boldsymbol{r}_1$ is the Green's function

$$G(\boldsymbol{r},\boldsymbol{r}_1) = \frac{1}{4\pi\epsilon_0}\left(\frac{1}{|\boldsymbol{r}-\boldsymbol{r}_1|} - \frac{R}{r_1}\frac{1}{|\boldsymbol{r}-\boldsymbol{r}_2|}\right), \tag{17.43}$$

for Poisson's equation $-\nabla^2\phi = q(\boldsymbol{r})/\epsilon_0$ inside the sphere with Dirichlet boundary conditions $\phi = f(\boldsymbol{r})$ given on the sphere. That Green's function satisfies (17.38)

$$-\nabla^2 G(\boldsymbol{r},\boldsymbol{r}_1) = \epsilon_0^{-1}\delta(\boldsymbol{r}-\boldsymbol{r}_1)$$

but now with $G(\boldsymbol{r},\boldsymbol{r}_1) = 0$ on $|\boldsymbol{r}| = R$.

A general solution formula can then be obtained from *Green's identity* (16.148) for $\psi = G(\boldsymbol{r},\boldsymbol{r}_1)$,

$$\int_{\mathcal{V}} (\phi\nabla^2 G - G\nabla^2\phi)\,\mathrm{d}V = \oint_{\partial\mathcal{V}} (\phi\boldsymbol{\nabla}G - G\boldsymbol{\nabla}\phi)\cdot\mathrm{d}\boldsymbol{S}. \tag{17.44}$$

Using $\nabla^2 G = -\epsilon_0^{-1}\delta(\boldsymbol{r}-\boldsymbol{r}_1)$, $\int_{\mathcal{V}} \phi(\boldsymbol{r})\delta(\boldsymbol{r}-\boldsymbol{r}_1)\mathrm{d}V = \phi(\boldsymbol{r}_1)$, $\nabla^2\phi = -q(\boldsymbol{r})/\epsilon_0$ and $G(\boldsymbol{r},\boldsymbol{r}_1) = 0$ on $r = R$, (17.44) reduces to

$$\phi(\boldsymbol{r}_1) = \int_{\mathcal{V}} G(\boldsymbol{r},\boldsymbol{r}_1)\,q(\boldsymbol{r})\mathrm{d}V - \epsilon_0 \oint_{\partial\mathcal{V}} f(\boldsymbol{r})\boldsymbol{\nabla} G\cdot \mathrm{d}\boldsymbol{S}. \tag{17.45}$$

Since $\boldsymbol{r}_1$ is any point inside the sphere, this is a solution formula for $\phi(\boldsymbol{r}_1)$ in terms of the known $q(\boldsymbol{r})$, $f(\boldsymbol{r})$ and $G(\boldsymbol{r},\boldsymbol{r}_1)$. For the integral on the spherical surface $\partial\mathcal{V}$ where $|\boldsymbol{r}| = R$ and $|\boldsymbol{r}-\boldsymbol{r}_2| = (R/r_1)|\boldsymbol{r}-\boldsymbol{r}_1|$, the expression for $\boldsymbol{\nabla} G$ simplifies to

$$-\epsilon_0 \boldsymbol{\nabla} G = \frac{1}{4\pi}\left(\frac{\boldsymbol{r}-\boldsymbol{r}_1}{|\boldsymbol{r}-\boldsymbol{r}_1|^3} - \frac{R}{r_1}\frac{\boldsymbol{r}-\boldsymbol{r}_2}{|\boldsymbol{r}-\boldsymbol{r}_2|^3}\right) = \frac{R^2-r_1^2}{4\pi R^2}\frac{\boldsymbol{r}}{|\boldsymbol{r}-\boldsymbol{r}_1|^3}.$$

For Laplace's equation, $\nabla^2\phi = 0$, (17.45) with $q(\boldsymbol{r}) = 0$ yields *Poisson's integral formula*

$$\phi(\boldsymbol{r}) = \oint_{\mathcal{S}'} f(\boldsymbol{r}')\frac{R^2-r^2}{\left(r^2+R^2-2rR\cos\alpha\right)^{3/2}}\frac{\mathrm{d}S'}{4\pi R}, \tag{17.46}$$

after the substitution $(\boldsymbol{r}_1,\boldsymbol{r}) \to (\boldsymbol{r},\boldsymbol{r}')$, where $\mathcal{S}'$ is the sphere $|\boldsymbol{r}'| = R$ with area element $\mathrm{d}S'$, and $\boldsymbol{r}\cdot\boldsymbol{r}' = rR\cos\alpha$. This integral is best performed by expressing $\boldsymbol{r}'$ and $f(\boldsymbol{r}')$ in terms of polar angle α from $\boldsymbol{r}$ and azimuthal angle β about $\boldsymbol{r}$, similarly to integral (17.57) below; in that case, $\mathrm{d}S' = R^2\sin\alpha\,\mathrm{d}\alpha\,\mathrm{d}\beta$ and

$$\begin{aligned}&\oint_{\mathcal{S}'} \frac{R^2-r^2}{\left(r^2+R^2-2rR\cos\alpha\right)^{3/2}}\frac{\mathrm{d}S'}{4\pi R}\\ &= \frac{R}{2}\int_0^\pi \frac{(R^2-r^2)\sin\alpha\,\mathrm{d}\alpha}{\left(r^2+R^2-2rR\cos\alpha\right)^{3/2}} = \frac{R^2-r^2}{2r}\int_{R-r}^{R+r}\frac{\mathrm{d}u}{u^2} = 1,\end{aligned} \tag{17.47}$$

where the substitution $u^2 = R^2+r^2-2Rr\cos\alpha$, and $R > r$ have been used.[9] Thus if the boundary condtion is $\phi = f(\boldsymbol{r}) = \phi_0$ constant on the sphere $|\boldsymbol{r}| = R$, then $\phi(\boldsymbol{r}) = \phi_0$ for all $\boldsymbol{r}$ inside the sphere. For nonconstant $\phi = f(\boldsymbol{r})$ on $|\boldsymbol{r}| = R$, the *Poisson kernel*

[9] If $r > R$, the lower bound in the u-integral is $r - R$, and (17.47) $= -R/r$.

$$\frac{1}{4\pi R}\frac{R^2-r^2}{\left(R^2+r^2-2Rr\cos\alpha\right)^{3/2}} \to 0 \quad \forall\alpha \neq 0 \tag{17.48}$$

as $r \to R$, but goes to infinity at $\alpha = 0$ in such a way that its surface integral is always 1, as obtained in (17.47); thus, this Poisson kernel tends of a delta function for the sphere of radius R as $r \to R$, and (17.46) yields the boundary condition $\phi(\boldsymbol{r}) = f(\boldsymbol{r})$ as $|\boldsymbol{r}| \to R$.

Sphere in a uniform field

A similar problem is that of a perfectly conducting sphere placed in a uniform electric field, $\boldsymbol{E} = -E_\infty\hat{\mathbf{z}}$, say (Fig. 17.13). Poisson's equation reduces to *Laplace's equation* $\nabla^2\phi = 0$, that is,

$$\frac{\partial}{\partial r}\left(r^2\frac{\partial\phi}{\partial r}\right) + \frac{1}{\sin\theta}\frac{\partial}{\partial\theta}\left(\sin\theta\frac{\partial\phi}{\partial\theta}\right) + \frac{1}{\sin^2\theta}\frac{\partial^2\phi}{\partial\varphi^2} = 0, \tag{17.49}$$

in spherical coordinates (16.113), with boundary conditions $\phi \to E_\infty z = E_\infty r\cos\theta$ as $r \to \infty$ and $\phi = 0$ on $r = R$. Clearly, we are looking for a solution of the form

$$\phi = E_\infty\, f(r)\cos\theta,$$

with $f(R) = 0$ and $f(r) \to r$ as $r \to \infty$. Substituting such ϕ function into Laplace's equation gives a second order differential equation for $f(r)$:

$$r^2\frac{\mathrm{d}^2 f}{\mathrm{d}r^2} + 2r\frac{\mathrm{d}f}{\mathrm{d}r} - 2f = 0. \tag{17.50}$$

This is an *equidimensional* differential equation; it does not change if we replace $r \to ar$ for any a, that is changing dimensions from meters to feet, for example. Linear constant coefficient differential equations (DE) are the simplest since they have exponential solutions. Equidimensional DE are the next simplest; they can be transformed into constant coefficients by the substitution $x = \ln r$. Then the DE in x has solutions of the form e^{sx} for some constant s, meaning that the original equidimensional DE has solutions of the form $e^{s\ln r} = r^s$. Indeed, substituting $f(r) = r^s$ for some unknown constant s into (17.50) gives a quadratic equation for s,

$$s(s-1) + 2s - 2 = s^2 + s - 2 = (s-1)(s+2) = 0,$$

such that the general solution for $f(r)$ is $f(r) = Ar + B/r^2$ for some constants A and B. The boundary conditions $f(r) \to r$ as $r \to \infty$ and $f(R) = 0$ give $f(r) = r - R^3/r^2$ and our potential

$$\phi = E_\infty\left(r - \frac{R^3}{r^2}\right)\cos\theta, \tag{17.51a}$$

$$\begin{aligned} \boldsymbol{E} &= -\boldsymbol{\nabla}\phi \\ &= -E_\infty\left(\hat{\boldsymbol{r}}\left(1 + 2\frac{R^3}{r^3}\right)\cos\theta - \hat{\boldsymbol{\theta}}\left(1 - \frac{R^3}{r^3}\right)\sin\theta\right). \end{aligned} \tag{17.51b}$$

The electric field $\boldsymbol{E}$ at $r = R$, $\theta = 0$ is $3\boldsymbol{E}_\infty$ as the sphere "pulls in" the field. The equipotentials and electric field lines are shown in Fig. 17.14.

The fields shown inside the sphere are the mathematical continuation of the functions (17.51), which are defined everywhere except for a *dipole* singularity at $r = 0$, but those fields are not physical. The electric field must be zero inside the conducting sphere since the potential there would be the solution of Laplace's equation $\nabla^2\phi = 0$ in $r < R$ with $\phi = 0$ on $r = R$. There are an infinite number of solutions to that problem, but all are singular except for the trivial one $\phi(\boldsymbol{r}) = 0$. Another singular solution is

$$\phi = \left(r^2 - \frac{R^5}{r^3}\right)(3\cos^2\theta - 1), \tag{17.52}$$

with a *quadrupole* singularity at $r = 0$, illustrated in Fig. 17.15. More generally,

$$\phi = \left(r^\ell - \frac{R^{2\ell+1}}{r^{\ell+1}}\right)P_\ell(\cos\theta),$$

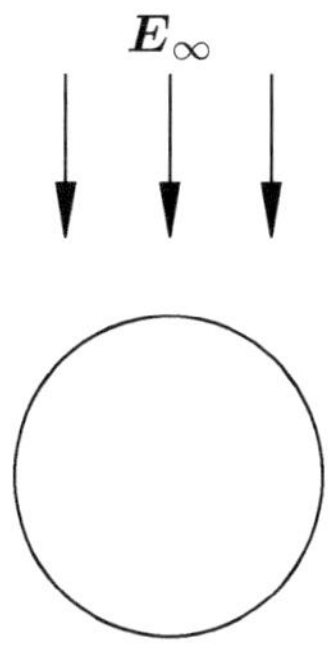

Fig. 17.13 Conducting sphere in a field uniform at infinity

Fig. 17.14 Dipole field (17.51).

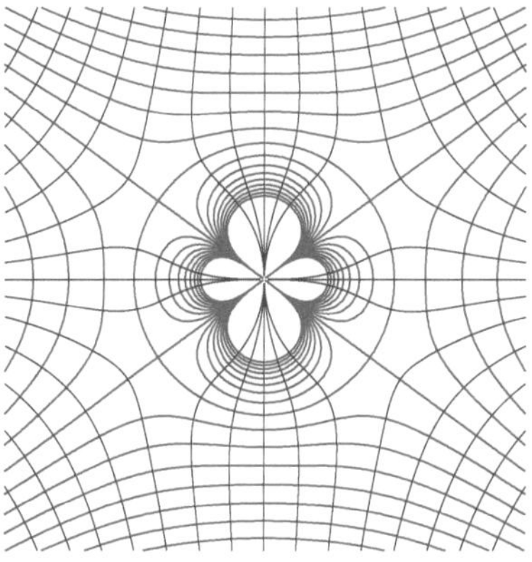
Fig. 17.15 Quadrupole field (17.52).

where $P_\ell(x)$ is the *Legendre polynomial* of degree $\ell \geq 0$, such that

$$\frac{1}{\sin\theta}\frac{\mathrm{d}}{\mathrm{d}\theta}\left(\sin\theta\frac{\mathrm{d}}{\mathrm{d}\theta}P_\ell(\cos\theta)\right) = -\ell(\ell+1)P_\ell(\cos\theta),$$

are all solutions of $\nabla^2\phi = 0$ with $\phi = 0$ on $r = R$, but each are increasingly more singular at $r = 0$. Requiring boundedness at $r = 0$ rejects all those solutions, except for the trivial one $\phi(\boldsymbol{r}) = 0$ for all $r \leq R$, and that is the physical solution.

Surface charge distribution

In the physical problem, a surface charge distribution $\sigma(\boldsymbol{r})$ on the sphere is triggered by the external field and the electric field has a jump discontinuity across that surface such that[10]

$$\lim_{h\to 0^+}(\boldsymbol{E}_+ - \boldsymbol{E}_-) = \frac{\sigma}{\epsilon_0}\hat{\boldsymbol{n}}, \tag{17.53}$$

[10] E.g. E. M. Purcell and D. J. Morin, *Electricity and Magnetism* (1.41), Cambridge University Press (2013).

where $\boldsymbol{E}_\pm = \boldsymbol{E}(\boldsymbol{r} \pm h\hat{\boldsymbol{n}})$ with $\boldsymbol{r}$ a point on the surface and $\hat{\boldsymbol{n}}$ the unit normal to the surface at that point. For (17.51b), the $\boldsymbol{E}$ field vanishes inside the conducting sphere; hence, the surface charge distribution on $|\boldsymbol{r}| = R$ is

$$\sigma(\theta,\varphi) = \epsilon_0\boldsymbol{E}\cdot\hat{\boldsymbol{r}} = -3\epsilon_0 E_\infty\cos\theta, \tag{17.54}$$

axisymmetric, with negative charge density on the top hemisphere $0 \leq \theta < \pi/2$ and positive charge density on the lower hemisphere $\pi/2 < \theta \leq \pi$. This surface charge provides the dipole part in (17.51), as we now verify explicitly.

We need to compute the potential due to the surface charge, that is

$$\phi_S(\boldsymbol{r}) = \frac{1}{4\pi\epsilon_0}\oint_{S'}\frac{\sigma(\boldsymbol{r}')\,\mathrm{d}S'}{|\boldsymbol{r}-\boldsymbol{r}'|}, \tag{17.55}$$

where the integral is over all $\boldsymbol{r}'$ on the sphere $S' : |\boldsymbol{r}'| = R$ and $\mathrm{d}S'$ is the area element for that sphere. The integral is easy—provided we use the proper spherical coordinates with $\hat{\boldsymbol{r}}$ as polar axis, not $\hat{\mathbf{z}}$ (Fig. 17.16).

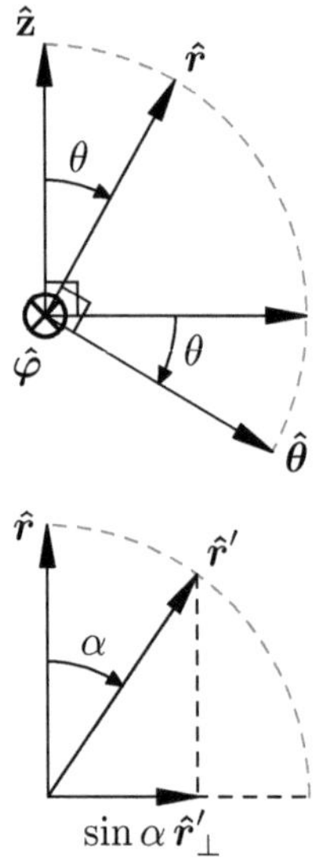

Fig. 17.16 Spherical coordinates with $\hat{\boldsymbol{r}}$ as polar axis.

The direction $\hat{\mathbf{z}}$ is determined by the electric field at infinity $\boldsymbol{E} \to -E_\infty\hat{\mathbf{z}}$. The integral (17.55) is over $\boldsymbol{r}'$ with $\boldsymbol{r}$ fixed at a polar angle θ from $\hat{\mathbf{z}}$, with azimuthal direction $\hat{\boldsymbol{\varphi}}$ such that $\hat{\mathbf{z}} \times \hat{\boldsymbol{r}} = \hat{\boldsymbol{\varphi}}\sin\theta$ and polar direction $\hat{\boldsymbol{\theta}} = \hat{\boldsymbol{\varphi}} \times \hat{\boldsymbol{r}}$. Then $\boldsymbol{r}'$ can be expressed in spherical coordinates with respect to the $\{\hat{\boldsymbol{r}}, \hat{\boldsymbol{\theta}}, \hat{\boldsymbol{\varphi}}\}$ basis as

$$\boldsymbol{r}'(\alpha,\beta) = R\Big(\hat{\boldsymbol{r}}\cos\alpha + \sin\alpha\big(\hat{\boldsymbol{\theta}}\cos\beta + \hat{\boldsymbol{\varphi}}\,\sin\beta\big)\Big),$$

where α is the polar angle from $\hat{\boldsymbol{r}}$ and β is the azimuthal angle about $\hat{\boldsymbol{r}}$. In those spherical coordinates, the area element $\mathrm{d}S'$ and distance $|\boldsymbol{r}-\boldsymbol{r}'|$ are

$$\mathrm{d}S' = R^2\sin\alpha\,\mathrm{d}\alpha\,\mathrm{d}\beta,$$
$$|\boldsymbol{r}-\boldsymbol{r}'|^2 = r^2 + R^2 - 2rR\cos\alpha,$$

and the surface density distribution (17.54) $\sigma = \sigma_0 \cos\theta'$, with $\sigma_0 = -3\epsilon_0 E_\infty$, is

$$\sigma(\alpha,\beta) = \sigma_0 \cos\theta' = \sigma_0 \hat{\mathbf{z}}\cdot\hat{\boldsymbol{r}}' = \sigma_0(\cos\alpha\cos\theta - \sin\alpha\cos\beta\sin\theta) \quad (17.56)$$

since $\hat{\mathbf{z}}\cdot\hat{\boldsymbol{r}} = \cos\theta$, $\hat{\mathbf{z}}\cdot\hat{\boldsymbol{\theta}} = -\sin\theta$ and $\hat{\mathbf{z}}\cdot\hat{\boldsymbol{\varphi}} = 0$. The integral (17.55) expressed in terms of these (α,β) spherical coordinates reads and evaluates as

$$\begin{aligned}\phi_S(\boldsymbol{r}) &= \frac{\sigma_0}{4\pi\epsilon_0}R^2\int_0^\pi\int_0^{2\pi}\frac{\cos\alpha\cos\theta-\sin\alpha\cos\beta\sin\theta}{\sqrt{r^2+R^2-2rR\cos\alpha}}\sin\alpha\,\mathrm{d}\beta\mathrm{d}\alpha\\ &= \frac{\sigma_0}{2\epsilon_0}R^2\cos\theta\int_0^\pi\frac{\sin\alpha\cos\alpha}{\sqrt{r^2+R^2-2rR\cos\alpha}}\mathrm{d}\alpha\\ &= \frac{\sigma_0}{4\epsilon_0}\frac{\cos\theta}{r^2}\int_{|r-R|}^{r+R}\left(r^2+R^2-u^2\right)\mathrm{d}u = \frac{\sigma_0}{3\epsilon_0}\frac{(\min(r,R))^3}{r^2}\cos\theta,\end{aligned} \quad (17.57)$$

where the substitution $u^2 = r^2+R^2-2rR\cos\alpha$ with $u\,\mathrm{d}u = rR\sin\alpha\,\mathrm{d}\alpha$ has been used. Since $\sigma_0 = -3\epsilon_0 E_\infty$, this surface charge integral indeed provides the dipole part of the potential (17.51a) for $r > R$, and is $-E_\infty r\cos\theta = -E_\infty z$ for $r < R$, where it cancels out the external field potential $\phi_\infty = E_\infty z$. The equipotentials of $\phi_S(\boldsymbol{r})$ are shown in meridional planes (ρ, z) in Fig. 17.17.

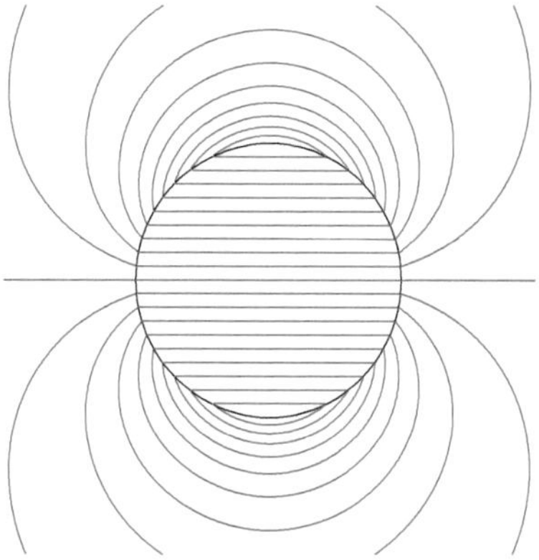

Fig. 17.17 Potential of a surface dipole distribution (17.57).

Well-posedness

We have encountered three fundamental partial differential equations: the heat equation $\partial_t T = \kappa\nabla^2 T$ (16.122), the wave equation $\partial_t^2\boldsymbol{E} = c^2\nabla^2\boldsymbol{E}$ (17.16), and Laplace's equation $\nabla^2\phi = 0$ (17.49). For the heat and wave equations, spatially periodic Fourier modes with real wavevector $\boldsymbol{k}$ are eigenmodes of the Laplacian

$$\nabla^2 e^{i\boldsymbol{k}\cdot\boldsymbol{r}} = -|\boldsymbol{k}|^2 e^{i\boldsymbol{k}\cdot\boldsymbol{r}}. \quad (17.58)$$

They yield fundamental *eigensolutions* decaying exponentially in time for the heat equation, and oscillating in time for the wave equation,

$$T(t,\boldsymbol{r}) = \widetilde{T}(\boldsymbol{k})\,e^{-\kappa|\boldsymbol{k}|^2 t}e^{i\boldsymbol{k}\cdot\boldsymbol{r}}, \quad (17.59)$$
$$\boldsymbol{E}(t,\boldsymbol{r}) = \widetilde{\boldsymbol{E}}_\pm(\boldsymbol{k})\,e^{\mp ic|\boldsymbol{k}|t}e^{i\boldsymbol{k}\cdot\boldsymbol{r}}, \quad (17.60)$$

which can be superposed to construct general solutions. Even when other functions are more convenient such as Bessel functions for cylindrical coordinates (16.132), the most useful representation of those *special functions* is superposition of Fourier modes (16.135). For Laplace's equation, however, Fourier mode solutions do not exist for *real* wavevectors $\boldsymbol{k}$. In cartesian coordinates with $\boldsymbol{k} = k_x\hat{\mathbf{x}} + k_y\hat{\mathbf{y}} + k_z\hat{\mathbf{z}}$, we need at least one of those components to be *imaginary* such that $|\boldsymbol{k}|^2 = k_x^2 + k_y^2 + k_z^2$ vanishes in (17.58).

For example, a solution $\phi(x, y)$ of Laplace's equation $\nabla^2\phi = 0$ with $\phi = \cos kx$ on $y = 0$ can be sought in the form $\phi = f(y)\cos kx$. Laplace's equation yields

$$\frac{d^2 f}{dy^2} = k^2 f, \tag{17.61}$$

whose solutions are $f(y) = e^{\pm ky}$ and the general solution is the linear combination $f(y) = Ae^{ky} + Be^{-ky}$ for arbitrary constants A and B. Those constants could be determined from initial conditions; for example, $f(0) = 1$, $\mathrm{d}f/\mathrm{d}y(0) = 0$ yields $f(y) = (e^{ky} + e^{-ky})/2 = \cosh ky$ corresponding to

$$\phi(x, y) = \cos kx \, \cosh ky. \tag{17.62}$$

Although these are valid initial conditions for the second order differential equation (17.61), yielding a unique solution (17.62) for Laplace's equation $\nabla^2\phi = 0$ with the double boundary condition $\phi = \cos kx$ and $\partial\phi/\partial y = 0$ for all x at $y = 0$. This is not a *well-posed* problem for Laplace's equation.

For the partial differential equation (PDE), any wavenumber k is allowed. Thus, if we consider the slightly different boundary conditions,

$$\phi(x, 0) = \cos kx + \epsilon \cos lx, \quad \frac{\partial\phi}{\partial y}(x, 0) = 0, \tag{17.63}$$

the unique solution to $\nabla^2\phi = 0$ with those boundary conditions is

$$\phi(x, y) = \cos kx \, \cosh ky + \epsilon \cos lx \cosh ly. \tag{17.64}$$

The problem with that solution is that no matter how small ϵ is, we can always pick l sufficiently large to make $\epsilon \cosh ly$ as large as we want for any $y \neq 0$, since $\cosh ly = (e^{ly} + e^{-ly})/2$ grows exponentially from 1 at $y = 0$. That means that the solution to the PDE is not continuously dependent on the initial conditions; a small change in those conditions can yield as big a change as we want in the solution. Laplace's equation, and Poisson's equation, cannot be solved as an initial value problem; we need to specify boundary conditions on all sides of the domain.

For example, $\phi = \cos kx + \epsilon \cos lx$ on $y = 0$ with ϕ bounded as $|y| \to \infty$ yields the unique solution

$$\phi(x, y) = \cos kx \; e^{-|ky|} + \epsilon \cos lx \; e^{-|ly|}, \tag{17.65}$$

and now the small perturbation proportional to ϵ cannot grow as large as we want for $y \neq 0$. In fact, it decays away from $y = 0$ and goes to 0 with ϵ. The problem is now well-posed with one boundary condition at $y = 0$ and another (boundedness) as $|y| \to \infty$.

We leave it as an exercise to derive the general solution to Laplace's equation $\nabla^2\phi = 0$ with known $\phi(x, 0) = f(x)$ and boundedness at infinity as

$$\phi(x, y) = \int_{-\infty}^{\infty} \frac{f(x')}{\pi} \frac{|y|}{(x - x')^2 + y^2} \, \mathrm{d}x', \tag{17.66}$$

using a 2D Green's function $-(2\pi)^{-1} \ln |\boldsymbol{r} - \boldsymbol{r}'|$ for $\boldsymbol{r} = x\hat{\mathbf{x}} + y\hat{\mathbf{y}}$, with an image across $y = 0$, or a Fourier expansion $f(x) = \int_{-\infty}^{\infty} f(k) e^{ikx} \mathrm{d}k$. The Poisson kernel in (17.66) is another function that tends to the Dirac delta $\delta(x - x')$ as $y \to 0$, such that $\phi(x, 0) = f(x)$.

Exercises

This entire chapter consists of solved examples. The student is encouraged to go through each example carefully and fill in the details, and study electromagnetism where many more interesting problems will be found.

(17.1) Verify that (17.52) solves Laplace's equation $\nabla^2\phi = 0$ with $\phi = 0$ on $|\boldsymbol{r}| = R$.

(17.2) Find the bounded spherically symmetric solution to $-\nabla^2\phi = \epsilon_0^{-1}D(\boldsymbol{r}; R)$ that vanishes at infinity, where $D(\boldsymbol{r}; R) = 3/(4\pi R^3)$ for $|\boldsymbol{r}| < R$ and $D(\boldsymbol{r}; R) = 0$ for $|\boldsymbol{r}| > R$, such that $\lim_{R\to 0} D(\boldsymbol{r}; R) = \delta(\boldsymbol{r})$ in an integral sense (17.39). Solve in both regions $|\boldsymbol{r}| \gtrless R$ separately, then match the potential ϕ and the electric field $\boldsymbol{E} = -\boldsymbol{\nabla}\phi$ at $r = R$. Sketch $\phi(r)$ and $\hat{\boldsymbol{r}} \cdot \boldsymbol{E}$.

(17.3) Derive (17.45) without the delta function but considering instead Green's identity (16.148) in a two-boundary domain $\mathcal{V}$ inside the sphere $|\boldsymbol{r}| < R$ but excluding a small sphere $|\boldsymbol{r} - \boldsymbol{r}_1| < \epsilon$, similar to the derivation of the mean value theorem (16.149), and taking the limit $\epsilon \to 0$.

(17.4) Consider a point charge Q_1 at $\boldsymbol{r} = \boldsymbol{r}_1$ and an infinite conducting plane at $z = 0$. Find the potential ϕ and electric field $\boldsymbol{E}$ using an image charge. What is the actual surface charge distribution filling the role of the fictitious image charge?

(17.5) Show that the bounded solution to $\nabla^2\phi = 0$ inside the disk of radius R in the $(x, y) = \rho(\cos\varphi, \sin\varphi)$ plane, with $\phi(R, \varphi) = f(\varphi)$ on its boundary, is given by Poisson's integral formula

$$\phi(\rho, \varphi) = \int_0^{2\pi} \frac{f(\theta)}{2\pi} \frac{R^2 - \rho^2}{R^2 + \rho^2 - 2\rho R\cos(\theta - \varphi)}\, \mathrm{d}\theta. \quad (17.67)$$

Use a Green's function with the method of images for a line charge with potential $-(2\pi)^{-1}\ln|\boldsymbol{r} - \boldsymbol{r}_1|$ instead of $1/(4\pi|\boldsymbol{r} - \boldsymbol{r}_1|)$), or Fourier series

$$\phi(\rho, \varphi) = \sum_{m=-\infty}^{\infty} F_m(\rho) e^{im\varphi}$$

that can be summed using Euler's formula and geometric series (19.45).

(17.6) Find $\phi(x, y)$ such that $\nabla^2\phi = 0$ with $\phi(x, 0) = f(x)$ and ϕ bounded as $|y| \to \infty$. (i) Using a Green's function, and (ii) using Fourier integrals $f(x) = \int_{-\infty}^{\infty} \tilde{f}(k) e^{ikx} \mathrm{d}k$.

Part III

Complex Calculus

Complex Algebra and Geometry

18

18.1 The cubic formula

To find the roots of the quadratic equation $ax^2+bx+c=0$, we complete the square

$$ax^2+bx+c = a\left(x+\frac{b}{2a}\right)^2 + c - \frac{b^2}{4a} = 0,$$

yielding $z^2 = (b^2-4ac)/(4a^2)$ for $z = x + b/(2a)$, then the roots

$$x = \frac{-b \pm \sqrt{b^2-4ac}}{2a}. \tag{18.1}$$

This is the *quadratic formula* of elementary algebra. A real quadratic polynomial has two, one, or zero real roots depending on whether the *discriminant* b^2-4ac is >0, $=0$, or <0. The quadratic function $y = ax^2+bx+c$ has a single extremum, $y_* = (4ac-b^2)/(4a)$ at $x_* = -b/(2a)$, and intersects $y=0$ only if $4ac-b^2 \leq 0$, and we can see the *bifurcation* from two to zero roots as $4ac-b^2$ goes positive.

What about the roots of the cubic $ax^3+bx^2+cx+d=0$? Completing the cube to eliminate the x^2 term yields

$$z^3 + pz + q = 0 \tag{18.2}$$

for $z = x + b/(3a)$, where the expressions for p and q are given in exercise 18.1. The cubic $w = z^3+pz+q$ has extrema at $z = \pm\sqrt{-p/3}$ provided $p<0$. These extrema are $w = q \pm \sqrt{-4p^3/27}$ and the cubic $w = z^3+pz+q$ intercepts $w=0$ at three real z values provided $p<0$ and $|q| < \sqrt{-4p^3/27}$, that is, provided the *discriminant*

$$q^2 + \frac{4p^3}{27} < 0. \tag{18.3}$$

A tricky further substitution, $z = w - p/(3w)$, transforms the cubic into a quadratic equation for w^3 whose roots are

$$w^3 = \frac{1}{2}\left(-q \pm \sqrt{q^2 + \frac{4p^3}{27}}\right). \tag{18.4}$$

Now we have a problem ... when the cubic has three real roots according to (18.3), formula (18.4) requires the square root of a *negative* number.

How is that possible?! For example, $z^3 - z = z(z-1)(z+1)$ has three real roots with $p = -1$ and $q = 0$, but formula (18.4) gives $w^3 = \pm\sqrt{-1}/\sqrt{27}$. This is the conundrum that confronted Italian mathematicians, Del Ferro, Tartaglia, and Cardano in the early 1500s and led to *imagining* that $\sqrt{-1}$ exists such that $(\sqrt{-1})^2 = -1$. Then $w = (\sqrt{-1})^{1/3}/\sqrt{3}$ for $p = -1$, $q = 0$ yields

$$z = w - \frac{p}{3w} = \frac{3w^2 - p}{3w} = \frac{(\sqrt{-1})^{2/3} + 1}{3w} = \frac{(-1)^{1/3} + 1}{3w} = 0,$$

that is, indeed, one of the three real roots of $z^3 - z$. The symbol $\sqrt{-1}$ was used for over 200 years, until Euler introduced the notation

$$i \triangleq \sqrt{-1} \quad \text{such that} \quad i^2 = -1, \tag{18.5}$$

for that *imaginary unit.* Thus began *complex* numbers $z = x + iy$ composed of a *real part* $\text{Re}(z) = x$ and an *imaginary part* $\text{Im}(z) = y$, both of which are real numbers. For example,

$$z = 2 + 3i \;\Leftrightarrow\; x = \text{Re}(z) = 2, \; y = \text{Im}(z) = 3.$$

A real number is a complex number with zero imaginary part, and an imaginary number is a complex number with zero real part.

18.2 The complex plane

Euler used series in the early 1700s to define the exponential function $\exp(x) = e^x$ and derive his famous formula for the exponential of an imaginary number,

$$e^{i\varphi} = \cos\varphi + i\sin\varphi. \tag{18.6}$$

Since $x = \cos\varphi$ and $y = \sin\varphi$ are the (x, y) coordinates of a point on the unit circle, it is surprising that it was not until 1797 that Caspar Wessel, then Jean-Robert Argand independently in 1806,[1] proposed that complex numbers should be considered as *directed line segments*, with magnitude and direction, in the euclidean (x, y) plane such that i represents the direction $(0, 1)$ and multiplication by i corresponds to rotation of directed line segments by $\pi/2$ in the $(1, 0)$ to $(0, 1)$ direction. In mathematical notation, that is

[1] Wessel's work in Copenhagen was overlooked for a hundred years, Argand's in Paris for only seven years.

$$z = x + iy \equiv (x, y) \Rightarrow iz = i(x + iy) = -y + ix \equiv (-y, x).$$

A second multiplication by i corresponds to another $\pi/2$ rotation from $(-y, x)$ to $(-x, -y)$, resulting in a sign change after two successive rotations by $\pi/2$, $(x, y) \to (-x, -y) = -(x, y)$ and indeed $i^2 = -1$.

The vector space $\mathbb{R}^2$ consists of ordered pairs of real numbers (x, y) representing the coordinates of directed line segments with the usual vector addition, $(x_1, y_1) + (x_2, y_2) = (x_1 + x_2, y_1 + y_2)$. When the complex product operation defined as

$$(x_1,\, y_1)(x_2,\, y_2) \triangleq (x_1x_2 - y_1y_2,\, x_1y_2 + y_1x_2)$$

is added, the set of real (x, y) becomes the *complex plane* $\mathbb{C}$, also known as the Argand plane (Fig. 18.1). Argand introduced the geometric concepts of *modulus* $|z|$, *argument* $\arg(z)$, and complex conjugate z^*.

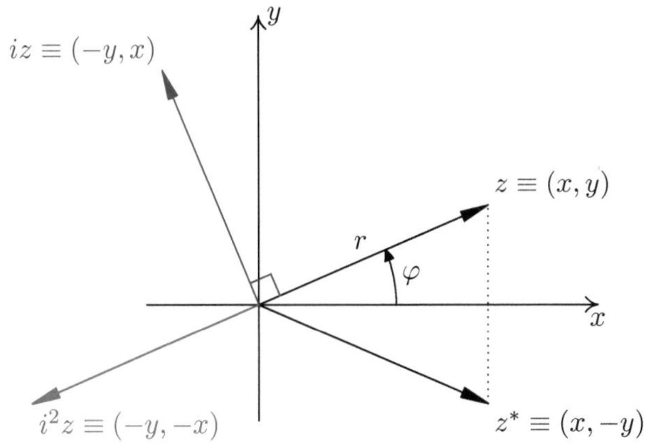

Fig. 18.1 The complex plane $\mathbb{C}$, also known as the Argand plane, interprets complex numbers $z = x + iy$ as vectors (x, y) in $\mathbb{R}^2$ with the additional operation of the complex product such that multiplication by i is rotation by $\pi/2$. The magnitude of z is $|z| = r = \sqrt{x^2 + y^2}$, and its *argument* $\arg(z) = \varphi$ with $x = r\cos\varphi$, $y = r\sin\varphi$. The complex conjugate $z^* = x - iy$ is the reflection about the real axis.

The *modulus*, or *norm*, or *magnitude*, of $z = x + iy$ is

$$|z| \triangleq \sqrt{x^2 + y^2} = r. \tag{18.7}$$

This modulus is equivalent to the euclidean norm of the 2D vector (x, y).

The *argument* or *angle* of $z = x + iy$ is

$$\arg(z) \equiv \text{angle}(z) = \varphi, \tag{18.8}$$

such that $(x, y) = (r\cos\varphi, r\sin\varphi)$. It is the angle between the real direction $(1, 0)$ and the vector (x, y) positive in the direction of the $\pi/2$ rotation from $(1, 0)$ to $(0, 1)$. The notation $\arg(z)$ is common but $\text{angle}(z)$ is clearer. The standard unique definition specifies the angle in $(-\pi, \pi]$

$$-\pi < \arg(z) \leq \pi. \tag{18.9}$$

Thus, $\arg(z) = \text{atan2}(y, x)$. The function $\text{atan2}(y, x)$ was introduced in the Fortran computer language and is now common in all computer languages. The $\text{atan2}(y, x)$ function is the arctangent function but returns an angle in $(-\pi, \pi]$, in contrast to $\arctan(y/x)$ that loses sign information and returns an angle in $[-\pi/2, \pi/2]$.

The *complex conjugate* of $z = x + iy$ is

$$z^* \triangleq x - iy, \tag{18.10}$$

is the reflection of z about the real axis. It is often written $\bar{z}$ in mathematics but the z^* notation is more common in applied mathematics, physics, and engineering where the overbar is commonly used to denote an average.

18.3 Complex algebra

The complex plane $\mathbb{C}$ consists of ordered pairs of real numbers (x, y) conveniently denoted $z = x + iy$, similar to our 2D vectors $\boldsymbol{r} = x\hat{\mathbf{x}} + y\hat{\mathbf{y}}$, with the following operations.

Addition

The sum of two complex numbers $z_1 + z_2$ is a complex number whose real part is the sum of the real parts, and imaginary part is the sum of the imaginary parts,

$$z_1 + z_2 = (x_1 + iy_1) + (x_2 + iy_2) \triangleq (x_1 + x_2) + i\,(y_1 + y_2). \qquad (18.11)$$

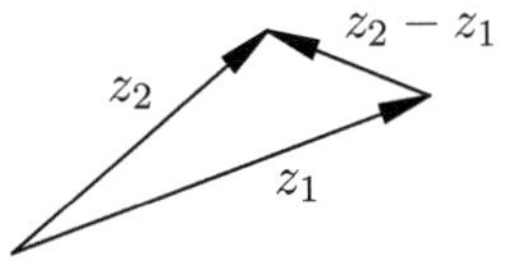

Fig. 18.2 Complex subtraction.

That is $(x_1, y_1) + (x_2, y_2) = (x_1 + x_2, y_1 + y_2)$ so complex numbers add like 2D cartesian vectors, and complex addition satisfies the *triangle inequality*

$$|z_1 + z_2| \leq |z_1| + |z_2|. \qquad (18.12)$$

Subtraction is the addition of the opposite vector $z_2 - z_1 = z_2 + (-z_1)$ and $z_2 - z_1$ is the vector from the tip of z_1 to that of z_2, just as for vectors $\boldsymbol{r}_2 - \boldsymbol{r}_1$ (Fig. 18.2).

Multiplication

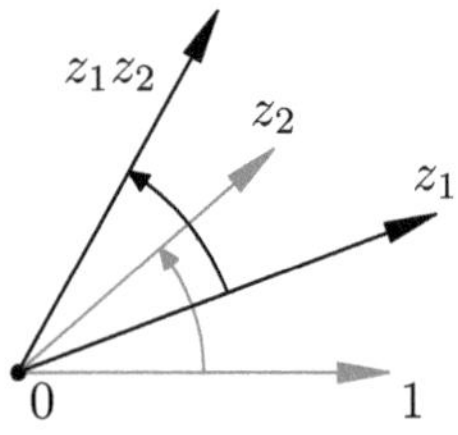

Fig. 18.3 Complex multiplication.

The product $z_1 z_2$ of *directed line segments* z_1 and z_2 is defined such that the triangle with sides $(z_1, z_1 z_2)$ is similar to the triangle with sides $(1, z_2)$ (Fig. 18.3). Thus, the magnitude $|z_1 z_2|$ is the product of the magnitudes

$$|z_1 z_2| = |z_1|\,|z_2|, \qquad (18.13)$$

and the angle between z_1 and $z_1 z_2$ is the same as that between $1 \equiv (1, 0)$ and z_2; that is, the angle of the product $z_1 z_2$ is the *sum* of the angles modulo 2π

$$\arg(z_1 z_2) = (\arg z_1 + \arg z_2) \qquad \bmod 2\pi, \qquad (18.14)$$

since $-\pi < \arg z \leq \pi$ for uniqueness of the angle. For instance,

$$\arg((-2)(-3)) = \arg(6) = 0,$$

but $\arg(-2) + \arg(-3) = \pi + \pi = 0 + 2\pi$, which is 0 *modulo* 2π.

This geometric definition of the product corresponds to the usual distributivity of multiplication with respect to addition but with the additional rule $i^2 = -1$, yielding

$$z_1 z_2 = (x_1 + iy_1)(x_2 + iy_2) = (x_1 x_2 - y_1 y_2) + i\,(x_1 y_2 + x_2 y_1).$$

Indeed, rotation by $\pi/2$ is multiplication by i with $i^2 = -1$, $i(x + iy) = -y + ix$; thus,

$$\begin{aligned} z_1 z_2 = (x_1 + iy_1)(x_2 + iy_2) &= x_1(x_2 + iy_2) + y_1(-y_2 + ix_2) \\ &= r_1\,(\cos\varphi_1(x_2 + iy_2) + \sin\varphi_1(-y_2 + ix_2)) \end{aligned}$$

is the vector $z_2 = x_2 + iy_2$ rotated by φ_1 and scaled by $r_1 = \sqrt{x_1^2 + y_1^2} = |z_1|$ since $iz_2 = -y_2 + ix_2$ is z_2 rotated by $\pi/2$.

The complex product is an operation between vectors $(x, y) \in \mathbb{C}$ that is not defined for vectors $(x, y) \in \mathbb{R}^2$. However, the product

$$z_1^* z_2 = (x_1 - iy_1)(x_2 + iy_2) = (x_1 x_2 + y_1 y_2) + i(x_1 y_2 - x_2 y_1)$$

has a real part that equals the dot product of the real vectors

$$(x_1, y_1) \cdot (x_2, y_2) = x_1 x_2 + y_1 y_2 = \mathrm{Re}(z_1^* z_2)$$

and an imaginary part that equals the $\mathbf{e}_3$ component of the cross product

$$(x_1, y_1, 0) \times (x_2, y_2, 0) = (0, 0, x_1 y_2 - x_2 y_1).$$

In particular,

$$z^* z = x^2 + y^2 = r^2 = |z|^2, \tag{18.15}$$

and this relation is used to compute complex magnitudes just as the dot product is used to compute vector magnitudes. For instance,

$$|a + b| = \sqrt{(a + b)(a + b)^*} = \sqrt{aa^* + ab^* + ba^* + bb^*}$$

for complex numbers a, b, is the complex equivalent of

$$|\boldsymbol{a} + \boldsymbol{b}| = \sqrt{(\boldsymbol{a} + \boldsymbol{b}) \cdot (\boldsymbol{a} + \boldsymbol{b})} = \sqrt{\boldsymbol{a} \cdot \boldsymbol{a} + 2\boldsymbol{a} \cdot \boldsymbol{b} + \boldsymbol{b} \cdot \boldsymbol{b}}$$

for real vectors $\boldsymbol{a}$, $\boldsymbol{b}$. However, while the dot product of two vectors is a *scalar* not a vector, and the cross product of *horizontal* vectors is a *vertical* vector, the product of complex numbers is a complex number. This yields complex division as the inverse of multiplication.

Division

Given z_1 and $z_2 \neq 0$ we can find a unique z such that $zz_2 = z_1$. That z is denoted z_1/z_2 and equals

$$\begin{aligned}\frac{z_1}{z_2} = \frac{z_1 z_2^*}{z_2 z_2^*} &= \frac{(x_1 + iy_1)(x_2 - iy_2)}{x_2^2 + y_2^2} \\ &= \left(\frac{x_1 x_2 + y_1 y_2}{x_2^2 + y_2^2}\right) + i\left(\frac{x_2 y_1 - x_1 y_2}{x_2^2 + y_2^2}\right).\end{aligned} \tag{18.16}$$

Thus, while division of vectors by vectors is not defined (what is north divided by northwest?!), we can divide complex numbers by complex numbers and the complex interpretation of 'north/northwest' is

$$\frac{i}{\frac{-1+i}{\sqrt{2}}} = \frac{\sqrt{2}\, i(-1 - i)}{(-1 + i)(-1 - i)} = \frac{1 - i}{\sqrt{2}},$$

which is "southeast"! Geometrically, it is left as an exercise to show that the magnitude of a ratio is the ratio of the magnitudes

$$\left|\frac{z_1}{z_2}\right| = \frac{|z_1|}{|z_2|}$$

and the angle of a ratio is the *difference* of the angles, modulo 2π

$$\arg\left(\frac{z_1}{z_2}\right) = \arg(z_1) - \arg(z_2) \quad \text{modulo } 2\pi.$$

Again, the "modulo 2π" is because of our definition $-\pi < \arg(z) \leq \pi$, for uniqueness of the $\arg(z)$ function, so we need to bring back the angle obtained from addition for $z_1 z_2$, or subtraction for z_1/z_2 to that range. For example, for $z_1 = -2$ and $z_2 = -i$,

$$\arg(z_1/z_2) = \arg(2/i) = \arg(-2i) = -\pi/2$$
$$\arg z_1 - \arg z_2 = \arg(-2) - \arg(-i) = \pi - (-\pi/2) = 3\pi/2$$

and $\arg(z_1/z_2) = \arg(z_1) - \arg(z_2) - 2\pi$ in this example.

All the well-known algebraic formula for real numbers hold for complex numbers, for instance,

$$z^2 - a^2 = (z-a)(z+a),$$

$$z^3 - a^3 = (z-a)(z^2 + az + a^2),$$

then for any positive integer n, we can show by induction that

$$z^{n+1} - a^{n+1} = (z-a)\sum_{k=0}^{n} z^{n-k}a^k. \tag{18.17}$$

That useful formula can also be derived by dividing by z^{n+1} or a^{n+1} in which case it becomes the *geometric sum*

$$1 + q + q^2 + \cdots + q^n = \frac{1-q^{n+1}}{1-q}, \tag{18.18}$$

for $q = a/z$ or $q = z/a$. This fundamental formula is best remembered by digesting its one-line proof: Let $S_n = 1 + q + \cdots + q^n$, then $qS_n = S_n + q^{n+1} - 1$. Solving for S_n yields (18.18).

Other basic algebraic formula are

$$(z+a)^2 = z^2 + 2za + a^2,$$
$$(z+a)^3 = z^3 + 3z^2a + 3za^2 + a^3,$$

and for any positive integer n we have the **binomial formula** (defining $0! = 1$)

$$(z+a)^n = \sum_{k=0}^{n} \binom{n}{k} z^{n-k}a^k = \sum_{k=0}^{n} \frac{n!}{k!(n-k)!} z^{n-k}a^k. \tag{18.19}$$

This can also be shown by induction. These classic algebraic formula are used to prove that $\mathrm{d}z^n/\mathrm{d}z = nz^{n-1}$ and that $\exp(z+a) = \exp(z)\exp(a)$.

18.4 Fundamental theorem of algebra

Argand also provided a constructive proof of the *fundamental theorem of algebra* that any polynomial of degree $n \geq 1$ has always at least one complex root, z_1 say. Since that root can be factored out, leaving a polynomial of degree $n-1$, this implies that a polynomial $P_n(z)$ of degree n in z can always be factored as a product of n polynomials of degree 1,

$$a_n z^n + \cdots + a_1 z + a_0 = a_n (z - z_1) \cdots (z - z_n), \tag{18.20}$$

where the $z_1, \ldots, z_n$ are the *zeroes* or *roots* of the polynomial, some of which can be repeated. For example, $z^2 + 1 = (z + i)(z - i)$ and $z^2 + 2z + 1 = (z + 1)(z + 1)$.

Argand starts form an arbitrary guess z_0 then takes a small step Δz such that

$$P(z_0 + \Delta z) = P(z_0) + Q_1 \, \Delta z + \cdots ,$$

where $P(z)$ is the polynomial of degree n and Q_1 is a polynomial of degree $n-1$ in z_0 that can be calculated by expanding the powers of $z = z_0 + \Delta z$ in $P(z)$ and collecting the Δz terms, and the dots $\cdots$ indicate terms of higher order in Δz. Argand then chooses $\Delta z = -\Delta t \, Q_1^* P(z_0)$ with Δt real, positive, and sufficiently small that the higher order terms are negligible and $0 \leq 1 - \Delta t |Q_1|^2 < 1$ such that

$$P(z_0 + \Delta z) = \left(1 - \Delta t |Q_1|^2\right) P(z_0) + \cdots$$

is smaller than $P(z_0)$ in magnitude. The process repeats from the new $z_0 := z_0 + \Delta z$, thereby converging directly toward a zero of $P(z)$. If $Q_1 = 0$, then $P(z_0 + \Delta z) = P(z_0) + Q_2 \, \Delta z^2 + \cdots$ and Argand picks $\Delta z^2 = -\Delta t^2 P(z_0) Q_2^*$, and similar to the next order if $Q_1 = Q_2 = 0$. In better notation, Q_1 is $\mathrm{d}P/\mathrm{d}z$ at z_0 and Argand proposed integrating the differential equation

$$\frac{\mathrm{d}z}{\mathrm{d}t} = -P(z) \left(\frac{\mathrm{d}P}{\mathrm{d}z} \right)^* \tag{18.21}$$

with t real, using Euler's method

$$z(t + \Delta t) = z(t) - \Delta t \, P(z(t)) \left(\frac{\mathrm{d}P}{\mathrm{d}z} \right)^*_{z=z(t)} .$$

This approach, with Argand's higher order modification, was studied by Hirsch and Smale[2] as a globally convergent Newton's method. Newton's method picks $\Delta t = 1/|Q_1|^2$ and converges quadratically near a root provided $Q_1 = \mathrm{d}P/\mathrm{d}z \neq 0$.

[2]Morris Hirsch and Stephen Smale, "On Algorithms for Solving $f(x) = 0$." *Communications on Pure and Applied Mathematics*, **32**, 281-312 (1979), §6: "A Sure-Fire Algorithm."

Exercises

(18.1) Complete the cube to transform the cubic equation $ax^3 + bx^2 + cx + d = 0$ into $z^3 + pz + q = 0$ for $z = x + b/(3a)$ with

$$p = \frac{3ac - b^2}{3a^2}, \quad q = \frac{27a^2 d - 9abc + 2b^2}{27a^3}.$$

(18.2) Show that $w = z^3 + pz + q$ with p, q real has three real roots when (18.3) applies.

(18.3) Show that the substitution $z = w - p/(3w)$ transforms $z^3 + pz + q = 0$ into a quadratic equation for w^3.

(18.4) Show/explain your work to evaluate

$$(2+3i)^*, \ |2+3i|, \ \mathrm{Im}(2+3i), \ \frac{1}{2+3i}, \ \arg(2+3i).$$

(18.5) Calculate $(1+i)/(2+3i)$. Find its magnitude and its angle.

(18.6) Consider two arbitrary complex numbers z_1 and z_2, not real, not imaginary, not 0. Sketch z_1, z_2, $z_1 + z_2$, $z_1 - z_2$, $z_1 + z_1^*$, and $z_1 - z_1^*$ in the complex plane.

(18.7) Consider $z_1 = -1 + i\,10^{-17}$ and $z_2 = -1 - i\,10^{-17}$. Calculate $|z_1 - z_2|$ and $\arg(z_1) - \arg(z_2)$. Are z_1 and z_2 close to each other? Are $\arg(z_1)$ and $\arg(z_2)$ close to each other? Show your work.

(18.8) Prove that $(z_1 + z_2)^* = z_1^* + z_2^*$ and $(z_1 z_2)^* = z_1^* z_2^*$ for any complex numbers z_1 and z_2.

(18.9) Prove that $|z_1 z_2| = |z_1||z_2|$ but $|z_1 + z_2| \le |z_1| + |z_2|$ for any complex numbers z_1 and z_2. When is $|z_1 + z_2| = |z_1| + |z_2|$?

(18.10) Show that $|z_1/z_2| = |z_1|/|z_2|$ and $\mathrm{angle}(z_1/z_2) = \mathrm{angle}(z_1) - \mathrm{angle}(z_2)$ modulo 2π. (Hint: use the corresponding results for product and $z_1/z_2 = z_1 z_2^* / |z_2|^2$.)

(18.11) Derive the formula (18.16) by solving the 2-by-2 linear system $z_2 z = z_1$ for (x, y) where $z = x + iy$.

(18.12) If a and b are arbitrary complex numbers, prove that $ab^* + a^*b$ is real and $ab^* - a^*b$ is imaginary.

(18.13) True or false: $(iz)^* z = 0$ since $i\,z$ is perpendicular to z. Explain.

(18.14) For vectors $\boldsymbol{a}$ and $\boldsymbol{b}$, the dot product $\boldsymbol{a} \cdot \boldsymbol{b} = 0$ when the vectors are perpendicular and the cross product $\boldsymbol{a} \times \boldsymbol{b} = 0$ when they are parallel. Show that two complex numbers a and b are parallel when $ab^* = a^*b$ and perpendicular when $ab^* = -a^*b$.

(18.15) Prove (18.17) and the binomial formula (18.19).

(18.16) Take a blank sheet of paper and derive the geometric sum formula in one line.

(18.17) Calculate $1 + z + z^2 + \cdots + z^{321}$ for $z = i$ and $z = 1 + i$. Explain your work.

(18.18) Calculate $1 - z + z^2 - z^3 + \cdots - z^{321}$ for $z = i$ and $z = 1 + i$. Explain your work.

(18.19) Let

$$a_n z^n + \cdots + a_1 z + a_0 = (z - z_1)(b_{n-1} z^{n-1} + \cdots + b_1 z + b_0)$$

where z is a variable and all a_k, b_l, z_1 are constants. Derive and solve the linear system of equations yielding the b_l's in terms of z_1 and the a_k's.

(18.20) Prove that if z is a complex root of a polynomial $a_n z^n + \cdots + a_1 z + a_0$ with real coefficients $a_0, \ldots, a_n$; then z^* is also a root.

(18.21) Newton's method to solve $f(z) = 0$ is to pick a guess z_0 then iterate

$$z_{n+1} = z_n - \frac{f(z_n)}{f'(z_n)}$$

for $n = 0, 1, \ldots$, where $f'(z) = \mathrm{d}f(z)/\mathrm{d}z$. Let z_* be a solution such that $f(z_*) = 0$. Use Taylor series to show quadratic convergence, that is

$$z_{n+1} - z_* = A(z_*)\,(z_n - z_*)^2 + \cdots,$$

where $(\cdots)$ are terms of higher order in $(z_n - z_*)$ that go to zero faster than $(z_n - z_*)^2$ as $z_n \to z_*$. Find $A(z_*)$.

(18.22) Try out Newton's and Argand's methods to find a root of $z^3 + 1 = 0$ starting from (a) $z_0 = i$, and (b) $z_0 = 0$. You will need to pick Δt and use Argand's modified Δz for the first step in (b). Discuss your choices and plot the iterates $z_0, z_1, z_2, \cdots$ in the complex plane $\mathbb{C}$.

Elementary Complex Functions

19

The algebraic operations defined for complex numbers, addition, subtraction, multiplication, and division, allow us to construct polynomial $f(z) = P_n(z)$ and rational $f(z) = P_n(z)/Q_m(z)$ functions of a complex variable $z = x + iy$, where $P_n(z)$ and $Q_m(z)$ are polynomials of degree n and m in z, respectively. Using the concept of series, we can take the limit $n \to \infty$ to define other "elementary" functions such as the exponential, the logarithm, roots, and trigonometric functions of a complex variable. These are the *elementary complex functions* considered in this chapter.

19.1 Limits and derivatives

The modulus $|z|$ allows for the definition of distance, then, limit, continuity, and differentiability. The *distance* between two complex numbers z_1 and z_2 is the modulus of their difference, $|z_1 - z_2|$, that is the euclidean distance between the complex numbers z_1 and z_2 in the complex plane. A complex number z_1 tends to a complex number z when $|z_1 - z| \to 0$.

Continuity

A function $f(z)$ is continuous at z if

$$\lim_{z_1 \to z} f(z_1) = f(z), \tag{19.1}$$

which means that $f(z_1)$ can be as close as we want to $f(z)$ by taking z_1 close enough to z, this is the "ϵ δ" definition of limit:

∀: for any; ∃: there exists.

$$\forall \epsilon > 0,\ \exists \delta > 0 \quad \text{s.t.} \quad |z_1 - z| < \delta \Rightarrow |f(z_1) - f(z)| < \epsilon.$$

For example, $f(z) = a_2 z^2 + a_1 z + a_0$ is continuous everywhere since from the triangle inequality

$$\begin{aligned} |f(z_1) - f(z)| &= |a_2(z_1^2 - z^2) + a_1(z_1 - z)| \\ &\le |a_2||z_1^2 - z^2| + |a_1||z_1 - z| = |z_1 - z|\,\big(|a_2||z_1 + z| + |a_1|\big). \end{aligned}$$

Thus, $|f(z_1) - f(z)| \to 0$ as $|z_1 - z| \to 0$. Find δ in exercise 19.2.

Differentiability

The *derivative* of a function $f(z)$ at z is

$$\frac{\mathrm{d}f(z)}{\mathrm{d}z} = \lim_{a\to 0} \frac{f(z+a) - f(z)}{a}, \tag{19.2}$$

where a is a complex number and $a \to 0$ means $|a| \to 0$. This limit must be the same no matter *how* $a \to 0$.

We can use the geometric sum formula (18.17) or the binomial formula (18.19) to deduce that

$$\frac{\mathrm{d}z^n}{\mathrm{d}z} = nz^{n-1}, \tag{19.3}$$

for any integer $n = 0, \pm 1, \pm 2, \ldots$, and we can define the antiderivative of z^n as $z^{n+1}/(n+1) + C$ for all integer $n \neq -1$. All the usual rules of differentiation—*derivative of a sum, product rule, quotient rule, chain rule,* etc.—hold for complex differentiation. The proofs proceed as for functions of one real variable.

So there is nothing special about complex derivatives, or is there? Consider the function $f(z) = \mathrm{Re}(z) = x$, the real part of z. What is its z-derivative? The differentiation rules are of no use here, so let's go back to the limit definition:

$$\frac{\mathrm{dRe}(z)}{\mathrm{d}z} = \lim_{a\to 0} \frac{\mathrm{Re}(z+a) - \mathrm{Re}(z)}{a} = \lim_{a\to 0} \frac{\mathrm{Re}(a)}{a} = ?! \tag{19.4}$$

What is that limit? If a is real, then $a = \mathrm{Re}(a)$, so the limit is 1, but if a is imaginary, then $\mathrm{Re}(a) = 0$ and the limit is 0. So there is no limit that holds for all $a \to 0$. The limit depends on *how* $a \to 0$, and we cannot define the z-derivative of $\mathrm{Re}(z)$. $\mathrm{Re}(z)$ is continuous everywhere, but nowhere z-differentiable! We will get back to this fundamental issue in Chapter 20. Here, (19.3) will suffice, together with series. We begin with the mother of all series.

19.2 Series

Geometric series

The *geometric series* is the limit of the geometric sum (18.18) as $n \to \infty$. It follows from (18.18), that the geometric series converges to $1/(1-q)$ if $|q| < 1$, and diverges if $|q| > 1$ (Fig. 19.1),

$$\sum_{n=0}^{\infty} q^n = 1 + q + q^2 + \cdots = \frac{1}{1-q}, \quad \text{iff} \quad |q| < 1. \tag{19.5}$$

Note that we have two different functions of q: on the left-hand side, we have the series $\sum_{n=0}^{\infty} q^n$, which only exists when $|q| < 1$; on the right-hand side, the function $1/(1-q)$, which is defined and differentiable everywhere except at $q = 1$. These two expressions, the geometric series and the function $1/(1-q)$, are identical in the disk $|q| < 1$, but they are not at all identical outside of that disk since the series does not make any sense (i.e. it diverges) outside of it. We leave it as an exercise to investigate what happens on the unit circle $|q| = 1$. Consider $q = 1$, $q = -1$, $q = i$, ...

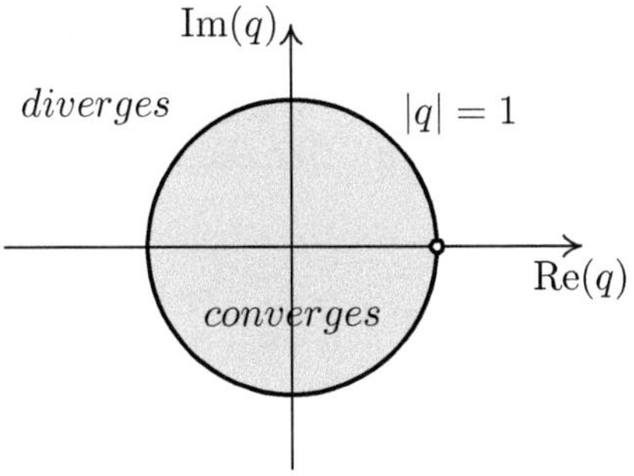

Fig. 19.1 The geometric series (18.18) $\sum_{n=0}^{\infty} q^n$ converges in the disk $|q| < 1$ where it equals $1/(1-q)$. The series diverges outside the unit disk although the function $1/(1-q)$ exists for any $q \neq 1$.

Ratio test

The geometric series leads to a useful test for convergence of the general series

$$\sum_{n=0}^{\infty} a_n = a_0 + a_1 + a_2 + \cdots, \tag{19.6}$$

understood as the limit of the partial sums $S_n = a_0 + a_1 + \cdots + a_n$ as $n \to \infty$. Any one of these finite partial sums exists but the infinite sum does not necessarily converge. For example, take $a_n = 1 \ \forall n$, then $S_n = n + 1$ but $S_n \to \infty$ as $n \to \infty$.

A *necessary* condition for convergence is that $a_n \to 0$ as $n \to \infty$ as you learned in calculus, but that is not sufficient. A *sufficient* condition for convergence is obtained by comparison to a geometric series. This leads to the *ratio rest*: the series (19.6) converges if

$$\lim_{n\to\infty} \frac{|a_{n+1}|}{|a_n|} = L < 1. \tag{19.7}$$

Indeed, if $L < 1$, then we can pick a q such that $L < q < 1$ and we can find a (sufficiently large) N such that $|a_{n+1}|/|a_n|$ is within $\epsilon = q - L$ of L, in other words such that $|a_{n+1}|/|a_n| < q$ for all $n \geq N$, so we can write

$$\begin{aligned} &|a_N| + |a_{N+1}| + |a_{N+2}| + |a_{N+3}| + \cdots \\ &\quad = |a_N| \left(1 + \frac{|a_{N+1}|}{|a_N|} + \frac{|a_{N+2}|}{|a_{N+1}|}\frac{|a_{N+1}|}{|a_N|} + \cdots\right) \\ &\quad < |a_N| \left(1 + q + q^2 + \cdots\right) = \frac{|a_N|}{1-q} < \infty. \end{aligned} \tag{19.8}$$

If $L > 1$, then we can reverse the proof (i.e. pick q with $1 < q < L$ and N such that $|a_{n+1}|/|a_n| > q \ \forall n \geq N$) to show that the series *diverges*. If $L = 1$, the ratio test does not determine convergence so the series may or may not converge.

Taylor series

A *power series* has the form

$$\sum_{n=0}^{\infty} c_n(z-a)^n = c_0 + c_1(z-a) + c_2(z-a)^2 + \cdots . \tag{19.9}$$

where the c_n's are complex coefficients and z and a are complex numbers. It is a series in powers of $(z-a)$. By the ratio test, the power series converges if

$$\lim_{n\to\infty}\left|\frac{c_{n+1}(z-a)^{n+1}}{c_n(z-a)^n}\right| = |z-a|\lim_{n\to\infty}\left|\frac{c_{n+1}}{c_n}\right| = \frac{|z-a|}{R} < 1, \tag{19.10}$$

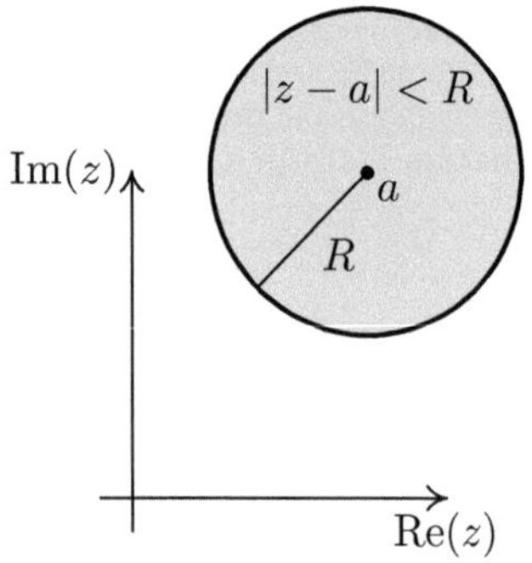

Fig. 19.2 A power series (19.9) converges in a disk $|z-a|<R$ and diverges outside of that disk. The radius of convergence R can be 0 or ∞.

where we have defined

$$\lim_{n\to\infty}\left|\frac{c_{n+1}}{c_n}\right| = \frac{1}{R}. \tag{19.11}$$

The power series converges if $|z-a|<R$. It diverges if $|z-a|>R$ (Fig. 19.2). Since $|z-a|=R$ is a circle of radius R centered at a, R is called the *radius of convergence* of the power series. R can be 0, ∞, or anything in between.

This geometric convergence inside a disk implies that power series can be differentiated (and integrated) term by term inside their disk of convergence. The disk of convergence of the derivative or integral series is the same as that of the original series. For instance, the geometric series $\sum_{n=0}^{\infty} z^n$ converges in $|z|<1$ and its term-by-term derivative $\sum_{n=0}^{\infty} nz^{n-1}$ does also, as shown by the ratio test.

If a power series converges to $f(z)$ say, then its derivatives are the term-by-term derivatives of the power series

$$f^{(k)}(z) = \frac{\mathrm{d}^k f(z)}{\mathrm{d}z^k} = \sum_{n=k}^{\infty} n(n-1)\cdots(n-k+1)\, c_n(z-a)^{n-k}.$$

Evaluating each of these series at $z=a$, only the first term $n=k$ survives and

$$\left.\frac{\mathrm{d}^k f(z)}{\mathrm{d}z^k}\right|_{z=a} = k(k-1)\cdots 1\ c_k = k!\ c_k,$$

yielding all the coefficients of the power series as

$$c_n = \frac{1}{n!}\left.\frac{\mathrm{d}^n f(z)}{\mathrm{d}z^n}\right|_{z=a} = \frac{f^{(n)}(a)}{n!}.$$

In other words, the convergent power series (19.9) can be written

$$\begin{aligned} f(z) &= f(a) + f'(a)(z-a) + \frac{f''(a)}{2}(z-a)^2 + \cdots \\ &= \sum_{n=0}^{\infty} \frac{f^{(n)}(a)}{n!}(z-a)^n, \end{aligned} \tag{19.12}$$

where $n! = n(n-1)\cdots 1$ is the factorial of n, with $0! = 1$ by convenient definition. The power series (19.12) is the *Taylor series* of $f(z)$ about $z = a$.

The equality between $f(z)$ and its Taylor series is only valid if the series converges. The geometric series

$$\frac{1}{1-z} = 1 + z + z^2 + \cdots = \sum_{n=0}^{\infty} z^n \tag{19.13}$$

is the Taylor series of $f(z) = 1/(1-z)$ about $z = 0$, but the function $1/(1-z)$ exists and is infinitely differentiable everywhere except at $z = 1$, while the series $\sum_{n=0}^{\infty} z^n$ only exists in the unit circle $|z| < 1$. The convergence of the series about $z = 0$ is limited by the singularity of $1/(1-z)$ at $z = 1$ since a power series always converges in a disk, Fig. 19.2.

Several useful Taylor series are more easily derived from the geometric series (19.5), (19.13) than from the general formula (19.12) (even if you really like calculating lots of derivatives!). For instance,

$$\frac{1}{1-z^2} = 1 + z^2 + z^4 + \cdots = \sum_{n=0}^{\infty} z^{2n} \tag{19.14}$$

$$\frac{1}{1+z} = 1 - z + z^2 - \cdots = \sum_{n=0}^{\infty} (-z)^n \tag{19.15}$$

$$\ln(1+z) = z - \frac{z^2}{2} + \cdots = \sum_{n=0}^{\infty} \frac{(-1)^n z^{n+1}}{n+1} \tag{19.16}$$

The last series is obtained by integrating both sides of the previous equation and matching at $z = 0$ to determine the constant of integration since $\ln 1 = 0$. These series converge only in $|z| < 1$ while the functions on the left-hand side exist for (much) larger domains of z.

19.3 Transcendentals

We have defined $i^2 = -1$ and made geometric sense of it. Now what is 2^i? i^i?! We can make sense of such complex powers but we first need to define the exponential function $\exp z$ and its inverse the (natural) logarithm, $\ln z$.

The complex versions of the Taylor series definition for the exponential, cosine, and sine functions

$$\exp z = 1 + z + \frac{z^2}{2} + \cdots = \sum_{n=0}^{\infty} \frac{z^n}{n!} \tag{19.17}$$

$$\cos z = 1 - \frac{z^2}{2} + \frac{z^4}{4!} \cdots = \sum_{n=0}^{\infty} (-1)^n \frac{z^{2n}}{(2n)!} \tag{19.18}$$

$$\sin z = z - \frac{z^3}{3!} + \frac{z^5}{5!} \cdots = \sum_{n=0}^{\infty} (-1)^n \frac{z^{2n+1}}{(2n+1)!} \tag{19.19}$$

[1]They are *entire* functions.

converge in the *entire* complex plane[1] for any z with $|z| < \infty$ as is readily checked from the ratio test. These convergent series serve as the definition of these functions for complex arguments.

We can verify the usual properties of these functions from the series expansions. In general, we can integrate and differentiate series term by term inside the disk of convergence of the power series. Doing so for $\exp z$ shows that it is equal to its derivative

$$\frac{\mathrm{d}\exp z}{\mathrm{d}z} = \frac{\mathrm{d}}{\mathrm{d}z}\left(\sum_{n=0}^{\infty}\frac{z^n}{n!}\right) = \sum_{n=1}^{\infty}\frac{z^{n-1}}{(n-1)!} = \exp z, \tag{19.20}$$

meaning that $\exp z$ is the solution of the complex differential equation

$$\frac{\mathrm{d}w}{\mathrm{d}z} = w \quad \text{with} \quad w(0) = 1.$$

Likewise the series (19.18) for $\cos z$ and (19.19) for $\sin z$ imply

$$\frac{\mathrm{d}\cos z}{\mathrm{d}z} = -\sin z, \qquad \frac{\mathrm{d}\sin z}{\mathrm{d}z} = \cos z. \tag{19.21}$$

Taking another derivative of both sides shows that $w_1(z) = \cos z$ and $w_2(z) = \sin z$ are solutions of the second order differential equation

$$\frac{\mathrm{d}^2 w}{\mathrm{d}z^2} = -w,$$

with $(w_1(0), w_1'(0)) = (1, 0)$ and $(w_2(0), w_2'(0)) = (0, 1)$.

Another slight *tour de force* with the series for $\exp(z)$ is to use the binomial formula (18.19) to obtain

$$\exp(z+a) = \sum_{n=0}^{\infty}\frac{(z+a)^n}{n!} = \sum_{n=0}^{\infty}\sum_{k=0}^{n}\binom{n}{k}\frac{z^k a^{n-k}}{n!}$$
$$= \sum_{n=0}^{\infty}\sum_{k=0}^{n}\frac{z^k a^{n-k}}{k!(n-k)!}. \tag{19.22}$$

The double sum is over the triangular region $0 \le n \le \infty$, $0 \le k \le n$ in n, k space. If we interchange the order of summation, we have to sum over $k = 0 \to \infty$ and $n = k \to \infty$ (sketch it!). Changing variables to k and $m = n - k$, instead of k and n, the range of m is 0 to ∞ as that of k, and the double sum reads

$$\exp(z+a) = \sum_{k=0}^{\infty}\sum_{m=0}^{\infty}\frac{z^k a^m}{k!m!} = \left(\sum_{k=0}^{\infty}\frac{z^k}{k!}\right)\left(\sum_{m=0}^{\infty}\frac{a^m}{m!}\right)$$
$$= \exp(z)\exp(a). \tag{19.23}$$

This is the principal property of the exponential function and we verified it from its series expansion (19.17) for general complex arguments z and a. It implies that if we define as before

$$e = \exp(1) = 1 + 1 + \frac{1}{2} + \frac{1}{6} + \frac{1}{24} + \frac{1}{120} + \cdots = 2.71828..., \tag{19.24}$$

then $\exp(n) = [\exp(1)]^n = e^n$ and $\exp(1) = [\exp(1/2)]^2$. Thus, $\exp(1/2) = e^{1/2}$ etc. so we can still identify $\exp(z)$ as the number e to the *complex power* z and (19.23) is the regular algebraic rule for exponents: $e^{z+a} = e^z e^a$. In particular,

$$\exp z = e^z = e^{x+iy} = e^x e^{iy}, \tag{19.25}$$

where e^x is the regular exponential function of a real variable x but e^{iy} is the exponential of a pure imaginary number. We can make sense of this from the series (19.17), (19.18) and (19.19) to obtain the very useful formula

$$\begin{aligned} e^{iz} &= \cos z + i \sin z, \\ e^{-iz} &= \cos z - i \sin z, \end{aligned} \tag{19.26}$$

which yield

$$\cos z = \frac{e^{iz} + e^{-iz}}{2}, \qquad \sin z = \frac{e^{iz} - e^{-iz}}{2i} \tag{19.27}$$

that could serve as the complex definitions of $\cos z$ and $\sin z$. These hold for any complex number z. For z real, this is *Euler's formula* usually written in terms of a real angle φ,

$$e^{i\varphi} = \cos\varphi + i \sin\varphi. \tag{19.28}$$

This is one of the most important formulas in all of mathematics! It reduces all of trigonometry to algebra among other things. For instance, $e^{i(\alpha+\beta)} = e^{i\alpha}e^{i\beta}$, implies that

$$\begin{aligned} \cos(\alpha+\beta) + i\sin(\alpha+\beta) &= (\cos\alpha + i\sin\alpha)(\cos\beta + i\sin\beta) \\ &= (\cos\alpha\cos\beta - \sin\alpha\sin\beta) \\ &\quad + i(\sin\alpha\cos\beta + \sin\beta\cos\alpha) \end{aligned} \tag{19.29}$$

Equating real and imaginary parts[2] yields two trigonometric identities in one swoop, the angle sum formulas from high school trigonometry.

[2] When α, β are real. More generally, equating parts even and odd in $(\alpha+\beta)$.

Polar representation

Introducing polar coordinates such that $(x, y) = (r\cos\varphi, r\sin\varphi)$, and using Euler's formula (19.28), any complex number can be written

$$z = x + iy = re^{i\varphi}, \tag{19.30}$$

where $r = |z| \geq 0$ and $\varphi = \arg(z) + 2k\pi$ is the *phase* of z, with $k = 0, \pm1, \pm2, \ldots$, an integer. The expression $z = x + iy$ is the cartesian or rectangular form of z, and $z = re^{i\varphi}$ is its *polar form.*

Replacing φ by $\varphi \pm 2\pi$ does not change z. However, the argument $\arg(z)$ is a function of z and is unique for every z. For instance, we can define $0 \leq \arg(z) < 2\pi$, or $-\pi < \arg(z) \leq \pi$. These are just two among an infinite number of possible definitions. Although computer functions (Fortran, C, Matlab, etc.) make a specific choice (typically the second one), that choice may not be suitable in some cases. The proper choice is problem dependent (see exercise 19.35 for example). The phase φ can be continued smoothly as we loop around $z = 0$, but $\arg(z)$ is necessarily discontinuous. For example, if we define $0 \le \arg(z) < 2\pi$, then a point moving about the unit circle at angular velocity ω will have a phase $\varphi = \omega t$ but $\arg(z) = \omega t \mod 2\pi$, which is discontinuous at $\omega t = 2k\pi$.

The cartesian representation $x + iy$ of a complex number z is perfect for addition/subtraction, but the polar representation $re^{i\varphi}$ is more convenient for multiplication and division since

$$z_1 z_2 = r_1 e^{i\varphi_1} r_2 e^{i\varphi_2} = r_1 r_2 e^{i(\varphi_1+\varphi_2)}, \tag{19.31}$$

$$\frac{z_1}{z_2} = \frac{r_1 e^{i\varphi_1}}{r_2 e^{i\varphi_2}} = \frac{r_1}{r_2} e^{i(\varphi_1-\varphi_2)}. \tag{19.32}$$

19.4 Logs and powers

The power series expansion of functions is remarkably powerful and closely tied to the theory of functions of a complex variable. Many complex functions beyond $\exp z$ can be defined by series. A priori, it doesn't seem very general. How, for instance, could we expand $f(z) = 1/z$ into a series in *positive* powers of z,

$$\frac{1}{z} = a_0 + a_1 z + a_2 z^2 + \cdots \quad ??$$

We cannot expand $1/z$ in powers of z but we can expand in powers of $z - a$ for any $a \neq 0$. That Taylor series is obtained easily using the geometric series, again,

$$\frac{1}{z} = \frac{1}{a + (z-a)} = \frac{1}{a} \frac{1}{1 + \left(\frac{z-a}{a}\right)} = \sum_{n=0}^{\infty} (-1)^n \frac{(z-a)^n}{a^{n+1}}. \tag{19.33}$$

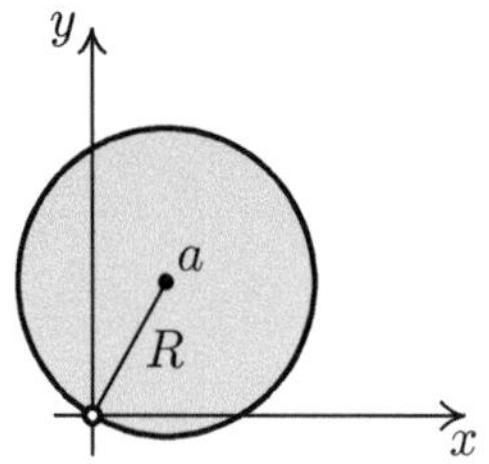

Fig. 19.3 The disk of convergence of the Taylor series (19.33) for $1/z$ is limited by the singularity at $z = 0$.

Thus, we can expand $1/z$ in powers of $z - a$ for any $a \neq 0$. That (geometric) series converges in the disk $|z - a| < |a|$. This is the disk of radius $|a|$ centered at a (Fig. 19.3). By taking a sufficiently far away from 0, that disk where the series converges can be made as big as one wants but it can never include the origin, which of course is the sole *singular point* of the function $1/z$. Integrating (19.33) for $a = 1$ term by term yields

$$\ln z = \sum_{n=0}^{\infty} (-1)^n \frac{(z-1)^{n+1}}{n+1} \tag{19.34}$$

as the antiderivative of $1/z$ that vanishes at $z = 1$. This looks nice; however, that series only converges for $|z - 1| < 1$. We need a better definition that works for a larger domain in the z-plane.

The Taylor series definition of the exponential $\exp(z) = \sum_{n=0}^{\infty} z^n/n!$ is very good. It converges for all z's; it led us to Euler's formula $e^{i\varphi} = \cos\varphi + i\sin\varphi$ and the key property of the exponential, namely $\exp(a + b) = \exp(a)\exp(b)$ (where a and b are any *complex* numbers), from which we deduced $\exp(z) \equiv e^z$ with $e = \exp(1) = 2.71828\ldots$, and $e^z = e^{x+iy} = e^x e^{iy}$.

Log as inverse of the exponential

Given z we want to define the function $\ln z$ as the inverse of the exponential. That is, we want to find a complex number $w = \ln z$ such that $e^w = z$. From the cartesian form $w = u + iv$, where $u = \mathrm{Re}(w)$ and $v = \mathrm{Im}(w)$ are real, we obtain

$$e^w = e^{u+iv} = e^u e^{iv} = z = |z|e^{i \arg z}, \tag{19.35}$$

yielding $e^u = |z|$ and $e^{iv} = e^{i \arg z}$; that is, $u = \ln |z|$ and $v = \arg z + 2k\pi$, where $|z| \geq 0$ is a *positive real number*, so $\ln |z|$ is the usual natural log of a positive real number, and $k = 0, \pm 1, \pm 2, \cdots$ is an integer. We have found the roots

$$e^w = z \Leftrightarrow w = \ln |z| + i \arg(z) + 2k\pi\, i,$$

but there is an infinite number of w's for each z, one for every k. We can take any *one* of those solutions as our definition of $\ln z$, in particular

$$w_0 = \ln z = \ln |z| + i \arg(z). \tag{19.36}$$

This definition is unique since we assume that $\arg z$ is uniquely defined in terms of z. However, different definitions of $\arg z$ lead to different definitions of $\ln z$.

For example, if $0 \leq \arg(z) < 2\pi$, then $\ln(-3i) = \ln 3 + i3\pi/2$, but if we define instead $-\pi < \arg(z) \leq \pi$, then $\ln(-3i) = \ln 3 - i\pi/2$. Note that we can now take logs of negative numbers, $\ln(-3) = \ln 3 + i \arg(-3)$, and the log of a product is the sum of the logs, modulo $2\pi i$,

$$\ln(ab) = \ln a + \ln b \mod 2\pi i \tag{19.37}$$

since $\exp \ln(ab) = ab = \exp(\ln a + \ln b + 2k\pi i)$. In particular,

$$\ln z = \ln(|z|e^{i \arg z}) = \ln |z| + \ln e^{i \arg z} = \ln |z| + i \arg z,$$

and $e^{\ln z} = z$ for any z but $\ln e^z = z \mod 2\pi i$.

Complex powers

As for functions of real variables, we can now define general complex powers such as 2^i in terms of the complex log and the complex exponential

$$a^b \triangleq e^{b \ln a} = e^{b \ln |a|} e^{ib \arg(a)}. \tag{19.38}$$

For example, $2^i = e^{i \ln 2} = \cos(\ln 2) + i \sin(\ln 2)$. Note that b is complex in general, so $e^{b \ln |a|}$ is not necessarily real. This definition involves $\arg(a)$ and different definitions for arg yield different values for a^b, in general.

We can now define the complex power functions

$$z^a = e^{a \ln z} = e^{a \ln |z|} e^{ia \arg(z)} \tag{19.39}$$

and the complex exponential functions

$$a^z = e^{z \ln a} = e^{z \ln |a|} e^{iz \arg(a)}. \tag{19.40}$$

These functions are well-defined once $\arg(\cdot)$ is defined. Different definitions for $\arg(a)$ imply definitions for a^b that do *not* simply differ by an *additive multiple* of $2\pi i$ as was the case for $\ln z$. For example

$$(-1)^i = e^{i \ln(-1)} = e^{-\arg(-1)} = e^{-\pi - 2k\pi}$$

for some k, so the various possible definitions of $(-1)^i$ will differ by a *multiplicative integer power* of $e^{-2\pi}$.

Roots

The **fundamental theorem of algebra** §18.4 states that an nth-order polynomial $a_n z^n + a_{n-1} z^{n-1} + \cdots + a_1 z + a_0$ with $a_n \neq 0$ can always be factored into a product of first-order polynomials as

$$a_n z^n + a_{n-1} z^{n-1} + \cdots + a_0 = a_n (z - z_1) \cdots (z - z_n). \tag{19.41}$$

The complex numbers $z_1, \ldots, z_n$ are the *roots* or *zeroes* of the polynomial. Some of these roots can be repeated as for the polynomial $2z^2 - 4z + 2 = 2(z-1)^2$.

For example, the equation $2z^2 - 2 = 0$ has two real roots $z = \pm 1$ and

$$2z^2 - 2 = 2(z-1)(z+1).$$

The equation $3z^2 + 3 = 0$ has no real roots; however, it has two imaginary roots $z = \pm i$ and

$$3z^2 + 3 = 3(z-i)(z+i).$$

In general, we cannot find the roots analytically, we need to use a numerical method such as Newton's or Argand's method (§18.4), except for very special polynomials such as $z^n - a$.

Roots of a The equation

$$z^n = a,$$

can be solved analytically for any complex a. For integer $n > 0$, it must have n roots. We might be tempted to write the solution as

$$z = a^{1/n} = e^{(\ln a)/n} = e^{(\ln |a|)/n} e^{i \arg(a)/n},$$

but this is only *one root* whose value depends on the definition of $\arg(a)$. The nth root function, $a^{1/n}$, must have a unique value for a given a but here we are looking for *all* the z's such that $z^n = a$. Using the polar representations $z = |z| e^{i \arg(z)}$ and $a = |a| e^{i \arg(a)}$ yields

$$z^n = a \;\Leftrightarrow\; |z|^n e^{i n \arg(z)} = |a| e^{i \arg(a)}.$$

Now $a = b$ implies that $|a| = |b|$, but

$$\arg(a) = \arg(b) + 2k\pi.$$

The magnitudes are equal $|a| = |b|$, but the angles are equal only up to a multiple of 2π. Indeed, $e^{i\pi/2} = i = e^{-i3\pi/2}$, for example, and $\pi/2 \neq -3\pi/2$ but $\pi/2 = -3\pi/2 + 2\pi$. The equation $z^n = a$ thus implies that

$$|z|^n = |a|, \qquad n \arg(z) = \arg(a) + 2k\pi,$$

where $k = 0, \pm 1, \pm 2, \ldots$ is any integer. Solving for the real $|z| \geq 0$ and $\arg(z)$ yields

$$|z| = |a|^{1/n}, \qquad \arg(z) = \frac{\arg(a)}{n} + k\frac{2\pi}{n}. \tag{19.42}$$

When n is a positive integer, this yields n distinct values for $\arg(z)$, yielding n distinct values for z. The roots are equispaced by angle $2\pi/n$ on the circle of radius $|a|^{1/n}$ in the complex plane.

For example,

$$z^3 = 1 \Leftrightarrow \begin{cases} z = 1, \\ z = e^{i2\pi/3} = (-1 + i\sqrt{3})/2, \\ z = e^{i4\pi/3} = (-1 - i\sqrt{3})/2 = e^{-i2\pi/3}, \end{cases}$$

the three roots are on the unit circle, equispaced by $2\pi/3$ (Fig. 19.4), and

$$z^3 - 1 = (z - 1)(z - e^{i2\pi/3})(z - e^{-i2\pi/3}).$$

Likewise, the five roots of $z^5 = e^{i\pi/3}$ are $z = e^{i\pi/15}e^{i2k\pi/5}$, for $k = 0, \pm 1, \pm 2, \ldots$; they lie on the unit circle equispaced by $2\pi/5$ (Fig. 19.5).

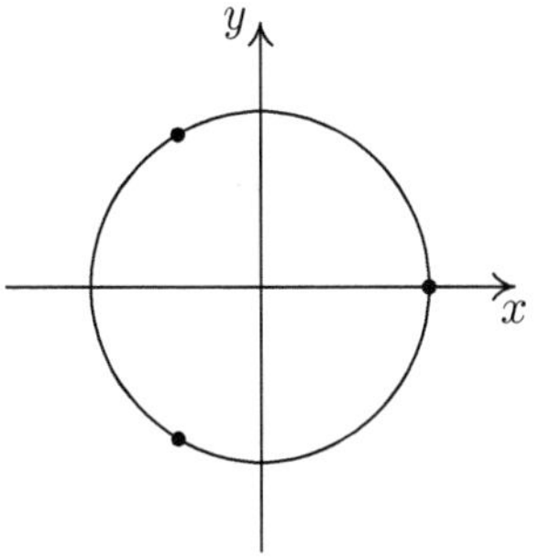

Fig. 19.4 The three roots of $z^3 = 1$.

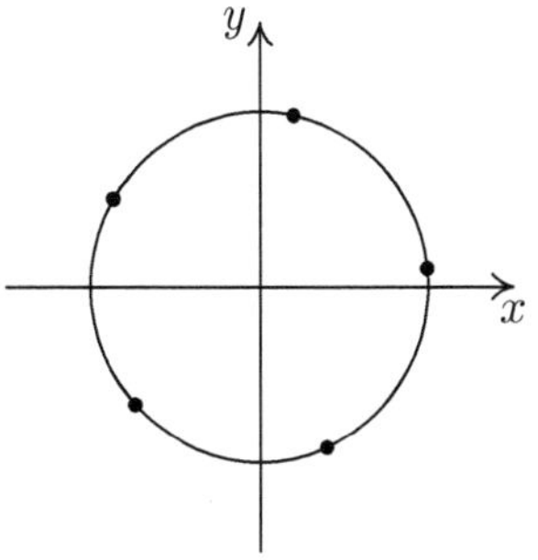

Fig. 19.5 The five roots of $z^5 = e^{i\pi/3}$.

Exercises

(19.1) Prove that the functions $f(z) = \mathrm{Re}(z)$ and $f(z) = z^*$ are continuous everywhere.

(19.2) For $f(z) = a_2 z^2 + a_1 z + a_0$ show that

$$|f(z_1) - f(z)| < \epsilon$$

when

$$|z_1 - z| < \delta = \frac{-|a_1| + \sqrt{|a_1|^2 + 4\epsilon|a_2|}}{2|a_2|}.$$

Does this hold also when $a_2 \to 0$? What $\delta(\epsilon)$ do you expect for $a_2 = 0$?

(19.3) Prove formula (19.3) from the limit definition of the derivative (a) using (18.17), (b) using (18.19).

(19.4) Prove that (19.3) also applies to negative integer powers $z^{-n} = 1/z^n$ from the limit definition of the derivative.

(19.5) Investigate the existence of $\mathrm{d}f/\mathrm{d}z$ from the limit definition for $f(z) = \mathrm{Im}(z)$ and $f(z) = |z|$.

(19.6) Explain why the domain of convergence of a power series is always a disk (possibly infinitely large), not an ellipse or a square or any other shape. (Anything can happen on the boundary of the disk: weak (algebraic) divergence or convergence, perpetual oscillations, etc.; recall the geometric series.)

(19.7) Show that if a function $f(z) = \sum_{n=0}^{\infty} c_n(z-a)^n$ for all z's within the (nonzero) disk of convergence of the power series, then the c_n's must have the form provided by formula (19.12).

(19.8) What is the Taylor series of $1/(1-z)$ about $z=0$? What is its radius of convergence? Does the series converge at $z=-2$? Why not?

(19.9) What is the Taylor series of the function $1/(1+z^2)$ about $z=0$? What is its radius of convergence? Use a computer or calculator to test the convergence of the series inside and outside its disk of convergence.

(19.10) What is the Taylor series of $1/z$ about $z=2$? What is its radius of convergence? (Hint: $z = a + (z-a)$.)

(19.11) What is the Taylor series of $1/(1+z)^2$ about $z=0$?

(19.12) Use series to compute the number e to four digits. How many terms do you need?

(19.13) Use series to compute $\exp(i)$, $\cos(i)$, and $\sin(i)$ to four digits.

(19.14) Express $\cos(2+3i)$ and $\sin(2+3i)$ in terms of cos, sin and exp of real numbers.

(19.15) The *hyperbolic* cosine and sine are defined as

$$\cosh z = 1 + \frac{z^2}{2} + \frac{z^4}{4!} \cdots = \sum_{n=0}^{\infty} \frac{z^{2n}}{(2n)!}$$
$$\sinh z = z + \frac{z^3}{3!} + \frac{z^5}{5!} \cdots = \sum_{n=0}^{\infty} \frac{z^{2n+1}}{(2n+1)!}.$$

(a) Use the series definition to show that

$$\frac{\mathrm{d}\cosh z}{\mathrm{d}z} = \sinh z, \qquad \frac{\mathrm{d}\sinh z}{\mathrm{d}z} = \cosh z,$$

and that $\cosh z$ and $\sinh z$ are both solutions of the differential equation

$$\frac{\mathrm{d}^2 f}{\mathrm{d}z^2} = f$$

with initial conditions $(f(0), f'(0)) = (1,0)$ for $\cosh z$, and $(f(0), f'(0)) = (0,1)$ for $\sinh z$ (here $f' = df/dz$).

(b) Show that

$$\begin{aligned} e^z &= \cosh z + \sinh z, \\ e^{-z} &= \cosh z - \sinh z, \\ \cosh z &= \frac{e^z + e^{-z}}{2} = \cos(iz), \\ \sinh z &= \frac{e^z - e^{-z}}{2} = \frac{\sin(iz)}{i}. \end{aligned} \qquad (19.43)$$

(c) Show that

$$\cos z = \cos(x+iy) = \cos x \cosh y - i \sin x \sinh y,$$
$$\sin z = \sin(x+iy) = \sin x \cosh y + i \sinh y \cos x.$$

(d) Sketch e^x, e^{-x}, $\cosh x$ and $\sinh x$ for real x.

(19.16) Prove that $\cos^2 z + \sin^2 z = 1$ for any complex number z.

(19.17) Prove that $\cosh^2 z - \sinh^2 z = 1$ for any complex number z.

(19.18) Prove that the following well-known identities hold for complex a and b also:

$$\cos(a+b) = \cos a \cos b - \sin a \sin b,$$
$$\sin(a+b) = \sin a \cos b + \sin b \cos a,$$
$$\cosh(a+b) = \cosh a \cosh b + \sinh a \sinh b,$$
$$\sinh(a+b) = \sinh a \cosh b + \sinh b \cosh a.$$

(19.19) Is $e^{-iz} = \left(e^{iz}\right)^*$?

(19.20) Show that

$$f_1(z) = 1 + \frac{z^3}{2\cdot 3} + \frac{z^6}{(2\cdot 3)\cdot(5\cdot 6)} + \cdots = \sum_{k=0}^{\infty} 3^k \left(\frac{1}{3}\right)_k \frac{z^{3k}}{(3k)!},$$
$$f_2(z) = z + \frac{z^4}{3\cdot 4} + \frac{z^7}{(3\cdot 4)\cdot(6\cdot 7)} + \cdots = \sum_{k=0}^{\infty} 3^k \left(\frac{2}{3}\right)_k \frac{z^{3k+1}}{(3k+1)!},$$

are two linearly independent solutions of *Airy's equation*

$$\frac{\mathrm{d}^2 f}{\mathrm{d}z^2} = z\, f, \qquad (19.44)$$

where $(z)_k = z(z+1)(z+2)\cdots(z+k-1)$ with $(z)_0 = 1$, is the *Pochhammer symbol.* Prove that the series converge in the *entire* complex plane.

(19.21) Prove the identity $\dfrac{1}{1-e^{ix}} = \dfrac{ie^{-ix/2}}{2\sin(x/2)}$.

(19.22) Use Euler's formula and geometric sums to derive the identities

$$\frac{1}{2} + \cos x + \cos 2x + \cdots + \cos nx = \frac{\sin(n+\frac{1}{2})x}{2\sin\frac{1}{2}x},$$
$$\sin x + \sin 2x + \cdots + \sin nx = \frac{\cos\frac{1}{2}x - \cos(n+\frac{1}{2})x}{2\sin\frac{1}{2}x}.$$

The first sum appeared in our quick look at the heat equation (16.124), and these identities are important in the study of waves and Fourier series. Use computer graphing software to plot the sums for a few increasing n. What happens when $n \to \infty$? (Hint: derive both identities at once as in (19.29).)

(19.23) Show that if p is real with $|p| < 1$ then

$$1 + 2p\cos x + 2p^2 \cos 2x + 2p^3 \cos 3x + \cdots = \frac{1-p^2}{1+p^2-2p\cos x}. \quad (19.45)$$

Otherwise, the series diverges if $|p| > 1$. What happens when $p = 1$? This is the *Poisson kernel* for a disk in 2D, times a 2π factor.

(19.24) The formula (19.29) leads to the well-known *double* and *triple* angle formula

$$\begin{aligned} \cos 2\theta &= 2\cos^2\theta - 1, \\ \sin 2\theta &= 2\sin\theta\cos\theta, \\ \cos 3\theta &= 4\cos^3\theta - 3\cos\theta, \\ \sin 3\theta &= \sin\theta(4\cos^2\theta - 1). \end{aligned} \quad (19.46)$$

These formulas suggest that $\cos n\theta$ is a polynomial of degree n in $\cos\theta$ and that $\sin n\theta$ is $\sin\theta$ times a polynomial of degree $n-1$ in $\cos\theta$. The polynomial for $\cos n\theta$ in powers of $\cos\theta$ is the *Chebyshev* polynomial $T_n(x)$ of degree n in x such that

$$T_n(\cos\theta) = \cos n\theta.$$

Thus,

$$T_0(x) = 1,\ T_1(x) = x,\ T_2(x) = 2x^2 - 1,$$
$$T_3(x) = 4x^3 - 3x, \ldots.$$

Derive explicit formulas for those polynomials for any n. These Chebyshev polynomials are important in numerical calculations; your calculator uses them to evaluate the "elementary" functions such as e^x, $\cos x$, $\sin x$, etc. (Hint: use Euler's formula

$$e^{in\theta} = \cos n\theta + i\sin n\theta = (e^{i\theta})^n = (\cos\theta + i\sin\theta)^n$$

and the binomial formula.)

(19.25) Evaluate $\ln(-1)$ and $\ln i$ *without* a calculator.

(19.26) Analyze and evaluate e^i and i^e.

(19.27) Find all the roots of $z^5 = e^{i\pi/3}$. Do your roots match those in Fig. 19.5?

(19.28) Find all the roots, visualize and locate them in the complex plane, and factor the corresponding polynomial (i) $z^2 + 1 = 0$, (ii) $z^3 + 1 = 0$, (iii) $z^4 + 1 = 0$, (iv) $z^2 = i$, (v) $2z^2 + 5z + 2 = 0$, and (vi) $2\,z^5 + 1 + i = 0$.

(19.29) Investigate the solutions of the equation $z^b = 1$ when (i) b is a rational number (that is, $b = p/q$ with p, q integers), (ii) when b is irrational, e.g. $b = \pi$, and (iii) when b is complex, e.g. $b = 1 + i$. Make a sketch showing the solutions in the complex plane.

(19.30) Revisit roots of $z^3 + pz + q$ obtained from w^3 in (18.4) through $z = w - p/(3w)$. There are six w's, but only three z's?! Explain. Illustrate with $z^3 - z$.

(19.31) If a and b are complex numbers, what is wrong with saying that if $w = e^{a+ib} = e^a e^{ib}$, then $|w| = e^a$ and $\arg(w) = b + 2k\pi$? Isn't that what we did in (19.35)?

(19.32) For $z = x + iy = re^{i\varphi}$ with x, y, r, φ real, $r \ge 0$ and $-\pi < \varphi \le \pi$, show that

$$\sqrt{z} \triangleq \sqrt{r}\,e^{i\varphi/2} = \sqrt{\frac{x + \sqrt{x^2+y^2}}{2}} + i\,\mathrm{sgn}(y)\sqrt{\frac{-x + \sqrt{x^2+y^2}}{2}},$$

where $\mathrm{sgn}(y) = 1$ if $y \ge 0$, and $\mathrm{sgn}(y) = -1$ if $y < 0$.

(19.33) Consider $w = \sqrt{z}$ for $z = re^{i\varphi}$ with $r \ge 0$ fixed but varying φ. Sketch $u = \mathrm{Re}(w)$ and $v = \mathrm{Im}(w)$ as functions of φ from $\varphi = 0 \to 4\pi$ for

$$\text{(i)}\ \sqrt{z} \triangleq \sqrt{r}\,e^{i\varphi/2}, \qquad \text{(ii)}\ \sqrt{z} \triangleq \sqrt{r}\,e^{i\,\arg(z)/2}.$$

Show that (i) is continuous as a function of φ but doubly valued as a function of z, while (ii) is single valued in z but discontinuous in φ.

(19.34) Use Matlab, Python or your favorite complex calculator to evaluate $\sqrt{z^2-1}$ and $\sqrt{z-1}\sqrt{z+1}$ for $z = -1 + i$. Since $z^2 - 1 = (z-1)(z+1)$, shouldn't these square roots equal each other? Explain.

(19.35) Show that $\sqrt{w} \triangleq \sqrt{|w|}e^{i(\arg w)/2}$ for $w = z^2 - 1$ with $-\pi < \arg(\cdot) \le \pi$ as computed by Matlab, Python, etc. is discontinuous across $z = x$ with $|x| < 1$ and $z = iy$ with $|y| > 0$. However, the alternative definition

$$\begin{aligned} \sqrt{z^2-1} &\triangleq \sqrt{|z^2-1|}\,e^{i\varphi_1/2}e^{i\varphi_2/2} \\ &= \sqrt{z-1}\sqrt{z+1}, \end{aligned} \quad (19.47)$$

for $\varphi_1 = \arg(z-1)$, $\varphi_2 = \arg(z+1)$ is discontinuous across $z = x$ with $|x| < 1$ but continuous across $z = iy$ with $|y| > 0$, (x, y real as usual). In particular, $\sqrt{z^2-1} \to z$ as $|z| \to \infty$.

(19.36) The inverse cosine is *one* solution w to $\cos w = z$. Use Euler's formula $\cos w = (e^{iw} + e^{-iw})/2 = z$ to derive a quadratic equation for e^{iw} and obtain

$$w = -i \ln\left(z + \sqrt{z^2 - 1}\right).$$

Use a similar approach to derive or verify that the inverse sine, cosh, and sinh functions can be defined as

$$w = -i \ln\left(iz + \sqrt{1 - z^2}\right) \quad \Rightarrow \quad \sin w = z,$$

$$w = \ln\left(z + \sqrt{z^2 - 1}\right) \quad \Rightarrow \quad \cosh w = z,$$

$$w = \ln\left(z + \sqrt{z^2 + 1}\right) \quad \Rightarrow \quad \sinh w = z.$$

Functions of a Complex Variable

20

20.1 Visualization of complex functions

A function $w = f(z)$ of a complex variable $z = x + iy$ has complex values $w = u + iv$, where u, v are real. The real and imaginary parts of $w = f(z)$ are functions of the real variables x and y

$$f(z) = u(x, y) + i\,v(x, y). \tag{20.1}$$

For example, $w = z^2 = (x + iy)^2$ is

$$w = z^2 = (x^2 - y^2) + i\,2xy \tag{20.2}$$

with a real part $u = x^2 - y^2$ and an imaginary part $v = 2xy$.

For real functions of one real variable, $y = f(x)$, we made an xy plot to visualize the function. In complex calculus, x and y are independent variables and $w = f(z)$ corresponds to *two* real functions of *two* real variables $\text{Re}(f(z)) = u(x, y)$ and $\text{Im}(f(z)) = v(x, y)$. One way to visualize $f(z)$ is to make a 3D plot with u as the *height* above the (x, y) plane. We can do the same for $v(x, y)$. A prettier idea is to *color* the surface $u = u(x, y)$ in the 3D space (x, y, u) by the value of $v(x, y)$ as in Fig. 20.1.

The left-side graph in Fig. 20.1 shows $\text{Re}(w) = u = x^2 - y^2 = \text{Re}(z^2)$. It is the upright parabola $u = x^2$ along the real axis $y = 0$, but $u = -y^2$ along the imaginary axis, $x = 0$. The right-side graph of $w^2 = z$ shows the two roots of that quadratic equation, that is, $w = \pm\sqrt{z}$ computed as $z = re^{i\varphi}$ and $w = \sqrt{r}\,e^{i\varphi/2}$ but for a double cover of the z plane, that is, $\varphi = 0 \to 4\pi$, not just 2π. This shows that those two roots are smoothly connected as one circles around the *branch point* $z = 0$ where both roots collide. The standard $\sqrt{z}$ function is

$$\sqrt{z} = \sqrt{|z|}\,e^{i(\arg z)/2}$$

with $-\pi < \arg(z) \leq \pi$. This selects the root with positive real part $u \geq 0$, that is the upper surface on the $w^2 = z$ graph.

Such 3D visualizations yield beautiful pictures—and illustrate the concept of a *Riemann surface*—but are only useful for relatively simple functions. It is usually more manageable to use 2D *contour plots* of $u(x, y)$ and $v(x, y)$, as in Fig. 20.2 that shows contour plots of $u = \text{Re}(z^2)$ and $v = \text{Im}(z^2)$.

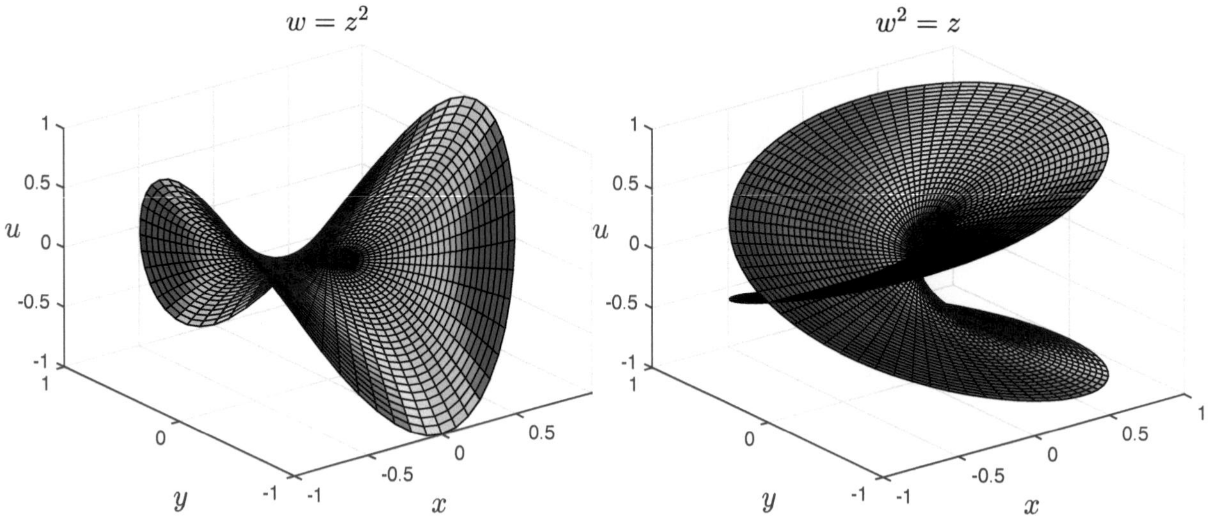

Fig. 20.1 3D visualization of $w = z^2$ and $w^2 = z$ with $u = \mathrm{Re}(w)$ as the height over the (x, y) plane and $v = \mathrm{Im}(w)$ as the color.

20.2 Cauchy–Riemann equations

A function $f(z)$ of a complex variable $z = x + iy$ is a special function $f(x + iy)$ of two real variables (x, y), and consists of two real functions $u(x, y)$ and $v(x, y)$ of two real variables

$$f(z) = f(x + iy) = u(x, y) + i\, v(x, y). \tag{20.3}$$

The example $z^2 = (x^2 - y^2) + i\,(2xy)$ was discussed and illustrated earlier. Other examples are

$$e^z = e^{x+iy} = e^x \cos y + i\, e^x \sin y$$

for which $u = e^x \cos y$ and $v = e^x \sin y$, and

$$\frac{1}{z} = \frac{1}{x + iy} = \frac{x}{x^2 + y^2} - i \frac{y}{x^2 + y^2}$$

that has $u = x/(x^2 + y^2)$ and $v = -y/(x^2 + y^2)$.

Now, if $f(z)$ is z-differentiable, then

$$\begin{aligned} \frac{\mathrm{d}f(z)}{\mathrm{d}z} &= \lim_{\Delta z \to 0} \frac{f(z + \Delta z) - f(z)}{\Delta z} \\ &= \lim_{\Delta z \to 0} \frac{f\,(x + \Delta x + i(y + \Delta y)) - f(x + iy)}{\Delta x + i\Delta y} \end{aligned}$$

has the same value no matter how $\Delta z = \Delta x + i\Delta y \to 0$. Picking $\Delta z = \Delta x$ with $\Delta y = 0$ gives

$$\frac{\mathrm{d}f}{\mathrm{d}z} = \frac{\partial f}{\partial x},$$

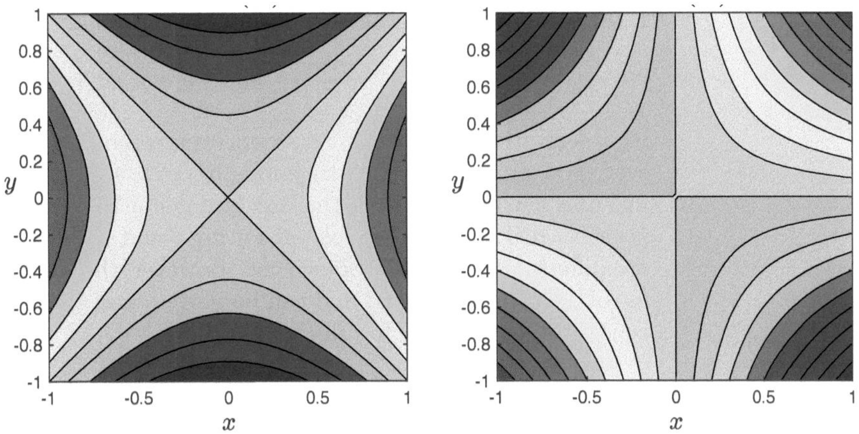

Fig. 20.2 2D visualization of $w = z^2$ with contour plots of $u = \mathrm{Re}(w) = x^2 - y^2$ on the left and $v = \mathrm{Im}(w) = 2xy$ on the right. Note the saddle structures in both u and v.

while picking $\Delta z = i\Delta y$ with $\Delta x = 0$ yields

$$\frac{\mathrm{d}f}{\mathrm{d}z} = \frac{1}{i}\frac{\partial f}{\partial y}.$$

Thus,

$$\frac{\mathrm{d}f}{\mathrm{d}z} = \frac{\partial f}{\partial x} = -i\,\frac{\partial f}{\partial y}, \tag{20.4}$$

and since $f = u + iv$ this is

$$\frac{\mathrm{d}f}{\mathrm{d}z} = \frac{\partial u}{\partial x} + i\frac{\partial v}{\partial x} = -i\frac{\partial u}{\partial y} + \frac{\partial v}{\partial y}. \tag{20.5}$$

Equating the real and imaginary parts of that last equation yields the *Cauchy–Riemann* equations

$$\begin{cases} \dfrac{\partial u}{\partial x} = \dfrac{\partial v}{\partial y}, \\ \dfrac{\partial u}{\partial y} = -\dfrac{\partial v}{\partial x}, \end{cases} \tag{20.6}$$

relating the partial derivatives of the real and imaginary parts of a differentiable function of a complex variable $f(z) = u(x,y) + i\,v(x,y)$. This derivation shows that the Cauchy–Riemann equations are *necessary* conditions on $u(x,y)$ and $v(x,y)$ if $f(z)$ is differentiable in a neighborhood of z. If $\mathrm{d}f/\mathrm{d}z$ exists, then the Cauchy–Riemann equations (20.6) *necessarily* hold. They are also *sufficient* as shown after some examples.

Example 1 The function $f(z) = z^2$ has $u = x^2 - y^2$ and $v = 2xy$. Its z-derivative $\mathrm{d}z^2/\mathrm{d}z = 2z$ exists everywhere and the Cauchy–Riemann equations (20.6) are satisfied everywhere since

$$\begin{cases} \dfrac{\partial u}{\partial x} = 2x = \dfrac{\partial v}{\partial y}, \\ \dfrac{\partial u}{\partial y} = -2y = -\dfrac{\partial v}{\partial x}. \end{cases}$$

Example 2 The function $f(z) = \ln z = \ln|z| + i \arg z$ has

$$u = \ln\sqrt{x^2 + y^2} \quad \text{and} \quad v = \text{atan2}(y, x).$$

Its z-derivative $\mathrm{d}\ln z/\mathrm{d}z = 1/z$ exists everywhere except at $z = 0$. However, $v = \arg z = \text{atan2}(y, x)$ jumps by $\pm 2\pi$ as z crosses the negative real axis $x < 0$, $y = 0$, with the standard definition $-\pi < \arg z \leq \pi$. The negative real axis is a *branch cut* for $\arg z$ and $\ln z$; those functions are not differentiable on that branch cut. However, that branch cut is a matter of definition of $\arg(z)$ and can be moved, and the Cauchy–Riemann equations are satisfied for all $(x, y) \neq (0, 0)$,

$$\begin{cases} \dfrac{\partial u}{\partial x} = \dfrac{x}{x^2 + y^2} = \dfrac{\partial v}{\partial y}, \\ \dfrac{\partial u}{\partial y} = \dfrac{y}{x^2 + y^2} = -\dfrac{\partial v}{\partial x}. \end{cases}$$

Example 3 The function $f(z) = z^* = x - iy$ has $u = x$, $v = -y$. The Cauchy–Riemann equations (20.6) do *not* hold anywhere for $f(z) = z^* = x - iy$ since

$$\frac{\partial u}{\partial x} = 1 \neq \frac{\partial v}{\partial y} = -1.$$

Its z-derivative $\mathrm{d}z^*/\mathrm{d}z$ does *not* exist anywhere. Indeed, from the limit definition of the derivative,

$$\frac{\mathrm{d}z^*}{\mathrm{d}z} = \lim_{a \to 0} \frac{(z^* + a^*) - z^*}{a} = \lim_{a \to 0} \frac{a^*}{a} = e^{-2i\alpha} \tag{20.7}$$

where $\Delta z = a = |a|e^{i\alpha}$, so the limit is different for every α. If a is real, then $\alpha = 0$ and the limit is 1, but if a is imaginary, then $\alpha = \pi/2$ and the limit is -1. If $|a| = e^{-\alpha}$, then $a \to 0$ in a *logarithmic* spiral as $\alpha \to \infty$, but there is no limit in that case since $e^{-2i\alpha}$ keeps spinning around the unit circle without ever converging to anything. We cannot define a unique limit as $a \to 0$, so z^* is not differentiable with respect to z.

Proof of sufficiency The Cauchy–Riemann equations (20.6) are also *sufficient* for $f(x, y) = u(x, y) + iv(x, y)$ to be differentiable with respect to $z = x + iy$. Functions $u(x, y)$ and $v(x, y)$ that satisfy the Cauchy–Riemann equations are called *conjugate harmonic functions*. They are the real and imaginary parts of a differentiable function of $x + iy$, $f(x + iy)$.

To prove this, we need to show that the z-derivative of the function $f(x, y) \triangleq u(x, y) + iv(x, y)$, that is,

$$\frac{\mathrm{d}f(x, y)}{\mathrm{d}z} \triangleq \lim_{\Delta z \to 0} \frac{f(x + \Delta x, y + \Delta y) - f(x, y)}{\Delta z},$$

exists independently of how the limit $\Delta z = \Delta x + i\Delta y \to 0$ is taken. Using differential notation to simplify the writing, with $w = f(x, y) = u + iv$, we have $\mathrm{d}w = \mathrm{d}u + i\mathrm{d}v$ with $\mathrm{d}z = \mathrm{d}x + i\mathrm{d}y$ and

$$\frac{\mathrm{d}w}{\mathrm{d}z} = \frac{\mathrm{d}u + i\mathrm{d}v}{\mathrm{d}x + i\mathrm{d}y}, \tag{20.8}$$

where $\mathrm{d}u = u(x+\mathrm{d}x, y+\mathrm{d}y) - u(x,y)$ and $\mathrm{d}v = v(x+\mathrm{d}x, y+\mathrm{d}y) - v(x,y)$. By the chain rule,[1]

$$\mathrm{d}u = \frac{\partial u}{\partial x}\mathrm{d}x + \frac{\partial u}{\partial y}\mathrm{d}y, \qquad \mathrm{d}v = \frac{\partial v}{\partial x}\mathrm{d}x + \frac{\partial v}{\partial y}\mathrm{d}y.$$

Substituting those differentials in (20.8) and rearranging terms yields

$$\frac{\mathrm{d}w}{\mathrm{d}z} = \left(\frac{\partial u}{\partial x} + i\frac{\partial v}{\partial x}\right)\frac{\mathrm{d}x}{\mathrm{d}x + i\mathrm{d}y} + \left(\frac{\partial u}{\partial y} + i\frac{\partial v}{\partial y}\right)\frac{\mathrm{d}y}{\mathrm{d}x + i\mathrm{d}y} \tag{20.9}$$

in the sense of limits, that is $\Delta z = \Delta x + i\Delta y \to 0$. For arbitrary $u(x,y)$ and $v(x,y)$, this limit will depend on how $\Delta z \to 0$. However, if $u(x,y)$ and $v(x,y)$ satisfy the Cauchy–Riemann equation (20.6), then we can use them to eliminate $v(x,y)$, and (20.9) becomes

$$\begin{aligned}\frac{\mathrm{d}w}{\mathrm{d}z} &= \left(\frac{\partial u}{\partial x} - i\frac{\partial u}{\partial y}\right)\frac{\mathrm{d}x}{\mathrm{d}x + i\mathrm{d}y} + \left(\frac{\partial u}{\partial y} + i\frac{\partial u}{\partial x}\right)\frac{\mathrm{d}y}{\mathrm{d}x + i\mathrm{d}y} \\ &= \left(\frac{\partial u}{\partial x} - i\frac{\partial u}{\partial y}\right)\frac{\mathrm{d}x + i\mathrm{d}y}{\mathrm{d}x + i\mathrm{d}y} = \frac{\partial u}{\partial x} - i\frac{\partial u}{\partial y},\end{aligned} \tag{20.10}$$

and the value of $\mathrm{d}w/\mathrm{d}z$ is thus independent of how $\Delta z \to 0$. □

[1] This requires continuity of the derivatives, for example,

$$\lim_{\Delta y \to 0} \frac{\partial u(x, y + \Delta y)}{\partial x} = \frac{\partial u(x,y)}{\partial x}.$$

Laplace's equation

Another important consequence of z-differentiability and the Cauchy–Riemann equations (20.6) is that the real and imaginary parts of a differentiable function $f(z) = u(x,y) + i\,v(x,y)$ both satisfy *Laplace's equation* (exercise 20.1)

$$\frac{\partial^2 u}{\partial x^2} + \frac{\partial^2 u}{\partial y^2} = 0 = \frac{\partial^2 v}{\partial x^2} + \frac{\partial^2 v}{\partial y^2}. \tag{20.11}$$

Since both the real and imaginary parts of $f(z) = u(x,y) + iv(x,y)$ satisfy Laplace's equation, this is also true for $f(z)$ seen as a function of (x,y) and

$$\nabla^2 f(x+iy) = \left(\frac{\partial^2}{\partial x^2} + \frac{\partial^2}{\partial y^2}\right) f(x+iy) = 0. \tag{20.12}$$

Differentiability of a function $f(z)$ of a complex variable z implies the Cauchy–Riemann and Laplace's equations. In fact, z-differentiability of $f(z)$ in a neighborhood of z (that is, $f(z)$ is *holomorphic*) implies that $f(z)$ is infinitely differentiable in the neighborhood of z and that its Taylor series converges in a disk in that neighborhood. This follows from Cauchy's formula as will be shown later. A function whose Taylor series converges in the neighborhood of a point is called *analytic* in that neighborhood. Since z-differentiability in a neighborhood implies analyticity in that neighborhood, the word *analytic* tends to be used interchangeably with *holomorphic*.

20.3 Conformal mapping

The Cauchy–Riemann equations connecting the real and imaginary parts of a complex differentiable function $f(z) = u(x, y) + i\,v(x, y)$ have great geometric implications illustrated in Fig. 20.3 and figures that follow.

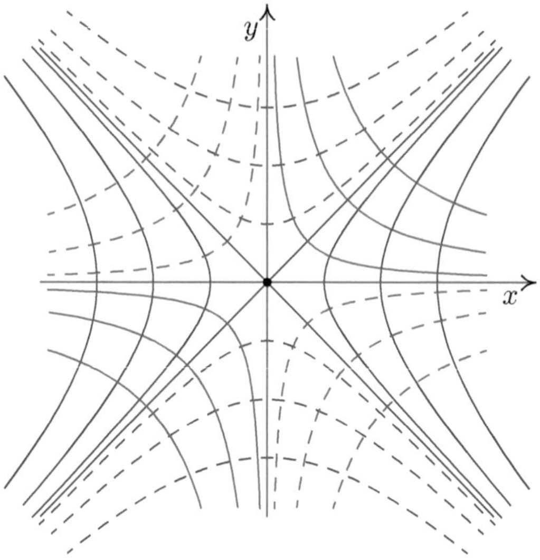

Fig. 20.3 For $z^2 = (x^2 - y^2) + i\,2xy$, the contours of $u(x, y) = x^2 - y^2 = 0, \pm1, \pm4, \pm9$ (blue) are *hyperbolas* with asymptotes $y = \pm x$. The contours of $v(x, y) = 2xy = 0, \pm1, \pm4, \pm9$ (red) are also *hyperbolas* but with asymptotes $x = 0$ and $y = 0$. Solid is positive, dashed is negative. The u and v contours intersect everywhere at 90°, except at $z = 0$ where $dz^2/dz = 2z = 0$.

Orthogonality

The Cauchy–Riemann equations (20.6) imply *orthogonality of the contours* of $u(x, y) = \text{Re}(f(z))$ and $v(x, y) = \text{Im}(f(z))$ wherever $\mathrm{d}f(z)/\mathrm{d}z$ exists but does not vanish. Indeed, consider the gradients ∇u and ∇v at a point (x, y). The Cauchy–Riemann equations (20.6) imply that those two gradients are perpendicular to each other,

$$\nabla u \cdot \nabla v = \frac{\partial u}{\partial x}\frac{\partial v}{\partial x} + \frac{\partial u}{\partial y}\frac{\partial v}{\partial y} = \frac{\partial v}{\partial y}\frac{\partial v}{\partial x} - \frac{\partial v}{\partial x}\frac{\partial v}{\partial y} = 0. \tag{20.13}$$

Since gradients are always perpendicular to their respective isocontours, ∇u is perpendicular to the u-contour and ∇v to the v-contour through that point, the orthogonality of ∇u and ∇v implies orthogonality of the contours (level curves) of u and v.

Orthogonality of the u and v contours for $u = \text{Re}(w)$ and $v = \text{Im}(w)$ with $w = f(z)$ holds wherever $\mathrm{d}w/\mathrm{d}z$ exists except at critical points where $\mathrm{d}w/\mathrm{d}z = 0$ and $\nabla u = \nabla v = 0$. For the example $w = z^2$ in Fig. 20.3, the contours are orthogonal everywhere except at $z = 0$ where $\mathrm{d}z^2/\mathrm{d}z = 2z = 0$. The gradients vanish at that point, which is a saddle point for both functions u and v.

The real and imaginary parts, $u(x, y)$ and $v(x, y)$, of any z-differentiable complex function $f(z)$ therefore provide orthogonal coordinates in the (x, y) plane. But the Cauchy–Riemann equations also imply that the gradients of u and v are not only orthogonal, but also have equal magnitudes. This implies that such (u, v) coordinates are not only orthogonal but also *conformal*, as discussed hereafter.

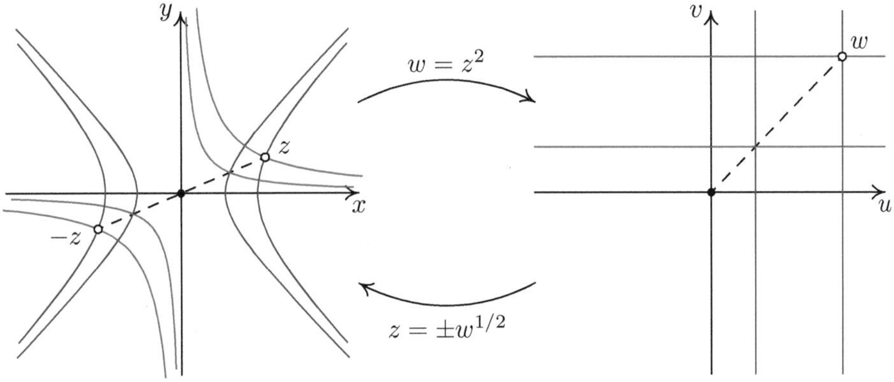

Fig. 20.4 $w = z^2$ as a mapping from the z-plane to the w-plane. There are two possible inverse maps $z = \pm\sqrt{w}$ since there are two z's for every w. The angles between corresponding curves are preserved. For example, the dashed line in the w-plane intersects the vertical and horizontal lines at $\pi/4$ and likewise for its pre-image in the z-plane intersecting the hyperbolas corresponding to vertical and horizontal lines in the w-plane.

Conformal mapping

We can visualize the function $w = f(z) = u(x, y) + iv(x, y)$ as a *map* from the complex plane $z = x + iy$ to the complex plane $w = u + iv$. For example, the map $z \to w = z^2$ is illustrated in Fig. 20.4.

In cartesian form, that map is $z = x + iy \to w = u + iv = (x^2 - y^2) + i(2xy)$. The vertical line $u = u_0$ in the w-plane is the image of the hyperbola $x^2 - y^2 = u_0$ in the z-plane. The horizontal line $v = v_0$ in the w-plane is the image of the hyperbola $2xy = v_0$ in the z-plane. Every point z in the z-plane has a single image w in the w-plane; however, the latter has two *pre-images* z and $-z$ in the z-plane. Indeed, the inverse functions are $z = \pm w^{1/2}$, as illustrated in Fig. 20.4.

The curves corresponding to constant u and constant v intersect at $90°$ in *both* planes, except at $z = w = 0$. That is the orthogonality of u and v contours, but the dotted radial line intersects the blue and red curves at $45°$ in *both* planes, for example. In fact *any angle* between any two curves in the z-plane is preserved in the w-plane *except at* $z = w = 0$ where they are *doubled* from the z- to the w-plane.

In polar form $z = re^{i\varphi} \to w = r^2 e^{i2\varphi}$. This means that every radial line from the origin with angle φ from the x-axis in the z-plane is mapped to a radial line from the origin with angle 2φ from the u-axis in the w-plane.

Angle preservation

That is the conformal map property of z-differentiable complex functions $f(z)$. If $f(z)$ is z-differentiable, then the mapping $w = f(z)$ *preserves all angles* at all z's where $\mathrm{d}f(z)/\mathrm{d}z \neq 0$ exists *and* does not vanish.

To show preservation of angles in general, consider three neighboring points in the z-plane: z, $z + \Delta z_1$, and $z + \Delta z_2$ (Fig. 20.5). We are interested in seeing what happens to the angle between the two vectors Δz_1 and Δz_2. If $\Delta z_1 = |\Delta z_1| e^{i\varphi_1}$ and $\Delta z_2 = |\Delta z_2| e^{i\varphi_2}$, then the angle between those two vectors is $\alpha = \varphi_2 - \varphi_1$ and this is the angle of the ratio

$$\frac{\Delta z_2}{\Delta z_1} = \frac{|\Delta z_2|}{|\Delta z_1|} e^{i(\varphi_2 - \varphi_1)} = \frac{|\Delta z_2|}{|\Delta z_1|} e^{i\alpha}.$$

The point z is mapped to the point $w = f(z)$. The points $z + \Delta z_1$ and $z + \Delta z_2$ are mapped to

$$\begin{aligned} w_1 &= f(z + \Delta z_1) \simeq f(z) + f'(z)\Delta z_1 + \frac{1}{2} f''(z)\Delta z_1^2 + \cdots \\ w_2 &= f(z + \Delta z_2) \simeq f(z) + f'(z)\Delta z_2 + \frac{1}{2} f''(z)\Delta z_2^2 + \cdots, \end{aligned} \tag{20.14}$$

respectively, where $f' = \mathrm{d}f/\mathrm{d}z$, $f'' = \mathrm{d}^2 f/\mathrm{d}z^2$, etc.

In general, a line segment $t \in [0, 1] \in \mathbb{R} \to z(t) = z + t\,\Delta z_1 \in \mathbb{C}$ in the z plane is mapped to a curve $w(t) = f(z(t))$ in the w plane. The angle between those curves is the angle between the tangent to those curves at their intersection point w. Thus, we are interested in the limit $\Delta z_1 \to 0$ and $\Delta z_2 \to 0$, with fixed ratio $|\Delta z_2|/|\Delta z_1|$ and fixed angles φ_1, φ_2.

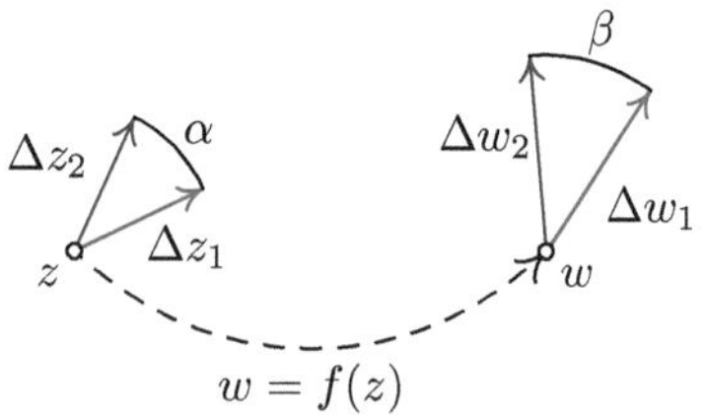

Fig. 20.5 $w = f(z)$ preserves angles wherever $0 < |\mathrm{d}f/\mathrm{d}z| < \infty$.

The angle between the tangents to the mapped curves at point w in the w plane is the limit of the angle between the secant vectors $\Delta w_1 = f(z + \Delta z_1) - f(z)$ and $\Delta w_2 = f(z + \Delta z_2) - f(z)$, which is the angle β of the ratio

$$\frac{\Delta w_2}{\Delta w_1} = \frac{|\Delta w_2|}{|\Delta w_1|} e^{i\beta}.$$

From the Taylor series expansions about z (20.14), this ratio

$$\frac{\Delta w_2}{\Delta w_1} = \frac{f'(z)\Delta z_2 + \frac{1}{2} f''(z)\Delta z_2^2 + \cdots}{f'(z)\Delta z_1 + \frac{1}{2} f''(z)\Delta z_1^2 + \cdots} \longrightarrow \frac{\Delta z_2}{\Delta z_1} \tag{20.15}$$

as Δz_1 and $\Delta z_2 \to 0$, with $|\Delta z_2|/|\Delta z_1|$ fixed, provided $f'(z) \neq 0$. Hence, in that limit,

$$\frac{\Delta w_2}{\Delta w_1} = \frac{|\Delta w_2|}{|\Delta w_1|} e^{i\beta} \to \frac{\Delta z_2}{\Delta z_1} = \frac{|\Delta z_2|}{|\Delta z_1|} e^{i\alpha} \tag{20.16}$$

implying that $|\Delta w_2|/|\Delta w_1| \to |\Delta z_2|/|\Delta z_1|$ and $\beta \to \alpha$, so angles are preserved.

Singular points

The analysis above requires that the Taylor approximations be valid in a neighborhood of point z, and that $f'(z)\Delta z$ dominates as $\Delta z \to 0$.

Hence, $f'(z)$ must exist and not vanish. If $f'(z)$ exists but $f'(z) = 0$, then angles are *not* preserved. At such points z, we need to go to the next order in the Taylor expansions to figure out what happens to angles. That is determined by the first nonzero term in the Taylor expansions.

If $f'(z) = 0$ but $f''(z) \neq 0$, then in the limit Δz_1, $\Delta z_2 \to 0$, with $|\Delta z_2|/|\Delta z_1|$ fixed, (20.15) yields

$$\frac{\Delta w_2}{\Delta w_1} = \frac{|\Delta w_2|}{|\Delta w_1|} e^{i\beta} \to \left(\frac{\Delta z_2}{\Delta z_1}\right)^2 = \frac{|\Delta z_2|^2}{|\Delta z_1|^2} e^{i2\alpha}$$

that is, $\beta \to 2\alpha$ so the angles are *doubled* at those z's. For example, $f(z) = z^2$ has $f'(z) = 2z$ that exists everywhere but vanishes at $z = 0$ but $f'' = 2 \neq 0$ for all z. Thus, the mapping $w = z^2$ preserves all angles—angles between any two curves in the z-plane will be the same as the angles between the image curves in the w-plane, *except* at $z = 0$ where the angles will be *doubled* in the w-plane.

If $f'(z) = f''(z) = 0$ but $f'''(z) \neq 0$, then in the same limit, (20.15) gives

$$\frac{\Delta w_2}{\Delta w_1} = \frac{|\Delta w_2|}{|\Delta w_1|} e^{i\beta} \to \left(\frac{\Delta z_2}{\Delta z_1}\right)^3 = \frac{|\Delta z_2|^3}{|\Delta z_1|^3} e^{i3\alpha}$$

so $\beta \to 3\alpha$ and the angles are *tripled* at those z's. For instance, $w = z^3$ has $w' = 3z^2$, $w' = 6z$, and $w''' = 6$, its derivative exists everywhere but vanishes together with its second derivative at $z = 0$. Thus $w = z^3$ preserves all angles except at $z = 0$ where the angles will be tripled in the w-plane since the third derivative does not vanish.

In general, if $f'(z) = f''(z) = \cdots = f^{(n-1)}(z) = 0$ but $f^{(n)}(z) \neq 0$, then angles at z will be multiplied by n at point $w = f(z)$.

A mapping that preserves all angles is called **conformal**. Analytic functions $f(z)$ provide conformal mappings between z and $w = f(z)$ at all points where $f'(z) \neq 0$. Reversing the role of z and w, analytic functions $z = g(w) = x(u, v) + iy(u, v)$ thus provide conformal coordinates (u, v) for domains in the (x, y)-plane. Figure 20.6 shows conformal coordinates for a Joukowski airfoil obtained by mapping the rectangular domain $(u, v) = [0, 1] \times [0, 2\pi)$ using a shifted and rotated exponential map $w = u + iv \to \zeta = \zeta_c + (1 - \zeta_c)e^w$ followed by a Joukowski map $\zeta \to z = \zeta + 1/\zeta$,

$$w = u + iv \;\to\; \zeta = \zeta_c + (1 - \zeta_c)e^w \;\to\; z = \zeta + \frac{1}{\zeta}. \tag{20.17}$$

The angles are preserved everywhere from $w = u + iv \to \zeta \to z = x + iy$ except at $\zeta = 1$ where $\mathrm{d}z/\mathrm{d}\zeta = 1 - \zeta^2 = 0$, corresponding to $w = 0$, and the sharp trailing edge of the Joukowski airfoil at $z = 2$. The other singular point, $\zeta = -1$, is inside the $u = 0$ circle $\zeta = \zeta_c + (1 - \zeta_c)e^{iv}$ and is mapped to $z = -2$, which is inside the Joukowski airfoil as long as $\mathrm{Re}(\zeta_c) < 0$.

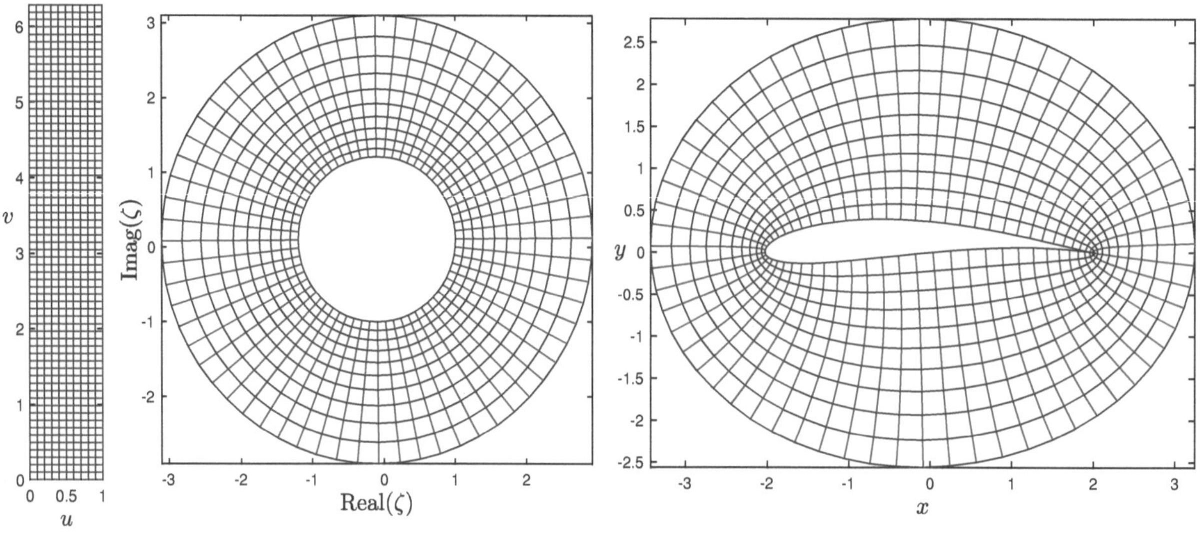

Fig. 20.6 Conformal map of $w = u + iv$ to $\zeta = \zeta_c + (1 - \zeta_c)e^w$ to a Joukowski airfoil $z = x + iy = \zeta + 1/\zeta$ for $\zeta_c = -0.1 + 0.1i$. Red: v contours. Blue: u contours. Innermost circle ($u = 0$) in the ζ-plane is centered at $\zeta = \zeta_c$ with radius $|1 - \zeta_c| > 1$.

Solving Laplace's equation by conformal mapping

Another application of conformal mapping is to solve Laplace's equation in a given z domain by conformally mapping that domain to another domain $z \to Z = f(z)$ where the corresponding solution of Laplace's equation is known, say $F(Z) = \Phi(X, Y) + i\Psi(X, Y)$. The solution(s) of Laplace's equation in the original domain is then simply $F(f(z))$.

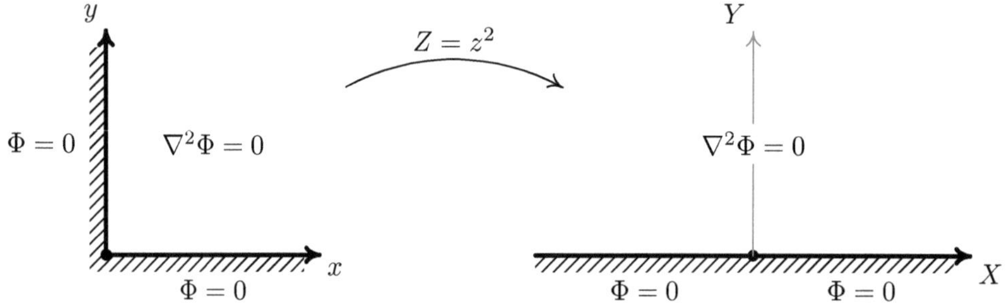

Fig. 20.7 Solving Laplace's equation in the first quadrant with $\Phi = 0$ on $xy = 0$ by mapping it to the upper half plane where the corresponding solution is trivial.

For example, to find a solution $\Phi(x, y)$ of Laplace's equation in the first quadrant $x \geq 0$, $y \geq 0$ with $\Phi = 0$ on $x = 0$ and on $y = 0$, we can map the $z = x + iy$ quadrant to the $Z = X + iY$ upper half plane with $Z = z^2$ (Fig. 20.7). The corresponding solution $\Phi(X, Y)$ in that upper half plane, that is, a solution of

$$\left(\frac{\partial^2}{\partial X^2} + \frac{\partial^2}{\partial Y^2}\right)\Phi = 0$$

with $\Phi = 0$ along $Y = 0$, is simply $\Phi = Y = \mathrm{Im}(Z)$. The solution in the original quadrant is then $\Phi = Y(x, y) = \mathrm{Im}(z^2) = 2xy$. Other examples are given in problems 20.11 – 20.14 in §20.2 and Fig. 20.14.

As another example, the complex velocity potential $w(z)$ for incompressible, irrotational flow around the cylinder of radius R centered at $z = 0$ in the complex z-plane is (exercise 20.9)

$$w(z) = V_\infty \left(z + \frac{R^2}{z} \right) + i\frac{\Gamma}{2\pi} \ln \frac{z}{R}, \qquad (20.18)$$

where the velocity $\boldsymbol{V} \to V_\infty \hat{\mathbf{x}}$ far from the cylinder, and Γ is the clockwise velocity circulation; that is, $\Gamma = \oint_{\mathcal{C}} \boldsymbol{V} \cdot \mathrm{d}\boldsymbol{r}$ over any simple clockwise closed loop $\mathcal{C}$ around the cylinder. The complex potential

$$w(z) = \phi(x, y) + i\,\psi(x, y)$$

has a real part $\phi(x, y)$ that is the velocity potential whose gradient is the velocity field $\boldsymbol{V} = \boldsymbol{\nabla}\phi$, guaranteeing irrotationality $\boldsymbol{\nabla} \times \boldsymbol{V} = 0$. The imaginary part $\psi(x, y)$ is the streamfunction such that the velocity $\boldsymbol{V} = \boldsymbol{\nabla} \times \psi \mathbf{e}_3$, guaranteeing incompressibility $\boldsymbol{\nabla} \cdot \boldsymbol{V} = 0$. The contours of the streamfunction $\psi(x, y)$ are the flow streamlines (Fig. 20.8).

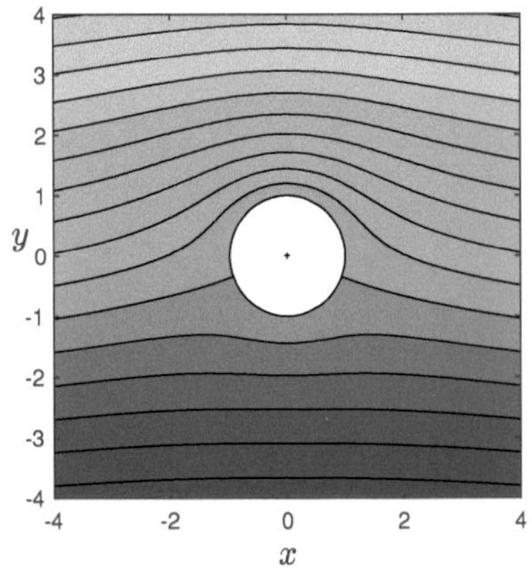

Fig. 20.8 Contours of $\psi = \mathrm{Im}(w)$ in eqn (20.18).

Incompressibility $\boldsymbol{\nabla} \cdot \boldsymbol{V} = 0$ and irrotationality $\boldsymbol{\nabla} \times \boldsymbol{V} = 0$ in two dimensions with $\boldsymbol{V} = V_x \hat{\mathbf{x}} + V_y \hat{\mathbf{y}} = \boldsymbol{\nabla}\phi = \boldsymbol{\nabla} \times \psi \mathbf{e}_3$ require that

$$V_x = \frac{\partial \phi}{\partial x} = \frac{\partial \psi}{\partial y}, \qquad V_y = \frac{\partial \phi}{\partial y} = -\frac{\partial \psi}{\partial x}, \qquad (20.19)$$

and we recognize the Cauchy–Riemann equations (20.6). Thus, finding a 2D incompressible, irrotational flow $\boldsymbol{V}(x, y)$ is reduced to finding an analytic complex function $w(z) = \phi(x, y) + i\psi(x, y)$ that satisfies the boundary conditions, and this can be done by conformal mapping of a known solution to the domain of interest. For 2D steady aerodynamics, this reduces to mapping a circle of radius R with its flow (20.18) to a desired airfoil, as illustrated in Fig. 20.6.

The complex potential for flow at angle of attack α around a cylinder of radius R centered at $z = z_c$ in the z plane is simply (20.18) with z replaced by $(z - z_c)e^{-i\alpha}$. Mapping that solution to $Z = z + 1/z$ for $R = |1 - z_c|$ yields flow around the Joukowski airfoil in the Z plane, as illustrated in Fig. 20.9. A nice introduction to complex variable methods applied to classical 2D airfoil theory can be found in chapter 4 of Acheson's book, for example.[2]

[2] D. J. Acheson, *Elementary Fluid Dynamics*, Oxford University Press, 1990.

20.4 Conformal mapping examples

Figures such as Figs. 20.3, 20.4, and 20.6, show contours of $u(x, y)$ and $v(x, y)$ with $w = u + iv = f(z)$ in the $z = x + iy$ plane. Hence, those figures visualize the *inverse* map $(u, v) \to (x, y)$, that is, $z = g(w)$ with contours of $u(x, y) = u_0$ and $v(x, y) = v_0$ corresponding to vertical lines and horizontal lines in the (u, v) plane, respectively. Given $w = f(z)$,

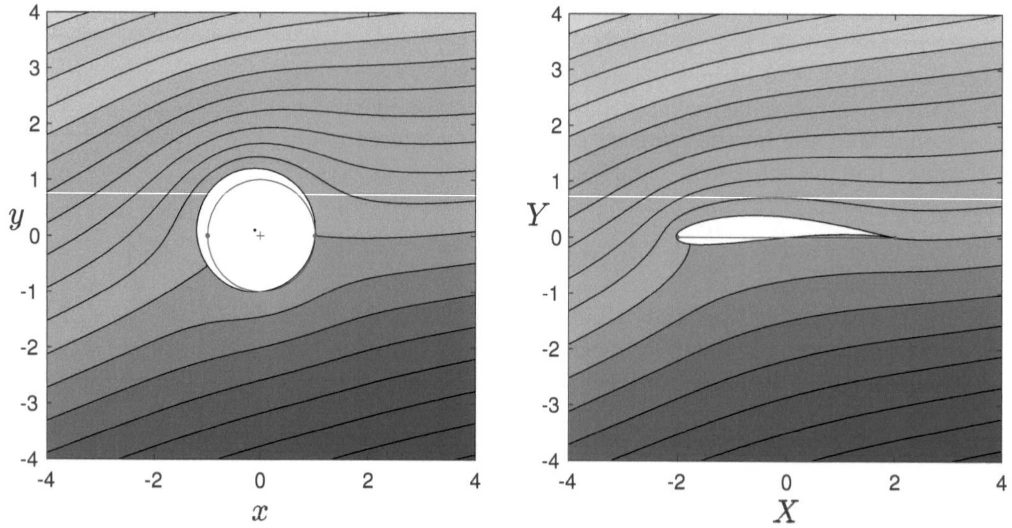

Fig. 20.9 Map of flow around a cylinder to flow around a Joukowski airfoil. The (red) unit circle in the z plane is mapped to the (red) line segment $[-2, 2]$ in the $Z = z + 1/z$ plane. The off-centered circle $z = z_c + (1 - z_c)e^{i\theta}$ is mapped to the Joukowski airfoil, here for $z_c = -0.1 + 0.1i$. Angle of attack $\alpha = \pi/12$.

contours of u and v are defined by implicit equations $u(x, y) = u_0$ and $v(x, y) = v_0$. The same curves are provided in explicit parametric forms $x = x(u_0, v), y = y(u_0, v)$, and $x = x(u, v_0), y = y(u, v_0)$, if we work with the inverse map $z = g(w)$, as in Fig. 20.6 and the following examples.

Quadratic map

The quadratic function

$$z = w^2 = (u^2 - v^2) + i\,2uv,$$

illustrated in Fig. 20.10, maps the right half w-plane to the entire z-plane. This map provides *conformal, confocal parabolic coordinates* (u, v) for the (x, y)-plane

$$x = u^2 - v^2, \qquad y = 2uv.$$

A vertical line with $u = u_0$ is mapped to the *parabola* $x = u_0^2 - v^2$, $y = 2x_0 v$, in the $z = (x, y)$ plane, that is,

$$x = u_0^2 - \frac{y^2}{4u_0^2}.$$

A horizontal line $v = v_0$ is mapped to the parabola $x = u^2 - v_0^2$, $y = 2uv_0$

$$x = \frac{y^2}{4v_0^2} - v_0^2.$$

All of those parabolas have $(x, y) = (0, 0)$ as their focus. The red and blue curves intersect at $90°$ in *both* planes.

Radial lines from the origin $w = \rho e^{i\varphi}$ in the w-plane are mapped to radials from the origin $z = \rho^2 e^{i2\varphi}$ in the z-plane. The angles between radials at the origin are doubled in the z-plane compared to angles in the w-plane.

Angles between the radial dashed lines and the red and blue curves are the same in *both* planes, *except at* $z = w = 0$ where the angles are doubled from w to z. The definition of the inverse function that was selected here is $z^{1/2} = |z|^{1/2} e^{i(\arg z)/2}$ with $-\pi < \arg(z) \leq \pi$ and corresponds to a one-to-one map between the right-half w-plane to the entire z-plane. The inverse function $z^{1/2}$ has a *branch cut* along the negative real axis, $x < 0$, $y = 0$. The $\arg z$ jumps by 2π across the cut and $z^{1/2}$ jumps from positive to negative imaginary, or vice versa. The definition $0 \leq \arg(z) < 2\pi$ would have a branch cut along the positive real axis in the z-plane and map the upper half w-plane to the entire z-plane, instead of the right-half w-plane.

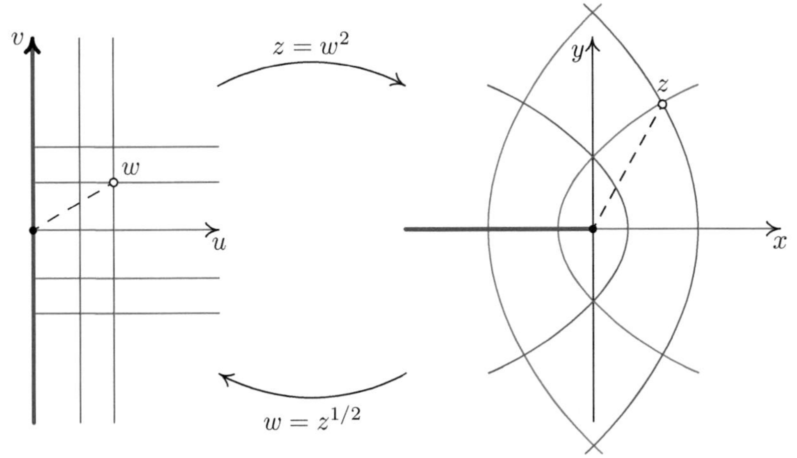

Fig. 20.10 $z = w^2$ and $w = \sqrt{z}$ with $-\pi < \arg z \leq \pi$.

Exponential map

The exponential function

$$z = e^w = e^u e^{iv}$$

maps the *strip* $-\infty < u < \infty$, $-\pi < v \leq \pi$ to the *entire* z-plane (Fig. 20.11). Indeed, $e^{w+2i\pi} = e^w$ is periodic of complex period $2\pi i$ in w, so e^w maps an infinite number of w's to the same z. This map provides conformal *log–polar coordinates* (u, v) for the (x, y) plane

$$x = e^u \cos v, \qquad y = e^u \sin v.$$

The vertical lines $w = u_0 + iv \to z = e^{u_0} e^{iv}$ are mapped to circles of radius $|z| = e^{v_0}$ in the z-plane. The horizontal lines $w = u + iv_0 \to z = e^u e^{iv_0}$ are mapped to radial lines with polar angle $\arg(z) = v_0$ in z-plane.

The radial from the origin $w = u + iau \to z = e^u e^{iau}$, with a real and fixed, is mapped to a *logarithmic spiral* in the z-plane since $|z| = e^u = e^{(\arg z)/a}$. The inverse function $w = \ln z = \ln|z| + i \arg z$ showed in this picture corresponds to the definition $-\pi < \arg z \leq \pi$. All angles are preserved, except at $z = 0$, which is the branch point.

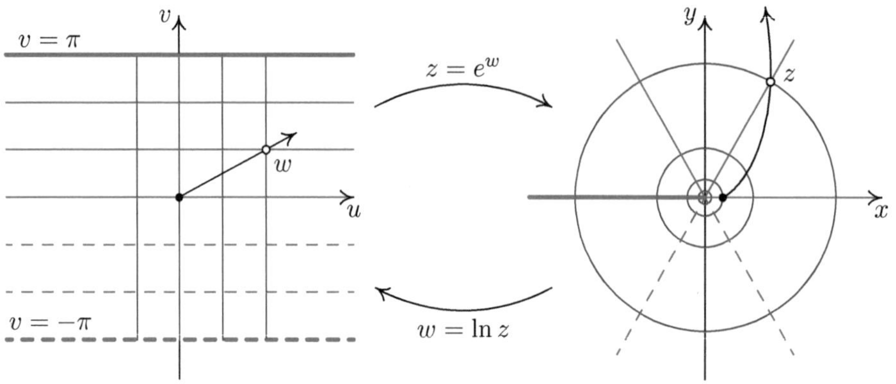

Fig. 20.11 $z = e^w \longleftrightarrow w = \ln z$.

Hyperbolic cosine map

The cosh function

$$z = \cosh w = \frac{e^w + e^{-w}}{2} = \frac{e^u e^{iv} + e^{-u} e^{-iv}}{2}$$
$$= \cosh u \cos v + i \sinh u \sin v$$

maps the *semi-infinite strip* $(u, v) \in [0, \infty) \times (-\pi, \pi]$ to the *entire* z-plane (Fig. 20.12). Indeed, $\cosh w = \cosh(-w)$ and $\cosh(w+2\pi i) = \cosh(w)$ so $\cosh w$ is even in w and periodic of period $2\pi i$. This map gives *conformal, confocal elliptic coordinates* (u, v) for the (x, y)-,plane

$$x = \cosh u \cos v, \qquad y = \sinh u \sin v$$

which are the right coordinates to solve Laplace's equation in an elliptical geometry. The vertical line segments $w = u_0 + iv$ with $u = u_0 \geq 0$ and $-\pi < v \leq \pi$ are mapped to *confocal ellipses* in the z-plane with

$$\frac{x^2}{\cosh^2 u_0} + \frac{y^2}{\sinh^2 u_0} = 1.$$

The semi-infinite horizontal lines $w = u + iv_0$ are mapped to *confocal hyperbolic arcs* in the z-plane with

$$\frac{x^2}{\cos^2 v_0} - \frac{y^2}{\sin^2 v_0} = 1.$$

All of those ellipses and hyperbolas have $(x, y) = (\pm 1, 0)$ as their foci.

The inverse map is $w = \ln(z + \sqrt{z^2 - 1})$ defined with $\varphi_{1,2} = \arg(z \pm 1)$ and

$$\sqrt{z^2 - 1} \triangleq \sqrt{|z^2 - 1|}\, e^{i\varphi_1/2} e^{i\varphi_2/2},$$

as in (19.47) such that $\sqrt{z^2 - 1} = \pm i\sqrt{|z^2 - 1|}$ on the top side or bottom side of $(-1, 1)$, respectively, and $\ln(z + \sqrt{z^2 - 1})$ has the branch cut $z \in (-\infty, 1)$.

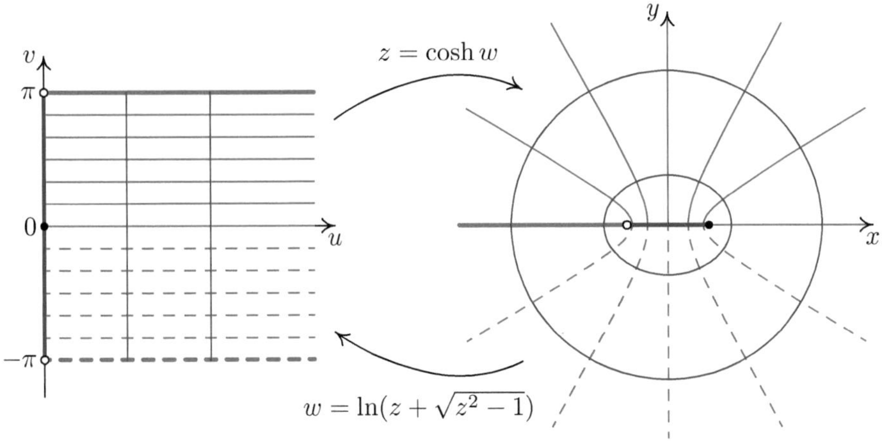

Fig. 20.12 $z = \cosh w \longleftrightarrow w = \ln(z + \sqrt{z^2 - 1})$.

Joukowski map

The Joukowski map,

$$\begin{aligned} w = z + \frac{1}{z} &= re^{i\varphi} + \frac{e^{-i\varphi}}{r} \\ &= \left(r + \frac{1}{r}\right)\cos\varphi + i\left(r - \frac{1}{r}\right)\sin\varphi, \end{aligned} \tag{20.20}$$

sends circles $|z| = r = r_0$ fixed to ellipses in the w-plane with (Fig. 20.13)

$$\frac{u^2}{\left(r_0 + \frac{1}{r_0}\right)^2} + \frac{v^2}{\left(r_0 - \frac{1}{r_0}\right)^2} = 1.$$

In particular, the unit circle $z = e^{i\varphi}$ is mapped to the line segment $w = 2\cos\varphi$. Radials $z = re^{i\varphi_0}$ with φ_0 fixed are mapped to hyperbolic arcs with

$$\frac{u^2}{\cos^2\varphi_0} - \frac{v^2}{\sin^2\varphi_0} = 1.$$

The Joukowski map is useful in 2D electrostatics and fundamental in 2D aerodynamics. Its real and imaginary parts (Fig. 20.14),

$$\begin{aligned} u &= x + \frac{x}{x^2 + y^2} = \left(r + \frac{1}{r}\right)\cos\varphi, \\ v &= y - \frac{y}{x^2 + y^2} = \left(r - \frac{1}{r}\right)\sin\varphi, \end{aligned} \tag{20.21}$$

provide solutions to

$$\nabla^2 u = 0 \quad \text{for } r > 1 \quad \text{with} \quad \frac{\partial u}{\partial r} = 0 \quad \text{on } r = 1$$

and

$$\nabla^2 v = 0 \quad \text{for } r > 1 \quad \text{with} \quad v = 0 \quad \text{on } r = 1.$$

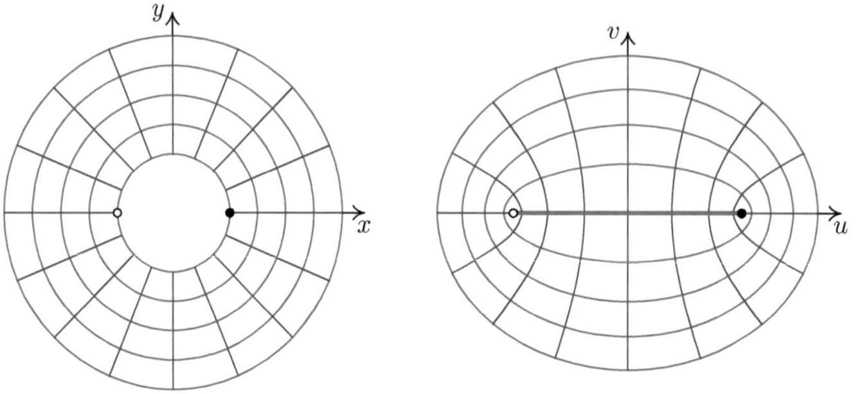

Fig. 20.13 The Joukowski map $w = z + 1/z$ maps circles centered at the origin in the z-plane to ellipses in the w plane. Radials from the origin in z are mapped to hyperbolic arcs in w.

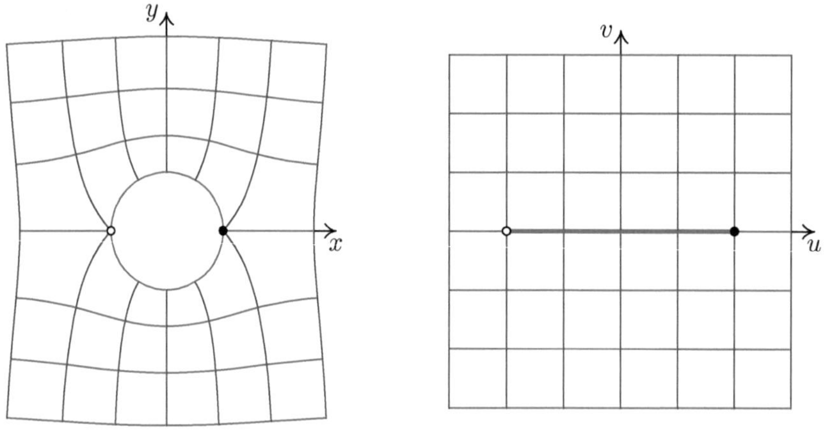

Fig. 20.14 The Joukowski map $w = z + 1/z$ maps the outside of the unit circle $|z| \geq 1$ to the entire w plane. The contours of $u(x, y) = \text{Re}(w)$ are the electric field lines (blue, vertical) when a perfect cylindrical conductor of radius 1 is placed in a uniform electric field. The contours of $v(x, y) = \text{Im}(w)$ are the streamlines (red, horizontal) for potential flow around the cylinder of radius 1.

Exercises

(20.1) Deduce (20.11) from the Cauchy–Riemann equations (20.6). Find $u(x,y)$ and $v(x,y)$ and verify (20.6) and (20.11) for (i) $f(z)=z^2$, (ii) z^3, (iii) e^z, and (iv) $1/z$.

(20.2) Given $u(x,y)$ find its conjugate harmonic function $v(x,y)$, if possible, such that $u(x,y)+iv(x,y)\equiv f(z)$ is z-differentiable, for (i) $u=y$, (ii) $u=x+y$, (iii) $u=\cos x\cosh y$, and (iv) $v=\ln\sqrt{x^2+y^2}$.

(20.3) Is $|z|$ differentiable with respect to z? Why? What about $\arg z$?

(20.4) Find $u(x,y)$ and $v(x,y)$ for $f(z)=\sqrt{z}=\sqrt{|z|}e^{i\arg(z)/2}$. Show that $f(z)$ is differentiable everywhere except across the negative real axis $x\le 0$, $y=0$ with the standard definition $-\pi<\arg(z)\le\pi$.

(20.5) Consider

$$f(z)=\sqrt{z^2-1}\triangleq\sqrt{|z^2-1|}\,e^{\frac{i}{2}\arg(z^2-1)}.$$

For the standard definition $-\pi<\arg(z^2-1)\le\pi$, show that $\arg(z^2-1)$ jumps by $\pm 2\pi$ when z crosses the imaginary axis *or* the real segment $[-1,1]$. Show that $f(z)$ is not differentiable for such z but is continuous and differentiable for all other z.

(20.6) Consider

$$f(z)=\sqrt{z^2-1}\triangleq\sqrt{z-1}\sqrt{z+1}$$
$$=\sqrt{r_1r_2}\,e^{i(\varphi_1+\varphi_2)/2},$$

where $r_{1,2}=|z\pm 1|$ and $\varphi_{1,2}=\arg(z\pm 1)$. For the standard definition $-\pi<\arg(\cdot)\le\pi$, show that $\varphi_1+\varphi_2$ jumps by $\pm 2\pi$ across $(-1,1)$ but by $\pm 4\pi$ across $(-\infty,-1)$. Deduce that $f(z)$ is not differentiable on $[-1,1]$ but is continuous and differentiable for all other z, including $x<-1$, $y=0$.

(20.7) Consider

$$f(z)=\ln\frac{z+a}{z-a}$$
$$=\ln\left|\frac{z+a}{z-a}\right|+i\,(\arg(z+a)-\arg(z-a))$$

where a is an arbitrary *positive real* number and $-\pi<\arg(z\pm a)\le\pi$. Show that $f(z)$ is continuous across the semi-infinite real line $(-\infty,-a)$ but jumps by $\pm 2\pi i$ when z crosses the real segment $(-a,a)$. Conclude that $f(z)$ is differentiable everywhere except along the *branch cut* $[-a,a]$. Show that the real and imaginary part of $f(z)$ correspond to the *bipolar coordinates* of chapter 1 and therefore that those functions satisfy Laplace's equation everywhere, except for $z=\pm a$ for the real part, and the *branch cut* $[-a,a]$ for the imaginary part.

(20.8) Complex numbers $z=x+iy$ correspond to vectors $\boldsymbol{r}=(x,y)$ in $\mathbb{R}^2$. The complex form of the gradient of a real function $\phi(x,y)$ is thus

$$\nabla\phi=\hat{\mathbf{x}}\frac{\partial\phi}{\partial x}+\hat{\mathbf{y}}\frac{\partial\phi}{\partial y}\equiv\frac{\partial\phi}{\partial x}+i\frac{\partial\phi}{\partial y}.$$

Show that if $w(z)=\phi(x,y)+i\psi(x,y)$ is analytic, with $\phi(x,y)$ and $\psi(x,y)$ real, then

$$\nabla\phi\equiv\left(\frac{\mathrm{d}w}{\mathrm{d}z}\right)^*. \qquad (20.22)$$

This is useful in 2D fluid dynamics where $w(z)$ represents a complex potential whose real part ϕ is a velocity potential and imaginary part ψ is a streamfunction (20.19). Expression (20.22) yields the velocity directly from the complex potential $w(z)$. For example, if $w(z)=z^3$, then the velocity is $(3z^2)^*$; that is, $\boldsymbol{V}=\nabla\phi=3(x^2-y^2)\,\hat{\mathbf{x}}-6xy\,\hat{\mathbf{y}}$.

(20.9) Find the potential $\phi(x,y)$ and the streamfunction $\psi(x,y)$ for the *lifting cylinder flow* $w(z)=\phi(x,y)+i\psi(x,y)$ in (20.18). Show that the cylinder $z(\varphi)=Re^{i\varphi}$ is a streamline for the flow; that is, $\psi(x,y)$ is constant on the cylinder. Show that the velocity field $\boldsymbol{V}(x,y)\to V_\infty\hat{\mathbf{x}}$ as $|z|=r=\sqrt{x^2+y^2}\to\infty$, but $\boldsymbol{V}\cdot\hat{\boldsymbol{n}}=0$ on the cylinder, where $\hat{\boldsymbol{n}}$ is the unit outward normal.

(20.10) The velocity in the Joukowski plane (Fig. 20.9) is the conjugate of $\mathrm{d}w/\mathrm{d}Z=(\mathrm{d}w/\mathrm{d}z)/(\mathrm{d}Z/\mathrm{d}z)$ by the chain rule. Show that $\mathrm{d}Z/\mathrm{d}z=0$ at the airfoil trailing edge at $z=1$. Find the circulation Γ in (20.18), with z replaced by $(z-z_c)e^{-i\alpha}$ and $R=|1-z_c|$, such that $\mathrm{d}w/\mathrm{d}z$ also vanishes at $z=1$, so the velocity is regular at the airfoil trailing edge. This is the *Kutta condition*.

(20.11) Show that $u=\cos(kx)e^{-ky}$ and $v=\sin(kx)e^{-ky}$ are solutions of Laplace's equation for any real k (i) by direct calculation and (ii) by finding a z

differentiable function $f(z) = u(x,y) + iv(x,y)$. These solutions occur in a variety of applications, e.g. surface gravity waves with the surface at $y = 0$ and $ky < 0$.

(20.12) Show that $w = z^3$ provides two solutions of Laplace's equation $\nabla^2\Phi = 0$ in the polar wedge $r \geq 0$, $0 < \varphi < \pi/3$, one solution to the *Dirichlet problem* with $\Phi = 0$ on the boundaries $\varphi = 0$ and $\varphi = \pi/3$, and another solution to the *Neumann problem* with $\partial\Phi/\partial\varphi = 0$ on $\varphi = 0, \pi/3$. Specify Φ for both problems in both cartesian coordinates (x,y) and polar coordinates (r,φ).

(20.13) Generalizing the previous problem, show that $w = z^{3n}$ with n a nonzero integer. Provide two solutions to Laplace's equation in the same polar wedge $r \geq 0$, $0 < \varphi < \pi/3$, one solution Φ_n with $\partial\Phi_n/\partial\varphi = 0$ for $\varphi = 0, \pi/3$ and one solution Ψ_n with $\Psi_n = 0$ for $\varphi = 0, \pi/3$. Specify Φ_n and Ψ_n in both cartesian coordinates (x,y) and polar coordinates (r,φ).

(20.14) Generalize the previous problems by finding a solution Φ of Laplace's equation in the wedge $r \geq 0$, $0 < \varphi < \alpha$ that vanishes on the boundaries of the wedge at $\varphi = 0$ and $\varphi = \alpha$, where α is a constant such that $0 < \alpha \leq 2\pi$. What is the behavior of the gradient of Φ as $r \to 0$ as a function of α? Can you find other nonzero solutions to the same problem?

(20.15) Consider the map $w = z^2$. Determine where the triangle with vertices (i) $(1,0)$, $(1,1)$, $(0,1)$ in the z-plane is mapped in the $w = u + iv$ plane; and (ii) same but for triangle $(0,0)$, $(1,0)$, $(1,1)$. Do not simply map the vertices; parametrize and map each edge and determine what happens to each angle.

(20.16) Analyze $w = 1/z$. Show that contours of u and v are circles in the z-plane. Determine the maps of constant (i) $x = \text{Re}(z)$, (ii) $y = \text{Im}(z)$, (iii) $\varphi = \arg(z)$, and (iv) $r = |z|$. Show that arbitrary lines $z(t) = a + ita$, for any complex constant a with t real, are mapped to circles.

(20.17) If (u,v) are curvilinear coordinates for the euclidean plane with cartesian coordinates $(x,y) \in \mathbb{R}^2$, then the del operator is (16.46)

$$\boldsymbol{\nabla} = \hat{\mathbf{x}}\frac{\partial}{\partial x} + \hat{\mathbf{y}}\frac{\partial}{\partial y} = (\boldsymbol{\nabla}u)\frac{\partial}{\partial u} + (\boldsymbol{\nabla}v)\frac{\partial}{\partial v}.$$

Show that the 2D Laplacian operator

$$\begin{aligned}\nabla^2 &= \boldsymbol{\nabla}\cdot\boldsymbol{\nabla} \\ &= |\boldsymbol{\nabla}u|^2\frac{\partial^2}{\partial u^2} + 2(\boldsymbol{\nabla}u)\cdot(\boldsymbol{\nabla}v)\frac{\partial^2}{\partial u\partial v} + |\boldsymbol{\nabla}v|^2\frac{\partial^2}{\partial v^2} \\ &\quad + (\nabla^2 u)\frac{\partial}{\partial u} + (\nabla^2 v)\frac{\partial}{\partial v}.\end{aligned}$$

Deduce that if the coordinates correspond to a conformal map, $w = f(z)$ with $z = x + iy$ and $w = u + iv$, then

$$\nabla^2 = \left|\frac{\mathrm{d}w}{\mathrm{d}z}\right|^2\left(\frac{\partial^2}{\partial u^2} + \frac{\partial^2}{\partial v^2}\right). \tag{20.23}$$

(20.18) Consider the change of variables from cartesian (x,y) to (u,v)

$$x = \cosh u \cos v, \qquad y = \sinh u \sin v.$$

Show that the coordinate curves with u fixed are confocal ellipses. Show that the coordinate curves with v fixed are confocal hyperbolas with the same foci as the ellipses. Identify the foci. Show that these (u,v) coordinates are conformal and that the Laplacian reads

$$\begin{aligned}\nabla^2 &= \frac{\partial^2}{\partial x^2} + \frac{\partial^2}{\partial y^2} \\ &= \frac{1}{\sinh^2 u + \sin^2 v}\left(\frac{\partial^2}{\partial u^2} + \frac{\partial^2}{\partial v^2}\right).\end{aligned}$$

(20.19) The Laplacian in cylindrical coordinates (ρ,φ,z) is derived in (16.110). Translating those results into polar coordinates $x = r\cos\varphi$, $y = r\sin\varphi$, the 2D Laplacian in polar coordinates is

$$\nabla^2 = \frac{\partial^2}{\partial r^2} + \frac{1}{r}\frac{\partial}{\partial r} + \frac{1}{r^2}\frac{\partial^2}{\partial\varphi^2}.$$

Show that the Laplacian in *log-polar coordinates* $x = e^u\cos v$, $y = e^u\sin v$ is

$$\nabla^2 = \frac{\partial^2}{\partial x^2} + \frac{\partial^2}{\partial y^2} = e^{-2u}\left(\frac{\partial^2}{\partial u^2} + \frac{\partial^2}{\partial v^2}\right).$$

Show that polar coordinates (r,φ) are orthogonal but not conformal, but log–polar coordinates (u,v) are orthogonal and conformal, with $r = e^u$ and $\varphi = v$.

Complex Integration

21

21.1 Complex integrals

What do we mean by $\int_a^b f(z)\mathrm{d}z$ when $f(z)$ is a complex function of the complex variable z and the bounds a and b are complex numbers in the z-plane? In general, we need to specify the *path* $\mathcal{C}$ in the complex plane to go from a to b and we need to write the integral as $\int_{\mathcal{C}} f(z)\mathrm{d}z$. Then if $z_0 = a$, z_1, $z_2, \ldots, z_N = b$ are successive points on the path from a to b, we can define the integral as usual as

$$\int_{\mathcal{C}} f(z)\,\mathrm{d}z \triangleq \lim_{\Delta z_n \to 0} \sum_{n=1}^{N} f_n\,\Delta z_n, \tag{21.1}$$

where $\Delta z_n = z_n - z_{n-1}$ and f_n is an approximation of $f(z)$ on the segment (z_{n-1}, z_n). The first-order Euler approximation selects $f_n = f(z_{n-1})$ or $f_n = f(z_n)$ while the second-order trapezoidal rule picks the average between those two values, that is, $f_n = \frac{1}{2}(f(z_{n-1}) + f(z_n))$. The Riemann sum definition also provides a practical way for estimating the integral as a sum of finite differences.

Bound If $|f(z)| \le M$ along the curve, then

$$\left|\int_{\mathcal{C}} f(z)\mathrm{d}z\right| \le M \int_{\mathcal{C}} |\mathrm{d}z| = ML, \tag{21.2}$$

where $L = \int_{\mathcal{C}} |\mathrm{d}z| \ge 0$ is the length of the curve $\mathcal{C}$ from a to b. Note also that the integral from a to b along $\mathcal{C}$ is minus that from b to a along the same curve since all the Δz_n change sign for that reversed curve. If $\mathcal{C}$ is from a to b, we sometimes use $-\mathcal{C}$ to denote the same path but with the opposite orientation, from b to a. If we have a parametrization for the curve, say $z(t)$ with t real and $z(t_a) = a$, $z(t_b) = b$ then the partition $z_0, \ldots, z_N$ can be obtained from a partition of the real t interval $t_0 = t_a, \ldots, t_N = t_b$, and in the limit the integral can be expressed as

$$\int_{\mathcal{C}} f(z)\mathrm{d}z = \int_{t_a}^{t_b} f\big(z(t)\big)\,\frac{\mathrm{d}z}{\mathrm{d}t}\mathrm{d}t, \tag{21.3}$$

which can be separated into real and imaginary parts to end up with a complex combination of real integrals over the real variable t.

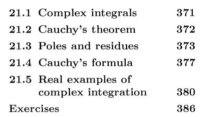

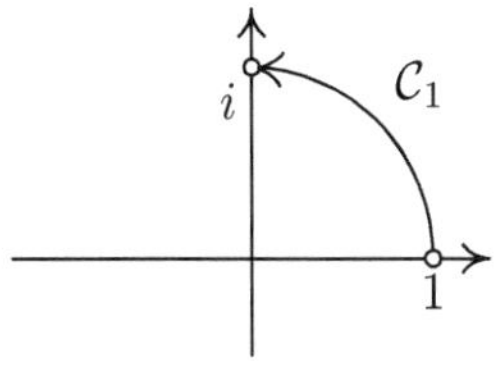

Fig. 21.1 Counterclockwise contour from $z = 1 \to i$.

Example 1 To compute the integral of $1/z$ along the path $\mathcal{C}_1$ that consists of the unit circle counterclockwise from $a = 1$ to $b = i$ (Fig. 21.1),

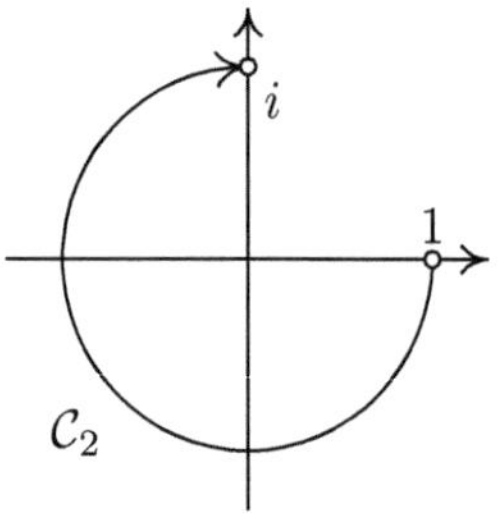

Fig. 21.2 Clockwise from $z = 1 \to i$.

we can parametrize the circle as $z(\varphi) = e^{i\varphi}$ with $\mathrm{d}z = ie^{i\varphi}\mathrm{d}\theta$ for $\varphi = 0 \to \pi/2$. Then

$$\int_{\mathcal{C}_1} \frac{1}{z}\mathrm{d}z = \int_0^{\pi/2} \frac{1}{e^{i\varphi}} ie^{i\varphi}\mathrm{d}\theta = i\frac{\pi}{2}.$$

Example 2 For the path $\mathcal{C}_2$ which consists of the portion of the unit circle *clockwise* from $a = 1$ to $b = i$ (Fig. 21.2), the parametrization is again $z = e^{i\varphi}$; however, now $\varphi = 0 \to -3\pi/2$ and

$$\int_{\mathcal{C}_2} \frac{1}{z}\mathrm{d}z = \int_0^{-3\pi/2} \frac{1}{e^{i\varphi}} ie^{i\varphi}\mathrm{d}\theta = -i\frac{3\pi}{2} \neq i\frac{\pi}{2}.$$

Clearly the integral of $1/z$ from $a = 1$ to $b = i$ depends on the path. Note that these two integrals differ by $2\pi i$.

Example 3 However, for the function z^2 over the same two paths with $z = e^{i\varphi}$, $z^2 = e^{i2\varphi}$ and $\mathrm{d}z = ie^{i\varphi}\mathrm{d}\theta$, we find that

$$\int_{\mathcal{C}_1} z^2\mathrm{d}z = \int_0^{\pi/2} ie^{i3\varphi}\mathrm{d}\theta = \frac{1}{3}\left(e^{i3\pi/2} - 1\right) = \frac{-i-1}{3} = \frac{b^3 - a^3}{3},$$

and

$$\int_{\mathcal{C}_2} z^2\mathrm{d}z = \int_0^{-3\pi/2} ie^{i3\varphi}\mathrm{d}\theta = \frac{1}{3}\left(e^{-i9\pi/2} - 1\right) = \frac{-i-1}{3} = \frac{b^3 - a^3}{3}.$$

Thus, for z^2 it appears that we obtain the expected result $\int_a^b z^2\mathrm{d}z = (b^3 - a^3)/3$, independent of the path of integration.

21.2 Cauchy's theorem

The integral of a complex function is independent of the path of integration if the integral over a *closed* contour vanishes. With $z = x + iy$ and $f(z) = u(x, y) + iv(x, y)$ the complex integral around a closed curve $\mathcal{C}$ can be written as

$$\oint_{\mathcal{C}} f(z)\mathrm{d}z = \oint_{\mathcal{C}} (u + iv)(\mathrm{d}x + i\mathrm{d}y) = \oint_{\mathcal{C}} (u\mathrm{d}x - v\mathrm{d}y) + i\oint_{\mathcal{C}} (v\mathrm{d}x + u\mathrm{d}y).$$

Hence, the real and imaginary parts of the integral are real *line integrals*. These line integrals can be turned into area integrals using Green's theorem:

$$\begin{aligned}\oint_{\mathcal{C}} f(z)\mathrm{d}z = &\oint_{\mathcal{C}} (u\mathrm{d}x - v\mathrm{d}y) \quad + i\oint_{\mathcal{C}} (v\mathrm{d}x + u\mathrm{d}y) \\ &= \int_A \left(-\frac{\partial v}{\partial x} - \frac{\partial u}{\partial y}\right)\mathrm{d}A + i\int_A \left(\frac{\partial u}{\partial x} - \frac{\partial v}{\partial y}\right)\mathrm{d}A,\end{aligned}$$

where A is the interior domain bounded by the closed curve $\mathcal{C}$. But the Cauchy–Riemann equations (20.6) give

$$\frac{\partial u}{\partial x} - \frac{\partial v}{\partial y} = 0, \quad \frac{\partial v}{\partial x} + \frac{\partial u}{\partial y} = 0, \tag{21.4}$$

whenever the function $f(z)$ is differentiable in the neighborhood of the point $z = x + iy$. Thus, both integrals vanish if $f(z)$ is z-differentiable at *all* points of A. This is *Cauchy's theorem*,

$$\oint_{\mathcal{C}} f(z)\mathrm{d}z = 0, \tag{21.5}$$

if $\mathrm{d}f(z)/\mathrm{d}z$ *exists everywhere inside and on the closed curve* $\mathcal{C}$.

Functions like e^z, $\cos z$, $\sin z$, and z^n with $n \geq 0$ are *entire* functions—functions that are complex differentiable in the *entire* complex plane. Hence, the integral of such functions around *any* closed contour $\mathcal{C}$ vanishes.

21.3 Poles and residues

Functions such as

$$f(z) = \frac{1}{z-a}$$

are differentiable everywhere except at the isolated singular point $z = a$. A function $f(z)$ has an *isolated singularity* at $z = a$ if it is differentiable in a neighborhood of $z = a$ but $f(z)$ is singular at $z = a$. The singularity is called a *pole of order* $n > 0$ if

$$\lim_{z \to a} (z-a)^n f(z) = C \neq 0,$$

that is, if

$$f(z) \sim \frac{C}{(z-a)^n}$$

near $z = a$. For example, $f(z) = (z-1)^{-1}$ and $f(z) = e^z/(z-1)$ have poles of order 1 (*simple* poles) at $z = 1$, while $f(z) = (z-i)^{-2}$ and $f(z) = (\cos z)/(z-i)^2$ have poles of order 2 (*double* poles) at $z = i$.

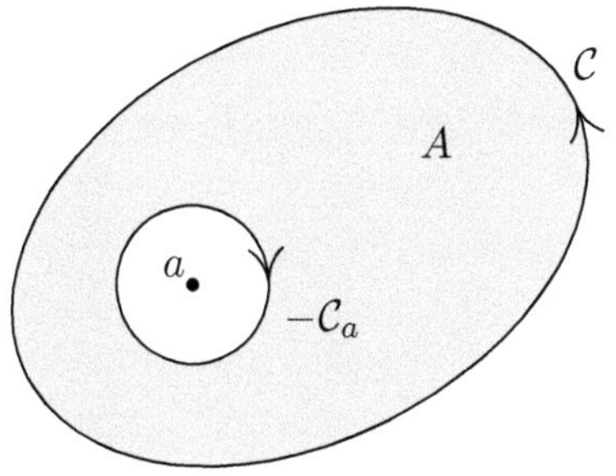

Fig. 21.3 Isolating the singularity at a.

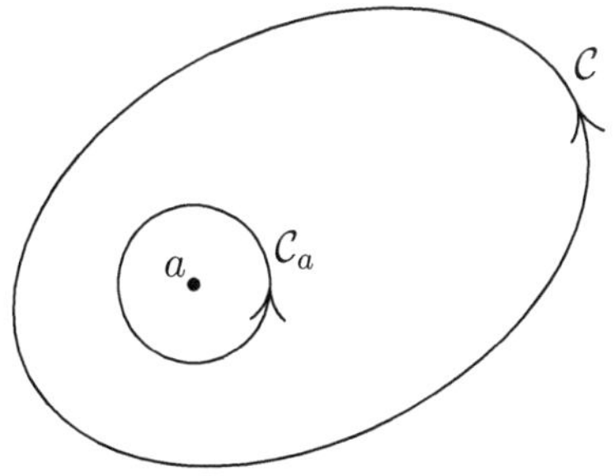

Fig. 21.4 Deforming the contour.

The integral of $1/(z-a)^n$ around any closed contour that does *not* include $z = a$ vanishes, by Cauchy's theorem. We can figure out the integral of $(z-a)^{-n}$, with n a positive integer, about any closed curve $\mathcal{C}$ enclosing the pole a by considering a small circle $\mathcal{C}_a$ centered at a of radius $\epsilon > 0$ as small as needed to be inside the outer closed curve $\mathcal{C}$. Now, the function $(z-a)^{-n}$ is analytic everywhere inside the domain A bounded by the *counterclockwise* outer boundary $\mathcal{C}$ and the *clockwise* inner circle boundary $-\mathcal{C}_a$ (emphasized here by the minus sign), so the interior A is always to the *left* when traveling on the boundary (Fig. 21.3).

By Cauchy's theorem, this implies that the integral over the closed contour of A, which consists of the sum $\mathcal{C} + (-\mathcal{C}_a)$ of the outer *counterclockwise* curve $\mathcal{C}$ and the inner *clockwise* circle $-\mathcal{C}_a$, vanishes,

$$\oint_{\mathcal{C}+(-\mathcal{C}_a)} \frac{1}{(z-a)^n}\mathrm{d}z = 0 = \oint_{\mathcal{C}} \frac{1}{(z-a)^n}\mathrm{d}z + \oint_{-\mathcal{C}_a} \frac{1}{(z-a)^n}\mathrm{d}z.$$

Therefore,

$$\oint_{\mathcal{C}} \frac{1}{(z-a)^n}\mathrm{d}z = \oint_{\mathcal{C}_a} \frac{1}{(z-a)^n}\mathrm{d}z,$$

since $\oint_{-\mathcal{C}_a} = -\oint_{\mathcal{C}_a}$. In other words the integral about the closed contour $\mathcal{C}$ equals the integral about the closed inner circle $\mathcal{C}_a$, now with the *same* orientation as the outer contour (Fig. 21.4).[1] Thus, the integral is invariant under deformation of the loop $\mathcal{C}$ as long as such deformation does not cross the *pole* at a. The loop can be shrunk to an arbitrary small circle surrounding the pole without changing the value of the integral. That remaining integral about the arbitrarily small circle is the *residue* integral.

[1]We used this singularity isolation technique in conjunction with the divergence theorem to evaluate the flux of $\hat{\boldsymbol{r}}/r^2 = \boldsymbol{r}/r^3$ (the inverse square law of gravity and electrostatics) through any closed surface enclosing the origin at $r = 0$, as well as in conjunction with Stokes' theorem for the circulation of a line current $\boldsymbol{B} = (\hat{\mathbf{z}} \times \boldsymbol{r})/|\hat{\mathbf{z}} \times \boldsymbol{r}|^2 = \hat{\boldsymbol{\varphi}}/\rho = \boldsymbol{\nabla}\varphi$ around a loop enclosing the z-axis at $\rho = 0$.

That *residue* integral can be calculated explicitly with the circle parameterization $z = a + \epsilon e^{i\varphi}$ and $\mathrm{d}z = i\epsilon\, e^{i\varphi}\mathrm{d}\varphi$, yielding the simple result that

$$\oint_{\mathcal{C}_a} \frac{\mathrm{d}z}{(z-a)^n} = \int_0^{2\pi} \frac{i\epsilon e^{i\varphi}\,\mathrm{d}\varphi}{\epsilon^n e^{in\varphi}} = \frac{i\epsilon}{\epsilon^n}\int_0^{2\pi} e^{i(1-n)\varphi}\mathrm{d}\varphi = \begin{cases} 2\pi i & \text{if } n = 1, \\ 0 & \text{if } n \neq 1. \end{cases}$$

We can collect all these various results in the following useful formula that for any integer $n = 0, \pm 1, \pm 2, \ldots$ and a closed contour $\mathcal{C}$ oriented counterclockwise

$$\oint_{\mathcal{C}} \frac{\mathrm{d}z}{(z-a)^n} = \begin{cases} 2\pi i & \text{if } n = 1 \text{ and } \mathcal{C} \text{ encloses } a, \\ 0 & \text{otherwise.} \end{cases} \tag{21.6}$$

For $n \leq 0$, that is for $n = -|n|$, this follows directly from Cauchy's theorem since $(z-a)^{-n} = (z-a)^{|n|}$ is complex differentiable for all z. For $n > 0$ this also follows from Cauchy's theorem that allows the deformation of the contour $\mathcal{C}$ to a small circle $\mathcal{C}_a$ around the pole, the integral *residue*, which can be calculated explicitly, as done above. If n is *not* an integer, then $z = a$ is a *branch point* from which a *branch cut* emanates to define the function $(z-a)^\alpha$ uniquely. The result (21.6) does not apply for noninteger exponent n; the integral depends on the choice of branch cut and the exponent.

If the function $f(z)$ has several poles, at a_1, a_2, and a_3 for example, we can use the same procedure to isolate all the poles inside the contour, a_1 and a_2 for example (Fig. 21.5). The poles outside the contour $\mathcal{C}$ (a_3 on the side figure) do not contribute to the integral, and Cauchy's theorem yields

$$\oint_{\mathcal{C}} f(z)\mathrm{d}z = \oint_{\mathcal{C}_1} f(z)\mathrm{d}z + \oint_{\mathcal{C}_2} f(z)\mathrm{d}z. \tag{21.7}$$

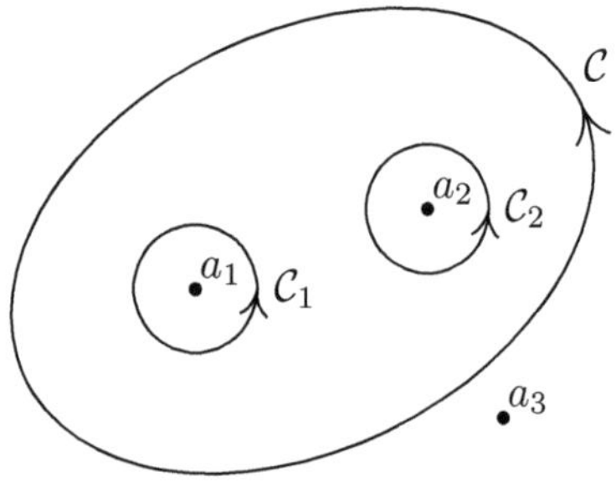

Fig. 21.5 Residues from the poles inside the contour.

The integral is then the sum of the residue integrals, where the sum is over all isolated singularities inside the contour. Each of these residue integrals can be calculated as above. The procedure is called the *calculus of residues*. In general, if $f(z)$ is differentiable everywhere inside and on the simple closed curve $\mathcal{C}$, except at the isolated singularities at $z = a_1, \ldots, a_N$, then

$$\oint_{\mathcal{C}} f(z)\mathrm{d}z = \sum_{k=1}^{N} \oint_{\mathcal{C}_k} f(z)\mathrm{d}z, \tag{21.8}$$

where $\mathcal{C}_k$ is a sufficiently small, simple closed loop that encloses only a_k. The residue integrals $\oint_{\mathcal{C}_k} f(z)\mathrm{d}z$ can be computed explicitly using (21.6) together with partial fractions or Taylor series. We call $\oint_{\mathcal{C}_k} f(z)\mathrm{d}z$ a *residue integral.* That residue integral divided by $2\pi i$ is called the *residue* of $f(z)$ at a_k, often denoted $\mathrm{Res}(f, a_k)$ in the literature.[2] We prefer to work with residue integrals instead of residues to avoid unnecessary formula and factors of $2\pi i$.

[2] So, $\oint_{C_k} f(z)\mathrm{d}z = 2\pi i\ \mathrm{Res}(f, a_k)$.

Example 1 *Simple poles.* Consider

$$\oint_C \frac{1}{z^2+1}\,\mathrm{d}z.$$

The function $f(z) = (z^2+1)^{-1}$ has simple poles at $z = \pm i$ since $z^2+1 = (z-i)(z+i)$. The integral can be calculated using partial fractions since

$$\frac{1}{z^2+1} = \frac{1}{2i}\left(\frac{1}{z-i} - \frac{1}{z+i}\right)$$

and

$$\oint_C \frac{\mathrm{d}z}{z^2+1} = \frac{1}{2i}\oint_C \frac{\mathrm{d}z}{z-i} - \frac{1}{2i}\oint_C \frac{\mathrm{d}z}{z+i}.$$

The integral is reduced to a sum of simple integrals such as (21.6), here with $a = \pm i$ and $n = 1$. The integrals are $2\pi i$ or 0, depending on whether the poles i and $-i$ are inside C. If C_1 is a circle centered at i sufficiently small to not include $-i$, the *residue* at i is

$$\oint_{C_1} \frac{1}{z^2+1}\,\mathrm{d}z = \frac{1}{2i}\oint_{C_1} \frac{1}{z-i}\mathrm{d}z = \pi,$$

and likewise if C_2 is a circle centered at $-i$ sufficiently small to not include i, the *residue* at $-i$ is

$$\oint_{C_2} \frac{1}{z^2+1}\,\mathrm{d}z = -\frac{1}{2i}\oint_{C_2} \frac{1}{z+i}\mathrm{d}z = -\pi.$$

In summary,

$$\oint_C \frac{1}{z^2+1}\,\mathrm{d}z = \begin{cases} \pi & \text{if } C \text{ encloses } i \text{ but not } -i, \\ -\pi & \text{if } C \text{ encloses } -i \text{ but not } i, \\ 0 & \text{if } C \text{ encloses both } \pm i \text{ or neither.} \end{cases} \tag{21.9}$$

This assumes that C is a *simple closed curve* (no self-intersections) and is oriented counterclockwise (otherwise, all the signs would be reversed). If the contour passes through a singularity, then the integral is not defined. It may have to be defined as a suitable limit (for instance, as the *Cauchy principal value*) (see example (21.38)). A more general approach is to use Taylor series expansions about the respective pole, instead of continued fraction expansions. For C_1 around pole i, for instance, we have

$$\frac{1}{z^2+1} = \frac{1}{(z-i)}\frac{1}{(z+i)} = \frac{1}{(z-i)}\frac{1}{2i+(z-i)} = \frac{1}{2i}\frac{1}{z-i}\sum_{k=0}^{\infty}\frac{(z-i)^k}{(2i)^k},$$

where the geometric series (19.5) has been used as a shortcut to the requisite Taylor series. Using (21.6) term by term then yields $\oint_{C_1} f(z)\,dz =$

π. A similar Taylor series expansion about $z = -i$ yields the other residue. For simple poles, only the first term of the Taylor series contributes, but that is not the case for higher order poles, as illustrated in the next examples.

Example 2 *Double pole.* Consider $f(z) = z^2/(z-1)^2$. That function is differentiable for all z except at $z = 1$ where there is a second order pole since $f(z) \sim 1/(z-1)^2$, near $z = 1$. This asymptotic behavior as $z \to 1$ might suggests that its residue around that pole vanishes if we apply (21.6) too quickly, but we have to be a bit more careful about integrating functions that diverge around vanishing circles. That is the story of calculus; $\infty \times 0$ requires further analysis. The Taylor series of z^2 about $z = 1$ is easily obtained as $z^2 = (z-1+1)^2 = (z-1)^2+2(z-1)+1$. Then, the residue integral is, using (21.6) term by term,

$$\oint_{C_1} \frac{z^2}{(z-1)^2}\,\mathrm{d}z = \oint_{C_1} \left(1 + \frac{2}{z-1} + \frac{1}{(z-1)^2}\right) \mathrm{d}z = 4\pi i.$$

Example 3 *Essential singularity.* Consider

$$\oint_C e^{1/z} \mathrm{d}z. \tag{21.10}$$

The function $e^{1/z}$ has an *infinite order pole*—an *essential singularity*–at $z = 0$ but Taylor series for e^t with $t = 1/z$ and the simple (21.6) makes this calculation straightforward,

$$\oint_C e^{1/z} \mathrm{d}z = \oint_C \sum_{n=0}^{\infty} \frac{1}{n!\, z^n} \mathrm{d}z = \sum_{n=0}^{\infty} \frac{1}{n!} \oint_C \frac{1}{z^n} \mathrm{d}z = 2\pi i, \tag{21.11}$$

since all integrals vanish except the $n = 1$ term from (21.6), if C encloses $z = 0$; otherwise, the integral is zero by Cauchy's theorem. Note that this "Taylor" series for $e^{1/z}$ is in powers of $t = 1/z$ expanding about $t = 0$ that corresponds to $z = \infty$! In terms of z, this is an example of a *Laurent series*, a power series that includes negative powers to capture poles and essential singularities.

Connection with $\ln z$

The integral of $1/z$ is, of course, directly related to $\ln z$, the natural log of z which can be defined as the antiderivative of $1/z$ that vanishes at $z = 1$; that is,

$$\ln z \equiv \int_1^z \frac{\mathrm{d}\zeta}{\zeta}.$$

We use ζ as the *dummy* variable of integration since z is the upper limit of integration.

But along what path from 1 to z? Here is the $2\pi i$ multiplicity again. We saw earlier that the integral of $1/z$ from a to b depends on how we go around the origin. If we get one result along one path, we can get the same result $+\ 2\pi i$ if we use a path that loops around the origin one more time counterclockwise than the original path. Or $-2\pi i$ if it loops clockwise, etc. (exercise 21.7). We find that

$$\int_1^z \frac{\mathrm{d}\zeta}{\zeta} = \ln|z| + i\arg(z) + 2ik\pi \tag{21.12}$$

for some specific k that depends on the actual path taken from 1 to z *and* our definition of $\arg(z)$. The notation $\int_1^z$ is not complete for this integral. The integral is *path-dependent* and it is necessary to specify that path in more details; however, all possible paths give the same answer *modulo* $2\pi i$ for this simple $f(z) = 1/z$.

21.4 Cauchy's formula

The combination of (21.5) with (21.6) and partial fraction and/or Taylor series expansions is quite powerful as shown in some examples, but there is another fundamental result that can be derived from them. This is *Cauchy's formula*,

$$\oint_{\mathcal{C}} \frac{f(z)}{z-a}\mathrm{d}z = 2\pi i f(a), \tag{21.13}$$

which holds for *any* closed counterclockwise contour $\mathcal{C}$ that encloses a provided that $\mathrm{d}f(z)/\mathrm{d}z$ exists everywhere inside and on $\mathcal{C}$.

The proof of this result follows the approach we used to calculate $\oint_{\mathcal{C}} \mathrm{d}z/(z-a)$ in section 21.3. Using Cauchy's theorem (21.5), the integral over $\mathcal{C}$ is equal to the integral over a small counterclockwise circle $\mathcal{C}_a$ of radius ϵ centered at a. That's because the function $f(z)/(z-a)$ is continuously differentiable in the domain between $\mathcal{C}$ and the circle $\mathcal{C}_a : z = a + \epsilon e^{i\varphi}$ with $\varphi = 0 \to 2\pi$, so

$$\oint_{\mathcal{C}} \frac{f(z)}{z-a}\mathrm{d}z = \oint_{\mathcal{C}_a} \frac{f(z)}{z-a}\mathrm{d}z = \int_0^{2\pi} f(a+\epsilon e^{i\varphi})\, i\mathrm{d}\varphi = 2\pi i f(a). \tag{21.14}$$

The final step follows from the fact that the integral has the same value no matter what $\epsilon > 0$ we pick. Then taking the limit $\epsilon \to 0^+$, the function $f(a + \epsilon e^{i\varphi}) \to f(a)$ because $f(z)$ is a nice continuous and differentiable function everywhere inside $\mathcal{C}$, and in particular at $z = a$.

Cauchy's formula applies for *any* a inside $\mathcal{C}$. To emphasize that, let us rewrite it with z in place of a, using ζ as the dummy variable of integration

$$2\pi i f(z) = \oint_{\mathcal{C}} \frac{f(\zeta)}{\zeta - z}\mathrm{d}\zeta. \tag{21.15}$$

This provides an integral formula for $f(z)$ at *any* z inside $\mathcal{C}$ in terms of its values on $\mathcal{C}$. Thus, knowing $f(z)$ on $\mathcal{C}$ completely determines $f(z)$ everywhere inside the contour.

Mean value theorem

Since (21.15) holds for any closed contour $\mathcal{C}$ as long as $f(z)$ is differentiable inside and on that contour, we can write it for a circle of radius r

centered at z, $\zeta = z + re^{i\varphi}$, where $\mathrm{d}\zeta = ire^{i\varphi}\mathrm{d}\varphi$ and (21.15) yields

$$f(z) = \frac{1}{2\pi}\int_0^{2\pi} f(z + re^{i\varphi})\mathrm{d}\varphi, \tag{21.16}$$

which states that $f(z)$ is equal to its average over a circle centered at z. This is true as long as $f(z)$ is differentiable at all points inside the circle of radius r. This *mean value theorem* also applies to the real and imaginary parts of $f(z) = u(x, y) + iv(x, y)$. It implies that $u(x, y)$, $v(x, y)$ and $|f(z)|$ do not have extrema inside a domain where $f(z)$ is differentiable. Points where $f'(z) = 0$ and therefore $\partial u/\partial x = \partial u/\partial y = \partial v/\partial x = \partial v/\partial y = 0$ are *saddle* points, not local maxima or minima.

Generalized Cauchy formula and Taylor series

Cauchy's formula also implies that if $f(z)$ is differentiable in a neighborhood of a point a, then $f(z)$ is *infinitely differentiable in that neighborhood.* Furthermore, $f(z)$ can be expanded in a Taylor series about a that converges inside a disk whose radius is equal to the distance between a and the nearest singularity of $f(z)$.

► To show that $f(z)$ is infinitely differentiable, we can show that the derivative of the right-hand side of (21.15) with respect to z exists by using the limit definition of the derivative and being careful to justify existence of the integrals and the limit. The final result is the same as that obtained by differentiating with respect to z under the integral sign, yielding

$$2\pi i f'(z) = \oint_C \frac{f(\zeta)}{(\zeta - z)^2}\mathrm{d}\zeta. \tag{21.17}$$

Doing this repeatedly we obtain

$$2\pi i f^{(n)}(z) = n! \oint_C \frac{f(\zeta)}{(\zeta - z)^{n+1}}\mathrm{d}\zeta, \tag{21.18}$$

where $f^{(n)}(z)$ is the nth derivative of $f(z)$ and $n! = n(n-1)\cdots 1$ is the factorial of n. Since all the integrals exist, all the derivatives exist. Formula (21.18) is the *generalized Cauchy formula* which we can rewrite in the form

$$\oint_C \frac{f(z)}{(z - a)^{n+1}}\,\mathrm{d}z = 2\pi i\,\frac{f^{(n)}(a)}{n!}. \tag{21.19}$$

□

► Another derivation of these results that establishes convergence of the Taylor series expansion at the same time is to use the geometric series (19.5) and the trick that we used in (19.33) to write

$$\frac{1}{\zeta - z} = \frac{1}{(\zeta - a) - (z - a)} = \frac{1}{\zeta - a}\,\frac{1}{1 - \dfrac{z - a}{\zeta - a}} = \sum_{n=0}^{\infty} \frac{(z - a)^n}{(\zeta - a)^{n+1}}, \tag{21.20}$$

where the geometric series converges provided $|z-a| < |\zeta-a|$. Cauchy's formula (21.15) then becomes

$$2\pi i f(z) = \oint_{C_a} \frac{f(\zeta)}{\zeta - z} d\zeta = \oint_{C_a} \sum_{n=0}^{\infty} f(\zeta) \frac{(z-a)^n}{(\zeta-a)^{n+1}}\, d\zeta$$
$$= \sum_{n=0}^{\infty} (z-a)^n \oint_{C_a} \frac{f(\zeta)}{(\zeta-a)^{n+1}}\, d\zeta, \qquad (21.21)$$

where C_a is a circle centered at a whose radius is as large as desired provided $f(z)$ is differentiable inside and on that circle. For instance, if $f(z) = 1/z$, then the radius of the circle must be less then $|a|$ since $f(z)$ has a singularity at $z = 0$ but is nice everywhere else. If $f(z) = 1/(z+i)$, then the radius must be less than $|a+i|$ which is the distance between a and $-i$ since $f(z)$ has a singularity at $-i$. In general, the radius of the circle must be less than the distance between a and the nearest singularity of $f(z)$. To justify interchanging the integral and the series we need to show that each integral exists and that the series of the integrals converges. If $|f(\zeta)| \le M$ on C_a and $|z-a|/|\zeta-a| \le q < 1$ since C_a is a circle of radius r centered at a and z is inside that circle while ζ is on the circle so $\zeta - a = re^{i\varphi}$, $d\zeta = ire^{i\varphi}d\varphi$ and

$$\left| \oint_{C_a} \frac{(z-a)^n f(\zeta)}{(\zeta-a)^{n+1}}\, d\zeta \right| \le 2\pi M q^n, \qquad (21.22)$$

showing that all integrals converge and the series of integrals also converges since $q < 1$.

The series (21.21) provides a power series expansion for $f(z)$,

$$2\pi i f(z) = \sum_{n=0}^{\infty} (z-a)^n \oint_{C_a} \frac{f(\zeta)}{(\zeta-a)^{n+1}}\, d\zeta = \sum_{n=0}^{\infty} c_n(a)(z-a)^n \quad (21.23)$$

that converges inside a disk centered at a with radius equal to the distance between a and the nearest singularity of $f(z)$. The series can be differentiated term by term and the derivative series also converges in the same disk. Hence, all derivatives of $f(z)$ exist in that disk. In particular we find that

$$c_n(a) = 2\pi i \frac{f^{(n)}(a)}{n!} = \oint_{C_a} \frac{f(\zeta)}{(\zeta-a)^{n+1}}\, d\zeta, \qquad (21.24)$$

which is the generalized Cauchy formula (21.18), (21.19), and the series (21.21) is none other than the familiar Taylor series

$$f(z) = f(a) + f'(a)(z-a) + \frac{f''(a)}{2}(z-a)^2 + \cdots = \sum_{n=0}^{\infty} \frac{f^{(n)}(a)}{n!}(z-a)^n. \qquad (21.25)$$

Finally, Cauchy's theorem tells us that the integral on the right of (21.24) has the same value on any closed contour (counterclockwise) enclosing a but no other singularities of $f(z)$, so the formula holds for any such closed contour as written in (21.18). However, convergence of the Taylor series only occurs inside a disk centered at a and of radius equal to the distance between a and the nearest singularity of $f(z)$.

21.5 Real examples of complex integration

One application of complex, or *contour*, integration is to turn difficult real integrals into simple complex integrals. We consider a few basic examples.

Example 1 Find the average of the function

$$f(\theta) = \frac{3}{5 + 4\cos\theta}.$$

Its minimum is $1/3$ and maximum is 3. It is periodic of period 2π so its average is

$$\frac{1}{2\pi}\int_0^{2\pi} f(\theta)\mathrm{d}\theta.$$

To compute that integral we think *integral over the unit circle in the complex plane!* Indeed, the unit circle with $|z| = 1$ has the simple parametrization

$$z = e^{i\theta} \to \mathrm{d}z = ie^{i\theta}\mathrm{d}\theta \Leftrightarrow \mathrm{d}\theta = \frac{\mathrm{d}z}{iz}. \tag{21.26}$$

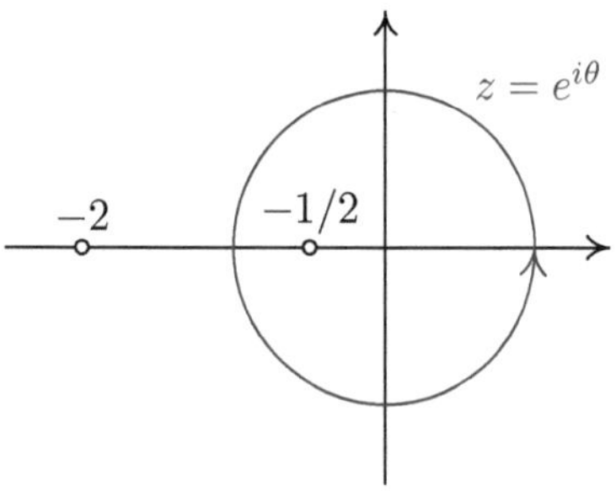

Fig. 21.6 Unit circle $z = e^{i\theta}$, poles and residues for (21.27).

Furthermore

$$\cos\theta = \frac{e^{i\theta} + e^{-i\theta}}{2} = \frac{z + 1/z}{2},$$

so we obtain (Fig. 21.6)

$$\int_0^{2\pi} \frac{3}{5 + 4\cos\theta}\mathrm{d}\theta = \oint_{|z|=1} \frac{3}{5 + 2(z + 1/z)}\frac{\mathrm{d}z}{iz}$$
$$= \frac{3}{2i}\oint_{|z|=1} \frac{\mathrm{d}z}{(z + \frac{1}{2})(z + 2)} = \frac{3}{2i}\left(\frac{2\pi i}{z + 2}\right)_{z=-\frac{1}{2}} = 2\pi. \tag{21.27}$$

We turned the integral of a real periodic function $f(\theta)$ over its period from $\theta = 0$ to 2π into a complex z integral over the unit circle $z = e^{i\theta}$. This is a general idea that applies for the integral of any periodic function over its period. That led us to the integral over a closed curve of a simple rational function (that is not always the case; it depends on the actual $f(\theta)$).

In this example, our complex function $(2z^2 + 5z + 2)^{-1}$ has two simple poles, at $-1/2$ and -2, and the denominator factors as

$$2z^2 + 5z + 2 = 2(z + 1/2)(z + 2).$$

Since -2 is outside the unit circle, it does not contribute to the integral, but the simple pole at $-1/2$ does. So the integrand has the form $g(z)/(z - a)$ with $a = -1/2$ inside our domain and $g(z) = 1/(z + 2)$, is a good analytic function inside the unit circle. So one application of Cauchy's formula, *et voilà*. The function $3/(5 + 4\cos\theta)$ which oscillates between $1/3$ and 3 has an average of 1.

An extended version of this example is to calculate

$$a_n = \frac{2-\delta_{n,0}}{\pi} \int_0^{\pi} \frac{3\cos n\theta}{5+4\cos\theta} \mathrm{d}\theta, \tag{21.28}$$

where n is an integer. These integrals give the Fourier coefficients of $f(\theta) = 3/(5+4\cos\theta)$. In general, an even function $f(\theta) = f(-\theta)$, periodic of period 2π can be expanded in a *Fourier cosine series*

$$f(\theta) = a_0 + a_1 \cos\theta + a_2 \cos 2\theta + a_3 \cos 3\theta + \cdots . \tag{21.29}$$

This Fourier expansion is useful in many applications: partial differential equations, numerical calculations, signal processing, etc. The coefficient a_0 is the average of $f(\theta)$. The other coefficients for $n \neq 0$ are given by

$$a_n = \frac{2}{\pi} \int_0^{\pi} f(\theta) \cos n\theta \, \mathrm{d}\theta,$$

as follows from (21.29) and orthogonality of the cosines

$$\int_0^{\pi} \cos m\theta \cos n\theta \, \mathrm{d}\theta = \frac{\pi}{(2-\delta_{m,0})} \delta_{m,n}.$$

For $f(\theta) = 3/(5+4\cos\theta)$, the Fourier coefficients are (21.28). To calculate those a_n's, we turn the integral into 1/2 of the integral from $-\pi \to \pi$ since $\cos n\theta = \cos(-n\theta)$ and write (21.28) as

$$a_n = \frac{2-\delta_{n,0}}{2\pi} \int_{-\pi}^{\pi} \frac{3e^{in\theta}}{5+4\cos\theta} \mathrm{d}\theta, \tag{21.30}$$

using $e^{in\theta} = \cos n\theta + i \sin n\theta$ and since $\sin n\theta$ is odd, the integral of that term is 0. This preparation of the integral is well worth it as it avoids having to deal with an nth-order pole at $z = 0$ when the substitution $z = e^{i\theta}$ is used. Proceeding as in example 1, we find that

$$a_n = \frac{2-\delta_{n,0}}{2\pi i} \oint_{|z|=1} \frac{3z^n}{2z^2+5z+2} \mathrm{d}z = (2-\delta_{n,0}) \left(\frac{-1}{2}\right)^n, \tag{21.31}$$

such that the Fourier series

$$\frac{3}{5+4\cos\theta} = 1 - \frac{\cos\theta}{2} + \frac{\cos 2\theta}{2^2} - \frac{\cos 3\theta}{2^3} + \cdots \tag{21.32}$$

converges exponentially.

Example 2

$$\int_{-\infty}^{\infty} \frac{\mathrm{d}x}{1+x^2} = \pi. \tag{21.33}$$

This integral is easily done with the fundamental theorem of calculus since $1/(1+x^2) = \mathrm{d}(\arctan x)/\mathrm{d}x$, but we use contour integration to demonstrate the method. The integral is equal to the integral of $1/(1+z^2)$ over the real line $z = x$ with $x = -\infty \to \infty$. That complex function has two simple poles at $z = \pm i$ since $z^2 + 1 = (z+i)(z-i)$.

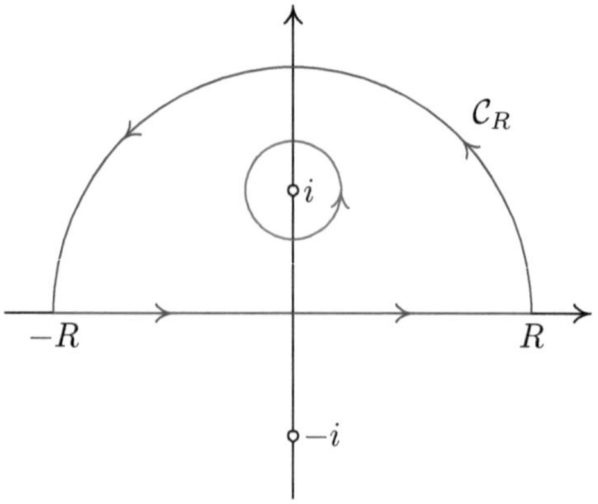

Fig. 21.7 Closing the contour and calculating the residue for (21.33).

So we turn this into a contour integration by considering the closed path $\mathcal{C}$ consisting of the real interval $z = x$ with $x = -R \to R$ together with the semi-circle $\mathcal{C}_R : z = Re^{i\theta}$ with $\theta = 0 \to \pi$ (Fig. 21.7). Since $z = i$ is the only simple pole inside our closed contour $\mathcal{C}$, Cauchy's formula gives

$$\oint_{\mathcal{C}} \frac{\mathrm{d}z}{z^2+1} = \oint_{\mathcal{C}} \frac{(z+i)^{-1}}{z-i}\mathrm{d}z = 2\pi i \left(\frac{1}{z+i}\right)_{z=i} = \pi.$$

To get the integral we want, we need to take $R \to \infty$ and figure out the $\mathcal{C}_R$ integral. The latter goes to zero as $R \to \infty$ since $|\mathrm{d}z| = R\mathrm{d}\theta$ and $R^2 - 1 \leq |z^2+1| \leq R^2+1$ on $z = Re^{i\theta}$. Thus,

$$\left|\int_{\mathcal{C}_R} \frac{\mathrm{d}z}{z^2+1}\right| \leq \int_{\mathcal{C}_R} \frac{|\mathrm{d}z|}{|z^2+1|} < \int_0^{\pi} \frac{R\mathrm{d}\theta}{R^2-1} = \frac{\pi R}{R^2-1} \to 0.$$

Example 3 We use the same technique for

$$\int_{-\infty}^{\infty} \frac{\mathrm{d}x}{1+x^4} = \int_{\mathbb{R}} \frac{\mathrm{d}z}{1+z^4} \tag{21.34}$$

This integrand has four simple poles at the simple roots of $z^4+1=0$. Those roots are $z_k = e^{i(2k-1)\pi/4}$ for $k = 1,2,3,4$; they are on the unit circle, equispaced by angle $\pi/2$.

We use the same closed contour $\mathcal{C}$ consisting of the real interval $[-R, R]$ and the semi-circle $\mathcal{C}_R$, as in the previous example, but now there are *two* simple poles inside that contour (Fig. 21.8). We need to isolate both singularities leading to

$$\oint_{\mathcal{C}} = \oint_{\mathcal{C}_1} + \oint_{\mathcal{C}_2}.$$

Note that $z^4+1 = (z^2-i)(z^2+i) = (z^2-z_1^2)(z^2-z_2^2)$, where $z_1^2 = i$ and $z_2^2 = -i$. Then for $\mathcal{C}_1$ that includes only z_1, we write $z^4+1 = (z-z_1)(z+z_1)(z^2+i)$ and Cauchy's formula gives

$$\oint_{\mathcal{C}_1} \frac{\mathrm{d}z}{z^4+1} = 2\pi i \left(\frac{1}{2z_1(z_1^2-z_2^2)}\right) = \frac{\pi}{2z_1}.$$

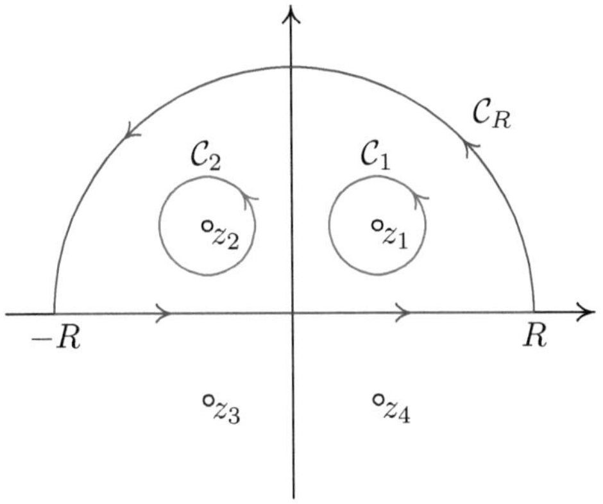

Fig. 21.8 Contour and residues for (21.34) with $z_1 = e^{i\pi/4}$ and $z_2 = -e^{-i\pi/4}$.

Likewise for $\mathcal{C}_2$ that includes only z_2, we write $z^4 + 1 = (z - z_2)(z + z_2)(z^2 - z_1^2)$ and Cauchy's formula yields

$$\oint_{\mathcal{C}_2} \frac{dz}{z^4+1} = 2\pi i \left(\frac{1}{2z_2(z_2^2 - z_1^2)} \right) = \frac{\pi}{2(-z_2)}.$$

Adding both residues and recalling $z_1 = e^{i\pi/4}$ and $z_2 = -e^{-i\pi/4}$ gives

$$\oint_{\mathcal{C}} \frac{\mathrm{d}z}{z^4+1} = \pi \frac{e^{-i\pi/4} + e^{i\pi/4}}{2} = \pi \cos\frac{\pi}{4} = \frac{\pi}{\sqrt{2}}.$$

As before we need to take $R \to \infty$ and figure out the $\mathcal{C}_R$ part. That part goes to zero as $R \to \infty$ since

$$\left| \int_{\mathcal{C}_R} \frac{\mathrm{d}z}{z^4+1} \right| \leq \int_{\mathcal{C}_R} \frac{|\mathrm{d}z|}{|z^4+1|} < \int_0^{\pi} \frac{R\mathrm{d}\theta}{R^4-1} = \frac{\pi R}{R^4-1}.$$

We could extend the same method to

$$\int_{-\infty}^{\infty} \frac{x^2}{1+x^8} \mathrm{d}x. \tag{21.35}$$

We would use the same closed contour again, but now there would be four simple poles inside it and therefore four separate contributions.

Example 4

$$\int_{-\infty}^{\infty} \frac{\mathrm{d}x}{(1+x^2)^2} = \int_{\mathbb{R}} \frac{\mathrm{d}z}{(z^2+1)^2} = \int_{\mathbb{R}} \frac{\mathrm{d}z}{(z-i)^2(z+i)^2}. \tag{21.36}$$

We use the same closed contour once more, but now we have a *double pole* inside the contour at $z = i$. We can figure out the contribution from that double pole by using the generalized form of Cauchy's formula (21.18). The integral over $\mathcal{C}_R$ can be shown to vanish as $R \to \infty$ and

$$\int_{-\infty}^{\infty} \frac{\mathrm{d}x}{(1+x^2)^2} = 2\pi i \left(\frac{\mathrm{d}}{\mathrm{d}z}(z+i)^{-2} \right)_{z=i} = \frac{\pi}{2}. \tag{21.37}$$

Example 5

$$\int_{-\infty}^{\infty} \frac{\sin x}{x} \mathrm{d}x = \pi. \tag{21.38}$$

This is a trickier problem. Our impulse is to consider $\int_{\mathbb{R}} (\sin z)/z \, \mathrm{d}z$ but that integrand is a good function! Indeed, $(\sin z)/z = 1 - z^2/3! + z^4/5! - \cdots$ is analytic in the entire plane; its Taylor series converges in the entire plane. There are no poles for $\sin z/z$ and no opportunity to evaluate the integral through residues. So we consider instead

$$\int_{\mathbb{R}} \frac{e^{iz}}{z} \mathrm{d}z = \int_{-\infty}^{\infty} \frac{e^{ix}}{x} \mathrm{d}x = \int_{-\infty}^{\infty} \frac{\cos x + i \sin x}{x} \, \mathrm{d}x, \tag{21.39}$$

with a simple pole at $z = 0$, but that is another problem since the pole is *on* the contour now! We added an integral of $(\cos x)/x$ that has a nonintegrable singularity at $x = 0$. It needs to be interpreted as the *Cauchy principal value*, that is, for $a < 0 < b$,

$$\int_a^b \frac{f(x)}{x} \, \mathrm{d}x = \lim_{\epsilon \to 0^+} \left(\int_a^{-\epsilon} \frac{f(x)}{x} \, \mathrm{d}x + \int_{\epsilon}^b \frac{f(x)}{x} \, \mathrm{d}x \right).$$

For $f(x) = \cos x$ and $b = -a$, that is 0 since $\cos x$ is even but x is odd.

Thus, we have to modify our favorite contour a bit to avoid the pole by going over or below it. If we go below and close along the upper half circle C_R as before, then we'll have a pole inside our contour. If we go over it, we won't have any pole inside the closed contour. We get the same result either way (luckily!), but the algebra is a bit simpler if we leave the pole out.

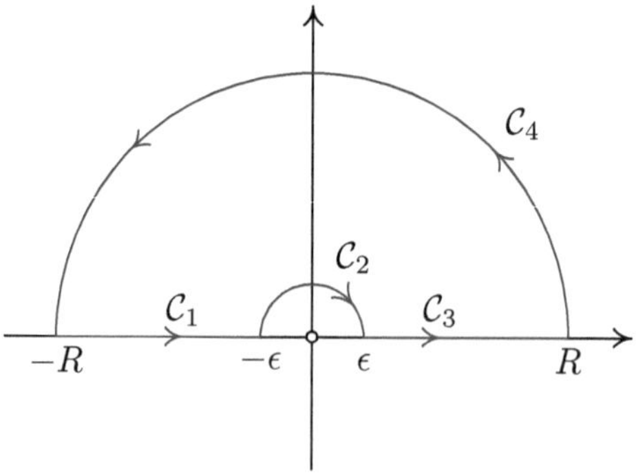

Fig. 21.9 Closing the contour for (21.39).

So we consider the closed contour $C = C_1 + C_2 + C_3 + C_4$, where C_1 is the real axis from $-R$ to $-\epsilon$, C_2 is the semi-circle from $-\epsilon$ to ϵ in the top half-plane, C_3 is the real axis from ϵ to R, and C_4 is our good old semi-circle of radius R (Fig. 21.9). The integrand e^{iz}/z is analytic everywhere except at $z = 0$ where it has a simple pole, but since that pole is outside our closed contour, Cauchy's theorem gives $\oint_C = 0$ or

$$\int_{C_1 + C_3} = -\int_{C_2} - \int_{C_4}$$

The integral over the semi-circle $\mathcal{C}_2 : z = \epsilon e^{i\theta}$, $dz = i\epsilon e^{i\theta}\mathrm{d}\theta$, is

$$-\int_{\mathcal{C}_2} \frac{e^{iz}}{z}\mathrm{d}z = i\int_0^{\pi} e^{i\epsilon e^{i\theta}}\mathrm{d}\theta \to \pi i \quad \text{as} \quad \epsilon \to 0.$$

As before we'd like to show that the $\int_{\mathcal{C}_4} \to 0$ as $R \to \infty$. This is trickier than the previous cases we've encountered. On the semi-circle $z = Re^{i\theta}$ and $\mathrm{d}z = iRe^{i\theta}\mathrm{d}\theta$, as we have used multiple times already, so

$$\int_{\mathcal{C}_4} \frac{e^{iz}}{z}\mathrm{d}z = i\int_0^{\pi} e^{iRe^{i\theta}}\mathrm{d}\theta = i\int_0^{\pi} e^{iR\cos\theta}e^{-R\sin\theta}\mathrm{d}\theta. \tag{21.40}$$

This is a scary integral, but we want the limit as $R \to \infty$. The integrand has two factors, $e^{iR\cos\theta}$ that oscillates rapidly as $R \to \infty$ but whose norm is always 1 and $e^{-R\sin\theta}$ which is real and exponentially small for all θ in $0 < \theta < \pi$, except at 0 and π where it is exactly 1. Sketch $e^{-R\sin\theta}$ in $0 \le \theta \le \pi$ and it will be clear that the integral should go to zero as $R \to \infty$. To show this rigorously, let's consider its modulus (norm) as we did in the previous cases. Then since (i) the modulus of a sum is less or equal to the sum of the moduli (triangle inequality), (ii) the modulus of a product is the product of the moduli, and (iii) $|e^{iR\cos\theta}| = 1$ when R and θ are real (which they are)

$$0 \le \left|\int_0^{\pi} e^{iR\cos\theta}e^{-R\sin\theta}\mathrm{d}\theta\right| < \int_0^{\pi} e^{-R\sin\theta}\mathrm{d}\theta. \tag{21.41}$$

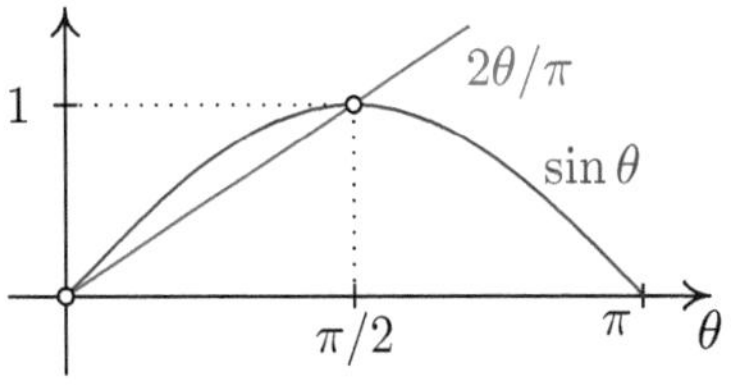

Fig. 21.10 Jordan's lemma (21.42).

We still cannot calculate that last integral but we don't need to. We just need to show that it is smaller than something that goes to zero as $R \to \infty$, so our integral will be *squeezed* to zero.

The plot of $\sin\theta$ for $0 \le \theta \le \pi$ illustrates that it is symmetric with respect to $\pi/2$ and that $2\theta/\pi \le \sin\theta$ when $0 \le \theta \le \pi/2$ (Fig. 21.10), or changing the signs $-2\theta/\pi \ge -\sin\theta$ and since e^x increases monotonically with x,

$$e^{-R\sin\theta} < e^{-2R\theta/\pi}$$

in $0 \le \theta \le \pi/2$. This yields **Jordan's lemma**

$$\begin{aligned}\left|\int_0^{\pi} e^{iRe^{i\theta}}\mathrm{d}\theta\right| &< \int_0^{\pi} e^{-R\sin\theta}\mathrm{d}\theta = 2\int_0^{\pi/2} e^{-R\sin\theta}\mathrm{d}\theta \\ &< 2\int_0^{\pi/2} e^{-2R\theta/\pi}\mathrm{d}\theta = \pi\frac{1-e^{-R}}{R}\end{aligned} \tag{21.42}$$

so $\int_{\mathcal{C}_4} \to 0$ as $R \to \infty$ and collecting our results we obtain (21.38).

The exercises contain several other important integrals evaluated using contour integration. More examples and results can be found in books devoted to complex analysis such as the following:

- R. V. Churchill and J. W. Brown, *Complex Variables and Applications*, McGraw Hill, 2009;
- M. J. Ablowitz and A. S. Fokas, *Complex Variables—Introduction and Applications*, Cambridge University Press, 2003; and
- L. V. Alfhors, *Complex Analysis*, McGraw Hill, 1979.

Exercises

Closed paths are simple and oriented *counterclockwise* unless specified otherwise.

(21.1) Calculate the integral of $f(z) = z + 2/z$ along the path C that goes once around the circle $|z| = R > 0$. How does your result depend on R?

(21.2) Calculate the integral of $f(z) = az + b/z + c/(z+1)$, where a, b, and c are complex constants, around (i) the circle of radius $R > 0$ centered at $z = 0$, (ii) the circle of radius 2 centered at $z = 0$, and (iii) the triangle $-1/2$, $-2+i$, $-1-2i$.

(21.3) Calculate the integral of $f(z) = 1/(z^2-4)$ around (i) the unit circle, (ii) the parallelogram 0, $2-i$, 4, $2+i$. (Hint: use partial fractions.)

(21.4) Calculate the integral of $f(z) = 1/(z^4-1)$ along the circle of radius 1 centered at i.

(21.5) Calculate the integral of $\sin(1/(3z))$ over the square 1, i, -1, $-i$. (Hint: use the Taylor series for $\sin z$.)

(21.6) Calculate the integral of $1/z$ from $z = 1$ to $z = 2e^{i\pi/4}$ (i) along the path $1 \to 2$ along the real line then $2 \to 2e^{i\pi/4}$ along the circle of radius 2, and (ii) along $1 \to 2$ on the real line, followed by $2 \to 2e^{i\pi/4}$ along the circle of radius 2, *clockwise*.

(21.7) If a is an arbitrary complex number, show that the integral of $1/z$ along the straight line from 1 to a is equal to the integral of $1/z$ from 1 to $|a|$ along the real line + the integral of $1/z$ along a circular path of radius $|a|$ from $|a|$ to a. *Draw a sketch of the three paths from 1 to a.* Calculate the integral using both a clockwise and a counterclockwise circular arc from $z = |a|$ to $z = a$. What happens if a is real but negative?

(21.8) Does the integral of $1/z^2$ from $z = a$ to $z = b$ (with a and b complex) depend on the path? Explain.

(21.9) Does the integral of $z^* = \text{conj}(z)$ depend on the path? Calculate $\int_a^b z^* \mathrm{d}z$ (i) along the straight line from a to b, and (ii) along the real direction from $\text{Re}(a)$ to $\text{Re}(b)$ then up in the imaginary direction from $\text{Im}(a)$ to $\text{Im}(b)$. Sketch the paths and compare the answers.

(21.10) The expansion (19.33) with $a = 1$ gives $1/z = \sum_{n=0}^{\infty}(1-z)^n$. Using this expansion together with (21.6) we find that

$$\oint_{|z|=1} \frac{\mathrm{d}z}{z} = \sum_{n=0}^{\infty} \oint_{|z|=1} (1-z)^n \mathrm{d}z = 0.$$

This does not match $\oint_{|z|=1} \mathrm{d}z/z = 2\pi i$. Why not?

(21.11) Why can we take $(z-a)^n$ outside of the integrals in (21.21)?

(21.12) Verify the estimate (21.22). Why does that estimate imply that the series of integrals converges?

(21.13) Consider the integral of $f(z)/(z-a)^2$ about a small circle C_a of radius ϵ centered at a: $z = a + \epsilon e^{i\varphi}$, $0 \le \varphi < 2\pi$. Study the limit of the φ-integral as $\epsilon \to 0^+$. Does your limit agree with the generalized Cauchy formula (21.18), (21.24)?

(21.14) Derive (21.19) by deforming the contour C to a small circle C_a about a, expanding $f(z)$ in a Taylor series about a (assuming that it exists) and using (21.6) term by term. In applications, this is a practical and more general approach that works for essential singularities as well.

(21.15) Use Cauchy's generalized formula (21.19) to show **Liouville's theorem** that if an entire function $f(z)$ (that is, differentiable in the entire complex plane) is bounded for all $z \in \mathbb{C}$, then it must be a constant.

(21.16) Let $f(z) = z^2 + 1$, $f'(z) = \mathrm{d}f(z)/\mathrm{d}z$, and $g(z) = f'(z)/f(z)$. (a) Calculate all possible values of $\oint g(z)\,\mathrm{d}z$ where the integral is over $|z| = R$. (b) Find the Taylor series $g(z) = a_0 + a_1 z + a_2 z^2 + \cdots$ and its radius of convergence. (c) Find the *Laurent series* $g(z) = b_0 + b_1/z + b_2/z^2 + \cdots$ that converges *outside* a disk of what radius? (d) Show how to answer (a) using (b) and (c).

(21.17) Let $P_n(z) = a_n z^n + \cdots + a_0$ be a polynomial of degree n in z, with $a_n \neq 0$, and derivative $P_n'(z) = \mathrm{d}P_n(z)/\mathrm{d}z$. Show that

$$\frac{1}{2\pi i} \oint_C \frac{P_n'(z)}{P_n(z)} \mathrm{d}z = n,$$

where C is a circle $|z| = R$ of sufficiently large radius R. Showing this without using the zeroes of $P_n(z)$ by expanding in a Laurent series, $P_n'(z)/P_n(z) = n/z + c_2/z^2 + \cdots$, yields a proof of the fundamental theorem of algebra.

(21.18) Let $f(z)$ be a *meromorphic* function, that is, a differentiable function except for isolated poles. Use residues to show (the *argument principle*)

$$\frac{1}{2\pi i}\oint_C \frac{f'(z)}{f(z)}\,\mathrm{d}z = n_0 - n_p,$$

where n_0 are the number of zeroes and n_p the number of poles, both counted with multiplicities, inside $\mathcal{C}$. Residues break up the integral as a sum of integrals over small contours surrounding a single pole a of order n_a where $f(z) = g(z)/(z-a)^{n_a}$ with $g(a) \neq 0$, or around a single zero b of order n_b where $f(z) = (z-b)^{n_b} h(z)$ with $h(b) \neq 0$, and $g(z)$ and $h(z)$ are analytic inside and on their residue contour.

(21.19) Calculate the integrals of $\cos(z)/z^n$ and $\sin(z)/z^n$ over the unit circle, where n is a positive integer.

(21.20) Sketch $f(x) = 1/(a + b\sin x)$. Calculate $\int_0^{2\pi} f(x)\mathrm{d}x$ where a and b are real numbers with $|a| > |b|$. Why does $|a| > |b|$ matter?

(21.21) Sketch $f(x) = 1/(1+\cos^2 x)$ and $g(x) = 1/(1+\sin^2 x)$. Calculate $\int_0^{2\pi} f(x)\mathrm{d}x$. Why is it equal to $\int_0^{2\pi} g(x)\mathrm{d}x$?

(21.22) Sketch $f(\theta) = e^{R\cos\theta}$ and $g(\theta) = e^{R\sin\theta}$ where R is a positive constant. Calculate $\int_0^{2\pi} e^{R\cos\theta}\mathrm{d}\theta$.

(21.23) Calculate $c_n \triangleq \int_0^{2\pi} e^{R\cos\theta}\cos n\theta\,\mathrm{d}\theta$, where n is an integer.

(21.24) Calculate $\int_{-\infty}^{\infty} \mathrm{d}x/(1+x^2+x^4)$. Can you provide an an priori upper bound for this integral based on integrals calculated earlier?

(21.25) Show that

$$F(k) = \int_{-\infty}^{\infty} \frac{e^{ikx}}{x^2+a^2}\,\mathrm{d}x = \frac{\pi e^{-|ka|}}{|a|},$$

the Fourier transform of the *Lorentzian*, where a and k are real. Show that $F(k) = F(|k|)$ and use that to calculate $F(k)$ without trouble as $|z| \to \infty$.

(21.26) Given the Poisson integral $\int_{-\infty}^{\infty} e^{-x^2}\mathrm{d}x = \sqrt{\pi}$, calculated in an earlier chapter with a classic trick ($x^2 \to x^2 + y^2$ then (x, y) to polar), show that

$$\int_{-\infty}^{\infty} e^{-x^2/a^2} e^{ikx}\mathrm{d}x = \sqrt{\pi}\,|a|\,e^{-k^2a^2/4},$$

where a and k are arbitrary *real* numbers. This is the *Fourier transform* of the Gaussian e^{-x^2/a^2}, and it is also a Gaussian in k, but if a is large then the Gaussian in x is wide, while that in k is very narrow, and vice versa if a is small. (Hint: complete the square, then integrate over the rectangle $-R \to R \to R + ika^2/2 \to -R + ika^2/2 \to -R$. Justify.)

(21.27) Calculate

$$\int_{-\infty}^{\infty} \frac{e^{ax}}{1+e^x}\mathrm{d}x$$

for $0 < a < 1$ using the rectangular contour $-R \to R \to R + 2\pi i \to -R + 2\pi i \to -R$.

(21.28) Show that

$$\int_0^{\infty} \frac{\mathrm{d}x}{1+x^3} = \frac{2\pi}{3\sqrt{3}}$$

using the straight line $z = 0 \to R$, the circular arc $z = R \to Re^{i2\pi/3}$ and the straight line $z = Re^{2i\pi/3} \to 0$.

(21.29) Show how to use contour integration to calculate

$$\int_0^{\infty} \frac{\mathrm{d}x}{1+x^5}.$$

(21.30) The *Fresnel* integrals come up in optics and quantum mechanics. They are the integrals of $\cos x^2$ and $\sin x^2$ from $0 \to \infty$. Calculate both at once by showing that

$$\int_0^{\infty} e^{ix^2}\mathrm{d}x = \frac{\sqrt{\pi}}{2}e^{i\pi/4}.$$

Consider the closed path that goes from 0 to R on the real axis, then on the circle of radius R to $Re^{i\pi/4}$ (or the vertical line $R \to R + iR$), then back along the diagonal $z = re^{i\pi/4}$ with r real.

(21.31) Show that the *Glauert integrals* of aerodynamics

$$\int_0^{2\pi} \frac{\cos n\theta}{\cos\theta - \cos\theta_0}\mathrm{d}\theta = \pi\frac{\sin n\theta_0}{\sin\theta_0},$$

where n is an integer, θ_0 is real, and the integral has to be interpreted as the *Cauchy principal value* (that is, with omitting a $\theta_0 \pm \epsilon$ interval around the singularity, then taking the limit $\epsilon \to 0$, as in Fig. 21.9).

Index